Business Foundations
A Changing World 12

Nassau Community Student Value Pack

create.mheducation.com

ISBN-13: 9781307421965

ISBN-10: 1307421962

Contents

Bonus Chapters

Online Supplements

Dedication

To James Ferrell

To Linda Hirt

To George Ferrell

Authors

O.C. FERRELL

O.C. Ferrell is the James T. Pursell Sr. Eminent Scholar in Ethics and Director of the Center for Ethical Organizational Cultures in the Raymond J. Harbert College of Business, Auburn University. He was formerly Distinguished Professor of Leadership and Business Ethics at Belmont University and University Distinguished Professor at the University of New Mexico. He has also been on the faculties of the University of Wyoming, Colorado State University, University of Memphis, Texas A&M University, Illinois State University, and Southern Illinois University. He received his PhD in marketing from Louisiana State University.

Dr. Ferrell is president of the Academy of Marketing Science. He is past president of the Academic Council of the American Marketing Association and chaired the American Marketing Association Ethics Committee. Under his leadership, the committee developed the AMA Code of Ethics and the AMA Code of Ethics for Marketing on the Internet. In addition, he is a former member of the Academy of Marketing Science Board of Governors and is a Society of Marketing Advances and Southwestern Marketing Association Fellow and an Academy of Marketing Science Distinguished Fellow. He has served for nine years as the vice president of publications for the Academy of Marketing Science. In 2010, he received a Lifetime Achievement Award from the Macromarketing Society and a special award for service to doctoral students from the Southeast Doctoral Consortium. He received the Harold Berkman Lifetime Service Award from the Academy of Marketing Science and, more recently, the Cutco Vector Distinguished Marketing Educator Award from the Academy of Marketing Science.

Dr. Ferrell has been involved in entrepreneurial engagements, co-founding Print Avenue in 1981, providing a solution-based printing company. He has been a consultant and served as an expert witness in legal cases related to marketing and business ethics litigation. He has conducted training for a number of global firms, including General Motors. His involvement with direct selling companies includes serving on the Academic Advisory Committee and as a fellow for the Direct Selling Education Foundation.

Dr. Ferrell is the co-author of 20 books and more than 100 published articles and papers. His articles have been published in the *Journal of Marketing Research, Journal of Marketing, Journal of Business Ethics, Journal of Business Research, Journal of the Academy of Marketing Science, AMS Review,* and the *Journal of Public Policy & Marketing,* as well as other journals.

GEOFFREY A. HIRT

Geoffrey A. Hirt of DePaul University previously taught at Texas Christian University and Illinois State University, where he was chairman of the Department of Finance and Law. At DePaul, he was chairman of the Finance Department from 1987 to 1997 and held the title of Mesirow Financial Fellow. He developed the MBA program in Hong Kong and served as director of international initiatives for the College of Business, supervising overseas programs in Hong Kong, Prague, and Bahrain, and was awarded the Spirit of St. Vincent DePaul award for his contributions to the university. Dr. Hirt directed the Chartered Financial Analysts (CFA) study program for the Investment Analysts Society of Chicago from 1987 to 2003. He has been a visiting professor at the University of Urbino in Italy, where he still maintains a relationship with the economics department. He received his PhD in finance from the University of Illinois at Champaign-Urbana, his MBA at Miami University of Ohio, and his BA from Ohio Wesleyan University.

Dr. Hirt is currently on the Dean's Advisory Board and Executive Committee of DePaul's School of Music. The Tyree Foundation funds innovative education programs in Chicago, and Dr. Hirt also serves on the Grant Committee. Dr. Hirt is past president and a current member of the Midwest Finance Association, a former editor of the *Journal of Financial Education,* and also a member of the Financial Management Association. He belongs to the Pacific Pension Institute, an organization of public pension funds, private equity firms, and international organizations such as the Asian Development Bank, the IMF, and the European Bank for Reconstruction and Development.

Dr. Hirt is widely known for his textbook *Foundations of Financial Management,* published by McGraw-Hill/Irwin. This book, in its sixteenth edition, has been used in more than 31 countries and translated into more than 14 different languages. Additionally, Dr. Hirt is well known for his textbook *Fundamentals of Investment Management,* also published by McGraw-Hill/Irwin and now in its tenth edition. Dr. Hirt enjoys golf, swimming, music, and traveling with his wife, who is a pianist and opera coach.

LINDA FERRELL

Linda Ferrell is Professor and Chair of the Marketing Department in the Raymond J. Harbert College of Business, Auburn University. She was formerly Distinguished Professor of Leadership and Business Ethics at Belmont University. She completed her PhD in business administration, with a concentration in management, at the University of Memphis. She has taught at the University of Tampa, Colorado State University, University of Northern Colorado, University of Memphis, University of Wyoming, and the University of New Mexico. She has also team-taught classes at Thammasat University in Bangkok, Thailand.

Her work experience as an account executive for McDonald's and Pizza Hut's advertising agencies supports her teaching of advertising, marketing management, marketing ethics, and marketing principles. She has published in the *Journal of Public Policy & Marketing, Journal of Business Research, Journal of the Academy of Marketing Science, Journal of Business Ethics, AMS Review, Journal of Academic Ethics, Journal of Marketing Education, Marketing Education Review, Journal of Teaching Business Ethics, Marketing Management Journal,* and *Case Research Journal,* and she is co-author of *Business Ethics: Ethical Decision Making and Cases* (eleventh edition), *Management* (third edition), and *Business and Society* (sixth edition).

Dr. Ferrell is the immediate past president of the Academy of Marketing Science and a past president for the Marketing Management Association. She is a member of the NASBA Center for the Public Trust Board, on the Mannatech Board of Directors, and on the college advisory board for Cutco/Vector. She is also on the Board, Executive Committee, and Academic Advisory Committee of the Direct Selling Education Foundation. She has served as an expert witness in cases related to advertising, business ethics, and consumer protection.

Welcome

The twelfth edition represents a complete and comprehensive revision. This is because so many events and changes in the environment relate to the foundational concepts in business. This means that an introduction to business product has to provide adequate coverage of dynamic changes in the economy as they relate to business decisions. We have listened to your feedback and incorporated needed changes in content, boxes, cases, exercises, support, online resources, and other features.

This is our fifth edition with a chapter on digital marketing and social networking in business. Since launching this chapter in the eighth edition, this dynamic area continues to change the face of business. Entrepreneurs and small businesses have to be able to increase sales and reduce costs by using social networking to communicate and develop relationships with customers. The sharing, or "gig," economy is transforming entrepreneurial opportunities for employees. For example, the number of independent contractors in our economy has increased to almost one-third of the workforce. The Internet is providing opportunities for peer-to-peer relationships for companies such as Uber, Lyft, TaskRabbit, as well as health care services like Dose. Digital marketing has helped many entrepreneurs launch successful businesses. The increase in independent contractors has contributed to a new trend called co-working. For example, WeWork, discussed in Chapter 10, is capitalizing on these co-working opportunities by offering flexible, agile workspaces to businesses ranging from freelancers to *Fortune* 500 companies.

Throughout the product, we recognize the importance of sustainability and "green" business. By using the philosophy *reduce, reuse, and recycle,* we believe every business can be more profitable and contribute to a better world through green initiatives. There is a "Going Green" box in each chapter that covers these environmental changes. Our "Entrepreneurship in Action" boxes also discuss many innovations and opportunities to use sustainability for business success. Sustainability is not only a goal of many businesses, but it is also providing career opportunities for many of our students.

We have been careful to continue our coverage of global business, ethics and social responsibility, and information technology as they relate to the foundations important in an introduction to business course. Our co-author team has a diversity of expertise in these important areas. O.C. Ferrell and Linda Ferrell have been recognized as leaders in business ethics education, and their insights are reflected in every chapter and in the "Business Integrity" boxes. Geoff Hirt has a strong background in global business development, especially world financial markets and trade relationships.

The foundational areas of introduction to business, entrepreneurship, small business management, marketing, accounting, and finance have been completely revised. Examples have been provided to which students can easily relate. An understanding of core functional areas of business is presented so students get a holistic view of the world of business. Box examples related to "Responding to Business Challenges,"

"Entrepreneurship in Action," "Going Green," and "Business Integrity" help provide real-world examples in these areas.

Our goal is to make sure that the content and teaching package for this book are of the highest quality possible. We wish to seize this opportunity to gain your trust, and we appreciate any feedback to help us continually improve these materials. We hope that the real beneficiary of all of our work will be well-informed students who appreciate the role of business in society and take advantage of the opportunity to play a significant role in improving our world. In this new edition, we have additional content to help our students understand how our free enterprise system operates and how we fit into the global competitive environment. This course is an opportunity for students to understand how they can create their own success and improve their quality of life.

O.C. Ferrell

Geoffrey A. Hirt

Linda Ferrell

Focused, Exciting, Applicable, Happening

Business Foundations: A Changing World, twelfth edition, offers faculty and students a **focused** resource that is **exciting, applicable,** and **happening!** What sets this learning program apart from the competition? An unrivaled mixture of exciting content and resources blended with application focused text and activities, and fresh topics and examples that show students what is happening in the world of business today!

Our product contains all of the essentials that most students should learn in a semester. *Business Foundations* has, since its inception, delivered a focused presentation of the essential material needed to teach introduction to business. An unrivaled mixture of exciting content and resources, application-focused content and activities, and fresh topics and examples that show students what is happening in the world of business today set this text apart!

Focused!

It's easy for students taking their first steps into business to become overwhelmed. Longer products try to solve this problem by chopping out examples or topics to make ad hoc shorter editions. *Business Foundations* carefully builds just the right mix of coverage and applications to give your students a firm grounding in business principles. Where other products have you sprinting through the semester to get everything in, Ferrell/Hirt/Ferrell allows you the breathing space to explore topics and incorporate other activities that are important to you and your students. The exceptional resources and the *Active Classroom Resource Manual* support you in this effort every step of the way.

Exciting

It's exciting to see students succeed! It's exciting to see more As and Bs in a course without grade inflation. Ferrell/Hirt/Ferrell makes these results possible for your course with its integrated learning package that is proven effective, tailored to each individual student, and easy to use.

Applicable

When students see how content applies to them, their life, their career, and the world around them, they are more engaged in the course. *Business Foundations* helps students maximize their learning efforts by setting clear objectives; delivering interesting cases and examples; focusing on core issues; and providing engaging activities to apply concepts, build skills, and solve problems.

Happening!

Because it isn't tied to the revision cycle of a larger book, *Business Foundations* inherits no outdated or irrelevant examples or coverage. Everything in the twelfth edition reflects the very latest developments in the business world—from the recent recession, high unemployment rates, and the financial instability in Europe to the growth of digital marketing and social networking. In addition, ethics continues to be a key issue, and Ferrell/Hirt/Ferrell use "Business Integrity" boxes to instill in students the importance of ethical conduct in business. To ensure you always know what's happening, join the author-led Facebook group page supporting this text.

New to This Edition

As always, when revising this material for the current edition, all examples, figures, and statistics have been updated to incorporate any recent developments that affect the world of business. Additionally, content was updated to ensure the most pertinent topical coverage is provided. We now provide bonus chapters in the text—Bonus Chapter A, The Legal and Regulatory Environment, and Bonus Chapter B, Personal and Financial Planning—to meet market demands. In addition, we have added a new online Appendix C, which provides the basics of risk management. Both insurable and noninsurable risk are covered in this appendix.

Here are the highlights for each chapter:

Chapter 1: The Dynamics of Business and Economics

- New boxed features describing real-world business issues
- Updated unemployment statistics
- New chart on online retailing

Chapter 2: Business Ethics and Social Responsibility

- New boxed features describing issues in business ethics and social responsibility
- New data on global trust in different industries
- New examples about ethical issues in the sharing economy
- New content about aggressive financial or business objectives
- New example of a bribery scandal

Chapter 3: Business in a Borderless World

- New boxed features describing issues in international business
- Updated list of top 10 countries with which the U.S. has trade deficits/surpluses
- New content on U.S. aluminum and steel tariffs
- Updated Euro Zone details
- New details on the EU's General Data Protection Regulation (GDPR)

Chapter 4: Options for Organizing Business

- New boxed features describing real-world business issues
- New chart of world's biggest dividend payers
- Updated table of America's largest private companies

Chapter 5: Small Business, Entrepreneurship, and Franchising

- New boxed features describing current business issues
- Examples of innovative small businesses
- New information on artificial intelligence
- Updated table of the fastest growing franchises
- Updated table of the most business friendly states

Chapter 6: The Nature of Management

- New boxed features describing current business issues
- New content about business models

- New table of compensation packages of CEOs
- New content on gender equality
- New *See for Yourself Videocase*—JCF Fitness

Chapter 7: Organization, Teamwork, and Communication

- New boxed features describing current business issues
- New examples of organizational culture
- New content on artificial intelligence
- New stats on email usage in the workplace
- New *See for Yourself Videocase*—Freshii

Chapter 8: Managing Operations and Supply Chains

- New boxed features describing current business operational issues
- New content on marketing research and artificial intelligence
- New section on blockchain technology
- New content on drone technology
- Extensive overhaul of Managing the Supply Chain section

Chapter 9: Motivating the Workforce

- New boxed features describing current business issues
- New examples of organizational culture

Chapter 10: Managing Human Resources

- New boxed features describing current HR issues
- Updated common job interview questions
- New content on wage gap
- New example of how soft benefits inspire loyalty

Chapter 11: Customer-Driven Marketing

- New boxed features describing current marketing issues
- New content on marketing orientation
- New content on supply chain management
- New content on marketing analytics dashboards

Chapter 12: Dimensions of Marketing Strategy

- New boxed features describing current marketing issues
- Logistics added as key term
- New definition for physical distribution key term
- New *See for Yourself Videocase*—Zappos

Chapter 13: Digital Marketing and Social Media

- New boxed features describing current digital marketing issues
- New stats on social media use by platform
- New stats on mobile app activities

Chapter 14: Accounting and Financial Statements

- New boxed features describing current accounting issues
- Updated rankings of accounting firms in the U.S.
- Updated financial information for Microsoft

Chapter 15: Money and the Financial System

- New boxed features describing current financial issues
- Updated life expectancy of money
- Updated cost to produce coins
- New content on cryptocurrency
- New *See for Yourself Videocase*—Kiva

Chapter 16: Financial Management and Securities Markets

- New boxed features describing current financial issues
- Updated short-term investment possibilities
- Updated U.S. corporate bond quotes
- New content on electronic markets

Acknowledgments

The twelfth edition of *Business Foundations: A Changing World* would not have been possible without the commitment, dedication, and patience of Jennifer Sawayda and Kelsey Reddick. Kelsey Reddick provided oversight for editing text content, and Jennifer Sawayda developed the cases, boxes, and the supplements. Anke Weekes, Executive Brand Manager, provided leadership and creativity in planning and implementing all aspects of the twelfth edition. Haley Burmeister, Product Developer, did an outstanding job of coordinating all aspects of the development and production process. Kathryn Wright was the Content Project Manager. Bruce Gin managed the technical aspects of Connect. Others important in this edition include Gabe Fedota (Marketing Manager) and Jessica Cuevas (Designer). Michael Hartline developed the Personal Career Plan in Appendix B. Vickie Bajtelsmit developed Bonus Chapter B on personal financial planning. Eric Sandberg of Interactive Learning assisted in developing the interactive exercises. Many others have assisted us with their helpful comments, recommendations, and support throughout this and previous editions. Thank you for all of your insight and feedback. We'd like to express our sincere thanks to the reviewers who helped us shape the twelfth edition. Your time and thoughtful feedback has helped us greatly make this another great revision:

Michael Bento
Owens Community College

Patty Boyle
Lane Community College

Dennis Brode
Sinclair Community College

Angela Casler
California State University

Steven M. Dunphy
Indiana University Northwest

Terri Gonzalez-Kreisman
Delgado Community College

Chad Grooms
Gateway Community and Technical College

Ivan Franklin Harber Jr.
Indian River State College

Dan Jones
Ball State University

Stephen Konrad
Portland State University

Hui Pate
Skyline College

Daniel Pfaltzgraf
University of Toledo

Linda L. Ridley
CUNY Hostos Community College

Michael Rose
Butler Community College

Amanda Stocklein
State Fair Community College

Rhonda K. Thomas
Butler Community College

Bruce Yuille
Mid Michigan Community College

Brenda Anthony, *Tallahassee Community College*

NaRita Gail Anderson, *University of Central Oklahoma*

Phyllis Alderdice, *Jefferson Community College*

Vondra Armstrong, *Pulaski Tech College*

John Bajkowski, *American Association of Individual Investors*

Gene Baker, *University of North Florida*

Lia Barone, *Norwalk Community College*

Ellen Benowitz, *Mercer County Community College*

Stephanie Bibb, *Chicago State University*

Gene Blackmun, *Rio Hondo College*

Susan Blumen, *Montgomery College*

Barbara Boyington, *Monmouth–Ocean County Small Business Development Center*

Suzanne Bradford, *Angelina College*

Alka Bramhandkar, *Ithaca College*

Dennis Brode, *Sinclair Community College*

Harvey S. Bronstein, *Oakland Community College*

Colin Brooks, *University of New Orleans*

Eric Brooks, *Orange County Community College*

Nicky Buenger, *Texas A&M University*

Anthony Buono, *Bentley College*

Tricia Burns, *Boise State University*

Diana Carmel, *Golden West College*

William Chittenden, *Texas State University*

Michael Cicero, *Highline Community College*

Margaret Clark, *Cincinnati State Tech & Community College*

Mark Lee Clark, *Collin College*

Debbie Collins, *Anne Arundel Community College–Arnold*

Karen Collins, *Lehigh University*

Katherine Conway, *Borough of Manhattan Community College*

Rex Cutshall, *Indiana University*

Dana D'Angelo, *Drexel University*

Laurie Dahlin, *Worcester State College*

Deshaun H. Davis, *Northern Virginia Community College*

Yalonda Ross Davis, *Grand Valley State University*

Peter Dawson, *Collin County Community College–Plano*

John DeNisco, *Buffalo State College*

Tom Diamante, *Corporate Consulting Associates, Inc.*

Joyce Domke, *DePaul University*

Glenn Doolittle, *Santa Ana College*

Michael Drafke, *College of DuPage*

John Eagan, *Erie Community College/City Campus SUNY*

Glenda Eckert, *Oklahoma State University*

Thomas Enerva, *University of Maine–Fort Kent*

Robert Ericksen, *Business Growth Center*

Donna Everett, *Santa Rosa Junior College*

Joe Farinella, *University of North Carolina–Wilmington*

Bob Farris, *Mt. San Antonio College*

Gil Feiertag, *Columbus State Community College*

James Ferrell, *R. G. Taylor, P.C.*

Cheryl Fetterman, *Cape Fear Community College*

Art Fischer, *Pittsburg State University*

Jackie Flom, *University of Toledo*

Anthony D. Fontes III, *Bunker Hill Community College*

Jennifer Friestad, *Anoka–Ramsey Community College*

Chris Gilbert, *Tacoma Community College/University of Washington*

Ross Gittell, *University of New Hampshire*

Connie Golden, *Lakeland Community College*

Terri Gonzales-Kreisman, *Phoenix College*

Kris Gossett, *Ivy Tech Community College of Indiana*

Carol Gottuso, *Metropolitan Community College*

Bob Grau, *Cuyahoga Community College–Western Campus*

Gary Grau, *Northeast State Tech Community College*

Jack K. Gray, *Attorney-at-Law, Houston, Texas*

Catherine Green, *University of Memphis*

Claudia Green, *Pace University*

Maurice P. Greene, *Monroe College*

Phil Greenwood, *University of Wisconsin–Madison*

David Gribbin, *East Georgia College*

Selina Andrea Griswold, *University of Toledo*

John P. Guess, *Delgado Community College*

Peggy Hager, *Winthrop University*

Michael Hartline, *Florida State University*

Paul Harvey, *University of New Hampshire*

Neil Herndon, *University of Missouri*

James Hoffman, *Borough of Manhattan Community College*

MaryAnne Holcomb, *Antelope Valley College*

Timothy D. Hovet, *Lane Community College*

Joseph Hrebenak, *Community College of Allegheny County–Allegheny Campus*

Stephen Huntley, *Florida Community College*

Rebecca Hurtz, *State Farm Insurance Co.*

Donald C. Hurwitz, *Austin Community College*

Scott Inks, *Ball State University*

Steven Jennings, *Highland Community College*

Carol Jones, *Cuyahoga Community College–Eastern Campus*

Sandra Kana, *Mid-Michigan Community College*
Norm Karl, *Johnson County Community College*
Janice Karlan, *LaGuardia Community College*
Eileen Kearney, *Montgomery County Community College*
Craig Kelley, *California State University–Sacramento*
Susan Kendall, *Arapahoe Community College*
Ina Midkiff Kennedy, *Austin Community College*
Kathleen Kerstetter, *Kalamazoo Valley Community College*
Arbrie King, *Baton Rouge Community College*
John Knappenberger, *Mesa State College*
Gail Knell, *Cape Cod Community College*
Anthony Koh, *University of Toledo*
Regina Korossy, *Pepperdine University*
Velvet Landingham, *Kent State University–Geauga*
Jeffrey Lavake, *University of Wisconsin–Oshkosh*
Daniel LeClair, *AACSB*
Chad T. Lewis, *Everett Community College*
Richard Lewis, *East Texas Baptist College*
Corinn Linton, *Valencia Community College*
Corrine Livesay, *Mississippi College*
Thomas Lloyd, *Westmoreland Community College*
Terry Loe, *Kennerow University*
Terry Lowe, *Illinois State University*
Kent Lutz, *University of Cincinnati*
Scott Lyman, *Winthrop University*
Dorinda Lynn, *Pensacola Junior College*
Isabelle Maignan, *ING*
Larry Martin, *Community College of Southern Nevada–West Charles*
Therese Maskulka, *Youngstown State University*
Theresa Mastrianni, *Kingsborough Community College*
Kristina Mazurak, *Albertson College of Idaho*
Debbie Thorne McAlister, *Texas State University–San Marcos*
Noel McDeon, *Florida Community College*
John McDonough, *Menlo College*
Tom McInish, *University of Memphis*
Mark McLean, *Delgado Community College*
Chris Mcnamara, *Fingers Lake Community College*
Kimberly Mencken, *Baylor University*
Mary Meredith, *University of Louisiana at Lafayette*
Michelle Meyer, *Joliet Junior College*
George Milne, *University of Massachusetts–Amherst*
Daniel Montez, *South Texas College*
Glynna Morse, *Augusta College*
Suzanne Murray, *Piedmont Technical College*

Stephanie Narvell, *Wilmington College–New Castle*
Fred Nerone, *International College of Naples*
Laura Nicholson, *Northern Oklahoma College*
Stef Nicovich, *Lynchburg College*
Michael Nugent, *SUNY–Stony Brook University New York*
Mark Nygren, *Brigham Young University–Idaho*
Lauren Paisley, *Genesee Community College*
James Patterson, *Paradise Valley Community College*
Wes Payne, *Southwest Tennessee Community College*
Dyan Pease, *Sacramento City College*
Constantine G. Petrides, *Borough of Manhattan Community College*
John Pharr, *Cedar Valley College*
Shirley Polejewski, *University of St. Thomas*
Daniel Powroznik, *Chesapeake College*
Krista Price, *Heald College*
Larry Prober, *Rider University*
Vincent Quan, *Fashion Institute Technology*
Michael Quinn, *Penn State University*
Stephen Pruitt, *University of Missouri–Kansas City*
Victoria Rabb, *College of the Desert*
Gregory J. Rapp, *Portland Community College*
Tom Reading, *Ivy Tech State College*
Delores Reha, *Fullerton College*
David Reiman, *Monroe County Community College*
Susan Roach, *Georgia Southern University*
Dave Robinson, *University of California–Berkeley*
Carol Rowey, *Surry Community College*
Marsha Rule, *Florida Public Utilities Commission*
Carol A. Rustad, *Sylvan Learning*
Cyndy Ruszkowski, *Illinois State University*
Martin St. John, *Westmoreland Community College*
Don Sandlin, *East Los Angeles College*
Nick Sarantakes, *Austin Community College*
Andy Saucedo, *Dona Ana Community College–Las Cruces*
Dana Schubert, *Colorado Springs Zoo*
Marianne Sebok, *Community College of Southern Nevada–West Charles*
Jeffery L. Seglin, *Seglin Associates*
Daniel Sherrell, *University of Memphis*
Morgan Shepherd, *University of Colorado Elaine Simmons, Guilford Technical Community College*
Greg Simpson, *Blinn College*
Nicholas Siropolis, *Cuyahoga Community College*
Robyn Smith, *Pouder Valley Hospital*
Kurt Stanberry, *University of Houston Downtown*
Cheryl Stansfield, *North Hennepin Community College*

Ron Stolle, *Kent State University–Kent*
Edith Strickland, *Tallahassee Community College*
Jeff Strom, *Virginia Western Community College*
Lisa Strusowski, *Tallahassee Community College*
Scott Taylor, *Moberly Area Community College*
Wayne Taylor, *Trinity Valley Community College*
Ray Tewell, *American River College*
Rodney Thirion, *Pikes Peak Community College*
Evelyn Thrasher, *University of
    Massachusetts–Dartmouth*
Steve Tilley, *Gainesville College*
Amy Thomas, *Roger Williams University*
Kristin Trask, *Butler Community College*
Allen D. Truell, *Ball State University*

George Valcho, *Bossier Parish Community College*
Ted Valvoda, *Lakeland Community College*
Gunnar Voltz, *Northern Arizona University–Flagstaff*
Sue Vondram, *Loyola University*
Elizabeth Wark, *Springfield College*
Emma Watson, *Arizona State University–West*
Ruth White, *Bowling Green State University*
Elisabeth Wicker, *Bossier Parish Community College*
Frederik Williams, *North Texas State University*
Richard Williams, *Santa Clara University*
Pat Wright, *University of South Carolina*
Lawrence Yax, *Pensacola Junior College–Warrington*
Bruce Yuille, *Cornell University–Ithaca*

Business in a Changing World

PART 1

Business in a Changing World

1 The Dynamics of Business and Economics

©Jagadeesh Nv/EPA/REX/Shutterstock

Chapter Outline

Learning Objectives

After reading this chapter, you will be able to:

LO 1-1 Define basic concepts such as business, product, profit, and economics.

LO 1-2 Identify the main participants and activities of business.

LO 1-3 Explain why studying business is important.

LO 1-4 Compare the four types of economic systems.

LO 1-5 Describe the role of supply, demand, and competition in a free-enterprise system.

LO 1-6 Specify why and how the health of the economy is measured.

LO 1-7 Outline the evolution of the American economy.

LO 1-8 Explain the role of the entrepreneur in the economy.

LO 1-9 Evaluate a small-business owner's situation and propose a course of action.

Enter the World of Business

Warren Buffet: The Oracle of Omaha

Warren Buffett did not change his life plans when he was rejected by Harvard. An avid reader with a photographic memory, he persevered. He researched other universities and discovered that Benjamin Graham, the author of the *The Intelligent Investor,* was a professor at Columbia University. He immediately enrolled there, was accepted, and became the star pupil and eventual partner of the famous Graham. After graduating, Buffett started a lucrative career, taking ownership in companies that he believed would do well in business.

Buffett's company, Berkshire Hathaway, has become a conglomerate with ownership in well-performing companies, including Geico, Heinz, Benjamin Moore, and See's Candies. Because Berkshire Hathaway owns companies in so many different industries, it reduces the risk that the failure of any one industry will significantly affect the company. Most importantly, Buffett believes in these firms and their value, a strategy that has made him one of the 10 richest people in the world.

Leading such a variety of companies comes with challenges as well. Companies require leaders who know the business and have specialized expertise—something nearly impossible for one person to do with so many different business areas. For this reason, Buffett depends on his managers to lead the various companies. He believes giving his managers autonomy allows them to achieve their highest performance. He wants his managers to "own" their job. By hiring knowledgeable managers and empowering them to run the companies as they believe best, Buffett is able to lead his vast business conglomerate successfully.

Buffett's organizational leadership philosophy includes focusing on the business. At his headquarters, 25 people run the organization. He encourages entrepreneurs "to focus on the business and not growing a large staff." Buffett has earned the moniker "Oracle of Omaha" because he has lived there most of his life and many investors follow his advice and decisions.[1]

Define basic concepts such as business, product, profit, and economics.

Introduction

We begin our study of business in this chapter by examining the fundamentals of business and economics. First, we introduce the nature of business, including its goals, activities, and participants. Next, we describe the basics of economics and apply them to the U.S. economy. Finally, we establish a framework for studying business in this text.

The Nature of Business

business
individuals or organizations who try to earn a profit by providing products that satisfy people's needs.

product
a good or service with tangible and intangible characteristics that provide satisfaction and benefits.

A **business** tries to earn a profit by providing products that satisfy people's needs. The outcomes of its efforts are **products** that have both tangible and intangible characteristics that provide satisfaction and benefits. When you purchase a product, you are buying the benefits and satisfaction you think the product will provide. A Subway sandwich, for example, may be purchased to satisfy hunger, while a Honda Accord may be purchased to satisfy the need for transportation and the desire to present a certain image.

Most people associate the word *product* with tangible goods—an automobile, smartphone, jeans, or some other tangible item. However, a product can also be a service, which occurs when people or machines provide or process something of value to customers. Dry cleaning, a checkup by a doctor, a movie or sports event—these are examples of services. An Uber ride satisfies the need for transportation and is therefore a service. A product can also be an idea. Accountants and attorneys, for example, provide ideas for solving problems.

The Goal of Business

profit
the difference between what it costs to make and sell a product and what a customer pays for it.

nonprofit organizations
organizations that may provide goods or services but do not have the fundamental purpose of earning profits.

The primary goal of all businesses is to earn a **profit**, the difference between what it costs to make and sell a product and what a customer pays for it. In addition, a business has to pay for all expenses necessary to operate. If a company spends $8 to produce, finance, promote, and distribute a product that it sells for $10, the business earns a profit of $2 on each product sold. Businesses have the right to keep and use their profits as they choose—within legal limits—because profit is the reward for their efforts and for the risks they take in providing products. Earning profits contributes to society by creating resources that support our social institutions and government. Businesses that create profits, pay taxes, and create jobs are the foundation of our economy. In addition, profits must be earned in a responsible manner. Not all organizations are businesses, however. **Nonprofit organizations**—such as National Public Radio (NPR), Habitat for Humanity, and other charities and social causes—do not have the fundamental purpose of earning profits, although they may provide goods or services and engage in fund-raising. They also utilize skills related to management, marketing, and finance. Profits earned by businesses support nonprofit organizations through donations from employees.

To earn a profit, a person or organization needs management skills to plan, organize, and control the activities of the business and to find and develop employees so that it can make products consumers will buy. A business also needs marketing expertise to learn what products consumers need and want and to develop, manufacture, price, promote, and distribute those products. Additionally, a business needs financial resources and skills to fund, maintain, and expand its operations. A business must cover the cost of labor, operate facilities, pay taxes, and provide management. Other challenges for businesspeople include abiding by laws and government regulations, and adapting to economic, technological, political, and social changes. Even nonprofit organizations engage in management, marketing, and finance activities to help reach their goals.

To achieve and maintain profitability, businesses have found that they must produce quality products, operate efficiently, and be socially responsible and ethical in dealing with customers, employees, investors, government regulators, and the community. Because these groups have a stake in the success and outcomes of a business, they are sometimes called **stakeholders.** Many businesses, for example, are concerned about how the production and distribution of their products affect the environment. New fuel requirements are forcing automakers to invest in smaller, lightweight cars. Electric vehicles may be a solution, but only about 1 percent of new car sales are plug-in-electric.[2] Other businesses are concerned with promoting science, engineering, and mathematics careers among women. Traditionally, these careers have been male dominated. A global survey found that when the number of men and women were evenly matched, the team was 23 percent more likely to have an increase in profit over teams dominated by one gender.[3] Nonprofit organizations, such as the American Red Cross, use business activities to support natural-disaster victims, relief efforts, and a national blood supply.

stakeholders
groups that have a stake in the success and outcomes of a business.

The People and Activities of Business

Figure 1.1 shows the people and activities involved in business. At the center of the figure are owners, employees, and customers; the outer circle includes the primary business activities—management, marketing, and finance. Owners have to put up resources—money or credit—to start a business. Employees are responsible for the work that goes on within a business. Owners can manage the business themselves or hire employees to accomplish this task. The president and chief executive officer (CEO) of Procter & Gamble, David S. Taylor, does not own P&G but is an employee who is responsible for managing all the other employees in a way that earns a profit for investors, who are the real owners. Finally, and most importantly, a business's major role is to satisfy the customers who buy its goods or services. Note also that forces beyond an organization's control—such as legal and regulatory forces, the economy, competition, technology, the political environment, and ethical and social concerns—all have an impact on the daily operations of businesses. You will learn more about these participants in business activities throughout this book. Next, we will examine the major activities of business.

LO 1-2

Identify the main participants and activities of business.

FIGURE 1.1
Overview of the Business
World

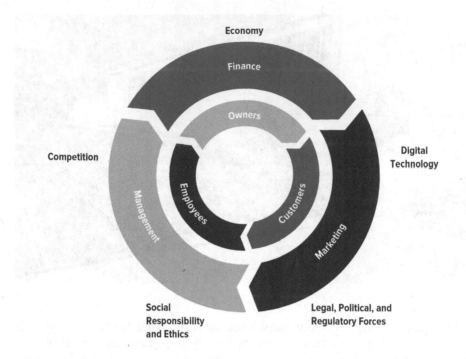

Management. Notice that in Figure 1.1, management and employees are in the same segment of the circle. This is because management involves developing plans, coordinating employees' actions to achieve the firm's goals, organizing people to work efficiently, and motivating them to achieve the business's goals. Management involves the functions of planning, organizing, leading, and controlling. Effective managers who are skilled in these functions display effective leadership, decision making, and delegation of work tasks. Management is also concerned with acquiring, developing, and using resources (including people) effectively and efficiently.

Operations is another element of management. Managers must oversee the firm's operations to ensure that resources are successfully transformed into goods and services. Although most people associate operations with the development of goods, operations management applies just as strongly to services. Managers at the Ritz-Carlton, for instance, are concerned with transforming resources such as employee actions and hotel amenities into a quality customer service experience. In essence, managers plan, organize, staff, and control the tasks required to carry out the work of the company or nonprofit organization. We take a closer look at management activities in Parts 3 and 4 of this text.

Marketing. Marketing and customers are in the same segment of Figure 1.1 because the focus of all marketing activities is satisfying customers. Marketing includes all the activities designed to provide goods and services that satisfy consumers' needs and wants. Marketers gather information and conduct research to determine what customers want. Using information gathered from marketing research, marketers plan and develop products and make decisions about how much to charge for their products and when and where to make them available. They also analyze the marketing environment to understand changes in competition and consumers. The retail environment is changing based on competition from online retailing such as Amazon. This has caused some retail stores and malls to close.[4] Marketing focuses on the four P's—product, price, place (or distribution), and promotion—also known as the marketing mix. Product management involves such key management decisions as product adoption, development, branding, and product positioning. Selecting the right price for the product is essential to the organization as it relates directly to profitability. Distribution is an important management concern because it involves making sure products are available to consumers in the right place at the right time. Supply chain management involves purchasing and logistics as well as operations to

M&Ms uses humor in its advertising to promote the chocolate candies.
©Kisler Creations/Alamy Stock Photo

coordinate suppliers, producers, and distributors to create value for customers. Marketers use promotion—advertising, personal selling, sales promotion (coupons, games, sweepstakes, movie tie-ins), and publicity—to communicate the benefits and advantages of their products to consumers and to increase sales. We will examine marketing activities in Part 5 of this text.

Finance.   Owners and finance are in the same part of Figure 1.1 because, although management and marketing have to deal with financial considerations, it is the primary responsibility of the owners to provide financial resources for the operation of the business. Moreover, the owners have the most to lose if the business fails to make a profit. Finance refers to all activities concerned with obtaining money and using it effectively. People who work as accountants, stockbrokers, investment advisors, or bankers are all part of the financial world. Owners sometimes have to borrow money from banks to get started or attract additional investors who become partners or stockholders. Owners of small businesses in particular often rely on bank loans for funding. Part 6 of this text discusses financial management.

Why Study Business?

LO 1-3

Studying business can help you develop skills and acquire knowledge to prepare for your future career, regardless of whether you plan to work for a multinational *Fortune* 500 firm, start your own business, work for a government agency, or manage or volunteer at a nonprofit organization. The field of business offers a variety of interesting and challenging career opportunities throughout the world, such as marketing, human resources management, information technology, finance, production, accounting, data analytics, and many more.

Explain why studying business is important.

Studying business can also help you better understand the many business activities that are necessary to provide satisfying goods and services. Most businesses charge a reasonable price for their products to ensure that they cover their production costs, pay their employees, provide their owners with a return on their investment, and perhaps give something back to their local communities and societies. Habitat for Humanity is an international nonprofit organization building housing for those who cannot afford simple, decent housing. Habitat operates like a business relying on volunteer labor and

Many companies engage in socially responsible behavior to give back to their communities. Bank of America partners with Habitat for Humanity to build homes for disadvantaged families.

©asiseeit/Getty Images

offers no-interest mortgages for repayment. Habitat ReStore is a retail unit that sells new and used building materials that are donated. The Home Depot Foundation has provided grants to remodel and renovate homes of U.S. military veterans.[5] Thus, learning about business can help you become a well-informed consumer and member of society.

Business activities help generate the profits that are essential not only to individual businesses and local economies, but also to the health of the global economy. Without profits, businesses find it difficult, if not impossible, to buy more raw materials, hire more employees, attract more capital, and create additional products that, in turn, make more profits and fuel the world economy. Understanding how our free-enterprise economic system allocates resources and provides incentives for industry and the workplace is important to everyone.

The Economic Foundations of Business

LO 1-4

Compare the four types of economic systems.

economics
the study of how resources are distributed for the production of goods and services within a social system.

natural resources
land, forests, minerals, water, and other things that are not made by people.

human resources (labor)
the physical and mental abilities that people use to produce goods and services.

financial resources (capital)
the funds used to acquire the natural and human resources needed to provide products.

economic system
a description of how a particular society distributes its resources to produce goods and services.

It is useful to explore the economic environment in which business is conducted. In this section, we examine economic systems, the free-enterprise system, the concepts of supply and demand, and the role of competition. These concepts play important roles in determining how businesses operate in a particular society.

Economics is the study of how resources are distributed for the production of goods and services within a social system. You are already familiar with the types of resources available. Land, forests, minerals, water, and other things that are not made by people are **natural resources. Human resources,** or **labor,** refer to the physical and mental abilities that people use to produce goods and services. **Financial resources,** or **capital,** are the funds used to acquire the natural and human resources needed to provide products. These resources are related to the *factors of production,* consisting of land, labor, capital, and enterprise used to produce goods and services. The firm can also have intangible resources such as a good reputation for quality products or being socially responsible. The goal is to turn the factors of production and intangible resources into a competitive advantage.

Economic Systems

An **economic system** describes how a particular society distributes its resources to produce goods and services. A central issue of economics is how to fulfill an unlimited demand for goods and services in a world with a limited supply of resources. Different economic systems attempt to resolve this central issue in numerous ways, as we shall see.

Although economic systems handle the distribution of resources in different ways, all economic systems must address three important issues:

1. What goods and services, and how much of each, will satisfy consumers' needs?
2. How will goods and services be produced, who will produce them, and with what resources will they be produced?
3. How are the goods and services to be distributed to consumers?

	Communism	Socialism	Capitalism
Business ownership	Most businesses are owned and operated by the government.	The government owns and operates some basic industries; individuals own small businesses.	Individuals own and operate all businesses.
Competition	Government controls competition and the economy.	Restricted in basic industries; encouraged in small business.	Encouraged by market forces and government regulations.
Profits	Excess income goes to the government. The government supports social and economic institutions.	Profits earned by small businesses may be reinvested in the business; profits from government-owned industries go to the government.	Individuals and businesses are free to keep profits after paying taxes.
Product availability and price	Consumers have a limited choice of goods and services; prices are usually high.	Consumers have some choice of goods and services; prices are determined by supply and demand.	Consumers have a wide choice of goods and services; prices are determined by supply and demand.
Employment options	Little choice in choosing a career; most people work for government-owned industries or farms.	More choice of careers; many people work in government jobs.	Unlimited choice of careers.

TABLE 1.1
Comparison of Communism, Socialism, and Capitalism

communism
first described by Karl Marx as a society in which the people, without regard to class, own all the nation's resources.

Communism, socialism, and capitalism, the basic economic systems found in the world today (Table 1.1), have fundamental differences in the way they address these issues. The factors of production in command economies are controlled by government planning. In many cases, the government owns or controls the production of goods and services. Communism and socialism are, therefore, considered command economies.

Communism. Karl Marx (1818–1883) first described communism as a society in which the people, without regard to class, own all the nation's resources. In his ideal political-economic system, everyone contributes according to ability and receives benefits according to need. In a communist economy, the people (through the government) own and operate all businesses and factors of production. Central government planning determines what goods and services satisfy citizens' needs, how the goods and services are produced, and how they are distributed. However, no true communist economy exists today that satisfies Marx's ideal.

On paper, communism appears to be efficient and equitable, producing less of a gap between rich and poor. In practice, however, communist economies have been marked by low standards of living, critical shortages of consumer goods, high prices, corruption, and little freedom. Russia, Poland, Hungary, and other eastern European nations have

Karl Marx, the founder of communism, described a society in which the people own all of the nation's resources and contribute according to their ability.

©Everett Historical/Shutterstock

socialism
an economic system in which the government owns and operates basic industries but individuals own most businesses.

capitalism (free enterprise)
an economic system in which individuals own and operate the majority of businesses that provide goods and services.

free-market system
pure capitalism, in which all economic decisions are made without government intervention.

turned away from communism and toward economic systems governed by supply and demand rather than by central planning. However, their experiments with alternative economic systems have been fraught with difficulty and hardship. Countries such as Venezuela have tried to incorporate communist economic principles without success. Even Cuba is experiencing changes to its predominately communist system. Massive government layoffs required many Cubans to turn toward the private sector, opening up more opportunities for entrepreneurship. The U.S. government has reestablished diplomatic relations with Cuba. Americans have more opportunities to visit Cuba than they have had for the past 50 years. Similarly, China has become the first communist country to make strong economic gains by adopting capitalist approaches to business. Economic prosperity has advanced in China with the government claiming to ensure market openness, equality, and fairness through state capitalism.[6] As a result of economic challenges, communism is declining and its future as an economic system is uncertain.

Socialism.   Socialism is an economic system in which the government owns and operates basic industries—postal service, telephone, utilities, transportation, health care, banking, and some manufacturing—but individuals own most businesses. For example, in France the postal service industry La Poste is fully owned by the French government and makes a profit. Central planning determines what basic goods and services are produced, how they are produced, and how they are distributed. Individuals and small businesses provide other goods and services based on consumer demand and the availability of resources. Citizens are dependent on the government for many goods and services.

Most socialist nations, such as Norway, India, and Israel, are democratic and recognize basic individual freedoms. Citizens can vote for political offices, but central government planners usually make many decisions about what is best for the nation. People are free to go into the occupation of their choice, but they often work in government-operated organizations. Socialists believe their system permits a higher standard of living than other economic systems, but the difference often applies to the nation as a whole rather than to its individual citizens. Socialist economies profess egalitarianism—equal distribution of income and social services. They believe their economies are more stable than those of other nations. Although this may be true, taxes and unemployment are generally higher in socialist countries. However, countries like Denmark have a high standard of living and they rate high in being happy.

Capitalism.   Capitalism, or free enterprise, is an economic system in which individuals own and operate the majority of businesses that provide goods and services. Competition, supply, and demand determine which goods and services are produced, how they are produced, and how they are distributed. The United States, Canada, Japan, and Australia are examples of economic systems based on capitalism.

There are two forms of capitalism: pure capitalism and modified capitalism. In pure capitalism, also called a **free-market system,** all economic decisions are made without government intervention.

Cuba and the United States restored diplomatic relations, and as a result, there are more opportunities to visit and do business with Cuba.

©Maurizio De Mattei/Shutterstock

This economic system was first described by Adam Smith in *The Wealth of Nations* (1776). Smith, often called the father of capitalism, believed that the "invisible hand of competition" best regulates the economy. He argued that competition should determine what goods and services people need. Smith's system is also called *laissez-faire* ("let it be") *capitalism* because the government does not interfere in business.

Modified capitalism differs from pure capitalism in that the government intervenes and regulates business to some extent. One of the ways in which the United States and Canadian governments regulate business is through laws. Laws such as the Federal Trade Commission Act, which created the Federal Trade Commission to enforce antitrust laws, illustrate the importance of the government's role in the economy. In the United States, states have leeway to regulate business. For example, the state of Washington requires employers to provide paid sick leave, and a number of states have legalized cannabis.[7]

Mixed Economies.   No country practices a pure form of communism, socialism, or capitalism, although most tend to favor one system over the others. Most nations operate as **mixed economies,** which have elements from more than one economic system. In socialist Sweden, most businesses are owned and operated by private individuals. In capitalist United States, an independent federal agency operates the postal service and another independent agency operates the Tennessee Valley Authority, an electric utility. In Great Britain and Mexico, the governments are attempting to sell many state-run businesses to private individuals and companies. In Germany, the Deutsche Post is privatized and trades on the stock market. In once-communist Russia, Hungary, Poland, and other eastern European nations, capitalist ideas have been implemented, including private ownership of businesses.

Countries such as China and Russia have used state capitalism to advance the economy. State capitalism tries to integrate the powers of the state with the advantages of capitalism. It is led by the government but uses capitalistic tools such as listing state-owned companies on the stock market and embracing globalization.[8] State capitalism includes some of the world's largest companies such as Russia's Gazprom, which has the largest reserves of natural gas. China's ability to make huge investments to the point of creating entirely new industries puts many private industries at a disadvantage.[9]

mixed economies
economies made up of elements from more than one economic system.

The Free-Enterprise System

Many economies—including those of the United States, Canada, and Japan—are based on free enterprise, and many communist and socialist countries, such as China and Russia, are applying more principles of free enterprise to their own economic systems. Free enterprise provides an opportunity for a business to succeed or fail on the basis of market demand. In a free-enterprise system, companies that can efficiently manufacture and sell products that consumers desire will probably succeed. Inefficient businesses and those that sell products that do not offer needed benefits will likely fail as consumers take their business to firms that have more competitive products.

A number of basic individual and business rights must exist for free enterprise to work. These rights are the goals of many countries that have recently embraced free enterprise.

1. Individuals must have the right to own property and to pass this property on to their heirs. This right motivates people to work hard and save to buy property.

2. Individuals and businesses must have the right to earn profits and to use the profits as they wish, within the constraints of their society's laws, principles, and values.

Two entrepreneurs present an idea for a new product. Entrepreneurs are more productive in free-enterprise systems.

©Cultura Creative (RF)/Alamy Stock Photo

demand
the number of goods and services that consumers are willing to buy at different prices at a specific time.

LO 1-5

Describe the role of supply, demand, and competition in a free-enterprise system.

supply
the number of products—goods and services—that businesses are willing to sell at different prices at a specific time.

 connect

▶ Need help understanding **supply and demand?** Visit your Connect ebook video tab for a brief animated explanation.

3. Individuals and businesses must have the right to make decisions that determine the way the business operates. Although there is government regulation, the philosophy in countries like the United States and Australia is to permit maximum freedom within a set of rules of fairness.

4. Individuals must have the right to choose what career to pursue, where to live, what goods and services to purchase, and more. Businesses must have the right to choose where to locate, what goods and services to produce, what resources to use in the production process, and so on.

Without these rights, businesses cannot function effectively because they are not motivated to succeed. Thus, these rights make possible the open exchange of goods and services. In the countries that favor free enterprise, such as the United States, citizens have the freedom to make many decisions about the employment they choose and create their own productivity systems. Many entrepreneurs are more productive in free-enterprise societies because personal and financial incentives are available that can aid in entrepreneurial success. For many entrepreneurs, their work becomes a part of their system of goals, values, and lifestyle. Consider the panelists ("sharks") on the ABC program *Shark Tank* who give entrepreneurs a chance to receive funding to realize their dreams by deciding whether to invest in their projects. They include Barbara Corcoran, who built one of New York's largest real estate companies; Mark Cuban, founder of Broadcast.com and MicroSolutions; and Daymond John, founder of clothing company FUBU, as well as others.[10]

The Forces of Supply and Demand

In the United States and in other free-enterprise systems, the distribution of resources and products is determined by supply and demand. **Demand** is the number of goods and services that consumers are willing to buy at different prices at a specific time. From your own experience, you probably recognize that consumers are usually willing to buy more of an item as its price falls because they want to save money. Consider handmade rugs, for example. Consumers may be willing to buy six rugs at $350 each, four at $500 each, but only two at $650 each. The relationship between the price and the number of rugs consumers are willing to buy can be shown graphically with a *demand curve* (see Figure 1.2).

Supply is the number of products that businesses are willing to sell at different prices at a specific time. In general, because the potential for profits is higher, businesses are willing to supply more of a good or service at higher prices. For example, a company that sells rugs may be willing to sell six at $650 each, four at $500 each, but just two at $350 each. The relationship between the price of rugs and the quantity the company is willing to supply can be shown graphically with a *supply curve* (see Figure 1.2).

In Figure 1.2, the supply and demand curves intersect at the point where supply and demand are equal. The price at which the number of products that businesses are willing to supply equals the amount of products that consumers are willing to buy at

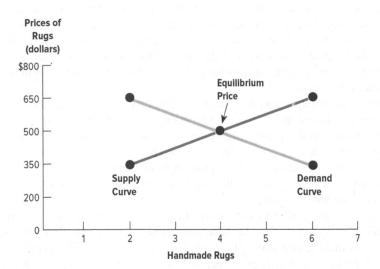

FIGURE 1.2
Equilibrium Price of
Handmade Rugs

a specific point in time is the **equilibrium price.** In our rug example, the company is willing to supply four rugs at $500 each, and consumers are willing to buy four rugs at $500 each. Therefore, $500 is the equilibrium price for a rug at that point in time, and most rug companies will price their rugs at $500. As you might imagine, a business that charges more than $500 (or whatever the current equilibrium price is) for its rugs will not sell as many and might not earn a profit. On the other hand, a business that charges less than $500 accepts a lower profit per rug than could be made at the equilibrium price.

If the cost of making rugs goes up, businesses will not offer as many at the old price. Changing the price alters the supply curve, and a new equilibrium price results. This is an ongoing process, with supply and demand constantly changing in response to changes in economic conditions, availability of resources, and degree of competition. For example, the price of oil can change rapidly and has been between $30 and $113 a barrel over the last eight years. Prices for goods and services vary according to these changes in supply and demand. Supply and demand is the force that drives the distribution of resources (goods and services, labor, and money) in a free-enterprise economy.

Critics of supply and demand say the system does not distribute resources equally. The forces of supply and demand prevent sellers who have to sell at higher prices (because their costs are high) and buyers who cannot afford to buy goods at the equilibrium price from participating in the market. According to critics, the wealthy can afford to buy more than they need, but the poor may be unable to buy enough of what they need to survive.

The Nature of Competition

Competition, the rivalry among businesses for consumers' dollars, is another vital element in free enterprise. According to Adam Smith, competition fosters efficiency and low prices by forcing producers to offer the best products at the most reasonable price; those who fail to do so are not able to stay in business. Thus, competition should improve the quality of the goods and services available and reduce prices. Competition allows for open markets and provides opportunities for both individuals and businesses to successfully compete. Entrepreneurs can discover new technology,

equilibrium price
the price at which the number of products that businesses are willing to supply equals the amount of products that consumers are willing to buy at a specific point in time.

competition
the rivalry among businesses for consumers' dollars.

ways to lower prices, as well as methods for providing better distribution or services. Founder Jeff Bezos of Amazon is a prime example. Amazon was able to offer products online at competitive prices. Today, Amazon competes against such retail giants as Walmart in a number of industries, including cloud computing, entertainment, food, and most consumer products found in retail stores. Bezos is now the richest person in the world.

Within a free-enterprise system, there are four types of competitive environments: pure competition, monopolistic competition, oligopoly, and monopoly.

pure competition
the market structure that exists when there are many small businesses selling one standardized product.

Pure competition exists when there are many small businesses selling one standardized product, such as agricultural commodities like wheat, corn, and cotton. No one business sells enough of the product to influence the product's price. And, because there is no difference in the products, prices are determined solely by the forces of supply and demand.

monopolistic competition
the market structure that exists when there are fewer businesses than in a pure-competition environment and the differences among the goods they sell are small.

Monopolistic competition exists when there are fewer businesses than in a pure-competition environment and the differences among the goods they sell are small. Aspirin, soft drinks, and jeans are examples of such goods. These products differ slightly in packaging, warranty, name, and other characteristics, but all satisfy the same consumer need. Businesses have some power over the price they charge in monopolistic competition because they can make consumers aware of product differences through advertising. Consumers value some features more than others and are often willing to pay higher prices for a product with the features they want. For example, many consumers are willing to pay a higher price for organic fruits and vegetables rather than receive a bargain on nonorganic foods. The same holds true for non-genetically modified foods.

oligopoly
the market structure that exists when there are very few businesses selling a product.

An **oligopoly** exists when there are very few businesses selling a product. In an oligopoly, individual businesses have control over their products' price because each business supplies a large portion of the products sold in the marketplace. Nonetheless, the prices charged by different firms stay fairly close because a price cut or increase by one company will trigger a similar response from another company. In the airline industry, for example, when one airline cuts fares to boost sales, other airlines quickly follow with rate decreases to remain competitive. On the other hand, airlines often raise prices at the same time. Oligopolies exist when it is expensive for new firms to enter the marketplace. Not just anyone can acquire enough financial capital to build an automobile production facility or purchase enough airplanes and related resources to build an airline.

monopoly
the market structure that exists when there is only one business providing a product in a given market.

When there is one business providing a product in a given market, a **monopoly** exists. Utility companies that supply electricity, natural gas, and water are monopolies. The government permits such monopolies because the cost of creating the good or supplying the service is so great that new producers cannot compete for sales. Government-granted monopolies are subject to government-regulated prices. Some monopolies exist because of technological developments that are protected by patent laws. Patent laws grant the developer of new technology a period of time (17 or 20 years) during which no other producer can use the same technology without the agreement of the original developer. The United States granted its first patent in 1790. Now its patent office receives hundreds of thousands of patent applications a year, although China has surpassed the United States in patent applications.[11] This monopoly allows the developer to recover research, development, and production expenses and to earn a reasonable profit. A drug can receive a 17-year patent from the time it is discovered or the chemical is identified. For example, Tamiflu lost its patent, and now the generic version can be made by other firms.

Economic Cycles and Productivity

Expansion and Contraction.   Economies are not stagnant; they expand and contract. **Economic expansion** occurs when an economy is growing and people are spending more money. Their purchases stimulate the production of goods and services, which in turn stimulates employment. The standard of living rises because more people are employed and have money to spend. Rapid expansions of the economy, however, may result in **inflation,** a continuing rise in prices. Inflation can be harmful if individuals' incomes do not increase at the same pace as rising prices, reducing their buying power. The worst case of hyperinflation occurred in Hungary in 1946. At one point, prices were doubling every 15.6 hours. One of the most recent cases of hyperinflation occurred in Zimbabwe.[12] Zimbabwe suffered from hyperinflation so severe that its inflation percentage rate rose into the hundreds of millions. With the elimination of the Zimbabwean dollar and certain price controls, the inflation rate began to decrease, but not before the country's economy was virtually decimated.[13]

Economic contraction occurs when spending declines. Businesses cut back on production and lay off workers, and the economy as a whole slows down. Contractions of the economy lead to **recession**—a decline in production, employment, and income. Recessions are often characterized by rising levels of **unemployment,** which is measured as the percentage of the population that wants to work but is unable to find jobs. Figure 1.3 shows the overall unemployment rate in the civilian labor force over the past 75 years. Rising unemployment levels tend to stifle demand for goods and services, which can have the effect of forcing prices downward, a condition known as *deflation.* Deflation poses a serious economic problem because price decreases could result in consumers delaying purchases. If consumers wait for lower prices, the economy could fall into a recession.

The United States has experienced numerous recessions, the most recent ones occurring in 1990–1991, 2002–2003, and 2008–2011. The most recent recession (or economic slowdown) was caused by the collapse in housing prices and consumers'

economic expansion
the situation that occurs when an economy is growing and people are spending more money; their purchases stimulate the production of goods and services, which in turn stimulates employment.

inflation
a condition characterized by a continuing rise in prices

economic contraction
a slowdown of the economy characterized by a decline in spending and during which businesses cut back on production and lay off workers.

recession
a decline in production, employment, and income.

unemployment
the condition in which a percentage of the population wants to work but is unable to find jobs.

FIGURE 1.3

Annual Average Unemployment Rate, Civilian Labor Force, 16 Years and Over

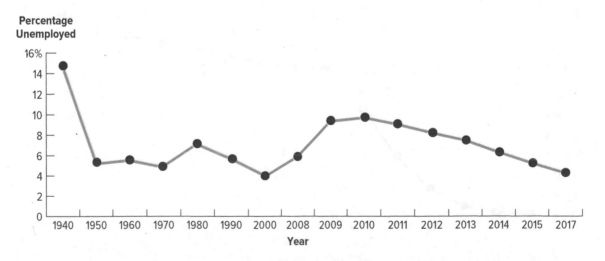

Sources: *Bureau of Labor Statistics,* "Labor Force Statistics from the Current Population Survey," http://data.bls.gov/timeseries/LNS14000000 (accessed March 25, 2018).

inability to stay current on their mortgage and credit card payments. This caused a crisis in the banking industry, with the government bailing out banks to keep them from failing. This in turn caused a slowdown in spending on consumer goods and a decrease in employment. Unemployment reached 10 percent of the labor force. Don't forget that personal consumption makes up almost 70 percent of gross domestic product, so consumer engagement is extremely important for economic activity. A severe recession may turn into a **depression,** in which unemployment is very high, consumer spending is low, and business output is sharply reduced, such as what occurred in the United States in the early 1930s. The most recent recession is often called the Great Recession because it was the longest and most severe economic decline since the Great Depression.

depression
a condition of the economy in which unemployment is very high, consumer spending is low, and business output is sharply reduced.

Economies expand and contract in response to changes in consumer, business, and government spending. War also can affect an economy, sometimes stimulating it (as in the United States during World Wars I and II) and sometimes stifling it (as during the Vietnam, Persian Gulf, and Iraq wars). Although fluctuations in the economy are inevitable and to a certain extent predictable, their effects—inflation and unemployment—disrupt lives and thus governments try to minimize them.

> LO 1-6

Specify why and how the health of the economy is measured.

gross domestic product (GDP)
the sum of all goods and services produced in a country during a year.

Measuring the Economy.   Countries measure the state of their economies to determine whether they are expanding or contracting and whether corrective action is necessary to minimize the fluctuations. One commonly used measure is **gross domestic product (GDP)**—the sum of all goods and services produced in a country during a year. GDP measures only those goods and services made within a country and therefore does not include profits from companies' overseas operations; it does include profits earned by foreign companies within the country being measured. However, it does not take into account the concept of GDP in relation to population (GDP per capita). Figure 1.4 shows the increase in U.S. GDP over several years, while Table 1.2 compares a number of economic statistics for a sampling of countries.

FIGURE 1.4
Growth in U.S. Gross Domestic Product

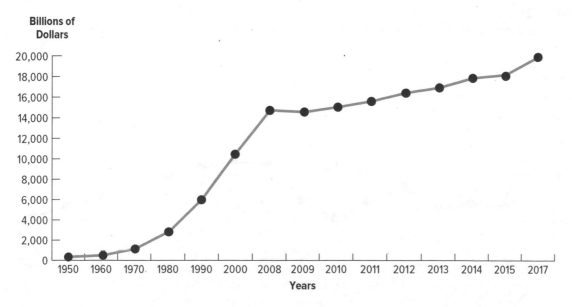

Source: *U.S. Department of Commerce Bureau of Economic Analysis,* "National Economic Accounts," www.bea.gov/national/index.htm#gdp (accessed March 25, 2018).

TABLE 1.2
Economic Indicators of Different Countries

Country	GDP (in billions of dollars)	GDP per Capita	Unemployment Rate (%)	Inflation Rate (%)
Argentina	911	20,700	8.1	26.9
Australia	1,235	49,900	5.6	2.0
Brazil	3,219	15,500	13.1	3.7
Canada	1,764	48,100	6.5	1.6
China	23,120	16,600	4	1.8
France	2,826	43,600	9.5	1.2
Germany	4,150	50,200	3.8	1.6
India	9,447	7,200	8.8	3.8
Israel	315	36,200	4.3	0.2
Japan	5,405	42,700	2.9	0.4
Mexico	2,406	19,500	3.6	5.9
Russia	4,000	27,900	5.5	4.2
South Africa	757	13,400	27.6	5.4
United Kingdom	2,880	43,600	4.4	2.6
United States	19,360	59,500	4.4	2.1

Source: CIA, *The World Fact Book*, https://www.cia.gov/library/publications/the-world-factbook/rankorder/rankorderguide.html (accessed March 25, 2018).

Another important indicator of a nation's economic health is the relationship between its spending and income (from taxes). When a nation spends more than it takes in from taxes, it has a **budget deficit.** In the 1990s, the U.S. government eliminated its long-standing budget deficit by balancing the money spent for social, defense, and other programs with the amount of money taken in from taxes.

In recent years, however, the budget deficit has reemerged and grown to record levels, partly due to defense spending in the aftermath of the terrorist attacks of September 11, 2001. Massive government stimulus spending during the most recent recession also increased the national debt. Because many Americans do not want their taxes increased and Congress has difficulty agreeing on appropriate tax rates, it is difficult to increase taxes and reduce the deficit. Like consumers and businesses, when the government needs money, it borrows from the public, banks, and even foreign investors. The national debt is more than $24 trillion.[14] This figure is especially worrisome because to reduce the debt to a manageable level, the government either has to increase its revenues (raise taxes) or reduce spending on social, defense, and legal programs, neither of which is politically popular. The United States recently enacted tax cuts that are projected to increase the national debt every year over the next 10 years.[15] The national debt figure changes daily and can be seen at the Department of the Treasury, Bureau of the Public Debt, website. Table 1.3 describes some of the other ways we evaluate our nation's economy.

budget deficit
the condition in which a nation spends more than it takes in from taxes.

TABLE 1.3
How Do We Evaluate Our
Nation's Economy?

Unit of Measure	Description
Trade balance	The difference between our exports and our imports. If the balance is negative, as it has been since the mid-1980s, it is called a trade deficit and is generally viewed as unhealthy for our economy.
Consumer Price Index	Measures changes in prices of goods and services purchased for consumption by typical urban households.
Per capita income	Indicates the income level of "average" Americans. Useful in determining how much "average" consumers spend and how much money Americans are earning.
Unemployment rate	Indicates how many working-age Americans are not working who otherwise want to work.
Inflation	Monitors price increases in consumer goods and services over specified periods of time. Used to determine if costs of goods and services are exceeding worker compensation over time.
Worker productivity	The amount of goods and services produced for each hour worked.

LO 1-7

Outline the evolution of the American economy.

standard of living
refers to the level of wealth and material comfort that people have available to them.

The American Economy

As we said previously, the United States is a mixed economy with a foundation based on capitalism. The answers to the three basic economic issues are determined primarily by competition and the forces of supply and demand, although the federal government does intervene in economic decisions to a certain extent. For instance, the federal government exerts oversight over the airline industry to make sure airlines remain economically viable as well as for safety and security purposes.

Standard of living refers to the level of wealth and material comfort that people have available to them. The United States, Germany, Australia, and Norway all have a high standard of living, meaning that most of their citizens are able to afford basic necessities and some degree of comfort. These nations are often characterized by a high GDP per capita. However, a higher GDP per capita does not automatically translate into a higher standard of living. Costs of goods and services are also factors. The European Union and Japan, for instance, tend to have higher costs of living than in the United States. Higher prices mean that it costs more to obtain a certain level of comfort than it does in other countries. Countries with low standards of living are usually characterized by poverty, higher unemployment, and lower education rates. To understand the current state of the American economy and its effect on business practices, it is helpful to examine its history and the roles of the entrepreneur and the government.

The Importance of the American Economy

open economy
an economy in which economic activities occur between the country and the international community.

The American economy is an **open economy,** or an economy in which economic activities occur between the country and the international community. As an open economy, the United States is a major player in international trade. Open economies tend to grow faster than economies that do not engage in international trade. This is because international trade is positively related to efficiency and productivity. Companies in the United States have greater access to a wider range of resources and knowledge, including technology. In today's global environment, the ability to harness technology is critical toward increased innovation.[16] In contrast, research indicates a negative relationship between regulatory actions and innovation in firms, suggesting that too much regulation hinders business activities and their contribution to the American economy.[17]

When looking at the American economy, growth in GDP and jobs are the two primary factors economists consider. A positive relationship exists between a country's employment rate and economic growth. A nation's output depends on the amount of labor used in the production process, so there is also a positive correlation between output and employment. In general, as the labor force and productivity increase, so does GDP. Profitable companies tend to hire more workers than those that are unprofitable. Therefore, companies that hire employees not only improve their profitability but also drive the economic well-being of the American economy.[18]

Government public policy also drives the economy through job creation. In order for any nation to ensure the social and economic health of the country, there must be a tax base to provide for the public interest. The vast majority of taxes come from individuals. It is estimated that the U.S. government obtains $2.3 trillion in individual income taxes annually.[19] Figure 1.5 shows the distribution of returns and income taxes paid by individuals based on their gross income. Unmarried individuals who earn more than $500,000 pay 37 percent of individual income taxes.[20]

Businesses are also an important form of tax revenue. Those that are classified as sole proprietorships, partnerships, and S corporations (discussed further in the chapter titled "Options for Organizing Business") pay taxes according to the individual income tax code. Corporations are taxed differently. Approximately 10.6 percent of the government's total revenues comes from corporate income taxes.[21] In 2017, the largest tax reform in the U.S. tax rate in over 30 years changed the corporate tax rate from 35 percent to 21 percent.[22] The average global corporate tax rate is 23 percent, while some countries such as Ireland have a corporate tax rate as low as 12.5 percent. The tax reform lowered the highest individual tax rate from 39.6 percent to 37 percent, but the reform capped the deduction of state and local taxes at $10,000. This could result in high-earners in states like New York and California with state and local taxes paying more federal taxes than before the tax reform.[23] The tax reform was designed to stimulate the economy but will increase the national debt.

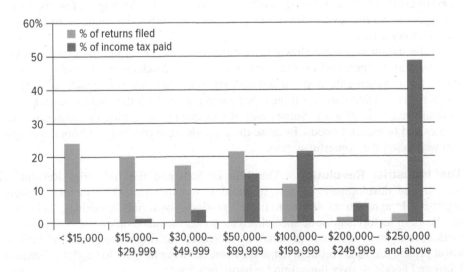

FIGURE 1.5

Individual Income Tax Statistics by Income Group

Source: Internal Revenue Service, Individual Income Tax Rates and Tax Shares, (accessed March 25, 2018).

Responding to Business Challenges

The Trix of the Trade: General Mills' Brand Strategy

In its 150-year history, General Mills has evolved from a flour mill to a packaged consumer goods company with revenues of $16 billion. More recently, General Mills has begun investing heavily in organic and natural foods with its acquisition of organic food brands Cascade Farms and Annie's. Sales of organic products are a $43.4 billion industry and growing. Demand for organic food is growing so rapidly that the supply has been unable to keep up. As a result, General Mills has begun underwriting the costs for farmers to convert their farms to organic crops.

The consumer goods market is characterized by monopolistic competition, meaning that General Mills has many competitors. This requires General Mills to adapt to changing consumer preferences. For example, consumers have been eating less cereal and looking for quick breakfast items like yogurt and breakfast bars. As a result, General Mills has expanded into more on-the-go products such as yogurt and granola. Perhaps as a way to meet the competition head on, General Mills has adopted a new brand strategy: "Consumers first."

General Mills wants its customer-centric focus to differentiate it from rival firms. In keeping with this customer emphasis, it has agreed to adopt genetically modified organism (GMO) labeling for some of its products. While General Mills believes GMO products are safe, it wants customers to know that it is listening to their concerns. In fact, General Mills became a first mover in this endeavor by reformulating its Cheerios to be GMO-free. However, as competitors also turn toward organic ingredients, it might not be long before other cereal manufacturers begin developing their own GMO-free cereals. General Mills must continue to innovate to maintain its competitive edge.[24]

Critical Thinking Questions

1. Why does the consumer goods industry operate in an environment of monopolistic competition? What must firms like General Mills do to succeed in this environment?
2. How is General Mills addressing the supply-demand problem that it is facing with organic food?
3. What are some of the ways in which General Mills is differentiating its products from the competition?

A Brief History of the American Economy

The Early Economy. Before the colonization of North America, Native Americans lived as hunter/gatherers and farmers, with some trade among tribes. The colonists who came later operated primarily as an *agricultural economy*. People were self-sufficient and produced everything they needed at home, including food, clothing, and furniture. Abundant natural resources and a moderate climate nourished industries such as farming, fishing, shipping, and fur trading. A few manufactured goods and money for the colonies' burgeoning industries came from England and other countries.

As the nation expanded slowly toward the West, people found natural resources such as coal, copper, and iron ore and used them to produce goods such as horseshoes, farm implements, and kitchen utensils. Farm families who produced surplus goods sold or traded them for things they could not produce themselves, such as fine furniture and window glass. Some families also spent time turning raw materials into clothes and household goods. Because these goods were produced at home, this system was called the domestic system.

The Industrial Revolution. The 19th century and the Industrial Revolution brought the development of new technology and factories. The factory brought together all the resources needed to make a product—materials, machines, and workers. Work in factories became specialized as workers focused on one or two tasks. As work became more efficient, productivity increased, making more goods available at lower prices. Railroads brought major changes, allowing farmers to send their surplus crops and goods all over the nation for barter or for sale.

Factories began to spring up along the railways to manufacture farm equipment and a variety of other goods to be shipped by rail. Samuel Slater set up the first American textile factory after he memorized the plans for an English factory and emigrated to the United States. Eli Whitney revolutionized the cotton industry with his cotton gin. Francis Cabot Lowell's factory organized all the steps in manufacturing cotton cloth for maximum efficiency and productivity. John Deere's farm equipment increased farm production and reduced the number of farmers required to feed the young nation. Farmers began to move to cities to find jobs in factories and a higher standard of living. Henry Ford developed the assembly-line system to produce automobiles. Workers focused on one part of an automobile and then pushed it to the next stage until it rolled off the assembly line as a finished automobile. Ford's assembly line could manufacture many automobiles efficiently, and the price of his cars was $200, making them affordable to many Americans.

The Manufacturing and Marketing Economies. Industrialization brought increased prosperity, and the United States gradually became a *manufacturing economy*—one devoted to manufacturing goods and providing services rather than producing agricultural products. The assembly line was applied to more industries, increasing the variety of goods available to the consumer. Businesses became more concerned with the needs of the consumer and entered the *marketing economy*. Expensive goods such as cars and appliances could be purchased on a time-payment plan. Companies conducted research to find out what products consumers needed and wanted. Advertising made consumers aware of products and important information about features, prices, and other competitive advantages.

Because these developments occurred in a free-enterprise system, consumers determined what goods and services were produced. They did this by purchasing the products they liked at prices they were willing to pay. The United States prospered, and American citizens had one of the highest standards of living in the world.

The Service and New Digital Economy. After World War II, with the increased standard of living, Americans had more money and more time. They began to pay others to perform services that made their lives easier. Beginning in the 1960s, more and more women entered the workforce. The United States began experiencing major shifts in the population. The U.S. population grew about 10 percent in the past decade to 327 million. This is the slowest pace of growth since the Great Depression, with the South leading the population gains. The United States is undergoing a baby bust, with record lows in the country's fertility rate.[25] While the birth rate in the United States is declining, new immigrants help with population gains.[26] The profile of the family is also changing: Today there are more single-parent families and individuals living alone, and in two-parent families, both parents often work.

One result of this trend is that time-pressed Americans are increasingly paying others to do tasks they used to do at home, like cooking, laundry, landscaping, and child care. These trends have gradually changed the United States to a *service economy*—one devoted to the production of services that make life easier for busy consumers. Businesses increased their demand for services, especially in the areas of finance and information technology. Service industries such as restaurants, banking, health care, child care, auto repair, leisure-related industries, and even education are growing rapidly and may account for as much as 80 percent of the U.S. economy. These trends continue with advanced technology contributing to

DID YOU KNOW? Approximately 57 percent of adult women are involved in the workforce.[27]

FIGURE 1.6

**Online Retailing and
E-Commerce**

Source: Aaron Smith and Monica Anderson, "Online Shopping and E-Commerce," Pew Research Center, December 19, 2016, http://www .pewinternet.org/2016/12/19/online-shopping-and-e-commerce/ (accessed March 25, 2018).

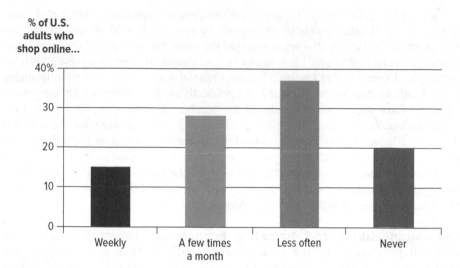

% of U.S. adults who shop online...

new service products based on technology and digital media that provide smartphones, social networking, and virtual worlds. This has led to the growth of e-commerce, or transactions involving goods and services over the Internet. E-commerce has led to firms that would have been unheard of a few decades ago, such as eBay, Shopify, Etsy, and Amazon. Figure 1.6 shows the percentage of adults who engage in e-commerce. More about the digital world, business, and new online social media can be found in the chapter titled "Digital Marketing and Social Networking."

LO 1-8

Explain the role of the entrepreneur in the economy.

entrepreneur
an individual who risks his or her wealth, time, and effort to develop for profit an innovative product or way of doing something.

The Role of the Entrepreneur

An **entrepreneur** is an individual who risks his or her wealth, time, and effort to develop for profit an innovative product or way of doing something. Heidi Ganahl is a true American entrepreneur. She took the unusual concept of a day care center for dogs and turned it into a successful $85 million franchise operation. Her business, Camp Bow Wow, which offers boarding, playtime, grooming, and other services for dogs, expanded to more than 140 locations and plans to grow to 1,000 over the next several years.[28]

The free-enterprise system provides the conditions necessary for entrepreneurs like Ganahl to succeed. In the past, entrepreneurs were often inventors who brought all the factors of production together to produce a new product. Thomas Edison, whose inventions include the record player and light bulb, was an early American entrepreneur. Henry Ford was one of the first persons to develop mass assembly methods in the automobile industry. Other entrepreneurs, so-called captains of industry, invested in the country's growth. John D. Rockefeller built Standard Oil out of the fledgling oil industry, and Andrew Carnegie invested in railroads and founded the United States Steel Corporation. Andrew Mellon built the Aluminum Company of America and Gulf Oil. J. P. Morgan started financial institutions to fund the business activities of other entrepreneurs. Although these entrepreneurs were born in another century, their legacy to the American economy lives on in the companies they started, many of which still operate today. Consider the history of Eli Lilly. Colonel Eli Lilly in Indianapolis, Indiana, was continually frustrated with the quality of pharmaceutical products sold at the time. As a pharmaceutical chemist, he decided to start his own

Entrepreneurship in Action

ATA: Engineering Good Teamwork

ATA Engineering
  Founder: Mary Baker
  Founded: 2000, in San Diego, CA
  Success: This 100 percent employee-owned firm has won numerous awards, including a quality award from NASA and a spot among *The Wall Street Journal's* Top Small Workplaces.

ATA Engineering is a unique firm because—although founder Mary Baker plays a role in the company as president and director—the company became entirely employee owned a mere four years after its founding. This gives the 120 employees at ATA Engineering greater motivation to make it profitable because they succeed when the company succeeds.

ATA Engineering Inc. is a provider of analysis and test-driven design solutions for structural, mechanical, electromechanical, and aerospace products. When it was founded, ATA Engineering was not the only engineering firm in the industry.

Competition was high. However, it was spun off from Structural Dynamics Research Corporation, a firm that had worked on NASA projects. As a spinoff, ATA Engineering had the expertise needed to continue this relationship. The company gained a competitive advantage with its role as a NASA contractor. More recently, the firm has helped to solve design issues on NASA's Mars Science Lab, which earned it the George M. Low Award, NASA's quality and performance award. Today, ATA Engineering operates in seven offices across the nation.[29]

Critical Thinking Questions

1. How does creating an employee-owned business give employees greater motivation to make the business profitable?
2. What are the disadvantages of an employee-owned business?
3. In what other ways does ATA stand apart from its competitors?

---

firm that would offer the highest-quality medicines. His firm, Eli Lilly and Company, would go on to make landmark achievements, including being one of the first pharmaceutical firms to mass-produce penicillin. Today, Eli Lilly is one of the largest pharmaceutical firms in the world.[30]

Entrepreneurs are constantly changing American business practices with new technology and innovative management techniques. Bill Gates, for example, built Microsoft, a software company whose products include Word and Windows, into a multibillion-dollar enterprise. Frederick Smith had an idea to deliver packages overnight, and now his FedEx Company plays an important role in getting documents and packages delivered all over the world for businesses and individuals. Steve Jobs co-founded Apple and turned the company into a successful consumer electronics firm that revolutionized many different industries, with products such as the iPod, iPhone, Mac computers, iPad, and Apple Watch. The company went from near bankruptcy in the 1990s to become one of the most valuable brands in the entire world. Entrepreneurs have been associated with such uniquely American concepts as Dell Computers, Ben & Jerry's, Levi's, McDonald's, Dr Pepper, SpaceX, Google, Facebook, and Walmart. Walmart, founded by entrepreneur Sam Walton, was the first retailer to reach $100 billion in sales in one year and now routinely passes that mark, with more than $453 billion in 2017.[31] We will examine the importance of entrepreneurship further in the chapter "Small Business, Entrepreneurship, and Franchising."

Apple Pay is a mobile payment system that allows users to store credit card or debit card information on their phones. When checking out at stores, users can bring up their credit card with two taps and use the information to pay for their purchases.

©How Hwee Young/Epa/REX/Shutterstock

Going Green

Rainforest Alliance Stands Out in a Forest of Nonprofits

When you think of the phrase "nonprofit organization," you likely think of a bare-bones operation run by people who are so dedicated that they do not mind working for peanuts. However, the Rainforest Alliance, a highly recognized nonprofit aiming to conserve biodiversity and promote sustainability, defies this mold. With thousands of members and supporters, the Rainforest Alliance has been fighting globally for wildlife and wild lands since 1987.

As more consumers and businesses become interested in preserving the world's resources, the Rainforest Alliance has gained widespread support. Interacting with key stakeholders is critical to the Rainforest Alliance's success. For instance, the Rainforest Alliance has many resources for businesses that partner with the organization. The Rainforest Alliance provides auditing and certification services for businesses—such as agricultural, tourism, and forestry organizations—that wish to display the Rainforest Alliance certification seal. The Rainforest Alliance blog describes how sustainability can positively affect a business, and its seal is one way businesses can inform

consumers that they have incorporated sustainability principles into their operations. The Rainforest Alliance even has its own business unit, called RA-Cert, involved with auditing, evaluation, and certification decisions. Additionally, the Rainforest Alliance offers resources to businesses that want to educate their own employees about sustainability.

Stakeholders generally have a positive image of the Rainforest Alliance. One survey noted that coffee drinkers were likely to spend extra on coffee certified by Rainforest Alliance. For individuals looking to aid the environment and support small growers, the Rainforest Alliance is making a positive difference with its strong mission and interaction with stakeholders.[32]

Critical Thinking Questions

1. Why might investing in sustainability and becoming certified by an organization like the Rainforest Alliance be a good business decision?
2. Who do you feel are Rainforest Alliance's primary customers: consumers or businesses?
3. As a nonprofit, how does the Rainforest Alliance differ from other organizations?

The Role of Government in the American Economy

The American economic system is best described as modified capitalism because the government regulates business to preserve competition and protect consumers and employees. Federal, state, and local governments intervene in the economy with laws and regulations designed to promote competition and to protect consumers, employees, and the environment. Many of these laws are discussed in Bonus Chapter A.

Additionally, government agencies such as the U.S. Department of Commerce measure the health of the economy (GDP, productivity, etc.) and, when necessary, take steps to minimize the disruptive effects of economic fluctuations and reduce unemployment. When the economy is contracting and unemployment is rising, the federal government through the Federal Reserve Board (see chapter titled "Money and the Financial System") tries to spur growth so that consumers will spend more money and businesses will hire more employees. To accomplish this, it may reduce interest rates or increase its own spending for goods and services. When the economy expands so fast that inflation results, the government may intervene to reduce inflation by slowing down economic growth. This can be accomplished by raising interest rates to discourage spending by businesses and consumers. Techniques used to control the economy are discussed in "Money and the Financial System."

The Role of Ethics and Social Responsibility in Business

In the past few years, you may have read about a number of scandals at well-known corporations, including Volkswagen, which cheated on emissions tests, and Wells Fargo. Wells Fargo employees opened more than 3.5 million accounts in customer names without permission and disclosed that the company had overcharged some auto loan and mortgage customers. CEO Tim Sloan told Congress he's "sorry for how the bank abused consumers."[33] In many cases, misconduct by individuals within these firms had an adverse effect on employees, investors, and others associated with these firms. These scandals undermined public confidence in corporate America and sparked a new debate about ethics in business. *Business ethics* generally refers to the standards and principles used by society to define appropriate and inappropriate conduct in the workplace. In many cases, these standards have been codified as laws prohibiting actions deemed unacceptable.

Society is increasingly demanding that businesspeople behave socially responsibly toward their stakeholders, including customers, employees, investors, government regulators, communities, and the natural environment. Diversity in the workforce is not only socially responsible, but also highly beneficial to the financial performance of companies. According to a McKinsey consulting firm study, organizations that have diverse leadership are more likely to report higher financial returns. This study defined diversity as women and minorities. Diversity creates increased employee satisfaction and improved decision making.[34] When actions are heavily criticized, a balance is usually required to support and protect various stakeholders.

While one view is that ethics and social responsibility are a good supplement to business activities, there is an alternative viewpoint. Ethical behavior can not only enhance a company's reputation, but can also drive profits.[35] The ethical and socially responsible conduct of companies such as Whole Foods, Starbucks, and the hotel chain Marriott provides evidence that good ethics is good business. There is growing recognition that the long-term value of conducting business in an ethical and socially responsible manner that considers the interests of all stakeholders creates superior financial performance.[36]

To promote socially responsible and ethical behavior while achieving organizational goals, businesses can monitor changes and trends in society's values. Businesses should determine what society wants and attempt to predict the long-term effects of their decisions. While it requires an effort to address the interests of all stakeholders, businesses can prioritize and attempt to balance conflicting demands. The goal is to develop a solid reputation of trust and avoid misconduct to develop effective workplace ethics.

Can You Learn Business in a Classroom?

Obviously, the answer is yes, or there would be no purpose for this textbook or course! To be successful in business, you need knowledge, skills, experience, and good judgment. The topics covered in this chapter and throughout this book provide some of the knowledge you need to understand the world of business. In addition, the "Build Your Skills" exercise near the end of each chapter will help you develop skills that may be useful in your future career. However, good judgment is based on knowledge and experience plus personal insight and understanding. Therefore, you need more courses in business, along with some practical experience in the business world, to help you develop the special insight necessary to put your personal stamp on knowledge as you apply it. The challenge in business is in the area of judgment, and judgment does not develop from memorizing an introductory business textbook. If you are observant in your daily experiences as an employee, as a student, and as a consumer, you will improve your ability to make good business judgments.

Whether you choose to work at an organization or become an entrepreneur, you will be required to know the basic concepts and principles in this book. You need to be prepared for changes in the way business will be conducted in the future. New business models or ways businesses create value will emerge based on the Internet, connected technologies, drones, driverless cars, and artificial intelligence. It should be exciting to think about your opportunities and the challenges of creating a successful career. Our society needs a strong economic foundation to help people develop a desired standard of living. Our world economy is becoming more digital and competitive, requiring new skills and job positions. Individuals like you can become leaders in business, nonprofits, and government to create a better life.

Figure 1.7 is an overview of how the chapters in this book are linked together and how the chapters relate to the participants, the activities, and the environmental factors found in the business world. The topics presented in the chapters that follow are those that will give you the best opportunity to begin the process of understanding the world of business.

FIGURE 1.7
The Organization of This Book

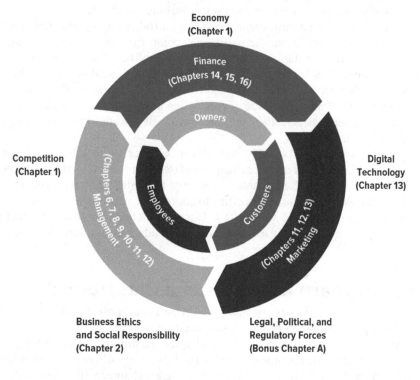

Special Topics:
Global Business (Chapter 3)
Forms of Ownership (Chapter 4)
Small Business, Entrepreneurship, and Franchising (Chapter 5)

So You Want a Job in the Business World

When most people think of a career in business, they see themselves entering the door to large companies and multinationals that they read about in the news and that are discussed in class. In a national survey, students indicated they would like to work for Google, Disney, Apple, and Ernst & Young. In fact, most jobs are not with large corporations, but are in small companies, nonprofit organizations, government, and even as self-employed individuals. There are nearly 27 million individuals in the United States who own their own businesses and have no employees. With more than 75 percent of the economy based on services, there are many jobs available in industries such as health care, finance, education, hospitality, and entertainment. E-commerce is creating the need for supply chain jobs related to purchasing, transportation, and operations. The world is changing quickly and large corporations replace the equivalent of their entire workforce every four years.

The fast pace of technology today means that you have to be prepared to take advantage of emerging job opportunities and markets. You must also become adaptive and recognize that business is becoming more global, with job opportunities around the world. If you want to obtain such a job, you shouldn't miss a chance to spend some time overseas. To get you started on the path to thinking about job opportunities, consider all of the changes in business today that might affect your possible long-term track and that could bring you lots of success. Companies are looking for employees with skills that can be used to address the changing business environment. For example, the demand for graduates that are good at analyzing data and navigating cloud-based computing is on the rise.

You're on the road to learning the key knowledge, skills, and trends that you can use to be a star in business. Business's impact on our society, especially in the area of sustainability and improvement of the environment, is a growing challenge and opportunity. Green businesses and green jobs in the business world are provided to give you a glimpse at the possibilities. Along the way, we will introduce you to some specific careers and offer advice on developing your own job opportunities. Research indicates that you won't be that happy with your job unless you enjoy your work and feel that it has a purpose. Because you spend most of your waking hours every day at work, you need to seriously think about what is important to you in a job.[37]

Review Your Understanding

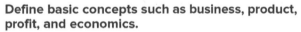

Define basic concepts such as business, product, profit, and economics.

A business is an organization or individual that seeks a profit by providing products that satisfy people's needs. A product is a good, service, or idea that has both tangible and intangible characteristics that provide satisfaction and benefits. Profit, the basic goal of business, is the difference between what it costs to make and sell a product and what a customer pays for it. Economics is the study of how resources are distributed for the production of goods and services within a social system.

Identify the main participants and activities of business.

The three main participants in business are owners, employees, and customers, but others—government regulators, suppliers, social groups, etc.—are also important. Management involves planning, organizing, and controlling the tasks required to carry out the work of the company. Marketing refers to those activities—research, product development, promotion, pricing, and distribution—designed to provide goods and services that satisfy customers. Finance refers to activities concerned with funding a business and using its funds effectively.

Explain why studying business is important.

Studying business can help you prepare for a career and become a better consumer.

Compare the four types of economic systems.

An economic system describes how a particular society distributes its resources. Communism is an economic system in which the people, without regard to class, own all the nation's resources. In a socialist system, the government owns and operates basic industries, but individuals own most businesses. Under capitalism, individuals own and operate the majority of businesses that provide goods and services. Mixed economies have elements from more than one economic system; most countries have mixed economies.

Describe the role of supply, demand, and competition in a free-enterprise system.

In a free-enterprise system, individuals own and operate the majority of businesses, and the distribution of resources is determined by competition, supply, and demand. Demand is the number of goods and services that consumers are willing to buy at different prices at a specific time. Supply is the number of goods or services that businesses are willing to sell at different prices at a specific time. The price at which the supply of a product equals demand at a specific point in time is the equilibrium price. Competition is the rivalry among businesses to convince consumers to buy goods or services. Four types of competitive environments are pure competition, monopolistic competition, oligopoly, and monopoly. These economic concepts determine how businesses may operate in a particular society and, often, how much they can charge for their products.

Specify why and how the health of the economy is measured.

A country measures the state of its economy to determine whether it is expanding or contracting and whether the country needs to take steps to minimize fluctuations. One commonly used measure is gross domestic product (GDP), the sum of all goods and services produced in a country during a year. A budget deficit occurs when a nation spends more than it takes in from taxes.

Outline the evolution of the American economy.

The American economy is an open economy that engages in significant international trade. Government public policy helps drive the economy through job creation, requiring a tax base to provide for the public interest. Much of the government's revenue comes from individual income taxes, but corporations pay a high corporate tax in the United States.

The American economy has evolved through several stages: the early economy, the Industrial Revolution, the manufacturing economy, the marketing economy, and the service and Internet-based economy of today.

Explain the role of the entrepreneur in the economy.

Entrepreneurs play an important role because they risk their time, wealth, and efforts to develop new goods, services, and ideas that fuel the growth of the American economy.

Evaluate a small-business owner's situation in order to propose a course of action.

"Solve the Dilemma" near the end of this chapter presents a problem for the owner of the firm. Should you, as the owner, raise prices, expand operations, or form a venture with a larger company to deal with demand? You should be able to apply your newfound understanding of the relationship between supply and demand to assess the situation and reach a decision about how to proceed.

Critical Thinking Questions

Enter the World of Business Questions

1. Describe Buffett's business strategy. Why do you think it was so successful?

2. What are some of the benefits of owning companies in several different industries? The risks?

3. Why is it important for Buffett to allow managers the ability to run the firms as they see fit?

Learn the Terms

budget deficit 17
business 4
capitalism (free enterprise) 10
communism 9
competition 13
demand 12
depression 16
economic contraction 15
economic expansion 15
economic system 8
economics 8
entrepreneur 22

equilibrium price 13
financial resources (capital) 8
free-market system 10
gross domestic product (GDP) 16
human resources (labor) 8
inflation 15
mixed economies 11
monopolistic competition 14
monopoly 14
natural resources 8
nonprofit organizations 4
oligopoly 14

open economy 18
product 4
profit 4
pure competition 14
recession 15
socialism 10
stakeholders 5
standard of living 18
supply 12
unemployment 15

Check Your Progress

1. What is the fundamental goal of business? Do all organizations share this goal?

2. Name the forms a product may take and give some examples of each.

3. Who are the main participants of business? What are the main activities? What other factors have an impact on the conduct of business in the United States?

4. What are four types of economic systems? Can you provide an example of a country using each type?

5. Explain the terms *supply, demand, equilibrium price,* and *competition.* How do these forces interact in the American economy?

6. List the four types of competitive environments and provide an example of a product of each environment.

7. List and define the various measures governments may use to gauge the state of their economies. If unemployment is high, will the growth of GDP be great or small?

8. Why are fluctuations in the economy harmful?

9. How did the Industrial Revolution influence the growth of the American economy? Why do we apply the term service economy to the United States today?

10. Explain the federal government's role in the American economy.

Get Involved

1. Discuss the economic changes occurring in Russia and eastern European countries, which once operated as communist economic systems. Why are these changes occurring? What do you think the result will be?

2. Why is it important for the government to measure the economy? What kinds of actions might it take to control the economy's growth?

3. Is the American economy currently expanding or contracting? Defend your answer with the latest statistics on GDP, inflation, unemployment, and so on. How is the federal government responding?

Build Your Skills

The Forces of Supply and Demand

Background

WagWumps are a new children's toy with the potential to be a highly successful product. WagWumps are cute and furry, and their eyes glow in the dark. Each family set consists of a mother, a father, and two children. Wee-Toys' manufacturing costs are about $6 per set, with $3 representing marketing and distribution costs. The wholesale price of a WagWump family for a retailer is $15.75, and the toy carries a suggested retail price of $26.99.

Task

Assume you are a decision maker at a retailer, such as Target or Walmart, that must determine the price the stores in your district should charge customers for the WagWump family set. From the information provided, you know that the SRP (suggested retail price) is $26.99 per set and that your company can purchase the toy set from your wholesaler for $15.75 each. Based on the following assumptions, plot your company's supply curve on the graph provided in Figure 1.8 and label it "supply curve."

Quantity	Price
3,000	$16.99
5,000	21.99
7,000	26.99

Using the following assumptions, plot your customers' demand curve on Figure 1.8 and label it "demand curve."

Quantity	Price
10,000	$16.99
6,000	21.99
2,000	26.99

FIGURE 1.8
Equilibrium Price of
WagWumps

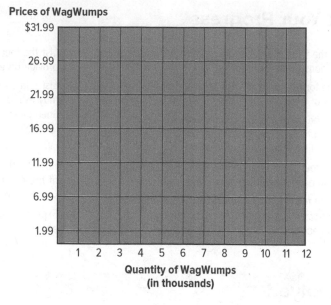

For this specific time, determine the point at which the quantity of toys your company is willing to supply equals the quantity of toys the customers in your sales district are willing to buy and label that point "equilibrium price."

Solve the Dilemma

Mrs. Acres Homemade Pies

Shelly Acres, whose grandmother gave her a family recipe for making pies, loved to cook, and she decided to start a business she called Mrs. Acres Homemade Pies. The company produces specialty pies and sells them in local supermarkets and select family restaurants. In each of the first six months, Shelly and three part-time employees sold 2,000 pies for $4.50 each, netting $1.50 profit per pie. The pies were quite successful and Shelly could not keep up with demand. The company's success results from a quality product and productive employees who are motivated by incentives and who enjoy being part of a successful new business.

To meet demand, Shelly expanded operations, borrowing money and increasing staff to four full-time employees. Production and sales increased to 8,000 pies per month, and profits soared to $12,000 per month. However, demand for Mrs. Acres Homemade Pies continues to accelerate beyond what Shelly can supply. She has several options: (1) maintain current production levels and raise prices; (2) expand the facility and staff while maintaining the current price; or (3) contract the production of the pies to a national restaurant chain, giving Shelly a percentage of profits with minimal involvement.

Evaluate a small-business owner's situation in order to propose a course of action.

Critical Thinking Questions

1. Explain and demonstrate the relationship between supply and demand for Mrs. Acres Homemade Pies.

2. What challenges does Shelly face as she considers the three options?

3. What would you do in Shelly's position?

Build Your Business Plan

The Dynamics of Business and Economics

 Have you ever thought about owning your business? If you have, how did your idea come about? Is it your experience with this particular field? Or might it be an idea that evolved from your desires for a particular good or service not being offered in your community? For example, perhaps you and your friends have yearned for a place to go have coffee, relax, and talk. Now is an opportunity to create the cafe bar you have been thinking of!

Whether you consider yourself a visionary or a practical thinker, think about your community. What needs are not being met? While it is tempting to suggest a new restaurant (maybe even one near campus), easier-to-implement business plans can range from a lawn care business or a designated driver business to a placement service agency for teenagers.

Once you have an idea for a business plan, think about how profitable this idea might be. Is there sufficient demand for this business? How large is the market for this particular business? What about competitors? How many are there?

To learn about your industry, you should do a thorough search of your initial ideas of a product on the Internet.

See for Yourself Videocase

Redbox Succeeds by Identifying Market Need

 Redbox's tell-tale bright red kiosks in stores and fast-food restaurants across the country have become an image of what a good business model can accomplish. The company's ability to offer customers a convenient and inexpensive DVD rental option has allowed it to succeed despite the widespread growth of streaming services such as Netflix and Amazon. However, the firm acknowledges the future of streaming, and Redbox has allegedly been discussing the launch of a new streaming service called "Redbox Variety." As one of the top rental companies in the United States, Redbox is a true entrepreneurial success story.

Building Redbox into a successful firm was not easy, however. It was fraught with challenges. Like most successful companies, Redbox started out by identifying a need. It recognized that consumers could not often find the movies they wanted in convenient locations. Like all good ideas, Redbox required funding to get started. This proved to be a major difficulty. Realizing that customers did not want to pay much for renting movies, Redbox decided to charge only one dollar. Yet the kiosks, which contain more than 800 components, required a large amount of capital. The combination of the capital-intensive nature of the business and the low prices was not an attractive recipe for venture capital funding.

However, Redbox was certain that demand for its product offerings would exceed the costs. The company finally found a partner in the more established Outerwall, formerly known as Coinstar, which already had partnerships with many different retailers. The alliance opened the way for Redbox to begin installing kiosks at the front of stores.

Redbox did not immediately expand across the country. Instead, it took a cautious approach toward its business model. It began by focusing its efforts on making one kiosk profitable, then replicating this way of thinking regionally and nationally. In this way, Redbox was able to test its concept without taking the risk of widespread failure.

Even though it was expanding, it was some time before Redbox was able to earn a profit. Like all entrepreneurs, the founders of Redbox had to take many risks if they wanted the company to succeed. "The risks for starting Redbox were significant," said Marc Achler, vice president of new business, strategy, and innovation. "The first couple years we had some red ink. It took us a while before we turned profitable." Yet with persistence and continual relationship building with retailers, Redbox has been able to secure more than 50 percent of the physical DVD-rental market.

One way that Redbox has been able to secure such a large share of the market is by meeting the needs of a variety of stakeholders. Redbox views its customers as its first priority and has developed its kiosks and database to meet their needs. For instance, customers can reserve movies online and pick them up at their nearest kiosk. If a kiosk happens to be out of a particular movie, customers can search the Redbox database to locate the movie at a nearby kiosk. This combination of convenience and low prices has attracted customers who desire a simplified process to renting movies.

Additionally, Redbox has created a process that also benefits the needs of its retail partners. Redbox kiosks help attract consumers to the store, where they may purchase additional products. Customers must come back the next day to return their movie, where they may once again purchase more products from the retailer. In this way, Redbox creates a win-win situation for both itself and its partners.

This is not to say that everything is easy for Redbox. For instance, it must continually safeguard against allowing underage children to rent inappropriate (rated-R) movies. And while Redbox has approached this changing and dynamic marketplace proactively, it must continue to do so in order to maintain its competitive position. With an increasing interest in streaming, Redbox has seen sales decline in recent years. Its first attempt to develop a streaming service with Verizon was unable to compete with Netflix and was discontinued. However, Redbox refuses to give up. It continues to look for other opportunities to gain advantages against the competition.[38]

Critical Thinking Questions

1. Why are consumers so willing to rent from Redbox?

2. How was Redbox able to overcome some of its earliest challenges?

3. What are some recommendations for ways that Redbox can maintain its high market share?

You can find the related video in the Video Library in Connect. Ask your instructor how you can access Connect.

Team Exercise

Major economic systems, including capitalism, socialism, and communism, as well as mixed economic systems were discussed in this chapter. Assuming that you want an economic system that is best for the majority, not just a few members of society, defend one of the economic systems as the best system. Form groups and try to reach an agreement on one economic system. Defend why you support the system that you advance.

Notes

1. Jena McGregor, "The Leadership Wisdom of Warren Buffett," *Washington Post,* March 2, 2015, https://www.washingtonpost.com/news/on-leadership/wp/2015/03/02/the-leadership-wisdom-in-warren-buffetts-letter/?utm_term=.4fcedcb25aff (accessed January 19, 2018); Robert Frank, "Warren Buffett Is the Most Charitable Billionaire," *CNBC,* September 21, 2017, https://www.cnbc.com/2017/09/21/warren-buffet-is-the-most-charitable-billionaire.html (accessed January 19, 2018); Investopedia Staff, "Warren Buffett: How He Does It," *Investopedia,* August 30, 2017, https://www.investopedia.com/articles/01/071801.asp (accessed January 19, 2018); CNBC LLC, "10 Top Brands Warren Buffett's Berkshire Hathaway Owns," *CNBC,* May 5, 2014, https://www.cnbc.com/2014/05/05/10-top-brands-warren-buffetts-berkshire-hathaway-owns.html?slide=7 (accessed January 19, 2018); BuffetBooks, "What Is Value Investing?" http://www.buffettsbooks.com/howtoinvestinstocks/course1/investing-for-beginners/what-is-value-investing.html#sthash.HPn7R7Gn.dpbs (accessed January 19, 2018); Joseph Chris, "9 Warren Buffett Leadership Style Doctrines," *Driving Business Connections* blog, September 15, 2015, http://www.josephchris.com/9-warren-buffett-leadership-style-doctrines (accessed January 4, 2018); Tanza Loudenback, "24 Mind-Blowing Facts about Warren Buffett and His $87 Billion Fortune," *Business Insider,* January 10, 2018, http://www.businessinsider.com/facts-about-warren-buffett-2016-12 (accessed January 19, 2018); Andrew Hill, "Buffet's Exceptional Style of Leadership," *Financial Times,* February 28, 2011, https://www.ft.com/content/73e667a8-436b-11e0-8f0d-00144feabdc0 (accessed January 3, 2018); Jory MacKay, "This Brilliant Strategy Used by Warren Buffett Will Help You Prioritize Your Time," *Inc.,* November 15, 2017, https://www.inc.com/jory-mackay/warren-buffetts-personal-pilot-reveals-billionaires-brilliant-method-for-prioritizing.html (accessed January 4, 2018); posted by Practical Wisdom, "Charlie Munger – Advice for the Young Generation—This Is What You Need to Know in Your Youth," *YouTube,* September 6, 2017, https://www.youtube.com/watch?v=Qf4trOzMy2I (accessed January 4, 2018); Juno Tay, "Warren Buffet's Speech to the University of Georgia Students Part 1 (Archive 2001)" Nasdaq, April 21, 2013, http://www.nasdaq.com/article/warren-buffett-speech-to-university-of-

georgia-students-part-1-cm238914 (accessed January 3, 2018); posted by Texas Business Hall of Fame, "Warren Buffett and Charles Munger Class of 2016," *YouTube,* January 23, 2017, https://www.youtube.com/watch?v=-ARcxChbVHQ (accessed January 19, 2018); posted by Motivation Madness, "Warren Buffett's Life Advice Will Change Your Future," *YouTube,* September 11, 2017, https://www.youtube.com/watch?v=PX5-XyBNi00 (accessed January 19, 2018).

2. Jay Young and Janet Loehrke, "Ward's Automotive Reports,"*US Bureau of Economic Analysis,* reported in *USA Today,* March 14, 2017, p. 1B.

3. Charisse Jones, "Here's the Secret to a Successful Company," *USA Today,* March 14, 2017, p. 1B.

4. Esther Fung, "Mall Owners Head for Exits as Retail Tenants Move Out," *The Wall Street Journal,* January 25, 2017, p. A1.

5. "Habitat for Humanity and Home Depot Foundation Partner to Repair Homes for U.S. Military Veterans and Families," Habitat.org, March 16, 2017.

6. James T. Areddy and Craig Karmin, "China Stocks Once Frothy, Fall by Half in Six Months," *The Wall Street Journal,* April 16, 2008, pp. 1, 7.

7. "American Politics: In Praise of State-ism," *The Economist,* January 6–12, 2018, p. 10.

8. "Special Report: The Visible Hand," *The Economist,* January 21, 2012, pp. 3–5.

9. "Special Report: The World in Their Hands," *The Economist,* January 21, 2012, pp. 15–17.

10. "The Shark Tank," ABC, http://abc.go.com/shows/shark-tank/bios (accessed April 2, 2014).

11. World Intellectual Property Organization, *World Intellectual Property Indicators,* 2014, http://www.wipo.int/edocs/pubdocs/en/wipo_pub_941_2014.pdf (accessed May 1, 2017).

12. Paul Toscano, "The Worst Hyperinflation Situations of All Time," *CNBC,* February 14, 2011, http://www.cnbc.com/id/41532451 (accessed May 1, 2017).

13. "Zimbabwe," CIA—*The World Factbook,* https://www.cia.gov/library/publications/the-world-factbook/geos/zi.html (accessed May 1, 2017).

14. "Federal Debt Clock," *U.S. Government Debt,* http://www.usgovernmentdebt,us/ (accessed March 25, 2018).

15. "The Great Experiment," *The Economist,* February 10-16, 2018, p. 23.

16. World Trade Organization, "The WTO can . . . stimulate economic growth and employment," https://www.wto.org/english/thewto_e/whatis_e/10thi_e/10thi03_e.htm (accessed May 1, 2017).

17. "The Criminalisation of American Business," *The Economist,* August 30–September 5, 2014, pp. 21–24; Tracy Gonzalez-Padron, G. Tomas M. Hult, and O.C. Ferrell, "Stakeholder Marketing Relationships to Social Responsibility and Firm Performance," Working paper, 2015.

18. Ryan C. Fuhrmann, "Okuns Law: Economic Growth and Unemployment," *Investopedia,* 2016, http://www.investopedia.com/articles/economics/12/okuns-law.asp (accessed May 1, 2017);World Trade Organization, "The WTO Can. . . Stimulate Economic Growth and Employment," https://www.wto.org/english/thewto_e/whatis_e/10thi_e/10thi03_e.htm (accessed May 1, 2017).

19. "U.S. Government Current Revenue History," *U.S. Government Revenue,* https://www.usgovernmentrevenue.com/current_revenue (accessed March 25, 2018).

20. Amir El-Sibaie, "2018 Tax Brackets," *Tax Foundation,* January 2, 2018, https://taxfoundation.org/2018-tax-brackets/ (accessed March 25, 2018).

21. Drew Desilver, "High-Income Americans Pay Most Income Taxes, but Enough to Be 'Fair'?" *Pew Research Center,* April 13, 2016, http://www.pewresearch.org/fact-tank/2016/04/13/high-income-americans-pay-most-income-taxes-but-enough-to-be-fair/ (accessed July 25, 2018).

22. "Corporate Tax in America: Let the Games Begin," *The Economist,* December 16–22, 2017, p. 58.

23. "High Tax States: Tax Replanning," *The Economist,* January 6, 2018, p. 17.

24. Sarah Elbert, "Food for Thought," *Delta Sky,* December 2016, pp. 66–70; Annie Gasparro, "General Mills Starts Making Some Cheerios Without GMOs," *The Wall Street Journal,* January 2, 2014, http://www.wsj.com/articles/SB10001424052702303370904579297211874270146 (accessed January 6, 2017); Hadley Malcolm, "General Mills to Label GMOs on Products across Country," *USA Today,* March 18, 2016, http://www.usatoday.com/story/money/2016/03/18/general-mills-to-label-gmos-on-products/81981314/ (accessed January 6, 2017); Stephanie Strom, "Paying Farmers to Go Organic, Even Before the Crops Come In," *The New York Times,* July 14, 2016, http://www.nytimes.com/2016/07/15/business/paying-farmers-to-go-organic-even-before-the-crops-come-in.html (accessed January 6, 2017); John Kell, "General Mills Reveals How It Plans to 'Renovate' Yogurt Products," *Fortune,* July 14, 2016, http://fortune.com/2016/07/14/general-mills-yogurt/ (accessed January 2, 2018); Nathan Bomey, "Slumping Yogurt, Cereal Sales Spoil General Mills' Performance," *USA Today,* September 20, 2017, https://www.usatoday.com/story/money/2017/09/20/general-mills-first-quarter-earnings/684218001/ (accessed January 2, 2018); Mary Ellen Shoup, "Chobani Beats Yoplait in Sales and Market Share as Dannon Takes No. 1 Spot in US Yogurt Market," *Dairy Reporter,* March 13, 2017, https://www.dairyreporter.com/Article/2017/03/13/Chobani-surpasses-Yoplait-in-sales-and-market-share (accessed January 2,

2018); Sean Rossman, "What Is French Yogurt and Is It the New Greek?" *USA Today,* July 6, 2017, https://www.usatoday.com/story/money/nation-now/2017/07/06/what-french-yogurt-and-new-greek/439935001/ (accessed January 2, 2018); Bruce Horovitz, "Cheerios Drop Genetically Modified Ingredients," *USA Today,* January 2, 2014, https://www.usatoday.com/story/money/business/2014/01/02/cheerios-gmos-cereals/4295739/ (accessed January 2, 2018).

25. Neil Shah, "Baby Bust Threatens Growth," *The Wall Street Journal,* December 4, 2014, p. A3.

26. U.S. Census Bureau, "State & County Quick Facts," http://quickfacts.census.gov/qfd/states/00000.html (accessed May 1, 2017); Haya El Nasser, Gregory Korte, and Paul Overberg, "308.7 Million," *USA Today,* December 22, 2010, p. 1A.

27. U.S. Department of Labor, "Data & Statistics," http://www.dol.gov/wb/stats/stats_data.htm (accessed May 1, 2017).

28. Dinah Eng and Heidi Ganahl, "Finding Her 'Wow' in Camp Bow Wow," *Fortune,* March 15, 2015, pp. 41–42; Camp Bow Wow, "About Us," http://www.campbowwow.com/about-us/our-history (accessed March 25, 2018).

29. Kelly K. Sports, "Top Small Workplaces 2008," *The Wall Street Journal,* February 22, 2009, https://www.wsj.com/articles/SB122347733961315417 (accessed January 1, 2018); "ATA Engineering Named as One of San Diego's 100 Fastest Growing Companies," ATA Engineering, October 2, 2008, http://www.ata-e.com/news/ata-engineering-named-one-san-diegos-100-fastest-growing-companies/ (accessed January 1, 2018); "This Year's Most Innovative Employee-Owned Companies Awarded in Chicago Last Week," The National Center for Employee Ownership and the Beyster Institute, http://www.ata-e.com/news/ata-wins-most-innovative-employee-owned-company-award/ (accessed January 1, 2018); Michelle Strulzenberger, "Employee Ownership Results in Better Workplace for ATA Engineering Inc.," January 16, 2009, http://axiomnews.ca/node/456 (accessed June 24, 2013); ATA Engineering, "ATA Engineering Recognized by Wall Street Journal as One of the Top Small Workplaces for 2008," October 13, 2008, http://www.ata-e.com/news/ata-engineering-recognized-wall-street-journal-one-top-small-workplaces-2008/ (accessed January 1, 2018); ATA Engineering website, http://www.ata-e.com/ (accessed January 1, 2018); ATA Engineering, "ATA Engineering Wins Top NASA Quality Award," June 26, 2013, http://www.ata-e.com/news/ata-engineering-wins-top-nasa-quality-award/ (accessed January 1, 2018); Bloomberg, "Company Overview of ATA Engineering, Inc.," February 17, 2018, https://www.bloomberg.com/research/stocks/private/snapshot.asp?privcapId=9636531 (accessed February 17, 2018); CalTech, "Mary Baker (MS '67, PHD '72)," *CalTech Alumni,* https://www.alumni.caltech.edu/mary-baker/ (accessed February 17, 2018).

30. Eli Lilly, "Heritage," www.lilly.com/about/heritage/Pages/heritage.aspx (accessed May 1, 2017).

31. Marcia Kaplan, "Walmart Ecommerce Sales Up 23 Percent in Fourth Quarter 2017," *Practical Ecommerce,* March 1, 2018, https://www.practicalecommerce.com/walmarts-ecommerce-sales-23-percent-fourth-quarter-2017 (accessed March 25, 2018).

32. Rainforest Alliance, "Annual Report 2012," 2012, http://www.rainforest-alliance.org/sites/default/files/about/annual_reports/AR2012_spreads-optimized.pdf (accessed June 24, 2013); Rainforest Alliance, "About Us," http://www.rainforest-alliance.org/about (accessed January 1, 2018); Kelly K. Spors, "Top Small Workplaces 2008," *The Wall Street Journal,* February 22, 2009, https://www.wsj.com/articles/SB122347733961315417 (accessed January 1, 2018); Richard Donovan, "Rainforest Alliance Launches TREES," Forest Stewardship Council, www.fscus.org/news/index.php?article=169 (accessed May 29, 2009); Rainforest Alliance, "Marketing Your Commitment to Sustainability," https://www.rainforest-alliance.org/business/marketing (accessed January 1, 2018); Rainforest Alliance, "Raising Awareness . . . Among Employees," https://www.rainforest-alliance.org/business/marketing/awareness/employees (accessed January 1, 2018); Rainforest Alliance, "What Does Rainforest Alliance CertifiedTM Mean?" October 25, 2016, https://www.rainforest-alliance.org/faqs/what-does-rainforest-alliance-certified-mean (accessed January 1, 2018).

33. Laura J. Keller and Shahien Nasiripour, "Wells Fargo's Uphill Battle," *Bloomberg Businessweek,* March 5, 2018, pp. 31–32.

34. Joann S. Lublin, "New Report Finds a Diversity Dividend at Work," *The Wall Street Journal,* January 20, 2015, http://blogs.wsj.com/atwork/2015/01/20/new-report-finds-a-diversity-dividend-at-work/ (accessed May 1, 2017).

35. "The 2011 World's Most Ethical Companies," *Ethisphere,* 2011, Q1, pp. 31–43.

36. Isabelle Maignon, Tracy L. Gonzalez-Padron, G. Tomas M. Hult, and O. C. Ferrell, "Stakeholder Orientation: Development and Testing of a Framework for Socially Responsible Marketing," *Journal of Strategic Marketing,* 19, no. 4 (July 2011), p. 313–338.

37. Small Business Administration Office of Advocacy, Frequently Asked Questions, 2012, www.sba.gov/sites/default/files/FAQ_Sept_2012.pdf (accessed May 1, 2017); Joel Holland, "Save the World, Make a Million," *Entrepreneur,* April 2010, http://www.entrepreneur.com/article/205556 (accessed May 1, 2017); iContact, www.icontact.com (accessed May 1, 2017); Leigh Buchanan, "The U.S. Now Has 27 Million Entrepreneurs," *Inc.,* http://www.inc.com/leigh-buchanan/us-entrepreneurshipreaches-record-highs.html (accessed May 1, 2017).

38. Small Business Administration Office of Advocacy, Frequently Asked Questions, 2012, www.sba.gov/sites/

default/files/FAQ_Sept_2012.pdf (accessed February 23, 2016); Joel Holland, "Save the World, Make a Million," *Entrepreneur,* April 2010, http://www.entrepreneur.com/article/205556 (accessed February 23, 2016); iContact, www.icontact.com (accessed February 23, 2016); Leigh Buchanan, "The U.S. Now Has 27 Million Entrepreneurs," *Inc.,* http://www.inc.com/leigh-buchanan/us-entrepreneurship-reaches-record-highs.html (accessed February 23, 2016).

Credits

Design elements: Part opener, ©Steve Allen/Getty Images; Consider Ethics and Social Responsibility icon, ©Design Pics/PunchStock; Think Globally icon, ©Sheff/Shutterstock; Responding to Business Challenges icon, ©Olivier LeMoal/Shutterstock; Going Green icon, ©Beboy/Shutterstock; Entrepreneurship in Action icon, ©Ruslan Grechka/Shutterstock; Test Prep icon, ©McGraw-Hill Education; Build Your Skills icon, ©Ilya Terentyev/Getty Images; Solve the Dilemma icon, ©Beautyimage/Shutterstock; Build Your Business Plan icon, ©ALMAGAMI/ Shutterstock; See for Yourself Videocase icon, ©MIKHAIL GRACHIKOV/Shutterstock.

2 Business Ethics and Social Responsibility

©JC Ridley/CSM/REX/Shutterstock

Chapter Outline

Learning Objectives

After reading this chapter, you will be able to:

LO 2-1 Describe the importance of business ethics and social responsibility.

LO 2-2 Detect some of the ethical issues that may arise in business.

LO 2-3 Specify how businesses can promote ethical behavior.

LO 2-4 Explain the four dimensions of social responsibility.

LO 2-5 Evaluate an organization's social responsibilities to owners, employees, consumers, the environment, and the community.

LO 2-6 Evaluate the ethics of a business's decision.

Enter the World of Business

NFL Tackles Safety Expectations

Playing professional football is not one of the "safest" jobs in the world, especially in the National Football League (NFL). Over the past few decades, retired NFL players have raised concerns about how repetitive head injuries/concussions sustained during games have affected them later in life. Many retired NFL players have faced neurological problems later in life, including permanent brain damage, dementia, and higher-than-average incidents of Alzheimer's disease and clinical depression.

Various retired NFL players brought lawsuits against the NFL, arguing that the NFL knew or should have known the risks and did not do enough to prevent these injuries. The NFL paid nearly $1 billion to retired NFL players who had suffered injuries in this regard and provided $10 million to fund brain injury research and education/safety programs. Researchers later discovered that 177 out of 202 brains from former football players who had played football sometime during their lives showed evidence of brain degenerative disease. These risks have convinced some NFL players to retire early. Parents are faced with an ethical issue about whether they should allow their children to play football in school at the risk of long-term injury. One poll found parents are 44 percent less likely to allow their children to play football.

Some observers note that football players accept the inherent dangers of the game. Seattle Seahawks star cornerback Richard Sherman has pointed out that NFL players have chosen their profession, know the risks, and have decided to play anyway. However, quarterback Brett Favre states that he has experienced memory loss. He believes concussions have had a negative impact on his life and fears he might develop a brain disease experienced by many retired NFL players.[1]

Introduction

Any organization, including nonprofits, has to manage the ethical behavior of employees and participants in the overall operations of the organization. Firms that are highly ethical tend to be more profitable with more satisfied employees and customers.[2] Therefore, there are no conflicts between profits and ethics—in fact, unethical conduct is more likely to lower profits than raise them. For instance, Volkswagen pleaded guilty to criminal charges for cheating on U.S. emissions tests, and the company could face $25 billion in the United States for fines, vehicle buybacks, and repairs.[3] Wrongdoing by some businesses has focused public attention and government involvement on encouraging more acceptable business conduct. Any organizational decision may be judged as right or wrong, ethical or unethical, legal or illegal.

In this chapter, we take a look at the role of ethics and social responsibility in business decision making. First, we define *business ethics* and examine why it is important to understand ethics' role in business. Next, we explore a number of business ethics issues to help you learn to recognize such issues when they arise. Finally, we consider steps businesses can take to improve ethical behavior in their organizations. The second half of the chapter focuses on social responsibility and unemployment. We describe some important issues and detail how companies have responded to them.

Describe the importance of business ethics and social responsibility.

business ethics
principles and standards that determine acceptable conduct in business.

Business Ethics and Social Responsibility

In this chapter, we define **business ethics** as the principles and standards that determine acceptable conduct in business organizations. Personal ethics, on the other hand, relates to an individual's values, principles, and standards of conduct. The acceptability of behavior in business is determined by not only the organization, but also stakeholders such as customers, competitors, government regulators, interest groups, and the public, as well as each individual's personal principles and values. The publicity and debate surrounding highly visible legal and ethical issues at a number of well-known firms, including Wells Fargo and Volkswagen, highlight the need for businesses to integrate ethics and responsibility into all business decisions. For instance, Wells Fargo provided incentives to its sales department that resulted in opening 3.5 million accounts without customer knowledge. This resulted in lowering customers' credit ratings and additional expenses for customers. Most unethical activities within organizations are supported by an organizational culture that encourages employees to bend the rules. On the other hand, trust in business is the glue that holds relationships together. In Figure 2.1, you can see that trust in financial services is lower than in other industries. While the majority of the population trusts business, a significant portion does not.

Organizations that exhibit a high ethical culture encourage employees to act with integrity and adhere to business values. For example, Illycaffé, an Italian family business, has been recognized as an ethical leader in quality, sustainability, and supply chain practices. The company is a leader in the science and technology of coffee and the world's most global coffee brand.[4] Many experts agree that ethical leadership, ethical values, and compliance

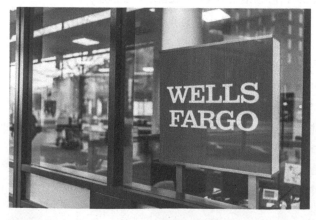

Wells Fargo opened 3.5 million fake accounts that hurt customers credit ratings.

©Kristi Blokhin/Shutterstock

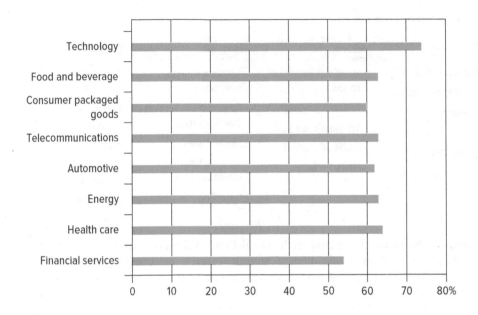

FIGURE 2.1
Global Trust in Different Industries

Source: Edelman, *2018 Edelman Trust Barometer Global Report,* http://cms.edelman.com/sites/default/files/2018-01/2018_Edelman_Trust_Barometer_Global_Report_Jan.PDF (accessed January 27, 2018).

are important in creating good business ethics. To truly create an ethical culture, however, managers must show a strong commitment to ethics and compliance. This "tone at the top" requires top managers to acknowledge their own role in supporting ethics and compliance, clearly communicate company expectations for ethical behavior to all employees, educate all managers and supervisors in the business about the company's ethics policies, and train managers and employees on what to do if an ethics crisis occurs.[5]

Businesses should not only make a profit, but also consider the social implications of their activities. For instance, Walmart to date has donated more than 1.2 billion pounds of food and, combined with the Walmart Foundation, has donated more than $122 million in grants to the charitable meal system.[6] However, profits permit businesses to contribute to society. The firms that are more well known for their strong social contributions tend to be those that are more profitable. We define **social responsibility** as a business's obligation to maximize its positive impact and minimize its negative impact on society. Although many people use the terms *social responsibility* and *ethics* interchangeably, they do not mean the same thing. Business ethics relates to an *individual's* or a *work group's* decisions that society evaluates as right or wrong, whereas social responsibility is a broader concept that concerns the impact of the *entire business's* activities on society. From an ethical perspective, for example, we may be concerned about a health care organization overcharging the government for Medicare services. From a social responsibility perspective, we might be concerned about the impact that this overcharging will have on the ability of the health care system to provide adequate services for all citizens.[7]

The most basic ethical and social responsibility concerns have been codified by laws and regulations that encourage businesses to conform to society's standards, values, and attitudes. Laws and regulations attempt to institutionalize ethical conduct and prevent harm to customers, the environment, and other stakeholders. Accounting, finance, and marketing professionals have to understand laws and regulations that

social responsibility
a business's obligation to maximize its positive impact and minimize its negative impact on society.

TABLE 2.1

Timeline of Ethical and Socially Responsible Activities

1960s	1970s	1980s	1990s	2000s
• Social issues	• Business ethics	• Standards for ethical conduct	• Corporate ethics programs	• Transparency in financial markets
• Consumer Bill of Rights	• Social responsibility	• Financial misconduct	• Regulation to support business ethics	• Cyber security
• Disadvantaged consumer	• Diversity	• Self-regulation	• Health issues	• Intellectual property
• Environmental issues	• Bribery	• Codes of conduct	• Safe working conditions	• Regulation of accounting and finance
• Product safety	• Discrimination	• Ethics training	• Detecting misconduct	• Executive compensation
	• Identifying ethical issues			• Identity theft

apply to their work. For example, the Dodd-Frank Act was passed to reform the financial industry and offer consumers protection against complex and/or deceptive financial products. At a minimum, managers are expected to obey all laws and regulations. Most legal issues arise as choices that society deems unethical, irresponsible, or otherwise unacceptable. However, all actions deemed unethical by society are not necessarily illegal, and both legal and ethical concerns change over time (see Table 2.1). More recently, identity theft has become the number-one consumer complaint with the Federal Trade Commission, and companies have an ethical responsibility to protect customer data. *Business law* refers to the laws and regulations that govern the conduct of business. Many problems and conflicts in business could be avoided if owners, managers, and employees knew more about business law and the legal system. Business ethics, social responsibility, and laws together act as a compliance system, requiring that businesses and employees act responsibly in society. In this chapter, we explore ethics and social responsibility. Bonus Chapter A provides an overview of the legal and regulatory environment.

The Role of Ethics in Business

You have only to pick up *The Wall Street Journal* or *USA Today* to see examples of the growing concern about legal and ethical issues in business. For example, the federal government accused Bristol-Myers Squibb of bribing state-owned hospitals in China to gain prescription sales. The company paid $14 million to settle the allegations.[8] Regardless of what an individual believes about a particular action, if society judges it to be unethical or wrong, whether correctly or not, that judgment directly affects the organization's ability to achieve its business goals.[9]

Many firms are recognized for their ethical conduct. 3M has been placed on the World's Most Ethical Companies list five years in a row. Furthermore, 3M's chief compliance officer states, "it is not enough to just win in business—it matters how you do it." 3M sees ethics as a competitive advantage.[10] The mass media frequently report about firms that engage in misconduct related to bribery, fraud, and unsafe products. However, the good ethical conduct of the vast majority of firms is not reported as often. Therefore, the public often gets the impression that misconduct is more widespread than it is in reality.

Often, misconduct starts as ethical conflicts but evolves into legal disputes when cooperative conflict resolution cannot be accomplished. This is because individuals may have different ethical beliefs and resort to legal activities to resolve issues. Also, there are many ethical gray areas, which occur when a new, undetermined or ambiguous situation arises. There may be no values, codes, or laws that answer the question about appropriate action. The sharing economy with peer-to-peer relationships like Uber, Lyft, and Airbnb provide new business models where existing regulations were inadequate, ambiguous, or in some cases blocking progress. For example, Uber has been accused of price gouging, endangering riders' safety, sexual harassment, stealing secret information on self-driving cars from Google, and violating local regulations on public transportation in some countries, states, and cities. To settle some of its legal fees, it has paid more than $500 million in the last two years.[11]

However, it is important to understand that business ethics goes beyond legal issues. Ethical conduct builds trust among individuals and in business relationships, which validates and promotes confidence in business relationships. Establishing trust and confidence is much more difficult in organizations that have reputations for acting unethically. If you were to discover, for example, that a manager had misled you about company benefits when you were hired, your trust and confidence in that company would probably diminish. And if you learned that a colleague had lied to you about something, you probably would not trust or rely on that person in the future.

Ethical issues are not limited to for-profit organizations either. Ethical issues include all areas of organizational activities. In government, politicians and some high-ranking officials have faced disciplinary actions over ethical indiscretions. There has been ethical misconduct in sports, and even ethical lapses in well-known nonprofits, such as the American Red Cross. Whether made in science, politics, sports, or business, most decisions are judged as right or wrong, ethical or unethical. Negative judgments can affect an organization's ability to build relationships with customers and suppliers, attract investors, and retain employees.[12]

Although we will not tell you in this chapter what you ought to do, others—your superiors, co-workers, and family—will make judgments about the ethics of your actions and decisions. Learning how to recognize and resolve ethical issues is a key step in evaluating ethical decisions in business.

Recognizing Ethical Issues in Business

Recognizing ethical issues is the most important step in understanding business ethics. An **ethical issue** is an identifiable problem, situation, or opportunity that requires a person to choose from among several actions that may be evaluated as right or wrong, ethical or unethical. Learning how to choose from alternatives and make a decision requires not only good personal values, but also knowledge competence in the business area of concern. Employees also need to know when to rely on their organizations' policies and codes of ethics or have discussions with co-workers or managers on appropriate conduct. Ethical decision making is not always easy because there are always gray areas that create dilemmas, no matter how decisions are made. For instance, should an employee report on a co-worker engaging in time theft? Or should you report a friend cheating on a test? Should a salesperson omit facts about a product's poor safety record in his presentation to a customer? Such questions require the decision maker to evaluate the ethics of his or her choice and decide whether to ask for guidance.

One of the principal causes of unethical behavior in organizations is rewards for overly aggressive financial or business objectives. It is not possible to discuss every issue, of course. However, a discussion of a few issues can help you begin to recognize

LO 2-2

Detect some of the ethical issues that may arise in business.

ethical issue
an identifiable problem, situation, or opportunity that requires a person to choose from among several actions that may be evaluated as right or wrong, ethical or unethical.

the ethical problems with which businesspersons must deal. Many ethical issues in business can be categorized in the context of their relation with abusive and intimidating behavior, conflicts of interest, fairness and honesty, communications, misuse of company resources, and business associations. The Global Business Ethics Survey found that workers witness many instances of ethical misconduct in their organizations and sometimes feel pressured to compromise standards (see Table 2.2). Overall, 47 percent of employees surveyed observed misconduct. Many observed multiple issues including abusive behavior (26 percent), lying (22 percent), and conflict of interest (15 percent).

To help you understand ethical issues that perplex businesspeople today, we will take a brief look at some of them in this section. Ethical issues can be more complex now than in the past. The vast number of news-format investigative programs has increased consumer and employee awareness of organizational misconduct. In addition, the multitude of cable channels and Internet resources has improved the awareness of ethical problems among the general public.

Bribery.   Many business issues seem straightforward and easy to resolve on the surface, but are, in reality, very complex. A person often needs several years of experience in business to understand what is acceptable or ethical. For example, it is considered improper to give or accept **bribes**, which are payments, gifts, or special favors intended to influence the outcome of a decision. A bribe benefits an individual or a company at the expense of other stakeholders. Companies that do business overseas should be aware that bribes are a significant ethical issue and are, in fact, illegal in many countries. In the United States, the Foreign Corrupt Practices Act imposes heavy penalties on companies found guilty of bribing foreign government officials.

Ethics is also related to the culture in which a business operates. In the United States, for example, it would be inappropriate for a businessperson to bring an elaborately wrapped gift to a prospective client on their first meeting—the gift could be viewed as a bribe. In Japan, however, it is considered impolite *not* to bring a gift. Experience with the culture in which a business operates is critical to understanding what is ethical or unethical. On the other hand, firms must also abide by the values and policies of global business.

bribes
payments, gifts, or special favors intended to influence the outcome of a decision.

TABLE 2.2
Organizational Misconduct in the United States

Misconduct Facts	Percentage
Observed misconduct	47
Abusive behavior	26
Lying to stakeholders	22
Conflict of interest	15
Internet abuse	16
Health violations	15
Pressure to compromise standards	16
Report observed misconduct	69
Experience retaliation for reporting	44

Source: Ethics and Compliance Initiative, 2018 Global Business Ethics Survey™: The State of Ethics and Compliance in the Workplace (Arlington, VA: Ethics and Compliance Initiative, 2018).

South Korea's president, Park Geun-hye was removed from office over a bribery scandal. The heir to the Samsung Group is standing trial in the investigation into bribery and embezzlement charges. There was a "donation" of $25 million to the National Pension Fund at the time of a Samsung merger. This case heightens the awareness of the political risks associated with bribery.[13] Such political scandals demonstrate that political ethical behavior must be proactively practiced at all levels of public service.

Misuse of Company Time. Theft of time is a common area of misconduct observed in the workplace.[14] One example of misusing time in the workplace is by engaging in activities that are not necessary for the job. For instance, many employees spend an average of one hour each day using social networking sites. Some companies have cho-

A former Siemens executive plead guilty in a $100 million Argentina bribery case. The engineering firm itself was involved in a decades long legal investigation.

©LUKAS BARTH/EPA-EFE/REX/Shutterstock

sen to block certain sites such as Facebook, YouTube, or Pandora from employees. In this case, the employee is misusing not only time, but also company resources by using the company's computer and Internet access for personal use.[15] Time theft costs can be difficult to measure but are estimated to cost companies hundreds of billions of dollars annually. It is widely believed that the average employee "steals" 4.5 hours a week with late arrivals, leaving early, long lunch breaks, inappropriate sick days, excessive socializing, and engaging in personal activities such as online shopping and watching sports while on the job. For example, on Cyber Monday, nearly 25 percent of employees say they shop online while at work.[16] All of these activities add up to lost productivity and profits for the employer–and relate to ethical issues in the area of time theft.

Abusive and Intimidating Behavior. Abusive or intimidating behavior is the most common ethical problem for employees. These concepts can mean anything from physical threats, false accusations, profanity, insults, yelling, harshness, and unreasonableness to ignoring someone or simply being annoying; and the meaning of these words can differ by person— you probably have some ideas of your own. Abusive behavior can be placed on a continuum from a minor distraction to a disruption of the workplace. For example, what one person may define as yelling might be another's definition of normal speech. Civility in our society is a concern, and the workplace is no exception. The productivity level of many organizations has been diminished by the time spent unraveling abusive relationships.

Abusive behavior is difficult to assess and manage because of diversity in culture and lifestyle. What does it mean to speak profanely? Is profanity only related to specific words or other such terms that are common in today's business world? If you

Misuse of company time through personal social media use and online shopping is very costly to businesses.

©Purestock/SuperStock

are using words that are normal in your language but that others consider to be profanity, have you just insulted, abused, or disrespected them?

Within the concept of abusive behavior, intent should be a consideration. If the employee was trying to convey a compliment but the comment was considered abusive, then it was probably a mistake. The way a word is said (voice inflection) can be important. Add to this the fact that we now live in a multicultural environment—doing business and working with many different cultural groups—and the businessperson soon realizes the depth of the ethical and legal issues that may arise. There are problems of word meanings by age and within cultures. For example, an expression such as "Did you guys hook up last night?" can have various meanings, including some that could be considered offensive in a work environment.

Bullying is associated with a hostile workplace when a person or group is targeted and is threatened, harassed, belittled, verbally abused, or overly criticized. Bullying may create what some consider a hostile environment, a term generally associated with sexual harassment. Recently there has been an explosion of concern about sexual harassment. Ethical misconduct has been unveiled with exposure of sexual harassment that has been part of the culture of the entertainment and hospitality industries, as well as other areas such as government and many corporations.[17] Although sexual harassment has legal recourse, bullying has little legal recourse at this time. Bullying is a widespread problem in the United States and can cause psychological damage that can result in health-endangering consequences to the target. Surveys reveal that bullying in the workplace is on the rise.[18] As Table 2.3 indicates, bullying can use a mix of verbal, nonverbal, and manipulative threatening expressions to damage workplace productivity. One may wonder why workers tolerate such activities. The problem is that 72 percent of bullies outrank their victims.[19]

Misuse of Company Resources.   Misuse of company resources has been identified by the Ethics Resource Center as a leading issue in observed misconduct in organizations. Issues might include spending an excessive amount of time on personal e-mails, submitting personal expenses on company expense reports, or using the company copier for personal use. Six Howard University employees were fired after they allegedly stole nearly $1 million via grants and tuition remission from the financial aid department.[20] While serious resource abuse can result in firing, some abuse can have legal

TABLE 2.3 **Actions Associated with Bullies**	
	1. Spreading rumors to damage others
	2. Blocking others' communication in the workplace
	3. Flaunting status or authority to take advantage of others
	4. Discrediting others' ideas and opinions
	5. Using e-mail to demean others
	6. Failing to communicate or return communication
	7. Insults, yelling, and shouting
	8. Using terminology to discriminate by gender, race, or age
	9. Using eye or body language to hurt others or their reputation
	10. Taking credit for others' work or ideas

repercussions. An Apple store employee was arrested for allegedly re-coding American Express and Visa gift cards and then using them to fraudulently purchase nearly $1 million in Apple gift cards.[21]

Employee internal theft or the misuse of the employer's assets is a major loss of resources for many firms, especially retailers. For example, employees may hide company items in a handbag, backpack, or briefcase. Customers may be overcharged while the employees keep the extra money. Employees may ship personal items using a firm's account number. Contract maintenance personnel may steal materials or office equipment. Food service employees may provide free drinks or food to friends or even customers hoping for an extra tip. Estheticians or hair stylists may pocket money when clients pay with cash. There are many others way to steal from the company. Firms need a good monitoring system and employee training to prevent the theft of resources.

Upset by alleged financial aid embezzlement by Howard University employees, students occupied the administration building to protest the misuse of university resources.

©Astrid Riecken For The Washington Post via Getty Images

The most common way that employees abuse resources is by using company computers for personal use. Typical examples of using a computer for personal use include shopping on the Internet, downloading music, doing personal banking, surfing the Internet for entertainment purposes, or visiting Facebook. Some companies choose to take a flexible approach to addressing this issue. For example, many have instituted policies that allow for some personal computer use as long as the use does not detract significantly from the workday.

No matter what approach a business chooses to take, it must have policies in place to prevent company resource abuse. Because misuse of company resources is such a widespread problem, many companies, like Coca-Cola, have implemented official policies delineating acceptable use of company resources. Coca-Cola's policy states that company assets should not be used for personal benefit but does allow employees some freedom in this area. The policy specifies that it is acceptable for employees to make the occasional personal phone call or e-mail, but they should use common sense to know when these activities become excessive.[22] This kind of policy is in line with that of many companies, particularly large ones that can easily lose millions of dollars and thousands of hours of productivity to these activities.

Conflict of Interest.   A conflict of interest, one of the most common ethical issues identified by employees, exists when an individual must choose whether to advance the individual's own personal interests or those of others. For example, a manager in a corporation is supposed to ensure that the company is profitable so that its stockholder-owners receive a return on their investment. In other words, the manager has a responsibility to investors. If she instead makes decisions that give her more power or money but do not help the company, then she has a conflict of interest—she is acting to benefit herself at the expense of her company and is not fulfilling her responsibilities as an employee. To avoid conflicts of interest, employees must be able to separate their personal financial interests from their business dealings. Conflict of interest has long been a serious problem in the financial industry. Asset management firm BlackRock paid $12 million to the U.S. Securities and Exchange Commission for not disclosing that its fund manager

The U.S. Securities and Exchange Commission is an independent government agency that protects investors and oversees securities transactions to prevent fraud.

©Kristi Blokhin/Shutterstock

had a major interest in a firm in which he deposited clients' money.[23] Conflict of interest can be particularly problematic in the finance industry because bad decisions can result in significant financial losses.

Insider trading is an example of a conflict of interest. Insider trading is the buying or selling of stocks by insiders who possess material that is still not public. Bribery can also be a conflict of interest. While bribery is an increasing issue in many countries, it is more prevalent in some countries than in others. Transparency International has developed a Corruption Perceptions Index (Table 2.4). Note that there are 17 countries perceived as less corrupt than the United States. The five countries rated by Transparency International as most corrupt include Yemen, Syria, North Korea, South Sudan, and Somalia.[24]

TABLE 2.4
Least Corrupt Countries

Rank	Country
1.	New Zealand
2.	Denmark
3.	Finland
3.	Norway
3.	Switzerland
6.	Singapore
6.	Sweden
8.	Canada
8.	Luxembourg
8.	Netherlands
8.	United Kingdom
12.	Germany
13.	Australia
13.	Hong Kong
13.	Iceland
16.	Austria
16.	Belgium
16.	United States
19.	Ireland
20.	Japan

Source: Corruption Perceptions Index 2017, Transparency https://www.transparency.org/news/feature/corruption_perceptions_index_2017 (accessed March 26, 2018).

Fairness and Honesty

Fairness and honesty are at the heart of business ethics and relate to the general values of decision makers. At a minimum, businesspersons are expected to follow all applicable laws and regulations. But beyond obeying the law, they are expected not to harm customers, employees, clients, or competitors knowingly through deception, misrepresentation, coercion, or discrimination. Honesty and fairness can relate to how the employees use the resources of the organization. In contrast, dishonesty is usually associated with a lack of integrity, lack of disclosure, and lying. One common example of dishonesty is theft of office supplies. Although the majority of office supply thefts involve small things such as pencils or Post-it Notes, some workers admit to stealing more expensive items or equipment such as computers or software. Employees should be aware of policies on stealing items and recognize how these decisions relate to ethical behavior.

One aspect of fairness relates to competition. For example, the former CEO of Uber was alleged to have conspired with a former Google engineer to steal trade secrets related to driverless cars.[25] Although numerous laws have been passed to foster competition and make monopolistic practices illegal, companies sometimes gain control over markets by using questionable practices that harm competition. For instance, the European Commission started an antitrust investigation into Google's practices to determine whether it was engaging in anticompetitive behavior. Several companies, including Microsoft, claimed that Google promoted its own search results over those of competitors in spite of their relevance. Because Google holds 90 percent of the search engine market in Europe, the controversy over how it is using its dominant position to remain ahead of competitors is not likely to die down.[26]

Another aspect of fairness and honesty relates to disclosure of potential harm caused by product use. For instance, the FDA has become increasingly concerned about food safety rules after a contamination crisis involving Blue Bell ice cream

---

Going Green

Chipotle Chips Away at Food Safety Advancements

Chipotle Mexican Grill has always done things differently from other restaurant chains. Steve Ells, founder and former co-CEO, established the first Chipotle restaurant as a high-end fast-casual restaurant chain. Fast-casual chains do not provide table service but offer higher-quality items than fast-food restaurants. Ells wanted to set Chipotle apart by creating a customer-service-oriented atmosphere where customers could see their food being prepared with fresh ingredients. Chipotle's "food with integrity" offerings use fresh food with ingredients grown naturally and sustainably from local farmers whenever possible. About 40 percent of beans used in its food offerings are organic, and Chipotle purchases about 10 million pounds of vegetables from local farms every year.

Despite its desire to be socially responsible, Chipotle encountered challenges in food safety. An outbreak of *E. coli* was traced to a small number of Chipotle stores. Ells reacted quickly, closing 43 restaurants in two states to completely rid the stores of any possible contaminated products, clean and sanitize their entire kitchens, and retrain staff on proper food handling procedures. The chain distributed free burrito coupons to win back customers. Ells overhauled Chipotle's managerial focus from career development back to day-to-day activities. The company's strong ethical reputation and quick response saved the firm.[27]

Critical Thinking Questions

1. How does Chipotle attempt to be socially responsible?
2. What actions did Chipotle take to assure stakeholders it was taking the E. coli outbreak seriously?
3. Why do you think customers are willing to do business with Chipotle even after the E. coli outbreak?

caused three deaths. The agency adopted new rules that now require food manufacturers to create and enforce detailed plans meant to prevent foodborne illness and contamination.[28]

Dishonesty has become a significant problem in the United States. In a survey of new students at Harvard, 23 percent admitted to cheating on schoolwork before attending Harvard, and 9 percent revealed that they had cheated on exams. Air Force officers at the Malmstrom Air Force Base in Montana were suspended after widespread cheating on monthly proficiency tests on operating warheads. Similarly, the U.S. Navy was criticized when sailors cheated on qualification exams for becoming nuclear reactor instructors. The Defense Secretary at the time believed the issue might be systematic, requiring an ethics overhaul in the military.[29]

Communications.   Communications is another area in which ethical concerns may arise. False and misleading advertising, as well as deceptive personal-selling tactics, anger consumers and can lead to the failure of a business. Truthfulness about product safety and quality is also important to consumers. Takata pleaded guilty to fraud and providing false data and agreed to pay $1 billion in a settlement to victims and car manufacturers for exploding airbags.[30]

Another important aspect of communications that may raise ethical concerns relates to product labeling. This becomes an even greater concern with potentially harmful products like cigarettes. The FDA warned three cigarette manufacturers against using "additive-free" or "natural" on their labeling out of concern that consumers would associate these terms as meaning that their products were healthier.[31] However, labeling of other products raises ethical questions when it threatens basic rights, such as freedom of speech and expression. This is the heart of the controversy surrounding the movement to require warning labels on movies and videogames, rating their content, language, and appropriate audience age. Although people in the entertainment industry claim that such labeling violates their First Amendment right to freedom of expression, other consumers—particularly parents—believe that labeling is needed to protect children from harmful influences. Internet regulation, particularly that designed to protect children and the elderly, is on the forefront in consumer protection legislation. Because of the debate surrounding the acceptability of these business activities, they remain major ethical issues.

Business Relationships.   The behavior of businesspersons toward customers, suppliers, and others in their workplace may also generate ethical concerns. Ethical behavior within a business involves keeping company secrets, meeting obligations and responsibilities, and avoiding undue pressure that may force others to act unethically.

Managers in particular, because of the authority of their position, have the opportunity to influence employees' actions. The National Business Ethics Survey found that employees who feel pressured to compromise ethical standards view top and middle managers as the greatest source of such pressure.[32]

It is the responsibility of managers to create a work environment that helps the organization achieve its objectives and fulfill its responsibilities. However, the methods that managers use to enforce these responsibilities should not compromise employee rights. Organizational pressures may encourage a person to engage in activities that he or she might otherwise view as unethical, such as invading others' privacy or stealing a competitor's secrets. The firm may provide only vague or lax supervision on ethical issues, creating the opportunity for misconduct. Managers who offer no ethical direction to employees create many opportunities for manipulation, dishonesty, and conflicts of interest. This happened to Wells Fargo in creating 3.5 million fake

Turnitin is an Internet service that allows teachers to determine if their students have plagiarized content.

©Daniel Acker/Bloomberg via Getty Images

accounts that hurt customers' credit ratings as well as other misconduct that damaged customers. The Federal Reserve restricted the bank's ability to grow until they could provide more oversight and reduce risks.[33]

Plagiarism—taking someone else's work and presenting it as your own without mentioning the source—is another ethical issue. As a student, you may be familiar with plagiarism in school—for example, copying someone else's term paper or quoting from a published work or Internet source without acknowledging it. In business, an ethical issue arises when employees copy reports or take the work or ideas of others and present it as their own. A manager attempting to take credit for a subordinate's ideas is engaging in another type of plagiarism.

> **plagiarism**
> the act of taking someone else's work and presenting it as your own without mentioning the source.

Making Decisions about Ethical Issues

It can be difficult to recognize specific ethical issues in practice. Managers, for example, tend to be more concerned about issues that affect those close to them, as well as issues that have immediate rather than long-term consequences. Thus, the perceived importance of an ethical issue substantially affects choices. However, only a few issues receive scrutiny, and most receive no attention at all.[34] Managers make intuitive decisions sometimes without recognizing the embedded ethical issue.

Table 2.5 lists some questions you may want to ask yourself and others when trying to determine whether an action is ethical. Open discussion of ethical issues does not eliminate ethical problems, but it does promote both trust and learning in an organization.[35] When people feel that they cannot discuss what they are doing with their co-workers or superiors, there is a good chance that an ethical issue exists. Once a

TABLE 2.5
Questions to Consider in Determining Whether an Action Is Ethical

Are there any potential legal restrictions or violations that could result from the action?
Does your company have a specific code of ethics or policy on the action?
Is this activity customary in your industry? Are there any industry trade groups that provide guidelines or codes of conduct that address this issue?
Would this activity be accepted by your co-workers? Will your decision or action withstand open discussion with co-workers and managers and survive untarnished?
How does this activity fit with your own beliefs and values?

person has recognized an ethical issue and can openly discuss it with others, he or she has begun the process of resolving that issue.

LO 2-3

Specify how businesses can promote ethical behavior.

Improving Ethical Behavior in Business

Understanding how people make ethical choices and what prompts a person to act unethically may result in better ethical decisions. Ethical decisions in an organization are influenced by three key factors: individual moral standards and values, the influence of managers and co-workers, and the opportunity to engage in misconduct (Figure 2.2). While you have great control over your personal ethics outside the workplace, your co-workers and superiors exert significant control over your choices at work through authority and example. In fact, the activities and examples set by co-workers, along with rules and policies established by the firm, are critical in gaining consistent ethical compliance in an organization. If the company fails to provide good examples and direction for appropriate conduct, confusion and conflict will develop and result in the opportunity for misconduct. If your boss or co-workers leave work early, you may be tempted to do so as well. If you see co-workers engaged in personal activities such as shopping online or if they ignore the misconduct of others, then you may be more likely to do so also. Having sound personal values is important because you will be responsible for your own conduct.

Because ethical issues often emerge from conflict, it is useful to examine the causes of ethical conflict. Business managers and employees often experience some tension between their own ethical beliefs and their obligations to the organizations in which they work. Many employees utilize different ethical standards at work than they do at home. This conflict increases when employees feel that their company is encouraging unethical conduct or exerting pressure on them to engage in it.

It is difficult for employees to determine what conduct is acceptable within a company if the firm does not have established ethics policies and standards. And without such policies and standards, employees may base decisions on how their peers and superiors behave. Professional **codes of ethics** are formalized rules and standards that

codes of ethics
formalized rules and standards that describe what a company expects of its employees.

FIGURE 2.2
Three Factors That Influence Business Ethics

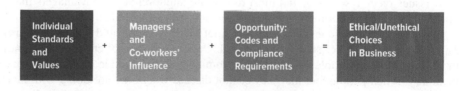

• Alerts employees about important issues and risks to address.
• Provides values such as integrity, transparency, honesty, and fairness that give the foundation for building an ethical culture.
• Gives guidance to employees when facing gray or ambiguous situations or ethical issues that they have never faced before.
• Alerts employees to systems for reporting or places to go for advice when facing an ethical issue.
• Helps establish uniform ethical conduct and values that provides a shared approach to dealing with ethical decisions.
• Serves as an important document for communicating to the public, suppliers, and regulatory authorities about the company's values and compliance.
• Provides the foundation for evaluation and improvement of ethical decision making.

TABLE 2.6
Why a Code of Ethics Is Important

describe what the company expects of its employees. Codes of ethics do not have to be so detailed that they take into account every situation, but they should provide guidelines and principles that can help employees achieve organizational objectives and address risks in an acceptable and ethical way. The development of a code of ethics should include not only a firm's executives and board of directors, but also legal staff and employees from all areas of a firm.[36] Table 2.6 lists why a code of ethics is important.

Codes of ethics, policies on ethics, and ethics training programs advance ethical behavior because they prescribe which activities are acceptable and which are not, and they limit the opportunity for misconduct by providing punishments for violations of the rules and standards. This creates compliance requirements to establish uniform behavior among all employees. Codes and policies on ethics encourage the creation of an ethical culture in the company. According to the National Business Ethics Survey (NBES), employees in organizations that have written codes of conduct and ethics training, ethics offices or hotlines, and systems for reporting are more likely to report misconduct when they observe it. The survey found that a company's ethical culture is the greatest determinant of future misconduct.[37]

The enforcement of ethical codes and policies through rewards and punishments increases the acceptance of ethical standards by employees. One of the most important components of an ethics program is a means through which employees can report observed misconduct anonymously. Although the risk of retaliation is still a major factor in whether an employee will report illegal conduct, the Global Business Ethics Survey found that whistleblowing has increased in the past few years. Approximately 76 percent of respondents said they reported misconduct when they observed it.[38] **Whistleblowing** occurs when an employee exposes an employer's wrongdoing to outsiders, such as the media or government regulatory agencies. However, more companies are establishing programs to encourage employees to report illegal or unethical practices internally so that they can take steps to remedy problems before they result in legal action or generate negative publicity. Unfortunately, whistleblowers are often treated negatively in organizations. The government, therefore, tries to encourage employees to report observed misconduct. Congress has also taken steps to close a legislative loophole in whistleblowing legislation that has led to the dismissal of many whistleblowers. Congress passed

whistleblowing
the act of an employee exposing an employer's wrongdoing to outsiders, such as the media or government regulatory agencies.

Consider Ethics and Social Responsibility

Eaton Maintains a Successful Ethics Program

Eaton Corporation is a power management company based in Cleveland that produces more than 900,000 different industrial components. Eaton, the maker of Golf Pride golf club grips, believes that high performance is only achieved by "doing business right." It is committed to social responsibility, which involves maximizing its positive impact on society while minimizing its negative impact.

At the heart of Eaton's values-based culture is a detailed code of ethics. Eaton's code of ethics contains standards meant to reduce the opportunity for misconduct. Every employee at Eaton has the responsibility to report any suspected violation of the code—also known as whistleblowing. All new employees go through ethics training immediately after they join so they can recognize ethical issues. The company also has tools to help employees monitor their own behavior before an ethical issue develops. For example, Eaton employees can record entertainment and gifts online, which fosters transparency.

Eaton modified a vision statement to focus on sustainability: "To improve the quality of life and the environment through the use of power management technologies and services." Eaton is undertaking three initiatives to increase its environmental impact: selling sustainable products, decreasing its environmental footprint, and reporting its progress toward its environmental goals. Eaton also incorporates sustainability into its community relations by supplying products that help customers reduce their energy consumption. Eaton Corporation has a companywide commitment to ethical business. Its ethical culture is the key to its reputation and global success.[39]

Critical Thinking Questions

1. What are some of the methods Eaton uses to encourage employees to be ethical?
2. How does Eaton incorporate sustainability into its business?
3. How does Eaton use sustainability to improve community relations?

Cynthia Cooper was a whistleblower who exposed accounting fraud at WorldCom.

©JOHN CHAPPLE/REX/Shutterstock

the Dodd-Frank Act, which includes a "whistleblower bounty program." The Securities and Exchange Commission can now award whistleblowers between 10 and 30 percent of monetary sanctions over $1 million. The hope is that incentives will encourage more people to come forward with information regarding corporate misconduct.

The current trend is to move away from legally based ethical initiatives in organizations to cultural- or integrity-based initiatives that make ethics a part of core organizational values. Organizations recognize that effective business ethics programs are good for business performance. Firms that develop higher levels of trust function more efficiently and effectively and avoid damaged company reputations and product images. Organizational ethics initiatives have been supportive of many positive and diverse organizational objectives, such as profitability, hiring, employee satisfaction, and customer loyalty.[40] Conversely, lack of organizational ethics initiatives and the absence of workplace values such as honesty, trust, and integrity can have a negative impact on organizational objectives and employee retention. According to one study, three of the most common factors that executives give for why turnover increases are employee loss of trust in the company, a lack of transparency among company leaders, and unfair employee treatment.[41]

The Nature of Social Responsibility

For our purposes, we classify four stages of social responsibility: financial, legal compliance, ethics, and philanthropy (Table 2.7). Another way of categorizing these four dimensions of social responsibility include economic, legal, ethical, and voluntary (including philanthropic).[42] Earning profits is the financial or economic foundation, and complying with the law is the next step. However, a business whose *sole* objective is to maximize profits is not likely to consider its social responsibility, although its activities will probably be legal. (We looked at ethical responsibilities in the first half of this chapter.) Voluntary responsibilities are additional activities that may not be required but which promote human welfare or goodwill. Legal and economic concerns have long been acknowledged in business, and voluntary and ethical issues are now being addressed by most firms.

Corporate citizenship is the extent to which businesses meet the legal, ethical, economic, and voluntary responsibilities placed on them by their various stakeholders. It involves the activities and organizational processes adopted by businesses to meet their social responsibilities. A commitment to corporate citizenship by a firm indicates a strategic focus on fulfilling the social responsibilities expected of it by its stakeholders. For example, CVS demonstrated corporate citizenship by eliminating tobacco products from its pharmacies. Although this cost the firm $2 billion in sales, CVS believed it was contradictory to market itself as a health care services business while still selling a dangerous product.[43] Corporate citizenship involves action and measurement of the extent to which a firm embraces the corporate citizenship philosophy and then follows through by implementing citizenship and social responsibility initiatives. One of the major corporate citizenship issues is the focus on preserving the environment. The majority of people agree that climate change is a global emergency, but there is no agreement on how to solve the problem.[44] Another example of a corporate citizenship issue might be animal rights—an issue that is important to many stakeholders. As the organic and local foods movements grow and become more profitable, more and more stakeholders are calling for more humane practices in factory farms as well.[45] Large factory farms are where most Americans get their meat, but some businesses are looking at more animal-friendly options in response to public outcry.

Part of the answer to climate change issues is alternative energy such as solar, wind, biofuels, and hydro applications. The drive for alternative fuels such as ethanol from corn has added new issues such as food price increases and food shortages. A survey

LO 2-4

Explain the four dimensions of social responsibility.

corporate citizenship the extent to which businesses meet the legal, ethical, economic, and voluntary responsibilities placed on them by their stakeholders.

 Need help understanding social responsibility? Visit your Connect ebook video tab for a brief animated explanation.

Stages	Examples
Stage I: Financial and economic viability	Starbucks offers investors a healthy return on investment, including paying dividends.
Stage 2: Compliance with legal and regulatory requirements	Starbucks specifies in its code of conduct that payments made to foreign government officials must be lawful according to the laws of the United States and the foreign country.
Stage 3: Ethics, principles, and values	Starbucks' mission and values create ethical culture with ethical leaders.
Stage 4: Philanthropic activities	Starbucks created the Starbucks College Achievement Plan that offers eligible employees full tuition to earn a bachelor's degree in partnership with Arizona State University.

TABLE 2.7
Social Responsibility Requirements

TIAA, a leading financial services provider led by Roger Ferguson, has been recognized for four consecutive years as one of the world's most ethical companies by the Ethisphere Institute.

©Ramin Talaie/EPA/REX/Shutterstock

revealed that 73 percent of American consumers feel that it is important for the government to invest in renewable energy.[46]

To respond to these developments, most companies are introducing eco-friendly products and marketing efforts. Lighting giant Philips is investigating ways to uses its LED lights for horticultural purposes. It is investing in vertical city farming, in which vegetables are grown in stacked layers under LED lighting. The company hopes this breakthrough will increase the yields of locally produced, fresh produce.[47] In one survey, 53 percent of respondents claimed they have boycotted or refused to purchase a company's products because it behaved socially irresponsibly.[48] This is because many businesses are promoting themselves as green-conscious and concerned about the environment without actually making the necessary commitments to environmental health.

The Ethisphere Institute selects an annual list of the world's most ethical companies based on the following criteria: corporate citizenship and responsibility; corporate governance; innovation that contributes to the public well-being; industry leadership; executive leadership and tone from the top; legal, regulatory, and reputation track record; and internal systems and ethics/compliance program.[49] Table 2.8 shows 26 from that list.

Although the concept of social responsibility is receiving more and more attention, it is still not universally accepted. Table 2.9 lists some of the arguments for and against social responsibility.

TABLE 2.8
A Selection of the World's Most Ethical Companies

L'Oréal	HASBRO Inc.
Starbucks	Intel
Marriott International	Xerox Corporation
3M Company	GE
T-Mobile US Inc.	Cummins Inc.
PepsiCo	Knights of Columbus
ManpowerGroup	LinkedIn
Colgate-Palmolive Company	Prudential
International Paper Co.	Texas Instruments
Visa Inc.	Whirlpool
USAA	Kellogg Company
Accenture	Aflac Incorporated
Wyndham Worldwide	Dell

Source: Ethisphere Institute, "The 2018 World's Most Ethical Companies® Honoree List," http://www.worldsmostethicalcompanies.com/honorees/ (accessed March 28, 2018).

For:

1. Social responsibility rests on stakeholder engagement and results in benefits to society and improved firm performance.

2. Businesses are responsible because they have the financial and technical resources to address sustainability, health, and education.

3. As members of society, businesses and their employees should support society through taxes and contributions to social causes.

4. Socially responsible decision making by businesses can prevent increased government regulation.

5. Social responsibility is necessary to ensure economic survival: If businesses want educated and healthy employees, customers with money to spend, and suppliers with quality goods and services in years to come, they must take steps to help solve the social and environmental problems that exist today.

Against:

1. It sidetracks managers from the primary goal of business—earning profit. The responsibility of business to society is to earn profits and create jobs.

2. Participation in social programs gives businesses greater power, perhaps at the expense of concerned stakeholders.

3. Does business have the expertise needed to assess and make decisions about social and economic issues?

4. Social problems are the responsibility of the government agencies and officials, who can be held accountable by voters.

5. Creation of nonprofits and contributions to them are the best ways to implement social responsibility.

TABLE 2.9
The Arguments For and Against Social Responsibility

Entrepreneurship in Action

A Step Ahead: Blake Mycoskie Provides Leadership for Social Entrepreneurship

Toms
Founder: Blake Mycoskie
Founded: 2006, in Santa Monica, California
Success: Toms' successful One for One® model is encouraging other entrepreneurs to develop similar companies.

On a trip to Argentina, entrepreneur Blake Mycoskie saw that many families could not afford to purchase shoes for their children. The situation inspired him to create Toms, a for-profit business with a socially focused mission. For each pair of shoes sold, it delivered a free pair of shoes to children in need. This concept became known as the One for One® model and has inspired other firms—such as eyeglass retailer Warby Parker—to adopt similar models.

After distributing its one-millionth pair of shoes, Toms began to consider other products that could be used for its model. Today, it has expanded into selling eyewear and coffee. For every pair of eyewear sold, the company provides treatment or prescription glasses for those in need. Purchasing a bag of coffee provides an entire week's supply of safe drinking water.

Mycoskie later sold half of the business to Bain Capital and used $100 million from the sale to start Toms Social Entrepreneurship Fund. This fund provides financial support to like-minded companies that want to use business as a way to improve society.[50]

Critical Thinking Questions

1. How did concern for social responsibility lead Mycoskie to create Toms?
2. How has Mycoskie expanding its One for One business model?
3. How has Toms paved the way for other companies?

Evaluate an organization's social responsibilities to owners, employees, consumers, the environment, and the community.

Social Responsibility Issues

Managers consider and make social responsibility decisions on a daily basis. Among the many social issues that managers must consider are their firms' relations with stakeholders, including owners and stockholders, employees, consumers, regulators, communities, and environmental advocates.

Social responsibility is a dynamic area with issues changing constantly in response to society's demands. There is much evidence that social responsibility is associated with improved business performance. Consumers are refusing to buy from businesses that receive publicity about misconduct. A number of studies have found a direct relationship between social responsibility and profitability, as well as a link that exists between employee commitment and customer loyalty—two major concerns of any firm trying to increase profits.[51] This section highlights a few of the many social responsibility issues that managers face; as managers become aware of and work toward the solution of current social problems, new ones will certainly emerge.

Relations with Owners and Stockholders.   Businesses must first be responsible to their owners, who are primarily concerned with earning a profit or a return on their investment in a company. In a small business, this responsibility is fairly easy to fulfill because the owner(s) personally manages the business or knows the managers well. In larger businesses, particularly corporations owned by thousands of stockholders, ensuring responsibility becomes a more difficult task.

A business's obligations to its owners and investors, as well as to the financial community at large, include maintaining proper accounting procedures, providing all relevant information to investors about the current and projected performance of the firm, and protecting the owners' rights and investments. In short, the business must maximize the owners' investments in the firm.

Employee Relations.   Another issue of importance to a business is its responsibilities to employees. Without employees, a business cannot carry out its goals. Employees expect businesses to provide a safe workplace, pay them adequately for their work, and keep them informed of what is happening in their company. They want employers to listen to their grievances and treat them fairly. Many firms have begun implementing extended parental leave for families with new babies. Credit Suisse announced it was giving its U.S. employees up to 20 weeks off after having a child. Facebook also extended its parental leave for up to four months. These types of benefits are becoming increasingly important as firms strive to attract top-quality employees.[52]

Congress has passed several laws regulating safety in the workplace, many of which are enforced by the Occupational Safety and Health Administration (OSHA). Labor unions have also made significant contributions to achieving safety in the workplace and improving wages and benefits. Most organizations now recognize that the safety and satisfaction of their employees are critical ingredients in their success, and many strive to go beyond what is legally expected of them. Healthy, satisfied employees also supply more than just labor

OSHA ordered its inspectors to crack down on employers who fail to file the necessary electronic paperwork to document injury reports.

©Herdik Herlambang/Shutterstock

to their employers. Employers are beginning to realize the importance of obtaining input from even the lowest-level employees to help the company reach its objectives.

A major social responsibility for business is providing equal opportunities for all employees regardless of their sex, age, race, religion, or nationality. Diversity is also helpful to a firm financially. Corporations that raised the share of female executives to 30 percent saw a 15 percent increase in profitability.[53] Also, it has been found that when men and women managers are evenly matched, there is a better chance of generating stronger profits. Thus, many firms are trying to become more inclusive, embracing diversity.[54] Yet, despite these benefits, women and minorities have been slighted in the past in terms of education, employment, and advancement opportunities; additionally, many of their needs have not been addressed by business. Women make up to 57 percent of undergraduate college students, but women account for only 19 percent of C-suites (top corporate offices). Women, who continue to bear most child-rearing responsibilities, often experience conflict between those responsibilities and their duties as employees. Consequently, day care has become a major employment issue for women, and more companies are providing day care facilities as part of their effort to recruit and advance women in the workforce. Many Americans today believe business has a social obligation to provide special opportunities for women and minorities to improve their standing in society.

Consumer Relations. A critical issue in business today is business's responsibility to customers, who look to business to provide them with satisfying, safe products and to respect their rights as consumers. The activities that independent individuals, groups, and organizations undertake to protect their rights as consumers are known as **consumerism**. To achieve their objectives, consumers and their advocates write letters to companies, lobby government agencies, make public service announcements, and boycott companies whose activities they deem irresponsible.

consumerism
the activities that independent individuals, groups, and organizations undertake to protect their rights as consumers.

Many of the desires of those involved in the consumer movement have a foundation in John F. Kennedy's 1962 consumer bill of rights, which highlighted four rights. The *right to safety* means that a business must not knowingly sell anything that could result in personal injury or harm to consumers. Defective or dangerous products erode public confidence in the ability of business to serve society. They also result in expensive litigation that ultimately increases the cost of products for all consumers. The right to safety also means businesses must provide a safe place for consumers to shop.

The *right to be informed* gives consumers the freedom to review complete information about a product before they buy it. This means that detailed information about risks and instructions for use are to be printed on labels and packages. The *right to choose* ensures that consumers have access to a variety of goods and services at competitive prices. The assurance of both satisfactory quality and service at a fair price is also a part of the consumer's right to choose. The *right to be heard* assures consumers that their interests will receive full and sympathetic consideration when the government formulates policy. It also ensures the fair treatment of consumers who voice complaints about a purchased product.

The role of the Federal Trade Commission's Bureau of Consumer Protection exists to protect consumers against unfair, deceptive, or fraudulent practices. The bureau, which enforces a variety of consumer protection laws, is divided into five divisions. The Division of Enforcement monitors legal compliance and investigates violations of laws, including unfulfilled holiday delivery promises by online shopping sites, employment opportunities fraud, scholarship scams, misleading advertising for health care products, and more.

Sustainability Issues.   Most people probably associate the term *environment* with nature, including wildlife, trees, oceans, and mountains. Until the 20th century, people generally thought of the environment solely in terms of how these resources could be harnessed to satisfy their needs for food, shelter, transportation, and recreation. As the earth's population swelled throughout the 20th century, however, humans began to use more and more of these resources and, with technological advancements, to do so with ever-greater efficiency. Although these conditions have resulted in a much-improved standard of living, they come with a cost. Plant and animal species, along with wildlife habitats, are disappearing at an accelerated rate. For example, the bumblebee population has suffered almost 90 percent decline in the past 20 years. Bees are important to pollinating most fruits and vegetables. The bumblebee was placed on the endangered species list and its habitats protected.[55]

Although the scope of the word *sustainability* is broad, in this book we discuss the term from a strategic business perspective. Thus, we define **sustainability** as conducting activities in such a way as to provide for the long-term well-being of the natural environment, including all biological entities. Sustainability involves the interaction among nature and individuals, organizations, and business strategies and includes the assessment and improvement of business strategies, economic sectors, work practices, technologies, and lifestyles so that they maintain the health of the natural environment. In recent years, business has played a significant role in adapting, using, and maintaining the quality of sustainability.

Environmental protection emerged as a major issue in the 20th century in the face of increasing evidence that pollution, uncontrolled use of natural resources, and population growth were putting increasing pressure on the long-term sustainability of these resources. In recent years, companies have been increasingly incorporating these issues into their overall business strategies. Some nonprofit organizations have stepped forward to provide leadership in gaining the cooperation of diverse groups in responsible environmental activities.

In the following sections, we examine some of the most significant sustainability and environmental health issues facing business and society today, including pollution and alternative energy.

Pollution.   A major issue in the area of environmental responsibility is pollution. Water pollution results from dumping toxic chemicals and raw sewage into rivers and oceans, oil spills, and the burial of industrial waste in the ground where it may filter into underground water supplies. Fertilizers and insecticides used in farming and grounds maintenance also run off into water supplies with each rainfall. Water pollution problems are especially notable in heavily industrialized areas. Society is demanding that water supplies be clean and healthful to reduce the potential danger from these substances.

Air pollution is usually the result of smoke and other pollutants emitted by manufacturing facilities, as well as carbon monoxide and hydrocarbons emitted by motor vehicles. In addition to the health risks posed by air pollution, when some chemical compounds emitted by manufacturing facilities react with air and rain, acid rain results. Acid rain has contributed to the deaths of many forests and lakes in North America as well as in Europe. Air pollution may also contribute to global warming; as carbon dioxide collects in the earth's atmosphere, it traps the sun's heat and prevents the earth's surface from cooling. It is indisputable that the global surface temperature has been increasing over the past 35 years. Worldwide passenger vehicle ownership has been growing due to rapid industrialization and consumer purchasing

---

sustainability
conducting activities in a way that allows for the long-term well-being of the natural environment, including all biological entities; involves the assessment and improvement of business strategies, economic sectors, work practices, technologies, and lifestyles so that they maintain the health of the natural environment.

FIGURE 2.3
Consumer Likelihood to Personally Address Social Responsibility Issues

Source: Cone Communications and Ebiquity, "2017 Cone Communications CSR Study," http://www.conecomm.com/2017-cone-communications-csr-study-pdf (accessed March 28, 2018).

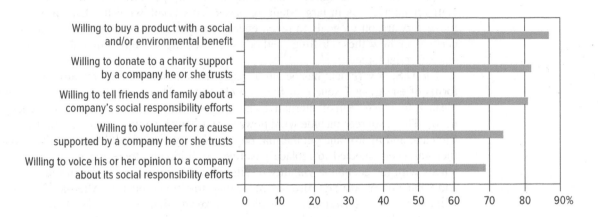

power in China, India, and other developing countries with large populations. The most important way to contain climate change is to control carbon emissions. For example, some utilities charge more for electricity in peak demand periods, which encourages behavioral changes that reduce consumption.

More and more consumers are recognizing the need to protect the planet. Figure 2.3 shows consumers' likelihood to personally address social responsibility and environmental issues. Although most consumers admit that sustainable products are important and that they bear responsibility for properly using and disposing of the product, many admit that they fail to do this.

Land pollution is tied directly to water pollution because many of the chemicals and toxic wastes that are dumped on the land eventually work their way into the water supply. A study conducted by the Environmental Protection Agency found residues of prescription drugs, soaps, and other contaminants in virtually every waterway in the United States. Effects of these pollutants on humans and wildlife are uncertain, but there is some evidence to suggest that fish and other water-dwellers are starting to suffer serious effects. Land pollution results from the dumping of residential and industrial waste, strip mining, forest fires, and poor forest conservation. In Brazil and other South American countries, rain forests are being destroyed to make way for farms and ranches, at a cost of the extinction of the many animals and plants (some endangered species) that call the rain forest home. For example, in Brazil trees were cleared over an area spanning seven times the territory of New York City. Brazil has committed to zero percent illegal deforestation by 2030.[56] Large-scale deforestation also depletes the oxygen supply available to humans and other animals.

Related to the problem of land pollution is the larger issue of how to dispose of waste in an environmentally responsible manner. Americans use approximately 100 billion plastic bags per year.[57] Some states and many other countries are also in the process of phasing out lightweight plastic bags.

Alternative Energy.   With ongoing plans to reduce global carbon emissions, countries and companies alike are looking toward alternative energy sources. Traditional fossil fuels are problematic because of their emissions, but also because stores have been greatly depleted. Foreign fossil fuels are often imported from politically and economically unstable regions, often making it unsafe to conduct business there. However, the United States is becoming an energy powerhouse with its ability to drill for natural gas in large shale reserves. This is allowing the United States to move forward on its goals to reach energy independence. On the other hand, concerns over how these drilling methods are affecting the environment make this a controversial topic.

The U.S. government has begun to recognize the need to look toward alternative forms of energy as a source of fuel and electricity. There have been many different ideas as to which form of alternative energy would best suit the United States' energy needs. These sources include wind power, solar power, nuclear power, biofuels, electric cars, and hydro- and geothermal power. As of yet, no "best" form of alternative fuel has been selected to replace gasoline. Additionally, there are numerous challenges with the economic viability of alternative energy sources. For instance, wind and solar power cost significantly more than traditional energy. Alternative energy will likely require government subsidies to make any significant strides. However, the news for solar power might be getting brighter. Electric cars are also gaining importance. Most automobile companies such as Tesla, BMW, General Motors, Nissan, and Toyota are introducing electric cars to help with sustainability.

Response to Environmental Issues.   Many firms are trying to eliminate wasteful practices, the emission of pollutants, and/or the use of harmful chemicals from their manufacturing processes. Other companies are seeking ways to improve their products. Utility providers, for example, are increasingly supplementing their services with alternative energy sources, including solar, wind, and geothermal power. Environmentalists are concerned that some companies are merely *greenwashing,* or "creating a positive association with environmental issues for an unsuitable product, service, or practice."

DID YOU KNOW? About 50 million tons of electronic waste is generated each year, including discarded laptops, mobile phones, and televisions.[59]

With the increasing popularity of electric vehicles, companies like BMW, Volkswagen, and Tesla have introduced new electric car models.

©Darren Brode/Shutterstock

Indeed, a growing number of businesses and consumers are choosing green power sources where available. New Belgium Brewing Company, the fourth-largest craft brewer in the United States, is the first all-wind-powered brewery in the country. Many businesses have turned to *recycling,* the reprocessing of materials—aluminum, paper, glass, and some plastic—for reuse. Such efforts to make products, packaging, and processes more environmentally friendly have been labeled "green" business or marketing by the public and media. For example, lumber products at The Home Depot may carry a seal from the Forest Stewardship Council to indicate that they were harvested from sustainable forests using environmentally friendly methods.[58]

It is important to recognize that, with current technology, environmental responsibility requires trade-offs. Society must weigh the huge costs of limiting or eliminating pollution against the health threat posed by the pollution. Environmental responsibility imposes costs on both business and the public. Managers must coordinate environmental goals with other social and economic ones.

Community Relations.   A final, yet very significant, issue for businesses concerns their responsibilities to the general welfare of the communities and societies in which they operate. Many businesses simply want to make their communities better places for everyone to live and work. The most common way that businesses exercise their community responsibility is through donations to local and national charitable organizations. For example, companies and their employees hold fundraising efforts to raise money for the United Way.[60] As a highly successful company, Adobe invests heavily in community development and sustainability. It invests 1 percent of its pretax profits in the Adobe Foundation, which partners with teams of employees to use the funds to improve local communities. The company donates software to more than 15,000 nonprofits.[61]

Unemployment

After realizing that the current pool of prospective employees lacks many basic skills necessary to work, many companies have become concerned about the quality of education in the United States. Unemployment reached 10 percent during the most recent recession but had fallen to 4.1 percent by 2018.

Although most would argue that unemployment is an economic issue, it also carries ethical implications. Protests often occur in areas where unemployment is high, particularly when there seems to be a large gap between rich and poor.

Factory closures are another ethical issue because factories usually employ hundreds of workers. Sometimes it is necessary to close a plant due to economic reasons. However, factory closures not only affect individual employees, but their communities as well. After years of withstanding closures, even as other factories around it closed, a factory in the small town of Hanover, Illinois, shuttered its doors and transferred to Mexico. Several of the factory workers felt betrayed, and about 100 lost their jobs. Factory closures also have repercussions on other businesses in the area because more unemployed people mean fewer sales.[62]

Another criticism levied against companies involves hiring standards. Some employers have been accused of having unreasonable hiring standards that most applicants cannot meet, often leaving these jobs unfilled. Critics have accused companies of not wanting to take the time to train employees.[63] Employers, however, believe there is a significant lack of skills needed among job applicants. A survey of employers conducted in Indiana revealed that 39 percent reported leaving positions unfilled because the applicants were not qualified.[64] With more companies requiring specialized knowledge and a strong educational background, jobs are becoming increasingly competitive among those looking for employment.

On the other hand, several businesses are working to reduce unemployment. After becoming frustrated with high unemployment rates, Starbucks

When Toys R Us and Babies R Us shuttered 740 U.S. stores, tens of thousands of employees lost their jobs.

©NorthWindPub/Alamy Stock Photo

founder and CEO Howard Schultz partnered with a national network of community lenders called Opportunity Network to develop Create Jobs for USA. The program provided funding for community businesses with the intent to reduce unemployment in their areas. Starbucks initially donated $5 million to the initiative. Other companies made significant contributions, including Banana Republic, Citi, Google Offers, and MasterCard. During the three years the initiative ran, it received donations of $15.2 million that it provided to community development financial institutions. These institutions were able to generate more funding from these donations, which helped to create or maintain more than 5,000 jobs.[65]

Additionally, businesses are beginning to take more responsibility for the hard-core unemployed. These are people who have never had a job or who have been unemployed for a long period of time. Some are mentally or physically handicapped; some are homeless. Organizations such as the National Alliance of Businessmen fund programs to train the hard-core unemployed so that they can find jobs and support themselves. Such commitment enhances self-esteem and helps people become productive members of society.

So You Want a Job in Business Ethics and Social Responsibility

In the words of Kermit the Frog, "It's not easy being green." It may not be easy, but green business opportunities abound. A popular catch phrase, "Green is the new black," indicates how fashionable green business is becoming. Consumers are more in tune with and concerned about green products, policies, and behaviors by companies than ever before. Companies are looking for new hires to help them see their business creatively and bring insights to all aspects of business operations.

The International Renewable Energy Industry estimates that the number of jobs in the renewable energy job market could rise to 24 million by 2030. Green business strategies not only give a firm a commercial advantage in the marketplace, but also help lead the way toward a greener world. The fight to reduce our carbon footprint in an attempt against climate change has opened up opportunities for renewable energy, recycling, conservation, and increasing overall efficiency in the way resources are used. New businesses that focus on hydro, wind, and solar power are on the rise and will need talented businesspeople to lead them. Carbon emissions' trading is gaining popularity as large corporations and individuals alike seek to lower their footprints. A job in this growing field could be similar to that of a stock trader, or you could lead the search for carbon-efficient companies in which to invest.

In the ethics arena, current trends in business governance strongly support the development of ethics and compliance departments to help guide organizational integrity. This alone is a billion-dollar business, and there are jobs in developing organizational ethics programs, developing company policies, and training employees and management. An entry-level position might be as a communication specialist or trainer for programs in a business ethics department. Eventually there's an opportunity to become an ethics officer that would have typical responsibilities of meeting with employees, the board of directors, and top management to discuss and provide advice about ethics issues in the industry, developing and distributing a code of ethics, creating and maintaining an anonymous, confidential service to answer questions about ethical issues, taking actions on possible ethics code violations, and reviewing and modifying the code of ethics of the organization.

There are also opportunities to help with initiatives to help companies relate social responsibility to stakeholder interests and needs. These jobs could involve coordinating and implementing philanthropic programs that give back to others important to the organization or developing a community volunteering program for employees. In addition to the human relations function, most companies develop programs to assist employees and their families to improve their quality of life. Companies have found that the healthier and happier employees are, the more productive they will be in the workforce.

Social responsibility, ethics, and sustainable business practices are not a trend; they are good for business and the bottom line. New industries are being created and old ones are adapting to the new market demands, opening up many varied job opportunities that will lead not only to a paycheck, but also to the satisfaction of making the world a better place.[66]

Review Your Understanding

Describe the importance of business ethics and social responsibility.

Business ethics refers to principles and standards that define acceptable business conduct. Acceptable business behavior is defined by customers, competitors, government regulators, interest groups, the public, and each individual's personal moral principles and values. Social responsibility is the obligation an organization assumes to maximize its positive impact and minimize its negative impact on society. Socially responsible businesses win the trust and respect of their employees, customers, and society and, in the long run, increase profits. Ethics is important in business because it builds trust and confidence in business relationships. Unethical actions may result in negative publicity, declining sales, and even legal action.

Detect some of the ethical issues that may arise in business.

An ethical issue is an identifiable problem, situation, or opportunity requiring a person or organization to choose from among several actions that must be evaluated as right or wrong. Ethical issues can be categorized in the context of their relation with conflicts of interest, fairness and honesty, communications, and business associations.

Specify how businesses can promote ethical behavior.

Businesses can promote ethical behavior by employees by limiting their opportunity to engage in misconduct. Formal codes of ethics, ethical policies, and ethics training programs reduce the incidence of unethical behavior by informing employees what is expected of them and providing punishments for those who fail to comply.

Explain the four dimensions of social responsibility.

The four dimensions of social responsibility are economic or financial viability (being profitable), legal (obeying the law), ethical (doing what is right, just, and fair), and philanthropic or voluntary (being a good corporate citizen).

Evaluate an organization's social responsibilities to owners, employees, consumers, the environment, and the community.

Businesses must maintain proper accounting procedures, provide all relevant information about the performance of the firm to investors, and protect the owners' rights and investments. In relations with employees, businesses are expected to provide a safe workplace, pay employees adequately for their work, and treat them fairly. Consumerism refers to the activities undertaken by independent individuals, groups, and organizations to protect their rights as consumers. Increasingly, society expects businesses to take greater responsibility for the environment, especially with regard to animal rights, as well as water, air, land, and noise pollution. Many businesses engage in activities to make the communities in which they operate better places for everyone to live and work.

Evaluate the ethics of a business's decision.

The "Solve the Dilemma" feature near the end of this chapter presents an ethical dilemma at Checkers Pizza. Using the material presented in this chapter, you should be able to analyze the ethical issues present in the dilemma, evaluate Barnard's plan, and develop a course of action for the firm.

Critical Thinking Questions

Enter the World of Business Questions

1. To what extent should there be something of a caveat emptor/buyer beware when someone chooses to professionally play football in the NFL?

2. Can the NFL ever really make professional football totally safe?

3. Is playing in the NFL different from being a NASCAR driver, a police officer, or an astronaut?

Learn the Terms

bribes 42
business ethics 38
codes of ethics 50
consumerism 57

corporate citizenship 53
ethical issue 41
plagiarism 49
social responsibility 39

sustainability 58
whistleblowing 51

Check Your Progress

1. Define *business ethics.* Who determines whether a business activity is ethical? Is unethical conduct always illegal?

2. Distinguish between ethics and social responsibility.

3. Why has ethics become so important in business?

4. What is an ethical issue? What are some of the ethical issues named in your text? Why are they ethical issues?

5. What is a code of ethics? How can one reduce unethical behavior in business?

6. List and discuss the arguments for and against social responsibility by business (Table 2-9). Can you think of any additional arguments (for or against)?

7. What responsibilities does a business have toward its employees?

8. What responsibilities does business have with regard to the environment? What steps have been taken by some responsible businesses to minimize the negative impact of their activities on the environment?

9. What are a business's responsibilities toward the community in which it operates?

Get Involved

1. Discuss some recent examples of businesses engaging in unethical practices. Classify these practices as issues of conflict of interest, fairness and honesty, communications, or business relationships. Why do you think the businesses chose to behave unethically? What actions might the businesses have taken?

2. Discuss with your class some possible methods of improving ethical standards in business. Do you think that business should regulate its own activities or that the federal government should establish and enforce ethical standards? How do you think businesspeople feel?

3. Find some examples of socially responsible businesses in newspapers or business journals. Explain why you believe their actions are socially responsible. Why do you think the companies chose to act as they did?

Build Your Skills

Making Decisions about Ethical Issues

Background

The merger of Lockheed and Martin Marietta created Lockheed Martin, the number-one company in the defense industry—an industry that includes such companies as Raytheon and Northrop Grumman.

You and the rest of the class are managers at Lockheed Martin Corporation, Orlando, Florida. You are getting ready to do the group exercise in an ethics training session. The training instructor announces you will be playing Gray Matters: The Ethics Game. You are told that Gray Matters, which was prepared for your company's employees, is also played at 41 universities, including Harvard University, and at 65 other companies. Although there are 55 scenarios in Gray Matters, you will have time during this session to complete only the four scenarios that your group draws from the stack of cards.[67]

Task

Form into groups of four to six managers and appoint a group leader who will lead a discussion of the case, obtain a consensus answer to the case, and be the one to report the group's answers to the instructor. You will have five minutes to reach each decision, after which time, the instructor will give the point values and rationale for each choice. Then you will have five minutes for the next case, etc., until all four cases have been completed. Keep track of your group's score for each case; the winning team will be the group scoring the most points.

Since this game is designed to reflect life, you may believe that some cases lack clarity or that some of your choices are not as precise as you would have liked. Also, some cases have only one solution, while others have more than one solution. Each choice is assessed to reflect which answer is the most correct. Your group's task is to select only one option in each case.

4

Mini-Case

For several months now, one of your colleagues has been slacking off, and you are getting stuck doing the work. You think it is unfair. What do you do?

Potential Answers

A. Recognize this as an opportunity for you to demonstrate how capable you are.

B. Go to your supervisor and complain about this unfair workload.

C. Discuss the problem with your colleague in an attempt to solve the problem without involving others.

D. Discuss the problem with the human resources department.

7

Mini-Case

You are aware that a fellow employee uses drugs on the job. Another friend encourages you to confront the person instead of informing the supervisor. What do you do?

Potential Answers

A. You speak to the alleged user and encourage him to get help.

B. You elect to tell your supervisor that you suspect an employee is using drugs on the job.

C. You confront the alleged user and tell him either to quit using drugs or you will "turn him in."

D. Report the matter to employee assistance.

36

Mini-Case

You work for a company that has implemented a policy of a smoke-free environment. You discover employees smoking in the restrooms of the building. You also smoke and don't like having to go outside to do it. What do you do?

Potential Answers

A. You ignore the situation.

B. You confront the employees and ask them to stop.

C. You join them, but only occasionally.

D. You contact your ethics or human resources representative and ask him or her to handle the situation.

40

Mini-Case

Your co-worker is copying company-purchased software and taking it home. You know a certain program costs $400, and you have been saving for a while to buy it. What do you do?

Potential Answers

A. You figure you can copy it too because nothing has ever happened to your co-worker.

B. You tell your co-worker he can't legally do this.

C. You report the matter to the ethics office.

D. You mention this to your supervisor.

Solve the Dilemma

Customer Privacy

 Checkers Pizza was one of the first to offer home delivery service, with overwhelming success. However, the major pizza chains soon followed suit, taking away Checkers' competitive edge. Jon Barnard, Checkers' founder and co-owner, needed a new gimmick to beat the competition. He decided to develop a computerized information database that would make Checkers the most efficient competitor and provide insight into consumer buying behavior at the same time. Under the system, telephone customers were asked their phone number; if they had ordered from Checkers before, their address and previous order information came up on the computer screen.

After successfully testing the new system, Barnard put the computerized order network in place in all Checkers outlets. After three months of success, he decided to give an award to the family that ate the most Checkers pizza. Through the tracking system, the company

 LO 2-6

Evaluate the ethics of a business's decision.

identified the biggest customer, who had ordered a pizza every weekday for the past three months (63 pizzas). The company put together a program to surprise the family with an award, free-food certificates, and a news story announcing the award. As Barnard began to plan for the event, however, he began to think that maybe the family might not want all the attention and publicity.

Critical Thinking Questions

1. What are some of the ethical issues in giving customers an award for consumption behavior without notifying them first?
2. Do you see this as a potential violation of privacy? Explain.
3. How would you handle the situation if you were Barnard?

Build Your Business Plan

🔲 connect

Business Ethics and Social Responsibility

 Think about which industry you are considering competing in with your good/service. Is there any kind of questionable practices in the way the product has been traditionally sold? Produced? Advertised? Have there been any recent accusations regarding safety within the industry? What about any environmental concerns?

For example, if you are thinking of opening a lawn care business, you need to be thinking about what possible effects the chemicals you are using will have on the client and the environment. You have a responsibility to keep your customers safe and healthy. You also have the social responsibility to let the community know of any damaging effect you may be directly or indirectly responsible for.

See for Yourself Videocase

Warby Parker: An Affordable World Vision

 Eyewear firm Warby Parker is known for more than just its affordable glasses. It has adopted a business model that helps advance entrepreneurship opportunities and economic development across the world. In 2008, co-founders Neil Blumenthal, David Gilboa, and three classmates embarked upon a plan to make glasses that would be affordable for the masses. The idea was developed for Wharton's business plan competition. Not only did their idea not win the competition, it did not even make it to the final round. Their plans for a company that sold glasses at a more affordable price seemed like a flop.

Fast track nearly a decade later, and Warby Parker has sold well over 1 million pairs of glasses. The premise behind the startup was simple. Blumenthal and Gilboa realized that one company had a near-monopoly over the optical industry, giving it the power to set steep prices. Blumenthal believed that by designing and manufacturing glasses in-house and selling them on the Internet, the company would be able to save on costs that it could then pass on to consumers. Warby Parker's unique business model enables the firm to sell designer-style eyeglasses for as little as $95 each.

From the get-go, the founders wanted to create a different kind of company centered on integrity. Blumenthal had previously worked with the nonprofit Visionspring, a charity that provides quality eyeglasses to those in need in developing countries. They decided to adopt a unique business model in which the organization would partner with Visionspring

to provide one pair of eyeglasses to someone in need for every pair of eyeglasses sold. Every month, Warby Parker tallies the number of eyeglasses sold and makes a donation to Visionspring. The donation covers the costs of sourcing glasses. At Warby Parker, this initiative is not simply philanthropy; rather, it has been incorporated into the firm's business model as a critical component of its operations.

On the surface, donating eyewear might not seem like it directly affects economic development. However, the founder of Visionspring, Jack Kassalow, describes just how important eyewear is for people who have vision impairment problems that could be solved with a pair of eyeglasses. "Vision is critical for work. If you can't see, you can't work. Vision is critical to learn. If you can't see, you can't learn. And [it's] critical as well to human security. If you can't see, it's hard to be safe as you move about the world," Kassawlow says.

However, Warby Parker and Visionspring are careful not to simply donate the glasses. While this might help individuals with their sight, it would miss out on the opportunity to encourage economic growth. Instead, Visionspring trains local women to be entrepreneurs and sell the glasses to tradespeople for $4 each. This model provides jobs for women and enables them to earn wages that they can then spend improving their communities. As for the tradespeople who purchase the glasses, it is estimated that their productivity increases by 35 percent and their earning power by 20 percent.

Warby Parker is living proof that a company can be socially responsible and profitable. Although it started out as an online-only firm, Warby Parker has since opened nearly 100 retail locations and is valued at $1.75 billion. Throughout its growth, Warby Parker has maintained a strong focus on the customer. For instance, it allows customers to try out five pairs of glasses for five days at no cost. Kasselow comments on how Warby Parker has been able to develop a highly efficient business model that is fair to both customers and those in developing nations. "The price is incredibly fair. But because they are able to be as efficient as they are in their supply chain, they're also able to bring it to market in a way that they can become a profitable company and scale their idea to provide affordable glasses to hopefully hundreds of millions of people, not only in the U.S. but all around the world," he says.[68]

Critical Thinking Questions

1. Describe Warby Parker's ethical vision in selling eyeglasses.

2. How does Warby Parker integrate social responsibility into its business?

3. Why does Warby Parker charge $4 to those in need of eyeglasses in developing countries?

You can find the related video in the Video Library in Connect. Ask your instructor how you can access Connect.

Team Exercise

Sam Walton, founder of Walmart, had an early strategy for growing his business related to pricing. The "Opening Price Point" strategy used by Walton involved offering the introductory product in a product line at the lowest point in the market. For example, a minimally equipped microwave oven would sell for less than anyone else in town could sell the same unit. The strategy was that if consumers saw a product, such as the microwave, and saw it as a good value, they would assume that all of the microwaves were good values. Walton also noted that most people don't buy the entry-level product; they want more features and capabilities and often trade up.

Form teams and assign the role of defending this strategy or casting this strategy as an unethical act. Present your thoughts on either side of the issue.

Notes

1. Ken Belson, Many Ex-Players May Be Ineligible for Payment in N.F.L. Concussion Settlement, *The New York Times,* October 17, 2013, http://www.nytimes.com/2013/10/18/sports/football/many-ex-players-may-be-ineligible-to-share-in-nfl-concussion-settlement.html?_r=0 (accessed December 27, 2017); Will Brinson, "Frontline PBS Doc 'League of Denial' Examines NFL Concussion Problem," *CBS Sports,* October 8, 2013, http://www.cbssports.com/nfl/eye-on-football/24051122/frontline-pbs-doc-league-o(af-denial-examines-nfl-concussion-problem (accessed December 27, 2017); Chris Chase, "NFL All-Pro Says if You Don't Like Concussions, Don't Watch Football," *USA Today,* October 23, 2013, http://ftw.usatoday.com/2013/10/richard-sherman-concussions/ (accessed December 27, 2017); Andrew M. Blecher, "NFL Concussion Litigation," http://NFLconcussionlitigation.com (accessed December 27, 2017); Travis Waldron, "What Does the NFL's Concussion Settlement Mean for the Future of Football?" *Think Progress,* August 29, 2013, https://thinkprogress.org/what-does-the-nfls-concussion-settlement-mean-for-the-future-of-football-d8650489ea7e/ (accessed December 27, 2017); Richard Sherman, "We Chose This Profession," *Sports Illustrated,* MMQB—In This Corner, October 24, 2013, http://mmqb.si.com/2013/10/23/richard-sherman-seahawks-concussions-in-the-nfl/ (accessed December 27, 2017); Jacque Wilson and Stephanie Smith, "Brett Favre: Memory Lapse 'Put a Little Fear in Me,'" *CNN,* October 28, 2013, http://www.cnn.com/2013/10/25/health/brett-favre-concussions/ (accessed December 27, 2017); Michael McCann, "Will New CTE Findings Doom the NFL Concussion Settlement?" *Sports Illustrated,* August 15, 2017, https://www.si.com/nfl/2017/08/15/new-cte-study-effect-nfl-concussion-settlement (accessed December 27, 2017); Tom Huddleston Jr., "The Football Industrial Complex Is in Big Trouble," *Fortune,* September 7, 2017, http://fortune.com/2017/09/07/nfl-ncaa-football-concussion-cte/ (accessed December 27, 2017).

2. Jacquelyn Smith, "The World's Most Ethical Companies," *Forbes,* March 6, 2013, http://www.forbes.com/sites/jacquelynsmith/2013/03/06/the-worlds-most-ethical-companies-in-2013/ (accessed May 3, 2017).

3. Mike Spector and Mike Colias, "Volkswagen Faces Up to Penalties," *The Wall Street Journal,* March 12, 2012, p. B1.

4. "Illycaffè Named as a World's Most Ethical Company by the Ethisphere Institute for the Fifth Consecutive Time," *PR Newswire,* March 15, 2017.

5. Kimberly Blanton, "Creating a Culture of Compliance," *CFO,* July/August 2011, pp. 19–21.

6. Walmart, "2016 Giving Report," https://cdn.corporate.walmart.com/64/66/4c15646a4b96909ccd0583edb609/wmt-065-giving-report-final.pdf (accessed March 26, 2018).

7. Kate Pickert, "Medicare Fraud Horror: Cancer Doctor Indicted for Billing Unnecessary Chemo," *Time,* August 15, 2013, http://nation.time.com/2013/08/15/medicare-fraud-horror-cancer-doctor-indicted-for-billing-unnecessary-chemo/ (accessed May 3, 2017).

8. Roger Yu, "Bristol-Myers Fined $14M over Bribe Claims," *USA Today,* October 6, 2015, p. 3B.

9. O.C. Ferrell, John Fraedrich, and Linda Ferrell, *Business Ethics: Ethical Decision Making and Cases,* 8th ed. (Mason, OH: South-Western Cengage Learning, 2011), p. 7.

10. "3M Named a 2017 World's Most Ethical Company," *Business Wire,* http://www.businesswire.com/news/home/20170313005407/en/ (accessed March 22, 2017).

11. Patrick Leger, "The Fall of Travis Kalanick," *Bloomberg Businessweek,* January 22, 2018, pp. 46–51.

12. Ferrell, Fraedrich, and Ferrell, *Business Ethics.*

13. Ben Dipietro, "The Morning Risk Report: South Korea Scandal Highlights Political Risks of Bribery," *The Wall Street Journal,* March 13, 2017, https://blogs.wsj.com/riskandcompliance/2017/03/13/the-morning-risk-report-south-korea-scandal-highlights-political-risks-of-bribery/ (accessed March 26, 2018).

14. Ethics Resource Center, *2011 National Business Ethics Survey®: Ethics in Transition* (Arlington, VA: Ethics Resource Center, 2012).

15. Bobby White, "The New Workplace Rules: No Video Watching," *The New York Times,* March 3, 2008, p. B1.

16. Brianna Bradley, "Nearly a Quarter of Staffers Will Shop Online Today," *Biz Women,* November 27, 2017, https://www.bizjournals.com/bizwomen/news/latest-news/2017/11/nearly-a-quarter-of-staffers-will-shop-online.html (accessed April 7, 2018).

17. "Sexual Harassment: #MeToo," *The Economist,* December 23, 2017–January 5, 2018, pp. 81–82.

18. Shana Lebowitz, "What's Behind a Rise in Workplace Bullying?" *USA Today,* October 8, 2013, www.usatoday.com/story/news/health/2013/10/08/hostile-workplace-less-productive/2945833/ (accessed May 3, 2017).

19. Carolyn Kinsey Goman, "Is Your Boss a Bully?" *Forbes,* April 6, 2014, http://www.forbes.com/sites/carolkinseygoman/2014/04/06/is-your-boss-a-bully/ (accessed May 3, 2017).

20. Abigail Hess, "Howard University Employees Fired Following Investigation into Stolen Financial Aid," *CNBC,* March 29, 2018, https://www.cnbc.com/2018/03/29/howard-university-fires-employees-who-allegedly-stole-financial-aid.html (accessed April 7, 2018).

21. NBC New York, "Apple Store Worker Charged with Using Bogus Credit Cards to Buy Nearly $1 Million in Apple Gift Cards," October 20, 2015, http://www.nbcnewyork.com/news/local/Apple-Worker-Arrest-Theft-Gift-Card-Fake-Credit-Information-Ruben-Profit-334735951.html (accessed May 3, 2017); Gregg Keizer, "Apple Employee Arrested, Charged with Buying $1M in Company Gift Cards with Fake Plastic," *Computerworld,* October 21, 2015, http://www.computerworld.com/article/2996000/technology-law-regulation/apple-employee-arrested-charged-with-buying-1m-in-company-gift-cards-with-fake-plastic.html (accessed May 3, 2017).

22. Coca-Cola Company, *Code of Business Conduct: Acting Around the Globe,* April 2009, p. 13, https://www.coca-colacompany.com/content/dam/journey/us/en/private/fileassets/pdf/2013/04/code-of-business-conduct-france-english.pdf (accessed May 3, 2017).

23. Reuters, "BlackRock to Pay $12 Million in SEC Conflict of Interest Case," *Fortune,* April 20, 2015, http://fortune.com/2015/04/20/blackrock-to-pay-12-million-in-sec-conflict-of-interest-case/ (accessed May 3, 2017).

24. *Corruption Perceptions Index 2016,* Transparency International, http://www.transparency.org/news/feature/corruption_perceptions_index_2016 (accessed April 9, 2017).

25. Joel Rosenblatt, "For Once, Uber Needs Travis Kalanick to Speak Up," *Bloomberg Businessweek,* February 5, 2018, pp. 22–23.

26. Holly Ellyatt, "EU Lawmakers Vote to Break Up Google," *CNBC,* November 27, 2014, http://www.cnbc.com/id/102222045# (accessed May 3, 2017).

27. FDA, "FDA Investigates Multistate Outbreak of E. Coli O26 Infections Linked to Chipotle Mexican Grill Restaurant," U.S. Food and Drug Administration, https://www.fda.gov/food/recallsoutbreaksemergencies/outbreaks/ucm470410.htm (accessed October 17, 2017); Diana Bradley, "Chris Arnold Guides Chipotle's Crisis Comeback," *PR Week,* https://www.prweek.com/article/1438230/chris-arnold-guides-chipotles-crisis-comeback#AB5RGAbJLBziCe3l.99 (accessed October 17, 2017); Erin Douglas, "Chipotle Shows Signs of Recovery with New Menu Items in Testing, but It Is Enough?" *Denver Post,* http://www.denverpost.com/2017/07/14/chipotle-new-menu-items-revenue-ecoli/ (accessed October 17, 2017); Joel Stein, "The Fast Food Ethicist," *Time,* July 23, 2012, pp. 39–44; Caitlin Dewey, "Analysts Predicted Chipotle Would Be the Death of McDonald's. They Were Wrong," *Washington Post,* October 26, 2017, https://www.washingtonpost.com/news/wonk/wp/2017/10/26/why-mcdonalds-is-beating-out-the-fresh-healthy-competition/?utm_term=.82e8dfcaa5c3 (accessed November 22, 2017); Sarah Begley, "Chipotle Customer Who Got E. Coli Asks for Free Burritos in Settlement," *Time,* September 14, 2016, http://time.com/4490672/chipotle-ecoli-free-burritos-settlement/ (accessed November 22, 2017); Jen Wieczner, "Chipotle's 'Free Burrito' Coupons Are Making People Less Scared to Eat There," *Fortune,* May 4, 2016, http://fortune.com/2016/03/04/chipotle-coupons-ecoli/ (accessed November 22, 2017).

28. Jesse Newman, "FDA Tightens Its Food-Safety Rules," *The Wall Street Journal,* September 11, 2015, p. B3.

29. Associated Press, "Hagel Orders Renewed Focus on Military Ethics amid Air Force, Navy Cheating Scandals," *Fox News,* February 5, 2014, http://www.foxnews.com/us/2014/02/05/hagel-orders-renewed-focus-on-military-ethics-amid-air-force-navy-cheating/ (accessed January 28, 2015); Helene

Cooper, "Cheating Accusations among Officers Overseeing Nuclear Arms," *The New York Times,* January 15, 2014, http://www.nytimes.com/2014/01/16/ us/politics/ air-force-suspends-34-at-nuclear-sites-over-test-cheating.html?_r=0 (accessed January 28, 2015); Associated Press, "Navy Kicks Out 34 for Cheating on Nuclear Training Tests," *CBS News,* August 20, 2014, http:// www.cbsnews.com/news/ navy-kicks-out-34-for-cheating-on-nuclear-training-tests/ (accessed January 28, 2015).

30. Kyle Stock, "Movers," *Bloomberg Businessweek,* March 6–12, 2017, p. 19.

31. Rachel Abrams, "F.D.A. Warns 3 Tobacco Makers about Language Used on Labels," *The New York Times,* August 27, 2015, http://www.nytimes.com/2015/08/28/business/ fda-warns-3-tobacco-makers-about-language-used-on-labels.html (accessed May 3, 2017).

32. Ethics Resource Center, *2005 National Business Ethics Survey* (Washington, DC: Ethics Resource Center, 2005), p. 43.

33. Erich Reimer, "The Federal Reserve's Extraordinary Wells Fargo Growth Restriction," *Seeking Alpha,* February 5, 2018, https://seekingalpha.com/article/4143027-federal-reserves-extraordinary-wells-fargo-growth-restriction (accessed March 28, 2018).

34. Thomas M. Jones, "Ethical Decision Making by Individuals in Organizations: An Issue-Contingent Model," *Academy of Management Review* 2 (April 1991), pp. 371–373.

35. Sir Adrian Cadbury, "Ethical Managers Make Their Own Rules," *Harvard Business Review* 65 (September–October 1987), p. 72.

36. Ferrell, Fraedrich, and Ferrell, *Business Ethics,* pp. 174–175.

37. Ethics Resource Center, *2009 National Business Ethics Survey* (Washington, DC: Ethics Resource Center, 2009), p. 41.

38. Ethics and Compliance Initiative, *2016 Global Business Ethics SurveyTM: Measuring Risk and Promoting Workplace Integrity* (Arlington, VA: Ethics and Compliance Initiative, 2016), p. 43.

39. CR Magazine, "CR's 100 Best Corporate Citizens," 2017, http://www.thecro.com/wp-content/uploads/2017/05/ CR_100Bestpages_digitalR.pdf (accessed September 22, 2017); "Eaton CEO Emphasizes Ethics," Carnegie Mellon Tepper School of Business, http://tepper.cmu.edu/news-multimedia/tepper-stories/eaton-ceo-emphasizes-ethics/ index.aspx (accessed June 1, 2011); Eaton Corporation, *2016 Sustainability Metrics* (Cleveland: Eaton Corporation, 2017), http://www.eaton.com/ecm/groups/public/@pub/@ sustainability/documents/content/pct_3099477.pdf (accessed September 22, 2017); Eaton Corporation, "About Us—Fast Facts," http://www.eaton.com/Eaton/OurCompany/AboutUs/ CorporateInformation/FastFacts/index.htm (accessed September 22, 2017); Eaton Corporation, "Accountability & Transparency," http://www.eaton.in/EatonIN/OurCompany/ Sustainability/AccountabilityTransparency/index.htm (accessed September 22, 2017); Eaton Corporation, *Code of Ethics,* 2015, http://www.eaton.com/ecm/groups/public/@pub/ @eaton/@corp/documents/content/pct_1223598.pdf (accessed September 22, 2017); Eaton Corporation, *Ethics: The Power of Doing Business Right* (Cleveland: Eaton Corporation, 2011), p. 6, web version; Eaton Corporation, *Global Ethics—Ethics Guide* (Cleveland: Eaton Corporation, 2011); Eaton Corporation, "Sustainability," http://www. eaton.com/Sustainability/index.htm (accessed September 22, 2017); Eaton Corporation, "Worldwide Gift and Entertainment Policy," http://www.eaton.com/ecm/ groups/public/@pub/@eaton/@corp/documents/content/ ct_251820.pdf (accessed September 22, 2017); Wayne Heilmen, "UCCS Speaker: Ethics Policies Must Be Enforced Consistently," *Colorado Springs Gazette,* March 15, 2011, http://gazette.com/uccs-speaker-ethics-policies-must-be-enforced-consistently/article/114612 (accessed September 22, 2017).

40. Ferrell, Fraedrich, and Ferrell, *Business Ethics,* p. 13.

41. "Trust in the Workplace: 2010 Ethics & Workplace Survey," Deloitte LLP (n.d.), www.deloitte.com/assets/Dcom-UnitedStates/Local%20 Assets/Documents/us_2010_Ethics_ and_Workplace_Survey_ report_071910.pdf (accessed May 3, 2017).

42. Archie B. Carroll, "The Pyramid of Corporate Social Responsibility: Toward the Moral Management of Organizational Stakeholders," *Business Horizons* 34 (July/ August 1991), p. 42.

43. Kelly Kennedy, "Pharmacies Look to Snuff Tobacco Sales," *USA Today,* February 6, 2014, p. 1A.

44. Bryan Walsh, "Why Green Is the New Red, White and Blue," *Time,* April 28, 2008, p. 46.

45. Adam Shriver, "Not Grass-Fed, but at Least Pain-Free," *The New York Times,* February 18, 2010, www.nytimes. com/2010/02/19/opinion/19shriver.html (accessed May 3, 2017).

46. "USA Snapshots: Energy Investment Shelton Group's Energy Please Survey," *USA Today,* March 22, 2017, p. A1.

47. Corrine Iozzio, "A Farm on Every Street Corner," *Fast Company,* April 2015, p. 68; Philips, "Philips Commercializing City Farming Solutions Based on LED 'Light' Recipes that Improve Crop Yield and Quality," 2016, http://www.lighting.philips.com/main/products/horticulture/ press-releases/Philips-commercializing-city-farming-solutions-based-on-LED-light-recipes.html (accessed May 3, 2017).

48. Cone Communications and Ebiquity, "2015 Cone Communications/Ebiquity Global CSR Study," http://www. conecomm.com/stuff/contentmgr/files/0/2482ff6f22fe47534 88d3fe37e948e3d/files/global_pdf_2015.pdf (accessed May 3, 2017).

49. "2017 World's Most Ethical Companies—Honorees," *Ethisphere,* http://ethisphere.com/worlds-most-ethical/wme-honorees/ (accessed May 3, 2017).

50. Blake Mycoskie, "The Founder of TOMS on Reimagining the Company's Mission," *Harvard Business Review,* January/February 2016, pp. 41–44; Ashley Fahey, "TOMS Founder Reflects on Conscious Capitalism,

Entrepreneurship Ahead of Talk in Charlotte," *Charlotte Business Journal,* November 22, 2017, https://www.bizjournals.com/charlotte/news/2017/11/22/toms-founder-reflects-on-conscious-capitalism.html (accessed November 30, 2017); Patrick Cole, "TOMS Free Shoe Plan, Boosted by Clinton, Reaches Million Mark," *Bloomberg,* September 15, 2010, http://www.bloomberg.com/news/2010-09-16/toms-shoe-giveaway-for-kids-boosted-by-bill-clinton-reaches-million-mark.html (accessed November 30, 2017); TOMS, "One for One," http://www.toms.com/one-for-one-en/ (accessed November 30, 2017); Booth Moore, "Toms Shoes' Model Is Sell a Pair, Give a Pair Away," *Los Angeles Times,* April 19, 2009, http://www.latimes.com/features/image/la-ig-greentoms19-2009apr19,0,3694310.story (accessed December 10, 2017); Stacy Perman, "Making a Do-Gooder's Business Model Work," *Bloomberg Businessweek,* January 23, 2009, http://www.businessweek.com/smallbiz/content/jan2009/sb20090123_264702.htm (accessed November 6, 2014); Michelle Prasad, "TOMS Shoes Always Feels Good," *KENTON Magazine,* March 19, 2011, http://kentonmagazine.com/toms-shoes-always-feel-good/ (accessed November 6, 2014); Craig Sharkton, "Toms Shoes—Philanthropy as a Business Model," sufac.com, August 23, 2008, http://sufac.com/2008/08/toms-shoes-philanthropy-as-a-business-model/ (accessed June 3, 2011); Mike Zimmerman, "The Business of Giving: TOMS Shoes," *Success Magazine,* September 30, 2009, http://www.successmagazine.com/the-business-of-giving/PARAMS/article/852 (accessed June 3, 2011); "TOMS Founder Shares Sole-ful Tale," *North Texas Daily,* April 14, 2011, http://www.ntdaily.com/?p=53882 (accessed March 5, 2012); Scott Gerber, "Exit Interview: Blake Mycoskie," *Inc.,* December 2014/January 2015, p. 144; "TOMS Is on a Mission to Brew Something Greater: TOMS Roasting Co. Launches as the Next One for One® Product," *PR Newswire,* March 12, 2014, https://www.prnewswire.com/news-releases/toms-is-on-a-mission-to-brew-something-greater-toms-roasting-co-launches-as-the-next-one-for-one-product-249740051.html (accessed December 10, 2017); Rick Tetzeli, "Behind Toms Founder Blake Mycoskie's Plan to Build an Army of Social Entrepreneurs," *Fast Company,* January 11, 2016, https://www.fastcompany.com/3054929/behind-toms-founder-blake-mycoskies-plans-to-build-an-army-of-social-entrepr (accessed December 10, 2017).

51. Ferrell, Fraedrich, and Ferrell, *Business Ethics,* pp. 13–19.

52. Rachel Emma Silverstein, "Wall Street Perk: Parental Leave," *The Wall Street Journal,* December 1, 2015, p. C3; David Cohen, "Facebook Offers All Employees 4 Months' Parental Leave," December 1, 2015, http://www.adweek.com/socialtimes/four-months-parental-leave/630794 (accessed May 3, 2017).

53. Sarah Nefter, "Closing the Gap," *Delta Sky,* March 2017, p. 78.

54. Charisse Jones, "Here's the Secret to a Successful Company," *USA Today,* March 14, 2017, p. B1.

55. Doyle Rice, "Bumblebee Lands Securely on Endangered List," *USA Today,* March 22, 2017, p. A2.

56. Jonathan Watts, "Amazon Deforestation Report Is Major Setback for Brazil Ahead of Climate Talks," *The Guardian,* November 27, 2015, http://www.theguardian.com/world/2015/nov/27/amazon-deforestation-report-brazil-paris-climate-talks (accessed May 3, 2017).

57. Janet Larsen and Savina Venkova, "Plan B Updates," April 22, 2014, http://www.earth-policy.org/plan_b_updates/2014/update122 (accessed May 3, 2017).

58. "Certification," Home Depot, https://corporate.homedepot.com/CorporateResponsibility/Environment/WoodPurchasing/Pages/Certification.aspx (accessed May 3, 2017).

59. Cody Boteler, "California, Home to Silicon Valley, Considers Controversial Right to Repair," *Waste Dive,* March 19, 2018, https://www.wastedive.com/news/silicon-valley-california-controversial-right-to-repair-legislation/519424/ (accessed March 28, 2018).

60. GE Foundation, "United Way," www.gefoundation.com/employee-programs/united-way/ (accessed May 3, 2017).

61. Adobe Systems, "Adobe Corporate Responsibility," http://www.adobe.com/corporate-responsibility.html (accessed January 29, 2015).

62. Chad Broughton, "Just Another Factory Closing," *The Atlantic,* September 23, 2015, http://www.theatlantic.com/business/archive/2015/09/factory-closure-privateequity/406264 (accessed May 3, 2017).

63. Peter Cappelli, "The Skills Gap Myth: Why Companies Can't Find Good People," *Time,* June 4, 2012, http://business.time.com/2012/06/04/the-skills-gap-myth-why-companies-cant-find-good-people/ (accessed May 3, 2017).

64. Peter Cappelli, "Why Companies Aren't Getting the Employees They Need," *The Wall Street Journal,* October 24, 2011, pp. R1, R6.

65. "Who Really Pays for CSR Initiatives," *Environmental Leader,* February 15, 2008, www.environmentalleader.com/2008/02/15/who-really-pays-for-csr-initiatives/ (accessed May 3, 2017); "Global Fund," www.joinred.com/globalfund (accessed April 7, 2014); Reena Jana, "The Business of Going Green," *BusinessWeek Online,* June 22, 2007, www.businessweek.com/stories/2007-06-22/the-business-benefits-of-going-green-businessweek-business-news-stock-market-and-financial-advice (accessed May 3, 2017); Emma Howard, "Green Jobs Boom: Meet the Frontline of the New Solar Economy," *The Guardian,* February 1, 2016, http://www.theguardian.com/global-development-professionals-network/2016/feb/01/solar-economy-renewable-energy-asia-africa (accessed May 3, 2017).

66. "Create Jobs for USA Supporters," http://createjobsforusa.org/supporters (accessed February 25, 2016); "Starbucks and Opportunity Finance Network: Taking Action to Reduce Unemployment in America," *Huffington Post,* February 5, 2013, www.huffingtonpost.com/create-jobs-for-usa/starbucks-and-opportunity_b_2622773.html (accessed May 3, 2017); Opportunity Finance Network, "Overview: Create

Jobs for USA," 2016, http://ofn.org/create-jobs-usa (accessed May 3, 2017).

67. George Sammet Jr., *Gray Matters:* The Ethics Game (Orlando, FL: Lockheed Martin Corporation, 1992).

68. McGraw-Hill video, http://www.viddler.com/embed/93 938820/?f=1&autoplay=0&player=full&disablebrand ing=0 (accessed April 11, 2016); Douglas MacMillan, "Warby Parker Adds Storefronts to Its Sales Strategy," *The Wall Street Journal,* November 17, 2014, http:// online.wsj.com/articles/warby-parker-adds-storefronts-to- its-sales-strategy-1416251866 (accessed November 17, 2014); Warby Parker, "Buy a Pair, Give a Pair," https:// www.warbyparker.com/buy-a-pair-give-a-pair (accessed November 17, 2014); "5 Minutes with Neil Blumenthal," *Delta Sky,* September 2014, p. 30; Warby Parker website, https://www.warbyparker.com/ (accessed April 11, 2016); Jessica Pressler, "20/30 Vision," *New York Magazine,* August 11, 2013, http://nymag.com/news/features/warby- parker-2013-8/ (accessed October 14, 2014); David Zax, "Fast Talk: How Warby Parker's Cofounders Disrupted the Eyewear Industry and Stayed Friends," *Fast Company,* February 22, 2012, http://www.fastcompany.com/1818215/ fast-talk-how-warby-parkers-cofounders-disrupted-eyewear- industry-and-stayed-friends (accessed October 14, 2014); Marcus Wohlsen, "Is Warby Parker Too Good to Last?" *Wired,* June 25, 2015, http://www.wired.com/2014/06/ warby-parkers-quest-to-prove-not-sucking-is-the-ultimate- innovation/ (accessed October 14, 2014); Max Chafkin, "Warby Parker Sees the Future of Retail," *Fast Company,* February 17, 2015, http://www.fastcompany.com/3041334/ most-innovative-companies-2015/warby-parker-sees-the- future-of-retail (accessed April 11, 2016); Chase Peterson- Withorn, "Warby Parker CEO Is Building a Brand that Gives Back," *Forbes,* September 8, 2014, http://www .forbes.com/sites/chasewithorn/2014/09/08/warby-parker- ceo-is-building-a-brand-that-gives-back/#c830c2069d3b (accessed April 11, 2016); L. V. Anderson, "Spectacular Advice," *Slate,* http:// www.slate.com/articles/business/ the_ladder/2016/03/career_and_productivity_advice_ from_warby_parker_co_founder_and_co_ceo_dave.html (accessed April 11, 2016); Jason Del Ray, "Warby Parker Is Valued at $1.75 Billion after a Pre-IPO Investment of $75 Million," *Recode,* March 14, 2018, https://www.recode .net/2018/3/14/17115230/warby-parker-75-million-funding-t -rowe-price-ipo (accessed October 6, 2018).

Credits

3 Business in a Borderless World

©Shutterstock

Chapter Outline

Learning Objectives

After reading this chapter, you will be able to:

LO 3-1 Explore some of the factors within the international trade environment that influence business.

LO 3-2 Assess some of the economic, legal, political, social, cultural, and technological barriers to international business.

LO 3-3 Specify some of the agreements, alliances, and organizations that may encourage trade across international boundaries.

LO 3-4 Summarize the different levels of organizational involvement in international trade.

LO 3-5 Contrast two basic strategies used in international business.

LO 3-6 Assess the opportunities and problems facing a small business that is considering expanding into international markets.

Enter the World of Business

Alibaba: China's National Treasure

Jack Ma founded Alibaba as an e-commerce business in 1999 after taking a trip to California's Silicon Valley. While there, he saw firsthand how the Internet was transforming the business world. He realized the commercial potential of pioneering an online marketing channel in China so small businesses could connect with local and international buyers.

Ma created an Internet marketplace he named "Alibaba" because of the association with opening doors and finding treasure. The initial business-to-business website did so well that Alibaba opened a second online marketplace for selling to consumers called Taobao, followed by an e-commerce mall for multinational retailers targeting Chinese consumers. In 2005, Ma formed a strategic alliance with Yahoo! Inc. co-founder Jerry Yang when he recognized the need for a search engine partner. The deal resulted in Yahoo! purchasing a 40 percent stake in Alibaba for $1 billion.

Ma's launch of a Chinese e-commerce business was not without risk. Obstacles existed in the economic and technological environments. Disposable incomes were still relatively low in China, and Internet connections were slow and expensive. However, as technology improved and disposable incomes increased, more Chinese consumers became interested in e-commerce.

Today, the company's many marketplaces and supporting businesses serve more than 600 million customers in 240 nations. Alibaba has held tightly to its top spot in China. With a 47 percent market share in China's online retail market, Alibaba has made it difficult for Amazon to compete on price, preventing the U.S. online retail giant from gaining traction. China, with approximately 890 million online shoppers, is overtaking the United States as the largest e-commerce market.[1]

Introduction

Consumers around the world can drink Coca-Cola and Pepsi, eat at McDonald's and Pizza Hut, buy an Apple phone made in China, and watch CNN and MTV on Samsung televisions. It may surprise you that German automaker BMW has manufacturing facilities in Mexico and South Africa that export many of their cars to the United States. In fact, one-third of all 3-series sold in the United States are built in South Africa.[2] The products you consume today are just as likely to have been made in China, India, or Germany as in the United States.[3] Likewise, consumers in other countries buy Western electrical equipment, clothing, rock music, cosmetics, and toiletries, as well as computers, robots, and household goods. Google's YouTube has more than 1 billion hours of global viewing of videos a day.[4]

Many U.S. firms are finding that international markets provide tremendous opportunities for growth. Accessing these markets can promote innovation, while intensifying global competition spurs companies to market better and less expensive products. Today, the more than 7 billion people who inhabit the earth comprise one tremendous marketplace.

In this chapter, we explore business in this exciting global marketplace. First, we look at the nature of international business, including barriers and promoters of trade across international boundaries. Next, we consider the levels of organizational involvement in international business. Finally, we briefly discuss strategies for trading across national borders.

The Role of International Business

LO 3-1

Explore some of the factors within the international trade environment that influence business.

international business
the buying, selling, and trading of goods and services across national boundaries.

International business refers to the buying, selling, and trading of goods and services across national boundaries. Falling political barriers and new technology are making it possible for more and more companies to sell their products overseas as well as at home. And, as differences among nations continue to narrow, the trend toward the globalization of business is becoming increasingly important. Starbucks serves millions of global customers at more than 24,000 locations in 75 markets.[5] The Internet and the ease by which mobile applications can be developed provide many companies with easier entry to access global markets than opening brick-and-mortar stores.[6]

Independent record labels are experiencing a surge in revenue from services like Spotify. Now through streaming, foreign markets are accessible and many are drawing almost half their listeners from outside the home country.[7] Amazon, an online retailer, has distribution centers from Nevada to Germany that fill millions of orders a day and ship them to customers in every corner of the world. Outside of the United States, China has become Apple's second largest market while Europe is the largest.[8] Indeed, most of the world's population and two-thirds of its total purchasing power are outside the United States.

When McDonald's sells a Big Mac in Moscow, Sony sells a television in Detroit, or a small Swiss medical supply company sells a shipment of orthopedic devices to a hospital in Monterrey, Mexico, the sale affects the economies of the countries involved.

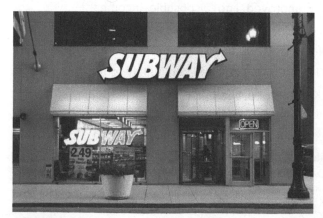

Subway has more restaurants around the world than any other fast-food chain.

©McGraw-Hill Education/Andrew Resek, photographer

The U.S. market, with 325 million consumers, makes up only 4.4 percent of the more than 7.6 billion people in the world to whom global companies must consider marketing. Global marketing requires balancing a firm's global brand with the needs of local consumers.[9] To begin our study of international business, we must first consider some economic issues: why nations trade, exporting and importing, and the balance of trade.

Why Nations Trade

Nations and businesses engage in international trade to obtain raw materials and goods that are otherwise unavailable to them or are available elsewhere at a lower price than that at which they themselves can produce. A nation, or individuals and organizations from a nation, sell surplus materials and goods to acquire funds to buy the goods, services, and ideas its people need. Countries like Ethiopia, Cameroon, and Kenya trade with Western nations in order to acquire technology and techniques to advance their economy. Which goods and services a nation sells depends on what resources it has available.

Many companies choose to outsource manufacturing to factories in Asia due to lower costs of labor.

©Roberto Westbrook/Blend Images

Some nations have a monopoly on the production of a particular resource or product. Such a monopoly, or **absolute advantage,** exists when a country is the only source of an item, the only producer of an item, or the most efficient producer of an item. An example would be an African mining company that possesses the only mine where a specialty diamond can be found. Russia has an absolute advantage in yuksporite, a rare and useful mineral that can be found only in Russia.

Most international trade is based on **comparative advantage,** which occurs when a country specializes in products that it can supply more efficiently or at a lower cost than it can produce other items. France has a comparative advantage in making wine because of its agricultural capabilities, reputation, and the experience of its vintners. The United States, having adopted new technological methods in hydraulic fracturing, has created a comparative advantage in the drilling and exporting of natural gas.[11] Other countries, particularly India and Ireland, are also gaining a comparative advantage over the United States in the provision of some services, such as call-center operations, engineering, and software programming. As a result, U.S. companies are increasingly **outsourcing,** or transferring manufacturing and other tasks to countries where labor and supplies are less expensive. Outsourcing has become a controversial practice in the United States because many jobs have moved overseas where those tasks can be accomplished for lower costs. Ireland has become a destination of choice for U.S. companies because of a well-educated workforce and technological capabilities.

absolute advantage
a monopoly that exists when a country is the only source of an item, the only producer of an item, or the most efficient producer of an item.

comparative advantage
the basis of most international trade, when a country specializes in products that it can supply more efficiently or at a lower cost than it can produce other items.

outsourcing
the transferring of manufacturing or other tasks—such as data processing—to countries where labor and supplies are less expensive.

Trade between Countries

To obtain needed goods and services and the funds to pay for them, nations trade by exporting and importing. **Exporting** is the sale of goods and services to foreign markets. The United States exports more than $2.3 trillion in goods and services annually.[12] U.S. businesses export many goods and services, particularly agricultural, entertainment (movies,

exporting
the sale of goods and services to foreign markets.

FIGURE 3.1

U.S. Exports to China
(millions of U.S. dollars)

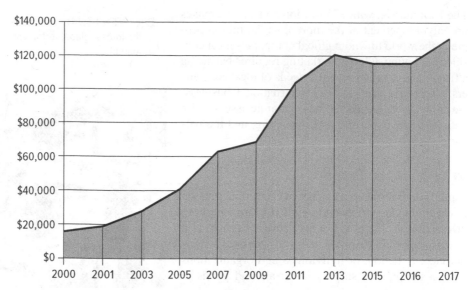

Source: U.S. Census Bureau, "Trade in Goods with China," https://www.census.gov/foreign-trade/balance/c5700.html (accessed April 7, 2018).

importing
the purchase of goods and
services from foreign sources.

television shows, etc.), and technological products. **Importing** is the purchase of goods and services from foreign sources. Many of the goods you buy in the United States are likely to be imports or to have some imported components. Sometimes, you may not even realize they are imports. The United States imports more than $2.9 trillion each year.[13]

 connect

▶ Need help understanding balance of trade? Visit your Connect ebook video tab for a brief animated explanation.

balance of trade
the difference in value
between a nation's exports
and its imports.

Balance of Trade

You have probably read or heard about the fact that the United States has a trade deficit, but what is a trade deficit? A nation's **balance of trade** is the difference in value between its exports and imports. Because the United States (and some other nations as well) imports more products than it exports, it has a negative balance of trade, or **trade deficit.** U.S. exports to China rapidly increased, as Figure 3.1 indicates, but not fast enough to offset the imports from China. Table 3.1 shows the overall trade deficit for the United States, which is currently around $502 million.[14] The trade deficit fluctuates according to such factors as the health of the United States and other economies, productivity, perceived quality, and exchange rates. Trade deficits are harmful because they can mean the failure of businesses, the loss of jobs, and a lowered standard of living.

TABLE 3.1

U.S. Trade Deficit (in billions of dollars)

	1990	2000	2010	2011	2012	2013	2014	2015	2016	2017
Exports	535.2	1,075.3	1,853.6	2,127.0	2,219.0	2,279.9	2,343.2	2,230.3	2,212.1	2,427
Imports	616.1	1,447.8	2,348.3	2,675.6	2,755.8	2,758.3	2,851.5	2,761.8	2,712.6	2,900
Trade surplus/deficit	−80.9	−372.5	−494.7	−548.6	−536.8	−478.4	−508.3	−531.5	−500.6	−568

Source: U.S. Bureau of the Census, Foreign Trade Division, "U.S. Trade in Goods and Services—Balance of Payments (BOP) Basis," March 7, 2018 www.census.gov/foreign-trade/statistics/historical/gands.pdf (accessed April 1, 2018).

TABLE 3.2
Top 10 Countries with which the United States Has Trade Deficits/Surpluses

Trade Deficit	Trade Surplus
1. China	Hong Kong
2. Japan	Netherlands
3. Germany	Australia
4. Mexico	Belgium
5. Canada	United Arab Emirates
6. Ireland	Singapore
7. Vietnam	Argentina
8. Italy	Panama
9. India	Dominican Republic
10. South Korea	Egypt

Sources: "Top Ten Countries with which the U.S. Has a Trade Deficit," January 2018, https://www.census.gov/foreign-trade/statistics/highlights/toppartners.html#def (accessed April 1, 2018).

Of course, when a nation exports more goods than it imports, it has a favorable balance of trade, or trade surplus. Until about 1970, the United States had a trade surplus due to an abundance of natural resources and the relative efficiency of its manufacturing systems. Table 3.2 shows the top 10 countries with which the United States has a trade deficit and a trade surplus.

The difference between the flow of money into and out of a country is called its **balance of payments.** A country's balance of trade, foreign investments, foreign aid, loans, military expenditures, and money spent by tourists comprise its balance of payments. As you might expect, a country with a trade surplus generally has a favorable balance of payments because it is receiving more money from trade with foreign countries than it is paying out. When a country has a trade deficit, more money flows out of the country than into it. If more money flows out of the country than into it from tourism and other sources, the country may experience declining production and higher unemployment because there is less money available for spending.

trade deficit
a nation's negative balance of trade, which exists when that country imports more products than it exports.

balance of payments
the difference between the flow of money into and out of a country.

International Trade Barriers

Completely free trade seldom exists. When a company decides to do business outside its own country, it will encounter a number of barriers to international trade. Any firm considering international business must research the other country's economic, legal, political, social, cultural, and technological background. Such research will help the company choose an appropriate level of involvement and operating strategies, as we will see later in this chapter.

Economic Barriers

When looking at doing business in another country, managers must consider a number of basic economic factors, such as economic development, infrastructure, and exchange rates.

LO 3-2

Assess some of the economic, legal, political, social, cultural, and technological barriers to international business.

Economic Development.   When considering doing business abroad, U.S. businesspeople need to recognize that they cannot take for granted that other countries offer the same things as are found in *industrialized nations*—economically advanced countries such as the United States, Japan, Great Britain, and Canada. Many countries in Africa, Asia, and South America, for example, are in general poorer and less economically advanced than those in North America and Europe; they are often called *less-developed countries* (LDCs). LDCs are characterized by low per-capita income (income generated by the nation's production of goods and services divided by the population), which means that consumers are less likely to purchase nonessential products. Nonetheless, LDCs represent a potentially huge and profitable market for many businesses because they may be buying technology to improve their infrastructures, and much of the population may desire consumer products. For example, automobile manufacturers are looking toward LDCs as a way to expand their customer base. The rising middle class has caused many consumers in India and China to desire their own vehicles. The automobile market in China is now larger than the market in the United States.

A country's level of development is determined in part by its **infrastructure**, the physical facilities that support its economic activities, such as railroads, highways, ports, airfields, utilities and power plants, schools, hospitals, communication systems, and commercial distribution systems. When doing business in LDCs, for example, a business may need to compensate for rudimentary distribution and communication systems, or even a lack of technology.

infrastructure
the physical facilities that support a country's economic activities, such as railroads, highways, ports, airfields, utilities and power plants, schools, hospitals, communication systems, and commercial distribution systems.

exchange rate
the ratio at which one nation's currency can be exchanged for another nation's currency.

Exchange Rates.   The ratio at which one nation's currency can be exchanged for another nation's currency is the **exchange rate**. Exchange rates vary daily and can be found in newspapers and through many sites on the Internet. Familiarity with exchange rates is important because they affect the cost of imports and exports. When the value of the U.S. dollar declines relative to other currencies, such as the euro, the price of imports becomes more economical for U.S. consumers. For example, if the exchange rate for the dollar moves from $1.40 per euro to $1.25 per euro, then imports from Europe will be less expensive. On the other hand, U.S. exports become relatively expensive for international markets—in this example, the European Union (EU). The U.S. dollar is most frequently used in international trade, with 81 percent of trade finance conducted in U.S. dollars.[15]

Occasionally, a government may intentionally alter the value of its currency through fiscal policy. Devaluation decreases the value of currency in relation to other currencies. If the U.S. government were to devalue the dollar, it would lower the cost of American goods abroad and make trips to the United States less expensive for foreign tourists. Thus, devaluation encourages the sale of domestic goods and tourism. On the other hand, when Switzerland's central bank let the value of the Swiss franc rise by 30 percent against the euro, it resulted in increasing the costs of exports. This made everything exported from Switzerland more expensive, including tourism. However, Swiss brands including expensive watches were offered at large discounts if bought using Swiss francs.[16] Revaluation, as in the Swiss example, increases the value of a currency in relation to other currencies, but occurs rarely.

Ethical, Legal, and Political Barriers

A company that decides to enter the international marketplace must contend with potentially complex relationships among the different laws of its own nation, international laws, and the laws of the nation with which it will be trading; various trade

Entrepreneurship in Action

Kenya Counts on Mobile Banking

M-Pesa

Founders: Nick Hughes and Susie Lonie

Founded: 2007, launched in Kenya through Safaricom

Success: Originally launched in Kenya, today this mobile-transfer payment system has 30 million users in 10 countries.

In 2007, mobile-transfer payment system M-Pesa was launched by British multinational Vodafone's Kenyan mobile operator Safaricom. Little did it know how revolutionary this innovation would be for the economy. M-Pesa used mobile services to leapfrog existing technological barriers and open up new opportunities for millions of Kenyans.

In Kenya, a less-developed country, banks tended to avoid catering to lower-income populations, and the distance to financial institutions could be far. The mobile-transfer payment service allowed Kenyan residents to quickly and securely transfer funds to merchants and family members. It also allowed

for a significant rise in entrepreneurship. It is estimated that M-Pesa extended financial services to 20 million Kenyans, prompting many to start their own small businesses. Unlike the United States, credit cards are rare in Kenya, as many Kenyans now use M-Pesa to pay for meals and other transactions.

Over the years, M-Pesa has expanded to nine other countries, including Tanzania, Romania, India, the Democratic Republic of the Congo, and Albania. While competition is rising for M-Pesa, the possibilities of worldwide expansion remain high as mobile payments gain traction globally.[17]

Critical Thinking Questions

1. What conditions in Kenya created the ideal conditions for a mobile banking service?
2. In what ways has M-Pesa paved the way for entrepreneurship in Kenya?
3. How can businesses in different industries profit from opportunities in less-developed countries?

---

restrictions imposed on international trade; changing political climates; and different ethical values. Legal and ethical requirements for successful business are increasing globally.

Laws and Regulations.   The United States has a number of laws and regulations that govern the activities of U.S. firms engaged in international trade and has a variety of friendship, commerce, and navigation treaties with other nations. These treaties allow business to be transacted between U.S. companies and citizens of the specified countries. We discuss some of these relationships in this chapter.

Once outside U.S. borders, businesspeople are likely to find that the laws of other nations differ from those of the United States. Many of the legal rights that Americans take for granted do not exist in other countries, and a firm doing business abroad must understand and obey the laws of the host country. Some countries have strict laws limiting the amount of local currency that can be taken out of the country and the amount of currency that can be brought in; others limit how foreign companies can operate within the country. While the United States has been slow to sell commercial drones globally due to regulations, Australia, Singapore, and Britain have regulations that have allowed drone deliveries to grow quickly.[18]

Some countries have copyright and patent laws that are less strict than those of the United States, and some countries fail to honor U.S. laws. Although countries such as China and Vietnam have and enforce their copyright laws, controlling the amount of counterfeit products sold in their countries is very challenging. China pays $8 billion a year for U.S. intellectual property, only slightly more than South Korea, an economy one-tenth the size of China.[19] Companies are angry because the counterfeits harm not only their sales, but also their reputations if the knock-offs are of poor quality. Such counterfeiting is not limited to China or Vietnam. It is estimated that nearly half of all software installed on personal computers worldwide is not properly licensed.[20]

In countries where these activities occur, laws against them may not be sufficiently enforced if counterfeiting is deemed illegal. Thus, businesses engaging in foreign trade may have to take extra steps to protect their products because local laws may be insufficient to do so.

Tariffs and Trade Restrictions.   Tariffs and other trade restrictions are part of a country's legal structure but may be established or removed for political reasons. An **import tariff** is a tax levied by a nation on goods imported into the country. A *fixed tariff* is a specific amount of money levied on each unit of a product brought into the country, while an *ad valorem tariff* is based on the value of the item. Most countries allow citizens traveling abroad to bring home a certain amount of merchandise without paying an import tariff. A U.S. citizen may bring $200, $800, or $1,600 worth of merchandise into the United States duty free depending on the country visited. After that, U.S. citizens must pay an ad valorem tariff based on the cost of the item and the country of origin. Thus, identical items purchased in different countries might have different tariffs.

Countries sometimes levy tariffs for political reasons, as when they impose sanctions against other countries to protest their actions. However, import tariffs are more commonly imposed to protect domestic products by raising the price of imported ones. Such protective tariffs have become controversial as Americans become increasingly concerned over the U.S. trade deficit. Protective tariffs allow more expensive domestic goods to compete with foreign ones. For example, the United States has imposed tariffs on steel imported into the United States because imports have caused many local steelworks to crash.[21] Other markets can produce steel more cheaply than the United States. The United States placed tariffs on aluminum and steel imports from all countries but Canada and Mexico in 2018. Small manufacturers that fabricate metal into parts for cars, appliances, and other components feared job loss and higher prices.[22] The United States indicated it would drop the 25 percent tariffs on steel and aluminum if other countries would make concessions on other trade issues.[23] In addition, tariffs were placed on some high-tech products from China. In response, China placed tariffs, up to 25 percent, on 128 U.S. products including fresh and dried fruit, soybeans, wine, pork, and some steel products. In 2017, the United States placed a 21 percent tariff on Canadian softwood, claiming the industry was unfairly subsidized. This tariff resulted in driving lumber prices to record highs in 2018 because Canadian producers passed the higher tariffs back to U.S. purchasers.[24]

The EU levies tariffs on many products, including some seafood imports and fruits. Critics of protective tariffs argue that their use inhibits free trade and competition. Supporters of protective tariffs say they insulate domestic industries, particularly new ones, against well-established foreign competitors. Once an industry matures, however, its advocates may be reluctant to let go of the tariff that protected it. Tariffs also help when, because of low labor costs and other advantages, foreign competitors can afford to sell their products at prices lower than those charged by domestic companies. Some Americans argue that

After the United States imposed a 21 percent tariff on Canadian softwood, lumber prices reached record highs.

©Spaces Images/Blend Images

import tariff
a tax levied by a nation on goods imported into the country.

tariffs should be used to keep domestic wages high and unemployment low. Recently, there are fears that a trade war could develop that damages the world economy.

Exchange controls restrict the amount of currency that can be bought or sold. Some countries control their foreign trade by forcing businesspeople to buy and sell foreign products through a central bank. If John Deere, for example, receives payments for its tractors in a foreign currency, it may be required to sell the currency to that nation's central bank. When foreign currency is in short supply, as it is in many LDCs, the government uses foreign currency to purchase necessities and capital goods and produces other products locally, thus limiting its need for foreign imports.

exchange controls
regulations that restrict the amount of currency that can be bought or sold.

A **quota** limits the number of units of a particular product that can be imported into a country. A quota may be established by voluntary agreement or by government decree. The United States imposes quotas on certain goods, such as garments produced in Vietnam and China. Quotas are designed to protect the industries and jobs of the country imposing the quota. Quotas help domestic suppliers but will lead to higher prices for consumers.

quota
a restriction on the number of units of a particular product that can be imported into a country.

An **embargo** prohibits trade in a particular product. Embargoes are generally directed at specific goods or countries and may be established for political, economic, health, or religious reasons. The United States currently maintains a trade embargo with Cuba. While the Obama administration reestablished trade and diplomatic relations between Cuba and the United States, the Trump administration tightened the embargo by restricting access to hotels and stores tied to the Cuban military from Americans. It is much easier to travel to Cuba than in previous decades, and U.S. citizens can bring back Cuban cigars and rum. The government has also approved the building of a U.S. factory in Cuba, the first time in more than 50 years.[25] It may be surprising to know that U.S. farmers export hundreds of millions of dollars' worth of commodities to Cuba each year, based on a 2000 law that provided permission for some trade to the embargoed country.[26] Health embargoes prevent the importing of various pharmaceuticals, animals, plants, and agricultural products. Muslim nations forbid the importation of alcoholic beverages on religious grounds.

embargo
a prohibition on trade in a particular product.

One common reason for setting quotas or tariffs is to prohibit **dumping,** which occurs when a country or business sells products at less than what it costs to produce them. For example, China accused the EU and Japan of dumping its stainless steel tubes, thus harming China's domestic industry.[27] A company may dump its products for several reasons. Dumping permits quick entry into a market. Sometimes, dumping occurs when the domestic market for a firm's product is too small to support an efficient level of production. In other cases, technologically obsolete products that are no longer salable in the country of origin are dumped overseas. Dumping is relatively difficult to prove, but even the suspicion of dumping can lead to the imposition of quotas or tariffs. China instituted anti-dumping duties on EU and Japanese imports.[28] As with other trade restrictions, dumping quotas or tariffs result in higher prices for consumers.

dumping
the act of a country or business selling products at less than what it costs to produce them.

Political Barriers. Unlike legal issues, political considerations are seldom written down and often change rapidly. Nations that have been subject to economic sanctions for political reasons in recent years

Due to the U.S. embargo against Cuba, many Cubans drive older automobiles.

©Horizon Images/Motion/Alamy Stock Photo

Responding to Business Challenges

Uber Attempts to Make the Right Turn

Uber provides ride-sharing services by connecting drivers and riders through an app. It has expanded its operations to 674 cities in 83 countries worldwide. As it expands, Uber is engaging in strategic partnerships with local companies. These alliances allow Uber to utilize the knowledge of domestic firms familiar with the country's culture. Uber has partnered with Times Internet in India, Baidu in China, and America Movil in Latin America.

Despite its success, Uber had faced problems in expanding internationally. In Spain, a judge ruled that Uber drivers are not legally authorized to transport passengers and that it unfairly competes against licensed taxi drivers. France and Germany instituted similar bans against unlicensed Uber drivers. Uber returned to Spain with UberX, which uses licensed drivers.

India is Uber's second-largest market after the United States. In New Delhi, the taxi industry banned app-based services without radio-taxi permits in the capital for safety reasons. Uber made changes to increase safety. Yet despite these changes, Uber continued to run afoul of Indian authorities. India asked Internet service providers to block Uber's websites because it continued to operate in the city despite being banned.

Uber has taken a global approach to expansion by applying the same practices in other countries as it does in the United States. However, it is realizing that it must take a more customized approach to achieve long-term international market success.[29]

Critical Thinking Questions

1. What are some of the major advantages Uber is experiencing by creating strategic partnerships with other companies?
2. What are some challenges Uber is facing as it expands globally?
3. What are some of the barriers that Uber is encountering in different countries?

---

cartel
a group of firms or nations that agrees to act as a monopoly and not compete with each other, in order to generate a competitive advantage in world markets.

include Cuba, Iran, Syria, and North Korea. While these were dramatic events, political considerations affect international business daily as governments enact tariffs, embargoes, or other types of trade restrictions in response to political events.

Businesses engaged in international trade must consider the relative instability of countries such as Iraq, Ukraine, and Venezuela. Political unrest in countries such as Pakistan, Somalia, and the Democratic Republic of the Congo may create a hostile or even dangerous environment for foreign businesses. Natural disasters can cripple a country's government, making the region even more unstable. Finally, a sudden change in power can result in a regime that is hostile to foreign investment. Some businesses have been forced out of a country altogether, as when Hugo Chavez conducted a socialist revolution in Venezuela to force out or take over American oil companies. Whether they like it or not, companies are often involved directly or indirectly in international politics. Today, Venezuela has a sinking economy with shortages of products and political unrest.

Political concerns may lead a group of nations to form a **cartel,** a group of firms or nations that agrees to act as a monopoly and not compete with each other, to generate a competitive advantage in world markets. Probably the most famous cartel is OPEC, the Organization of Petroleum Exporting Countries, founded in the 1960s to increase the price of petroleum throughout the world and to maintain

Political instability in many nations has led to an influx of refugees. The potential for political turmoil is a substantial risk businesses face when expanding overseas.

©Sk Hasan Ali/Shutterstock

high prices. By working to ensure stable oil prices, OPEC hopes to enhance the economies of its member nations. In 2018, OPEC and some non-OPEC members agreed to cut oil production to increase and stabilize prices.

Social and Cultural Barriers

Most businesspeople engaged in international trade underestimate the importance of social and cultural differences; but these differences can derail an important transaction. Tiffany & Co. learned that more attentive customer service was necessary in order to succeed in Japan, and bold marketing and advertising served as the recipe for success in China.[30] And in Europe, Starbucks took the unprecedented step of allowing its locations to be franchised in order to reach smaller markets that are unfamiliar.

Sociocultural differences can create challenges for businesses that want to invest in other countries.
©LEE SNIDER PHOTO IMAGES/Shutterstock

This way, Starbucks reduced some of the cultural and social risks involved in entering such markets.[31] For example, Starbucks opened its first store in Italy in 2018.[32] Starbucks waited to enter Italy because it needed to understand the coffee culture there. Unfortunately, cultural norms are rarely written down, and what is written down may well be inaccurate.

Cultural differences include differences in spoken and written language. Although it is certainly possible to translate words from one language to another, the true meaning is sometimes misinterpreted or lost. Consider some translations that went awry in foreign markets:

- Scandinavian vacuum manufacturer Electrolux used the following in an American campaign: "Nothing sucks like an Electrolux."
- The Coca-Cola name in China was first read as "Ke-kou-ke-la," meaning "bite the wax tadpole."
- In Italy, a campaign for Schweppes Tonic Water translated the name into Schweppes Toilet Water.[33]

Translators cannot just translate slogans, advertising campaigns, and website language; they must know the cultural differences that could affect a company's success.

Differences in body language and personal space also affect international trade. Body language is nonverbal, usually unconscious communication through gestures, posture, and facial expression. Personal space is the distance at which one person feels comfortable talking to another. Americans tend to stand a moderate distance away from the person with whom they are speaking. Arab businessmen tend to stand face-to-face with the object of their conversation. Additionally, gestures vary from culture to culture, and gestures considered acceptable in American society—pointing, for example—may be considered rude in others. Table 3.3 shows some of the behaviors considered rude or unacceptable in other countries. Such cultural differences may generate uncomfortable feelings or misunderstandings when businesspeople of different countries negotiate with each other.

Family roles also influence marketing activities. Many countries do not allow children to be used in advertising, for example. Advertising that features people in nontraditional social roles may or may not be successful either. Companies should

TABLE 3.3
Cultural Behavioral
Differences

	Gestures Viewed as Rude or Unacceptable
Japan, Hong Kong, Middle East	Summoning with the index finger
Middle and Far East	Pointing with index finger
Thailand, Japan, France	Sitting with soles of shoes showing
Brazil, Germany	Forming a circle with fingers (the "O.K." sign in the United States)
Japan	Winking means "I love you"
Buddhist countries	Patting someone on the head

Source: Adapted from Judie Haynes, "Communicating with Gestures," EverythingESL (n.d.), www.everythingesl.net/inservices/body_language.php (accessed April 7, 2017).

also guard against marketing that could be perceived as reinforcing negative stereotypes. Coca-Cola was forced to pull an online advertisement and issue an apology after releasing a Christmas ad showing fair-skinned people arriving at an indigenous village in Mexico bearing gifts of sodas and a Christmas tree. Mexican activists claimed the advertisement reinforced negative stereotypes of indigenous people and called for the government's antidiscrimination commission to issue sanctions against the company.[34]

The people of other nations quite often have a different perception of time as well. Americans value promptness; a business meeting scheduled for a specific time seldom starts more than a few minutes late. In Mexico and Spain, however, it is not unusual for a meeting to be delayed half an hour or more. Such a late start might produce resentment in an American negotiating in Spain for the first time.

Companies engaged in foreign trade must observe the national and religious holidays and local customs of the host country. In many Islamic countries, for example, workers expect to take a break at certain times of the day to observe religious rites. Companies also must monitor their advertising to guard against offending customers. In Thailand and many other countries, public displays of affection are unacceptable in advertising messages; in many Middle Eastern nations, it is unacceptable to show the soles of one's feet.[35] In Russia, smiling is considered appropriate only in private settings, not in business.

With the exception of the United States, most nations use the metric system. This lack of uniformity creates problems for both buyers and sellers in the international marketplace. American sellers, for instance, must package goods destined for foreign markets in liters or meters, and Japanese sellers must convert to the English system if they plan to sell a product in the United States. Tools also must be calibrated in the correct system if they are to function correctly. Hyundai and Honda service technicians need metric tools to make repairs on those cars.

The literature dealing with international business is filled with accounts of sometimes humorous but

After just two years in the U.S. market, Primark scaled back three of its eight retail locations. U.S. shoppers have yet to warm to the retailer's store-only shopping strategy.

©Helen89/Shutterstock

often costly mistakes that occurred because of a lack of understanding of the social and cultural differences between buyers and sellers. Such problems cannot always be avoided, but they can be minimized through research on the cultural and social differences of the host country.

Technological Barriers

Many countries lack the technological infrastructure found in the United States, and some marketers are viewing such barriers as opportunities. For instance, marketers are targeting many countries such as India and China and some African countries where there are few private phone lines. Citizens of these countries are turning instead to wireless communication through cell phones. Technological advances are creating additional global marketing opportunities. Along with opportunities, changing technologies also create new challenges and competition. The U.S. market share of the personal computer market is dropping as new competitors emerge that are challenging U.S. PC makers. In fact, out of the top five global PC companies—Lenovo, Hewlett-Packard, Dell, Asus, and Acer Group—three are from Asian countries. On the other hand, Apple Inc.'s iPad and other tablet computer makers have significantly eroded the market share of traditional personal computers, placing the industry in the maturity stage of the product life cycle.[36]

Trade Agreements, Alliances, and Organizations

LO 3-3

Specify some of the agreements, alliances, and organizations that may encourage trade across international boundaries.

Although these economic, political, legal, and sociocultural issues may seem like daunting barriers to international trade, there are also organizations and agreements—such as the General Agreement on Tariffs and Trade, the World Bank, and the International Monetary Fund—that foster international trade and can help companies get involved in and succeed in global markets. Various regional trade agreements, such as the North American Free Trade Agreement and the EU, also promote trade among member nations by eliminating tariffs and trade restrictions. In this section, we'll look briefly at these agreements and organizations.

General Agreement on Tariffs and Trade

During the Great Depression of the 1930s, nations established so many protective tariffs covering so many products that international trade became virtually impossible. By the end of World War II, there was considerable international momentum to liberalize trade and minimize the effects of tariffs. The **General Agreement on Tariffs and Trade (GATT)**, originally signed by 23 nations in 1947, provided a forum for tariff negotiations and a place where international trade problems could be discussed and resolved. More than 100 nations abided by its rules. GATT sponsored rounds of negotiations aimed at reducing trade restrictions. The most recent round, the Uruguay Round (1988–1994), further reduced trade barriers for most products and provided new rules to prevent dumping.

The **World Trade Organization (WTO)**, an international organization dealing with the rules of trade between nations, was created in 1995 by the Uruguay Round. Key to the World Trade Organization are the WTO agreements, which are the legal ground rules for international commerce. The agreements were negotiated and signed by most of the world's trading nations and ratified by their parliaments. The goal is to help producers of goods and services and exporters and importers conduct their business.

General Agreement on Tariffs and Trade (GATT)
a trade agreement, originally signed by 23 nations in 1947, that provided a forum for tariff negotiations and a place where international trade problems could be discussed and resolved.

World Trade Organization (WTO)
international organization dealing with the rules of trade between nations.

In addition to administering the WTO trade agreements, the WTO presents a forum for trade negotiations, monitors national trade policies, provides technical assistance and training for developing countries, and cooperates with other international organizations. Based in Geneva, Switzerland, the WTO has also adopted a leadership role in negotiating trade disputes among nations.[37] For example, the WTO ruled that China's antidumping measures taken against Japan and the EU violated trade rules because China had not adequately proven that the imports of the stainless steel tubes had harmed China's domestic industry.[38]

The North American Free Trade Agreement

North American Free Trade Agreement (NAFTA)
agreement that eliminates most tariffs and trade restrictions on agricultural and manufactured products to encourage trade among Canada, the United States, and Mexico.

The **North American Free Trade Agreement (NAFTA)**, which went into effect on January 1, 1994, effectively merged Canada, the United States, and Mexico into one market of nearly 450 million consumers. NAFTA virtually eliminated all tariffs on goods produced and traded among Canada, Mexico, and the United States to create a free trade area. The estimated annual output for this trade alliance is about $20.8 trillion.[39] NAFTA makes it easier for U.S. businesses to invest in Mexico and Canada; provides protection for intellectual property (of special interest to high-technology and entertainment industries); expands trade by requiring equal treatment of U.S. firms in both countries; and simplifies country-of-origin rules, hindering Japan's use of Mexico as a staging ground for further penetration into U.S. markets. The U.S., Canada, and Mexico reached an agreement that has been called NAFTA 2.0, The United States-Mexico-Canada Agreement, or USMCA. When the agreement is ratified in all three countries it will result in major changes.

Canada's nearly 37 million consumers are relatively affluent, with a per capita GDP of $48,100.[40] Trade between the United States and Canada totals exceeding $580 billion.[41] In fact, Canada is the single largest trading partner of the United States. NAFTA has also increased trade between Canada and Mexico. Mexico is Canada's fifth largest export market and third largest import market.[42]

With a per capita GDP of $19,500, Mexico's nearly 125 million consumers are less affluent than Canadian consumers.[43] However, trade between the United States and Mexico has tripled since NAFTA was initiated. Mexico purchases more than $242 billion in U.S. products annually.[44] Millions of Americans cite their heritage as Mexican, making them the most populous Hispanic group in the country. These individuals often have close ties to relatives in Mexico and assist in Mexican–U.S. economic development and trade. Mexico is on a course of a market economy, rule of law, respect for human rights, and responsible public policies. There is also a commitment to the environment and sustainable human development. Many U.S. companies have taken advantage of Mexico's low labor costs and proximity to the United States to set up production facilities, sometimes called *maquiladoras.* Mexico is also attracting major technological industries, including electronics, software, and aerospace. Investors see many growth opportunities in Mexico, particularly in light of recent reforms. For instance, Mexico passed legislation to open up its state-controlled oil reserves to foreign companies. Additionally, if the

NAFTA, which went into effect on January 1, 1994, has increased trade among Mexico, the United States, and Canada.

©JORGE NUNEZ/EPA-EFE/REX/Shutterstockage

United States does well economically, Mexico—its biggest customer—is also likely to do well.[45]

However, there is great disparity within Mexico. The country's southern states cannot seem to catch up with the more affluent northern states on almost any socio-economic indicator. The disparities are growing, as can be seen comparing the south to the northern industrial capital of Monterrey, which is beginning to seem like south Texas.[46] Drug gang wars threaten the economic stability of Mexico, especially in the northern states close to the U.S. border. However, this situation is improving as the economy is growing and violence is decreasing.

Despite its benefits, NAFTA has been controversial. While many Americans feared the agreement would erase jobs in the United States, Mexicans have been disappointed that the agreement failed to create more jobs. Moreover, Mexico's rising standard of living has increased the cost of doing business there; many hundreds of *maquiladoras* have closed their doors and transferred work to China and other nations where labor costs are cheaper. Indeed, China has become the United States' second-largest importer. USMCA, the renegotiated agreement, has the goal to have more car and truck parts made in North America. Cars or trucks with 75 percent of its components made in Canada, Mexico, or the U.S. will qualify for zero tariffs. It makes improvements to environmental and labor regulations that will benefit Mexico in particular. Additionally, the agreement would also create changes for agricultural products, intellectual property, copyrights, and digital trade between the three countries.[47]

The European Union

The **European Union (EU)**, also called the *European Community* or *Common Market,* was established in 1958 to promote trade among its members, which initially included Belgium, France, Italy, West Germany, Luxembourg, and the Netherlands.

East and West Germany united in 1991, and by 1995 the United Kingdom, Spain, Denmark, Greece, Portugal, Ireland, Austria, Finland, and Sweden had joined as well. The Czech Republic, Estonia, Hungary, Latvia, Lithuania, Poland, Slovakia, and Slovenia joined in 2004. In 2007, Bulgaria and Romania also became members, Cyprus and Malta joined in 2008, and Croatia joined in 2013. Today, the Euro Zone (countries that have adopted the euro as their currency) consists of 19 separate countries with varying political landscapes.[48] In 1991, East and West Germany united, and by 2015, the EU included the United Kingdom, Spain, Denmark, Greece, Portugal, Ireland, Austria, Finland, Sweden, Cyprus, Poland, Hungary, the Czech Republic, Slovenia, Estonia, Latvia, Lithuania, Slovakia, Malta, Romania, Bulgaria, Belgium, France, Germany, Italy, Luxembourg, The Netherlands, and Croatia. The Former Yugoslav Republic of Macedonia, Montenegro, Serbia, and Turkey are candidate countries that hope to join the European Union in the near future.[49] Until 1993, each nation functioned as a separate market, but at that time members officially unified into one of the largest single world markets, which today has nearly half a billion consumers with a GDP of more than $17.1 trillion.[50]

To facilitate free trade among members, the EU is working toward standardization of business regulations and requirements, import duties, and value-added taxes; the elimination of customs checks; and the creation of a standardized currency for use by all members. Many European nations (Austria, Belgium, Finland, France, Germany, Greece, Ireland, Italy, Luxembourg, the Netherlands, Portugal, Spain, and Slovenia) link their exchange rates to a common currency, the *euro;* however, some

European Union (EU)
a union of European nations established in 1958 to promote trade among its members; one of the largest single markets today.

EU members have rejected use of the euro in their countries. Although the common currency requires many marketers to modify their pricing strategies and will subject them to increased competition, the use of a single currency frees companies that sell goods among European countries from the nuisance of dealing with complex exchange rates.[51] The long-term goals are to eliminate all trade barriers within the EU, improve the economic efficiency of the EU nations, and stimulate economic growth, thus making the union's economy more competitive in global markets, particularly against Japan and other Pacific Rim nations, and in North America. However, several disputes and debates still divide the member-nations, and many barriers to completely free trade remain. Consequently, it may take many years before the EU is truly one deregulated market.

The EU has also enacted some of the world's strictest laws concerning antitrust issues, which have had unexpected consequences for some non-European firms. The European Parliament is also encouraging the breakup of Google's search engine business from its other businesses.[52] The EU passed General Data Protection Regulation (GDPR) privacy law after five years of debate. The very complex 2,000-page law attempts to make data privacy clear with one's personal data a fundamental human right.[53] Consent to collect and use personal data now has to be "unambiguous." Individuals have the right to demand to view their data, have their data deleted, and have to give permission to have data transferred to another service.

In 2016, the United Kingdom voted to exit the European Union. This decision to exit, called "Brexit," resulted in the value of the pound falling sharply. There remain many questions about the impact of the proposed exit on trade relationships with other countries.[54] The government's priority appears to be first reaching trade agreements with EU members and then with America. Once outside the EU, the UK could abolish tariffs that mainly existed to protect other countries' products, such as the tariff on oranges to protect Spanish growers.[55]

Asia-Pacific Economic Cooperation

Asia-Pacific Economic Cooperation (APEC)
an international trade alliance that promotes open trade and economic and technical cooperation among member nations.

The **Asia-Pacific Economic Cooperation (APEC)**, established in 1989, promotes open trade and economic and technical cooperation among member economies, which initially included Australia, Brunei Darussalam, Canada, Indonesia, Japan, South Korea, Malaysia, New Zealand, the Philippines, Singapore, Thailand, and the United States. Since then, the alliance has grown to include Chile; China; Hong Kong, China; Mexico; Papua New Guinea; Peru; Russia; Chinese Taipei; and Vietnam. The 21-member alliance represents approximately 40 percent of the world's population, 54 percent of the world's GDP, and nearly 44 percent of global trade.[56] APEC differs from other international trade alliances in its commitment to facilitating business and its practice of allowing the business/private sector to participate in a wide range of APEC activities.[57]

Companies of the APEC have become increasingly competitive and sophisticated in global business in the past three decades. The Japanese and South Koreans, in particular, have made tremendous inroads on world markets for automobiles, cameras, and audio and video equipment. Products from Samsung, Sony, Canon, Toyota, Daewoo, Mitsubishi, Suzuki, and Lenovo are sold all over the world and have set standards of quality by which other products are often judged. The People's Republic of China, a country of more than 1.3 billion people, has launched a program of economic reform to stimulate its economy by privatizing many industries, restructuring its banking system, and increasing public spending on infrastructure (including railways and telecommunications). For many years, China was a manufacturing powerhouse with 10 percent growth at its height. However, in recent years, growth has slowed to 6.8 percent.[58] China's export market has consistently outpaced its import growth in recent years and its GDP represents the world's second-largest economy, behind the United States. In fact, China has overtaken the United States as the world's largest trader.[59] The global Internet retailer Alibaba is one of the world's most valuable companies. Half of all Internet retail sales were done in China in 2018.[60]

Increased industrialization has also caused China to become the world's largest emitter of greenhouse gases. China has overtaken the United States to become the world's largest oil importer.[61] On the other hand, China has also begun a quest to become a world leader in green initiatives and renewable energy. This is an increasingly important quest as the country becomes more polluted.

Another risk area for China is the fact that the government owns or has stakes in so many enterprises. On the one hand, China's system of state-directed capitalism has benefited the country because reforms and decisions can be made more quickly. On the other hand, state-backed companies lack many of the competitors that private industries have. Remember that competition often spurs innovation and lowers costs. If China's firms lack sufficient competition, their costs may very likely increase.[62] China's growing debt liabilities have also caused concern among foreign investors.[63]

Less visible Pacific Rim regions, such as Thailand, Singapore, Taiwan, Vietnam, and Hong Kong, have also become major manufacturing and financial centers. Vietnam, with one of the world's most open economies, has bypassed its communist government with private firms moving ahead despite bureaucracy, corruption, and poor infrastructure. In a country of 96 million, Vietnamese firms now compete internationally due to an agricultural miracle, making the country one of the world's main providers of farm produce.[64] As China's labor costs continue to grow, more businesses are turning toward Vietnam to open factories.[65] More recently, the United States has expressed concern about fair competition in the region.

Association of Southeast Asian Nations

The **Association of Southeast Asian Nations (ASEAN)**, established in 1967, promotes trade and economic integration among member nations in Southeast Asia, including Malaysia, the Philippines, Singapore, Thailand, Brunei Darussalam, Vietnam, Laos, Indonesia, Myanmar, and Cambodia.[66] The 10-member alliance represents 600 million people with a GDP of $2.4 trillion.[67] ASEAN's goals include the promotion of free trade, peace, and collaboration between its members.[68]

However, ASEAN is facing challenges in becoming a unified trade bloc. Unlike members of the EU, the economic systems of ASEAN members are quite different, with political systems including democracies (Philippines and Malaysia), constitutional monarchies (Cambodia), and communism (Vietnam).[69] Major conflicts have also occurred between member-nations. In Thailand the military staged a coup and placed the country under martial law, a change that not only impacted Thailand but also ASEAN as a whole.[70] Unlike the EU, ASEAN

Association of Southeast Asian Nations (ASEAN) a trade alliance that promotes trade and economic integration among member nations in Southeast Asia.

Clothing, timber, rice, and fish are among Cambodia's major exports.
©Glowimages/Getty Images

will not have a common currency or fully free labor flows between member-nations. In this way, ASEAN plans to avoid some of the pitfalls that occurred among nations in the EU during the latest worldwide recession.[71]

World Bank

World Bank
an organization established by the industrialized nations in 1946 to loan money to underdeveloped and developing countries; formally known as the International Bank for Reconstruction and Development.

The **World Bank,** more formally known as the International Bank for Reconstruction and Development, was established by the industrialized nations, including the United States, in 1946 to loan money to underdeveloped and developing countries. It loans its own funds or borrows funds from member countries to finance projects ranging from road and factory construction to the building of medical and educational facilities. The World Bank and other multilateral development banks (banks with international support that provide loans to developing countries) are the largest source of advice and assistance for developing nations. The International Development Association and the International Finance Corporation are associated with the World Bank and provide loans to private businesses and member countries.

International Monetary Fund

International Monetary Fund (IMF)
an organization established in 1947 to promote trade among member nations by eliminating trade barriers and fostering financial cooperation.

The **International Monetary Fund (IMF)** was established in 1947 to promote trade among member-nations by eliminating trade barriers and fostering financial cooperation. It also makes short-term loans to member countries that have balance-of-payment deficits and provides foreign currencies to member nations. The IMF tries to avoid financial crises and panics by alerting the international community about countries that will not be able to repay their debts. The IMF's Internet site provides additional information about the organization, including news releases, frequently asked questions, and members.

The IMF is the closest thing the world has to an international central bank. If countries get into financial trouble, they can borrow from the World Bank. However, the global economic crisis created many challenges for the IMF as it was forced to significantly increase its loans to both emerging economies and more developed nations. The usefulness of the IMF for developed countries is limited because these countries use private markets as a major source of capital.[72] Yet the European debt crisis changed this somewhat. Portugal, Ireland, Greece, and Spain (often referred to with the acronym PIGS) required billions of dollars in bailouts from the IMF to keep their economies afloat.

Getting Involved in International Business

LO 3-4

Summarize the different levels of organizational involvement in international trade.

Businesses may get involved in international trade at many levels—from a small Kenyan firm that occasionally exports African crafts to a huge multinational corporation such as Shell Oil that sells products around the globe. The degree of commitment of resources and effort required increases according to the level at which a business involves itself in international trade. This section examines exporting and importing, trading companies, licensing and franchising, contract manufacturing, joint ventures, direct investment, and multinational corporations.

Exporting and Importing

Many companies first get involved in international trade when they import goods from other countries for resale in their own businesses. For example, a grocery store chain may import bananas from Honduras and coffee from Colombia. A business

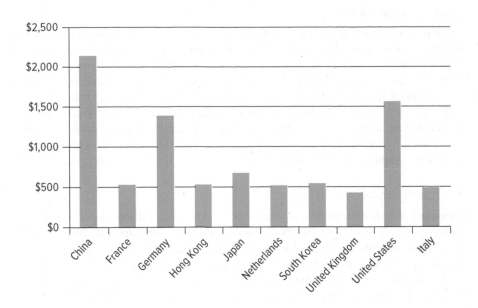

FIGURE 3.2
Top Exporting Countries (in billions of $)

Source: Central Intelligence Agency, "Country Comparison: Exports," https://www.cia.gov/library/publications/the-world-factbook/rankorder/2078rank.html (accessed April 2, 2017).

may get involved in exporting when it is called upon to supply a foreign company with a particular product. For example, the U.S. will be a net exporter of energy by 2022.[73] Such exporting enables enterprises of all sizes to participate in international business. Exporting to other countries becomes a necessity for established countries that seek to grow continually. Products often have higher sales growth potential in foreign countries than they have in the parent country. For instance, General Motors and YUM! Brands sell more of their products in China than in the United States. Cadillac's sales are growing rapidly in China where it has a long heritage and pedigree.[74] Figure 3.2 shows some of the world's largest exporting countries.

Exporting sometimes takes place through **countertrade agreements,** which involve bartering products for other products instead of for currency. Such arrangements are fairly common in international trade, especially between Western companies and eastern European nations. An estimated 40 percent or more of all international trade agreements contain countertrade provisions.

Although a company may export its wares overseas directly or import goods directly from their manufacturer, many choose to deal with an intermediary, commonly called an *export agent*. Export agents seldom produce goods themselves; instead, they usually handle international transactions for other firms. Export agents either purchase products outright or take them on consignment. If they purchase them outright, they generally mark up the price they have paid and attempt to sell the product in the international marketplace. They are also responsible for storage and transportation.

An advantage of trading through an agent instead of directly is that the company does not have to deal with foreign currencies or the red tape (paying tariffs and handling paperwork) of international business. A major disadvantage is that, because the export agent must make a profit, either the price of the product must be increased or the domestic company must provide a larger discount than it would in a domestic transaction.

countertrade agreements
foreign trade agreements that involve bartering products for other products instead of for currency.

Trading Companies

trading company
a firm that buys goods in one country and sells them to buyers in another country.

A **trading company** buys goods in one country and sells them to buyers in another country. Trading companies handle all activities required to move products from one country to another, including consulting, marketing research, advertising, insurance, product research and design, warehousing, and foreign exchange services to companies interested in selling their products in foreign markets. Trading companies are similar to export agents, but their role in international trade is larger. By linking sellers and buyers of goods in different countries, trading companies promote international trade. WTSC offers a 24-hour-per-day online world trade system that connects 20 million companies in 245 countries, offering more than 60 million products.[75]

Licensing and Franchising

licensing
a trade agreement in which one company—the licensor—allows another company—the licensee—to use its company name, products, patents, brands, trademarks, raw materials, and/or production processes in exchange for a fee or royalty.

Licensing is a trade arrangement in which one company—the *licensor*—allows another company—the *licensee*—to use its company name, products, patents, brands, trademarks, raw materials, and/or production processes in exchange for a fee or royalty. The Coca-Cola Company and PepsiCo frequently use licensing as a means to market their soft drinks, apparel, and other merchandise in other countries. Licensing is an attractive alternative to direct investment when the political stability of a foreign country is in doubt or when resources are unavailable for direct investment. Licensing is especially advantageous for small manufacturers wanting to launch a well-known brand internationally. Yoplait is a French yogurt that is licensed for production in the United States.

franchising
a form of licensing in which a company—the franchiser—agrees to provide a franchisee a name, logo, methods of operation, advertising, products, and other elements associated with a franchiser's business in return for a financial commitment and the agreement to conduct business in accordance with the franchiser's standard of operations.

Franchising is a form of licensing in which a company—the *franchiser*—agrees to provide a *franchisee* the name, logo, methods of operation, advertising, products, and other elements associated with the franchiser's business, in return for a financial commitment and the agreement to conduct business in accordance with the franchiser's standard of operations. Wendy's, McDonald's, H&R Block, and Holiday Inn are well-known franchisers with international visibility. Table 3.4 lists the top 10 global franchises.

Licensing and franchising enable a company to enter the international marketplace without spending large sums of money abroad or hiring or transferring personnel to

TABLE 3.4
Top 10 Global Franchises

Ranking	Franchise
1	McDonald's
2	KFC
3	Burger King
4	Pizza Hut
5	7 Eleven
6	Marriott International
7	RE/MAX
8	Dunkin' (Dunkin' Donuts)
9	InterContinental Hotels and Resorts
10	Subway

Source: "Top 100 Global Franchises - Rankings (2018)," Franchise Direct, https://www.franchisedirect.com/top100globalfranchises/rankings/ (accessed April 1, 2018).

handle overseas affairs. They also minimize problems associated with shipping costs, tariffs, and trade restrictions, and they allow the firm to establish goodwill for its products in a foreign market, which will help the company if it decides to produce or market its products directly in the foreign country at some future date. However, if the licensee (or franchisee) does not maintain high standards of quality, the product's image may be hurt; therefore, it is important for the licensor to monitor its products overseas and to enforce its quality standards.

Contract Manufacturing

Contract manufacturing occurs when a company hires a foreign company to produce a specified volume of the firm's product to specification; the final product carries the domestic firm's name. Foxconn Technology Group in Taiwan is the world's largest contract manufacturer and makes products for other companies. Its clients include Sony Corp, Apple Inc., and Nintendo Co. In fact, Foxconn is the largest iPhone assembler.[76]

contract manufacturing the hiring of a foreign company to produce a specified volume of the initiating company's product to specification; the final product carries the domestic firm's name.

Outsourcing

Earlier, we defined *outsourcing* as transferring manufacturing or other tasks (such as information technology operations) to companies in countries where labor and supplies are less expensive. Many U.S. firms have outsourced tasks to India, Ireland, Mexico, and the Philippines, where there are many well-educated workers and significantly lower labor costs. Services, such as taxes or customer service, can also be outsourced.

Although outsourcing has become politically controversial in recent years amid concerns over jobs lost to overseas workers, foreign companies transfer tasks and jobs to U.S. companies—sometimes called *insourcing*—far more often than U.S. companies outsource tasks and jobs abroad.[77] However, some firms are bringing their outsourced jobs back after concerns that foreign workers were not adding enough value. Companies such as General Electric and Caterpillar are returning to the United States due to increasing labor costs in places such as China, the expense of shipping products across the ocean, and fears of fraud or intellectual property theft. Companies from other countries have also been moving some of their production to the United States; Caterpillar and Ford brought production of some of their excavators and medium-duty commercial trucks back to the United States.[78]

offshoring the relocation of business processes by a company or subsidiary to another country; offshoring is different than outsourcing.

Offshoring

Offshoring is the relocation of a business process by a company, or a subsidiary, to another country. Offshoring is different than outsourcing: the company retains control of the process because it is not subcontracting to a different company. Companies may choose to offshore for a number of reasons, ranging from lower wages, skilled labor, or taking advantage of time zone differences in order to offer services around the clock. Some banks have chosen not to outsource because of concerns about data security in other countries. These institutions may, instead, engage in offshoring, which allows a company more control over international operations

IBM has more employees in India than in the United States.
©Ramesh Pathania/Mint via Getty Images

Going Green

Algae: A Biofuel Breakthrough

If Euglena has its way, algae will become the fuel of the future. The name of this Tokyo-based company comes from the product it sells: the algae *Euglena gracilis*. Euglena's business is focused on five key areas: food, fiber, feed, fertilizer, and fuel. The firm is also investigating the use of algae for pharmaceuticals. The company has been successful in developing food and cosmetics using algae and is now making its mark in the biofuel industry.

As carbon emissions become an increasing problem— particularly for countries like the United States, China, and India—governments are looking toward biofuels as an alternative fuel source. Research has shown that the algae in which Euglena specializes has strong potential for biofuel production. Euglena oversees all aspects of the process, including cultivation of the algae and the refining and marketing of the fuel. A strategic alliance with Isuzu Motors has allowed Euglena to successfully mix its algae biofuel with light oil to power city buses. However, Euglena is determined to create its own algae biofuel that will not require oil so it can completely eliminate carbon

emissions. The company is working with ANA Holdings to develop algae-based jet fuel. These partnerships have allowed Euglena to build the nation's first algae biofuel refinery, which it estimates will produce 33,000 gallons of jet fuel yearly.

Euglena has certainly attracted the attention of investors, but any promising breakthrough brings competition. ExxonMobil has partnered with Synthetic Genomics to investigate algae biofuels, and others are trying to create biofuels from garbage. The biofuel industry is growing globally, requiring firms like Euglena to remain innovative as competition heats up.[79]

Critical Thinking Questions

1. How has partnering with other businesses been crucial to Euglena's success? How would you recommend they approach expansion opportunities into other countries?
2. What are some type of barriers Euglena is likely to face as it grows?
3. What are some of the opportunities Euglena might have expanding into countries like the United States, China, and India? What are some potential barriers?

---

because the offshore office is an extension of the company. Shell, for example, opened a delivery center in India and moved its global IT jobs to that area.[80]

Joint Ventures and Alliances

Many countries, particularly LDCs, do not permit direct investment by foreign companies or individuals. A company may also lack sufficient resources or expertise to operate in another country. In such cases, a company that wants to do business in another country may set up a **joint venture** by finding a local partner (occasionally, the host nation itself) to share the costs and operation of the business. Qualcomm formed a joint venture with Chinese chip maker Semiconductor Manufacturing International Corporation to produce semiconductors. Qualcomm hopes this joint venture will help it to gain a foothold in selling chips in the Chinese market.[81]

In some industries, such as automobiles and computers, strategic alliances are becoming the predominant means of competing. A **strategic alliance** is a partnership formed to create competitive advantage on a worldwide basis. In such industries, international competition is so fierce and the costs of competing on a global basis are so high that few firms have the resources to go it alone, so they collaborate with other companies. An example of a strategic alliance is the partnership between LinkedIn and accounting firm Ernst & Young. The companies hope to use their combined expertise to assist other companies in using technology, social networking, and sales effectively.[82]

Direct Investment

Companies that want more control and are willing to invest considerable resources in international business may consider **direct investment,** the ownership of overseas facilities. Direct investment may involve the development and operation of new

joint venture
a partnership established for a specific project or for a limited time.

strategic alliance
a partnership formed to create competitive advantage on a worldwide basis.

direct investment
the ownership of overseas facilities.

facilities—such as when Starbucks opens a new coffee shop in Japan—or the purchase of all or part of an existing operation in a foreign country. General Motor's Cadillac brand was so successful in exporting to China, it opened a $1.2 billion plant in Shanghai to build the CT6 Sedan and XT5 SUV. Cadillac doubled its dealerships and avoided a 25 percent import tariff because the vehicles are locally built.[83]

The highest level of international business involvement is the **multinational corporation (MNC),** a corporation, such as IBM or ExxonMobil, that operates on a worldwide scale, without significant ties to any one nation or region. Table 3.5 lists 10 well-known multinational corporations. MNCs are more than simple corporations. They often have greater assets than some of the countries in which they do business. Nestlé, with headquarters in Switzerland, operates more than 400 factories around the world and receives

Under CEO Marillyn A. Hewson's leadership, Lockheed Martin's global business strategy is aimed at international expansion and strengthening national and economic security.

©Aflo/REX/Shutterstock

revenues from Europe; North, Central, and South America; Africa; and Asia.[84] The Royal Dutch/Shell Group, one of the world's major oil producers, is another MNC. Its main offices are located in The Hague and London. Other MNCs include BASF, British Petroleum, Matsushita, Mitsubishi, Siemens, Toyota, and Unilever. Many MNCs have been targeted by anti-globalization activists at global business forums, and some protests have turned violent. The activists contend that MNCs increase the gap between rich and poor nations, misuse and misallocate scarce resources, exploit the labor markets in LDCs, and harm their natural environments.[85]

multinational corporation (MNC)
a corporation that operates on a worldwide scale, without significant ties to any one nation or region.

Company	Country	Description
Royal Dutch Shell	Netherlands	Oil and gas; largest company in the world in terms of revenue
Toyota	Japan	Largest automobile manufacturer in the world
Walmart	United States	Largest retailer in the world; largest private employer in the world
Siemens	Germany	Engineering and electronics; largest engineering company in Europe
Nestlé	Switzerland	Nutritional, snack-food, and health-related consumer goods
Samsung	South Korea	Subsidiaries specializing in electronics, electronic components, telecommunications equipment, medical equipment, and more
Unilever	United Kingdom	Consumer goods including cleaning and personal care, foods, beverages
Boeing	United States	Aerospace and defense; largest U.S. exporter
Lenovo	China	Computer technology; highest share of PC market
Subway	United States	Largest fast-food chain; fastest-growing franchises in 105 countries

TABLE 3.5
Large Multinational Companies

LO 3-5

Contrast two basic strategies used in international business.

International Business Strategies

Planning in a global economy requires businesspeople to understand the economic, legal, political, and sociocultural realities of the countries in which they will operate. These factors will affect the strategy a business chooses to use outside its own borders.

Developing Strategies

multinational strategy
a plan, used by international companies, that involves customizing products, promotion, and distribution according to cultural, technological, regional, and national differences.

Companies doing business internationally have traditionally used a **multinational strategy,** customizing their products, promotion, and distribution according to cultural, technological, regional, and national differences. When McDonald's opened its first restaurant in Vietnam, it offered its traditional menu items as well as McPork sandwiches specifically targeted toward Vietnam consumers.[86] Many soap and detergent manufacturers have adapted their products to local water conditions, washing equipment, and washing habits. For customers in some LDCs, Colgate-Palmolive Co. has developed an inexpensive, plastic, hand-powered washing machine for use in households that have no electricity. Even when products are standardized, advertising often has to be modified to adapt to language and cultural differences. For example, Mars has been in the China pet food market for more than two decades with brands like Royal Canin, Whiskies, and Pedigree. U.S.-based Mars has competitive advantage because in China people are concerned about the safety of locally produced pet-food.

global strategy (globalization)
a strategy that involves standardizing products (and, as much as possible, their promotion and distribution) for the whole world, as if it were a single entity.

More and more companies are moving from this customization strategy to a **global strategy (globalization),** which involves standardizing products (and, as much as possible, their promotion and distribution) for the whole world, as if it were a single entity. Examples of globalized products are American clothing, movies, music, and cosmetics. Social media sites are important channels that brands are using to connect with their global customers. 3M, Pampers, and Corona had the highest engagement with their global followers on Twitter.[87]

Before moving outside their own borders, companies must conduct environmental analyses to evaluate the potential of and problems associated with various markets and to determine what strategy is best for doing business in those markets. Failure to do so may result in losses and even negative publicity. Some companies rely on local managers to gain greater insights and faster response to changes within a country. Astute businesspeople today "think globally, act locally." That is, while constantly being aware of the total picture, they adjust their firms' strategies to conform to local needs and tastes.

Managing the Challenges of Global Business

As we've pointed out in this chapter, many past political barriers to trade have fallen or been minimized, expanding and opening new market opportunities. Managers who can meet the challenges of creating and implementing effective and sensitive business strategies for the global marketplace can help lead their companies to success. For example, the Commercial Service is the global business solutions unit of the U.S. Department of Commerce that offers U.S. firms wide and deep practical knowledge of international markets and industries, a unique global network, inventive use of information technology, and a focus on small and mid-sized businesses. Another example is the benchmarking of best international practices that benefits U.S. firms, which is conducted by the network of CIBERs (Centers for International Business Education and Research) at leading business schools in the United States. These CIBERs are funded by the U.S. government to help U.S. firms become more competitive globally. A major element of the assistance that these governmental organizations can provide firms (especially for small and medium-sized firms) is knowledge of the internationalization process.[88]

Small businesses, too, can succeed in foreign markets when their managers have carefully studied those markets and prepared and implemented appropriate strategies. Being globally aware is therefore an important quality for today's managers and will remain a critical attribute for managers of the 21st century.

So You Want a Job in Global Business

Have you always dreamt of traveling the world? Whether backpacking your way through Central America or sipping espressos at five-star European restaurants is your style, the increasing globalization of business might just give you your chance to see what the world has to offer. Most new jobs will have at least some global component, even if located within the United States, so being globally aware and keeping an open mind to different cultures is vital in today's business world. Think about the 1.3 billion consumers in China who have already purchased mobile phones. In the future, some of the largest markets will be in Asia.

Many jobs discussed in chapters throughout this text tend to have strong international components. For example, product management and distribution management are discussed as marketing careers in the chapter "Dimensions of Marketing Strategy." As more and more companies sell products around the globe, their function, design, packaging, and promotions need to be culturally relevant to many different people in many different places. Products very often cross multiple borders before reaching the final consumer, both in their distribution and through the supply chain to produce the products.

Jobs exist in export and import management, product and pricing management, distribution and transportation, and advertising. Many "born global" companies such as Google operate virtually and consider all countries their market. Many companies sell their products through eBay and other Internet sites and never leave the United States. Today, communication and transportation facilitates selling and buying products worldwide with delivery in a few days. You may have sold or purchased a product on eBay outside the United States without thinking about how easy and accessible international markets are to business. If you have, welcome to the world of global business.

To be successful, you must have an idea not only of differing regulations from country to country, but of different language, ethics, and communication styles and varying needs and wants of international markets. From a regulatory side, you may need to be aware of laws related to intellectual property, copyrights, antitrust, advertising, and pricing in every country. Translating is never only about translating the language. Perhaps even more important is ensuring that your message gets through. Whether on a product label or in advertising or promotional materials, the use of images and words varies widely across the globe.

Review Your Understanding

Explore some of the factors within the international trade environment that influence business.

International business is the buying, selling, and trading of goods and services across national boundaries. Importing is the purchase of products and raw materials from another nation; exporting is the sale of domestic goods and materials to another nation. A nation's balance of trade is the difference in value between its exports and imports; a negative balance of trade is a trade deficit. The difference between the flow of money into a country and the flow of money out of it is called the balance of payments. An absolute or comparative

advantage in trade may determine what products a company from a particular nation will export.

Assess some of the economic, legal, political, social, cultural, and technological barriers to international business.

Companies engaged in international trade must consider the effects of economic, legal, political, social, and cultural differences among nations. Economic barriers are a country's level of development (infrastructure) and exchange rates. Wide-ranging legal and political barriers include differing laws (and enforcement), tariffs, exchange controls, quotas, embargoes, political instability, and war. Ambiguous cultural and social barriers involve differences in spoken and body language, time, holidays and other observances, and customs.

Specify some of the agreements, alliances, and organizations that may encourage trade across international boundaries.

Among the most important promoters of international business are the General Agreement on Tariffs and Trade, the World Trade Organization, the North American Free Trade Agreement, the European Union, the Asia-Pacific Economic Cooperation, the Association of Southeast Asian Nations, the World Bank, and the International Monetary Fund.

Summarize the different levels of organizational involvement in international trade.

A company may be involved in international trade at several levels, each requiring a greater commitment of resources and effort, ranging from importing/exporting to multinational corporations. Countertrade agreements occur at the import/export level and involve bartering products for other products instead of currency. At the next level, a trading company links buyers and sellers in different countries to foster trade. In licensing and franchising, one company agrees to allow a foreign company the use of its company name, products, patents, brands, trademarks, raw materials, and production processes in exchange for a flat fee or royalty. Contract manufacturing occurs when a company hires a foreign company to produce a specified volume of the firm's product to specification; the final product carries the domestic firm's name. A joint venture

is a partnership in which companies from different countries agree to share the costs and operation of the business. The purchase of overseas production and marketing facilities is direct investment. Outsourcing, a form of direct investment, involves transferring manufacturing to countries where labor and supplies are cheap. Offshoring is the relocation of business processes by a company or subsidiary to another country; it differs from outsourcing because the company retains control of the offshored processes. A multinational corporation is one that operates on a worldwide scale, without significant ties to any one nation or region.

Contrast two basic strategies used in international business.

Companies typically use one of two basic strategies in international business. A multinational strategy customizes products, promotion, and distribution according to cultural, technological, regional, and national differences. A global strategy (globalization) standardizes products (and, as much as possible, their promotion and distribution) for the whole world, as if it were a single entity.

Assess the opportunities and problems facing a small business that is considering expanding into international markets.

The "Solve the Dilemma" feature near the end of this chapter presents a small business considering expansion into international markets. Based on the material provided in the chapter, analyze the business's position, evaluating specific markets, anticipating problems, and exploring methods of international involvement.

Critical Thinking Questions

Enter the World of Business Questions

1. What barriers did Alibaba need to overcome in its home country?

2. How do you think Alibaba's strategic alliance with Yahoo! helped it to gain market share?

3. What are some of the barriers that firms such as Amazon and eBay face as they try to compete against Alibaba in China?

Learn the Terms

absolute advantage 75
Asia-Pacific Economic
  Cooperation (APEC) 88
Association of Southeast Asian Nations
  (ASEAN) 89
balance of payments 77
balance of trade 76
cartel 82
comparative advantage 75

contract manufacturing 93
countertrade agreements 91
direct investment 94
dumping 81
embargo 81
European Union (EU) 87
exchange controls 81
exchange rate 78
exporting 75

franchising 92
General Agreement on Tariffs and
  Trade (GATT) 85
global strategy (globalization) 96
import tariff 80
importing 76
infrastructure 78
international business 74
International Monetary Fund (IMF) 90

Check Your Progress

1. Distinguish between an absolute advantage and a comparative advantage. Cite an example of a country that has an absolute advantage and one with a comparative advantage.

2. What effect does devaluation have on a nation's currency? Can you think of a country that has devaluated or revaluated its currency? What have been the results?

3. What effect does a country's economic development have on international business?

4. How do political issues affect international business?

5. What is an import tariff? A quota? Dumping? How might a country use import tariffs and quotas to control its balance of trade and payments? Why can dumping result in the imposition of tariffs and quotas?

6. How do social and cultural differences create barriers to international trade? Can you think of any additional social or cultural barriers (other than those mentioned in this chapter) that might inhibit international business?

7. Explain how a countertrade agreement can be considered a trade promoter. How does the World Trade Organization encourage trade?

8. At what levels might a firm get involved in international business? What level requires the least commitment of resources? What level requires the most?

9. Compare and contrast licensing, franchising, contract manufacturing, and outsourcing.

10. Compare multinational and global strategies. Which is better? Under what circumstances might each be used?

Get Involved

1. If the United States were to impose additional tariffs on cars imported from Japan, what would happen to the price of Japanese cars sold in the United States? What would happen to the price of American cars? What action might Japan take to continue to compete in the U.S. automobile market?

2. Although NAFTA has been controversial, it has been a positive factor for U.S. firms desiring to engage in international business. What industries and specific companies have the greatest potential for opening stores in Canada and Mexico? What opportunities exist for

small businesses that cannot afford direct investment in Mexico and Canada?

3. Identify a local company that is active in international trade. What is its level of international business involvement and why? Analyze the threats and opportunities it faces in foreign markets, as well as its strengths and weaknesses in meeting those challenges. Based on your analysis, make some recommendations for the business's future involvement in international trade. (Your instructor may ask you to share your report with the class.

Build Your Skills

Global Awareness

Background

As American businesspeople travel the globe, they encounter and must quickly adapt to a variety of cultural norms quite different from the United States. When encountering individuals from other parts of the world, the best

attitude to adopt is, "Here is my way. Now what is yours?" The more you see that you are part of a complex world and that your culture is different from, not better than, others, the better you will communicate and the more effective you will be in a variety of situations. It takes time, energy, understanding, and tolerance to learn about and appreciate other cultures. Naturally you're more comfortable doing things the way you've always done them. Remember, however, that this fact will also be true of the people from other cultures with whom you are doing business.

Task

You will "travel the globe" by answering questions related to some of the cultural norms that are found in other countries. Form groups of four to six class members and determine the answers to the following questions. Your instructor has the answer key, which will allow you to determine your group's Global Awareness IQ, which is based on a maximum score of 100 points (10 points per question).

Match the country with the cultural descriptor provided:

A. Saudi Arabia   F. China
B. Japan   G. Greece
C. Great Britain   H. Korea
D. Germany   I. India
E. Venezuela   J. Mexico

_____ 1. When people in this country table a motion, they want to discuss it. In America, "to table a motion" means to put off discussion.

_____ 2. In this country, special forms of speech called *keigo* convey status among speakers. When talking with a person in this country, one should know the person's rank. People from this country typically will not initiate a conversation without a formal introduction.

_____ 3. People from this country often pride themselves on enhancing their image by keeping others waiting.

_____ 4. When writing a business letter, people in this country like to provide a great deal of background information and detail before presenting their main points.

_____ 5. For a man to inquire about another man's wife (even a general question about how she is doing) is considered very offensive in this country.

_____ 6. When in this country, you are expected to negotiate the price on goods you wish to purchase.

_____ 7. While North Americans want to decide the main points at a business meeting and leave the details for later, people in this country need to have all details decided before the meeting ends to avoid suspicion and distrust.

_____ 8. Children in this country learn from a very early age to look down respectfully when talking to those of higher status.

_____ 9. Until recently in this country, the eldest male was legally the ruler of the household, and the custom was to keep the women hidden.

_____ 10. Many businesspeople from the United States experience frustration because yes does not always mean the same thing in other cultures. For example, the word *yes* in this country means, "OK, I want to respect you and not offend you." It does not necessarily show agreement.

Solve the Dilemma

Global Expansion or Business as Usual?

Audiotech Electronics, founded in 1959 by a father and son, currently operates a 35,000-square-foot factory with 75 employees. The company produces control consoles for television and radio stations and recording studios. It is involved in every facet of production—designing the systems, installing the circuits in its computer boards, and even manufacturing and painting the metal cases housing the consoles. The company's products are used by all the major broadcast and cable networks. The firm's newest products allow television correspondents to simultaneously hear and communicate with their counterparts in different geographic locations. Audiotech has been very successful meeting its customers' needs efficiently.

Audiotech sales have historically been strong in the United States, but recently, growth is stagnating. Even though Audiotech is a small, family-owned firm, it believes it should evaluate and consider global expansion.

LO 3-6

Assess the opportunities and problems facing a small business that is considering expanding into international markets.

Critical Thinking Questions

1. What are the key issues that need to be considered in determining global expansion?

2. What are some of the unique problems that a small business might face in global expansion that larger firms would not?

3. Should Audiotech consider a joint venture? Should it hire a sales force of people native to the countries it enters?

Build Your Business Plan   ≡ connect

Business in a Borderless World

Think about the good/service you are contemplating for your business plan. If it is an already established good or service, try to find out if the product is currently being sold internationally. If not, can you identify opportunities to do so in the future? What countries do you think would respond most favorably to your product? What problems would you encounter if you attempted to export your product to those countries?

If you are thinking of creating a new good or service for your business plan, think about the possibility of eventually marketing that product in another country. What countries or areas of the world do you think would be most responsive to your product?

Are there countries the United States has trade agreements or alliances with that would make your entry into the market easier? What would be the economic, social, cultural, and technological barriers you would have to recognize before entering the prospective country(ies)? Think about the specific cultural differences that would have to be taken into consideration before entering the prospective country.

Visit Connect to practice building your business plan with the Business Plan Prep Exercises.

See for Yourself Videocase

Electra Bikes: Better, Cooler, Awesomer!

Twenty-three years ago, Swiss snowboard designer Benno Bänziger and his German business partner Jeano Erforth decided they wanted to make bike riding fun again. At the time, cruiser bikes were out of style and were in danger of disappearing. Bänziger and Erforth converted their T-shirt company in Vista, California, into a bicycle manufacturer called Electra Bicycle Company. However, they did not want to manufacture just any type of cruiser bike. They wanted their bikes to look hip with a vintage style, based off rockabilly culture and resembling the look of muscle cars. The shop went from selling a few hundred the first year it was in business to becoming a global sensation. Electra eventually caught the attention of Trek Bicycle Corporation, which acquired the firm in 2014.

As Electra began to expand globally, it found that it had a comparative advantage its bigger competitors did not have. First of all, many bike enthusiasts worldwide appreciated the vintage look combined with the most up-to-date technology of Electra bicycles. One particular advantage Electra has is its patented Flat Foot Technology®. This technology enables better leg extension, making the ride more comfortable. Because of its patent, Electra is currently the only company that can use the technology it developed. These advantages combined with its "genuine Americana message" proved to be a hit overseas.

Global expansion offered Electra the unique opportunity to reach new market niches of bicycle enthusiasts. It started slowly, expanding first into markets that were closer to the United States, including Canada and Australia. The company currently has distributors in 26 countries that sell its bikes. In addition to offices in the United States, Electra opened a European office that supports sales for 25 countries. Like many other firms, Electra outsources production to Taiwan and China. Using freight ships, the firm ships its products to markets such as Australia, the European Union, and the domestic American market. By outsourcing production, the firm is able to save more money than if it had the bikes produced in the United States.

Like most companies that expand globally, Electra is subject to import duties that can vary depending upon where the bicycles are manufactured. It is also subject to regulations that vary from country to country. Sometimes, Electra must have additional features due to safety laws. For instance, some bikes shipped to certain countries have front brakes, while others have lighting systems or fenders. While these different features add to costs, they ensure that Electra stays on the right side of the laws within the countries in which they do business.

Regulations can vary by type of product as well. Kevin Cox, president of Electra Bicycle, describes how Electra had to take this into consideration when designing a bicycle with a motor called the Townie Go. According to Cox, "There are certain speed regulations that exist, and those regulations are different between Europe and [the] U.S." The company chose to adopt the stricter U.S. standard, although it meant a slower speed because it would give Electra "one product platform that has global reach."

Electra also faces economic and geographical challenges as well. "Currency fluctuation in Asia, that's a big one because that in and of itself can skew the cost of our bicycle," Cox says. Additionally, certain areas such as Europe and many parts of the United States have highly seasonal weather that limits bike riding to certain times of the year. These weather variations cause fluctuations in demand based upon the season. However, Cox also points out that having a global reach helps to balance out these fluctuations "because while it's winter in the U.S., I'm enjoying a nice summer in Australia." In other words, there is a constant worldwide demand for bicycles year-round.

As Electra expanded globally, the infrastructure of operating a global network became more complex. This is why, in 2014, Electra made the decision to sell the company to Trek. The decision benefits Electra because it can now utilize Trek's extensive network to reach new markets. At the same time, tapping into Trek's distribution networks frees Electra to focus less on distribution concerns and more on its core competency: developing high-quality, durable, vintage bicycles.[89]

Critical Thinking Questions

1. Describe how Electra maintains a worldwide comparative advantage.

2. What are some global difficulties Electra had to overcome when it expanded into different countries?

3. Why did Electra—which markets itself with a "genuine Americana" message—decide to outsource production to Asia? Do you believe this is appropriate?

You can find the related video in the Video Library in Connect. Ask your instructor how you can access Connect.

Team Exercise

Visit Transparency International's Country Corruption Index website: https://www.transparency.org/research/cpi/overview. Form groups and select two countries. Research some of the economic, ethical, legal, regulatory, and political barriers that would have an impact on international trade. Be sure to pair a country with a high level of perceived corruption (lower scores) with a country that has a low level of perceived corruption (higher scores). Report your findings.

Ask your instructor about the role-play exercises available with this book to practice working with a business team.

Notes

1. Lulu Yilun Chen, "Online Giant Alibaba Aims beyond China and E-Commerce: QuickTake," *Washington Post,* November 13, 2017, https://www.washingtonpost.com/business/online-giant-alibaba-aims-beyond-china-and-e-commerce-quicktake/2017/11/13/9147da86-c846-11e7-b506-8a10ed11ecf5_story.html (accessed November 19, 2017); Katherine Rushton, "Alibaba Is Now the Biggest Retailer in the World," *Telegraph (U.K.),* October 28, 2014, www.telegraph.co.uk/finance/newsbysector/retailandconsumer/11193340/Alibaba-is-now-the-biggest-retailer-in-the-world.html (accessed November 25, 2017); Laura Lorenzetti, "Alibaba Heads to Hollywood with Its New Cash Stockpile," *Fortune,* October 28, 2014, http://fortune.com/2014/10/28/alibaba-heads-to-hollywood-with-its-new-cash-stockpile/ (accessed November 25, 2017); Bill George, "Jack Ma on Alibaba, Entrepreneurs, and the Role of Handstands," *The New York Times,* September 22, 2014, http://dealbook.nytimes.com/2014/09/22/jack-ma-on-alibaba-entrepreneurs-and-the-role-of-handstands/?_r=0 (accessed November 25, 2017); Frank Langfitt, "From a Chinese Apartment to Wall Street Darling: The Rise of Alibaba," *National Public Radio,* September 8, 2014, www.npr.org/blogs/parallels/2014/09/08/326930271/from-a-chinese-apartment-to-wall-street-darling-the-rise-of-alibaba (accessed November 25, 2017); Aaron Pressman and Adam Lashinsky, "Data Sheet–Alibaba's Vast and Growing Reach," *Fortune,* November 13, 2017, http://fortune.com/2017/11/13/data-sheet-alibaba-payments-shopping/ (accessed November 19, 2017); Daniel Keyes, "Amazon is Struggling to Find its Place China," *Business Insider,* August 30, 2017, www.businessinsider.com/amazon-is-struggling-to-find-its-place-china-2017-8 (accessed November 19, 2017); Kathy Chu, "Alibaba to Act Faster Against Counterfeits," *The Wall Street Journal,* May 15, 2014, p. B1; "E-commerce with Chinese characteristics," *The Economist,* November 15, 2007, www.economist.com/node/10125658 (accessed November 25, 2017); Eric Markowitz, "From Start-up to Billion-Dollar Company," *Inc.,* April 6, 2012, https://www.inc.com/eric-markowitz/alibaba-film-dawn-of-the-chinese-internet-revolution.html (accessed November 25, 2017).

2. Nick Parker, "BMWs Billion-Dollar Bet on Mexico," *CNN Money,* July 8, 2014, http://money.cnn.com/2014/07/08/news/companies/bmw-mexico/ (accessed May 3, 2017).

3. Deloitte, "2016 Global Manufacturing Competitiveness Index," 2016, http://www2.deloitte.com/global/en/pages/about-deloitte/articles/global-manufacturing-competitiveness-index.html (accessed May 3, 2017).

4. Jack Nicas, "YouTube Viewership Notches a Global Milestone," *The Wall Street Journal,* February 28, 2017, p. B1.

5. Starbucks, "Starbucks Coffee International," https://www.starbucks.com/business/international-stores (accessed April 1, 2018).

6. Elisabeth Sullivan, "Choose Your Words Wisely," *Marketing News,* February 15, 2008, p. 22.

7. Anne Steele, "Overseas Sales Jump for Indie Records," *The Wall Street Journal,* March 16, 2018, p. B5.

8. Rayna Hollander, "As Apple's Third Largest Market by Sales, China is an Attractive Growth Opportunity," *Business Insider,* December 5, 2017, https://www.businessinsider.com/china-is-apples-third-largest-market-2017-12 (accessed July 25, 2018).

9. Worldometers, "Current World Population," www.worldometers.info/world-population/ (accessed April 1, 2018).

10. "Subway," *Fortune,* www.forbes.com/companies/subway/ (accessed May 3, 2017).

11. Paul Davidson, "We Produce More at Home with New Drilling Methods," *USA Today,* February 11, 2014, p. 1B.

12. Chris Isidore, "These Are the Top U.S. Exports," *CNN Money,* March 7, 2018, http://money.cnn.com/2018/03/07/news/economy/top-us-exports/index.html (accessed April 1, 2018).

13. Kimberly Amadeo, "U.S. Imports and Exports: Components and Statistics," *The Balance,* March 3, 2018, https://www.thebalance.com/u-s-imports-and-exports-components-and-statistics-3306270 (accessed April 1, 2018).

14. Ibid.

15. Ian Bremmer, "Sea of Troubles," *Time* 185, no. 1 (2015), p. 18.

16. "Shaken, Not Stirred," *The Economist,* January 24, 2015, p. 48.

17. Daniel Runde, "M-Pesa and the Rise of the Global Mobile Money Market," August 12, 2015, https://www.forbes.com/sites/danielrunde/2015/08/12/m-pesa-and-the-rise-of-the-global-mobile-money-market/#64299fd65aec (accessed March 3, 2018); Kieron Monks, "M-Pesa: Kenya's Mobile Money Success Story Turns 10," *CNN,* February 24, 2017, https://www.cnn.com/2017/02/21/africa/mpesa-10th-anniversary/index.html (accessed March 3, 2018); BBC, "M-Pesa's Founders on Launching Mobile Wallet," *BBC News,* November 22, 2010, www.bbc.com/news/av/business-11793288/m-pesa-founders-on-launching-kenya-s-mobile-wallet (accessed March 3, 2018); Janelle Richards, "Kenyan Entrepreneurs Help Youth Thrive in Africa's Emerging 'Silicon Valley,'" *NBC News,* December 4, 2017, https://www.nbcnews.com/news/nbcblk/kenyan-entrepreneurs-help-youth-thrive-africa-s-emerging-silicon-valley-n826156 (accessed March 3, 2018); Charles Graeber, "Ten Days in Kenya with No Cash, Only a Phone," *Bloomberg Businessweek,* June 5, 2014, www.businessweek.com/articles/2014-06-05/safaricoms-m-pesa-turns-kenya-into-a-mobile-payment-paradise (accessed March 3, 2018); Vivienne Walt, "Is Africa's Rise for Real This Time?" *Fortune,* September 18, 2014, pp. 166–172.

18. Andy Pasztan, "U.S. Drone Deliveries Ready for Takeoff," *The Wall Street Journal,* March 12, 2018, p. B1.

19. Nathaniel Taplin, "Walling off China Won't Help U.S.," *The Wall Street Journal,* March 10–11, 2018, p. B12.

20. BSA, "Security Threats Rank as Top Reason Not to Use Unlicensed Software," http://globalstudy.bsa.org/2013/ (accessed May 3, 2017).

21. Sonja Elmquist, "U.S. Calls for 256% Tariff on Imports of Steel from China," *Bloomberg Business,* December 22, 2015, www.bloomberg.com/news/articles/2015-12-22/u-s-commerce-department-to-put-256-tariff-on-chinese-steel (accessed May 3, 2017).

22. Andrew Tangel, Bob Tita, and Josh Zumbrun, "Metal Users Harden of Levy," *The Wall Street Journal,* March 10–11, 2018, p. B5.

23. Andrea Thomas, Paul Vieira, David Winning, "U.S. Allies Weigh Response to Tariffs," *The Wall Street Journal,* March 13, 2018, p. 4-7.

24. Jen Skerritt, "Timber Tariffs Are Hammering U.S. Builders," *Bloomberg Businessweek,* March 12, 2018, p. 35.

25. Kitty Bean Yancey, "Back to Cuba: People-to-People Trips Get the Green Light," *USA Today,* August 4, 2011, p. 4A; Alan Gomez, "Feds Approve First U.S. Factory in Cuba," *USA Today,* February 16, 2016, p. 1B.

26. Kitty Bean Yancey and Laura Bly, "Door May Be Inching Open for Tourism," *USA Today,* February 20, 2008, p. A5; Sue Kirchhoff and Chris Woodyard, "Cuba Trade Gets New Opportunity,'" *USA Today,* February 20, 2008, p. B1; Gardiner Harris, "Trump Tightens Cuba Embargo, Restricting Access to Hotels and Businesses," *The New York Times,* November 8, 2017, https://www.nytimes.com/2017/11/08/us/politics/trump-tightens-cuba-embargo-restricting-access-to-hotels-businesses.html (accessed April 1, 2018).

27. European Commission, "EU Wins a WTO Dispute on Chinese Anti-Dumping Duties," February 13, 2015, http://trade.ec.europa.eu/doclib/press/index.cfm?id=1257 (accessed May 3, 2017).

28. European Commission, "EU Wins a WTO Dispute on Chinese Anti-Dumping Duties."

29. Rob Davies, "Uber Suffers Legal Setbacks in France and Germany," *The Guardian,* June 9, 2016, https://www.theguardian.com/technology/2016/jun/09/uber-suffers-legal-setbacks-in-france-and-germany (accessed April 15, 2017); "Uberworld," *The Economist,* September 3, 2016, p. 9; Jefferson Graham, "App Greases the Wheels," *USA Today,* May 27, 2015, p. 5B; Karun, "Times Internet and Uber Enter Into a Strategic Partnership, *Uber Blog,* March 22, 2015, http://blog.uber.com/times-internet (accessed April 15, 2017); R. Jai Krishna and Joanna Sugden, "India Asks Internet Service Providers to Block Uber Website in Delhi," *The Wall Street Journal,* May 14, 2015, www.wsj.com/articles/india-asks-internet-service-providers-to-block-uber-website-in-delhi-1431606032 (accessed April 15, 2017); Saritha Rai, "Uber Gets Serious about Passenger Safety In India, Introduces Panic Button," *Forbes,* February 12, 2015, www.forbes.com/sites/saritharai/2015/02/12/uber-gets-serious-about-passenger-safety-in-india-introduces-panic-button/ (accessed April 15, 2017); Sam Schechner and Tom Fairless, "Europe Steps Up Pressure on Tech Giants," *The Wall Street Journal,* April 2, 2015, www.wsj.com/articles/europe-steps-up-pressure-on-technology-giants-1428020273 (accessed April 15, 2017); Aditi Shrivastava, "Uber Resumes Operations in Delhi Post 1.5 Months Ban," *The Economic Times,* January 23, 2015, http://articles.economictimes.indiatimes.com/2015-01-23/news/58382689_1_indian-taxi-market-radio-taxi-scheme-uber-spokesman (accessed April 15, 2017); UNM Daniels Fund Ethics Initiative, "Truth, Transparency, and Trust: Uber Important in the Sharing Economy," PPT presentation, https://danielsethics.mgt.unm.edu/teaching-resources/presentations.asp (accessed April 15, 2017); Maria Vega Paul, "Uber Returns to Spanish Streets in Search of Regulatory U-Turn," *Reuters,* March 30, 2016, www.reuters.com/article/us-spain-uber-tech-idUSKCN0WW0AO (accessed April 15, 2017).

30. Laurie Burkitt, "Tiffany Finds Sparkle in Overseas Markets," *The Wall Street Journal,* December 26, 2013, p. B4.

31. Julie Jargon, "Starbucks Shifts in Europe," *The Wall Street Journal,* November 30–December 1, 2013, p. B3.

32. Kyle Stock, "Movers," *Bloomberg Businessweek,* March 6–12, 2017, p. 19.

33. "Slogans Gone Bad," Joe-ks, www.joe-ks.com/archives_apr2004/slogans_gone_bad.htm (accessed May 3, 2017).

34. Nina Lakhani, "Coca-Cola Apologizes for Indigenous People Ad Intended as 'Message of Unity,'" *The Guardian,* December 5, 2015, www.theguardian.com/world/2015/dec/05/coca-cola-mexico-ad-indigenous-people (accessed May 3, 2017).

35. J. Bonasia, "For Web, Global Reach Is Beauty—and Challenge," *Investor's Business Daily,* June 13, 2001, p. A6.

36. Gartner, "Gartner Says Worldwide PC Shipments Declined 9.5 Percent in Second Quarter of 2015," July 9, 2015, www.gartner.com/newsroom/id/3090817 (accessed May 3, 2017).

37. "What Is the WTO," World Trade Organization (n.d.), www.wto.org/english/thewto_e/whatis_e/whatis_e.htm (accessed May 3, 2017).

38. European Commission, "EU Wins a WTO Dispute on Chinese Anti-Dumping Duties."

39. "The North American Free Trade Agreement (NAFTA)," *export.gov,* http://export.gov/FTA/nafta/index.asp (accessed April 7, 2017); Kimberly Amadeo, "Why NAFTA's Six Advantages Outweigh its Six Disadvantages," *The Balance,* March 6, 2018, https://www.thebalance.com/nafta-pros-and-cons-3970481 (accessed April 1, 2018).

40. Central Intelligence Agency, "Guide to Country Comparisons," *World Factbook,* https://www.cia.gov/library/publications/the-world-factbook/rankorder/rankorderguide.html (accessed April 1, 2018); Worldometers, "Canada Population," www.worldometers.info/world-population/canada-population/ (accessed April 1, 2018).

41. U.S. Census Bureau, "Trade in Goods with Canada," https://www.census.gov/foreign-trade/balance/c1220.html (accessed April 1, 2018).

42. Statistics Canada, "Table 1 Merchandise Trade: Canada's Top 10 Principal Trading Partners—Seasonally Adjusted, Current Dollars," December 5, 2014, www.statcan.gc.ca/daily-quotidien/141205/t141205b001-eng.htm (accessed May 3, 2017).

43. Central Intelligence Agency, "Guide to Country Comparisons."

44. U.S. Census Bureau, "Trade in Goods with Mexico," *Foreign Trade,* https://www.census.gov/foreign-trade/balance/c2010.html (accessed April 1, 2018).

45. Jen Wieczner, "Why 2014 Could Be Mexico's Year," *Fortune,* January 13, 2014, p. 37–38.

46. "A Tale of Two Mexicos: North and South," *The Economist,* April 26, 2008, p. 53–54.

47. U.S. Census Bureau, "Top Trading Partners—December 2013: Year-to-Date Total Trade," www.census.gov/foreign-trade/statistics/highlights/top/top1312yr.html (accessed May 3, 2017); Heather Long, "U.S., Canada and Mexico Just Reached a Sweeping New NAFTA Deal. Here's What's in It," *The Washington Post,* October 1, 2018, https://www.washingtonpost.com/business/2018/10/01/us-canada-mexico-just-reached-sweeping-new-nafta-deal-heres-whats-it/?noredirect=on&utm_term=.a0f83040eba0 (accessed October 6, 2018).

48. "Crisis Revisited," *The Economist,* December 13, 2014, p. 17; "Euro Area," European Commission, https://ec.europa.eu/info/business-economy-euro/euro-area_en (accessed January 28, 2017).

49. "Euro Area 1999-2015," European Central Bank, https://www.ecb.europa.eu/euro/intro/html/map.en.html (accessed January 28, 2017); "Europe in 12 Lessons," *Europa,* http://europa.eu/abc/12lessons/lesson_2/index_en.htm (accessed January 28, 2017).

50. "IMF World Economic Outlook Database," International Monetary Fund, October 2017, www.imf.org/external/pubs/ft/weo/2017/02/weodata/weorept.aspx?pr.x=89&pr.y=6&sy=2017&ey=2017&scsm=1&ssd=1&sort=country&ds=.&br=1&c=998&s=NGDPD%2CPPPGDP%2CPPPPC&grp=1&a=1 (accessed February 11, 2018).

51. Stanley Reed, with Ariane Sains, David Fairlamb, and Carol Matlack, "The Euro: How Damaging a Hit?" *BusinessWeek,* September 29, 2003, p. 63; Irene Chapple, "How the Euro Became a Broken Dream," *CNN,* November 3, 2011, www.cnn.com/2011/09/23/business/europe-euro-creation-maastricht-chapple/ (accessed May 3, 2017).

52. Julia Fioretti, "EU Watchdogs to Apply 'Right to Be Forgotten' Rule on Web Worldwide," *Reuters,* November 26, 2014, www.reuters.com/article/2014/11/26/us-google-eu-privacy-idUSKCN0JA1HU20141126 (accessed May 3, 2017); "Drawing the Line," *The Economist,* October 4, 2014, www.economist.com/news/international/21621804-google-grapples-consequences-controversial-ruling-boundary-between (accessed May 3, 2017); Samuel Gibbs, "European Parliament Votes Yes on Google Breakup Motion," *The Guardian,* November 27, 2014, www.theguardian.com/technology/2014/nov/27/european-parliament-votes-yes-google-breakup-motion (accessed May 3, 2017).

53. Ludwig Siegele, "The Dodd-Frank of Data," *The Economist,* The World in 2018, p. 123.

54. Alex Hunt and Brian Wheeler, "Brexit: All Your Need to Know about the UK Leaving the EU," *BBC,* March 26, 2018, www.bbc.com/news/uk-politics-32810887 (accessed April 1, 2018).

55. "Negotiating Post-Brexit Deals: Trading Places," *The Economist,* February 4, 2017, p. 48.

56. Asia-Pacific Economic Cooperation, "About APEC," www.apec.org/About-Us/About-APEC.aspx (accessed November 21, 2014); "U.S.-APEC Trade Facts," Office of the Unites States Trade Representative, https://ustr.gov/trade-agreements/other-initiatives/asia-pacific-economic-cooperation-apec/us-apec-trade-facts (accessed January 31, 2017).

57. Asia-Pacific Economic Cooperation, "About APEC."

58. "China GDP Annual Growth Rate," *Trading Economics,* 2017, www.tradingeconomics.com/china/gdp-growth-annual (accessed April 1, 2018).

59. Charles Riley and Feng Ke, "China to Overtake U.S. as World's Top Trader," *CNN,* January 10, 2014, http://money.cnn.com/2014/01/10/news/economy/china-us-trade/ (accessed May 3, 2017).

60. "Retail Sales," *The Economist,* The World in 2018, p. 12.

61. U.S. Environmental Protection Agency, "Global Greenhouse Gas Emissions Data," www.epa.gov/climatechange/ghgemissions/global.html (accessed February 29, 2016); Joshua Keating, "China Passes U.S. as World's Largest Oil Importer," *Slate,* October 11, 2013, www.slate.com/blogs/the_world_/2013/10/11/china_now_world_s_largest_net_oil_importer_surpassing_united_states.html (accessed May 3, 2017).

62. "The Rise of Capitalism," *The Economist,* January 21, 2012, p. 11.

63. Dexter Roberts, "Corporate China's Black Hole of Debt," *Bloomberg Businessweek,* November 19–22, 2012, p. 15–16.

64. Central Intelligence Agency, "Guide to Country Comparisons."

65. Kathy Chu, "China Loses Edge on Labor Costs," *The Wall Street Journal,* December 3, 2015, p. B1, B4.

66. Association of Southeast Asian Nations, "Overview," www.aseansec.org/64.htm (accessed April 10, 2014).

67. "ASEAN Economic Community: 12 Things to Know," Asian Development Bank, December 29, 2015, www.adb.org/features/asean-economic-community-12-things-know (accessed January 6, 2016).

68. Simon Long, "Safety in Numbers," *The Economist,* The World in 2015 Edition, p. 68.

69. R.C., "No Brussels Sprouts in Bali," *The Economist,* November 18, 2011, www.economist.com/blogs/banyan/2011/11/asean-summits (accessed May 3, 2017).

70. "Thaksin Times," *The Economist,* January 31, 2015, p. 31.

71. Eric Bellman, "Asia Seeks Integration Despite EU s Woes," *The Wall Street Journal,* July 22, 2011, p. A9.

72. David J. Lynch, "The IMF Is . . . Tired Fund Struggles to Reinvent Itself," *USA Today,* April 19, 2006. p. B1.

73. Sarah McFarlane, "U.S. Oil-Export Trade Consolidates," *The Wall Street Journal,* March 26, 2018, p. B10.

74. David Welch and Yon Zhang, "Where Cadillac Is Still Praised," *Bloomberg Businessweek,* February 6–12, 2017, p. 16.

75. WTSC Industrial Group website, www.wtsc.eu/ (accessed April 1, 2018).

76. Takashi Mochizuki and Eva Don, "TV Snag Blunts FoxConn's Plan for Sharp," *The Wall Street Journal,* January 5, 2017.

77. Walter B. Wriston, "Ever Heard of Insourcing?" Commentary, *The Wall Street Journal,* March 24, 2004, p. A20.

78. James Hagerty and Mark Magnier, "Companies Tiptoe Back toward Made in U.S.A.,'" *The Wall Street Journal,* January 13, 2015, www.wsj.com/articles/companies-tiptoe-back-toward-made-in-the-u-s-a-1421206289 (accessed May 3, 2017).

79. Ted Redmond and Yuko Takeo, "Skimming Profits from Pond Scum," *Bloomberg Businessweek,* July 20–26, 2015, pp. 31–32; Chisaki Watanabe, "Japan's Isuzu, Euglena to Begin Biodiesel Development With Algae," *Bloomberg,* June 25, 2014, www.bloomberg.com/news/articles/2014-06-25/japan-s-isuzu-euglena-to-begin-biodiesel-development-with-algae (accessed March 3, 2018); Durga Madhab Mahapatra, H. N. Chanakya, and T. V. Ramachandra, "*Euglena* sp. as a Suitable Source of Lipids for Potential Use as Biofuel and Sustainable Wastewater Treatment," *Journal of Applied Psychology* 25, no. 3 (2013), pp. 855–865; Euglena, "Research and Business Strategy," www.euglena.jp/en/labo/research.html (accessed March 3, 2018); Sarah Karacs, "Food, Face Cream and Jet Fuel: Japanese Firm Finds Many Uses for Algae," *CNN Tech,* March 24, 2017, http://money.cnn.com/2017/03/24/technology/japan-algae-euglena/index.html (accessed March 3, 2018); Yoko Shoji, "Euglena Plans Japanese Refinery for Algae-Derived Biofuel," *Nikkei Asian Review,* December 2, 2015, https://asia.nikkei.com/Tech-Science/Tech/Euglena-plans-Japanese-refinery-for-algae-derived-jet-fuel (accessed March 3, 2018).

80. Sobia Khan, "Shell to Open Largest Offshore Delivery Centre Globally in Bengaluru," *Economic Times,* June 5, 2015, http://economictimes.indiatimes.com/jobs/shell-to-open-largest-offshore-delivery-centre-globally-in-bengaluru/articleshow/47548572.cms (accessed May 3, 2017).

81. Paul Mozer, "Qualcomm in Venture with Chinese Chip Maker," *The New York Times,* June 23, 2015, www.nytimes.com/2015/06/24/business/international/qualcomm-in-venture-with-chinese-chip-maker.html?_r=0 (accessed May 3, 2017).

82. Calum Fuller, "EY and LinkedIn Announce Strategic Alliance," *Accountancy Age,* October 30, 2015, www.accountancyage.com/aa/news/2432737/ey-and-linkedin-announce-strategic-alliance (accessed May 3, 2017).

83. David Welch and Yon Zhang, "Where Cadillac Is Still Praised," *Bloomberg Businessweek,* February 6–12, 2017, p. 17.

84. Guo Changdong and Ren Ruqin, "Nestle CEO visits Tianjin," *China Daily,* August 12, 2010, www.chinadaily.com.cn/m/tianjin/e/2010-08/12/content_11146560.htm (accessed May 3, 2017); Nestlé, "How Many Factories Do You Have," www.nestle.com/ask-nestle/our-company/answers/how-many-factories-do-you-have (accessed May 3, 2017).

85. O. C. Ferrell, John Fraedrich, and Linda Ferrell, *Business Ethics,* 6th ed. (Boston: Houghton Mifflin, 2005), p. 227–230.

86. Vu Trong Khanh, "Vietnam Gets Its First McDonald's" *The Wall Street Journal,* February 11, 2014, p. B4.

87. Kimberlee Morrison, "Who Are the Most (and Least) Engaged Brands on Twitter?" *Ad Week,* March 16, 2015, www.adweek.com/socialtimes/the-most-and-least-engaged-brands-on-twitter/617045 (accessed May 3, 2017).

88. Export.gov, www.export.gov/about/index.asp (accessed May 3, 2017); CIBER Web, http://CIBERWEB.msu.edu (accessed May 3, 2017).

89. McGraw-Hill video, www.viddler.com/embed/50051953/?f=1&autoplay=0&player=full&disablebranding=0 (accessed April 11, 2016); Brain Staff, "Trek Announces Acquisition of Electra," *Bicycle Retailer,* January 6, 2014, www.bicycleretailer.com/north-america/2014/01/06/trek-announces-acquisition-electra#.Vwu7TjArLIV (accessed April 11, 2016); Ron Callahan, "Electra Bicycle Company Opens Global Headquarters with Slowest Bicycle Race," *Bike World News,* November 10, 2012, www.bikeworldnews.com/2012/11/10/electra-bicycle-company-opens-global-headquarters-slowest-bicycle-race/ (accessed April 11, 2016); Electra Bicycle Company website, www.electrabike.com/ (accessed April 11, 2016).

Credits

Starting and Growing a Business

PART 2

Starting and Growing a Business

4 Options for Organizing Business

©imac/Alamy Stock Photo

Chapter Outline

Learning Objectives

After reading this chapter, you will be able to:

LO 4-1 Describe the advantages and disadvantages of the sole proprietorship form of organization.

LO 4-2 Describe the two types of business partnership and their advantages and disadvantages.

LO 4-3 Describe the corporate form of organization and its advantages and disadvantages.

LO 4-4 Assess the advantages and disadvantages of mergers, acquisitions, and leveraged buyouts.

LO 4-5 Propose an appropriate organizational form for a startup business.

Enter the World of Business

Louisville Slugger Hits the Ball Out of the Park

From small business to privately held company to public corporation—the legal form of ownership for the Louisville Slugger baseball bat has changed multiple times in its more than 130-year history. What has not changed is the popularity of the brand among Major League Baseball stars.

It started in 1884 when 11-year-old Bud Hillerich made a bat for Pete Browning, a star on the professional Louisville major league baseball team. Bud was an apprentice in his father's small woodworking shop who played amateur baseball. He wanted to create a business out of selling bats and finally convinced his father to pursue the endeavor. Bud renamed his bat the "Louisville Slugger" and registered with the U.S. Patent Office. The company was incorporated in 1897 as J.F. Hillerich & Son.

In 1905, the Pittsburgh Pirates' Honus Wagner, "the Flying Dutchman," signed an endorsement with Louisville Slugger, becoming the first athlete to endorse a product. In 1911, salesman Frank Bradsby became a partner of the company, lending his professional sales and marketing expertise. The company was renamed Hillerich & Bradsby Company Inc. (H&B). Among the baseball legends who used Louisville Slugger bats are Babe Ruth, Lou Gehrig, and Hank Aaron.

The Louisville Slugger has sold more than 100 million bats, making it the most popular bat brand in baseball history. Sixty percent of all Major League Baseball players use Louisville Slugger bats. In 2015, Wilson Sporting Goods, a subsidiary of a billion-dollar Finnish corporation and maker of the official football for the National Football League, announced it had acquired Louisville Slugger for $70 million. Although Wilson owns the brand, H&B remains its exclusive manufacturing partner and continues to use its expertise to manufacture the iconic baseball bat.[1]

Introduction

The legal form of ownership taken by a business is seldom of great concern to you as a customer. When you eat at a restaurant, you probably don't care whether the restaurant is owned by one person (a sole proprietorship), has two or more owners who share the business (a partnership), or is an entity owned by many stockholders (a corporation); all you want is good food. If you buy a foreign car, you probably don't care whether the company that made it has laws governing its form of organization that are different from those for businesses in the United States. All businesses must select a form of organization that is most appropriate for their owners and the scope of their business. A business's legal form of ownership affects how it operates, how much it pays in taxes, and how much control its owners have.

This chapter examines three primary forms of business ownership—sole proprietorship, partnership, and corporation—and weighs the advantages and disadvantages of each. These forms are the most often used whether the business is a traditional brick-and-mortar company, an online-only one, or a combination of both. We also take a look at S corporations, limited liability companies, and cooperatives; discuss some trends in business ownership; and touch on one of the most common forms of organizations for nonprofits. You may wish to refer to Table 4.1 to compare the various forms of business ownership mentioned in the chapter.

Sole Proprietorships

Sole proprietorships, businesses owned and operated by one individual, are the most common form of business organization in the United States. Common examples include many retailers such as restaurants, hair salons, flower shops, dog kennels, and independent grocery stores. For example, Chris Nolte, a veteran who served in Iraq, started his own electric-assisted bicycle business called Propel Electric Bikes.[2] Sole proprietors also include independent contractors who complete projects or engage in entrepreneurial activities for different organizations but who are not employees. These include drivers for Uber and those engaged in direct selling for firms such as Mary

connect

▶ Need help understanding forms of business ownership? Visit your Connect ebook video tab for a brief animated explanation.

LO 4-1

Describe the advantages and disadvantages of the sole proprietorship form of organization.

sole proprietorships businesses owned and operated by one individual; the most common form of business organization in the United States.

TABLE 4.1 Various Forms of Business Ownership

Structure	Ownership	Taxation	Liability	Use
Sole proprietorship	One owner	Individual income taxed	Unlimited	Owned by a single individual and is the easiest way to conduct business
Partnership	Two or more owners	Individual owners' income taxed	Somewhat limited	Easy way for two individuals to conduct business
Corporation	Any number of shareholders	Corporate and shareholder taxed	Limited	A legal entity with shareholders or stockholders
S corporation	Up to 100 shareholders	Taxed as a partnership	Limited	A legal entity with tax advantages for restricted number of shareholders
Limited liability company	Unlimited number of shareholders	Taxed as a partnership	Limited	Avoid personal lawsuits

FIGURE 4.1

Comparison of Sole
Proprietorships,
Partnerships,
S Corporations, and
C Corporations

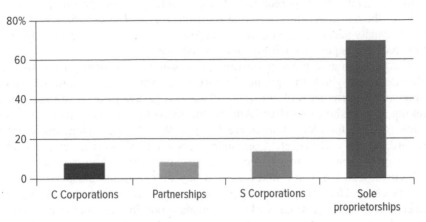

Source: Scott Greenberg, "Pass-Through Businesses: Data and Policy," *Tax Foundation,* January 17, 2017, https://taxfoundation.org/pass-through-businesses-data-and-policy/ (accessed April 10, 2018)

Kay, Avon, or Tupperware. Many sole proprietors focus on services—small retail stores, financial counseling, automobile repair, child care, and the like—rather than on the manufacture of goods, which often requires large sums of money not available to most small businesses. Amazon provides small proprietors the opportunity to set up online shops in the Amazon marketplace. Amazon takes a percentage of the sales and provides shipping and online services to more than 20,000 small companies that earn more than $100,000 a year selling through Amazon. This opens up opportunities for proprietors through the biggest online retailer that accounts for almost half of all Internet retail sales in America.[3] As you can see in Figure 4.1, proprietorships far outnumber corporations. However, they net far fewer sales and less income. Differences between S corporations and C corporations will be discussed later in this text.

Sole proprietorships are typically small businesses employing fewer than 50 people. (We'll look at small businesses in greater detail in the chapter "Small Business, Entrepreneurship, and Franchising.") Sole proprietorships constitute approximately three-fourths of all businesses in the United States. It is interesting to note that women business owners are less likely to get access to credit than their male counterparts.[4] In many areas, small businesses make up the vast majority of the economy.

Advantages of Sole Proprietorships

Sole proprietorships are generally managed by their owners. Because of this simple management structure, the owner/manager can make decisions quickly. This is just one of many advantages of the sole proprietorship form of business.

Ease and Cost of Formation Forming a sole proprietorship is relatively easy and inexpensive. In some states, creating a sole proprietorship involves merely announcing the new business in the local newspaper. Other proprietorships, such as barber

Sole proprietorships, the most common form of business organization in the United States, commonly include businesses such as florists, restaurants, and hair salons.

©Hero Images Inc./Alamy Stock Photo

shops and restaurants, may require state and local licenses and permits because of the nature of the business. The cost of these permits may run from $25 to $100. Lawyers are not usually needed to create such enterprises, and the owner can usually take care of the required paperwork without much assistance.

Of course, an entrepreneur starting a new sole proprietorship must find a suitable site from which to operate the business, even if it is an online business. Some sole proprietors look no farther than their garage or a spare bedroom when seeking a workshop or office. Among the more famous businesses that sprang to life in their founders' homes are Google, Walt Disney, Dell, eBay, Hewlett-Packard, Apple, and Mattel.[5] Computers, personal copiers, scanners, and websites have been a boon for home-based businesses, permitting them to interact quickly with customers, suppliers, and others. Many independent salespersons and contractors can perform their work using a smartphone or tablet computer as they travel. E-mail and social networks have made it possible for many proprietorships to develop in the services area. Internet connections also allow small businesses to establish websites to promote their products and even to make low-cost long-distance phone calls with voice-over Internet protocol (VoIP) technology. One of the most famous services using VoIP is Skype, which allows people to make free calls over the Internet.

Secrecy.   Sole proprietorships make possible the greatest degree of secrecy. The proprietor, unlike the owners of a partnership or corporation, does not have to discuss publicly his or her operating plans, minimizing the possibility that competitors can obtain trade secrets. Financial reports need not be disclosed, as do the financial reports of publicly owned corporations.

Distribution and Use of Profits.   All profits from a sole proprietorship belong exclusively to the owner. He or she does not have to share them with any partners or stockholders. The owner decides how to use the funds—for expansion of the business, for salary increases, for travel to purchase additional inventory, or to find new customers.

Flexibility and Control of the Business.   The sole proprietor has complete control over the business and can make decisions on the spot without anyone else's approval. This control allows the owner to respond quickly to competitive business conditions or to changes in the economy. The ability to quickly change prices or products can provide a competitive advantage for the business.

Government Regulation.   Sole proprietorships have the most freedom from government regulation. Many government regulations—federal, state, and local—apply only to businesses that have a certain number of employees, and securities laws apply only to corporations that issue stock. Nonetheless, sole proprietors must ensure that they follow all laws that do apply to their business. For example, sole proprietorships must be careful to obey employee and consumer protection regulation.

Taxation.   Profits from sole proprietorships are considered personal income and are taxed at individual tax rates. The owner, therefore, pays one income tax that includes the business and individual income. Another tax benefit is that a sole proprietor is allowed to establish a tax-exempt retirement account or a tax-exempt profit-sharing account. Such accounts are exempt from current income tax, but payments taken after retirement are taxed when they are received.

Closing the Business.   A sole proprietorship can be dissolved easily. No approval of co-owners or partners is necessary. The only legal condition is that all financial obligations must be paid or resolved. If a proprietor does a going-out-of-business sale, most states require that the business actually close.

Disadvantages of Sole Proprietorships

What may be seen as an advantage by one person may turn out to be a disadvantage to another. For profitable businesses managed by capable owners, many of the following factors do not cause problems. On the other hand, proprietors starting out with little management experience and little money are likely to encounter many of the disadvantages.

Unlimited Liability.   The sole proprietor has unlimited liability in meeting the debts of the business. In other words, if the business cannot pay its creditors, the owner may be forced to use personal, nonbusiness holdings such as a car or a home to pay off the debts. There are only a few states in which houses and homesteads cannot be taken by creditors, even if the proprietor declares bankruptcy. The more wealth an individual has, the greater is the disadvantage of unlimited liability.

Limited Sources of Funds.   Among the relatively few sources of money available to the sole proprietorship are banks, friends, family, the Small Business Administration, or his or her own funds. The owner's personal financial condition determines his or her credit standing. Additionally, sole proprietorships may have to pay higher interest rates on funds borrowed from banks than do large corporations because they are considered greater risks. More proprietors are using nonbank financial institutions for transactions that charge higher interest rates than banks. Often, the only way a sole proprietor can borrow for business purposes is to pledge a car, a house, other real estate, or other personal assets to guarantee the loan. If the business fails, the owner may lose the personal assets as well as the business. Publicly owned corporations, in contrast, can not only obtain funds from commercial banks, but can sell stocks and bonds to the public to raise money. If a public company goes out of business, the owners do not lose personal assets. However, they will lose the value of their stocks or bonds.

Limited Skills.   The sole proprietor must be able to perform many functions and possess skills in diverse fields such as management, marketing, finance, accounting, bookkeeping, and personnel management. Specialized professionals, such as accountants or attorneys, can be hired by businesses for help or advice. Sometimes, sole proprietors need assistance with certain business functions. For instance, Network Solutions offers web services for small and medium-sized businesses that want to grow their online presence. The company offers website hosting, or storage space and access for websites, as well as tools to help build a website and online marketing services.[6] In the end, however, it is up to the business owner to make the final decision in all areas of the business.

Lack of Continuity.   The life expectancy of a sole proprietorship is directly linked to that of the owner and his or her ability to work. The serious illness of the owner could result in failure of the business if competent help cannot be found. It is difficult to arrange for the sale of a proprietorship and, at the same time, assure customers that the business will continue to meet their needs. For instance, how does one sell a veterinary practice? A veterinarian's major asset is patients. If the vet dies suddenly, the equipment can be sold, but the patients will not necessarily remain loyal to the office. On the other hand, a veterinarian who wants to retire could take in a younger partner and sell the practice to the partner over time. One advantage to the partnership is that some of the customers are likely to stay with the business, even if ownership changes.

Lack of Qualified Employees.   It is sometimes difficult for a small sole proprietorship to match the wages and benefits offered by a large competing corporation because the proprietorship's profits may not be as high. In addition, there may be less room for advancement within a sole proprietorship, so the owner may have difficulty attracting and retaining qualified employees. On the other hand, the trend of large corporations downsizing and outsourcing tasks has created renewed opportunities for small businesses to acquire well-trained employees.

Taxation.   Although we listed taxation as an advantage for sole proprietorships, it can also be a disadvantage, depending on the proprietor's income. Under current tax rates, sole proprietors pay a higher tax rate than do small corporations on income of less than $75,000. However, sole proprietorships avoid the double taxation of corporate and personal taxes that occurs with corporations. The tax effect often determines whether a sole proprietor chooses to incorporate his or her business.

LO 4-2

Describe the two types of business partnership and their advantages and disadvantages.

partnership
a form of business organization defined by the Uniform Partnership Act as "an association of two or more persons who carry on as co-owners of a business for profit."

Partnerships

One way to minimize the disadvantages of a sole proprietorship and maximize its advantages is to have more than one owner. Most states have a model law governing partnerships based on the Uniform Partnership Act. This law defines a **partnership** as "an association of two or more persons who carry on as co-owners of a business for profit." For example, Rebecca Minkoff, a luxury handbag and accessories designer, partnered with her brother, Uri Minkoff, to establish her namesake brand.[7] Partnerships are the least used form of business. They are typically larger than sole proprietorships but smaller than corporations. Partnerships can be a fruitful form of business, as long as some basic keys to success, which are outlined in Table 4.2, are followed.

TABLE 4.2
Keys to Success in Business Partnerships

1. Keep profit sharing equitable based on contributions.
2. Partners should have different skill sets or resource contributions.
3. Ethics and compliance are required.
4. Must maintain effective communication skills.
5. Maintain transparency with stakeholders.
6. Must be realistic in resource and financial management.
7. Previous experience related to business is helpful.
8. Maintain life balance in time spent on business.
9. Focus on customer satisfaction and product quality.
10. Maintain resources in line with sales and growth expectations and planning.

Types of Partnership

There are two basic types of partnership: general partnership and limited partnership. A **general partnership** involves a complete sharing in the management of a business. In a general partnership, each partner has unlimited liability for the debts of the business. Professionals such as lawyers, accountants, and architects often join together in general partnerships.

A **limited partnership** has at least one general partner, who assumes unlimited liability, and at least one limited partner, whose liability is limited to his or her investment in the business. Limited partnerships exist for risky investment projects where the chance of loss is great. The general partners accept the risk of loss; the limited partners' losses are limited to their initial investment. Limited partners do not participate in the management of the business but share in the profits in accordance with the terms of a partnership agreement.

Colhoc Limited Partnership owns and operates the Columbus Blue Jackets NHL franchise. John P. McConnell is majority owner of the Blue Jackets.

©Aaron Doster/CSM/REX/Shutterstock

Usually, the general partner receives a larger share of the profits after the limited partners have received their initial investment back. A *master limited partnership* (MLP) is a limited partnership traded on securities exchanges. MLPs have the tax benefits of a limited partnership but the liquidity (ability to convert assets into cash) of a corporation. Popular examples of MLPs include oil and gas companies and pipeline operators.[8]

Articles of Partnership

Articles of partnership are legal documents that set forth the basic agreement between partners. Most states require articles of partnership, but even if they are not required, it makes good sense for partners to draw them up. Articles of partnership usually list the money or assets that each partner has contributed (called *partnership capital*), state each partner's individual management role or duty, specify how the profits and losses of the partnership will be divided among the partners, and describe how a partner may leave the partnership as well as any other restrictions that might apply to the agreement. Table 4.3 lists some of the issues and provisions that should be included in articles of partnership.

Advantages of Partnerships

Law firms, accounting firms, and investment firms with several hundred partners have partnership agreements that are quite complicated in comparison with the partnership agreement among two or three people owning a computer repair shop. The advantages must be compared with those offered by other forms of business organization, and not all apply to every partnership.

Ease of Organization. Starting a partnership requires little more than drawing up articles of partnership. No legal charters have to be granted, but the name of the business should be registered with the state.

Availability of Capital and Credit. When a business has several partners, it has the benefit of a combination of talents and skills and pooled financial resources. Partnerships tend to be larger than sole proprietorships and, therefore, have greater

general partnership
a partnership that involves a complete sharing in both the management and the liability of the business.

limited partnership
a business organization that has at least one general partner, who assumes unlimited liability, and at least one limited partner, whose liability is limited to his or her investment in the business.

articles of partnership
legal documents that set forth the basic agreement between partners.

TABLE 4.3
Issues and Provisions in
Articles of Partnership

1. Name, purpose, location
2. Duration of the agreement
3. Authority and responsibility of each partner
4. Character of partners (i.e., general or limited, active or silent)
5. Amount of contribution from each partner
6. Division of profits or losses
7. Salaries of each partner
8. How much each partner is allowed to withdraw
9. Death of partner
10. Sale of partnership interest
11. Arbitration of disputes
12. Required and prohibited actions
13. Absence and disability
14. Restrictive covenants
15. Buying and selling agreements

earning power and better credit ratings. Because many limited partnerships have been formed for tax purposes rather than for economic profits, the combined income of all U.S. partnerships is quite low. Nevertheless, the professional partnerships of many lawyers, accountants, and banking firms make quite large profits. For instance, the more than 700 partners at the international law firm Morgan, Lewis & Bockius LLP take home large incomes as the firm earns revenues of more than $1 billion a year.[9]

Combined Knowledge and Skills.   Partners in the most successful partnerships acknowledge each other's talents and avoid confusion and conflict by specializing in a particular area of expertise such as marketing, production, accounting, or service. The diversity of skills in a partnership makes it possible for the business to be run by a management team of specialists instead of by a generalist sole proprietor. Co-founders Justin Wetherill, Edward Trujillo, and David Reiff credit diversity as being a key component to the success of their company uBreakiFix, an iPhone repair service. In just nine years, the startup has grown to more than 375 locations, generating more than $147 million in annual revenue. They have also embarked upon a franchising strategy, which is sure to spur further growth.[10] Service-oriented partnerships in fields such as law, financial planning, and accounting may attract customers because clients may think that the service offered by a diverse team is of higher quality than that provided by one person. Larger law firms, for example, often have individual partners who specialize in certain areas of the law—such as family, bankruptcy, corporate, entertainment, and criminal law.

Decision Making.   Small partnerships can react more quickly to changes in the business environment than can large partnerships and corporations. Such fast reactions are possible because the partners are involved in day-to-day operations and can make decisions quickly after consultation. Large partnerships with hundreds of partners in many states are not common. In those that do exist, decision making is likely to be slow. However, some partnerships have been successful despite their large size. The accounting firm Ernst & Young is the second largest accounting and advisory firm in the United States. In one year, it promoted 753 individuals to the rank of partner, 30 percent of whom were women. With global revenues of more than $29.6 billion, some have attributed Ernst & Young's success to its strong approach to diversity and the innovation of its teams.[11]

Regulatory Controls.   Like a sole proprietorship, a partnership has fewer regulatory controls affecting its activities than does a corporation. A partnership does not have to file public financial statements with government agencies or send out quarterly financial statements to several thousand owners, as do corporations such as Apple and Ford Motor Co. A partnership does, however, have to abide by all laws relevant to the industry or profession in which it operates as well as state and federal laws relating to financial reports, employees, consumer protection, and environmental regulations, just as the sole proprietorship does.

Disadvantages of Partnerships

Partnerships have many advantages compared to sole proprietorships and corporations, but they also have some disadvantages. Limited partners have no voice in the management of the partnership, and they may bear most of the risk of the business, while the general partner reaps a larger share of the benefits. There may be a change in the goals and objectives of one partner but not the other, particularly when the partners are multinational organizations. This can cause friction, giving rise to an enterprise that fails to satisfy both parties or even forcing an end to the partnership. Many partnership disputes wind up in court or require outside mediation. A partnership can be jeopardized when two business partners cannot resolve disputes. For instance, two co-founders of photo-sharing mobile application Snapchat reached a financial settlement with their former fraternity brother, who sued the co-founders because the business was based around his idea.[12] In some cases, the ultimate solution may be dissolving the partnership. Major disadvantages of partnerships include the following.

Unlimited Liability.   In general partnerships, the general partners have unlimited liability for the debts incurred by the business, just as the sole proprietor has unlimited liability for his or her business. Such unlimited liability can be a distinct disadvantage to one partner if his or her personal financial resources are greater than those of the others. A potential partner should check to make sure that all partners have comparable resources to help the business in time of trouble. This disadvantage is eliminated for limited partners, who can lose only their initial investment.

Business Responsibility.   All partners are responsible for the business actions of all others. Partners may have the ability to commit the partnership to a contract without approval of the other partners. A bad decision by one partner may put the other partners' personal resources in jeopardy. Personal problems such as a divorce can eliminate a significant portion of one partner's financial resources and weaken the financial structure of the whole partnership.

Life of the Partnership.   A partnership is terminated when a partner dies or withdraws. In a two-person partnership, if one partner withdraws, the firm's liabilities would be paid off and the assets divided between the partners. Obviously, the partner who wishes to continue in the business would be at a serious disadvantage. The business could be disrupted, financing would be reduced, and the management skills of the departing partner would be lost. The remaining partner would have to find another or reorganize the business as a sole proprietorship. In very large partnerships such as those found in law firms and investment banks, the continuation of the partnership may be provided for in the articles of partnership. The provision may simply state the terms for a new partnership agreement among the remaining partners. In such cases, the disadvantage to the other partners is minimal.

Selling a partnership interest has the same effect as the death or withdrawal of a partner. It is difficult to place a value on a partner's share of the partnership. No public value is placed on the partnership, as there is on publicly owned corporations. What is a law firm worth? What is the local hardware store worth? Coming up with a fair value that all partners can agree to is not easy. Selling a partnership interest is easier if the articles of partnership specify a method of valuation. Even if there is not a procedure for selling one partner's interest, the old partnership must still be dissolved and a new one created. In contrast, in the corporate form of business, the departure of owners has little effect on the financial resources of the business, and the loss of managers does not cause long-term changes in the structure of the organization.

Distribution of Profits. Profits earned by the partnership are distributed to the partners in the proportions specified in the articles of partnership. This may be a disadvantage if the division of the profits does not reflect the work each partner puts into the business. You may have encountered this disadvantage while working on a student group project: You may have felt that you did most of the work and that the other students in the group received grades based on your efforts. Even the perception of an unfair profit-sharing agreement may cause tension between the partners, and unhappy partners can have a negative effect on the profitability of the business.

Limited Sources of Funds. As with a sole proprietorship, the sources of funds available to a partnership are limited. Because no public value is placed on the business (such as the current trading price of a corporation's stock), potential partners do not always know what one partnership share is worth, although third parties can access the value. Moreover, because partnership shares cannot be bought and sold easily in public markets, potential owners may not want to tie up their money in assets that cannot be readily sold on short notice. Accumulating enough funds to operate a national business, especially a business requiring intensive investments in facilities and equipment, can be difficult. Partnerships also may have to pay higher interest rates on funds borrowed from banks than do large corporations because partnerships may be considered greater risks.

Taxation of Partnerships

Partnerships are quasi-taxable organizations. This means that partnerships do not pay taxes when submitting the partnership tax return to the Internal Revenue Service. The tax return simply provides information about the profitability of the organization and the distribution of profits among the partners. Partners must report their share of profits on their individual tax returns and pay taxes at the income tax rate for individuals. Master limited partnerships require financial reports similar to corporations, which are discussed in the next section.

Beer Hound Brews the Perfect Partnership

Kenny Thacker always dreamed about brewing beer, so after he became unemployed, the former construction worker set out to make his dream a reality. Using a $25,000 loan from his father, Thacker opened Beer Hound Brewery in 2012. The brewery proved popular with the locals, but nonetheless, the business tanked financially in the three years it was open. There were not enough customers in the small town where the brewery was located for Thacker to turn a profit. As a sole business owner, he did not have all the skills needed to succeed financially.

Help came when Thacker partnered with Frank Becker, a restaurant franchise manager, and Rick Cash, a retired manager at Philip Morris. The three men reopened Beer Hound Brewery, this time in a more desirable location where the brewery could develop a stronger customer base. The various business responsibilities of running the brewery were split up among the three men. Thacker focused on perfecting his brewing recipes, while his two partners addressed the marketing, operations and financing aspects of the business. The brewery was a hit when it opened. Finding the right business partners helped transform the fledgling brewery into a successful business. Revenue at the brewery approached $500,000 in its first year, and Thacker, Becker, and Cash estimated revenues would surpass $1 million in 2017.[13]

Critical Thinking Questions

1. What made the brewery more successful after Thacker brought on his two business partners?
2. Why was it important for Thacker, Becker, and Cash to share in the responsibility of running the brewery?
3. How did the owners of Beer Hound Brewery grow its customer base?

Corporations

When you think of a business, you probably think of a huge corporation such as General Electric, Procter & Gamble, or Sony because a large portion of your consumer dollars go to such corporations. A **corporation** is a legal entity, created by the state, whose assets and liabilities are separate from its owners. As a legal entity, a corporation has many of the rights, duties, and powers of a person, such as the right to receive, own, and transfer property. Corporations can enter into contracts with individuals or with other legal entities, and they can sue and be sued in court.

Corporations account for the majority of all U.S. sales and income. Thus, most of the dollars you spend as a consumer probably go to incorporated businesses. Most corporations are not mega-companies like General Mills or Ford Motor Co.; even small businesses can incorporate. As we shall see later in the chapter, many smaller firms elect to incorporate as "S Corporations," which operate under slightly different rules and have greater flexibility than do traditional "C Corporations" like General Mills.

Corporations are typically owned by many individuals and organizations who own shares of the business, called **stock** (thus, corporate owners are often called *shareholders* or *stockholders*). Stockholders can buy, sell, give or receive as gifts, or inherit their shares of stock. As owners, the stockholders are entitled to all profits that are left after all the corporation's other obligations have been paid. These profits may be distributed in the form of cash payments called **dividends**. For example, if a corporation earns $100 million after expenses and taxes and decides to pay the owners $40 million in dividends, the stockholders receive 40 percent of the profits in cash dividends. In recent years, companies in the U.S. have paid shareholders more than $438 billion per year in dividends. Table 4.4 lists the world's biggest dividend payers. However, not all after-tax profits are paid to stockholders in dividends. Some corporations may retain profits to expand the business. For example, Alphabet (Google's parent company), Amazon, Biogen, and Tesla have not paid dividends in the past because they reinvest their earnings.

LO 4-3

Describe the corporate form of organization and its advantages and disadvantages.

corporation
a legal entity, created by the state, whose assets and liabilities are separate from its owners.

stock
shares of a corporation that may be bought or sold.

dividends
profits of a corporation that are distributed in the form of cash payments to stockholders.

TABLE 4.4
World's Biggest Dividend Payers

Rank	Company
1.	Royal Dutch Shell Plc
2.	China Mobile Limited
3.	Exxon Mobil Corp.
4.	Apple Inc
5.	Microsoft Corporation
6.	AT&T Inc.
7.	HSBC Holdings plc
8.	China Construction Bank Corp.
9.	Verizon Communications Inc.
10.	Johnson & Johnson

Source: Janus Henderson Global Dividend Index, Edition 17, February 2018, p. 12.

Creating a Corporation

A corporation is created, or incorporated, under the laws of the state in which it incorporates. The individuals creating the corporation are known as *incorporators*. Each state has a specific procedure, sometimes called *chartering the corporation,* for incorporating a business. Most states require a minimum of three incorporators; thus, many small businesses can be and are incorporated. Another requirement is that the new corporation's name cannot be similar to that of another business. In most states, a corporation's name must end in "company," "corporation," "incorporated," or "limited" to show that the owners have limited liability. (In this text, however, the word *company* means any organization engaged in a commercial enterprise and can refer to a sole proprietorship, a partnership, or a corporation.)

The incorporators must file legal documents generally referred to as *articles of incorporation* with the appropriate state office (often the secretary of state). The articles of incorporation contain basic information about the business. The following 10 items are found in the Model Business Corporation Act, issued by the American Bar Association, which is followed by most states:

1. Name and address of the corporation.
2. Objectives of the corporation.
3. Classes of stock (common, preferred, voting, nonvoting) and the number of shares for each class of stock to be issued.
4. Expected life of the corporation (corporations are usually created to last forever).
5. Financial capital required at the time of incorporation.
6. Provisions for transferring shares of stock between owners.
7. Provisions for the regulation of internal corporate affairs.
8. Address of the business office registered with the state of incorporation.
9. Names and addresses of the initial board of directors.
10. Names and addresses of the incorporators.

Based on the information in the articles of incorporation, the state issues a **corporate charter** to the company. After securing this charter, the owners hold an organizational meeting at which they establish the corporation's bylaws and elect a board of directors. The bylaws might set up committees of the board of directors and describe the rules and procedures for their operation.

corporate charter
a legal document that the state issues to a company based on information the company provides in the articles of incorporation.

Types of Corporations

If the corporation does business in the state in which it is chartered, it is known as a *domestic corporation.* In other states where the corporation does business, it is known as a *foreign corporation.* If a corporation does business outside the nation in which it is incorporated, it is called an *alien corporation.* A corporation may be privately or publicly owned.

A **private corporation** is owned by just one or a few people who are closely involved in managing the business. These people, often a family, own all the corporation's stock, and no stock is sold to the public. Many corporations are quite large, yet remain private, including Publix Super Markets. It is the nation's seventh largest privately held corporation, with annual revenues of more than $34 billion. Founded in 1930, today the company is run by the founder's grandson, who is the fourth family member to lead the company.[14] Table 4.5 lists the 10 largest private companies in the United States. Privately owned corporations are not required to disclose financial information publicly, but they must, of course, pay taxes.

private corporation
a corporation owned by just one or a few people who are closely involved in managing the business.

A **public corporation** is one whose stock anyone may buy, sell, or trade. A few thousand multinational firms influence what billions of consumers watch, wear, and eat.[15] Companies like IBM, McDonald's, Caterpillar, and Procter & Gamble earn more than half of their revenue outside the United Sates, as Table 4.6 indicates. Thousands of smaller public corporations in the United States have sales under $10 million. In large public corporations such as AT&T, the stockholders are often far removed from the

public corporation
a corporation whose stock anyone may buy, sell, or trade.

TABLE 4.5
America's Largest Private Companies

Rank	Company	Industry	Revenue (in billions)	Employees
1.	Cargill	Food, drink, and tobacco	$109.7	150,000
2.	Koch Industries	Multicompany	$100	100,000
3.	Albertsons	Food markets	$59.7	273,000
4.	Deloitte	Business services and supplies	$36.8	244,400
5.	PricewaterhouseCoopers	Business services and supplies	$35.9	223,000
6.	Mars	Food, drink, and tobacco	$35	80,000
7.	Publix Super Markets	Food markets	$34	191,000
8.	Bechtel	Construction	$32.9	55,000
9.	Ernst & Young	Business services and supplies	$29.6	231,000
10.	C&S Wholesale Grocers	Food, drink, and tobacco	$28.1	17,500

Source: "America's Largest Private Companies," Forbes, https://www.forbes.com/largest-private-companies/list/#tab:rank (accessed April 8, 2018).

TABLE 4.6

American Companies with More than Half of Their Revenues from Outside the United States

Company	Description
Caterpillar Inc.	Designs, manufactures, markets, and sells machinery, engines, and financial products
Dow Chemical	Manufactures chemicals, with products including plastics, oil, and crop technology
General Electric	Operates in the technology infrastructure, energy, capital finance, and consumer and industrial fields, with products including appliances, locomotives, weapons, lighting, and gas
General Motors	Sells automobiles with brands including Chevrolet, Buick, Cadillac, and Isuzu
IBM	Conducts technological research, develops intellectual property including software and hardware, and offers consulting services
Intel	Manufactures and develops semiconductor chips and microprocessors
McDonald's	Operates second-largest chain of fast-food restaurants worldwide after Subway
Nike	Designs, develops, markets, and sells athletic shoes and clothing
Procter & Gamble	Sells consumer goods with brands including Tide, Bounty, Crest, and Iams
Yum! Brands	Operates and licenses restaurants including Taco Bell, Kentucky Fried Chicken, and Pizza Hut

management of the company. In other public corporations, the managers are often the founders and the major shareholders. Facebook CEO Mark Zuckerberg, for example, holds 16.1 percent of Facebook stock. While he announced that he would donate 35 to 75 million of his shares to charity, he holds an 87 percent voting majority position in the firm.[16] Publicly owned corporations must disclose financial information to the public under specific laws that regulate the trade of stocks and other securities.

initial public offering (IPO) selling a corporation's stock on public markets for the first time.

A private corporation that needs more money to expand or to take advantage of opportunities may have to obtain financing by "going public" through an **initial public offering (IPO)**—that is, becoming a public corporation by selling stock so that it can be traded in public markets. Digital media companies are leading a surge in initial public offerings. Chinese e-commerce company Alibaba released the largest IPO globally at $25 billion.[17]

Also, privately owned firms are occasionally forced to go public with stock offerings when a major owner dies and the heirs have large estate taxes to pay. The tax payment may only be possible with the proceeds of the sale of stock. This happened to the brewer Adolph Coors Inc. After Adolph Coors died, the business went public and his family sold shares of stock to the public in order to pay the estate taxes.

On the other hand, public corporations can be "taken private" when one or a few individuals (perhaps the management of the firm) purchase all the firm's stock so that it can no longer be sold publicly. Taking a corporation private may be desirable when owners want to exert more control over the

Music streaming service Spotify went public on the New York Stock Exchange, earning a $26 billion valuation.

©JUSTIN LANE/EPA-EFE/REX/Shutterstock

firm or they want the flexibility to make decisions for restructuring operations. For example, Michael Dell took his company private in order to set a new direction as PC sales continue to decline. Becoming a private company again allows Dell to focus on the needs of the company more fully without having to worry about the stock price for investors.[18] Taking a corporation private is also one technique for avoiding a takeover by another corporation.

Quasi-public corporations and nonprofits are two types of public corporations. **Quasi-public corporations** are owned and operated by the federal, state, or local government. The focus of these entities is to provide a service to citizens, such as mail delivery, rather than earning a profit. Indeed, many quasi-public corporations operate at a loss. Examples of quasi-public corporations include the National Aeronautics and Space Administration (NASA) and the U.S. Postal Service.

quasi-public corporations
corporations owned and operated by the federal, state, or local government.

Like quasi-public corporations, **nonprofit corporations** focus on providing a service rather than earning a profit, but they are not owned by a government entity. Organizations such as the Sesame Workshop, the Elks Clubs, the American Lung Association, the American Red Cross, museums, and private schools provide services without a profit motive. United Way Worldwide is the largest charity in the U.S. with total revenue of $3.9 billion.[19] To fund their operations and services, nonprofit organizations solicit donations from individuals and companies, apply for grants from the government and other charitable foundations, and charge for their services. Habitat for Humanity, a nonprofit that builds affordable homes using volunteer labor for families in need, operates hundreds of Habitat for Humanity ReStore locations, outlets for discount furniture, appliances, building materials, and more. All of the profits from its stores benefit its home-building program.[20] Nonprofits do not have shareholders, and most are organized as 501(c)(3) organizations. A 501(c)(3) organization receives certain tax exemptions, and donors contributing to these organizations may reduce their tax deductibility for their donations. Organizations that have 501(c)(3) status include public charities (for example, the Leukemia & Lymphoma Society), private foundations (for example, the Daniels Fund), and private operating foundations that sponsor and fund their own programs (for example, day camp for underprivileged children).[21]

nonprofit corporations
corporations that focus on providing a service rather than earning a profit but are not owned by a government entity.

Elements of a Corporation

The Board of Directors. A **board of directors**, elected by the stockholders to oversee the general operation of the corporation, sets the long-range objectives of the corporation. It is the board's responsibility to ensure that the objectives are achieved on schedule. Board members have a duty of care and loyalty to oversee the management of the firm or for any misuse of funds. An important duty of the board of directors is to hire corporate officers, such as the president and the chief executive officer (CEO), who are responsible to the directors for the management and daily operations of the firm. The role and expectations of the board of directors took on greater significance after the accounting scandals of the early 2000s and the passage of the Sarbanes-Oxley Act.[22] As a result, most corporations have restructured how they compensate board directors for their time and expertise.

board of directors
a group of individuals, elected by the stockholders to oversee the general operation of the corporation, who set the corporation's long-range objectives.

However, some experts now speculate that Sarbanes-Oxley did little to motivate directors to increase company oversight. At the same time, the pay rate of directors is rising. On average, corporate directors are paid around $245,000, with compensation ranging from $0 to more than $1 million. Over the past several years, the trend of increasing directors' pay has continued, resulting in higher and higher pay levels. Although such pay is meant to attract top-quality directors, concerns exist over whether excessive pay will have unintended consequences. Some believe that this trend is contributing to the declining effectiveness in corporate governance.[23]

Directors can be employees of the company (*inside directors*) or people unaffiliated with the company (*outside directors*). Inside directors are usually the officers responsible for running the company. Outside directors are often top executives from other companies, lawyers, bankers, even professors. Directors today are increasingly chosen for their expertise, competence, and ability to bring diverse perspectives to strategic discussions. Outside directors are also thought to bring more independence to the monitoring function because they are not bound by past allegiances, friendships, a current role in the company, or some other issue that may create a conflict of interest. Many of the corporate scandals uncovered in recent years might have been prevented if each of the companies' boards of directors had been better qualified, more knowledgeable, and more independent.

There is a growing shortage of available and qualified board members. Boards are increasingly telling their own CEOs that they should be focused on serving their company, not serving on outside boards. Because of this, the average CEO sits on less than one outside board. This represents a decline from a decade ago when the average was two. Because many CEOs are turning down outside positions, many companies have taken steps to ensure that boards have experienced directors. They have increased the mandatory retirement age to 72 or older, and some have raised it to 75 or even older. Minimizing the amount of overlap between directors sitting on different boards helps to limit conflicts of interest and provides for independence in decision making.

Stock Ownership.   Corporations issue two types of stock: preferred and common. Owners of **preferred stock** are a special class of owners because, although they generally do not have any say in running the company, they have a claim to profits before any other stockholders. Other stockholders do not receive any dividends unless the preferred stockholders have already been paid. Dividend payments on preferred stock are usually a fixed percentage of the initial issuing price (set by the board of directors). For example, if a share of preferred stock originally cost $100 and the dividend rate was stated at 7.5 percent, the dividend payment will be $7.50 per share per year. Dividends are usually paid quarterly. Most preferred stock carries a cumulative claim to dividends. This means that if the company does not pay preferred-stock dividends in one year because of losses, the dividends accumulate to the next year. Such dividends unpaid from previous years must also be paid to preferred stockholders before other stockholders can receive any dividends.

Although owners of **common stock** do not get such preferential treatment with regard to dividends, they do get some say in the operation of the corporation. Their ownership gives them the right to vote for members of the board of directors and on other important issues. Common stock dividends may vary according to the profitability of the business, and some corporations do not issue dividends at all, but instead plow their profits back into the company to fund expansion.

Common stockholders are the voting owners of a corporation. They are usually entitled to one vote per share of common stock. During an annual stockholders' meeting, common stockholders elect a board of directors. Some boards find it easier than others to attract high-profile individuals. For example, in 2018 the board of Procter & Gamble consisted of Francis S. Blake, former CEO of Home Depot; Angela F. Braly, CEO of Anthem; Amy L. Chang, CEO of Accompany Inc.; Kenneth I. Chenault, former CEO of American Express; and others.[24] Because they can choose the board of

preferred stock
a special type of stock whose owners, though not generally having a say in running the company, have a claim to profits before other stockholders.

common stock
stock whose owners have voting rights in the corporation, yet do not receive preferential treatment regarding dividends.

directors, common stockholders have some say in how the company will operate. Common stockholders may vote by *proxy,* which is a written authorization by which stockholders assign their voting privilege to someone else, who then votes for his or her choice at the stockholders' meeting. It is a normal practice for management to request proxy statements from shareholders who are not planning to attend the annual meeting. Most owners do not attend annual meetings of the very large companies, such as Westinghouse or Boeing, unless they live in the city where the meeting is held.

Common stockholders have another advantage over preferred shareholders. In most states, when the corporation decides to sell new shares of common stock in the marketplace, common stockholders have the first right, called a *preemptive right,* to purchase new shares of the stock from the corporation. A preemptive right is often included in the articles of incorporation. This right is important because it allows stockholders to purchase new shares to maintain their original positions. For example, if a stockholder owns

Procter & Gamble's board of directors includes Kenneth I. Chenault, former CEO of American Express.

©lev radin/Shutterstock

10 percent of a corporation that decides to issue new shares, that stockholder has the right to buy enough of the new shares to retain the 10 percent ownership.

Advantages of Corporations

Because a corporation is a separate legal entity, it has some very specific advantages over other forms of ownership. The biggest advantage may be the limited liability of the owners.

Limited Liability.   Because the corporation's assets (money and resources) and liabilities (debts and other obligations) are separate from its owners', in most cases the stockholders are not held responsible for the firm's debts if it fails. Their liability or potential loss is limited to the amount of their original investment. Although a creditor can sue a corporation for not paying its debts, even forcing the corporation into bankruptcy, it cannot make the stockholders pay the corporation's debts out of their personal assets. Occasionally, the owners of a private corporation may pledge personal assets to secure a loan for the corporation; this would be most unusual for a public corporation.

Ease of Transfer of Ownership.   Stockholders can sell or trade shares of stock to other people without causing the termination of the corporation, and they can do this without the prior approval of other shareholders. The transfer of ownership (unless it is a majority position) does not affect the daily or long-term operations of the corporation.

Perpetual Life.   A corporation usually is chartered to last forever unless its articles of incorporation stipulate otherwise. The existence of the corporation is unaffected by the death or withdrawal of any of its stockholders. It survives until the owners sell it

Caterpillar pays foreign taxes as well as U.S. taxes when profits are brought back into the country.

©Paul Brady/123RF

or liquidate its assets. However, in some cases, bankruptcy ends a corporation's life. Bankruptcies occur when companies are unable to operate and earn profits. Eventually, uncompetitive businesses must close or seek protection from creditors in bankruptcy court while the business tries to reorganize.

External Sources of Funds.   Of all the forms of business organization, the public corporation finds it easiest to raise money. When a large or public corporation needs to raise more money, it can sell more stock shares or issue bonds (corporate "IOUs," which pledge to repay debt), attracting funds from anywhere in the United States and even overseas. The larger a corporation becomes, the more sources of financing are available to it. We take a closer look at some of these in the chapter "Money and the Financial System."

Expansion Potential.   Because large public corporations can find long-term financing readily, they can easily expand into national and international markets. And, as a legal entity, a corporation can enter into contracts without as much difficulty as a partnership.

Disadvantages of Corporations

Corporations have some distinct disadvantages resulting from tax laws and government regulation.

Double Taxation.   As a legal entity, the corporation must pay taxes on its income just like you do. The United States has a tax rate of 21 percent. The global average is 23 percent.[25] Global companies such as Apple and Caterpillar have to pay foreign taxes as well as U.S. taxes when profits are brought back into the country. When after-tax corporate profits are paid out as dividends to the stockholders, the dividends are taxed a second time as part of the individual owner's income. This process creates double taxation for the stockholders of dividend-paying corporations. Double taxation does not occur with the other forms of business organization.

Forming a Corporation.   The formation of a corporation can be costly. A charter must be obtained, and this usually requires the services of an attorney and payment of legal fees. Filing fees ranging from $25 to $150 must be paid to the state that awards the corporate charter, and certain states require that an annual fee be paid to maintain the charter. Today, a number of Internet services such as LegalZoom.com and Business.com make it easier, quicker, and less costly to form a corporation. However, in making it easier for people to form businesses without expert consultation, these services have increased the risk that people will not choose the kind of organizational form that is right for them. Sometimes, one form works better than another. The business's founders may fail to take into account disadvantages, such as double taxation with corporations.

DID YOU KNOW? The first corporation with a net income of more than $1 billion in one year was General Motors, with a net income in 1955 of $1,189,477,082.[26]

To B or Not to B: Cascade Engineering Adopts B Corporation Certification

B Corporations are for-profit companies that meet very high standards of social and environmental performance, accountability, and transparency. Cascade Engineering makes an unlikely B corporation in that it manufactures plastic products, operating in an industry that tends to be seen as environmentally unfriendly. However, Cascade's founder, Fred Keller, has a different view. He welcomes regulation, believing it motivates companies to improve processes. Under his leadership, Cascade has adopted the triple bottom line approach of people, planet, and profits.

B corporations are not your typical corporation. The B stands for beneficial. It is a certification awarded by the nonprofit B Lab to signal that member companies conform to a set of transparent and comprehensive social and environmental performance standards. These businesses are purpose driven and are designed to give back to communities, the environment, and employees.

From an environmental standpoint, Cascade has spent more than a decade helping customers reduce oil use and eliminate waste. It has developed a cradle-to-cradle certified product line based on a design concept that stresses reusable and safe materials, renewable energy, water quality, social fairness, and continuous improvement. Unlike many manufacturers, Cascade supports a switch to renewable energy. Company decisions focus on sustainable solutions as the best alternative for the continued success of the company and society.

Cascade is aware of ramifications caused by plastic. When evaluating alternatives, therefore, sustainability is a primary consideration. Although many believe companies should focus on profits first, Keller believes businesses should be analyzed not only by profit but also by how they benefit society and the environment.[27]

Critical Thinking Questions

1. How does Cascade use B corporation certification as a way to improve both company processes as well as the environment?
2. How does B corporation certification encourage corporations to be socially responsible?
3. Do you think it is possible for a plastics company to make a positive contribution toward sustainability?

Disclosure of Information. Corporations must make information available to their owners, usually through an annual report to shareholders. The annual report contains financial information about the firm's profits, sales, facilities and equipment, and debts, as well as descriptions of the company's operations, products, and plans for the future. Public corporations must also file reports with the Securities and Exchange Commission (SEC), the government regulatory agency that regulates securities such as stocks and bonds. The larger the firm, the more data the SEC requires. Because all reports filed with the SEC are available to the public, competitors can access them. Additionally, complying with securities laws takes time.

Employee-Owner Separation. Many employees are not stockholders of the company for which they work. This separation of owners and employees may cause employees to feel that their work benefits only the owners. Employees without an ownership stake do not always see how they fit into the corporate picture and may not understand the importance of profits to the health of the organization. If managers are part owners but other employees are not, management–labor relations take on a different, sometimes difficult, aspect from those in partnerships and sole proprietorships. However, this situation is changing as more corporations establish employee stock ownership plans (ESOPs), which give shares of the company's stock to its employees. Such plans build a partnership between employee and employer and can boost productivity because they motivate employees to work harder so that they can earn dividends from their hard work as well as from their regular wages.

Other Types of Ownership

In this section, we take a brief look at joint ventures, S corporations, limited liability companies, and cooperatives–businesses formed for special purposes.

Joint Ventures

joint venture
a partnership established for a specific project or for a limited time.

A **joint venture** is a partnership established for a specific project or for a limited time. The partners in a joint venture may be individuals or organizations, as in the case of the international joint ventures discussed in the chapter "Business in a Borderless World." Control of a joint venture may be shared equally, or one partner may control decision making. Joint ventures are especially popular in situations that call for large investments, such as extraction of natural resources and the development of new products. Joint ventures can even take place between businesses and governments. For example, Sony set up joint ventures to make and sell PlayStation consoles and games in China.[28]

S Corporations

S corporation
corporation taxed as though it were a partnership with restrictions on shareholders.

An **S corporation** is a form of business ownership that is taxed as though it were a partnership. Net profits or losses of the corporation pass to the owners, thus eliminating double taxation. The benefit of limited liability is retained. Formally known as Subchapter S Corporations, they have become a popular form of business ownership for entrepreneurs and represent almost half of all corporate filings.[29] The owners of an S corporation get the benefits of tax advantages and limited liability. Advantages of S corporations include the simple method of taxation, the limited liability of shareholders, perpetual life, and the ability to shift income and appreciation to others. Avoiding double taxation is reason enough for approximately 4.6 million U.S. companies to operate as S corporations.[30] Disadvantages include restrictions on the number (100) and types (individuals, estates, and certain trusts) of shareholders and the difficulty of formation and operation. S corporations are mainly small businesses

limited liability company (LLC)
form of ownership that provides limited liability and taxation like a partnership but places fewer restrictions on members.

Fiat Chrysler Automobiles (FCA) US LLC is the automaker that designs, manufactures, sells, and distributes brands like Jeep, Dodge, and Ram. FCA US is an example of a limited liability company.

©Steve Lagreca/Shutterstock

Limited Liability Companies

A **limited liability company (LLC)** is a form of business ownership that provides limited liability, as in a corporation, but is taxed like a partnership. Although relatively new in the United States, LLCs have existed for many years abroad. Professionals such as lawyers, doctors, and engineers often use the LLC form of ownership. Many consider the LLC a blend of the best characteristics of corporations, partnerships, and sole proprietorships. One of the major reasons for the LLC form of ownership is to protect the members' personal assets in case of lawsuits. LLCs are flexible, simple to run, and do not require the members to hold meetings, keep

Moving on Up: The VIP Moving & Storage LLC Experience

VIP Moving & Storage LLC
Founder: Marshall Powell Ledbetter Sr.
Founded: 1932, in Columbia, Tennessee
Success: VIP Moving & Storage, the official mover for the Tennessee Titans, recently expanded to Long Beach, California.

For families and football teams, VIP Moving & Storage offers customized services to meet their moving and storage needs. This limited liability company offers a wide variety of moving services, including corporate moves, residential moves, packing, unpacking, furniture assembly, and storage for valuables in its 50,000-square-foot storage facility.

Currently owned by Marshall Powell Ledbetter III, VIP Moving was started by his grandfather under a different name in 1932. Ledbetter purchased the company in 2000, changing the name to VIP Moving & Storage. He was only able to

purchase the warehouse, requiring him to buy new trucks and trailers. By owning its own fleet, the company is able to offer the ultimate in customization.

For more than 20 years, VIP Moving has been the official mover for the Tennessee Titans and has also worked with the New York Jets, Los Angeles Rams, Seattle Seahawks, and New England Patriots. The company's movers meet the teams at the airport to transport their equipment, set up the locker rooms, and tear them down. VIP is committed to offering the best service for moving and storage—one box at a time.[31]

Critical Thinking Questions

1. What are the advantages for VIP Moving & Storage of operating as an LLC?
2. How is an LLC form of ownership beneficial to owner Marshall Powell Ledbetter III?
3. How has VIP Moving & Storage leveraged partnerships with NFL teams to expand its business?

---

minutes, or make resolutions, all of which are necessary in corporations. Mrs. Fields Famous Brands LLC—known for its cookies and brownies–is an example of a limited liability company.[32]

Cooperatives

Another form of organization in business is the **cooperative** or **co-op**, an organization composed of individuals or small businesses that have banded together to reap the benefits of belonging to a larger organization. Berkshire Co-op Market, for example, is a grocery store cooperative based in Massachusetts; Ocean Spray is a cooperative of cranberry farmers.[33] REI operates a bit differently because it is owned by consumers rather than farmers or small businesses. A co-op is set up not to make money as an entity. It exists so that its members can become more profitable or save money. Co-ops are generally expected to operate without profit or to create only enough profit to maintain the co-op organization.

Many cooperatives exist in small farming communities. The co-op stores and markets grain; orders large quantities of fertilizer, seed, and other supplies at discounted prices; and reduces costs and increases efficiency with good management. A co-op can purchase supplies in large quantities and pass the savings on to its members. It also can help distribute the products of its members more efficiently than each could on an individual basis. A cooperative can advertise its members' products and thus generate demand. Ace Hardware, a cooperative of independent hardware store owners, allows its members to share in the savings that result from buying supplies in large quantities; it also provides advertising, which individual members might not be able to afford on their own.

cooperative (co-op)
an organization composed of individuals or small businesses that have banded together to reap the benefits of belonging to a larger organization.

LO 4-4

Assess the advantages and disadvantages of mergers, acquisitions, and leveraged buyouts.

Trends in Business Ownership: Mergers and Acquisitions

Companies large and small achieve growth and improve profitability by expanding their operations, often by developing and selling new products or selling current products to new groups of customers in different geographic areas. Such growth, when carefully planned and controlled, is usually beneficial to the firm and ultimately helps it reach its goal of enhanced profitability. But companies also grow by merging with or purchasing other companies.

merger
the combination of two companies (usually corporations) to form a new company.

A **merger** occurs when two companies (usually corporations) combine to form a new company. An **acquisition** occurs when one company purchases another, generally by buying most of its stock. The acquired company may become a subsidiary of the buyer, or its operations and assets may be merged with those of the buyer. The government sometimes scrutinizes mergers and acquisitions in an attempt to protect customers from monopolistic practices. For example, the decision to authorize American Airlines' acquisition of U.S. Airways was carefully analyzed. Google paid $3.2 billion for smart-home company Nest Labs.[34] The company was just one of many that Google acquired during the year. While these acquisitions have the potential to diversify Google's service offerings and benefit it financially, some believe that Google might be investing in companies of which it has little knowledge. In these cases, acquisitions could end up harming the acquiring company. Perhaps partially for this reason, Google restructured to become a holding firm called Alphabet Inc. Divisions such as Google Nest are now operated as semi-independent businesses.[35] Acquisitions sometimes involve the purchase of a division or some other part of a company rather than the entire company. Table 4.7 highlights the largest mergers of all time.

acquisition
the purchase of one company by another, usually by buying its stock.

When firms that make and sell similar products to the same customers merge, it is known as a *horizontal merger,* as when Martin Marietta and Lockheed, both defense contractors, merged to form Lockheed Martin. Horizontal mergers, however, reduce the number of corporations competing within an industry, and for this reason they are usually reviewed carefully by federal regulators before the merger is allowed to proceed.

When companies operating at different but related levels of an industry merge, it is known as a *vertical merger.* In many instances, a vertical merger results when one

TABLE 4.7
The Largest Mergers of All Time

Rank	Acquirer	Target	Transaction Value (in billions)
1.	Vodafone	Mannesmann	$180
2.	Time Warner	America Online (AOL)	$165
3.	Verizon Communications	Vodafone	$129
4.	RFS Holdings	ABN Amro	$98
5.	AB InBev	SABMiller	$90
6.	Pfizer	Warner-Lambert	$89
7.	AT&T	BellSouth	$86

corporation merges with one of its customers or suppliers. For example, if Burger King were to purchase a large Idaho potato farm—to ensure a ready supply of potatoes for its french fries—a vertical merger would result.

A *conglomerate merger* results when two firms in unrelated industries merge. For example, the purchase of Sterling Drug, a pharmaceutical firm, by Eastman Kodak, best known for its films and cameras, represented a conglomerate merger because the two companies were of different industries. (Kodak later sold Sterling Drug to a pharmaceutical company.)

When a company (or an individual), sometimes called a *corporate raider,* wants to acquire or take over another company, it first offers to buy some or all of the other company's stock at a premium over its current price in a *tender offer.* Most such offers are "friendly," with both groups agreeing to the proposed deal, but some are "hostile," when the second company does not want to be taken over.

Walmart acquired women's clothing retailer ModCloth to expands its online shopping footprint.

©Casimiro PT/Shutterstock

To head off a hostile takeover attempt, a threatened company's managers may use one or more of several techniques. They may ask stockholders not to sell to the raider, file a lawsuit in an effort to abort the takeover, institute a *poison pill* as Energizer did (in which the firm allows stockholders to buy more shares of stock at prices lower than the current market value) or *shark repellant* (in which management requires a large majority of stockholders to approve the takeover), or seek a *white knight* (a more acceptable firm that is willing to acquire the threatened company). In some cases, management may take the company private or even take on more debt so that the heavy debt obligation will "scare off" the raider.

In a **leveraged buyout (LBO)**, a group of investors borrows money from banks and other institutions to acquire a company (or a division of one), using the assets of the purchased company to guarantee repayment of the loan. In some LBOs, as much as 95 percent of the buyout price is paid with borrowed money, which eventually must be repaid.

leveraged buyout (LBO) a purchase in which a group of investors borrows money from banks and other institutions to acquire a company (or a division of one), using the assets of the purchased company to guarantee repayment of the loan.

Because of the explosion of mergers, acquisitions, and leveraged buyouts in the 1980s and 1990s, financial journalists coined the term *merger mania.* Many companies joined the merger mania simply to enhance their own operations by consolidating them with the operations of other firms. Mergers and acquisitions enabled these companies to gain a larger market share in their industries, acquire valuable assets such as new products or plants and equipment, and lower their costs. Mergers also represent a means of making profits quickly, as was the case during the 1980s when many companies' stock was undervalued. Quite simply, such companies represent a bargain to other companies that can afford to buy them. Additionally, deregulation of some industries has permitted consolidation of firms within those industries for the first time, as is the case in the banking and airline industries.

Some people view mergers and acquisitions favorably, pointing out that they boost corporations' stock prices and market value, to the benefit of their stockholders. In many instances, mergers enhance a company's ability to meet foreign competition

If you have a good idea and want to turn it into a business, you are not alone. Small businesses are popping up all over the United States, and the concept of entrepreneurship is hot. Entrepreneurs seek opportunities and creative ways to make profits. Business emerges in a number of different organizational forms, each with its own advantages and disadvantages. Sole proprietorships are the most common form of business organization in the United States. They tend to be small businesses and can take pretty much any form—anything from a hair salon to a scuba shop, from an organic produce provider to a financial advisor. Proprietorships are everywhere serving consumers' wants and needs. Proprietorships have a big advantage in that they tend to be simple to manage—decisions get made quickly when the owner and the manager are the same person and they are fairly simple and inexpensive to set up. Rules vary by state, but at most all you will need is a license from the state.

Many people have been part of a partnership at some point in their life. Group work in school is an example of a partnership. If you ever worked as a DJ on the weekend with your friend and split the profits, then you have experienced a partnership. Partnerships can be either general or limited. General partners have unlimited liability and share completely in the management, debts, and profits of the business. Limited partners, on the other hand, consist of at least one general partner and one or more limited partners who do not participate in the management of the company but share in the profits. This form of partnership is used more often in risky investments where the limited partner stands only to lose his or her initial investment. Real estate limited partnerships are an example of how investors can minimize their financial exposure, given the poor performance of the real estate market in recent years. Although it has its advantages, partnership is the least utilized form of business. Part of the reason is that all partners are responsible for the actions and decisions of all other partners, whether or not all of the partners were involved. Usually, partners will have to write up articles of partnership that outline respective responsibilities in the business. Even in states where it is not required, it is a good idea to draw up this document as a way to cement each partner's role and hopefully minimize conflict. Unlike a corporation, proprietorships and partnerships both expire upon the death of one or more of those involved.

Corporations tend to be larger businesses, but do not need to be. A corporation can consist of nothing more than a small group of family members. In order to become a corporation, you will have to file in the state under which you wish to incorporate. Each state has its own procedure for incorporation, meaning there are no general guidelines to follow. You can make your corporation private or public, meaning the company issues stocks, and shareholders are the owners. While incorporating is a popular form of organization because it gives the company an unlimited lifespan and limited liability (meaning that if your business fails, you cannot lose personal funds to make up for losses), there is a downside. You will be taxed as a corporation and as an individual, resulting in double taxation. No matter what form of organization suits your business idea best, there is a world of options out there for you if you want to be or experiment with being an entrepreneur.

in an increasingly global marketplace. Additionally, companies that are victims of hostile takeovers generally streamline their operations, reduce unnecessary staff, cut costs, and otherwise become more efficient with their operations, which benefits their stockholders whether or not the takeover succeeds.

Critics, however, argue that mergers hurt companies because they force managers to focus their efforts on avoiding takeovers rather than managing effectively and profitably. Some companies have taken on a heavy debt burden to stave off a takeover, later to be forced into bankruptcy when economic downturns left them unable to handle the debt. Mergers and acquisitions also can damage employee morale and productivity, as well as the quality of the companies' products.

Many mergers have been beneficial for all involved; others have had damaging effects for the companies, their employees, and customers. No one can say whether mergers will continue to slow, but many experts say the utilities, telecommunications, financial services, natural resources, computer hardware and software, gaming, managed health care, and technology industries are likely targets.

Review Your Understanding

Describe the advantages and disadvantages of the sole proprietorship form of organization.

Sole proprietorships—businesses owned and managed by one person—are the most common form of organization. Their major advantages are the following: (1) They are easy and inexpensive to form, (2) they allow a high level of secrecy, (3) all profits belong to the owner, (4) the owner has complete control over the business, (5) government regulation is minimal, (6) taxes are paid only once, and (7) the business can be closed easily. The disadvantages include the following: (1) The owner may have to use personal assets to borrow money, (2) sources of external funds are difficult to find, (3) the owner must have many diverse skills, (4) the survival of the business is tied to the life of the owner and his or her ability to work, (5) qualified employees are hard to find, and (6) wealthy sole proprietors pay a higher tax than they would under the corporate form of business.

Describe the two types of business partnership and their advantages and disadvantages.

A partnership is a business formed by several individuals; a partnership may be general or limited. Partnerships offer the following advantages: (1) They are easy to organize, (2) they may have higher credit ratings because the partners possibly have more combined wealth, (3) partners can specialize, (4) partnerships can make decisions faster than larger businesses, and (5) government regulations are few. Partnerships also have several disadvantages: (1) General partners have unlimited liability for the debts of the partnership, (2) partners are responsible for each other's decisions, (3) the termination of one partner requires a new partnership agreement to be drawn up, (4) it is difficult to sell a partnership interest at a fair price, (5) the distribution of profits may not correctly reflect the amount of work done by each partner, and (6) partnerships cannot find external sources of funds as easily as can large corporations.

Describe the corporate form of organization and its advantages and disadvantages.

A corporation is a legal entity created by the state, whose assets and liabilities are separate from those of its owners. Corporations are chartered by a state through articles of incorporation. They have a board of directors made up of corporate officers or people from outside the company. Corporations, whether private or public, are owned by stockholders. Common stockholders have the right to elect the board of directors. Preferred stockholders do not have a vote but get preferential dividend treatment over common stockholders. Advantages of the corporate form of business include the following: (1) The owners have limited liability, (2) ownership (stock) can be easily transferred, (3) corporations usually last forever, (4) raising money is easier than for other forms of business, and (5) expansion into new businesses is simpler because of the ability of the company to enter into contracts. Corporations also have disadvantages: (1) The company is taxed on its income, and owners pay a second tax on any profits received as dividends; (2) forming a corporation can be expensive; (3) keeping trade secrets is difficult because so much information must be made available to the public and to government agencies; and (4) owners and managers are not always the same and can have different goals.

Assess the advantages and disadvantages of mergers, acquisitions, and leveraged buyouts.

A merger occurs when two companies (usually corporations) combine to form a new company. An acquisition occurs when one company buys most of another company's stock. In a leveraged buyout, a group of investors borrows money to acquire a company, using the assets of the purchased company to guarantee the loan. They can help merging firms to gain a larger market share in their industries, acquire valuable assets such as new products or plants and equipment, and lower their costs. Consequently, they can benefit stockholders by improving the companies' market value and stock prices. However, they also can hurt companies if they force managers to focus on avoiding takeovers at the expense of productivity and profits. They may lead a company to take on too much debt and can harm employee morale and productivity.

Propose an appropriate organizational form for a startup business.

After reading the facts in the "Solve the Dilemma" feature near the end of this chapter and considering the advantages and disadvantages of the various forms of business organization described in this chapter, you should be able to suggest an appropriate form for the startup nursery.

Critical Thinking

Enter the World of Business Questions

1. Why might a small business like J.F. Hillerich's choose to incorporate as a private company?

2. How do you think adding a partner with marketing expertise allowed the Louisville Slugger brand to become so popular?

3. What are some of the benefits of a private corporation like Hillerich & Bradsby compared to a public corporation like Wilson Sporting Goods? What are the benefits of a public corporation?

Learn the Terms

acquisition 130
articles of partnership 115
board of directors 123
common stock 124
cooperative (co-op) 129
corporate charter 121
corporation 119
dividends 119

general partnership 115
initial public offering (IPO) 122
joint venture 128
leveraged buyout (LBO) 131
limited liability company (LLC) 128
limited partnership 115
merger 130
nonprofit corporations 123

partnership 114
preferred stock 124
private corporation 121
public corporation 121
quasi-public corporations 123
S corporation 128
sole proprietorships 110
stock 119

Check Your Progress

1. Name five advantages of a sole proprietorship.

2. List two different types of partnerships and describe each.

3. Differentiate among the different types of corporations. Can you supply an example of each type?

4. Would you rather own preferred stock or common stock? Why?

5. Contrast how profits are distributed in sole proprietorships, partnerships, and corporations.

6. Which form of business organization has the least government regulation? Which has the most?

7. Compare the liability of the owners of partnerships, sole proprietorships, and corporations.

8. Why would secrecy in operating a business be important to an owner? What form of organization would be most appropriate for a business requiring great secrecy?

9. Which form of business requires the most specialization of skills? Which requires the least? Why?

10. The most common example of a cooperative is a farm co-op. Explain the reasons for this and the benefits that result for members of cooperatives.

Get Involved

1. Select a publicly owned corporation and bring to class a list of its subsidiaries. These data should be available in the firm's corporate annual report, *Standard & Poor's Corporate Records,* or *Moody Corporate Manuals.* Ask your librarian for help in finding these resources.

2. Select a publicly owned corporation and make a list of its outside directors. Information of this nature can be found in several places in your library: the company's annual report, its list of corporate directors, and various financial sources. If possible, include each director's title and the name of the company that employs him or her on a full-time basis.

Build Your Skills

Selecting a Form of Business

Background

Ali Bush sees an opportunity to start her own website development business. Ali has just graduated from the University of Mississippi with a master's degree in computer science. Although she has many job opportunities outside the Oxford area, she wishes to remain there to care for her aging parents. She already has most of the computer equipment necessary to start the business, but she needs additional software. She is considering the purchase of a server to maintain websites for small businesses. Ali feels she has the ability to take this startup firm and create a long-term career opportunity for herself and others. She knows she can hire Ole Miss students to work on a part-time basis to support her business. For now, as she starts the business, she can work out of the extra bedroom of her apartment. As the business grows, she'll hire the additional full- and/or part-time help needed and reassess the location of the business.

LO 4-5

Propose an appropriate organizational form for a startup business.

Task

Using what you've learned in this chapter, decide which form of business ownership is most appropriate for Ali. Use the tables provided to assist you in evaluating the advantages and disadvantages of each decision.

| Sole Proprietorships | |
Advantages	Disadvantages
•	•
•	•
•	•
•	•
•	•

| Corporation | |
Advantages	Disadvantages
•	•
•	•
•	•
•	•
•	

| Limited Liability Company | |
Advantages	Disadvantages
•	•
•	•
•	•
•	•
•	•

Solve the Dilemma

To Incorporate or Not to Incorporate

Thomas O'Grady and Bryan Rossisky have decided to start a small business buying flowers, shrubs, and trees wholesale and reselling them to the general public. They plan to contribute $5,000 each in startup capital and lease a 2.5-acre tract of land with a small, portable sales office.

Thomas and Bryan are trying to decide what form of organization would be appropriate. Bryan thinks they should create a corporation because they would have limited liability and the image of a large organization. Thomas thinks a partnership would be easier to start and would allow them to rely on the combination of their talents and financial resources. In addition, there might be fewer reports and regulatory controls to cope with.

Critical Thinking Questions

1. What are some of the advantages and disadvantages of Thomas and Bryan forming a corporation?

2. What are the advantages and disadvantages of their forming a partnership?

3. Which organizational form do you think would be best for Thomas and Bryan's company and why?

Build Your Business Plan

Options for Organizing Business

Your team needs to think about how you should organize yourselves that would be most efficient and effective for your business plan. The benefits of having partners include having others to share responsibilities with and to toss ideas off of each other. As your business evolves, you will have to decide whether one or two members will manage the business while the other members are silent partners. Or perhaps you will all decide on working in the business to keep costs down, at least initially. However you decide on team member involvement in the business, it is imperative to have a written agreement so that all team members understand what their responsibilities are and what will happen if the partnership dissolves.

It is not too soon for you and your partners to start thinking about how you might want to find additional funding for your business. Later on in the development of your business plan, you might want to show your business plan to family members. Together, you and your partners will want to develop a list of potential investors in your business.

Visit Connect to practice building your business plan with the Business Plan Prep Exercises.

See for Yourself Videocase

PODS Excels at Organizing a Business

What happens when homeowners need to store their belongings temporarily? Before 1998, people would choose to either rent storage space, which can be costly and inconvenient, or store their belongings in their front yards. Yet, starting in 1998, another option was introduced: PODS.

PODS, short for Portable On Demand Storage, was founded after a group of firemen noticed the difficulties that many people faced when they needed to store their belongings for a short period. PODS delivers storage containers and leaves them in front of a house or business. A specially made hydraulic lift called PODZILLA is able to place the container on ground level, which makes it easier for owners to store their belongings inside the container. PODS will then pick up the containers and move them to either its warehouses for storage or to anywhere else in the country.

When PODS was first started, banks and financial institutions were uncertain about how successful the moving and storage services company would be. Another issue was the expense of the actual containers. PODS containers, made of plywood over steel frames, cost between $2,200 and $2,500 each. If the company failed, the banks could repossess the containers. However, because PODS was a first-mover and there were no other comparable companies around, the banks feared that they would not be able to resell the containers. The risk for banks and financial institutions was high. This meant that PODS initially depended on venture capitalists for funding.

Once the company got started, however, PODS proved it was well worth the investment. More than a decade later, PODS can be found on three continents and has more than 200 million customers across the United States, Canada, the United Kingdom, and Australia. More franchises are planned for France, South Africa, Europe, Asia, Latin America, and South America. In addition to its convenience, PODS has also become known for its high-quality services and social responsibility. For instance, the company provided PODS containers to help recovery efforts in Hurricane Katrina.

As PODS has expanded, its business organization has also undergone changes. Originally, PODS was formed as a sole proprietorship. Many businesses start off as sole proprietorships because of the benefits involved, such as ease of formation and greater control over operations. PODS's initial name was PODS LLC, meaning that it was a limited liability company. Limited liability companies provide more protection to owners so that their personal belongings will not be seized to pay the company's debts. It also frees owners from some of the restrictions that exist for corporations.

However, as businesses grow nationally, they become much more difficult for one or two individuals to handle. PODS soon realized that its rapid expansion required a new form of business organization. PODS decided to become a private corporation and renamed itself PODS Enterprises Inc. Although more people became involved in the ownership of the company, a small group of individuals maintains control over much of the general corporate operations. PODS stock is not issued publicly.

Eventually, PODS also decided to adopt a franchise model. The corporation began to allow other entrepreneurs, called franchisees, to license its name and products for a fee. This provided PODS with additional funding as well as the opportunity to expand into more areas. Because PODS is already successful, franchisees have a lower failure rate than starting their own businesses from scratch. Franchisees also understand their particular markets better than a corporation can. "Franchisees bring another advantage, though, and that is their knowledge and connections in the local market, so they can take advantage of particularities in a market," said the senior vice president of Franchise Operations. This increases PODS's adaptability when it expands into other areas.

On the other hand, because corporations are able to bring together several knowledgeable individuals, PODS corporate headquarters finds that it is better able to handle larger markets such as Los Angeles and Chicago. With this business model, PODS has figured out how to meet the needs of both local markets, through franchisees, and larger markets. Because of PODS's ability to understand the best ways of organizing its business, the company has been able to reap the benefits from all types of market sizes.[36]

Critical Thinking Questions

1. What are some advantages of sole proprietorships for PODS? What are some disadvantages?

2. What are some advantages of private corporations for PODS? What are some disadvantages?

3. How has adopting a franchise model made PODS more adaptable?

You can find the related video in the Video Library in Connect. Ask your instructor how you can access Connect.

Team Exercise

Form groups and find examples of mergers and acquisitions. Mergers can be broken down into traditional mergers, horizontal mergers, and conglomerate mergers. When companies are found, note how long the merger or acquisition took, if there were any requirements by the government before approval of the merger or acquisition, and if any failed mergers or acquisitions were found that did not achieve government approval. Report your findings to the class, and explain what the companies hoped to gain from the merger or acquisition.

Ask your instructor about the role-play exercises available with this book to practice working with a business team.

Notes

1. Louisville Slugger Museum and Factory, "About Us—FAQs," 2017, https://www.sluggermuseum.com/about-us/faqs (accessed December 26, 2017); Louisville Slugger, "Our History," 2017, http://www.slugger.com/en-us/our-history (accessed December 26, 2017); "Baseball Bat," How Products Are Made website, http://www.madehow.com/Volume-2/Baseball-Bat.html (accessed December 26, 2017); David Mielach, "Louisville Slugger's Unlikely Home Run," *Business News Daily,* July 8, 2012, https://www.businessnewsdaily.com/2806-louisville-slugger-story.html (accessed December 26, 2017); Darren Rovell, "Wilson Buys Louisville Slugger," *ESPN,* March 23, 2015, http://www.espn.com/mlb/story/_/id/12544727/wilson-parent-company-buys-louisville-slugger-70-million (accessed March 17, 2018); Sam Sanders, "Wilson Sporting Goods Acquires Louisville Slugger Brand," *NPR,* March 23, 2015, https://www.npr.org/sections/thetwo-way/2015/03/23/394887115/wilson-sporting-goods-acquires-louisville-slugger-brand (accessed March 17, 2018); Funding Universe, "Hillerich & Bradsby Company, Inc. History," http://www.fundinguniverse.com/company-histories/hillerich-bradsby-company-inc-history/ (accessed March 17, 2018); Company-Histories.com, "Hillerich & Bradsby Company, Inc.," http://www.company-histories.com/Hillerich-Bradsby-Company-Inc-Company-History.html (accessed March 17, 2018).

2. Kara Stiles, "Veteran-Turned-Entrepreneur Navigates Young and Evolving Electric Bike Industry," *Forbes,* October 4, 2017, https://www.forbes.com/sites/karastiles/2017/10/04/veteran-turned-entrepreneur-navigates-a-young-and-evolving-electric-bike-industry/#3ef10fbf2e24 (accessed April 11, 2018).

3. "Retail," *The Economist,* The World in 2018, p. 121.

4. The Entrepreneurs Help Page, www.tannedfeet.com/sole_proprietorship.htm (accessed May 4, 2017); Kent Hoover, "Startups Down for Women Entrepreneurs, Up for Men," *San Francisco Business Times,* May 2, 2008, http://sanfrancisco.bizjournals.com/sanfrancisco/stories/2008/05/05/smallb2.html (accessed May 4, 2017).

5. Maggie Overfelt, "Start-Me-Up: How the Garage Became a Legendary Place to Rev Up Ideas," *Fortune Small Business,* September 1, 2003, http://money.cnn.com/magazines/fsb/fsb_archive/2003/09/01/350784/index.htm (accessed May 4, 2017).

6. Network Solutions LLC, "Hosting Options," 2015, http://www.networksolutions.com/web-hosting/index-v5.jsp (accessed May 4, 2017); Network Solutions LLC, Network Solutions website, http://www.networksolutions.com/web-hosting/index-v5.jsp (accessed February 23, 2015).

7. Chris Beier, "How These Famous Siblings Worked Out Their Co-Founding Differences," *Inc.,* October 25, 2017,

https://www.inc.com/video/rebecca-minkoff/inside-the-worlds-smartest-dressing-room.html (accessed April 11, 2018).

8. "Master Limited Partnership—MLP," *Investopedia,* http://www.investopedia.com/terms/m/mlp.asp (accessed May 4, 2017).

9. "100 Largest Law Firms in the World," *Ranker,* 2016, http://www.ranker.com/list/100-largest-law-firms-in-the-world/business-and-company-info (accessed March 1, 2016); "Morgan, Lewis & Bockius LLP," Chambers & Partners, 2016, http://www.chambersandpartners.com/usa/firm/64563/morgan-lewis-bockius (accessed March 1, 2016).

10. Bill Orben, "uBreakiFix Launches Franchise Unit, Sets Sights on 125 Stores This Year," *Orlando Business Journal,* February 20, 2013, http://www.bizjournals.com/orlando/blog/2013/02/ubreakifix-launches-franchise-unit.html (accessed May 4, 2017); "About uBreakiFix," uBreakiFix Careers website, http://careers.ubreakifix.com/about-ubreakifix/#.VOuVr3nTBVI (accessed May 4, 2017); Matthew Richardson, "Orlando's uBreakiFix Gets $20M Biz Loan, Plans Expansion in U.S., Canada," *Orlando Business Journal,* January 12, 2018, https://www.bizjournals.com/orlando/news/2018/01/12/orlandos-ubreakifix-gets-20m-biz-loan-plans-to.html (accessed April 8, 2018); "uBreakiFix Expands in Los Angeles Area, Opens in Culver City," *Global Newswire,* April 2, 2018, https://globenewswire.com/news-release/2018/04/02/1458616/0/en/uBreakiFix-Expands-in-Los-Angeles-Area-Opens-in-Culver-City.html (accessed April 8, 2018).

11. "EY Reports Record Global Revenues in 2016—Up by 9%," Ernst & Young, October 6, 2016, http://www.ey.com/GL/en/Newsroom/News-releases/News-EY-promotes-753-new-partners-worldwide (accessed April 7, 2017).

12. Sarah Griffiths, "Snapchat Settles Lengthy Lawsuit with Former University Classmate—and Admit the App WAS His Idea," *Daily Mail,* September 10, 2014, http://www.dailymail.co.uk/sciencetech/article-2750807/Snapchat-founders-settle-lengthy-lawsuit-former-classmate-admit-app-WAS-idea.html (accessed March 1, 2016).

13. Kerry Hannon, "How an Unemployed Construction Worker Became a Craft Beer Entrepreneur," *Time,* June 12, 2015, http://time.com/money/3919129/beer-houndbrewery-kenny-thacker/ (accessed March 17, 2018); Alicia Adamczyk, "Advice from 7 Baby Boomers Who Reinvented Their Careers," *Time,* August 13, 2015, http://time.com/money/3992427/second-act-career-baby-boomers/ (accessed March 17, 2018); Allison Brophy Champion, "Culpeper-Crafted Olde Yella Beer Wins Gold," *Culpeper Star Exponent,* August 24, 2015, http://www.dailyprogress.com/starexponent/culpeper-crafted-olde-yella-beer-winsgold/article_27097080-4a8f-11c5-9c64-53ca5f4e6b78.html (accessed November 5, 2015); Kenny Thacker, "Beer Hound Brewery-Grand Opening," *CvilleCalendar.com,* October 9, 2012, http://www.cvillecalendar.com/calendar/Beer_Hound_BreweryGrand_Openi_10132012 (accessed November 5, 2015); Wally Bunker, "Beer Hound Brewery

Plan to Open...Again," *Culpepper Times,* November 3, 2014, http://culpepertimes.com/2014/11/03/beer-hound-brewery-plans-to-open-again/153412/ (accessed March 17, 2018).

14. "America's Largest Private Companies," *Forbes,* https://www.forbes.com/largest-private-companies/list/#tab:rank (accessed April 8, 2018).

15. "In Retreat," *The Economist,* January 28, 2017, p. 11.

16. Vindu Goel and Nick Wingfield, "Mark Zuckerberg Vows to Donate 99% of His Facebook Shares for Charity," *The New York Times,* December 1, 2015, http://www.nytimes.com/2015/12/02/technology/mark-zuckerberg-facebook-charity.html (accessed May 4, 2017); John Shinal, "Mark Zuckerberg Has Sold $357 Million of Stock This Month, Part of Plans to Unload up to $12 Billion," *CNBC,* February 26, 2018, https://www.cnbc.com/2018/02/26/mark-zuckerberg-has-sold-357-million-in-facebook-shares-in-february.html (accessed April 8 ,2018); U.S. Securities and Exchange Commission, Schedule 13G, Facebook, Inc., December 31, 2017, https://www.sec.gov/Archives/edgar/data/1326801/000119312518038194/0001193125-18-038194-index.htm (accessed April 8, 2018).

17. Leslie Picker and Lulu Yilun Chen, "Alibaba's Banks Boost IPO Size to Record of $25 Billion," *Bloomberg,* September 22, 2014, http://www.bloomberg.com/news/articles/2014-09-22/alibaba-s-banks-said-to-increase-ipo-size-to-record-25-billion (accessed May 4, 2017).

18. Brendan Marasco, "3 Reasons Dell Went Private," *The Motley Fool,* November 1, 2013, www.fool.com/investing/general/2013/11/01/3-reasons-dell-went-private.aspx (accessed May 4, 2017); Katherine Noyes, "As a Private Company, Dell-EMC Will Enjoy a Freedom HP Can Only Dream of," *CIO,* October 12, 2015, http://www.cio.com/article/2991551/as-a-private-company-dell-emc-will-enjoy-a-freedom-hp-can-only-dream-of.html (accessed May 4, 2017).

19. "The 100 Largest U.S. Charities," *Forbes,* 2017, https://www.forbes.com/top-charities/list/ (accessed April 11, 2018).

20. "Habitat for Humanity ReStore," Habitat for Humanity, https://www.habitat.org/restores (accessed April 11, 2018).

21. "Private Operating Foundations," Elko & Associates Limited, February 23, 2011, http://blog.elkocpa.com/nonprofit-tax-exempt/private-operating-foundations (accessed February 23, 2015); Foundation Group, "What Is a 501(c)(3)?" http://www.501c3.org/what-is-a-501c3/ (accessed May 4, 2017).

22. O. C. Ferrell, John Fraedrich, and Linda Ferrell, *Business Ethics: Ethical Decision Making and Cases,* 8th ed. (Mason, OH: South-Western Cengage Learning, 2011), p. 109.

23. Chris Morris, "Pay Raised for Corporate Board Members Far Outpace Average Americans,'" *Fortune,* October 18, 2017, http://fortune.com/2017/10/18/board-of-directors-pay/ (accessed April 8, 2018); Theo Francis and Joann S. Lublin, "Corporate Directors' Pay Ratchets Higher as Risks Grow," *The Wall Street Journal,* February 24, 2015, http://www.wsj.com/articles/corporate-directors-pay-ratchets-higher-as-risks-grow-1456279452 (accessed May 4, 2017).

24. "Board Composition," Procter & Gamble, https://us.pg.com/who-we-are/structure-governance/corporate-governance/board-composition (accessed April 8, 2018).

25. "High Stakes, High Expectations as Earnings Season Heats Up," *Yahoo! Finance,* April 8, 2018, https://finance.yahoo.com/news/high-stakes-high-expectations-earnings-110444292.html (accessed April 8, 2018); Diana Furchtgott-Roth, "Make Domestic Growth Great Again," *U.S. News,* December 20, 2017, https://www.usnews.com/opinion/economic-intelligence/articles/2016-12-20/donald-trump-and-republicans-are-right-to-lower-the-corporate-tax-rate (accessed April 8, 2018).

26. Kane, Joseph Nathan, *Famous First Facts,* 4th ed. (New York: H.W. Wilson Company, 1981), p. 202.

27. Adam Bluestein, "Regulate Me. Please." *Inc.,* May 2011, pp. 72–80; Cascade Engineering website, www.cascadeng.com (accessed December 23, 2017); B Corporation website,www.bcorporation.net (accessed December 23, 2017); "The Nonprofit Behind B Corps," http://www.bcorporation.net/what-are-b-corps/the-non-profit-behind-b-corps (accessed December 23, 2017); Make It Right, "Cradle to Cradle®," 2017, http://makeitright.org/c2c/ (accessed December 23, 2017); Lynn Golodner, "Welfare to Career: Plastics Company Helps People Break Barriers to Success," *Corp Magazine,* December 23, 2015, http://www.corpmagazine.com/welfare-career-plastics-company-helps-people-break-barriers-success/ (accessed December 23, 2017); B Lab, "Cascade Engineering," B Corporation website, https://www.bcorporation.net/community/cascade-engineering (accessed December 23, 2017).

28. Shi Jing, "Gaming Giants in the Winning Track," *China Daily,* June 21, 2017, http://www.chinadaily.com.cn/business/2017-06/21/content_29825940.htm (accessed April 8, 2018).

29. Robert D. Hisrich and Michael P. Peters, *Entrepreneurship,* 5th ed. (Boston: McGraw-Hill, 2002), pp. 315–16.

30. "The History and Challenges of America's Dominant Business Structure," S-CORP, http://s-corp.org/our-history/ (accessed April 8, 2018).

31. Conversation with Marshall Powell Ledbetter III; VIP Moving & Storage website, http://www.vipmovingandstorage.com/ (accessed July 14, 2015); "VIP Moving & Storage Inc.," Better Business Bureau, https://www.bbb.org/losangelessiliconvalley/business-reviews/moving-and-storage-company/vip-moving-and-storage-in-pacoima-ca-1023815 (accessed January 28, 2018).

32. "Company Overview of Mrs. Fields Famous Brands, LLC," *Bloomberg Business,* February 23, 2015, http://www.bloomberg.com/research/stocks/private/snapshot.asp?privcapId=3553769 (accessed February 23, 2015).

33. "Coop Directory Service Listing," Coop Directory, www.coopdirectory.org/directory.htm#Massachusetts (accessed March 2, 2016).

34. Aaron Tilley, "Google Acquires Smart Thermostat Maker Nest for $3.2 Billion," *Forbes,* January 13, 2014, www.forbes.com/sites/aarontilley/2014/01/13/google-acquires-nest-for-3-2-billion/ (accessed May 4, 2017).

35. Eric Savitz, "Did Google Buy a Lemon? Motorola Mobility Whiffs Q4," *Forbes,* January 8, 2012, www.forbes.com/sites/ericsavitz/2012/01/08/did-google-buy-a-lemon-motorola-mobility-whiffs-q4/ (accessed May 4, 2017); Richard Nieva, "Alphabet? Google? Either Way, It's Ready to Rumble," *CNET,* January 29, 2016, http://www.cnet.com/news/larry-page-sergey-brin-google-alphabet/ (accessed May 4, 2017).

36. "PODS LLC: Company Profile," *Bloomberg,* http://www.bloomberg.com/profiles/companies/751647Z:US-pods-llc (accessed March 29, 2016); "PODS Taking a Chunk Out of Moving and Storage Market," *USA Today,* August 4, 2006, www.usatoday.com/money/smallbusiness/2006-08-04-pods_x.htm (accessed March 29, 2016); Consortium Media Services, "PODS Recognized as One of the Nation's Best Service Providers by Professional Organizing Association," *Yahoo! Voices,* February 28, 2011, http://voices.yahoo.com/pods-recognized-as-one-nationsbest-service-7701664.html (accessed July 26, 2012); Soti, "Portable On Demand Storage (PODS) Case Study," http://www.soti.net/PDF/PODS-SOTI-Casestudy.pdf (accessed March 29, 2016); World Franchise Associates, "PODS Multi-unit Franchise," 2015, http://www.worldfranchisecentre.com/p-detail.php?bid=91 (accessed March 29, 2016).

Credits

5 Small Business, Entrepreneurship, and Franchising

©Rob Wilson/Shutterstock

Chapter Outline

Learning Objectives

After reading this chapter, you will be able to:

LO 5-1 Define *entrepreneurship* and *small business*.

LO 5-2 Explain the importance of small business in the U.S. economy and why certain fields attract small business.

LO 5-3 Specify the advantages of small-business ownership.

LO 5-4 Analyze the disadvantages of small-business ownership and the reasons why many small businesses fail.

LO 5-5 Describe how to start a small business and what resources are needed.

LO 5-6 Evaluate the demographic, technological, and economic trends that are affecting the future of small business.

LO 5-7 Explain why many large businesses are trying to "think small."

LO 5-8 Assess two entrepreneurs' plans for starting a small business.

Enter the World of Business

Chick-fil-A Earns Grade A for Quality

From the day it was founded, Chick-fil-A has served communities differently than other fast-food franchises. The first Chick-fil-A was established in 1967 after the founder, Truett Cathy, had started in the restaurant business in 1946. The firm remains a nationwide family-owned restaurant. Chick-fil-A is not about just providing chicken but wants to be part of the community.

Chick-fil-A takes a different approach to the prep work method of their food. Its motto is "Cook less, more often." Quality is emphasized over quantity. This allows its chicken sandwiches to stand out from the competition. The restaurant's name has a hidden meaning to show they are the best in quality. Its name is a take on "chicken fillet" with a capital "A" at the end. The "A" stands for top quality.

Because Chick-fil-A has more than 2,000 restaurants run by different franchised operators, it makes certain that operators are aware of the chain's high standards of quality. Opening a franchise requires a $10,000 initial fee, and the individual must not be involved in other business ventures. Chick-fil-A wants to make certain the operator can run the franchise hands-on and full-time. Training for new operators lasts multiple weeks to equip them with the expertise to make the day-to-day decisions of running a successful business.

Chick-fil-A's emphasis on quality and training has enabled its franchise operators to work independently as business owners while continuing to advance the company's values, goals, and reputation for quality. This successful operation of its franchise has contributed to Chick-fil-A's current status as the eighth-largest fast food chain in the nation. Despite its success, it's unlikely anyone will see the company traded on the New York Stock Exchange. Before Cathy passed away, he created a contract that allows his children to sell the privately owned corporation but not go public.[1]

Introduction

Although many business students go to work for large corporations upon graduation, others may choose to start their own business or find employment opportunities in small organizations. Small businesses employ about half of all private-sector employees.[2] Each small business represents the vision of its owners to succeed through providing new or better products. Small businesses are the heart of the U.S. economic and social system because they offer opportunities and demonstrate the freedom of people to make their own destinies. Today, the entrepreneurial spirit is growing around the world, from Russia and China to India, Germany, Brazil, and Mexico. Countries with the healthiest "entrepreneurship ecosystems" include the United States, Canada, Australia, Denmark, and Sweden.[3] This chapter surveys the world of entrepreneurship and small business. First, we define entrepreneurship and small business and examine the role of small business in the American economy. Then, we explore the advantages and disadvantages of small-business ownership and analyze why small businesses succeed or fail. Next, we discuss how an entrepreneur goes about starting a business and the challenges facing small businesses today. Finally, we look at entrepreneurship in larger organizations.

The Nature of Entrepreneurship and Small Business

In the chapter "The Dynamics of Business and Economics," we defined an *entrepreneur* as a person who risks his or her wealth, time, and effort to develop for profit an innovative product or way of doing something. **Entrepreneurship** is the process of creating and managing a business to achieve desired objectives. Many large businesses you may recognize (Levi Strauss and Co., Procter & Gamble, McDonald's, Dell Computers, Microsoft, and Google) all began as small businesses based on the visions of their founders. Some entrepreneurs who start small businesses have the ability to see emerging trends; in response, they create a company to provide a product that serves customer needs. For example, rather than inventing a major new technology, an innovative company may take advantage of technology to create new markets, such as Amazon. Or it may offer a familiar product that has been improved or placed in a unique retail environment, such as Starbucks and its coffee shops. A company may innovate by focusing on a particular market segment and delivering a combination of features that consumers in that segment could not find anywhere else. The sharing economy, or gig economy, can use technology to connect service providers or homeowners. Porch.com was founded as a way to connect homeowners with contractors. Founder Matt Ehrlichman conceived of the idea after becoming frustrated with problems in building his own house. Porch.com's software provides a more transparent way to link homeowners with licensed professionals. Approximately $1.5 trillion worth of home remodeling projects have been featured through the site. The software is so effective that home improvement store Lowe's installed it in 1,700 retail locations.[4]

Success requires creativity, innovation, and entrepreneurship, and it requires more than a formal education. It requires the ability to learn and develop skills and knowledge and be an independent thinker. Consider that Steve Jobs (Apple), Richard Branson (Virgin Group), Larry Allison (former CEO of Oracle), and Michael Dell (Dell Computers) did not graduate from college. Mark Zuckerberg dropped out of Harvard after his sophomore year to focus on Facebook. Bill Gates, co-founder of Microsoft, also dropped out of Harvard after his sophomore year. Of course, smaller businesses do not have to evolve into such highly visible companies to be successful,

Company	Entrepreneur
Hewlett-Packard	Bill Hewlett, David Packard
Walt Disney Productions	Walt Disney
Starbucks	Howard Schultz
Amazon	Jeff Bezos
Dell	Michael Dell
Microsoft	Bill Gates
Apple	Steve Jobs
Walmart	Sam Walton
Google	Larry Page, Sergey Brin
Ben & Jerry's	Ben Cohen, Jerry Greenfield
Ford	Henry Ford
General Electric	Thomas Edison

TABLE 5.1
Great Entrepreneurs of Innovative Companies

but those entrepreneurial efforts that result in rapidly growing businesses gain visibility along with success. Table 5.1 lists some of the greatest entrepreneurs of the past century.

The entrepreneurship movement is accelerating, and many new, smaller businesses are emerging. Many entrepreneurs with five or fewer employees are considered **microentrepreneurs.** Technology once available only to the largest firms can now be obtained by a small business. Websites, podcasts, online videos, social media, cellular phones, and even expedited delivery services enable small businesses to be more competitive with today's giant corporations. Small businesses can also form alliances with other companies to produce and sell products in domestic and global markets.

Another growing trend among small businesses is social entrepreneurship. **Social entrepreneurs** are individuals who use entrepreneurship to address social problems. They operate by the same principles as other entrepreneurs but view their organizations as vehicles to create social change. Although these entrepreneurs often start their own nonprofit organizations, they can also operate for-profit organizations committed to solving social issues. Blake Mycoskie, the founder of Toms Shoes, is an example of a social entrepreneur. He founded the firm with the purpose of donating one pair of shoes to a child in need for every pair of shoes sold to consumers. The firm has since expanded this one-for-one model with other products and has developed a fund to invest in other social entrepreneurial startups.[5] Shamayim Harris is restoring rundown homes in Michigan's Highland Park, an impoverished city of 10,000 in the Detroit metro area. Through seed grants and crowdfunding, she is helping to develop a self-sustaining community.[6]

What Is a Small Business?

This question is difficult to answer because smallness is relative. In this book, we will define a **small business** as any independently owned and operated business that is not dominant in its competitive area and does not employ more than 500 people.

microentrepreneur entrepreneurs who develop businesses with five or fewer employees.

social entrepreneurs individuals who use entrepreneurship to address social problems.

small business any independently owned and operated business that is not dominant in its competitive area and does not employ more than 500 people.

Small Business Administration (SBA) an independent agency of the federal government that offers managerial and financial assistance to small businesses.

Microentrepreneurs, sometimes called micropreneurs, that employ five or fewer employees are growing rapidly. A local Mexican restaurant may be the most patronized Mexican restaurant in your community, but because it does not dominate the restaurant industry as a whole, the restaurant can be considered a small business. This definition is similar to the one used by the **Small Business Administration (SBA)**, an independent agency of the federal government that offers managerial and financial assistance to small businesses. On its website, the SBA outlines the first steps in starting a small business and offers a wealth of information to current and potential small-business owners.

The Role of Small Business in the American Economy

No matter how you define a small business, one fact is clear: They are vital to the American economy. As you can see in Table 5.2, more than 99 percent of all U.S. firms are classified as small businesses, and they employ about half of private workers. Small firms are also important as exporters, representing 98 percent of U.S. exporters of goods and contributing 33 percent of the value of exported goods.[7] In addition, small businesses are largely responsible for fueling job creation and innovation. Small businesses also provide opportunities for minorities and women to succeed in business. Women own more than 9 million businesses nationwide, with great success in the professional services, retail, communication, and administrative services areas.[8] Minority-owned businesses have been growing faster than other classifiable firms as well, representing 28.6 percent of all small businesses.[9] For example, Mexican-born José de Jesus Legaspi went into the real estate business and focused his market niche on inner-city areas with a high percentage of Hispanic consumers. When Legaspi decided to begin investing in struggling malls, he refashioned the malls he acquired as cultural centers appealing to Hispanic consumers of all generations. One of his malls, renamed La Gran Plaza, went from being 20 percent occupied to 80 percent.[10]

Job Creation. The energy, creativity, and innovative abilities of small-business owners have resulted in jobs for many people. About 63 percent of net new jobs annually are created by small businesses.[11] Table 5.3 indicates that 82.2 percent of all businesses employ fewer than 500 people. Businesses employing 19 or fewer people account for 66.8 percent of all businesses.[12]

Many small businesses today are being started because of encouragement from larger ones. Many new jobs are also created by big-company/small-company alliances. Whether through formal joint ventures, supplier relationships, or product or

TABLE 5.2
Importance of Small Businesses to Our Economy

Small firms represent 99.7 percent of all employer firms.
Small firms have generated 62 percent of net new jobs.
Small firms hire approximately 37 percent of high-tech workers (such as scientists, engineers, computer programmers, and others).
Small firms produce 16 times more patents per employee than large patenting firms.
Small firms employ nearly half of all private-sector employees.
Small firms pay 42 percent of the total U.S. private payroll.

Source: "How Many Jobs Do U.S. Small Businesses Create? 2017," Small Business, December 1, 2017, https://smallbusiness.com/about-small-businesses/how-many-jobs-small-business-create/ (accessed April 21, 2018); Small Business Administration Department of Advocacy, "Frequently Asked Questions," March 2014, www.sba.gov/sites/default/files/FAQ_March_2014_0.pdf (accessed April 21, 2018).

Firm Size	Number of Firms	Percentage of All Firms
0–19 employees	4,524,688	66.6
20–99 employees	689,685	10.1
100–499 employees	358,217	5.3
500+ employees	1,222,611	18.0

TABLE 5.3
Number of Firms by
Employment Size

Source: "SUSB Employment Change Data Tables," Statistics of U.S. Businesses, https://www.census.gov/data/tables/2015/econ/susb/2015-susb-employment.html (accessed April 21, 2018).

marketing cooperative projects, the rewards of collaborative relationships are creating numerous jobs for small-business owners and their employees. In India, for example, many small information technology (IT) firms provide IT services to global markets. Because of lower costs, international companies can often find Indian businesses to provide their information-processing solutions.[13]

Innovation.   Perhaps one of the most significant strengths of small businesses is their ability to innovate, bringing significant benefits to customers. Consider Stitch Fix, which offers fashion clothing customized for each individual. It delivers a personalized shopping experience for each customer—the small firm claims two people getting the same "fix" or clothing is nearly zero, but for those who shop big retailers it is a possibility.[14] Similarly, Andrew Blackmon co-founded the online tuxedo rental service The Black Tux. There was such extreme interest in the service after a single mention in *GQ Magazine* that it quickly ran into supply chain issues. The company was able to decrease turnaround times with manufacturers and improve its supply chain. Both of these businesses provide an important service that many people want.

Small firms produce more than half of all innovations. Among the important 20th-century innovations by U.S. small firms are the airplane, the audio tape recorder, fiber-optic examining equipment, the heart valve, the optical scanner, the pacemaker, the personal computer, soft contact lenses, the Internet, and the zipper. Artificial intelligence will create new opportunities for innovation. The ability of computers to perform tasks that normally require human intelligence will provide the ability to create new businesses. The simulation of human behavior will permit facial recognition, speech recognition, decision making, and the control of robots to do tasks associated with people. Small businesses will have improved ability to compete with large businesses because the technology is becoming available and affordable. Customer service, marketing, human resources, supply chains, and manufacturing will all be more efficient to develop and manage. Artificial intelligence will provide the opportunity for new business models for startups. Companies in the early stage of development will have the ability to be highly independent and maximize all the elements that create value. The barrier to enter and advance a new business will be easier.

Katrina Lake, founder and CEO of Stitch Fix, developed her personal shopping subscription service to provide women personal styling service at an affordable price point.

©Griffin Lipson/BFA/REX/Shutterstock

Shark Tank Takes a Bite of Success

The television show *Shark Tank* features entrepreneurs who pitch their business ideas to a panel of investors. These investors include real estate mogul Barbara Corcoran, founder of FUBU clothing line Daymond John, and owner of the National Basketball Association's Dallas Mavericks Mark Cuban. The investors decide whether to invest their own money with these entrepreneurs, based upon whether they believe their ideas hold promise. Pitches range from ideas not yet realized to those that have already achieved sales and need more money for growth.

Some believe this show is having a positive impact on entrepreneurial-minded individuals. Clear articulation of the business opportunity, understanding of the business market, knowledge of cost and profitability metrics, and idea protection are lessons that many viewers have learned from watching the show. These lessons are important for potential entrepreneurs when they present their ideas to venture capitalists.

The "sharks" have invested more than $66 million into business ventures that were pitched on the show. They offer advice

on steps entrepreneurs should take to make their business idea more viable:

- Pursue profitability. Do what works.
- Take ownership for your success.
- Keep persisting.
- Refuse to be intimidated.
- Work hard to be successful.

Shark Tank has given people a better and more optimistic view of the nature of American business. This reality show offers insight into regular people trying to create something they are passionate about.[15]

Critical Thinking Questions

1. What type of financing are entrepreneurs seeking when they are guests on *Shark Tank?*
2. Why do you think the advice of the "sharks" is important for entrepreneurial success, particularly given the high failure rate of businesses?
3. What are some alternative ways entrepreneurs featured on the show can get help or financing if the sharks choose not to invest?

The innovation of successful firms takes many forms. For instance, franchises make up approximately 2 percent of all small businesses. Many of today's largest businesses started off as small firms that used innovation to achieve success.[16] Small businessman Ray Kroc found a new way to sell hamburgers and turned his ideas into one of the most successful fast-food franchises in the world—McDonald's. David Galboa co-founded the successful company Warby Parker, an online retailer that sells stylish glasses at lower prices. Similar to Toms Shoes, Warby Parker is a social enterprise that gives a pair of glasses to seeing-impaired individuals in developing countries for every pair of glasses sold. The company has sold more than 1 million pairs of glasses.[17] Entrepreneurs provide fresh ideas and usually have greater flexibility to change than do large companies.

LO 5-2

Explain the importance of small business in the U.S. economy and why certain fields attract small business.

Industries That Attract Small Business

Small businesses are found in nearly every industry, but retailing and wholesaling, services, manufacturing, and high technology are especially attractive to entrepreneurs. These fields are relatively easy to enter and require low initial financing. Small-business owners in these industries also find it easier to focus on specific groups of consumers; new firms in these industries initially suffer less from heavy competition than do established firms.

Retailing.   Retailers acquire goods from producers or wholesalers and sell them to consumers. Main streets, shopping centers, and malls are generally lined with independent music stores, sporting-goods stores, dry cleaners, boutiques, drugstores, restaurants, caterers, service stations, and hardware stores that sell directly to consumers. Retailing attracts entrepreneurs because gaining experience and exposure in retailing is relatively easy. Additionally, an entrepreneur opening a new retail store or establishing a new website does not have to spend the large sums of money for the equipment and distribution systems that a manufacturing business requires. All that a new retailer needs is the ability to understand the market and provide a product that satisfies a need. However, it is important for entrepreneurs to anticipate the costs of opening a retail or wholesale business beforehand. Plenty Grocery & Deli in Chicago, for instance, raised $15,000 on Kickstarter, a crowdfunding site, to open its small neighborhood grocery store.[18]

The retailing industry is particularly attractive to entrepreneurs.

©Lynne Neuman/Shutterstock

Many opportunities exist for nonstore retailing as well. Nonstore retailing involves selling products outside of a retail facility. There are two types of nonstore retailing: direct marketing—which uses the telephone, catalogs, and other media to give consumers an opportunity to place orders by mail, telephone, or the Internet—and direct selling. Nonstore retailing is an area that provides great opportunity for entrepreneurs because of a lower cost of entry. JCPenney also found that it significantly affects sales. The organization decided to engage in more direct marketing by resurrecting its catalog—which it had discontinued in 2010—based on market research findings suggesting that catalog users are more inspired to purchase items online.[19] Smaller businesses can engage in a form of direct marketing by featuring their products on eBay, Amazon, or Etsy.

Direct selling involves the marketing of products to ultimate consumers through face-to-face sales presentations at home, in the workplace, and in party environments. Well-known direct selling companies include Amway, Avon, Herbalife, and Mary Kay. The cost of getting involved in direct selling is low and often involves buying enough inventory to get started. Many people view direct selling as a part-time business opportunity. Often, those who become independent contractors for direct selling companies are enthusiastic about the product and have the opportunity to recruit other distributors and receive commissions on their sales.

Wholesaling.   Wholesalers provide both goods and services to producers and retailers. They can assist their customers with almost every business function. Wholesalers supply products to industrial, retail, and institutional users for resale or for use in making other products. Wholesaling activities range from planning and negotiating for supplies, promoting, and distributing (warehousing and transporting) to providing management and merchandising assistance to clients. Wholesalers are extremely important for many products, especially consumer goods, because of the marketing activities they perform. Although it is true that wholesalers themselves can be eliminated, their functions must be passed on to some other organization such as the producer or another intermediary, often a small business. Frequently, small businesses are closer to the final customers and know what it takes to keep them satisfied. Some smaller businesses find their real niche as a service provider or distributor of other firms' products. John Hinnen, a tinkerer, spent his early life in his parent's garage designing toys and novelties. With a lot of failure, he moved on, but, later in life, after watching the movie *Elf* and seeing a snowball fight sequence, he took a plastic bat and created the snow slugger that could make snowballs and be used to defend against snowball attacks. The toy company Wham-O picked the product up to manufacture for Hinnen.[20]

Services. The service sector includes businesses that do not actually produce tangible goods. Services include intangible products that involve a performance, inauguration, or any effort to provide something of value that cannot be physically possessed. Services can also be part of the wholesale market and involve any product that is intangible and therefore cannot be touched. The service sector accounts for 80 percent of U.S. jobs, excluding farmworkers. Real estate, insurance and personnel agencies, barbershops, banks, television and computer repair shops, copy centers, dry cleaners, and accounting firms are all service businesses. Services also attract individuals—such as beauticians, morticians, jewelers, doctors, and veterinarians—whose skills are not usually required by large firms. Many of these service providers are retailers who provide their services to ultimate consumers. José Neves founded his 10-year-old company to assist retail luxury boutiques in putting their inventory online. The services provide an opportunity for stores to track their fashion brands from payment and shipping, to customer service and in-store returns.[21]

Manufacturing. Manufacturing goods can provide unique opportunities for small businesses. For example, Irene Rhodes founded Consumer Fire Products by using her background in engineering to create a system that sprays a biodegradable protective foam on a house when a wildfire is nearby. She has revenues of $2 million a year.[22] Small businesses sometimes have an advantage over large firms because they can customize products to meet specific customer needs and wants. Such products include custom artwork, jewelry, clothing, and furniture.

Technology. *High technology* is a broad term used to describe businesses that depend heavily on advanced scientific and engineering knowledge. People who were able to innovate or identify new markets in the fields of computers, biotechnology, genetic engineering, robotics, and other markets have become today's high-tech giants. One innovative technology was developed by a teenager interested in virtual reality. Only a few years ago, virtual reality was considered a dead technology past its prime. However, when 19-year-old Palmer Luckey developed a virtual gaming headset, it caught the attention of programmer John Carmack. Together, they brought virtual reality to a new level for gamers. The company, Oculus Rift, was sold to Facebook for $2 billion.[23] In general, high-technology businesses require greater capital and have higher initial startup costs than do other small businesses. Many of the biggest, nonetheless, started out in garages, basements, kitchens, and dorm rooms.

sharing economy
an economic model involving the sharing of underutilized resources.

Sharing Economy. The past few years have seen a rise in the **sharing economy,** an economic model involving the sharing of underutilized resources. Under this model, entrepreneurs earn income by renting out an underutilized resource such as lodging or vehicles.[24] The ride-sharing service Uber is the company most associated with the sharing economy. Rather than employing people outright, Uber acts more as a "labor broker," providing a mobile app that connects buyers (passengers) with sellers (drivers).[25] Although Uber does maintain control over variables such as rates, drivers act as independent contractors taking on jobs whenever or wherever they desire. Airbnb is another well-known company operating in the sharing economy. Its website connects those in need of lodging with sellers of those services.

The sharing economy is often referred to as the "gig economy" because independent contractors earn income going from job to job.[26] The sharing economy offers opportunities for those who want to be their own entrepreneurs or want additional income even though they have another job. Elise Benun, fired from her second job out of college, started Marketing Mentor to help creative professionals find clients with large budgets. Elise is a gig worker operating her own business. Currently 34 percent of the U.S. workforce consists of independent gig workers that are self-employed.[27] Services offered through this model often cost less than more traditional services. It is even taking market share away from established firms. For instance, Airbnb is now twice as valuable as Hilton. The company is buying related companies to move into a wider tourism market.[28]

Despite the opportunities in the sharing economy, Uber and similar firms have been experiencing pressure over whether workers are independent contractors or employees. In addition, Uber and Airbnb often are in conflict with local regulatory rules about lodging or ride sharing. As independent contractors, workers act as their own bosses and pay their own taxes and benefits, unlike employees of a firm. In spite of the controversy, however, overall perception of the sharing economy appears to be high. According to one study, approximately 86 percent of respondents believe the sharing economy makes life more affordable, while 78 percent believe the sharing of underutilized resources reduces waste.[29]

Advantages of Small-Business Ownership

There are many advantages to establishing and running a small business. These can be categorized into personal advantages and business advantages. Table 5.4 lists some of the traits that can help entrepreneurs succeed.

LO 5-3

Specify the advantages of small-business ownership.

Independence

Independence is probably one of the leading reasons that entrepreneurs choose to go into business for themselves. Being a small-business owner means being your own boss. Many people start their own businesses because they believe they will do better for themselves than they could do by remaining with their current employer or

TABLE 5.4
Successful Traits of Young Entrepreneurs

Trait	Definition	Trait	Definition
Intuitive	Using one's intuition to derive what's true without conscious reasoning	Innovative	Being able to come up with new and creative ideas
Productive	Being able to produce large amounts of something during a specific time period	Risk-taker	Having the ability to pursue risky endeavors despite the possibility of failure
Resourceful	Understanding how to use and spend resources wisely	Persistent	Continuing in a certain action in spite of obstacles
Charismatic	Having the ability to inspire others behind a central vision	Friendly	Being able to have mutually beneficial interactions with people

by changing jobs. They may feel stuck on the corporate ladder and that no business would take them seriously enough to fund their ideas. Sometimes people who venture forth to start their own small business are those who simply cannot work for someone else. Such people may say that they just do not fit the "corporate mold." In a survey of top entrepreneurs, ambition, ability, and a timeless work ethic were important to success.[30]

More often, small-business owners just want the freedom to choose whom they work with, the flexibility to pick where and when to work, and the option of working in a family setting. The availability of a company website, social media, and other Internet resources make it easy to start a business and work from home.

Costs

As already mentioned, small businesses often require less money to start and maintain than do large ones. Of top entrepreneurs, 78 percent used personal savings to start their business, with 29 percent using their credit card and 22 percent using family loans.[31] Obviously, a firm with just 14 people spends less money on wages and salaries, rent, utilities, and other expenses than does a firm employing tens of thousands of people in several large facilities. Rather than maintain the expense of keeping separate departments for accounting, advertising, and legal counseling, small businesses often hire other firms (sometimes small businesses themselves) to supply these services as they are needed. Additionally, small-business owners can sometimes rely on friends and family members to help them save money by volunteering to work on a difficult project.

Flexibility

With small size comes the flexibility to adapt to changing market demands. Small businesses usually have only one layer of management–the owners. Decisions therefore can be made and executed quickly. In larger firms, decisions about even routine matters can take weeks because they must pass through multiple levels of management before action is authorized. When Taco Bell introduces a new product, for example, it must first research what consumers want, then develop the product and test it before introducing it nationwide—a process that sometimes takes years. An independent snack shop, however, can develop and introduce a new product (perhaps to meet a customer's request) in a much shorter time. In fact, 56 percent of successful small-business CEOs say they use social media to collect feedback on products.[32]

Focus

Small firms can focus their efforts on a precisely defined market niche—that is, a specific group of customers. Many large corporations must compete in the mass market or for large market segments. Smaller firms can develop products for particular groups of customers or to satisfy a need that other companies have not addressed. For example, Megan Tamte launched a chain of boutiques called Evereve targeted toward young mothers. As a young mother herself, she recognized the many problems mothers faced when they take their young children out shopping. Evereve stores were built with double-wide aisles to accommodate strollers, large dressing rooms, and play areas for children. It has since opened 59 stores and experienced a 30 percent annual growth rate.[33] By targeting small niches or product needs, businesses can sometimes avoid competition from larger firms, helping them to grow into stronger companies.

Entrepreneurship in Action

Sseko Helps Women Get a Step Ahead

Sseko Designs
Founder: Liz Forkin Bohannon
Founded: 2008, in Portland, Oregon
Success: Sseko Designs has successfully used social entrepreneurship to develop a for-profit company that improves the lives of women in Uganda and East Africa.

Social entrepreneurship involves creating social value or solving social problems using a business model. Liz Forkin Bohannon, founder of Sseko Designs, designed a for-profit business that uses entrepreneurial principles to create educational opportunities for Ugandan women. When Liz traveled to Uganda, she discovered that high school girls had to return to their villages and work to save money for tuition. Most of these girls did not continue their education because the families needed the money for their subsistence.

Liz developed a work-study model for Ugandan women during the nine-month period they have to earn revenues for college. The women began by making ribbon sandals to sell to U.S. consumers. In the process, they learn skills and have the chance to earn wages for college. The sandals became a success, and Sseko has expanded into other products.

To continue the company's growth, the company sought funding via the popular reality show *Shark Tank*. Although it was denied funding, Sseko's exposure on the show spread its message across the nation. Today, Sseko is Uganda's largest footwear exporter and has enabled 87 female employees to get a university education.[34]

Critical Thinking Questions

1. How does Sseko's social entrepreneurship model differ from traditional charitable initiatives that other companies might take?
2. Do you believe the Bohannans are too focused on their social mission? Should they compromise on the values in favor of profit?
3. What are the disadvantage small-business ownership Liz Forkin Bohannon faces?

Reputation

Reputation, or how a firm is perceived by its various stakeholders, is highly significant to an organization's success. Small firms, because of their capacity to focus on narrow niches, can develop enviable reputations for quality and service. A good example of a small business with a formidable reputation is W. Atlee Burpee and Co., which has the country's premier bulb and seed catalog. Burpee has an unqualified returns policy (complete satisfaction or your money back) that demonstrates a strong commitment to customer satisfaction.

LO 5-4

Analyze the disadvantages of small-business ownership and the reasons why many small businesses fail.

Disadvantages of Small-Business Ownership

The rewards associated with running a small business are so enticing that it's no wonder many people dream of it. However, as with any undertaking, small-business ownership has its disadvantages.

High Stress Level

A small business is likely to provide a living for its owner, but not much more (although there are exceptions as some examples in this chapter have shown). There are ongoing worries about

Some entrepreneurs choose to start their businesses from scratch so they can run their businesses as they see fit. While they might start off small and struggle to attract customers, they are not limited by the restrictions of a franchise agreement.

©Arina P Habich/Shutterstock

competition, employee problems, new equipment, expanding inventory, rent increases, or changing market demand. In addition to other stresses, small-business owners tend to be victims of physical and psychological stress. The small-business person is often the owner, manager, sales force, shipping and receiving clerk, bookkeeper, and custodian. Having to multitask can result in long hours for most small-business owners. Many creative persons fail, not because of their business concepts, but rather because of difficulties in managing their business. Fear of failure is the most common concern when starting a business.[35]

High Failure Rate

Despite the importance of small businesses to our economy, there is no guarantee of success. Half of all businesses fail within the first five years.[36] Restaurants are a case in point. Look around your own neighborhood, and you can probably spot the locations of several restaurants that are no longer in business.

Small businesses fail for many reasons (see Table 5.5). A poor business concept—such as insecticides for garbage cans (research found that consumers are not concerned with insects in their garbage)—will produce disaster nearly every time. Expanding a hobby into a business may work if a genuine market niche exists, but all too often people start such a business without identifying a real need for the good or service. Other notable causes of small-business failure include the burdens imposed by government regulation, insufficient funds to withstand slow sales, and vulnerability to competition from larger companies. Three major causes of small-business failure deserve a close look: undercapitalization, managerial inexperience or incompetence, and inability to cope with growth.

DID YOU KNOW? About 20 percent of small businesses make it past the first year of operation.[37]

undercapitalization
the lack of funds to operate a business normally.

Undercapitalization.   The shortest path to failure in business is **undercapitalization,** the lack of funds to operate a business normally. Too many entrepreneurs think that all they need is enough money to get started, that the business can survive on cash generated from sales soon thereafter. But almost all businesses suffer from seasonal variations in sales, which make cash tight, and few businesses make money from the start. Many small rural operations cannot obtain financing within their own communities because small rural banks often lack the necessary financing expertise or assets

TABLE 5.5
Challenges in Starting a New Business

1. Underfunded (not providing adequate startup capital)
2. Not understanding your competitive niche
3. Lack of effective utilization of websites and social media
4. Lack of a marketing and business plan
5. If operating a retail store, poor site selection
6. Pricing mistakes–too high or too low
7. Underestimating the time commitment for success
8. Not finding complementary partners to bring in additional experience
9. Not hiring the right employees and/or not training them properly
10. Not understanding legal and ethical responsibilities

sizable enough to counter the risks involved with small-business loans. That is why personal savings and loans from family members are top sources for financing a small business. For startups, personal bank loans represent less than 10 percent of the sources of starting capital.[38] Without sufficient funds, the best small-business idea in the world will fail. Consider Necco Wafers, the heart-shaped pressed sugar Valentine's Day candy bearing messages including "Kiss Me" and "Be Mine." The firm was started by an English Immigrant in 1847 and is the oldest confectionary company in America. Today, the still small company makes 630 million wafers a year. In 2018, the company was looking for a new owner to continue its operations due to capitalization. Without a new owner to bring in more capital, the company could fail.[39]

Managerial Inexperience or Incompetence.   Poor management is the cause of many business failures. Just because an entrepreneur has a brilliant vision for a small business does not mean he or she has the knowledge or experience to manage a growing business effectively. A person who is good at creating great product ideas and marketing them may lack the skills and experience to make good management decisions in hiring, negotiating, finance, and control. Moreover, entrepreneurs may neglect those areas of management they know little about or find tedious, at the expense of the business's success.

Zynga Inc., a social game developer and creator of mobile games like FarmVille and Words with Friends, expanded too quickly and had to let go hundreds of employees.

©JHVEPhoto/Shutterstock

Inability to Cope with Growth.   Sometimes, the very factors that are advantages for a small business turn into serious disadvantages when the time comes to grow. Growth often requires the owner to give up a certain amount of direct authority, and it is frequently hard for someone who has called all the shots to give up control. It has often been said that the greatest impediment to the success of a business is the entrepreneur. Similarly, growth requires specialized management skills in areas such as credit analysis and promotion—skills that the founder may lack or not have time to apply. The founders of many small businesses, including Dell Computers, found that they needed to bring in more experienced managers to help manage their companies through growing pains.

Poorly managed growth probably affects a company's reputation more than anything else, at least initially. And products that do not arrive on time or goods that are poorly made can quickly reverse a success. The principal immediate threats to small and mid-sized businesses include rising inflation, energy and other supply shortages or cost escalations, and excessive household and/or corporate debt. For this reason, some small-business owners choose to stay small and are not interested in wide-scale growth. These micropreneurs operate small-scale businesses with no more than five employees. It is estimated that 95 percent of small businesses are microbusinesses.[40]

Starting a Small Business

LO 5-5

Describe how to start a small business and what resources are needed.

We've told you how important small businesses are and why they succeed and fail, but *how do you go about* starting your own business in the first place? To start any business, large or small, you must have some kind of general idea. Sam Walton, founder of Walmart stores, had a vision of a discount retailing enterprise that spawned the world's largest retailing empire and changed the way companies look at business. Next, you need to devise a strategy to guide planning and development in the business. Finally, you must make decisions about form of ownership; the financial resources needed; and whether to acquire an existing business, start a new one, or buy a franchise.

The Business Plan

A key element of business success is a **business plan**—a precise statement of the rationale for the business and a step-by-step explanation of how it will achieve its goals. The business plan should include an explanation of the business, an analysis of the competition, estimates of income and expenses, and other information. It should also establish a strategy for acquiring sufficient funds to keep the business going. The U.S. SBA website provides an overview of a plan for small businesses to use to gain financing. Many financial institutions decide whether to loan a small business money based on its business plan. A good business plan should act as a guide and reference document—not a shackle that limits the business's flexibility and decision-making ability. The business plan must be revised periodically to ensure that the firm's goals and strategies adapt to changes in the environment. Business plans allow companies to assess market potential, determine price and manufacturing requirements, identify optimal distribution channels, and refine product selection. It is also important to evaluate and update the business plan to account for changes in the company. Salem, Oregon-based Rich Duncan Construction learned this the hard way. The company rewrote part of its business plan after 14 years to account for high growth rates and identify weaknesses that needed to be addressed.[41]

Forms of Business Ownership

After developing a business plan, the entrepreneur has to decide on an appropriate legal form of business ownership—whether it is best to operate as a sole proprietorship, partnership, or corporation—and to examine the many factors that affect that decision, which we explored in "Options for Organizing Business."

Financial Resources

The expression "it takes money to make money" holds especially true in developing a business enterprise. To make money from a small business, the owner must first provide or obtain money (capital) to get started and to keep it running smoothly. Even a small retail store will probably need at least $50,000 in initial financing to rent space, purchase or lease necessary equipment and furnishings, buy the initial inventory, and provide working capital. Of all startups, 90 percent start with less than $100,000, with almost 50 percent of these business ventures starting with less than $5,000.[42] Often, the small-business owner has to put up a significant percentage of the necessary capital. Few new business owners have a large amount of their own capital and must look to other sources for additional financing. Open Door Labs Inc. buys and resells houses. It borrows about 90 percent of the purchase price of a home, and they plan to reach 20 metro areas in the coming years. Therefore, they have to find investors to loan funds or borrow from banks.[43]

Equity Financing.   The most important source of funds for any new business is the owner. Many owners include among their personal resources ownership of a home, the accumulated value in a life insurance policy, or a savings account. A new business owner may sell or borrow against the value of such assets to obtain funds to operate a business. Additionally, the owner may bring useful personal assets—such as a computer, desks and other furniture, a car or truck—as part of his or her ownership interest in the firm. Such financing is referred to as *equity financing* because the owner uses real personal assets rather than borrowing funds from outside sources to get started in a new business. The owner can also provide working capital by reinvesting profits into the business or simply by not drawing a full salary.

Small businesses can also obtain equity financing by finding investors for their operations. They may sell stock in the business to family members, friends, employees, or other investors. **Venture capitalists** are persons or organizations that agree to provide some funds for a new business in exchange for an ownership interest or stock. Venture capitalists hope to purchase the stock of a small business at a low price and then sell the stock for a profit after the business has grown successful. Although these forms of equity financing have helped many small businesses, they require that the small-business owner share the profits of the business—and sometimes control, as well—with the investors.

Small-business owners often use debt financing from banks or the Small Business Administration to start their own organizations.

©Vico Collective/Erik Palmer/Blend Images LLC

venture capitalists persons or organizations that agree to provide some funds for a new business in exchange for an ownership interest or stock.

Debt Financing.   New businesses can borrow more as they become established. Banks are the main suppliers of external financing to small businesses. On the federal level, the SBA offers financial assistance to qualifying businesses. Five Star Guitars, out of Hillsboro, Oregon, was named among the Persons of the Year by the Oregon Small Business Association (SBA). SBA loans allow companies such as Five Star Guitars to acquire retail space, employ salespeople, and serve specific target markets.[44] They can also look to family and friends as sources for long-term loans or other assets, such as computers or an automobile, that are exchanged for an ownership interest in a business. In such cases, the business owner can usually structure a favorable repayment schedule and sometimes negotiate an interest rate below current bank rates. If the business goes bad, however, the emotional losses for all concerned may greatly exceed the money involved. Anyone lending a friend or family member money for a venture should state the agreement clearly in writing before any money changes hands.

The amount a bank or other institution is willing to loan depends on its assessment of the venture's likelihood of success and of the entrepreneur's ability to repay the loan. The bank will often require the entrepreneur to put up *collateral,* a financial interest in the property or fixtures of the business, to guarantee payment of the debt. Additionally, the small-business owner may have to provide personal property as collateral, such as his or her home, in which case the loan is called a *mortgage.* If the small business fails to repay the loan, the lending institution may eventually claim and sell the collateral or mortgage to recover its loss.

Banks and other financial institutions can also grant a small business a *line of credit*—an agreement by which a financial institution promises to lend a business a predetermined sum on demand. A line of credit permits an entrepreneur to take quick advantage of opportunities that require external funding. Small businesses may obtain funding from their suppliers in the form of a *trade credit*—that is, suppliers allow the business to take possession of the needed goods and services and pay for them at a later date or in installments. Occasionally, small businesses engage in *bartering*—trading their own products for the goods and services offered by other businesses.

For example, an accountant may offer accounting services to an office supply firm in exchange for office supplies and equipment.

Additionally, some community groups sponsor loan funds to encourage the development of particular types of businesses. State and local agencies may guarantee loans, especially to minority business people or for development in certain areas.

Approaches to Starting a Small Business

Starting from Scratch versus Buying an Existing Business.   Although entrepreneurs often start new small businesses from scratch much the way we have discussed in this section, they may elect instead to buy an existing business. This has the advantage of providing a built-in network of customers, suppliers, and distributors and reducing some of the guesswork inherent in starting a new business from the ground up. Actor Sarah Michelle Gellar co-founded Foodstirs, a baking brand, after being dismissed by countless investors. Gellar leveraged her brand based on her ability to get on shows like the *Harry Connick Jr. Show* and promote the baking mix and prepared foods with high-quality ingredients. Their one-minute mug cake was picked up by 8,000 Starbucks stores.[45] However, an entrepreneur who buys an existing business also takes on any problems the business already has.

Franchising.   Many small-business owners find entry into the business world through franchising. A license to sell another's products or to use another's name in business, or both, is a **franchise.** The company that sells a franchise is the **franchiser.** Dunkin' Donuts, Subway, and Jiffy Lube are well-known franchisers with national visibility. The purchaser of a franchise is called a **franchisee.**

The franchisee acquires the rights to a name, logo, methods of operation, national advertising, products, and other elements associated with the franchiser's business in return for a financial commitment and the agreement to conduct business in accordance with the franchiser's standard of operations. The initial fee to join a franchise varies greatly. In addition, franchisees buy equipment, pay for training, and obtain a mortgage or lease. The franchisee also pays the franchiser a monthly or annual fee based on a percentage of sales or profits. In return, the franchisee often receives building specifications and designs, site recommendations, management and accounting support, and perhaps most importantly, immediate name recognition. Visit the website of the International Franchise Association to learn more on this topic.

The practice of franchising first began in the United States in the 19th century when Singer used it to sell sewing machines. This method of goods distribution soon became commonplace in the automobile, gasoline, soft drink, and hotel industries. The concept of franchising grew especially rapidly during the 1960s, when it expanded to diverse industries. Table 5.6 shows the 10 fastest growing franchises.

The entrepreneur will find that franchising has both advantages and disadvantages. Franchising allows a franchisee the opportunity to set up a small business relatively quickly, and because of its association with an established brand, a franchise outlet often reaches the break-even point faster than an independent business would. Franchisees commonly report the following advantages:

- Management training and support.
- Brand-name appeal.
- Standardized quality of goods and services.
- National and local advertising programs.
- Financial assistance.

franchise
a license to sell another's products or to use another's name in business, or both.

franchiser
the company that sells a franchise.

franchisee
the purchaser of a franchise.

Top 10 Fastest Growing Franchises
Dunkin' (Dunkin' Donuts)
7-Eleven Inc.
Planet Fitness
Jan-Pro Franchising International Inc.
Taco Bell
Orangetheory Fitness
Great Clips
Mac Tools
Cruise Planners
Jazzercise Inc.

TABLE 5.6
Fastest Growing
Franchises

Source: "2018 Fastest-Growing Franchises Ranking," Entrepreneur, https://www.entrepreneur.com/franchises/fastestgrowing (accessed Aril 21, 2018).

- Proven products and business formats.
- Centralized buying power.
- Site selection and territorial protection.
- Greater chance for success.[46]

However, the franchisee must sacrifice some freedom to the franchiser. Some shortcomings experienced by franchisees include:

- Franchise fees and profit sharing with the franchiser.
- Strict adherence to standardized operations.
- Restrictions on purchasing.
- Limited product line.
- Possible market saturation.
- Less freedom in business decisions.[47]

Strict uniformity is the rule rather than the exception. Entrepreneurs who want to be their own bosses are often frustrated with the restrictions of a franchise.

Help for Small-Business Managers

Because of the crucial role that small business and entrepreneurs play in the U.S. economy, a number of organizations offer programs to improve the small-business owner's ability to compete. These include entrepreneurial training programs and programs sponsored by the SBA. Such programs provide small-business owners with invaluable assistance in managing their businesses, often at little or no cost to the owner.

Entrepreneurs can learn critical marketing, management, and finance skills in seminars and college courses. In addition, knowledge, experience, and judgment are necessary for success in a new business. While knowledge can be communicated and some experiences can be simulated in the classroom, good judgment must be developed by the entrepreneur. Local chambers of commerce and the U.S. Department of Commerce offer information and assistance helpful in operating a small business.

Growing a Family Business: Haney's Appledale Farm

While many small businesses fail, others last for generations under the management of hard-working, entrepreneurial-minded owners. Haney's Appledale Farm, in Nancy, Kentucky, has been in operation since 1870. Currently, Haney's is run by brothers Don and Mark, the fifth generation of the Haney family to operate the business.

Haney's Appledale Farm is a sustainable 450-acre family farm that grows 35 varieties of apples and 15 varieties of peaches. The Haneys employ conservation and sustainable practices in their farm operation. They use Integrated Pest Management, which includes special traps so they can identify pests, take immediate action to eliminate the invasive intruders, and minimize risks to human health and the environment.

In season, customers pick their own apples or peaches on the farm's orchards. Their company sells apple-related items such as cookies, cakes, jams, jellies, preserves, apple cider, caramel apples, and Haney's Apple Slush. The bakery has created a signature "Fried Apple Pie" made from scratch every day. Although the Haneys have had offers to go national with this product, they have declined the opportunity so product integrity can be maintained.

The Haney brothers run the farm simply but effectively. One of their most effective tools is Facebook and a website to let customers know when the harvest is ready. The 150-year history of the Haney family farm illustrates how a small business can prosper through many generations.[48]

Critical Thinking Questions

1. What are the advantages the Haney family has in running their own small business? What about disadvantages?
2. What are some ways Haney's owners have used innovative techniques to keep the farm operating so successfully?
3. What are some of the advantages of keeping Haney's Appledale Farm a small business rather than expanding nationally?

National publications such as *Inc.* and *Entrepreneur* share statistics, advice, tips, and success/failure stories. Additionally, most urban areas have weekly business journals/newspapers that provide stories on local businesses as well as on business techniques that a manager or small business can use.

The SBA offers many types of management assistance to small businesses, including counseling for firms in difficulty, consulting on improving operations, and training for owner/managers and their employees. Among its many programs, the SBA funds Small Business Development Centers (SBDCs). These are business clinics, usually located on college campuses, that provide counseling at no charge and training at only a nominal charge. SBDCs are often the SBA's principal means of providing direct management assistance.

The Service Corps of Retired Executives (SCORE) and the Active Corps of Executives (ACE) are volunteer agencies funded by the SBA to provide advice for owners of small firms. Both are staffed by experienced managers whose talents and knowledge the small firms could not ordinarily afford. SCORE has more than 11,000 volunteers at over 320 locations across the country and shares mentor expertise from more than 60 industries.[49] The SBA also has organized Small Business Institutes (SBIs) on almost 500 university and college campuses in the United States. Seniors, graduate students, and faculty at each SBI provide onsite management counseling.

Finally, the small-business owner can obtain advice from other small-business owners, suppliers, and even customers. A customer may approach a small business it frequents with a request for a new product, for example, or a supplier may offer suggestions for improving a manufacturing process. Networking—building relationships and sharing information with colleagues—is vital for any businessperson, whether you work for a huge corporation or run your own small business. Incubators, or organizations created to accelerate the development and success of startup organizations, often provide network opportunities and potential capital to jumpstart a business.[50] Communicating with other business owners is a great way to find ideas for dealing with employees and government regulation, improving processes, or solving problems. New technology is making it easier to network. For example, some states are establishing social networking sites for the use of their businesses to network and share ideas.

The Future for Small Business

LO 5-6

Evaluate the demographic, technological, and economic trends that are affecting the future of small business.

Although small businesses are crucial to the economy, their size and limited resources can make them more vulnerable to turbulence and change in the marketplace than large businesses.[51] Next, we take a brief look at the demographic, technological, and economic trends that will have the most impact on small business in the future.

Demographic Trends

America's baby boom started in 1946 and ended in 1964. The baby boomer generation consists of 75 million Americans.[52] This segment of the population is wealthy, but many small businesses do not actively pursue it. Some exceptions, however, include Gold Violin, which sells designer canes and other products online and through a catalog, and LifeSpring Nutrition, which delivers nutritional meals and snacks directly to the customer. Industries such as travel, financial planning, and health care will continue to grow as boomers age. Many experts believe that the boomer demographic is the market of the future.

Other consumers, comprising a market with a huge potential for small business, that have surpassed baby boomers as the United States' largest living generation are called millennials or Generation Y.[53] Millennials number around 83 million and possess a number of unique characteristics.[54] Born between 1981 and 1997, this cohort is not solely concerned about money. Those that fall into this group tend to be concerned with advancement, recognition, and improved capabilities. They tend to need direct, timely feedback and frequent encouragement and recognition. Millennials do well when training sessions combine entertainment with learning. Working remotely is more acceptable to this group than previous generations, and virtual communication may become as important as face-to-face meetings.[55]

Also to be considered are the growing number of immigrants living in the United States, who now represent more than 17 percent, or 27.6 million, of the total U.S. workforce. The largest employer of legal immigrants is retail, followed by educational services and non-hospital health care services.[56]

This vast group provides still another greatly untapped market for small businesses. Retailers who specialize in ethnic products, and service providers who offer bi- or multilingual employees, will find a large amount of business potential in this market. Table 5.7 ranks top states in the United States for small businesses and startups.

The Latino population is the biggest and fastest growing minority segment in the U.S.—and a lucrative market for businesses looking for ways to meet the segment's many needs.

©Vico Collective/Erik Palmer/Blend Images LLC

Technological and Economic Trends

Advances in technology have opened up many new markets to small businesses. Undoubtedly, the Internet will continue to provide new opportunities for small businesses. Slack is a messaging app that

TABLE 5.7
Most Business-Friendly States

1. North Carolina
2. Texas
3. Utah
4. Nebraska
5. Virginia
6. Georgia
7. Florida
8. Colorado
9. North Dakota
10. Indiana

Source: "Best States for Business," Forbes, www.forbes.com/best-states-for-business/list/#tab:overall (accessed April 21, 2018).

allows teams to engage in efficient collaboration through cloud-based computing. In 2017, the company was valued at $3.8 billion.[57] Technology has also enabled the substantial growth of entrepreneurs working out of their houses, known as home-based businesses. Many of today's largest businesses started out in homes, including Mary Kay, Ford, and Apple. Approximately 52 percent of small businesses are based out of the home.[58] Technological advancements have increased the ability of home-based businesses to interact with customers and operate effectively.[59]

Technological advances and an increase in service exports have created new opportunities for small companies to expand their operations abroad. Changes in communications and technology can allow small companies to customize their services quickly for international customers. Also, free trade agreements and trade alliances are helping to create an environment in which small businesses have fewer regulatory and legal barriers.

In recent years, economic turbulence has provided both opportunities and threats for small businesses. As large information technology companies such as Cisco, Oracle, and Sun Microsystems had to recover from an economic slowdown and an oversupply of Internet infrastructure products, some smaller firms found new niche markets. Smaller companies can react quickly to change and can stay close to their customers. While well-funded dot-coms were failing, many small businesses were learning how to use the Internet to promote themselves and sell products online. For example, arts and crafts dealers and makers of specialty products found they could sell their wares on existing websites, such as eBay and Etsy. Service providers related to tourism, real estate, and construction also found they could reach customers through their own or existing websites.

Deregulation of the energy market and interest in alternative fuels and in fuel conservation have spawned many small businesses. Southwest Windpower Inc. manufactures and markets small wind turbines for producing electric power for homes, sailboats, and telecommunications. As entrepreneurs begin to realize that worldwide energy markets are valued in the hundreds of billions of dollars, the number of innovative companies entering this market will increase. In addition, many small businesses have the desire and employee commitment to purchase such environmentally friendly products. New Belgium Brewing Company received the U.S. Environmental Protection Agency and Department of Energy Award for leadership in conservation for making a 10-year commitment to purchase wind energy.

The future for small business remains promising. The opportunities to apply creativity and entrepreneurship to serve customers are unlimited. Large companies such as Walmart, which employ roughly 1 percent of the U.S. population, or 2.1 million employees, often adapt slowly to changes in the environment, whereas small businesses can adapt immediately.[60] This flexibility provides small businesses with a definite advantage over large companies.

Making Big Businesses Act "Small"

LO 5-7

Explain why many large businesses are trying to "think small."

The continuing success and competitiveness of small businesses through rapidly changing conditions in the business world have led many large corporations to take a closer look at what makes their smaller rivals tick. More and more firms are emulating small businesses in an effort to improve their own bottom line. Beginning in the 1980s and continuing through the present, the buzzword in business has been to *downsize* or *rightsize* to reduce management layers, corporate staff, and work tasks in order to make the firm more flexible, resourceful, and innovative. Many well-known U.S. companies—including IBM, Ford, Apple, General Electric, Xerox, and 3M—have downsized to improve their competitiveness, as have German, British, and Japanese firms. Other firms have sought to make their businesses "smaller" by making their operating units function more like independent small businesses, each responsible for its profits, losses, and resources. Of course, some large corporations, such as Southwest Airlines, have acted like small businesses from their inception, with great success.

Trying to capitalize on small-business success in introducing innovative new products, more and more companies are attempting to instill a spirit of entrepreneurship into even the largest firms. In major corporations, **intrapreneurs,** like entrepreneurs, take responsibility for, or "champion," the development of innovations of any kind *within* the larger organization.[61] Often, they use company resources and time to develop a new product for the company.

intrapreneurs
individuals in large firms who take responsibility for the development of innovations within the organizations.

So You Want to Be an Entrepreneur or Small-Business Owner

In times when jobs are scarce, many people turn to entrepreneurship as a way to find employment. As long as there are unfulfilled needs from consumers, there will be a demand for entrepreneurs and small businesses. Entrepreneurs and small-business owners have been, and will continue to be, a vital part of the U.S. economy, whether in retailing, wholesaling, manufacturing, technology, or services. Creating a business around your idea has a lot of advantages. For many people, independence is the biggest advantage of forming their own small business, especially for those who do not work well in a corporate setting and like to call their own shots. Smaller businesses are also cheaper to start up than large ones in terms of salaries, infrastructure, and equipment. Smallness also provides a lot of flexibility to change with the times. If consumers suddenly start demanding new and different products, a small business is more likely to deliver quickly.

Starting your own business is not easy, especially in slow economic times. Even in a good economy, taking an idea and turning it into a business has a very high failure rate. The possibility of failure can increase even more when money is tight. Reduced revenues and expensive materials can hurt a small business more than a large one because small businesses have fewer resources. When people are feeling the pinch from rising food and fuel prices, they tend to cut back on other expenditures—which could potentially harm your small business. The increased cost of materials will also affect your bottom line. However, several techniques can help your company survive:

- Set clear payment schedules for all clients. Small businesses tend to be worse about collecting payments than large ones, especially if the clients are

acquaintances. However, you need to keep cash flowing into the company in order to keep business going.

- Take the time to learn about tax breaks. A lot of people do not realize all of the deductions they can claim on items such as equipment and health insurance.

- Focus on your current customers, and don't spend a lot of time looking for new ones. It is far less expensive for a company to keep its existing customers happy.

- Although entrepreneurs and small-business owners are more likely to be friends with their customers, do not let this be a temptation to give things away for free. Make it clear to your customers what the basic price is for what you are selling and charge for extra features, extra services, etc.

- Make sure the office has the conveniences employees need—like a good coffee maker and other drinks and snacks. This will not only make your employees happy, but it will also help maintain productivity by keeping employees closer to their desks.

- Use your actions to set an example. If money is tight, show your commitment to cutting costs and making the business work by doing simple things like taking the bus to work or bringing a sack lunch every day.

- Don't forget to increase productivity in addition to cutting costs. Try not to focus so much attention on cost cutting that you don't try to increase sales.

In unsure economic times, these measures should help new entrepreneurs and small-business owners sustain their businesses. Learning how to run a business on a shoestring is a great opportunity to cut the fat and to establish lean, efficient operations.[62]

Review Your Understanding

Define *entrepreneurship* and *small business*.

An entrepreneur is a person who creates a business or product and manages his or her resources and takes risks to gain a profit; entrepreneurship is the process of creating and managing a business to achieve desired objectives. A small business is one that is not dominant in its competitive area and does not employ more than 500 people.

Explain the importance of small business in the U.S. economy and why certain fields attract small business.

Small businesses are vital to the American economy because they provide products, jobs, innovation, and opportunities. Retailing, wholesaling, services, manufacturing, high technology, and the sharing economy attract small businesses because these industries are relatively easy to enter, require relatively low initial financing, and may experience less heavy competition.

Specify the advantages of small-business ownership.

Small-business ownership offers some personal advantages, including independence, freedom of choice, and the option of working at home. Business advantages include flexibility, the ability to focus on a few key customers, and the chance to develop a reputation for quality and service.

Analyze the disadvantages of small-business ownership and the reasons why many small businesses fail.

Small businesses have many disadvantages for their owners such as expense, physical and psychological stress, and a high failure rate. Small businesses fail for many reasons: undercapitalization, management inexperience or incompetence, neglect, disproportionate burdens imposed by government regulation, and vulnerability to competition from larger companies.

Describe how to start a small business and what resources are needed.

First, you must have an idea for developing a small business. Next, you need to devise a business plan to guide planning and development of the business. Then, you must decide what form of business ownership to use: sole proprietorship, partnership,

or corporation. Small-business owners are expected to provide some of the funds required to start their businesses, but funds also can be obtained from friends and family, financial institutions, other businesses in the form of trade credit, investors (venture capitalists), state and local organizations, and the Small Business Administration. In addition to loans, the Small Business Administration and other organizations offer counseling, consulting, and training services. Finally, you must decide whether to start a new business from scratch, buy an existing one, or buy a franchise operation.

Evaluate the demographic, technological, and economic trends that are affecting the future of small business.

Changing demographic trends that represent areas of opportunity for small businesses include more elderly people as baby boomers age, millennials (or Generation Y), and an increasing number of immigrants to the United States. Technological advances and an increase in service exports have created new opportunities for small companies to expand their operations abroad, while trade agreements and alliances have created an environment in which small business has fewer regulatory and legal barriers. Economic turbulence presents both opportunities and threats to the survival of small businesses.

Explain why many large businesses are trying to "think small."

More large companies are copying small businesses in an effort to make their firms more flexible, resourceful, and innovative and, generally, to improve their bottom line. This effort often involves downsizing (reducing management layers, laying off employees, and reducing work tasks) and intrapreneurship, when an employee takes responsibility for (champions) developing innovations of any kind within the larger organization.

Assess two entrepreneurs' plans for starting a small business.

Based on the facts given in the "Solve the Dilemma" feature near the end of this chapter and the material presented in this chapter, you should be able to assess the feasibility and potential success of Gray and McVay's idea for starting a small business.

Critical Thinking Questions

Enter the World of Business Questions

1. What are some of the benefits of operating Chick-fil-A as a franchise?

2. What are some of the disadvantages of operating Chick-fil-A as a franchise?

3. Looking to the future, how can Chick-fil-A stay ahead of demographic, technological, and economic trends?

Learn the Terms

business plan 154
entrepreneurship 142
franchise 156
franchisee 156
franchiser 156

intrapreneurs 161
microentrepreneurs 143
sharing economy 148
small business 143
Small Business Administration (SBA) 144

social entrepreneurs 143
undercapitalization 152
venture capitalists 155

Check Your Progress

1. Why are small businesses so important to the U.S. economy?

2. Which fields tend to attract entrepreneurs the most? Why?

3. What are the advantages of starting a small business? The disadvantages?

4. What are the principal reasons for the high failure rate among small businesses?

5. What decisions must an entrepreneur make when starting a small business?

6. What types of financing do small entrepreneurs typically use? What are some of the pros and cons of each?

7. List the types of management and financial assistance that the Small Business Administration offers.

8. Describe the franchising relationship.

9. What demographic, technological, and economic trends are influencing the future of small business?

10. Why do large corporations want to become more like small businesses?

Get Involved

1. Interview a local small-business owner. Why did he or she start the business? What factors have led to the business's success? What problems has the owner experienced? What advice would he or she offer a potential entrepreneur?

2. Using business journals, find an example of a company that is trying to emulate the factors that make small businesses flexible and more responsive. Describe and

evaluate the company's activities. Have they been successful? Why or why not?

3. Using the business plan outline in online Appendix A, create a business plan for a business idea that you have. (A man named Fred Smith once did a similar project for a business class at Yale. His paper became the basis for the business he later founded: Federal Express!)

Build Your Skills

Creativity

Background

The entrepreneurial success stories in this chapter are about people who used their creative abilities to develop innovative products or ways of doing something that became the basis of a new business. Of course, being creative is not just for entrepreneurs or inventors; creativity is an important tool to help you find the optimal solutions to the problems you face on a daily basis. Employees rely heavily on their creativity skills to help them solve daily workplace problems.

According to brain experts, the right-brain hemisphere is the source of creative thinking; and the creative part of the brain can "atrophy" from lack of use. Let's see how much "exercise" you're giving your right-brain hemisphere.

Task

1. Take the following self-test to check your Creativity Quotient.[63]

2. Write the appropriate number in the box next to each statement according to whether the statement describes your behavior always (3), sometimes (2), once in a while (1), or never (0).

	Always 3	Sometimes 2	Once in a While 1	Never 0
1. I am a curious person who is interested in other people's opinions.				
2. I look for opportunities to solve problems.				
3. I respond to changes in my life creatively by using them to redefine my goals and revising plans to reach them.				
4. I am willing to develop and experiment with ideas of my own.				
5. I rely on my hunches and insights.				
6. I can reduce complex decisions to a few simple questions by seeing the "big picture."				
7. I am good at promoting and gathering support for my ideas.				
8. I think further ahead than most people I associate with by thinking long term and sharing my vision with others.				
9. I dig out research and information to support my ideas.				
10. I am supportive of the creative ideas from my peers and subordinates and welcome "better ideas" from others.				
11. I read books and magazine articles to stay on the "cutting edge" in my areas of interest. I am fascinated by the future.				
12. I believe I am creative and have faith in my good ideas.				
Subtotal for each column				
Grand Total				

3. **Check your score using the following scale:**

   30–36 High creativity. You are giving your right-brain hemisphere a regular workout.

   20–29 Average creativity. You could use your creativity capacity more regularly to ensure against "creativity atrophy."

   10–19 Low creativity. You could benefit by reviewing the questions you answered "never" in the above assessment and selecting one or two of the behaviors that you could start practicing.

   0–9 Undiscovered creativity. You have yet to uncover your creative potential.

Solve the Dilemma

The Small-Business Challenge

 Jack Gray and his best friend, Bruce McVay, decided to start their own small business. Jack had developed recipes for fat-free and low-fat cookies and muffins in an effort to satisfy his personal health needs. Bruce had extensive experience in managing food-service establishments. They knew that a startup company needs a quality product, adequate funds, a written business plan, some outside financial support, and a good promotion program. Jack and Bruce felt they had all of this and more and were ready to embark on their new low-fat cookie/muffin store. Each had $35,000 to invest and with their homes and other resources, they had borrowing power of an additional $125,000.

However, they still have many decisions to make, including what form or organization to use, how to market their product, and how to determine exactly what products to sell—whether just cookies and muffins or additional products.

Critical Thinking Questions

1. Evaluate the idea of a low-fat cookie and muffin retail store.

2. Are there any concerns in connection with starting a small business that Jack and Bruce have not considered?

3. What advice would you give Jack and Bruce as they start up their business?

LO 5-8

Assess two entrepreneurs' plans for starting a small business.

Build Your Business Plan

connect

Small Business, Entrepreneurship, and Franchising

Now you can get started writing your business plan! Refer to Guidelines for the Development of the Business Plan following "The Dynamics of Business and Economics," which provides you with an outline for your business plan. As you are developing your business plan, keep in mind that potential investors might be reviewing it. Or you might have plans to go to your local Small Business Development Center for an SBA loan.

At this point in the process, you should think about collecting information from a variety of (free) resources. For example, if you are developing a business plan for a local business, good, or service, you might want to check out any of the following sources for demographic information: your local Chamber of Commerce, Economic Development office, census bureau, or City Planning office.

Go on the Internet and see if there have been any recent studies done or articles on your specific type of business, especially in your area. Remember, you always want to explore any secondary data before trying to conduct your own research.

Visit Connect to practice building your business plan with the Business Plan Prep Exercises.

See for Yourself Videocase

Sonic—A Successful Franchise with an Old-Fashioned Drive-In Experience

For those who are nostalgic for the classic drive-in diner experience, the Sonic fast-food chain helps fill that need. Sonic offers customers a dose of nostalgia with its 1950s-style curbside speakers and carhop service. As the United States' largest drive-in fast-food chain, Sonic offers a unique and diverse menu selection that helps set it apart from a highly competitive fast-food franchise market. Founder Troy Smith launched the first Sonic Drive-In (known then as Top Hat Drive-In) in Shawnee, Oklahoma, in 1953 as a sole proprietorship. He later added a partner, Charlie Pappe, and eventually turned the business into a franchise.

Despite its traditional feel, the company has seized upon new trends and opportunities to secure more business. Customers at Sonic frequently eat in their cars or at tables outside the restaurant. However, Sonic has begun building indoor dining prototypes in colder areas to test whether this will entice more customers to eat at its locations. The prototype still makes use of the restaurant's traditional patio but encloses it to protect customers from the elements. Each of these restaurants maintains its carhop and drive-thru features in order to retain the "Sonic experience."

Today, Sonic is a publicly traded company and ranks among the top 30 restaurants among *Franchise Times'* Top 200+. Franchising is an appealing option for entrepreneurs looking to begin businesses without creating them from scratch. In the case of Sonic, when a franchisee purchases a franchise, he or she is getting a business that already has a national reputation and a national advertising campaign. The company also offers its franchisees tremendous support and training. As a pioneer, Troy Smith was required to innovate; as a Sonic franchisee, one steps into an already proven system.

That being said, successfully running a franchise is not easy. One entrepreneur who owns 22 Sonic franchises said the franchisee's job is to ensure that each customer has the best experience possible, thereby making repeat visits more likely. To accomplish this, a franchisee must build his or her location(s); purchase equipment; hire excellent employees; make certain the products live up to Sonic's reputation; maintain a clean, inviting facility; and much more. In order to run 22 franchises, the entrepreneur runs his locations as limited partnerships, ensuring that a managing partner is on site at each location to keep day-to-day operations running smoothly.

Some of Sonic's success may be attributed to its stringent requirements for selecting franchisees. Although franchisees must have excellent financial credentials and prior restaurant/entrepreneurial experience, the most important factor is that each franchisee fit into the Sonic culture. Sonic offers two types of franchises. The traditional franchise, which includes the full restaurant set-up, requires an initial investment of between $1.1 million and $2 million. Franchisees are required to pay 5 percent in ongoing royalty fees and a franchise fee of $45,000. A Sonic in a travel plaza, a mall food court, or a college campus are all examples of the nontraditional model. Because these set-ups do not include the drive-in and carhop features, initial investment is less. However, royalty and advertising fees still apply.

For entrepreneurs looking for limited risk, franchises like Sonic are great options. The advantages are abundant, as discussed earlier. There is a high failure rate among small businesses. Entering into a successful franchise significantly cuts down on the risk of failure, although a franchisee does have to watch for market saturation, poor location choice, and other determining factors. However, there are also disadvantages; chiefly, franchisees are often required to follow a strict model set by the franchiser. For instance, in addition to prior restaurant experience, Sonic requires its franchisees to be financially and operationally able to open two or more drive-ins. These types of requirements may make it difficult for entrepreneurs who want to set their own terms. However, with Sonic's successful business model and brand equity, there is no shortage of individuals who would like to operate a Sonic franchise.[64]

Critical Thinking Questions

1. What is Sonic's competitive advantage over other fast-food franchises?

2. What are the advantages of becoming a Sonic franchisee?

3. What are the disadvantages of buying into the Sonic franchise?

You can find the related video in the Video Library in Connect. Ask your instructor how you can access Connect.

Team Exercise

Explore successful global franchises. Go to the companies' websites and find the requirements for applying for three franchises. The chapter provides examples of successful franchises. What do the companies provide, and what is expected to be provided by the franchiser? Compare and contrast each group's findings for the franchises researched. For example, at Subway, the franchisee is responsible for the initial franchise fee, finding locations, leasehold improvements and equipment, hiring employees and operating restaurants, and paying an 8 percent royalty to the company and a fee into the advertising fund. The company provides access to formulas and operational systems, store design and equipment ordering guidance, a training program, an operations manual, a representative on site during opening, periodic evaluations and ongoing support, and informative publications.

Ask your instructor about the role-play exercises available with this book to practice working with a business team.

Notes

1. Kate Taylor, "The Incredible Story of How Chick-fil-A Took over Fast Food," *Business Insider,* January 25, 2016, http://www.businessinsider.com/chick-fil-a-history-and-facts-2016-1/#chick-fil-a-has-its-roots-in-a-restaurant-called-the-dwarf-grill-opened-by-founder-truett-cathy-in-1946-1 (accessed November 2, 2017); "Who We Are," Chick-fil-A, https://www.chick-fil-a.com/About/Who-We-Are (accessed November 2, 2017); Hayley Peterson, "Why Chick-fil-A's Restaurants Sell 4 times as Much as KFC's," *Business Insider,* August 1, 2017, http://www.businessinsider.com/why-chick-fil-a-is-so-successful-2017-8 (accessed November 2, 2017); "Franchise Opportunities," Chick-fil-A, https://www.chick-fil-a.com/Franchising/Franchise (accessed March 11, 2018); Kate Taylor, "Why Chick-fil-A Will Never Go Public," *Business Insider,* January 28, 2016, http://www.businessinsider.com/chick-fil-a-will-never-go-public-2016-1 (accessed April 11, 2018).

2. "Frequently Asked Questions," Small Business Administration Department of Advocacy, March 2014, www.sba.gov/sites/default/files/FAQ_March_2014_0.pdf (accessed May 4, 2017).

3. "Global Entrepreneurship Index," Global Entrepreneurship Development Institute, 2016, http://thegedi.org/global-entrepreneurship-and-development-index/ (accessed March 4, 2016).

4. Marco della Cava, "USA Today Entrepreneur of the Year," *USA Today,* December 11, 2014, pp. 1B–2B.

5. Ben Schiller, "TOMS Founder Blake Mycoskie Announces a New Fund for Social Good Startups," *Fast Company,* November 11, 2015, http://www.fastcompany.com/3053526/the-fast-company-innovation-festival/toms-founder-blake-mycoskie-announces-a-new-fund-for-so (accessed May 4, 2017).

6. "Taking Back the Block," *FastCompany.com,* April 2017, p. 79.

7. "Frequently Asked Questions," Small Business Administration Department of Advocacy.

8. "About Us," National Association of Women Business Owners, https://nawbo.org/about (accessed February 13, 2015).

9. Valentina Zarya, "Women-Owned Businesses Are Trailing in Size and Revenue," *Fortune,* September 2, 2015, http://fortune.com/2015/09/02/women-business-size-revenue/ (accessed May 4, 2017).

10. Sam Frizell, "Mercado of America," *Time,* April 28, 2015, 42–45; "About Us," The Legaspi Company website, http://www.thelegaspi.com/jos-de-jes-s-legaspi/ (accessed March 3, 2016).

11. "Frequently Asked Questions," Small Business Administration Department of Advocacy.

12. "SUSB Employment Change Data Tables," *Statistics of U.S. Businesses,* www.census.gov/data/tables/2014/econ/susb/2014-susb-employment.htlm (accessed April 12, 2017).

13. "Bittersweet Synergy: Domestic Outsourcing in India," *The Economist,* October 22, 2009, p. 74.

14. Ray A. Smith, "I Love My Unique, Personalized Shirt! Oh You Have One, Too," *The Wall Street Journal,* April 7–8, 2018, pp. A1, A9.

15. Andrew Medal, "5 Important Startup Lessons from 'Shark Tank'," *Entrepreneur,* May 29, 2015, https://www.entrepreneur.com/article/246716 (accessed January 1, 2018); Gary Golden, "This Student Entrepreneur's Pitch Impressed Daymond John So Much that He Got a 'Shark Tank' Invite," *Inc.,* August 31, 2017, https://www.inc.com/gary-golden/how-a-20-year-old-college-entrepreneur-perfected-h.html (accessed January 1, 2018); Laura Woods, "The Best Advice 'Shark Tank' Investors Have Given Entrepreneurs," *Business Insider,* January 20, 2017, http://www.businessinsider.com/the-best-advice-shark-tank-investors-have-given-entrepreneurs-2017-1/#-1 (accessed January 1, 2018); Steven Key, "3 Lies You Heard on 'Shark Tank'," *Entrepreneur,* September 3, 2013, https://www.entrepreneur.com/article/228162 (accessed January 1, 2018); Naomi Schaefer Riley, "'Shark Tank' Offers Valuable Business Lessons," *The New York Post,* November 26, 2013, http://nypost.com/2013/11/26/shark-tank-offers-valuable-lessons-about-american-business/ (accessed January 1, 2018); Alice Daniel, "Inside the Shark Tank," *Success,* http://www.success.com/article/inside-the-shark-tank (accessed January 1, 2018); Alison Griswold, "Successful Companies That Didn't Get a 'Shark Tank' Deal", *Business Insider,* November 18, 2013, http://www.businessinsider.com/successful-companies-that-didnt-get-a-shark-tank-deal-2013-11 (accessed January 1, 2018); Carol Tice, "The Shark Tank Effect: Top Success Stories from the First Three Seasons," *Entrepreneur,* September 13, 2012, http://www.entrepreneur.com/slideshow/224405#0 (accessed January 1, 2018).

16. "Small Biz Stats & Trends," SCORE Association, https://www.score.org/node/148155 (accessed March 3, 2016).

17. Matthew Diebel, "A Visionary Approach to Selling Eyewear," *USA Today,* December 1, 2014, p. 3B.

18. Alina Dizik, "Where Your Favorite Condiment Is Never Out of Stock," *The Wall Street Journal,* March 4, 2015, p. D3.

19. Suzanne Kapner, "J.C. Penney Resurrects Its Catalog," *The Wall Street Journal,* January 19, 2015, http://www.wsj.com/articles/j-c-penney-resurrects-its-catalog-1421695574 (accessed May 4, 2017).

20. Joe Keohane, "How One of America's Most Beloved Toy Makers Rebounded from Near Death," *Entrepreneur,* February 20, 2018, https://www.entrepreneur.com/article/308693 (accessed April 21, 2018).

21. Claire Dodson, "Future Forward," *Fastcompany.com,* April 2017, p. 20.

22. "Fight Fire with Foam—and a Resume That Includes the Space Shuttle," *Inc.,* September 2016, p. 51.

23. Lev Grossman, "Head Trip," *Time,* April 7, 2014, p. 36–41.

24. Rachel Botsman presentation, "The Shared Economy Lacks a Shared Definition," *Fast Company,* November 21, 2013, http://www.fastcoexist.com/3022028/the-sharing-economy-lacks-a-shared-definition (accessed May 4, 2017); Natasha Singer, "In the Sharing Economy, Workers Find Both Freedom and Uncertainty," *The New York Times,* August 16, 2014, http://www.nytimes.com/2014/08/17/technology/in-the-sharing-economy-workers-find-both-freedom-and-uncertainty.html?_r=1 (accessed May 4, 2017).

25. Singer, "In the Sharing Economy, Workers Find Both Freedom and Uncertainty."

26. Arun Sundararajan, "The Gig Economy Is Coming. What Will It Mean for Work?" *The Guardian,* July 25, 2015, http://www.theguardian.com/commentisfree/2015/jul/26/will-we-get-by-gig-economy (accessed May 4, 2017).

27. Hal Conick, "How to Make It as a Creative in the Gig Economy," *Marketing News,* April 2018, pp. 53–54.

28. Brian Chesky and Jo Gebbia, "Why Airbnb Is Now Almost Twice as Valuable as Hilton," *Vanity Fair,* March 10, 2017, p. 12.

29. PricewaterhouseCoopers, *The Sharing Economy: Consumer Intelligence Series,* https://www.pwc.com/us/en/industry/entertainment-media/publications/consumer-intelligence-series/assets/pwc-cis-sharing-economy.pdf

30. "How Dreamers Become Doers," *Inc.* September 2016, p. 44.

31. Ibid.

32. Ibid.

33. David Whitford, "How a Terrible, Horrible, No Good, Very Bad Day Spawned a $70 Million Business," *Inc.,* November 2015, http://www.inc.com/magazine/201511/david-whitford/not-a-spectator-anymore.html (accessed August 9, 2016).

34. "Sseko Designs: Empowering Ugandan Women," University of New Mexico Daniels Fund Ethics Initiative, https://danielsethics.mgt.unm.edu/pdf/Sseko.pdf (accessed November 24, 2017); Dianne Kroncke, "Sseko Designs Company Growth," Phone Interview with Kayla Joy Asbury, Portland, Oregon, October 13, 2017; Eugene Rowe, "Sseko Designs—Brianna Leever," Community Manager and Sseko Fellows Program," Radio Interview, http://www.hombabiz.com/sseko-designs-brianna-leever/ (accessed on October 12, 2017); Malia Spencer, "Life After 'Shark Tank' Is Pretty Good for One Portland Startup," *Portland Biz Journals,* February 17, 2015, https://www.bizjournals.com/portland/blog/techflash/2015/02/life-after-shark-tank-is-pretty-good-for-one.html (accessed on *October 12, 2017);* Sseko Designs website, https://ssekodesigns.com/ (accessed October 12, 2017); U.S. Mission Uganda, "Like Sseko Designs, You Too Can Take Advantage of the African Growth and Opportunity Act," United States Embassy in Uganda, June 27, 2016, https://ug.usembassy.gov/learn-ugandans-taking-advantage-african-growth-opportunity-act/ (accessed October 12, 2017); "Sseko Designs, Uganda: An AGOA Success Story," United States AID Video, https://www.usaid.gov/news-information/videos/node/218021 (accessed October 12, 2017); Pete Williams, "Liz Forkin Bohannon from Sseko

and the Future of Retail," *YouTube,* https://www.youtube.com/watch?v=X11t0vy6wiY (accessed October 12, 2017).

35. "How Dreamers Become Doers," p. 44.

36. "Frequently Asked Questions," Small Business Administration Department of Advocacy.

37. Keith Speights, "Success Rate: What Percentage of Business Fail in Their First Year?" *USA Today,* May 21, 2017, https://www.usatoday.com/story/money/business/small-business-central/2017/05/21/what-percentage-of-businesses-fail-in-their-first-year/101260716/ (accessed April 21, 2018).

38. "How Dreamers Become Doers," p. 43.

39. John Clarke, "Necco Wafers, Taste of Childhood and Chalk, Face the Final Crunch," *The Wall Street Journal,* April 9, 2018, p. A1.

40. Susan Payton, "Attention, Micropreneurs: You're Not Alone in Small Business," *Forbes,* May 12, 2014, http://www.forbes.com/sites/allbusiness/2014/05/12/attention-micropreneurs-youre-not-alone-in-small-business/ (accessed May 4, 2017).

41. Richard Duncan, "How I Blew It with My Business Plan," *The Wall Street Journal,* May 29, 2015, http://blogs.wsj.com/experts/2015/05/29/how-i-blew-it-with-my-business-plan/ (accessed May 4, 2017.

42. "How Dreamers Become Doers," p. 45.

43. Rolfe Winkler, "Home Reseller Seeks Funds," *The Wall Street Journal,* March 30, 2018, p. B2.

44. "Portland District Office: Success Stories," U.S. Small Business Administration, https://www.sba.gov/offices/district/or/portland/success-stories (accessed April 7, 2017).

45. Jason Feifer, "Turning Rejection into Triumph: How Sarah Michelle Gellar and Her Co-Founders Built a New Baking Brand," *Entrepreneur,* March 27, 2018, https://www.entrepreneur.com/article/310091 (accessed April 21, 2018).

46. Thomas W. Zimmerer and Norman M. Scarborough, *Essentials of Entrepreneurship and Small Business Management,* 6th ed. (Upper Saddle River, NJ: Pearson Prentice Hall, 2005), pp. 118–124.

47. Ibid.

48. Haney's Appledale Farm, http://www.haneysappledalefarm.com/ (accessed December 8, 2017); Mark Haney, "President's Column," *Kentucky Farm Bureau Magazine* 16, no. 8 (October 2017), p. 3, https://www.kyfb.com/federation/newsroom/kentucky-farm-bureau-news/kfbn-2017/october-2017/ (accessed December 8, 2017); "Power of the Neighborhood Market," Kentucky Department of Agriculture, https://www.kyagr.com/marketing/documents/EDU_Power-of-the-Neighborhood-Market.pdf (accessed December 8, 2017); "Kentucky Farm Bureau, "Voice of Kentucky Agriculture," Kentucky Farm Bureau, July 31, 2017, https://www.kyfb.com/federation/newsroom/kentucky-farm-bureau-the-voice-of-kentucky-agriculture/ (accessed December 8, 2017); "Kentucky Farm Bureau, KFB President Mark Haney Announces Formation of Water Management Working Group," *Kentucky Farm Bureau Magazine,* December 9, 2014, https://www.kyfb.com/federation/newsroom/haney-announces-water-management-working-group/ (accessed December 8, 2017); "KFB President Mark Haney Emphasizes KFB Loves KY During Annual Address," *Kentucky Farm Bureau Newsroom,* December, 2, 2016, https://www.kyfb.com/federation/newsroom/kfb-president-mark-haney-emphasizes-kfb-loves-ky-during-annual-address/ (accessed December 8, 2017); "Haney's Appledale Farm," Kentucky Department of Travel, 2017, https://www.kentuckytourism.com/haneys-appledale-farm/2267/ (accessed December 8, 2017); Dianne Kroncke, "Give More Than You're Taking—Appledale Farm Interview with Don Haney," Phone Interview, Auburn University, Alabama, November 13, 2017; Diane Kroncke, "Personal Interview, Be the Best," Appledale Farm Interview with Mark Haney," Phone Interview, Auburn University, Alabama, November 29, 2017; Dianne Kroncke, "Personal Interview with Kaycee Rader of Nancy, Kentucky," at Grand Floridian, Orlando, Florida, November 20, 2017; "Agritourism," Lake Cumberland, Somerset Polaski County Tourism, 2015, http://www.lakecumberlandtourism.com/explore/agritourism (accessed December 8, 2017); Mark Haney, "Mark Haney Association Plans Should Be Part of Health Care Reform; Would Provide Bargaining Power," *Northern Kentucky Tribune,* September 26, 2017, http://www.nkytribune.com/2017/09/mark-haney-association-plans-should-be-part-of-health-care-reform-would-provide-bargaining-power/ (accessed December 8, 2017); "Historical Data on Integrated Pest Management Programs," University of Kentucky, Agriculture, Integrated Pest Management Programs, College of Agriculture, Food and Environment, https://ipm.ca.uky.edu/trapdata (accessed November 27, 2017); "Integrated Pest Management," *Wikipedia,* November 22, 2017, https://en.wikipedia.org/wiki/Integrated_pest_management (accessed December 8, 2017); "Haney's Appledale Farm," *Words to Live By,* 2017, https://www.wordstoliveby.com/blogs/words-we-live-by/haneys-appledale-farm (accessed December 8, 2017).

49. "About SCORE," SCORE, https://www.score.org/about-score (accessed April 4, 2017).

50. "Getting Started with Business Incubators," *Entrepreneur,* http://www.entrepreneur.com/article/52802 (accessed May 4, 2017).

51. Adapted from "Tomorrow's Entrepreneur," *Inc. State of Small Business* 23, no. 7 (2001), pp. 80–104.

52. Cheryl Corley, "Millennials Now Out Number Baby Boomers, Census Bureau Says," *NPR,* July 7, 2015, http://www.npr.org/2015/06/25/417349199/millenials-now-out-number-baby-boomers-census-bureau-says (accessed May 4, 2017).

53. Richard Fry, "Millennials Match Baby Boomers as Largest Generation in U.S. Electorate, But Will They Vote?" Pew Research Center, May 16, 2016, http://www.pewresearch.org/fact tank/2016/05/16/millennials-match-baby-boomers-as-largest-generation-in-u-s-electorate-but-will-they-vote/ (accessed April 20, 2017).

54. Corley, "Millennials Now Out Number Baby Boomers, Census Bureau Says."

55. Molly Smith, "Managing Generation Y as They Change the Workforce," *Reuters,* January 8, 2008, www.reuters.com/article/2008/01/08/idUS129795=08-Jan-2008=BW20080108 (accessed March 4, 2016).

56. Drew Desilver, "Immigrants Don't Make up a Majority of Workers in Any U.S. Industry," Pew Research Center, March 16, 2017, http://www.pewresearch.org/fact-tank/2017/03/16/immigrants-dont-make-up-a-majority-of-workers-in-any-u-s-industry/ (accessed April 6, 2017).

57. Matt Weinberger, "$3.8 Billion Slack Is Finally Launching Its Long-Awaited Version for Big Businesses Next Week," *Business Insider,* January 25, 2017, http://www.businessinsider.com/slack-to-launch-enterprise-edition-2017-1 (accessed April 6, 2017).

58. "Small Business Facts & Data," Small Business & Entrepreneurship Council, http://sbecouncil.org/about-us/facts-and-data/ (accessed April 6, 2017).

59. Jason Nazar, "16 Surprising Statistics about Small Businesses," *Forbes,* September 9, 2013, www.forbes.com/sites/jasonnazar/2013/09/09/16-surprising-statistics-about-small-businesses/ (accessed May 4, 2017); "Home-Based Businesses," U.S. Small Business Administration, www.sba.gov/content/home-based-businesses (accessed February 13, 2015).

60. Henry Blodget, "Walmart Employs 1% of America. Should It Be Forced to Pay Its Employees More?" *Business Insider,* September 20, 2010, http://www.businessinsider.com/walmart-employees-pay (accessed April 6, 2017).

61. Gifford Pinchott III, *Intrapreneuring* (New York: Harper & Row, 1985), p. 34.

62. Paul Brown, "How to Cope with Hard Times," *The New York Times,* June 10, 2008, www.nytimes.com/2008/06/10/business/smallbusiness/10toolkit.html?r%205%201&ref%205%20smallbusiness&orefslogin&gwh=A256B42494736F9E2C604851BF6451DC&gwt=regi (accessed April 22, 2014).

63. Adapted from Carol Kinsey Gorman, *Creativity in Business: A Practical Guide for Creative Thinking,* Crisp Publications Inc., 1989, pp. 5–6. © Crisp Publications Inc., 1200 Hamilton Court, Menlo Park, CA 94025.

64. Sonic Beach website, http://www.sonicbeach.com/ (accessed March 29, 2016); Sonic website, www.sonicdrivein.com (accessed March 29, 2016); "Strictly Speaking," Sonic website, www.sonicdrivein.com/business/franchise/faq.jsp (accessed July 27, 2012); "Awards," Sonic, https://www.sonicdrivein.com/corporate/awards (accessed March 29, 2016); "Sonic Drive-In Restaurants," *Entrepreneur,* 2016, https://www.sonicdrivein.com/corporate/awards (accessed March 29, 2016).

Credits

Managing for Quality and Competitiveness

PART 3

Managing for Quality and Competitiveness

6 The Nature of Management

©QualityHD/Shutterstock

Chapter Outline

Learning Objectives

After reading this chapter, you will be able to:

LO 6-1 Explain management's role in the achievement of organizational objectives.

LO 6-2 Describe the major functions of management.

LO 6-3 Distinguish among three levels of management and the concerns of managers at each level.

LO 6-4 Specify the skills managers need in order to be successful.

LO 6-5 Summarize the systematic approach to decision making used by many business managers.

LO 6-6 Recommend a new strategy to revive a struggling business.

Enter the World of Business

Home Depot Builds Effective Management System

When Home Depot was founded in 1979, the founders built a strong culture placing customers and employees at the top and executives at the bottom. However, after Robert Nardelli, a high-level executive at General Electric, took over as CEO, the style of leadership at Home Depot abruptly changed. Nardelli took a top-down approach to running Home Depot: executives at the top and customers and employees at the bottom.

This new management style was disastrous, and customer satisfaction hit an all-time low. After Nardelli was ousted, new CEO Frank Blake quickly refocused on customers and employees. As an authentic leader, Blake led by example. He quickly admitted to the customer service problems the company faced, apologized for the inconvenience it caused, and encouraged customers to leave feedback so it could make improvements. Each one of the complaints was addressed; some angry followers were appeased by phone calls from managers and personal e-mails responding to their specific issues. Both customer satisfaction and company morale rose once more.

After Blake stepped down as CEO, he was replaced by Craig Menear. Menear proved that he would continue managing the company in a way that would honor the original culture of Home Depot. This is being put to the test as brick-and-mortar retailers are experiencing major changes because of online competition. While many companies have struggled because of online retailing, Home Depot has continued to see revenues increase, partly because of the resolve of leadership to invest in a strong e-commerce strategy. The company's online sales have increased 21.5 percent since Menear became CEO. Thanks to the transformational leadership of its committed managers, Home Depot is effectively competing in the "Amazon-era" with more than $78 billion in revenues.[1]

Introduction

For any organization—small or large, for profit or nonprofit—to achieve its objectives, it must have resources to support operations; employees to make and sell the products; and financial resources to purchase additional goods and services, pay employees, and generally operate the business. To accomplish this, it must also have one or more managers to plan, organize, staff, direct, and control the work that goes on.

This chapter introduces the field of management. It examines and surveys the various functions, levels, and areas of management in business. The skills that managers need for success and the steps that lead to effective decision making are also discussed.

management
a process designed to achieve an organization's objectives by using its resources effectively and efficiently in a changing environment.

LO 6-1

Explain management's role in the achievement of organizational objectives.

managers
those individuals in organizations who make decisions about the use of resources and who are concerned with planning, organizing, staffing, directing, and controlling the organization's activities to reach its objectives.

staffing
the hiring of people to carry out the work of the organization.

downsizing
the elimination of a significant number of employees from an organization.

The Importance of Management

Management is a process designed to achieve an organization's objectives by using its resources effectively and efficiently in a changing environment. *Effectively* means having the intended result; *efficiently* means accomplishing the objectives with a minimum of resources. **Managers** make decisions about the use of the organization's resources and are concerned with planning, organizing, directing, and controlling the organization's activities so as to reach its objectives. Consider Waze, Google's crowdsourcing map app that taps data from 90 million users to deliver information on traffic patterns and infrastructure problems. Management of Waze sells location-based advertising and traffic data such as the location of a local McDonald's restaurant. The firm helped plan the launch of carpool or ride-sharing programs to help commuters, provided organizational structure for the launch, and was involved in directing and controlling implementation of the service.[2] Management is universal. It takes place not only in business, but also in government, the military, labor unions, hospitals, schools, and religious groups—any organization requiring the coordination of resources.

Every organization must acquire resources (people, services, raw materials, equipment, finances, and information) to effectively pursue its objectives and coordinate their use to turn out a final good or service. Employees are one of the most important resources in helping a business attain its objectives. Hiring people to carry out the work of the organization is known as **staffing.** Beyond recruiting people for positions within the firm, managers must determine what skills are needed for specific jobs, how to motivate and train employees, how much to pay, what benefits to provide, and how to prepare employees for higher-level jobs in the firm at a later date. Sometimes, they must also make the difficult decision to reduce the workforce. This is known as **downsizing,** the elimination of significant numbers of employees from an organization. After a downsizing situation, an effective manager will promote optimism and positive thinking and minimize criticism and fault-finding. These elements of staffing will be explored in detail in the "Motivating the Workforce" and "Managing Human Resources" chapters.

Acquiring suppliers is another important part of managing resources and ensuring that products are made available to customers. As firms reach global markets, companies such as PepsiCo, Corning, and Charles Schwab enlist hundreds of diverse suppliers

As a result of over-hiring, Snap Inc., the parent company of Snapchat, cut more than 220 workers, including 120 engineers.[3]

©dennizn/Shutterstock

that provide goods and services to support operations. A good supplier maximizes efficiencies and provides creative solutions to help the company reduce expenses and reach its objectives. Finally, the manager needs adequate financial resources to pay for essential activities. Primary funding comes from owners and shareholders, as well as banks and other financial institutions. All these resources and activities must be coordinated and controlled if the company is to earn a profit. Organizations must also have adequate supplies of resources of all types, and managers must carefully coordinate their use if they are to achieve the organization's objectives.

Management Functions

To harmonize the use of resources so that the business can develop, produce, and sell products, managers engage in a series of activities: planning, organizing, directing, and controlling (Figure 6.1). Although this book discusses each of the four functions separately, they are interrelated; managers may perform two or more of them at the same time.

Planning

Planning, the process of determining the organization's objectives and deciding how to accomplish them, is the first function of management. Planning is a crucial activity because it designs the map that lays the groundwork for the other functions. It involves forecasting events and determining the best course of action from a set of options or choices. The plan itself specifies what should be done, by whom, where, when, and how. For some managers, one major decision that requires extensive planning is selecting the right type of automation for warehouses and distribution facilities. Data gathering is a major phase of the planning process to determine what the facilities need and which automation can maximize order efficiency. Potential pitfalls in this process that managers should plan for include being swayed by advanced technology that is not needed, under-automating the facility, or over-automating the facility.[4] All businesses—from the smallest restaurant to the largest multinational corporation—need to develop plans for achieving success. But before an organization can plan a course of action, it must first determine what it wants to achieve.

Mission.   A **mission,** or mission statement, is a declaration of an organization's fundamental purpose and basic philosophy. It seeks to answer the question: "What business are we in?" Good mission statements are clear and concise statements that

LO 6-2

Describe the major functions of management.

planning
the process of determining the organization's objectives and deciding how to accomplish them; the first function of management.

mission
the statement of an organization's fundamental purpose and basic philosophy.

FIGURE 6.1
The Functions of Management

WeWork, a leader in the shared co-working space industry, has declared its mission to be to create a world where people work to make a life, not just a living.

©Yonhap/Epa/REX/Shutterstock

explain the organization's reason for existence. A well-developed mission statement, no matter what the industry or size of business, will answer five basic questions:

1. Who are we?
2. Who are our customers?
3. What is our operating philosophy (basic beliefs, values, ethics, etc.)?
4. What are our core competencies and competitive advantages?
5. What are our responsibilities with respect to being a good steward of environmental, financial, and human resources?

A mission statement that delivers a clear answer to these questions provides the foundation for the development of a strong organizational culture, a good marketing plan, and a coherent business strategy. Tesla's mission is "to accelerate the world's transition to sustainable energy."[5]

Goals.   A goal is the result that a firm wishes to achieve. A company almost always has multiple goals, which illustrates the complex nature of business. A goal has three key components: an attribute sought, such as profits, customer satisfaction, or product quality; a target to be achieved, such as the volume of sales or extent of management training to be completed; and a time frame, which is the time period in which the goal is to be attained. Sometimes goals have unintended consequences. When Wells Fargo set goals for salespeople to generate new accounts, they did not intend for their employees to fraudulently set up new accounts without the customer's knowledge. Incidents such as this do tremendous damage to reputation, brand, and customer attraction and retention. To be successful at achieving goals, it is necessary to know what is to be achieved, how much, when, and how succeeding at a goal is to be determined.

Objectives.   Objectives, the ends or results desired by an organization, derive from the organization's mission. A business's objectives may be elaborate or simple. Common objectives relate to profit, competitive advantage, efficiency, and growth.

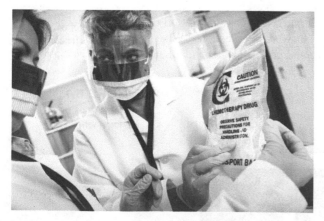

One of the FDA's recent annual goals was to ensure the quality of medications by issuing new regulations.

©fstop123/Getty Images

The principal difference between goals and objectives is that objectives are generally stated in such a way that they are measurable. Organizations with profit as an objective want to have money and assets left over after paying off business expenses. Objectives regarding competitive advantage are generally stated in terms of percentage of sales increase and market share, with the goal of increasing those figures. Efficiency objectives involve making the best use of the organization's resources. Growth objectives relate to an organization's ability to adapt and to get new products to the marketplace in a timely fashion. One of the most important objectives for businesses is sales. In the United States, the consumption of bottled water surpassed soda for the first time. Water marketers took advantage of the trend, and to boost sales, they advertised in Super Bowl LI.

Rebecca Ray Designs Has It in the Bag

Rebecca Ray Designs

Founder: Rebecca Yuhasz Smith

Founded: 1998, in Chagrin Falls, Ohio

Success: Due to her reputation for quality, the Kentucky Derby used Rebecca Ray-designed gift bags for its VIP guests.

Rebecca Yuhasz Smith's retail and wholesale company, Rebecca Ray Designs, grew out of her love of animals and the sporting lifestyle. Rebecca partnered with Amish communities in Ohio and Pennsylvania to create and sell equestrian-themed luxury handbags. As both manager and founder, Rebecca had to plan the organization's goals, organize different job functions, lead others to develop and sell the products, and control activities needed for achieving objectives. Luxury bags are hand-sewn by Amish women with pedal-powered sewing machines. Amish harness-makers hand-make the company's leather products, and the firm employs a sales team who ride on horseback.

In 2007 Rebecca Ray Designs won the category of *Country Living Magazine*'s Pitch Your Product competition. Rebecca wanted to expand, but she underestimated the human resources skills needed. After hiring sales representatives, the number of boutiques selling Rebecca Ray Designs went from 35 to about 50. Today Rebecca's luxury products include handbags, accessories, and home décor sold at 300 retailers as well as online. In her role as manager, Rebecca turned Rebecca Ray Designs into a successful high-fashion business.[6]

Critical Thinking Questions

1. What management skills did Rebecca Yuhasz Smith need to found her successful business?
2. How does Rebecca Yuhasz Smith use the four functions of management to operate her business?
3. What types of decisions did Rebecca Yuhasz Smith need to make as her business grew?

Pepsi launched LIFEWTR, and Fiji advertised its water to an audience of more than 111 million.[7] Objectives provide direction for all managerial decisions; additionally, they establish criteria by which performance can be evaluated.

Plans. There are three general types of plans for meeting objectives—strategic, tactical, and operational. A firm's highest managers develop its **strategic plans,** which establish the long-range objectives and overall strategy or course of action by which the firm fulfills its mission. Strategic plans generally cover periods of one year or longer. They include plans to add products, purchase companies, sell unprofitable segments of the business, issue stock, and move into international markets. For example, Combekk, a Dutch firm that makes knives and tools, set an objective to increase sales by introducing a heavyweight pot called a Dutch Oven that cost $450. The 100 percent recycled iron collected from bridges and former train rails has an added value to help create an incredible sales success.[8] Strategic plans must take into account the organization's capabilities and resources, the changing business environment, and organizational objectives. Plans should be market-driven, matching customers' desire for value with operational capabilities, processes, and human resources.[9]

Tactical plans are short range and designed to implement the activities and objectives specified in the strategic plan. These plans, which usually cover a period of one year or less, help keep the organization on the course established in the strategic plan. General Motors, for instance, developed tactical plans to release redesigned versions of its vehicles that target millennials as part of its strategic plan to grow market share and reduce rental deliveries.[10] Because tactical plans allow the organization to react to changes in the environment while continuing to focus on the company's overall strategy, management must periodically review and update them. Declining performance or failure to meet objectives set out in tactical plans are possible reasons for revising them. The differences between the two types of planning result in different

strategic plans
those plans that establish the long-range objectives and overall strategy or course of action by which a firm fulfills its mission.

tactical plans
short-range plans designed to implement the activities and objectives specified in the strategic plan.

operational plans
very short-term plans that specify what actions individuals, work groups, or departments need to accomplish in order to achieve the tactical plan and ultimately the strategic plan.

crisis management
an element in planning that deals with potential disasters such as product tampering, oil spills, fire, earthquake, computer virus, or airplane crash.

contingency planning
an element in planning that deals with potential disasters such as product tampering, oil spills, fire, earthquake, computer virus, or airplane crash.

activities in the short term versus the long term. For instance, a strategic plan might include the use of social media to reach consumers. A tactical plan could involve finding ways to increase traffic to the site or promoting premium content to those who visit the site. A fast-paced and ever-changing market requires companies to develop short-run or tactical plans to deal with the changing environment.

Many traditional retailers have been under enormous pressure because of online providers such as Amazon. A retailing organization may have a five-year strategic plan that calls for store closures and revamping their sales structure. The tactical part of the plan would address which stores to close and the timeline, how to boost their online presence, and how to handle employee separation and layoffs. Tactical plans are designed to execute the overall strategic plan. Because of their short-term nature, they are easier to adjust or abandon if changes in the environment or the company's performance so warrant.

Operational plans are very short term and specify what actions specific individuals, work groups, or departments need to accomplish in order to achieve the tactical plan and, ultimately, the strategic plan. They apply to details in executing activities in one month, week, or even day. For example, a work group may be assigned a weekly production quota to ensure there are sufficient products available to elevate market share (tactical goal) and ultimately help the firm be number one in its product category (strategic goal). Returning to our retail store example, operational plans may specify the schedule for opening one new store, hiring and training new employees, obtaining merchandise, and opening for actual business.

Another element of planning is **crisis management** or **contingency planning,** which deals with potential disasters such as product tampering, oil spills, fire, earthquake, computer viruses, or even a reputation crisis due to unethical or illegal conduct by one or more employees. Unfortunately, many businesses do not have updated contingency plans to handle the types of crises that their companies might encounter. According to the Federal Emergency Management Agency (FEMA), approximately 40 percent of small businesses do not reopen after a disaster.[11] Businesses that have well-thought-out contingency plans tend to respond more effectively when problems occur than do businesses that lack such planning.

Many companies—including Ashland Oil, H. J. Heinz, and Johnson & Johnson—have crisis management teams to deal specifically with problems, permitting other managers to continue to focus on their regular duties. Some companies even hold periodic disaster drills to ensure that their employees know how to respond when a crisis does occur. After the horrific earthquake in Japan, many companies in U.S. earthquake zones reevaluated their crisis management plans. Crisis management plans generally cover maintaining business operations throughout a crisis and communicating with the public, employees, and officials about the nature of and the company's response to the problem. Communication is especially important to minimize panic and damaging rumors; it also demonstrates that the company is aware of the problem and plans to respond.

WE'RE SORRY

A chicken restaurant without any chicken. It's not ideal. Huge apologies to our customers, especially those who travelled out of their way to find we were closed. And endless thanks to our KFC team members and our franchise partners for working tirelessly to improve the situation. It's been a hell of a week, but we're making progress, and every day more and more fresh chicken is being delivered to our restaurants. Thank you for hearing with us.

Visit kfc.co.uk/crossed-the-road for details about your local restaurant.

After KFC's chicken shortage in the UK went viral on Twitter, the fast-food company acknowledged the issue in a timely manner, addressed guest concerns on social media, and published a series of ads apologizing for the supply chain error.
©Ray Tang/REX/Shutterstock

Sometimes, disasters occur that no one can anticipate, but companies can still plan for how to react to the disaster. There can be a major crisis when supply that is time sensitive is disrupted. That's what happened to KFC in the UK when it pared back its logistics network to cut expenses. The move resulted in two-thirds of its outlets without chicken for several days.[12] Incidents such as this highlight the importance of planning for crises and the need to respond publicly and quickly when a disaster occurs.

Organizing

Rarely are individuals in an organization able to achieve common goals without some form of structure. **Organizing** is the structuring of resources and activities to accomplish objectives in an efficient and effective manner. Managers organize by reviewing plans and determining what activities are necessary to implement them; then, they divide the work into small units and assign it to specific individuals, groups, or departments. As companies reorganize for greater efficiency, more often than not, they are organizing work into teams to handle core processes such as new product development instead of organizing around traditional departments such as marketing and production. Organizing occurs continuously because change is inevitable.

Organizing is important for several reasons. It helps create synergy, whereby the effect of a whole system equals more than that of its parts. It also establishes lines of authority, improves communication, helps avoid duplication of resources, and can improve competitiveness by speeding up decision making. Procter & Gamble underwent a major, multiyear reorganization cutting more than 20,000 jobs, and the pending sale of Duracell Batteries and merger with Coty will result in more loss of jobs for P&G. Major changes in staffing and structure present challenges and opportunities for companies such as P&G.[13]

A business model relates to how a firm is organized to operate and provide value to stakeholders. It is the map or blueprint for running a business—a conceptual tool for organizing how a business operates.[14] Examples of business models include a Subway franchise, Avon direct selling, Amazon and online retailing, and Netflix's subscription business model that provides access to entertainment. General business models relate to manufacturing to create a product or a distribution that resells to retailers. Today many businesses are trying to create new business models that focus on business sectors such as the sharing economy. Uber and Airbnb provide access but not ownership of products. In the future, many business models will emerge related to the digital economy, driverless vehicles, robotics, drones, and artificial intelligence. Artificial intelligence will allow managers to gain extraordinary control over their workers as well as forecasting demand, developing customer relationships, and managing the supply chain. New business models will develop around artificial intelligence systems.[15] Because organizing is so important, we'll take a closer look at it in the chapter titled "Organization, Teamwork, and Communication."

organizing
the structuring of resources and activities to accomplish objectives in an efficient and effective manner.

Directing

During planning and organizing, staffing occurs and management must direct the employees. **Directing** is motivating and leading employees to achieve organizational objectives. Good directing involves telling employees what to do and when to do it through the implementation of deadlines and then encouraging them to do their work. For example, as a sales manager, you would need to learn how to motivate

directing
motivating and leading employees to achieve organizational objectives.

salespersons, provide leadership, teach sales teams to be responsive to customer needs, and manage organizational issues as well as evaluate sales results. Finally, directing also involves determining and administering appropriate rewards and recognition. All managers are involved in directing, but it is especially important for lower-level managers who interact daily with the employees operating the organization. For example, an assembly-line supervisor for Frito-Lay must ensure that her workers know how to use their equipment properly and have the resources needed to carry out their jobs safely and efficiently, and she must motivate her workers to achieve their expected output of packaged snacks.

Managers may motivate employees by providing incentives—such as the promise of a raise or promotion—for them to do a good job. But most workers want more than money from their jobs: They need to know that their employer values their ideas and input. Managers should give younger employees some decision-making authority as soon as possible. Smart managers, therefore, ask workers to contribute ideas for reducing costs, making equipment more efficient, improving customer service, or even developing new products. This participation also serves to increase employee morale. Recognition and appreciation are often the best motivators. Employees who understand more about their effect on the financial success of the company may be induced to work harder for that success, and managers who understand the needs and desires of workers can encourage their employees to work harder and more productively. The motivation of employees is discussed in detail in the chapter titled "Motivating the Workforce."

Controlling

controlling
the process of evaluating and correcting activities to keep the organization on course.

Planning, organizing, staffing, and directing are all important to the success of an organization, whether its objective is earning a profit or something else. But what happens when a firm fails to reach its goals despite a strong planning effort? **Controlling** is the process of evaluating and correcting activities to keep the organization on course. Control involves five activities: (1) measuring performance, (2) comparing present performance with standards or objectives, (3) identifying deviations from the standards, (4) investigating the causes of deviations, and (5) taking corrective action when necessary.

Controlling and planning are closely linked. Planning establishes goals and standards. By monitoring performance and comparing it with standards, managers can determine whether performance is on target. When performance is substandard, management must determine why and take appropriate actions to get the firm back on course. In short, the control function helps managers assess the success of their plans. You might relate this to your performance in this class. If you did not perform as well on early projects or exams, you must take corrective action such as increasing studying or using website resources to achieve your overall objective of getting an A or B in the course. When the outcomes of plans do not meet expectations, the control process facilitates revision of the plans. Control can take many forms such as visual inspections, testing, and statistical modeling processes. The basic idea is to ensure that operations meet requirements and are satisfactory to reach objectives.

The control process also helps managers deal with problems arising outside the firm. For example, if a firm is the subject of negative publicity, management should use the control process to determine why, and to guide the firm's response.

Types of Management

LO 6-3

Distinguish among three levels of management and the concerns of managers at each level.

All managers—whether the sole proprietor of a jewelry store or the hundreds of managers of a large company such as Home Depot—perform the four functions just discussed. In the case of the jewelry store, the owner handles all the functions, but in a large company with more than one manager, responsibilities must be divided and delegated. This division of responsibility is generally achieved by establishing levels of management and areas of specialization—finance, marketing, and so on.

Levels of Management

As we have hinted, many organizations have multiple levels of management—top management, middle management, and first-line (or supervisory) management. These levels form a pyramid, as shown in Figure 6.2. As the pyramid shape implies, there are generally more middle managers than top managers and still more first-line managers. Very small organizations may have only one manager (typically, the owner), who assumes the responsibilities of all three levels. Large businesses have many managers at each level to coordinate the use of the organization's resources. Managers at all three levels perform all four management functions, but the amount of time they spend on each function varies, as we shall see (Figure 6.3).

Top Management. In businesses, **top managers** include the president and other top executives, such as the chief executive officer (CEO), chief financial officer (CFO), and chief operations officer (COO), who have overall responsibility for the organization. For example, the CEO of a company manages the overall strategic direction of the company and plays a key role in representing the company to stakeholders. The COO is responsible for daily operations of the company. The COO reports to the CEO and is often considered the number two in command. In public corporations, even the CEO has a boss, which is the board of directors. With technological advances accelerating and privacy concerns increasing, many companies are adding a new executive in the form of a chief privacy officer (CPO). The position of privacy officer has grown and the International Association of Privacy Professionals now have more than 20,000 members in 83 countries.[16] In government, top management

top managers
the president and other top executives of a business, such as the chief executive officer (CEO), chief financial officer (CFO), and chief operations officer (COO), who have overall responsibility for the organization.

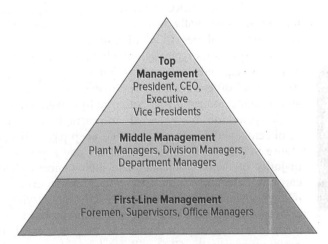

FIGURE 6.2
Levels of Management Planning

FIGURE 6.3

Importance of
Management Functions to
Managers in Each Level

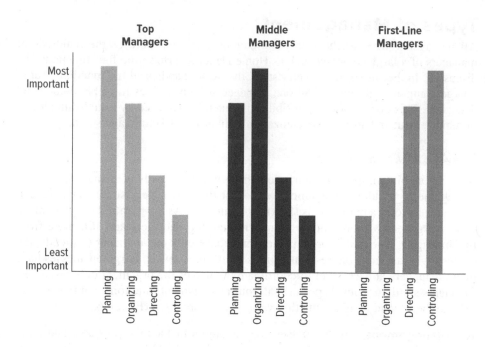

JCPenney CEO Marvin Ellison earns about $10.8
million annually.[18]

©MediaPunch/REX/Shutterstock

refers to the president, a governor, or a mayor or city manager; in education, a chancellor of a university or a superintendent of education.

Top-level managers spend most of their time planning. They make the organization's strategic decisions, decisions that focus on an overall scheme or key idea for using resources to take advantage of opportunities. They decide whether to add products, acquire companies, sell unprofitable business segments, and move into foreign markets. Top managers also represent their company to the public and to government regulators.

Given the importance and range of top management's decisions, top managers generally have many years of varied experience and command top salaries. In addition to salaries, top managers' compensation packages typically include bonuses, long-term incentive awards, stock, and stock options. Table 6.1 lists the compensation packages of different CEOs. Top management may also get perks and special treatment that is criticized by stakeholders.

Compensation committees are increasingly working with boards of directors and CEOs to attempt to keep pay in line with performance in order to benefit stockholders and key stakeholders. The majority of major companies cite their concern about attracting capable leadership for the CEO and other top executive positions in their organizations. However, many firms are trying to curb criticism of excessive executive compensation by trying to align CEO compensation with performance. In other words, if the company performs poorly, the CEO will not be paid as well. This type of compensation method is making a difference.[19] Chipotle continues

CEO	Company	Total Compensation
Ginni Rometty	IBM	$96 million
Tim Cook	Apple	$150 million
Elon Musk	Tesla	$99 million
John D. Wren	Omnicom Group	$26 million
Sundar Pichai	Google	$106 million
Mark G. Parker	Nike	$47 million

TABLE 6.1
Compensation Packages of CEOs

Sources: Emmie Martin, "The 5 Highest-Paid CEOs in the U.S.," CNBC, May 11, 2017, https://www.cnbc.com/2017/05/11/highest-paid-ceos-in-america.html (accessed April 16, 2018); Samuel Stebbins, "Highest Paid CEOs in 2017," 24/7 Wall St., December 28, 2017, https://247wallst.com/special-report/2017/12/28/highest-paid-ceos-in-2017/ (accessed April 16, 2018).

to struggle after attempting to recover from *E. coli* outbreaks in its restaurants. The co-CEOs had their pay cut in half, and the stock continues to underperform, losing 30 percent of its value in one year. Unsuccessful management has negative consequences for leaders, whereas successful management translates into happy stockholders who are willing to compensate their top executives fairly and in line with performance.[20]

Workforce diversity is an important issue in today's corporations. Effective managers at enlightened corporations have found that diversity is good for workers and for the bottom line. Putting together different kinds of people to solve problems often results in better solutions. Kaiser Permanente topped Diversity Inc's list with its culture of diversity and inclusiveness. The company has minorities in top management positions at a level 66.6 percent higher than the other top 10 companies on the list. Executive compensation is tied to diversity, diversity metrics, and progress in growing supplier diversity.[21] A diverse workforce is better at making decisions regarding issues related to consumer diversity. Reaching fast-growing demographic groups such as Hispanics, African Americans, Asian Americans, and others will be beneficial to large companies as they begin to target these markets.[22] Managers from companies devoted to workforce diversity devised five rules that make diversity recruiting work (see Table 6.2).

Diversity concerns also relate to gender equality. Gender equality relates to pay, promotions, respect, and how one is treated in the workplace. The higher ranks in management often have significantly fewer women than men, and the women often receive lower pay, although Starbucks has achieved equal pay and career advancement for women. In addition, women occupy only 15 percent of board of director seats globally.[23] Diversity is explored in greater detail in the chapter titled "Managing Human Resources."

Middle Management.   Rather than making strategic decisions about the whole organization, **middle managers** are responsible for tactical and operational planning that will implement the general guidelines established by top management. Thus, their responsibility is more narrowly focused than that of top managers. Middle managers are involved in the specific operations of the organization and spend more time organizing than other managers. In business, plant managers, division managers, and department managers make up middle management. The product manager for laundry detergent at a consumer products manufacturer, the department chairperson in a

middle managers
those members of an organization responsible for the tactical planning that implements the general guidelines established by top management.

TABLE 6.2
Five Rules of Successful
Diversity Recruiting

Rule	Action
1. Involve employees	Educate all employees on the tangible benefits of diversity recruiting to garner support and enthusiasm for those initiatives.
2. Communicate diversity	Prospective employees are not likely to become excited about joining your company just because you say that your company is diversity-friendly; they need to see it.
3. Support diversity initiatives and activities	By supporting community-based diversity organizations, your company will generate the priceless word-of-mouth publicity that will lead qualified diversity candidates to your company.
4. Delegate resources	If you are serious about diversity recruiting, you will need to spend some money getting your message out to the right places.
5. Promote your diversity initiatives	Employers need to sell their company to prospective diversity employees and present them with a convincing case as to why their company is a good fit for the diversity candidate.

Source: Adapted from Juan Rodriguez, "The Five Rules of Successful Diversity Recruiting," Diversityjobs.com, www.diversityjobs.com/Rules-of-Successful Diversity-Recruiting (accessed February 25, 2010).

university, and the head of a state public health department are all middle managers. The ranks of middle managers have been shrinking as more and more companies downsize to be more productive.

First-Line Management.   Most people get their first managerial experience as **first-line managers,** those who supervise workers and the daily operations of the organization. They are responsible for implementing the plans established by middle management and directing workers' daily performance on the job. They spend most of their time directing and controlling. Common titles for first-line managers are foreman, supervisor, and office manager.

first-line managers
those who supervise
both workers and the
daily operations of an
organization.

Areas of Management

At each level, there are managers who specialize in the basic functional areas of business: finance, production and operations, human resources (personnel), marketing, IT, and administration.

Each of these management areas is important to a business's success. For instance, a firm cannot survive without someone obtaining needed financial resources (financial managers) or staff (human resources managers). While larger firms will most likely have all of these managers, and even more depending upon that particular firm's needs, in smaller firms, these important tasks may fall onto the owner or a few employees. Yet whether or not companies have managers for specific areas, every company must have someone responsible for obtaining financial resources, transforming resources into finished products for the marketplace, hiring and/or dealing with staff, marketing goods and services, handling the firm's information technology resources, and managing a business segment or the overall business. These different types of managers are discussed in more detail in Table 6.3.

TABLE 6.3   Areas of Management

Manager	Function
Financial manager	Focus on obtaining the money needed for the successful operation of the organization and using that money in accordance with organizational goals.
Production and operations manager	Develop and administer the activities involved in transforming resources into goods, services, and ideas ready for the marketplace.
Human resources manager	Handle the staffing function and deal with employees in a formalized manner.
Marketing manager	Responsible for planning, pricing, and promoting products and making them available to customers through distribution.
Information technology (IT) manager	Responsible for implementing, maintaining, and controlling technology applications in business, such as computer networks.
Administrative manager	Manage an entire business or a major segment of a business; do not specialize in a particular function.

financial managers
those who focus on obtaining needed funds for the successful operation of an organization and using those funds to further organizational goals.

production and operations managers
those who develop and administer the activities involved in transforming resources into goods, services, and ideas ready for the marketplace.

human resources managers
those who handle the staffing function and deal with employees in a formalized manner.

Skills Needed by Managers

Managers are typically evaluated using the metrics of how effective and efficient they are. Managing effectively and efficiently requires certain skills—technical expertise, conceptual skills, analytical skills, human relations skills, and leadership. Table 6.4 describes some of the roles managers may fulfill.

LO 6-4
Specify the skills managers need in order to be successful.

marketing managers
those who are responsible for planning, pricing, and promoting products and making them available to customers.

information technology (IT) managers
those who are responsible for implementing, maintaining, and controlling technology applications in business, such as computer networks.

administrative managers
those who manage an entire business or a major segment of a business; they are not specialists but coordinate the activities of specialized managers.

TABLE 6.4   Managerial Roles

General Role Category	Specific Role	Example Activity
Interpersonal	Figure	Attending award banquet
	Liaison	Coordinating production schedule with supply manager
	Leadership	Conducting performance appraisal for subordinates
Informational	Monitor	Contacting government regulatory agencies
	Disseminator	Conducting meetings with subordinates to pass along safety policy
	Spokesperson	Meeting with consumer group to discuss product safety
Decisional	Entrepreneur	Changing work process
	Disturbance handler	Deciding which unit moves into new facilities
	Resource allocator	Deciding who receives new computer equipment
	Negotiator	Settling union grievance

Source: Roles developed by management professor Henry Mintzberg.

IT managers are responsible for implementing, maintaining, and controlling technology applications in business, such as computer networks.

©Huntstock/Getty Images

technical expertise
the specialized knowledge and training needed to perform jobs that are related to particular areas of management.

conceptual skills
the ability to think in abstract terms and to see how parts fit together to form the whole.

Technical Expertise

Managers need **technical expertise,** the specialized knowledge and training required to perform jobs related to their area of management. Accounting managers need to be able to perform accounting jobs, and production managers need to be able to perform production jobs. Although a production manager may not actually perform a job, he or she needs technical expertise to train employees, answer questions, provide guidance, and solve problems. Technical skills are most needed by first-line managers and are least critical to top-level managers.

Conceptual Skills

Conceptual skills, the ability to think in abstract terms, and to see how parts fit together to form the whole, are needed by all managers, but particularly top-level managers. Top management must be able to evaluate continually where the company will be in the future. Conceptual skills also involve the ability to think creatively. Recent scientific research has revealed that creative thinking, which is behind the development of many innovative products and ideas, can be learned. As a result, IBM, AT&T, GE, Hewlett-Packard, Intel, and other top U.S. firms hire creative consultants to teach their managers how to think creatively.

Analytical Skills

Analytical skills refer to the ability to identify relevant issues and recognize their importance, understand the relationships between them, and perceive the underlying causes of a situation. When managers have identified critical factors and causes, they can take appropriate action. All managers need to think logically, but this skill is probably most important to the success of top-level managers. To be analytical, it is necessary to think about a broad range of issues and to weigh different options before taking action. Because analytical skills are so important, questions that require analytical skills are often a part of job interviews. Questions such as "Tell me how you would resolve a problem at work if you had access to a large amount of data?" may be part of the interview process. The answer would require the interviewee to try to explain how to sort data to find relevant facts that could resolve the issue. Analytical thinking is required in complex or difficult situations where the solution is often not clear. Resolving ethical issues often requires analytical skills.

Financial managers are responsible for obtaining the necessary funding for organizations to succeed, both in the short term and in the long term.

©pixdeluxe/Getty Images

Human Relations Skills

People skills, or **human relations skills,** are the ability to deal with people, both inside and outside the organization. Those who can relate to others, communicate well with others, understand the needs of others, and show a true appreciation for others are generally more successful than managers who lack such skills. People skills are especially important in hospitals, airline companies, banks, and other organizations that provide services. For example, Southwest Airlines places great value on its employees. New hires go through extensive training to teach employees about the airline and its reputation for impeccable customer service. All employees in management positions at Southwest take mandatory leadership classes that address skills related to listening, staying in touch with employees, and handling change without compromising values.

Leadership

Leadership is the ability to influence employees to work toward organizational goals. Strong leaders manage and pay attention to the culture of their organizations and the needs of their employees. Table 6.5 offers some requirements for successful leadership.

Managers often can be classified into three types based on their leadership style. *Autocratic leaders* make all the decisions and then tell employees what must be done and how to do it. They generally use their authority and economic rewards to get employees to comply with their directions. Martha Stewart is an example of an autocratic leader. She built up her media empire by paying close attention to every detail.[24] *Democratic leaders* involve their employees in decisions. The manager presents a situation and encourages his or her subordinates to express opinions and contribute ideas. The manager then considers the employees' points of view and makes the decision. Herb Kelleher, co-founder of Southwest Airlines, had a democratic leadership style. Under his leadership, employees were encouraged to discuss concerns and provide input.[25] *Free-rein leaders* let their employees work without much interference. The manager sets performance standards and allows employees to find their own ways to meet them. For this style to be effective, employees must know what the standards are, and they must be motivated to attain them. The free-rein style of leadership can be a powerful motivator because it demonstrates a great deal of trust and confidence in the employee. Warren Buffett, CEO of Berkshire Hathaway, exhibits free-rein leadership among the managers who run the company's various businesses.

The effectiveness of the autocratic, democratic, and free-rein styles depends on several factors. One consideration is the type of employees. An autocratic style of leadership is generally best for stimulating unskilled, unmotivated employees; highly skilled, trained, and motivated employees may respond better to democratic or free-rein leadership styles. Employees who have been involved in decision

• Communicate objectives and expectations.
• Gain the respect and trust of stakeholders.
• Develop shared values.
• Acquire and share knowledge.
• Empower employees to make decisions.
• Be a role model for appropriate behavior.
• Provide rewards and take corrective action to achieve goals.

TABLE 6.5
Requirements for Successful Leadership

analytical skills
the ability to identify relevant issues, recognize their importance, understand the relationships between them, and perceive the underlying causes of a situation.

human relations skills
the ability to deal with people, both inside and outside the organization.

LO 6-5

Summarize the systematic approach to decision making used by many business managers.

leadership
the ability to influence employees to work toward organizational goals.

 connect

Need help understanding **leaders vs. managers?** Visit your Connect ebook video tab for a brief animated explanation.

Twitter CEO Jack Dorsey believes in a democratic leadership style.

©ARoger Askew/REX/Shutterstock

making generally require less supervision than those not similarly involved. Other considerations are the manager's abilities and the situation itself. When a situation requires quick decisions, an autocratic style of leadership may be best because the manager does not have to consider input from a lot of people. If a special task force must be set up to solve a quality-control problem, a normally democratic manager may give free rein to the task force.

Many managers, however, are unable to use more than one style of leadership. Some are incapable of allowing their subordinates to participate in decision making, let alone make any decisions. What leadership style is "best" depends on specific circumstances, and effective managers will strive to adapt their leadership style as circumstances warrant. Many organizations offer programs to develop good leadership skills. Now in his late 80s, Warren Buffet, chair of Berkshire Hathaway, has developed dozens of exemplary leaders with proven track records. Leadership continuity is important to all firms.[26] When plans fail, very often leaders are held responsible for what goes wrong. For example, Wells Fargo CEO John Stumpf resigned before probably being fired for oversight of a bank that opened more than 3.5 million fake accounts for customers.[27]

Another type of leadership style that has been gaining in popularity is *authentic leadership*. Authentic leadership is a bit different from the other three leadership styles because it is not exclusive. Both democratic and free-rein leaders could qualify as authentic leaders depending upon how they conduct themselves among stakeholders. Authentic leaders are passionate about the goals and mission of the company, display corporate values in the workplace, and form long-term relationships with stakeholders.[28] Chobani founder and CEO Hamdi Ulukaya feels a deep responsibility to consumers and employees. He believes in equitable polices for employees and embraces diversity, especially related to immigrants and refugees. He wants to stand for something more than profit.[29]

Former Wells Fargo CEO John Stumpf, who testified before the Senate Banking Committee after the company opened 3.5 million fake accounts for customers, resigned and was replaced by Tim Sloan.

©Michael Reynolds/Epa/REX/Shutterstock

While leaders might incorporate different leadership styles depending on the business and the situation, all leaders must be able to align employees behind a common vision to be effective.[30] Strong leaders also realize the value that employees can provide by participating in the firm's corporate culture. It is important that companies develop leadership training programs for employees. Because managers cannot oversee everything that goes on in the company, empowering employees to take more responsibility for their decisions can aid in organizational growth and productivity. Leaders often change directions when they see opportunities. Founder and CEO Michael Preysman of online clothing retailer Everlane said he would open no brick-and-mortar stores. But, based on demand, he opened a minimalist glass storefront in Manhattan and plans to roll out additional stores around the country.[31]

Going Green

New Belgium Brewing Brews Up "Green" Management Style

Through the implementation of effective management, New Belgium Brewing (NBB) has grown from operating out of a basement to having two state-of-the-art facilities and more than 800 employees. NBB was founded after its co-founder took a trip to Belgium and was inspired to produce high-quality beers in his hometown of Fort Collins, Colorado. Over time, NBB developed a strong customer base that has propelled it to the fourth-largest craft brewery in the United States.

Since the beginning, managers made sustainability a top corporate value. NBB offers an onsite recycling center and gives employees fat-tired cruiser bikes to ride to work after one year of employment. NBB strives to produce a cost- and energy-efficient product that reduces its impact on the environment. It became the first fully wind-powered brewery in the United States after employees and owners unanimously agreed to invest in a wind turbine. NBB also installed a photovoltaic system that produces 3 percent of the company's electricity and uses sun tubes to provide natural lighting.

From planning goals to organizing and assigning responsibilities, managers at NBB assumed a leadership role. Controlling and monitoring activities and correcting for mistakes enabled them to operate profitably while providing a socially responsible product. Under strong leadership, employees are encouraged to actively participate in the business. With a consistent and positive company culture, NBB has provided exceptional leadership for employees to continue the vision, direction, and values the company initially developed.[32]

Critical Thinking Questions

1. How do managers carry out the management functions at NBB?
2. What type of leadership style do you believe NBB's managers practice? Why?
3. How would you describe NBB's organizational culture?

Employee Empowerment

Businesses are increasingly realizing the benefits of participative corporate cultures characterized by employee empowerment. **Employee empowerment** occurs when employees are provided with the ability to take on responsibilities and make decisions about their jobs. Employee empowerment does not mean that managers are not needed. Managers are important for guiding employees, setting goals, making major decisions, and other responsibilities emphasized throughout this chapter. However, a participative corporate culture has been found to be beneficial because employees in these companies feel like they are taking an active role in the firm's success.

Leaders who wish to empower employees adopt systems that support an employee's ability to provide input and feedback on company decisions. *Participative decision making,* a type of decision making that involves both manager and employee input, supports employee empowerment within the organization. One of the best ways to encourage participative decision making is through employee and managerial training. As mentioned earlier, employees should be trained in leadership skills, including teamwork, conflict resolution, and decision making. Managers should also be trained in ways to empower employees to make decisions while also guiding employees through challenging situations in which the right decision might not be so clear.[33]

A section on leadership would not be complete without a discussion of leadership in teams. In today's business world, decisions made by teams are becoming the norm. Employees at Zappos, for instance, often work in teams and are encouraged to make decisions that they believe will reinforce the company's mission and values. Teamwork has often been an effective way for encouraging employee empowerment. Although decision making in teams is collective, the most effective teams are those in which all employees are encouraged to contribute their ideas and recommendations. Because each employee can bring in his or her own unique insights, teams often result

employee empowerment when employees are provided with the ability to take on responsibilities and make decisions about their jobs.

Rubrik CEO Bipul Sinha invites all 900 of his employees to board meetings to empower employees with information and embrace transparency.[35]

©David Paul Morris/Bloomberg via Getty Images

in innovative ideas or decisions that would not have been reached by only one or two people. Michelle Peluso, IBM's chief marketing officer, tries to pull designers and data scientists, as well as marketers, together in teams. For example, she has teams attend sessions where they review and refine their work using only mobile devices to better understand how consumers use their products.[34] However, truly empowering employees in team decision making can be difficult. It is quite common for more outspoken employees to dominate the team and others to engage in groupthink, in which team members go with the majority rather than what they think is the right decision. Training employees to listen to one another and then provide relevant feedback can help to prevent these common challenges. Another way is to rotate the team leader so that no one person can assume dominancy.[36]

Decision Making

Managers make many different kinds of decisions, such as the hours in a workday, which employees to hire, what products to introduce, and what price to charge for a product. Decision making is important in all management functions and at all levels, whether the decisions are on a strategic, tactical, or operational level. A systematic approach using the following six steps usually leads to more effective decision making: (1) recognizing and defining the decision situation, (2) developing options to resolve the situation, (3) analyzing the options, (4) selecting the best option, (5) implementing the decision, and (6) monitoring the consequences of the decision (Figure 6.4).

Recognizing and Defining the Decision Situation

The first step in decision making is recognizing and defining the situation. The situation may be negative—for example, huge losses on a particular product—or positive—for example, an opportunity to increase sales.

Situations calling for small-scale decisions often occur without warning. Situations requiring large-scale decisions, however, generally occur after some warning signs. Effective managers pay attention to such signals. Declining profits, small-scale losses in previous years, inventory buildup, and retailers' unwillingness to stock a product

FIGURE 6.4

Steps in the Decision-Making Process

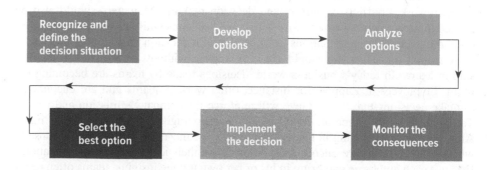

are signals that may foreshadow huge losses to come. If managers pay attention to such signals, problems can be contained.

Once a situation has been recognized, management must define it. Losses reveal a problem—for example, a failing product. One manager may define the situation as a product quality problem; another may define it as a change in consumer preference. These two viewpoints may lead to vastly different solutions. The first manager, for example, may seek new sources of raw materials of better quality. The second manager may believe that the product has reached the end of its lifespan and decide to discontinue it. This example emphasizes the importance of carefully defining the problem rather than jumping to conclusions.

Developing Options

Once the decision situation has been recognized and defined, the next step is to develop a list of possible courses of action. The best lists include both standard and creative plans. **Brainstorming,** a technique in which group members spontaneously suggest ideas to solve a problem, is an effective way to encourage creativity and explore a variety of options. As a general rule, more time and expertise are devoted to the development stage of decision making when the decision is of major importance. For example, after seven years of losses, Sears Holding Corp. has raised doubts about its ability to keep operating. Sears lost $2.22 billion in 2016. Besides reducing costs and selling its Craftsman brand, new options are needed to develop a turnaround plan.[37] When the decision is of less importance, less time and expertise will be spent on this stage. Options may be developed individually, by teams, or through analysis of similar situations in comparable organizations. Creativity is a very important part of selecting the most viable option. Creativity depends on new and useful ideas, regardless of where they originate or the method used to create them. The best option can range from a required solution to an identified problem or a volunteered solution to an observed problem by an outside work group member.[38]

brainstorming
a technique in which group members spontaneously suggest ideas to solve a problem.

Analyzing Options

After developing a list of possible courses of action, management should analyze the practicality and appropriateness of each option. An option may be deemed impractical because of a lack of financial resources, legal restrictions, ethical and social responsibility considerations, authority constraints, technological constraints, economic limitations, or simply a lack of information and expertise. For example, after experiencing the greatest loss since the U.S. government bailed out American International Group Inc., the company's directors looked at options related to then-CEO Peter Hancock, who did not achieve desired performance.[39]

When assessing appropriateness, the decision maker should consider whether the proposed option adequately addresses the situation. When analyzing the consequences of an option, managers should consider its impact on the situation and on the organization as a whole. For example, when considering a price cut to boost sales, management must think about the consequences of the action on the organization's cash flow and consumers' reaction to the price change.

Selecting the Best Option

When all courses of action have been analyzed, management must select the best one. Selection is often a subjective procedure because many situations do not lend

Zappos Puts the Right Foot Forward

The online shoe and clothing retailer Zappos is famous for its strong customer service and fun corporate culture. Under the leadership of CEO Tony Hsieh, exceptional customer service is a key value of the company. Employees are empowered to engage in Zappos' strong participative culture by identifying problems and working together to come up with solutions.

In 2015, Hsieh changed Zappos from a more traditional top-down structure into what is called a *holacracy*. Traditional managerial roles were eliminated, and employees became their own leaders with their own roles. Employee teams have tactical meetings to discuss key issues and next actions. Hsieh made this decision because he believes it will help Zappos grow without losing productivity.

Hsieh recognized this change may not be welcome among employees—especially among managers—so he agreed to provide six months of severance pay for employees who decided to leave. Approximately 18 percent accepted the offer.

One concern is that nonmanagerial employees may not fully understand the risks and uncertainty that will affect company operations. Some employees are struggling to adapt as Zappos continues to work out the kinks of such a massive structural change. In 2016, Zappos fell off *Fortune* magazine's 100 Best Companies to Work For. Despite these challenges, Hsieh believes that in the long term, the change will empower employees to become leaders and allow Zappos to achieve both growth and greater productivity.[40]

Critical Thinking Questions

1. Since Zappos eliminated traditional manager positions, does this mean the managerial functions of planning, organizing, leading, and controlling have also been eliminated? Why or why not?
2. What do you think some of the benefits are of employee-centered leadership? What might be some disadvantages?
3. Why is it important for Hsieh to take a long-term perspective of the company, even if the decision he makes might hurt the firm in the short run?

---

themselves to quantitative analysis. The best option always relates to analyzing risks and trade-offs. For example, how Amazon uses its Alexa virtual assistant involves many alternatives. A decision had to be made whether the voice-activated device could store bank accounts data and make payments with all the risk associated with this service. Nearly all options create dilemmas that create an assistant of risks and rewards.[41] Of course, it is not always necessary to select only one option and reject all others; it may be possible to select and use a combination of several options.

Implementing the Decision

To deal with the situation at hand, the selected option or options must be put into action. Implementation can be fairly simple or very complex, depending on the nature of the decision. For example, China is a country where almost everything is purchased online. Walmart meets this challenge by crowdsourcing delivery drivers, such as Uber drivers with cell phones and scooters to provide one-hour delivery within about a two-mile radius from 161 Walmart supermarkets. This is an example of making changes to adjust to current consumer behavior in purchasing groceries.[42] Effective implementation of a decision to abandon a product, close a plant, purchase a new business, or something similar requires planning. For example, when a product is dropped, managers must decide how to handle distributors and customers and what to do with the idle production facility. Additionally, they should anticipate resistance from people within the organization. (People tend to resist change because they fear the unknown.) Finally, management should be ready to deal with the unexpected consequences. No matter how well planned implementation is, unforeseen problems will arise. Management must be ready to address these situations when they occur.

Monitoring the Consequences

After managers have implemented the decision, they must determine whether it has accomplished the desired result. Without proper monitoring, the consequences of decisions may not be known quickly enough to make efficient changes. If the desired result is achieved, management can reasonably conclude that it made a good choice. If the desired result is not achieved, further analysis is warranted. Was the decision simply wrong, or did the situation change? Should some other option have been implemented?

If the desired result is not achieved, management may discover that the situation was incorrectly defined from the beginning. That may require starting the decision-making process all over again. Finally, management may determine that the decision was good even though the desired results have not yet shown up, or it may determine a flaw in the decision's implementation. In the latter case, management would not change the decision but would change the way in which it is implemented.

Management in Practice

Management is not exact and calculated. There is no mathematical formula for managing an organization and achieving organizational goals, although many managers passionately wish for one! Managers plan, organize, direct, and control, but management expert John P. Kotter says even these functions can be boiled down to two basic activities:

1. Figuring out what to do despite uncertainty, great diversity, and an enormous amount of potentially relevant information.
2. Getting things done through a large and diverse set of people despite having little direct control over most of them.[43]

Managers spend as much as 75 percent of their time working with others—not only with subordinates but with bosses, people outside their hierarchy at work, and people outside the organization itself. In these interactions, they discuss anything and everything remotely connected with their business.

Managers spend a lot of time establishing and updating an agenda of goals and plans for carrying out their responsibilities. An **agenda** contains both specific and vague items, covering short-term goals and long-term objectives. Like a calendar, an agenda helps the manager figure out what must be done and how to get it done to meet the objectives set by the organization. Technology tools such as smartphones can help managers manage their agendas, contacts, communications, and time.

Managers also spend a lot of time **networking**—building relationships and sharing information with colleagues who can help them achieve the items on their agendas. Managers spend much of their time communicating with a variety of people and participating in activities that, on the surface, do not seem to have much to do with the goals of their organization. Nevertheless, these activities are crucial to getting the job done. Networks are not limited to immediate subordinates and bosses; they include other people in the company as well as customers, suppliers, and friends. These contacts provide managers with information and advice on diverse topics. Managers ask, persuade, and push members of their network in order to get information and to get things done. Networking helps managers carry out their responsibilities. Social media sites have increased the ability of both managers and subordinates to network. Internal social networks such as Yammer allow employees to connect with one another, while

agenda
a calendar, containing both specific and vague items, that covers short-term goals and long-term objectives.

networking
the building of relationships and sharing of information with colleagues who can help managers achieve the items on their agendas.

Websites like LinkedIn help managers and employees network to achieve their professional goals.

©dennizn/Shutterstock

social networks such as Facebook or Twitter enable managers to connect with customers. Sales managers are even using social networks to communicate with their distributors. LinkedIn has been used for job networking and is gaining in popularity among the younger generation as an alternative to traditional job hunting. Some speculate that social networks might eventually replace traditional résumés and job boards.[44]

Finally, managers spend a great deal of time confronting the complex and difficult challenges of the business world today. Some of these challenges relate to rapidly changing technology (especially in production and information processing), increased scrutiny of individual and corporate ethics and social responsibility, the impact of social media, the changing nature of the workforce, new laws and regulations, increased global competition and more challenging foreign markets, declining educational standards (which may limit the skills and knowledge of the future labor and customer pool), and time itself—that is, making the best use of it. But such diverse issues cannot simply be plugged into a computer program that supplies correct, easy-to-apply solutions. It is only through creativity and imagination that managers can make effective decisions that benefit their organizations.

So You Want to Be a Manager

Managers are needed in a wide variety of organizations. Experts suggest that employment will increase by millions of jobs in upcoming years. But the requirements for the jobs become more demanding with every passing year—with the speed of technology and communication increasing by the day, and the stress of global commerce increasing pressures to perform. However, if you like a challenge and if you have the right kind of personality, management remains a viable field. Even as companies are forced to restructure, management remains a vital role in business. In fact, the Bureau of Labor Statistics predicts that management positions in public relations, marketing, and advertising will increase 9 percent overall between 2014 and 2024. Demand for financial managers is estimated to increase 7 percent in the same time period. Computer and IT managers will continue to be in strong demand, with the number of jobs increasing 15 percent between 2014 and 2024.[45]

Salaries for managerial positions remain strong overall. While pay can vary significantly depending on your level of experience, the firm where you work, and the region of the country where you live, below is a list of the nationwide average incomes for a variety of different managers:

Chief executive: $180,700
Computer and IT manager: $136,280
Marketing manager: $137,400
Financial manager: $130,230
General and operations manager: $117,200
Medical/health services manager: $103,680
Administrative services manager: $92,250
Human resources manager: $114,140
Sales manager: $126,040[45]

In short, if you want to be a manager, there are opportunities in almost every field. There may be fewer middle management positions available in firms, but managers remain a vital part of most industries and will continue to be long into the future—especially as navigating global business becomes ever more complex.

Review Your Understanding

Explain management's role in the achievement of organizational objectives.

Management is a process designed to achieve an organization's objectives by using its resources effectively and efficiently in a changing environment. Managers make decisions about the use of the organization's resources and are concerned with planning, organizing, directing, and controlling the organization's activities so as to reach its objectives.

Describe the major functions of management.

Planning is the process of determining the organization's objectives and deciding how to accomplish them. Organizing is the structuring of resources and activities to accomplish those objectives efficiently and effectively. Directing is motivating and leading employees to achieve organizational objectives. Controlling is the process of evaluating and correcting activities to keep the organization on course.

Distinguish among three levels of management and the concerns of managers at each level.

Top management is responsible for the whole organization and focuses primarily on strategic planning. Middle management develops plans for specific operating areas and carries out the general guidelines set by top management. First-line, or supervisory, management supervises the workers and day-to-day operations. Managers can also be categorized as to their area of responsibility: finance, production and operations, human resources, marketing, IT, or administration.

Specify the skills managers need in order to be successful.

To be successful, managers need leadership skills (the ability to influence employees to work toward organizational goals), technical expertise (the specialized knowledge and training needed to perform a job), conceptual skills (the ability to think in abstract terms and see how parts fit together to form the whole), analytical skills (the ability to identify relevant issues and recognize their importance, understand the relationships between issues, and perceive the underlying causes of a situation), and human relations (people) skills.

Summarize the systematic approach to decision making used by many business managers.

A systematic approach to decision making follows these steps: recognizing and defining the situation, developing options, analyzing options, selecting the best option, implementing the decision, and monitoring the consequences.

Recommend a new strategy to revive a struggling business.

Using the decision-making process described in this chapter, analyze the struggling company's problems described in "Solve the Dilemma" feature near the end of this chapter and formulate a strategy to turn the company around and aim it toward future success.

Critical Thinking Questions

Enter the World of Business Questions

1. What type of leader was Robert Nardelli? What type of leaders were CEOs Frank Blake and Craig Menear?

2. What are some ways that Blake used the management functions of leading and controlling to change the culture of Home Depot?

3. As an effective manager, Craig Menear must successfully engage in planning, organizing, leading, and controlling. Why is this especially important for Home Depot's ability to compete against online retailers?

Learn the Terms

administrative managers   185
agenda   193
analytical skills   187
brainstorming   191
conceptual skills   186
controlling   180

crisis management   178
contingency planning   178
directing   179
downsizing   174
employee empowerment   189
financial managers   185

first-line managers   184
human relations skills   187
human resources managers   185
information technology (IT) managers   185
leadership   187

Check Your Progress

1. Why is management so important, and what is its purpose?

2. Explain why the American Heart Association would need management, even though its goal is not profit related.

3. Why must a company have financial resources before it can use human and physical resources?

4. Name the four functions of management and briefly describe each function.

5. Identify the three levels of management. What is the focus of managers at each level?

6. In what areas can managers specialize? From what area do top managers typically come?

7. What skills do managers need? Give examples of how managers use these skills to do their jobs.

8. What are three styles of leadership? Describe situations in which each style would be appropriate.

9. Explain the steps in the decision-making process.

10. What is the mathematical formula for perfect management? What do managers spend most of their time doing?

Get Involved

1. Give examples of the activities that each of the following managers might be involved in if he or she worked for the Coca-Cola Company:

   Financial manager

   Production and operations manager

   Personnel manager

   Marketing manager

   Administrative manager

   Information technology manager

   Foreman

2. Interview a small sample of managers, attempting to include representatives from all three levels and all areas of management. Discuss their daily activities and relate these activities to the management functions of planning, organizing, directing, and controlling. What skills do the managers say they need to carry out their tasks?

3. You are a manager of a firm that manufactures conventional ovens. Over the past several years, sales of many of your products have declined; this year, your losses may be quite large. Using the steps of the decision-making process, briefly describe how you arrive at a strategy for correcting the situation.

Build Your Skills

Functions of Management

Background

Although the text describes each of the four management functions separately, you learned that these four functions are interrelated, and managers sometimes perform two or more of them at the same time. Here you will broaden your perspective of how these functions occur simultaneously in management activities.

Task

1. Imagine that you are the manager in each scenario described in the following table and you have to decide which management function(s) to use in each.

2. Mark your answers using the following codes:

Codes	Management Functions
P	Planning
O	Organizing
D	Directing
C	Controlling

No.	Scenario	Answer(s)
1	Your group's work is centered on a project that is due in two months. Although everyone is working on the project, you have observed your employees involved in what you believe is excessive socializing and other time-filling behaviors. You decide to meet with the group to have them help you break down the project into smaller subprojects with mini-deadlines. You believe this will help keep the group members focused on the project and that the quality of the finished project will then reflect the true capabilities of your group.	
2	Your first impression of the new group you'll be managing is not too great. You tell your friend at dinner after your first day on the job: "Looks like I got a baby sitting job instead of a management job."	
3	You call a meeting of your work group and begin it by letting them know that a major procedure used by the work group for the past two years is being significantly revamped, and your department will have to phase in the change during the next six weeks. You proceed by explaining to them the reasoning your boss gave you for this change. You then say, "Let's take the next 5 to 10 minutes to let you voice your reactions to this change." After 10 minutes elapse with the majority of comments being critical of the change, you say: "I appreciate each of you sharing your reactions; and I, too, recognize that *all* change creates problems. "The way I see it, however, is that we can spend the remaining 45 minutes of our meeting focusing on why we don't want the change and why we don't think it's necessary, or we can work together to come up with viable solutions to solve the problems that implementing this change will most likely create." After about five more minutes of comments being exchanged, the consensus of the group is that the remainder of the meeting needs to be focused on how to deal with the potential problems the group anticipates having to deal with as the new procedure is implemented.	
4	You are preparing for the annual budget allocation meetings to be held in the plant manager's office next week. You are determined to present a strong case to support your department getting money for some high-tech equipment that will help your employees do their jobs better. You will stand firm against any suggestions of budget cuts in your area.	
5	Early in your career, you learned an important lesson about employee selection. One of the nurses on your floor unexpectedly quit. The other nurses were putting pressure on you to fill the position quickly because they were overworked even before the nurse left, and then things were really bad. After a hasty recruitment effort, you made a decision based on insufficient information. You ended up regretting your quick decision during the three months of problems that followed until you finally had to discharge the new hire. Since then, you have never let anybody pressure you into making a quick hiring decision.	

Solve the Dilemma

Making Infinity Computers Competitive

 Infinity Computers Inc. produces notebook computers, which it sells through direct mail catalog companies under the Infinity name and in some retail computer stores under their private brand names. Infinity's products are not significantly different from competitors', nor do they have extra product-enhancing features, although they are very price competitive. The strength of the company has been its CEO and president, George Anderson, and a highly motivated, loyal workforce. The firm's weakness is having too many employees and too great a reliance on one product. The firm switched to computers with the Intel Core i7 processors after it saw a decline in its netbook computer sales.

Recognizing that the strategies that initially made the firm successful are no longer working effectively, Anderson wants to reorganize the company to make it more responsive and competitive and to cut costs. The threat of new technological developments and current competitive conditions could eliminate Infinity.

Recommend a new strategy to revive a struggling business.

Critical Thinking Questions

1. Evaluate Infinity's current situation and analyze its strengths and weaknesses.

2. Evaluate the opportunities for Infinity, including using its current strategy, and propose alternative strategies.

3. Suggest a plan for Infinity to compete successfully over the next 10 years.

Build Your Business Plan

The Nature of Management

The first thing you need to be thinking about is "What is the mission of your business? What is the shared vision your team members have for this business? How do you know if there is demand for this particular business?" Remember, you need to think about the customer's *ability and willingness* to try this particular product.

Think about the various processes or stages of your business in the creation and selling of your good or service. What functions need to be performed for these processes to be completed? These functions might include buying, receiving, selling, customer service, and/or merchandising.

Operationally, if you are opening up a retail establishment, how do you plan to provide your customers with superior customer service? What hours will your customers expect you to be open? At this point in time, how many employees are you thinking you will need to run your business? Do you (or one of your partners) need to be there all the time to supervise?

Visit Connect to practice building your business plan with the Business Plan Prep Exercises.

See for Yourself Videocase

Building a Strong and Healthy Business

JCF Health & Fitness is a total body-training program created for women. The program utilizes personalized group fitness in a team-like atmosphere. Daris Wilson, founder of JCF Health & Fitness discovered that many women feel unsure about how to use gym equipment beyond cardio machines and basic free weights. Wilson made it his mission to eliminate the intimidation that women feel in a traditional gym setting by building his company using a group fitness model.

Wilson started his career as an accountant in New Orleans. After evacuating to Jersey City during Hurricane Katrina, Wilson decided to stay in the area and work on the New York Stock Exchange. Though he was successful in his career, the job soon drained him, and he found he had very little time to spend with his family. As a hobby, Wilson had already started working as a personal trainer and fitness instructor while maintaining his full-time career. His desire to help clients meet their personal goals and improve their bodies and lives drove his entrepreneurial spirit and helped him grow JCF Health & Fitness.

After Hurricane Sandy devastated the Jersey shore, New York, and coastal areas of New England, Wilson decided to restructure JCF Health & Fitness and pursue the business full-time. The company's mission is to empower individuals toward happiness through a culture of fitness, leadership, camaraderie, and results. At JCF Health & Fitness, Wilson learns about the limitations and fitness levels of his clients through a questionnaire and a one-on-one meeting designed to developed a rapport and create goals. This makes people enjoy working out and then motivates them to continue. "It's how we speak. How we think. How we train."

Wilson turned to mentoring programs to improve his abilities as a business owner and manager. Wilson's mentors put a focus on organization, marketing, and sales. He realized he needed to structure his resources and activities to accomplish the objectives he set in an efficient and effective manner. He also learned how important it was to market his business. JCF Health & Fitness often offers 21-day boot camp programs at a discounted rate on Groupon to attract new clients. And, to highlight its services, JCF Health & Fitness released a series of spotlight videos featuring real clients.

Wilson continues to set new goals for JCF Health & Fitness, including opening a second gym location to expand the company's physical presence. Wilson believes his business is successful because he had a clear sense of what he wanted to offer from the very beginning. His ability to meet the needs of his clients has created a sense of trust and community.

The growth of Wilson's business has created a need to reevaluate the organization of JCF Health & Fitness. Instead of offering exclusively month-to-month memberships, Wilson introduced yearly memberships. He moved his outdoor training program inside a gym and hired several additional employees to help him train and run the operational side of his business. Because of this, Wilson's role has changed, and instead of giving all of his time to training clients, he has to focus more on the strategy, marketing, and the development of his team.[46]

Critical Thinking Questions

1. Why is it important for JCF Health & Fitness employees to share the same mission as Wilson?

2. How do the goals of JCF Health & Fitness set them apart from other fitness companies?

3. How can Wilson continue to use an entrepreneurial mindset to grow his company?

You can find the related video in the Video Library in Connect. Ask your instructor how you can access Connect.

Team Exercise

Form groups and assign the responsibility of locating examples of crisis management implementation for companies dealing with natural disasters (explosions, fires, earthquakes, etc.), technology disasters (viruses, plane crashes, compromised customer data, etc.), or ethical or legal disasters. How did these companies communicate with key stakeholders? What measures did the company take to provide support to those involved in the crisis? Report your findings to the class.

Ask your instructor about the role play exercises available with this book to practice working with a business team.

Notes

1. Brad Tuttle, "Why Home Depot Is Immune to the 'Amazon Effect,'" *Time: Money,* August 16, 2016, http://time.com/money/4453962/home-depot-amazon-effect-sales/ (accessed September 17, 2017); Greenleaf, "How Home Depot Overcame a Difficult Cultural Shift: A Q&A with CFO Carol Tome," Greenleaf Center for Servant Leadership, July 10, 2015, https://www.greenleaf.org/how-home-depot-overcame-a-difficult-cultural-shift-a-qa-with-cfo-carol-tome/ (accessed September 27, 2017); Heidi N. Moore, "Chrysler: The End of Bob Nardelli. Again," *The Wall Street Journal,* April 21, 2009, https://blogs.wsj.com/deals/2009/04/21/chrysler-the-end-of-bob-nardelli-again/ (accessed September 27, 2017); "CEO Craig Menear Talks Innovation: Follow the Consumer," Home Depot, October 19, 2016, https://corporate.homedepot.com/newsroom/ceo-craig-menear-talks-innovation-aspen-institute (accessed September 27, 2017); "Home Depot Builds Out Its Online Customer Service," *Internet Retailer,* June 4, 2010, https://www.digitalcommerce360.com/2010/06/04/home-depot-builds-out-its-online-customer-service/ (accessed September 27, 2017); Joann Lublin, Matt Murray, and Rick Brooks, "Home Depot Names GE's Nardelli as New CEO in a Surprise Move," *The Wall Street Journal,* December 6, 2000, https://www.wsj.com/articles/SB976051062408860254 (accessed September 27, 2017); John Kell, "Home Depot's Former CEO Frank Blake to Retire as Chairman," *Fortune,* January 16, 2015, http://fortune.com/2015/01/16/home-depot-former-ceo-retires-as-chairman/ (accessed September 27, 2017); Julie Creswell and Michael Barbaro, "Home Depot Ousts Highly Paid Chief," *The New York Times,* January 4, 2007, http://www.nytimes.com/2007/01/04/business/04home.html?mcubz=3 (accessed September 27, 2017); Louis Uchitelle, "Home Depot Girds for Continued Weakness," *The New York Times,* May 18, 2009, http://www.nytimes.com/2009/05/19/business/19depot.html (accessed September 27, 2017); Nathan Owen Rosenberg, "The Key to Home Depot's Success Is Transformational Leadership," *Insigniam,* http://insigniam.com/blog/the-key-to-home-depots-success-is-transformational-leadership/ (accessed September 27, 2017); Parija B. Kavilanz, "Nardelli Out at Home Depot," *CNN Money,* January 3, 2007, http://money.cnn.com/2007/01/03/news/companies/home_depot/ (accessed September 27, 2017); "Home Unimprovement: Was Nardelli's Tenure at Home Depot a Blueprint for Failure?" *Knowledge[[commat]]Wharton,* http://knowledge.wharton. upenn.edu/article/home-unimprovement-was-nardellis-tenure-at-home-depot-a-blueprint-for-failure/ (accessed September 27, 2017); Rachel Tobin, "Frank Blake Is Home Depot's 'Calmer-in-Chief,'" *Seattle Times,* September, 4, 2010, https://www.seattletimes.com/business/frank-blake-is-home-depots-calmer-in-chief/ (accessed September 27, 2017); Rachel Tobin Ramos, "Home Depot Laying Off 1,000 Nationwide," *Atlanta Journal-Constitution,* January 26, 2010, http://www.ajc.com/business/home-depot-laying-off-000-nationwide/ADq8GoBrxpX5h37LIxBxyM/ (accessed September 27, 2017); "Home Depot Builds Out Its Online Customer Service," *Internet Retailer,* June 4, 2010, https://www.digitalcommerce360.com/2010/06/04/home-depot-builds-out-its-online-customer-service/ (accessed September 27, 2017).

2. "Waze: For Clearing the Roads," *Fast Company,* March/April 2018, p. 67.

3. Sarah Frier, "Snap Cuts about 100 Employees in Latest Round of Downsizing," *Bloomberg,* March 29, 2018, https://www.bloomberg.com/news/articles/2018-03-29/snap-cuts-about-100-employees-in-latest-round-of-downsizing (accessed April 16, 2018).

4. Suzanne Heyn, "Sorting through Options," *Inbound Logistics,* May 2014, pp. 48–52.

5. "About Tesla," Tesla, https://www.tesla.com/about (accessed April 7, 2017).

6. Kathyrn Kroll, "Rebecca Ray Designs Grew after Owner Let Others into Business," Cleveland.com, June 21, 2009, http://www.cleveland.com/business/index.ssf/2009/06/rebecca_ray_designs_grew_after.html (accessed November 11, 2017); "About Us," Rebecca Ray Designs, http://rebecca-ray-designs.myshopify.com/pages/about-us (accessed July 22, 2013); Karen Ammond, "Rebecca Ray Designs," December 13, 2012, http://eliteprofessionals.org/2014/10/05/rebecca-ray-designs/ (accessed November 11, 2017); Holly Phillips, "Horse Love: Rebecca Ray Designs," *The English Room,* September 1, 2015, http://www.theenglishroom.biz/2015/09/01/horse-love-rebecca-ray-designs/ (accessed November 11, 2017).

7. Kristina Monllos, "As Soda Sales Suffer, Beverage Marketers Are Shifting to a New Stream of Income: Water," *Adweek,* March 20, 2017, http://www.adweek.com/brand-marketing/as-soda-sales-suffer-beverage-marketers-

are-shifting-to-a-new-stream-of-income-water/ (accessed April 7, 2017).

8. "The Dutch Oven, Disrupted," *Bloomberg Businessweek,* March 19, 2018, p. 75.

9. G. Tomas, M. Hult, David W. Cravens, and Jagdish Sheth, "Competitive Advantage in the Global Marketplace: A Focus on Marketing Strategy," *Journal of Business Research* 51 (January 2001), p. 1.

10. "Chevrolet Remains the Industry's Fastest-Growing Full-Line Brand, with 11 Consecutive Months of Growth," General Motors, March 1, 2016, https://www.gm.com/investors/sales/us-sales-production.html (accessed May 4, 2017).

11. "Protecting Your Businesses," Federal Emergency Management Agency, https://www.fema.gov/protecting-your-businesses (accessed April 7, 2017).

12. James E. Ellis, "At KFC, A Bucketful of Trouble," *Bloomberg Businessweek,* March 5, 2018, pp. 20–21.

13. Alexander Coolidge, "P&G to Cut $10 Billion More, Execs Say," *USA Today,* February 19, 2016, https://www.usatoday.com/story/money/nation-now/2016/02/18/procter-gamble-cuts-wall-street/80563832/ (accessed April 7, 2017).

14. Anna-Greta Nystrom and Mila Mustonen, "The Dynamic Approach to Business Models," *AMS Review,* 2017, p. 123.

15. "AI-Spy," *The Economist,* March 31–April 6, 2018, p. 13.

16. "About the IAPP," IAPP, 2017, https://iapp.org/about/information-privacy-professionals-credentials-from-the-iapp-receive-ansi-accreditation/ (accessed April 7, 2017).

17. Jena McGregor, "The Number of Women CEOs in the Fortune 500 Is at an All-Time High—of 32," *The Washington Post,* June 7, 2017, https://www.washingtonpost.com/news/on-leadership/wp/2017/06/07/the-number-of-women-ceos-in-the-fortune-500-is-at-an-all-time-high-of-32/ (accessed April 16, 2018).

18. Evan Clark, "Marvin Ellison's Pay Tops $10 Million at J.C. Penney," *WWD,* April 9, 2018, http://wwd.com/fashion-news/fashion-scoops/marvin-ellisons-pay-tops-10-million-at-j-c-penney-1202646280/ (accessed April 16, 2018).

19. Ross Kerber, "Growth in Compensation for U.S. CEOs May Have Slowed," *Reuters,* March 17, 2014, www.reuters.com/article/2014/03/17/us-compensation-ceos-2013-insight-idUSBREA2G05520140317 (accessed March 29, 2014).

20. Melissa Behrend, "More Companies Are Reducing CEO Compensation Due to Poor Quarterly Results," *Chief Executive,* March 21, 2016, http://chiefexecutive.net/more-companies-are-reducing-ceo-compensation-due-to-poor-quarterly-results/ (accessed April 7, 2017).

21. "No. 1 Kaiser Permanente, DiversityInc Top 50," *Diversity Inc,* 2017, http://www.diversityinc.com/kaiser-permanente/ (accessed April 7, 2017).

22. Laura Nichols, "Agencies Called to Step Up the Pace on Diversity Efforts," *PRWeek,* February 7, 2014, www.prweekus.com/article/agencies-called-step-pace-diversity-efforts/1283550 (accessed March 29, 2014).

23. Linda Eling Lee, Ric Marshall, Damon Rallis, and Matt Moscardi, "Women on Boards: Global Trends in Gender Diversity on Corporate Boards," *MCSI,* November 2015, p. 3.

24. Del Jones, "Autocratic Leadership Works—Until It Fails," *USA Today,* June 5, 2003, www.usatoday.com/news/nation/2003-06-05-raines-usat_x.htm (accessed May 4, 2017).

25. George Manning and Kent Curtis, *The Art of Leadership* (New York: McGraw-Hill, 2003), p. 125.

26. "Mr. Buffet Has Still Got It," *The Economist,* March 24–30, 2018, p. 16.

27. Richard Gonzales, "Wells Fargo CEO John Stumpf Resigns amid Scandal," *NPR,* October 12, 2016, https://www.npr.org/sections/thetwo-way/2016/10/12/497729371/wells-fargo-ceo-john-stumpf-resigns-amid-scandal (accessed April 12, 2018).

28. Bruce J. Avolio and William L. Gardner, "Authentic Leadership Development: Getting to the Root of Positive Forms of Leadership," *The Leadership Quarterly,* 2005, pp. 315–338.

29. Robert Sation, "A New Model of Leadership," *Fact Company,* April 2017, p. 18.

30. John P. Kotter, "What Leaders Really Do," *Harvard Business Review,* December 2001, http://fs.ncaa.org/Docs/DIII/What%20Leaders%20Really%20Do.pdf (accessed May 4, 2017).

31. "For Building the Next-Gen Clothing Brand," *Fast Company,* March–April 2018, p. 79.

32. Corporate Sustainability Report, New Belgium Brewing website, http://www.newbelgium.com/docs/default-source/sustainability/2017sustainabilitybrochure.pdf?pdf=sustainabilityreport (accessed September 18, 2017); Kelly K. Spors, "Top Small Workplaces 2008," *The Wall Street Journal,* February 22, 2009, http://online.wsj.com/article/SB122347733961315417.html (accessed April 16, 2013); "We Are 100% Owned," New Belgium Brewing website, http://www.newbelgium.com/community/Blog/13-01-16/ We-are-100-Employee-Owned.aspx (accessed July 23, 2013); The facts of this case are from Peter Asmus, "Goodbye Coal, Hello Wind," *Business Ethics,* 13 (July/August 1999), pp. 10–11; Darren Dahl, "New Belgium Brewing's Ownership Culture Wins Again," *Forbes,* November 19, 2016, https://www.forbes.com/sites/darrendahl/2016/11/19/new-belgium-brewings-ownership-culture-wins-again/#7c19ef382b6a (accessed September 18, 2017); Susan Adams, "New Belgium Brewing Hires a New CEO from the Liquor Industry," *Forbes,* July 17, 2017, https://www.forbes.com/sites/susanadams/2017/07/17/new-belgium-brewing-hires-a-new-ceo-from-the-liquor-industry/#4414498e5e7a (accessed November 12, 2017).

33. C. L. Pearce and C. C. Manz, "The New Silver Bullets of Leadership: The Importance of Self and Shared Leadership in Knowledge Work," *Organizational Dynamics* 34, no. 2 (2005), pp. 130–140.

34. Michelle Peluseo, "Watson's New Champion," *Fast Company,* April 2017, p. 24.

35. Bipul Sinha, "Why I Invite all 900 of My Employees to Board Meetings," *Quartz at Work,* March 21, 2018, https://work.qz.com/1232424/why-i-invite-all-900-of-my-employees-to-board-meetings/ (accessed April 16, 2018).

36. Deborah Harrington-Mackin, *The Team Building Tool Kit* (New York: New Directions Management, 1994); Joseph P. Folger, Marshall Scott Poole, and Randall K. Stutman, *Working through Conflict: Strategies for Relationships, Groups, and Organizations,* 6th ed. (Upper Saddle River, NJ: Pearson Education, 2009).

37. Anne Steele, "Sears Creates Stir as It Casts Doubt about Its Future," *The Wall Street Journal,* March 2017, p. B1.

38. Kerrie Unsworth, "Unpacking Creativity," *Academy of Management Review,* 26 (April 2001), pp. 289–297.

39. Joann S. Lublin, Leslie Seism, and David Benoit, "AIG's Bound to Weigh Ouster of CEO," *The Wall Street Journal,* February 28, 2017, p. B1.

40. Adapted from "Zappos: Delivering Happiness to Stakeholders," Daniels Fund Ethics Initiative, http://danielsethics.mgt.unm.edu/pdf/Zappos%20Case.pdf (accessed November 25, 2017); Richard Feloni, "Inside Zappos CEO Tony Hsieh's Radical Management Experiment That Prompted 14% of Employees to Quit," *Business Insider,* May 16, 2015, http://www.businessinsider.com/tony-hsieh-zappos-holacracy-managementexperiment-2015-5 (accessed May 12, 2017); Rebecca Greenfield, "Zappos CEO Tony Hsieh: Adopt Holacracy or Leave," *Fast Company,* March 30, 2015, https://www.fastcompany.com/3044417/zappos-ceo-tony-hsieh-adopt-holacracy-or-leave (accessed May 12, 2017); Zack Guzman, "Zappos CEO Tony Hsieh on Getting Rid of Managers: What I Wish I'd Done Differently," *CNBC,* September 13, 2016, http://www.cnbc.com/2016/09/13/zappos-ceo-tony-hsieh-the-thing-i-regret-aboutgetting-rid-of-managers.html (accessed May 12, 2017); "How It Works," HolacracyOne LLC, http://holacracy.org/how-it-work (accessed May 12, 2017); Rachel Emma Silverman, "At Zappos, Some Employees Find Offer to Leave Too Good to Refuse," *The Wall Street Journal,* May 7, 2014, http://www.wsj.com/articles/at-zappos-someemployees-find-offer-to-leave-too-good-to-refuse-1431047917 (accessed

May 12, 2017); "Holacracy," Zappos Insights, http://www.zapposinsights.com/about/holacracy (accessed May 12, 2017); Zappos.com, "Company Statement from Zappos.com," YouTube, April 1, 2016, https://www.youtube.com/watch?v=3zieP6NUWL8 (accessed May 12, 2017).

41. Anna Maria Andriotos and Lara Stevens, "Amazon Voices Payment Strategy," *The Wall Street Journal,* April 7–8, 2018, pp. B1, B2.

42. James E. Ellis, "China Doesn't Want to Go to the Store for Groceries, Either," *Bloomberg Businessweek,* December 4, 2017, pp. 23–24.

43. *Harvard Business Review* 60 (November–December 1982), p. 160.

44. Dan Schwabel, "5 Reasons Why Your Online Presence Will Replace Your Resume in 10 Years," *Forbes,* February 21, 2012, www.forbes.com/sites/danschawbel/2011/02/21/5-reasons-why-your-online-presence-will-replace-your-resume-in-10-years/ (accessed May 4, 2017).

45. U.S. Bureau of Labor Statistics, "Occupational Outlook Handbook," December 17, 2015, http://www.bls.gov/ooh/ (accessed March 7, 2016).

46. JCF Health & Fitness, "The JCF Difference," https://jcf.fitness/ (accessed June 3, 2018); "JCF Health & Fitness Bootcamp," Groupon, https://www.groupon.com/deals/jcf-health-fitness-bootcamp (accessed June 3, 2018); "JCF Camper Spotlight—Julie," YouTube, July 26, 2016, https://www.youtube.com/watch?v=wF4s7glPPvY (accessed June 3, 2018); Stephen McMillian, "Boot Camp's Focus Is Women's Fitness, Nutrition," NJ.com, May 13, 2016, http://www.nj.com/jjournal-news/index.ssf/2016/05/boot_camps_focus_is_womens_fit.html (accessed June 3, 2018); Rushion McDonald, "Episode 26: Brian Dobbins, Daris Wilson, and Abigail Gonzalez," *Money Making Conversations,* November 27, 2017, https://www.rushionmcdonald.com/episode-26-brian-dobbins-daris-wilson-and-abigail-gonzalez/ (accessed June 3, 2018); William G. Nickels, James M. McHugh, and Susan M. McHugh, *Understanding Business* (New York: McGraw-Hill, 2019), p. 165.

Credits

7 Organization, Teamwork, and Communication

©Terry Putman/Shutterstock

Learning Objectives

After reading this chapter, you will be able to:

LO 7-1   Explain the importance of organizational culture.

LO 7-2   Describe how organizational structures develop.

LO 7-3   Describe how specialization and departmentalization help an organization achieve its goals.

LO 7-4   Determine how organizations assign responsibility for tasks and delegate authority.

LO 7-5   Compare and contrast some common forms of organizational structure.

LO 7-6   Distinguish between groups and teams.

LO 7-7   Identify the types of groups that exist in organizations.

LO 7-8   Describe how communication occurs in organizations.

LO 7-9   Analyze a business's use of teams.

Enter the World of Business

Keurig Green Mountain Brews Effective Communication

Keurig Green Mountain is a leader in the specialty coffee industry. The company sells coffee and beverage selections through a coordinated, multichannel distribution network of wholesale and consumer direct operations. It also sells Keurig single-pack coffee packets and Keurig brewers, which are single-cup brewing systems.

Keurig Green Mountain is decentralized with few layers of management. Although it has functional departments that vary across the company, an openness of communication allows employees access to all levels of the organization. The company uses digital communication channels such as e-mail to inform groups of decisions. Verbal communication is used at meetings, and employees are encouraged to share their views. Written forms of communication, such as agendas, outline the results of meetings and serve to guide efficient decision making. This communication across channels ensures that the collaborative nature of getting things done is spread equally throughout the company.

Corporate social responsibility is a major objective of Keurig Green Mountain. Keurig Green Mountain invests in sustainably grown coffee initiatives, and the company is developing K-Cup pods that are increasingly recyclable. This is important as one major criticism of the company is the large amount of waste its pods create over time. Keurig Green Mountain announced the pods would be 100 percent recyclable by 2020. To ensure employees are on board, Keurig Green Mountain communicates this objective across its various communication channels.

In 2016, JAB Holding Company purchased Keurig Green Mountain for $13.9 billion. The company was also taken private. A privately held company with fewer owners might make it easier for management to strengthen communication channels both internally and externally. The company must continue to display effective communication skills to ensure a collaborative corporate culture focused on developing specialty coffees sustainably and responsibly.[1]

Introduction

An organization's structure determines how well it makes decisions and responds to problems, and it influences employees' attitudes toward their work. A suitable structure can minimize a business's costs and maximize its efficiency. Even companies that operate within the same industry may utilize different organizational structures. For example, in the consumer electronics industry, Samsung is organized as a conglomerate with separate business units or divisions. Samsung is largely decentralized. Apple, under CEO Tim Cook, has moved from a hierarchical structure to a more collaborative approach among divisions.[2] On the other hand, Ford Motor Co.'s board views the CEO, Jim Hackett, as being in charge of the firm's strategy in a more hierarchical structure.[3] A manufacturing firm, like Ford, may require more leadership and control from the top.

Because a business's structure can so profoundly affect its success, this chapter will examine organizational structure in detail. First, we discuss how an organization's culture affects its operations. Then we consider the development of structure, including how tasks and responsibilities are organized through specialization and departmentalization. Next, we explore some of the forms organizational structure may take. Finally, we consider communications within business.

LO 7-1

Explain the importance of organizational culture.

organizational culture
a firm's shared values, beliefs, traditions, philosophies, rules, and role models for behavior.

Organizational Culture

One of the most important aspects of organizing a business is determining its **organizational culture,** a firm's shared values, beliefs, traditions, philosophies, rules, and role models for behavior. Also called corporate culture, an organizational culture exists in every organization, regardless of size, organizational type, product, or profit objective. Sometimes behaviors, programs, and policies enhance and support the organizational culture. For instance, the sixth-largest accounting firm, Grant Thornton, established an unlimited vacation policy to give its employees more freedom. Less than 1 percent of American firms have this policy, but it seems to be growing in companies like Netflix, where employees have greater autonomy. Some speculate, however, that these policies will only work at firms with employees who are already highly motivated to work hard and are less likely to take vacations in the first place.[4] A firm's culture may be expressed formally through its mission statement, codes of ethics, memos, manuals, and ceremonies, but it is more commonly expressed informally. Examples of informal expressions of culture include dress codes (or the lack thereof), work habits, extracurricular activities, and stories. Employees often learn the accepted standards through discussions with co-workers.

McDonald's has organizational cultures focused on quality, service, cleanliness, and value. Nordstrom stresses a culture of excellent customer service. As a result, employees are empowered to use their best judgment in delivering the best services.[5] When such values are shared by all members of an organization, they will be expressed in its relationships with customers. Zappos stresses

Google embraces a corporate culture that focuses on employee happiness. The tech company offers perks and benefits like in-office gyms and fitness areas, free meals, and more.

©Uladzik Kryhin/Shutterstock

the importance of fitting into the organizational culture. When trainees have completed training, the company offers them $3,000 to leave the company. If the employee leaves for the money, they are not the right fit for Zappos, so the company is happy to see them go.[6] However, organizational cultures that fail to understand the values of their customers may experience rejection. The values and integrity of customers must always be considered. Google focused on online advertising that is effective and easily measured. The Google culture was focused more on metrics and sales of advertising than customer values. Companies such as AT&T have pulled advertisements from YouTube because of concerns over the possibility of appearing next to offensive material.[7]

The Zappos tagline "Powered by Service" emphasizes the company's focus on its customers.

©Jonathan Weiss/Alamy

Organizational culture helps ensure that all members of a company share values and suggests rules for how to behave and deal with problems within the organization. Kevin Johnson took over as CEO of Starbucks, so he has to embrace the corporate culture that founder and former CEO Howard Schultz built. Employees like to work at Starbucks because of the firm's culture. It was troubling to executives when a Starbucks manager in Philadelphia had two minority customers removed from the restaurant. The company's move to support their ethical organizational culture was swift and decisive. They offered a half day of racial sensitivity training in the vast majority of their company-owned restaurants.[8] The key to success in any organization is satisfying stakeholders, especially customers. Establishing a positive organizational culture sets the tone for all other decisions, including building an efficient organizational structure.

Developing Organizational Structure

Structure is the arrangement or relationship of positions within an organization. Rarely is an organization, or any group of individuals working together, able to achieve common objectives without some form of structure, whether that structure is explicitly defined or only implied. A professional baseball team such as the Colorado Rockies is a business organization with an explicit formal structure that guides the team's activities so that it can increase game attendance, win games, and sell souvenirs such as T-shirts. But even an informal group playing softball for fun has an organization that specifies who will pitch, catch, bat, coach, and so on. Governments and nonprofit organizations also have formal organizational structures to facilitate the achievement of their objectives. Getting people to work together efficiently and coordinating the skills of diverse individuals require careful planning. Developing appropriate organizational structures is, therefore, a major challenge for managers in both large and small organizations.

An organization's structure develops when managers assign work tasks and activities to specific individuals or work groups and coordinate the diverse activities required to reach the firm's objectives. When Best Buy, for example, has a sale, the store manager must work with the advertising department to make the public aware of the sale, with department managers to ensure that extra salespeople are scheduled

LO 7-2

Describe how organizational structures develop.

structure
the arrangement or relationship of positions within an organization.

to handle the increased customer traffic, and with merchandise buyers to ensure that enough sale merchandise is available to meet expected consumer demand. All the people occupying these positions must work together to achieve the store's objectives.

The best way to begin to understand how organizational structure develops is to consider the evolution of a new business such as a clothing store. At first, the business is a sole proprietorship in which the owner does everything—buys, prices, and displays the merchandise; does the accounting and tax records; and assists customers. As the business grows, the owner hires a salesperson and perhaps a merchandise buyer to help run the store. As the business continues to grow, the owner hires more salespeople. The growth and success of the business now require the owner to be away from the store frequently, meeting with suppliers, engaging in public relations, and attending trade shows. Thus, the owner must designate someone to manage the salespeople and maintain the accounting, payroll, and tax functions. If the owner decides to expand by opening more stores, still more managers will be needed. Figure 7.1 shows these stages of growth with three **organizational charts** (visual displays of organizational structure, chain of command, and other relationships).

Growth requires organizing—the structuring of human, physical, and financial resources to achieve objectives in an effective and efficient manner. Growth necessitates hiring people who have specialized skills. With more people and greater specialization, the organization needs to develop a formal structure to function efficiently. Often organizations undergo structural changes when the current structure is no longer deemed effective. For instance, Zappos, an online shoe and clothing retailer, adopted an organizational structure called a *holacracy,* a structure in which job titles are abandoned, traditional managers are eliminated, and authority is distributed to teams. Although these constitute massive changes for the firm, CEO Tony Hsieh believes this more fluid organizational structure will allow Zappos to grow in size while simultaneously increasing in productivity.[9] Zappos has a major focus on service, but its structure might not work for a manufacturing firm like Ford that is more hierarchical in its structure. As we shall see, structuring an organization requires that management assign work tasks to specific individuals and departments and assign responsibility for the achievement of specific organizational objectives.

organizational chart
a visual display of the organizational structure, lines of authority (chain of command), staff relationships, permanent committee arrangements, and lines of communication.

FIGURE 7.1

The Evolution of a Clothing Store, Phases 1, 2, and 3

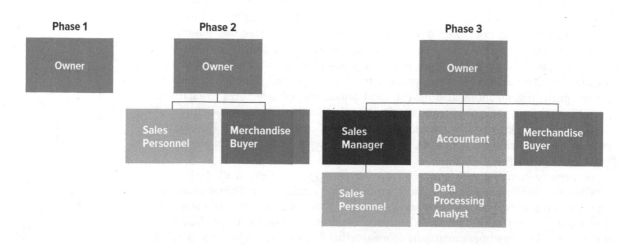

Assigning Tasks

LO 7-3

For a business to earn profits from the sale of its products, its managers must first determine what activities are required to achieve its objectives. At Celestial Seasonings, for example, employees must purchase herbs from suppliers, dry the herbs and place them in tea bags, package and label the tea, and then ship the packages to grocery stores around the country. Other necessary activities include negotiating with supermarkets and other retailers for display space, developing new products, planning advertising, managing finances, and managing employees. All these activities must be coordinated, assigned to work groups, and controlled. Two important aspects of assigning these work activities are specialization and departmentalization.

Describe how specialization and departmentalization help an organization achieve its goals.

Specialization

After identifying all activities that must be accomplished, managers then break these activities down into specific tasks that can be handled by individual employees. This division of labor into small, specific tasks and the assignment of employees to do a single task is called **specialization.**

The rationale for specialization is efficiency. People can perform more efficiently if they master just one task rather than all tasks. In *The Wealth of Nations,* 18th-century economist Adam Smith discussed specialization, using the manufacture of straight pins as an example. Individually, workers could produce 20 pins a day when each employee produced complete pins. Thus, 10 employees working independently of each other could produce 200 pins a day. However, when one worker drew the wire, another straightened it, a third cut it, and a fourth ground the point, 10 workers could produce 48,000 pins per day.[10] To save money and achieve the benefits of specialization, some companies outsource and hire temporary workers to provide key skills. Many highly skilled, diverse, experienced workers are available through temp agencies.

Specialization means workers do not waste time shifting from one job to another, and training is easier. However, efficiency is not the only motivation for specialization. Specialization also occurs when the activities that must be performed within an organization are too numerous for one person to handle. Recall the example of the clothing store. When the business was young and small, the owner could do everything; but when the business grew, the owner needed help waiting on customers, keeping the books, and managing other business activities.

Overspecialization can have negative consequences. Employees may become bored and dissatisfied with their jobs, and the result of their unhappiness is likely to be poor quality work, more injuries, and high employee turnover. In extreme cases, employees in crowded specialized electronic plants are unable to form working relationships with one another. In some factories in Asia, workers are cramped together and overworked. Fourteen global vehicle manufacturers pledged to increase their oversight of the factories in their supply chain to ensure human rights and healthy working conditions. However, the task is monumental for these global companies because their supply chains encompass many different countries with different labor practices, and it can be difficult to oversee the operations of dozens of supplier and subcontractors.[11] This is why some manufacturing firms allow job rotation so that employees do not become dissatisfied and leave. Although some degree of specialization is necessary for efficiency, because of differences in skills, abilities, and interests, all people are not equally suited for all jobs. We examine some strategies to overcome these issues in the chapter titled "Motivating the Workforce."

specialization
the division of labor into small, specific tasks and the assignment of employees to do a single task.

Departmentalization

After assigning specialized tasks to individuals, managers next organize workers doing similar jobs into groups to make them easier to manage. **Departmentalization** is the grouping of jobs into working units usually called departments, units, groups, or divisions. As we shall see, departments are commonly organized by function, product, geographic region, or customer (Figure 7.2). Most companies use more than

FIGURE 7.2
Departmentalization

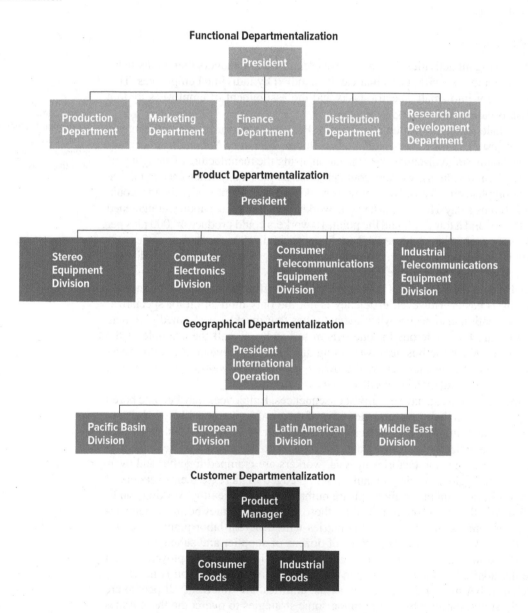

one departmentalization plan to enhance productivity. For instance, many consumer goods manufacturers have departments for specific product lines (beverages, frozen dinners, canned goods, and so on) as well as departments dealing with legal, purchasing, finance, human resources, and other business functions. For smaller companies, accounting can be set up online, almost as an automated department. Accounting software can handle electronic transfers so you never have to worry about a late bill. Many city governments also have departments for specific services (for example, police, fire, waste disposal) as well as departments for legal, human resources, and other business functions.

Functional Departmentalization.   Functional departmentalization groups jobs that perform similar functional activities, such as finance, manufacturing, marketing, and human resources. Each of these functions is managed by an expert in the work done by the department—an engineer supervises the production department; a financial executive supervises the finance department. This approach is common in small organizations. Tesla uses functional departmentalization with Finance, Technology, Global Sales and Service, Engineering, and Legal departments in addition to the chairman and CEO. The company also has two divisions: automotive and energy generation and storage.[12] A weakness of functional departmentalization is that because it tends to emphasize departmental units rather than the organization as a whole, decision making that involves more than one department may be slow, and it requires greater coordination. Thus, as businesses grow, they tend to adopt other approaches to organizing jobs.

> **functional departmentalization**
> the grouping of jobs that perform similar functional activities, such as finance, manufacturing, marketing, and human resources.

Product Departmentalization.   Product departmentalization, as you might guess, organizes jobs around the products of the firm. Unilever has global units, including personal care, foods, refreshment, and home care.[13] Each division develops and implements its own product plans, monitors the results, and takes corrective action as necessary. Functional activities—production, finance, marketing, and others—are located within each product division. Consequently, organizing by products duplicates functions and resources and emphasizes the product rather than achievement of the organization's overall objectives. However, it simplifies decision making and helps coordinate all activities related to a product or product group. PepsiCo Inc. is organized into six business units: (1) North America Beverages; (2) Frito-Lay North America; (3) Quaker Foods North America; (4) Latin America; (5) Europe Sub-Saharan Africa; and (6) Asia, Middle East & North Africa. PepsiCo has actually adopted a combination of two types of departmentalization. While it clearly uses product departmentalization in North America, the company also chooses to divide its segments into geographic regions—a type of geographic departmentalization.[14]

> **product departmentalization**
> the organization of jobs in relation to the products of the firm.

Geographical Departmentalization.   Geographical departmentalization groups jobs according to geographic location, such as a state, region, country, or continent. Diageo, the premium beverage company known for brands such as Johnnie Walker and Tanqueray, is organized into five geographic regions, allowing the company to get closer to its customers and respond more quickly and efficiently to regional competitors.[15] Multinational corporations often use a geographical approach because of vast differences between different regions. Coca-Cola, General Motors, and

> **geographical departmentalization**
> the grouping of jobs according to geographic location, such as state, region, country, or continent.

Grouping jobs into departments like marketing, customer service, or purchasing can enhance productivity.

©Adam Hester/Blend Images

Caterpillar are organized by region. However, organizing by region requires a large administrative staff and control system to coordinate operations, and tasks are duplicated among the different regions.

Customer Departmentalization. Customer departmentalization arranges jobs around the needs of various types of customers. This allows companies to address the unique requirements of each group. Airlines, such as British Airways and Southwest Airlines, provide prices and services customized for either business/frequent travelers or infrequent/vacationing customers. Customer departmentalization, like geographical departmentalization, does not focus on the organization as a whole and therefore requires a large administrative staff to coordinate the operations of the various groups.

customer departmentalization
the arrangement of jobs around the needs of various types of customers.

Assigning Responsibility

After all workers and work groups have been assigned their tasks, they must be given the responsibility to carry them out. Management must determine to what extent it will delegate responsibility throughout the organization and how many employees will report to each manager.

LO 7-4

Determine how organizations assign responsibility for tasks and delegate authority.

Delegation of Authority

Delegation of authority means not only giving tasks to employees, but also empowering them to make commitments, use resources, and take whatever actions are necessary to carry out those tasks. Let's say a marketing manager at Nestlé has assigned an employee to design a new package that is less wasteful (more environmentally responsible) than the current package for one of the company's frozen dinner lines. To carry out the assignment, the employee needs access to information and the authority to make certain decisions on packaging materials, costs, and so on. Without the authority to carry out the assigned task, the employee would have to get the approval of others for every decision and every request for materials.

As a business grows, so do the number and complexity of decisions that must be made; no one manager can handle them all. 3M delegates authority to its employees by encouraging them to share ideas and make decisions. 3M believes employee ideas can have such an impact on the firm that it encourages them to spend 15 percent of their time working on and sharing their own projects. This "15 percent culture" has created the collaboration that drives innovation in the company.[16] Delegation of authority frees a manager to concentrate on larger issues, such as planning or dealing with problems and opportunities.

Delegation also gives a **responsibility,** or obligation, to employees to carry out assigned tasks satisfactorily and holds them accountable for the proper execution of their assigned work. The principle of **accountability** means that employees who accept an assignment and the authority to carry it out are answerable to a superior for the outcome. While there can be delegation of authority with employee responsibility,

delegation of authority
giving employees not only tasks, but also the power to make commitments, use resources, and take whatever actions are necessary to carry out those tasks.

responsibility
the obligation, placed on employees through delegation, to perform assigned tasks satisfactorily and be held accountable for the proper execution of work.

accountability
the principle that employees who accept an assignment and the authority to carry it out are answerable to a superior for the outcome.

the manager delegating still is accountable for oversight of the final result. Returning to the Nestlé example, if the packaging design prepared by the employee is unacceptable or late, the employee must accept the blame. If the new design is innovative, attractive, and cost-efficient, as well as environmentally responsible, or is completed ahead of schedule, the employee will accept the credit.

The process of delegating authority establishes a pattern of relationships and accountability between a superior and his or her subordinates. The president of a firm delegates responsibility for all marketing activities to the vice president of marketing. The vice president accepts this responsibility and has the authority to obtain all relevant information, make certain decisions, and delegate any or all activities to his or her subordinates. The vice president, in turn, delegates all advertising activities to the advertising manager, all sales activities to the sales manager, and so on. These managers then delegate specific tasks to their subordinates. However, the act of delegating authority to a subordinate does not relieve the superior of accountability for the delegated job. Even though the vice president of marketing delegates work to subordinates, he or she is still ultimately accountable to the president for all marketing activities.

Degree of Centralization

The extent to which authority is delegated throughout an organization determines its degree of centralization.

Centralized Organizations.   In a **centralized organization**, authority is concentrated at the top, and very little decision-making authority is delegated to lower levels. Although decision-making authority in centralized organizations rests with top levels of management, a vast amount of responsibility for carrying out daily and routine procedures is delegated to even the lowest levels of the organization. Many government organizations, including the U.S. Army, the Postal Service, and the IRS, are centralized.

Businesses tend to be more centralized when the decisions to be made are risky and when low-level managers are not highly skilled in decision making. In the banking industry, for example, authority to make routine car loans is given to all loan managers, while the authority to make high-risk loans, such as for a large residential development, may be restricted to upper-level loan officers.

Overcentralization can cause serious problems for a company, in part because it may take longer for the organization as a whole to implement decisions and to respond to changes and problems on a regional or national scale. For example, McDonald's tested fresh beef in its Quarter Pounder regionally in the United States. The success of the tests allowed McDonald's to engage in a national rollout. The move was in response to competitors promoting their fresh beef and casting McDonald's frozen beef in a negative light.[17] Centralized decision making can also prevent front-line service employees from providing insights and recommendations to improve the customer experience as well as reporting and resolving problems.

Decentralized Organizations.   A **decentralized organization** is one in which decision-making authority is delegated as far down the chain of command as possible. Decentralization is characteristic of organizations that operate in complex, unpredictable environments. Businesses that face intense competition often decentralize to improve responsiveness and enhance creativity. Lower-level managers who interact with the external environment often develop a good understanding of it and thus are

centralized organization
a structure in which authority is concentrated at the top and very little decision-making authority is delegated to lower levels.

decentralized organization
an organization in which decision-making authority is delegated as far down the chain of command as possible.

able to react quickly to changes. Johnson & Johnson has a very decentralized, flat organizational structure.

Delegating authority to lower levels of managers may increase the organization's productivity. Decentralization requires that lower-level managers have strong decision-making skills. In recent years, the trend has been toward more decentralized organizations, and some of the largest and most successful companies, including GE, IBM, Google, and Nike, have decentralized decision-making authority. Decentralization can be a key to being better, not just bigger. Nonprofit organizations can benefit from decentralization as well.

Span of Management

How many subordinates should a manager manage? There is no simple answer. Experts generally agree, however, that top managers should not directly supervise more than four to eight people, while lower-level managers who supervise routine tasks are capable of managing a much larger number of subordinates. For example, the manager of the finance department may supervise 25 employees, whereas the vice president of finance may supervise only five managers. **Span of management** (also called *span of control*) refers to the number of subordinates who report to a particular manager. A *wide span of management or control* exists when a manager directly supervises a very large number of employees. A *narrow span of management or control* exists when a manager directly supervises only a few subordinates (Figure 7.3). At Whole Foods, the best employees are recruited and placed in small teams. Employees are empowered to discount, give away, and sample products, as well as to assist in creating a respectful workplace where goals are achieved, individual employees succeed, and customers are core in business decisions. Whole Foods teams get to vote on new employee hires as well. This approach allows Whole Foods to offer unique and "local market" experiences in each of its stores. This level of customization is in contrast to more centralized national supermarket chains such as Kroger, Safeway, and Publix.[18]

Should the span of management be wide or narrow? To answer this question, several factors need to be considered. A narrow span of management is appropriate when superiors and subordinates are not in close proximity, the manager has many responsibilities in addition to the supervision, the interaction between superiors and subordinates is frequent, and problems are common. However, when superiors and subordinates are located close to one another, the manager has few responsibilities other than supervision, the level of interaction between superiors and subordinates is low, few problems arise, subordinates are highly competent, and a set of specific

span of management
the number of subordinates who report to a particular manager.

FIGURE 7.3
Span of Management: Wide Span and Narrow Span

Wide Span: Flat Organization

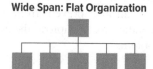

Narrow Span: Tall Organization

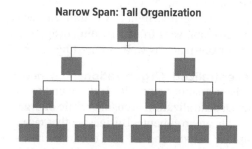

Whole Foods Focuses on the Whole Team

Since its founding in 1978, Whole Foods has been committed to great customer service and selling the highest-quality natural and organic products. Employees are essential toward achieving these goals as they daily interact with customers. Employees are therefore highly valued at Whole Foods and are labeled team members to empower them through their everyday contributions.

At each store, individuals are divided into 8 to 10 self-directed work teams. Initially, when candidates are hired, they are hired on a provisional basis. Before candidates are hired on a provisional basis, they undergo a 60-day process of interviews on the phone, with team members, and with leaders. If two-thirds of the team members vote in favor of the candidate, the candidate becomes part of the team. The team approach has been adopted throughout the entire chain of command—even among regional leaders. Through its use of teams, Whole Foods has been able to turn its workers into significant contributors of value for the company.

Despite Whole Foods' success, the company has struggled to get rid of its "whole paycheck" reputation. Due to financial struggles, Amazon acquired Whole Foods for $13.7 billion. Such a cultural shift might take away some of the team autonomy practiced at individual stores. Amazon and Whole Foods will need to balance these changes with the successful team approach practiced at individual stores.[19]

Critical Thinking Questions

1. What are some of the ways Whole Foods encourages teamwork among employees?
2. Why does the use of teams allow Whole Foods to tap into a variety of talent?
3. How does Whole Foods make use of self-directed work teams?

---

operating procedures governs the activities of managers and their subordinates, a wide span of management will be more appropriate. Narrow spans of management are typical in centralized organizations, while wide spans of management are more common in decentralized firms.

Organizational Layers

Complementing the concept of span of management is **organizational layers,** the levels of management in an organization. Organizational layers relate to a description of the layers and number of layers in the organizational structure. Span of management or control covered in the last section is the number of subordinates, functions, or people that have need to be managed. A company with many layers of managers is considered tall; in a tall organization, the span of management is narrow (see Figure 7.3). Because each manager supervises only a few subordinates, many layers of management are necessary to carry out the operations of the business. McDonald's, for example, has a tall organization with many layers, including store managers, district managers, regional managers, and functional managers (finance, marketing, and so on), as well as a chief executive officer and many vice presidents. Because there are more managers in tall organizations than in flat organizations, administrative costs are usually higher. Communication is slower because information must pass through many layers.

Organizations with few layers are flat and have wide spans of management. When managers supervise a large number of employees, fewer management layers are needed to conduct the organization's activities. Managers in flat organizations typically perform more administrative duties than managers in tall organizations because there are fewer of them. They also spend more time supervising and working with subordinates.

Many of the companies that have decentralized also flattened their structures and widened their spans of management, often by eliminating layers of middle

organizational layers
the levels of management in an organization.

Sugar Bowl Bakery Hits the Sweet Spot

Sugar Bowl Bakery
Founders: Tom, Binh, Andrew, Sam, and Paul Ly
Founded: 1984, in San Francisco, California
Success: The bakery has expanded to a 120,000-square-foot production facility and has become one of America's largest family-owned, national, minority bakery manufacturers.

In the 1980s, a family of refugees from Vietnam came to the United States to start a new life. Brothers Tom, Binh, Andrew, Sam, and Paul Ly had a deeply ingrained spirit of teamwork. Despite limited baking experience, the brothers decided to purchase a struggling coffee shop. To purchase the shop, they worked any jobs they could find and pooled together their money. After the bakery was opened, they shared operational responsibility. Andrew Ly took English language night courses so he could build business connections and promote the company.

While Sugar Bowl Bakery began as a coffee and donut shop, today its dessert products are manufactured with modern technology and can be found in nationwide retailers, including Costco and Walgreens. It is still very much a team endeavor, with Andrew Ly as CEO and 8 of 12 second-generation children working at the bakery. Even now, the brothers hold monthly meetings to discuss conflicts and new ideas. This culture of shared responsibility has led to continual success for more than 30 years, with double-digit growth each year.[20]

Critical Thinking Questions

1. How do the founders of Sugar Bowl Bakery use teamwork to make their bakery a success?
2. Why is it important for the brothers to continue to hold monthly meetings to discuss new ideas?
3. Do you think Sugar Bowl has a centralized or decentralized structure? Why?

---

management. As mentioned earlier in this chapter, Johnson & Johnson has both a decentralized and flat organizational structure.

LO 7-5

Compare and contrast some common forms of organizational structure.

Forms of Organizational Structure

Along with assigning tasks and the responsibility for carrying them out, managers must consider how to structure their authority relationships—that is, what structure the organization itself will have and how it will appear on the organizational chart. Common forms of organization include line structure, line-and-staff structure, multidivisional structure, and matrix structure.

Line Structure

line structure
the simplest organizational structure, in which direct lines of authority extend from the top manager to the lowest level of the organization.

The simplest organizational structure, **line structure,** has direct lines of authority that extend from the top manager to employees at the lowest level of the organization. For example, a convenience store employee at 7-Eleven may report to an assistant manager, who reports to the store manager, who reports to a regional manager, or, in an independent store, directly to the owner (Figure 7.4). This structure has a clear chain of command, which enables managers to make decisions quickly.

A mid-level manager facing a decision must consult only one person, his or her immediate supervisor. However, this structure requires that managers possess a wide range of knowledge and skills. They are responsible for a variety of activities and must be knowledgeable about them all. Line structures are most common in small businesses.

Line-and-Staff Structure

line-and-staff structure
a structure having a traditional line relationship between superiors and subordinates and also specialized managers—called staff managers—who are available to assist line managers.

The **line-and-staff structure** has a traditional line relationship between superiors and subordinates, and specialized managers—called staff managers—are available to

FIGURE 7.4
Line Structure

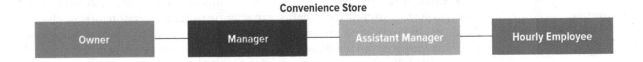

Convenience Store

| Owner | Manager | Assistant Manager | Hourly Employee |

assist line managers (Figure 7.5). Line managers can focus on their area of expertise in the operation of the business, while staff managers provide advice and support to line departments on specialized matters such as finance, engineering, human resources, and the law. Staff managers do not have direct authority over line managers or over the line manager's subordinates, but they do have direct authority over subordinates in their own departments. However, line-and-staff organizations may experience problems with overstaffing and ambiguous lines of communication. Additionally, employees may become frustrated because they lack the authority to carry out certain decisions.

connect

▶ Need help understanding line vs. staff employees? Visit your Connect ebook video tab for a brief animated explanation.

FIGURE 7.5
Line-and-Staff Structure

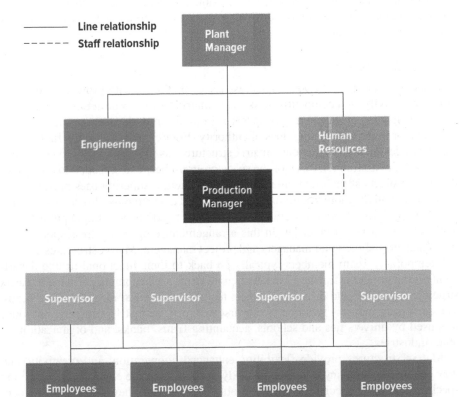

——— Line relationship
- - - - Staff relationship

Plant Manager

Engineering   Human Resources

Production Manager

Supervisor   Supervisor   Supervisor   Supervisor

Employees   Employees   Employees   Employees

Multidivisional Structure

restructure
to change the basic structure of an organization.

multidivisional structure
a structure that organizes departments into larger groups called divisions.

As companies grow and diversify, traditional line structures become difficult to coordinate, making communication difficult and decision making slow. When the weaknesses of the structure—the "turf wars," miscommunication, and working at cross-purposes—exceed the benefits, growing firms tend to **restructure,** or change the basic structure of an organization. Growing firms tend to restructure into the divisionalized form. A **multidivisional structure** organizes departments into larger groups called divisions. Just as departments might be formed on the basis of geography, customer, product, or a combination of these, so too divisions can be formed based on any of these methods of organizing. Within each of these divisions, departments may be organized by product, geographic region, function, or some combination of all three. Indra Nooyi, CEO of PepsiCo, rearranged the company's organizational structure after taking the helm. Prior to her tenure, PepsiCo was organized geographically. She created new units that span international boundaries and make it easier for employees in different geographic regions to share business practices.[21]

Multidivisional structures permit delegation of decision-making authority, allowing divisional and department managers to specialize. They allow those closest to the action to make the decisions that will affect them. Delegation of authority and divisionalized work also mean that better decisions are made faster, and they tend to be more innovative. Most importantly, by focusing each division on a common region, product, or customer, each is more likely to provide products that meet the needs of its particular customers. However, the divisional structure inevitably creates work duplication, which makes it more difficult to realize the economies of scale that result from grouping functions together.

Matrix Structure

matrix structure
a structure that sets up teams from different departments, thereby creating two or more intersecting lines of authority; also called a project management structure.

Another structure that attempts to address issues that arise with growth, diversification, productivity, and competitiveness, is the matrix. A **matrix structure,** also called a project management structure, sets up teams from different departments, thereby creating two or more intersecting lines of authority (Figure 7.6). One of the first organizations to design and implement a matrix structure was the National Aeronautics and Space Administration (NASA) for the space program because it needed to coordinate different projects at the same time. The matrix structure superimposes project-based departments on the more traditional, function-based departments. Project teams bring together specialists from a variety of areas to work together on a single project, such as developing a new fighter jet. In this arrangement, employees are responsible to two managers—functional managers and project managers. Matrix structures are usually temporary: Team members typically go back to their functional or line department after a project is finished. However, more firms are becoming permanent matrix structures, creating and dissolving project teams as needed to meet customer needs. The aerospace industry was one of the first to apply the matrix structure, but today it is used by universities and schools, accounting firms, banks, and organizations in other industries.

Matrix structures provide flexibility, enhanced cooperation, and creativity, and they enable the company to respond quickly to changes in the environment by giving special attention to specific projects or problems. However, they are generally expensive and quite complex, and employees may be confused as to whose authority has priority—the project manager's or the immediate supervisor's.

FIGURE 7.6
Matrix Structure

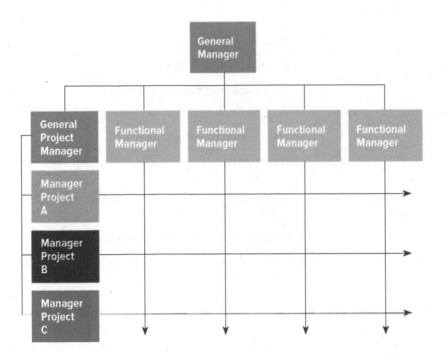

The Role of Groups and Teams in Organizations

Regardless of how they are organized, most of the essential work of business occurs in individual work groups and teams, so we'll take a closer look at them now. There has been a gradual shift toward an emphasis on teams and managing them to enhance individual and organizational success. Some experts now believe that highest productivity results only when groups become teams.[22]

Traditionally, a **group** has been defined as two or more individuals who communicate with one another, share a common identity, and have a common goal. A **team** is a small group whose members have complementary skills; have a common purpose, goals, and approach; and hold themselves mutually accountable.[23] Think of a team like a sports team. Members of a basketball team have different skill sets and work together to score and win the game. All teams are groups, but not all groups are teams. Table 7.1 points out some important differences between them. Work groups emphasize individual work products, individual accountability, and even individual leadership. Your class is a group that can further be separated into teams of two or three classmates. Work teams share leadership roles, have both individual and mutual accountability, and create collective work products. In other words, a work group's performance depends on what its members do as individuals, while a team's performance is based on creating a knowledge center and a competency to work together to accomplish a goal. On the other hand, it is also important for team members to retain their individuality and avoid becoming just "another face in the crowd." The purpose of teams should be toward collaboration versus collectivism. Although the team is working toward a common goal, it is important that all team members actively contribute their ideas and work together to achieve this common goal.[24]

LO 7-6

Distinguish between groups and teams.

group
two or more individuals who communicate with one another, share a common identity, and have a common goal.

team
a small group whose members have complementary skills; have a common purpose, goals, and approach; and hold themselves mutually accountable.

TABLE 7.1   Differences between Groups and Teams

Working Group	Team
Has strong, clearly focused leader	Has shared leadership roles
Has individual accountability	Has individual and group accountability
Has the same purpose as the broader organizational mission	Has a specific purpose that the team itself delivers
Creates individual work products	Creates collective work products
Runs efficient meetings	Encourages open-ended discussion and active problem-solving meetings
Measures its effectiveness indirectly by its effects on others (e.g., financial performance of the business)	Measures performance directly by assessing collective work products
Discusses, decides, and delegates	Discusses, decides, and does real work together

Source: Robert Gatewood, Robert Taylor, and O. C. Ferrell, Management: Comprehension Analysis and Application, (New York: McGraw-Hill Education, 1995), p. 427.

LO 7-7

Identify the types of groups that exist in organizations.

The type of groups an organization establishes depends on the tasks it needs to accomplish and the situation it faces. Some specific kinds of groups and teams include committees, task forces, project teams, product-development teams, quality-assurance teams, and self-directed work teams. All of these can be *virtual teams*—employees in different locations who rely on e-mail, audio conferencing, fax, Internet, videoconferencing, or other technological tools to accomplish their goals. Virtual teams are becoming a part of everyday business, with the number of employees working remotely from their employer increasing more than 80 percent in the last several years.[25] Virtual teams have also opened up opportunities for different companies. For instance, inside salespeople use virtual technology such as e-mail and social media to connect with prospects and clients.[26]

Committees

committee
a permanent, formal group that performs a specific task.

A **committee** is usually a permanent, formal group that does some specific task. For example, many firms have a compensation or finance committee to examine the effectiveness of these areas of operation as well as the need for possible changes. Ethics committees are formed to develop and revise codes of ethics, suggest methods for implementing ethical standards, and review specific issues and concerns.

Task Forces

task force
a temporary group of employees responsible for bringing about a particular change.

A **task force** is a temporary group of employees responsible for bringing about a particular change. They typically come from across all departments and levels of an organization. Task force membership is usually based on expertise rather than organizational position. Occasionally, a task force may be formed from individuals outside a company. Coca-Cola has often used task forces to address problems and provide recommendations for improving company practices or products. While some task forces might last a few months, others last for years. When Coca-Cola faced lawsuits alleging discrimination practices in hiring and promotion, it developed a five-year task force to examine pay and promotion practices among minority employees. Its experiences helped Coca-Cola realize the advantages of having a cross-functional task force made up of employees from different departments, and it continued to use task forces to tackle major company issues. Other companies that have also recognized the benefits of task forces include IBM, Prudential, and General Electric.[27]

Teams

Teams are becoming far more common in the U.S. workplace as businesses strive to enhance productivity and global competitiveness. In general, teams have the benefit of being able to pool members' knowledge and skills and make greater use of them than can individuals working alone. Team building is becoming increasingly popular in organizations, with around half of executives indicating their companies had team-building training. Teams require harmony, cooperation, synchronized effort, and

Florida State University formed a temporary Brand Development Task Force comprised of campus leaders to refine and develop the university's brand.

©Mario Houben/CSM/REX/Shutterstock

flexibility to maximize their contribution.[28] Teams can also create more solutions to problems than can individuals. Furthermore, team participation enhances employee acceptance of, understanding of, and commitment to team goals. Teams motivate workers by providing internal rewards in the form of an enhanced sense of accomplishment for employees as they achieve more, and external rewards in the form of praise and certain perks. Consequently, they can help get workers more involved. They can help companies be more innovative, and they can boost productivity and cut costs.

According to psychologist Ivan Steiner, team productivity peaks at about five team members. People become less motivated and group coordination becomes more difficult after this size. Jeff Bezos, Amazon CEO, says that he has a "two-pizza rule": If a team cannot be fed by two pizzas, it is too large. Keep teams small enough where everyone gets a piece of the action.[29]

Project Teams.   Project teams are similar to task forces, but normally they run their operation and have total control of a specific work project. Like task forces, their membership is likely to cut across the firm's hierarchy and be composed of people from different functional areas. They are almost always temporary, although a large project, such as designing and building a new airplane at Boeing Corporation, may last for years.

Product-development teams are a special type of project team formed to devise, design, and implement a new product. Sometimes product-development teams exist within a functional area—research and development—but now they more frequently include people from numerous functional areas and may even include customers to help ensure that the end product meets the customers' needs. Intel informs its product development process through indirect input from customers. It has a social scientist on staff who leads a research team on how customers actually use products. This is done mainly by observation and asking questions. Once enough information is gathered, it is relayed to the product-development team and incorporated into Intel's designs.[30]

Quality-Assurance Teams.   Quality-assurance teams, sometimes called quality circles, are fairly small groups of workers brought together from throughout the

project teams
groups similar to task forces that normally run their operation and have total control of a specific work project.

product-development teams
a specific type of project team formed to devise, design, and implement a new product.

quality-assurance teams
(or quality circles)
small groups of workers brought together from throughout the organization to solve specific quality, productivity, or service problems.

Connection and Collaboration: How Timberland Works as a Team

From its product development team to its Global Stewards, Timberland is committed to using teamwork to advance its goals of sustainable product offerings. In addition to teamwork among its employees, Timberland has also partnered with suppliers and even competitors to create solutions for positively affecting the environment.

Timberland grew out of the Abington Shoe Company founded in 1955, becoming Timberland Company in 1978. The firm began by manufacturing footwear and later expanded into clothing and accessories. Timberland is known for its eco-friendly products. One-third of the company's footwear is made at least partially from recycled materials.

Timberland uses teamwork to positively affect the sustainability movement. In 2001, the company started a cross-functional team made up of employees from different areas in the company to determine how it could contribute to sustainability.

Timberland formed its Global Stewards Program, teams of passionate volunteer employees who take the time to participate as civic leaders within their communities, empower employees, and lead social responsibility projects.

Timberland has also partnered with other important stakeholders. It partnered with its leather suppliers to decrease energy and water usage at its leather tanneries. Timberland also teamed with competitors on sustainability initiatives. The company worked with Nike and many other firms to create the Higg Index, a green index for apparel, and with Adidas on a green footwear index. Timberland recognizes that by developing teams among various stakeholders, it can create a major difference within its industry.[31]

Critical Thinking Questions

1. How does Timberland use teamwork to advance its goals?
2. Why would Timberland partner with the competition?
3. How do you think Timberland's use of teams empowers employees?

---

organization to solve specific quality, productivity, or service problems. Although the *quality circle* term is not as popular as it once was, the concern about quality is stronger than ever. Companies such as IBM and Xerox as well as companies in the automobile industry have used quality circles to shift the organization to a more participative culture. The use of teams to address quality issues will no doubt continue to increase throughout the business world.

self-directed work team (SDWT)
a group of employees responsible for an entire work process or segment that delivers a product to an internal or external customer.

Self-Directed Work Teams. A **self-directed work team (SDWT)** is a group of employees responsible for an entire work process or segment that delivers a product to an internal or external customer.[32] SDWTs permit the flexibility to change rapidly to meet the competition or respond to customer needs. The defining characteristic of an SDWT is the extent to which it is empowered or given authority to make and implement work decisions. Thus, SDWTs are designed to give employees a feeling of "ownership" of a whole job. Employees at 3M as well as an increasing number of companies encourage employees to be active to perform a function or operational task. With shared team responsibility for work outcomes, team members often have broader job assignments and cross-train to master other jobs, thus permitting greater team flexibility.

DID YOU KNOW? A survey of employees revealed that approximately 53 percent consider how senior management communicates with them to be a key factor in their job satisfaction.[33]

LO 7-8

Describe how communication occurs in organizations.

Communicating in Organizations

Communication within an organization can flow in a variety of directions and from a number of sources, each using both oral and written forms of communication. The success of communication systems within the organization has a tremendous effect on the overall success of the firm. Communication mistakes can lower productivity and morale.

Alternatives to face-to-face communications—such as meetings—are growing, thanks to technology such as voice-mail, e-mail, social media, and online newsletters. Many companies use internal networks called intranets to share information with employees. Intranets increase communication across different departments and levels of management and help with the flow of everyday business activities. Another innovative approach is cloud computing. Rather than using physical products, companies using cloud computing technology can access computing resources and information over a network. Cloud computing allows companies to have more control over computing resources and can be less expensive than hardware or software. Salesforce. com uses cloud computing in its customer relationship management solutions.[34] Companies can even integrate aspects of social media into their intranets, allowing employees to post comments and pictures, participate in polls, and create group calendars. However, increased access to the Internet at work has also created many problems, including employee abuse of company e-mail and Internet access.[35] The increasing use of e-mail as a communication tool also inundates employees and managers with e-mails, making it easier to overlook individual communications. For this reason, it is advised that employees place a specific subject in the subject line, keep e-mails brief, and avoid using e-mail if a problem would be better solved through telephone contact or face-to-face interaction.[36]

Cloud-based tools like Dropbox, iCloud, Google Drive, Salesforce, and more are transforming internal communication at companies. They allow employees to share documents, collaborate with their teams, and even work from home.

©Krisztian Bocsi/Bloomberg via Getty Images

Formal and Informal Communication

Formal channels of communication are intentionally defined and designed by the organization. They represent the flow of communication within the formal organizational structure, as shown on organizational charts. Table 7.2 describes the different forms of formal communication. Traditionally, formal communication patterns were classified as vertical and horizontal, but with the increased use of teams and matrix structures, formal communication may occur in a number of patterns (Figure 7.7).

Along with the formal channels of communication shown on an organizational chart, all firms communicate informally as well. Communication between friends, for instance, cuts across department, division, and even management-subordinate boundaries. Such friendships and other nonwork social relationships comprise the *informal organization* of a firm, and their impact can be great.

The most significant informal communication occurs through the **grapevine,** an informal channel of communication, separate from management's formal, official communication channels. Grapevines exist in all organizations. Information passed along the grapevine may relate to the job or organization, or it may be gossip and rumors unrelated to either. The accuracy of grapevine information has been of great concern to managers.

grapevine
an informal channel of communication, separate from management's formal, official communication channels.

TABLE 7.2   Types of Formal Communication

Type	Definition	Examples
Upward	Flows from lower to higher levels of the organization	Progress reports, suggestions for improvement, inquiries, grievances
Downward	Traditional flow of communication from upper organizational levels to lower organizational levels	Directions, assignments of tasks and responsibilities, performance feedback, details about strategies and goals, speeches, employee handbooks, job descriptions
Horizontal	Exchange of information among colleagues and peers on the same organizational level, such as across or within departments, who inform, support, and early coordinate activities both within the department and between other departments	Task forces, project teams, communication from the finance department to the marketing department concerning budget requirements
Diagonal	When individuals from different levels and different departments communicate	A manager from the finance department communicates with a lower-level manager from the marketing department

FIGURE 7.7
The Flow of Communication in an Organizational Hierarchy

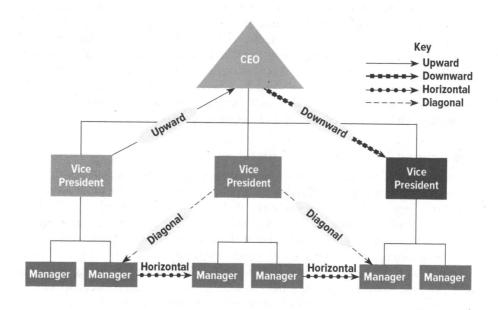

Managers can turn the grapevine to their advantage. Using it as a "sounding device" for possible new policies is one example. Managers can obtain valuable information from the grapevine that could improve decision making. Some organizations use the grapevine to their advantage by floating ideas, soliciting feedback, and reacting accordingly. People love to gossip, and managers need to be aware that grapevines exist in every organization. Managers who understand how the grapevine works also can use it to their advantage by feeding it facts to squelch rumors and incorrect information. For instance, rather than confronting employees about gossip and placing them on the defense, some employers ask employees—especially those who are the spreaders of gossip—for assistance in squelching the untrue rumors. This tactic turns employees into advocates for sharing truthful information.[37]

Monitoring Communications

Technological advances and the increased use of electronic communication in the workplace have made monitoring its use necessary for most companies. Failing to monitor employees' use of e-mail, social media, and the Internet can be costly. Many companies require that employees sign and follow a policy on appropriate Internet use. These agreements often require that employees will use corporate computers only for work-related activities. Additionally, some companies use software programs to monitor employee computer usage.[38] Instituting practices that show respect for employee privacy but do not abdicate employer responsibility are increasingly necessary in today's workplace. Merck, for instance, has a section on employee privacy in its code of conduct that reassures both current and former employees that their information will be protected and used only for legitimate business purposes.[39]

Artificial intelligence (AI) is having a significant impact on workplace monitoring, benchmarking, and understanding how employees "feel" about their jobs. More than 40 percent of employers globally have implemented AI processes within their organization. One such tool called Xander can determine an employee's optimism, confusion, anger, or happiness. With such an understanding of employee attitudes, you can correct or improve the negative experiences in the workplace and enhance the positive ones. Software companies have developed artificial intelligence tools that can make routine human resource decisions such as hiring, firing, and raises. There are concerns that AI technology might incorporate biases that can lead to discriminatory decision making.[40]

Improving Communication Effectiveness

Without effective communication, the activities and overall productivity of projects, groups, teams, and individuals will be diminished. Communication is an important area for a firm to address at all levels of management. Apple supplier Foxconn is one example of how essential communication is to a firm. Despite criticisms of unfair labor conditions, the Fair Labor Association determined that Foxconn had formal procedures in place at its factories to prevent many major accidents. However, it concluded that the firm had a communication problem. These procedures were not being communicated to the factory workers, contributing to unsafe practices and two tragic explosions.[41]

One of the major issues of effective communication is in obtaining feedback. If feedback is not provided, then communication will be ineffective and can drag down overall performance. Managers should always encourage feedback, including concerns and challenges about issues. Listening is a skill that involves hearing, and most employees listen much more than they actively communicate to others. Therefore, managers should encourage employees to provide feedback—even if it is negative. It is interesting to note that employees list a failure to listen to their concerns as a top complaint in the workplace.[42] Employees often notice issues that managers overlook, and employee feedback can alert managers to these issues. This will allow the organization to identify strengths and weaknesses and make adjustments when needed. At the same time, strong feedback mechanisms help to empower employees as they feel that their voices are being heard.

Interruptions can be a serious threat to effective communication. Various activities can interrupt the message. For example, interjecting a remark can create discontinuance in the communication process or disrupt the uniformity of the message. Even small interruptions can be a problem if the messenger cannot adequately understand or interpret the communicator's message. One suggestion is to give the communicator space or time to make another statement rather than quickly responding or making your own comment.

Strong and effective communication channels are a requirement for companies to distribute information to different levels of the company. Businesses have several channels for communication, including face-to-face, e-mail, phone, and written communication (for example, memos). Each channel has advantages and disadvantages, and some are more appropriate to use than others. For instance, a small task requiring little instruction might be communicated through a short memo or e-mail. An in-depth task would most likely require a phone conversation or face-to-face contact. E-mail has become especially helpful for businesses, and both employees and managers are increasingly using e-mail rather than memos or phone conversations. However, it is important that employees use e-mail correctly. Inappropriate use of e-mail can include forwarding sexually explicit or otherwise offensive material. Additionally, many employees have used work e-mail accounts to send personal information. This may be against company policy. It is important for employees to remember that e-mail sent from corporate accounts is the property of the firm, so they should exert caution in making sure their e-mail messages contain appropriate content. It is, therefore, important for companies to communicate their e-mail policies throughout the organization. Communicators using e-mail, whether managers or employees, must exert caution before pushing that "Send" button.

E-mail is a dominant form of workplace communication in many offices. Employees need to be coached on best practices in using their e-mail so that it's not a distraction or disruptive to productivity. On average, users check their e-mail every 11 minutes, and 84 percent have their e-mail up at all times. On average, we receive 87 e-mails each day, with 70 percent being opened within six seconds of their receipt. It takes us just over a minute to transition to our original tasks interrupted by e-mail. Some guidance on best practices in managing workplace productivity suggests that it's best to check your e-mail every 45 minutes. Also, responding too quickly and during evenings and weekends can increase stress and diminish productivity.[43]

Communication is necessary in helping every organizational member understand what is expected of him or her. Many business problems can be avoided if clear communication exists within the company. Even the best business strategies are of little use if those who will oversee them cannot understand what is intended. Communication might not seem to be as big of a concern to management as finances, human resources, and marketing, but in reality, it can make the difference between successful implementation of business activities or failure.

So You Want a Job in Managing Organizational Culture, Teamwork, and Communication

Jobs dealing with organizational culture and structure are usually at the top of the organization. If you want to be a CEO or high-level manager, you will help shape these areas of business. On the other hand, if you are an entrepreneur or small-business person, you will need to make decisions about assigning tasks, departmentalization, and assigning responsibility. Even managers in small organizations have to make decisions about decentralization, span of management, and forms of organizational structure. Micro-entrepreneurs with five or fewer employees still have to make decisions about assigning tasks and whether to work as a group or a team. While these decisions may be part of your job, there are usually no job titles dealing with these specific areas. Specific jobs that attempt to improve organizational culture could include ethics and compliance positions as well as those who are in charge of communicating memos, manuals, and policies that help establish the culture. These positions will be in communications, human resources, and positions that assist top organizational managers.

Teams are becoming more common in the workplace, and it is possible to become a member of a product development group or quality assurance team. There are also human resource positions that encourage teamwork through training

activities. The area of corporate communications provides lots of opportunities for specific jobs that facilitate communication systems. Thanks to technology, there are job positions to help disseminate information through online newsletters, intranets, or internal computer networks to share information to increase collaboration. In addition to the many advances using electronic communications, there are technology concerns that create new job opportunities. Monitoring workplace communications such as the use of e-mail and the Internet has created new industries. There have to be internal controls in the organization to make sure that the organization does not engage in any copyright infringement. If this is an area of interest, there are specific jobs that provide an opportunity to use your technological skills to assist in maintaining appropriate standards in communicating and using technology.

If you go to work for a large company with many divisions, you can expect a number of positions dealing with the tasks discussed here. If you go to work for a small company, you will probably engage in most of these tasks as a part of your position. Organizational flexibility requires individual flexibility, and those employees willing to take on new domains and challenges will be the employees who survive and prosper in the future.

Review Your Understanding

Explain the importance of organizational culture.

Organizational culture is the firm's shared values, beliefs, traditions, philosophies, and role models for behavior. It helps ensure that all members of a company share values and suggests rules for how to behave and deal with problems within the organization.

Describe how organizational structures develop.

Structure is the arrangement or relationship of positions within an organization; it develops when managers assign work activities to work groups and specific individuals and coordinate the diverse activities required to attain organizational objectives. Organizational structure evolves to accommodate growth, which requires people with specialized skills.

Describe how specialization and departmentalization help an organization achieve its goals.

Structuring an organization requires that management assign work tasks to specific individuals and groups. Under

specialization, managers break labor into small, specialized tasks and assign employees to do a single task, fostering efficiency. Departmentalization is the grouping of jobs into working units (departments, units, groups, or divisions). Businesses may departmentalize by function, product, geographic region, or customer, or they may combine two or more of these.

Determine how organizations assign responsibility for tasks and delegate authority.

Delegation of authority means assigning tasks to employees and giving them the power to make commitments, use resources, and take whatever actions are necessary to accomplish the tasks. It lays responsibility on employees to carry out assigned tasks satisfactorily and holds them accountable to a superior for the proper execution of their assigned work. The extent to which authority is delegated throughout an organization determines its degree of centralization. Span of management refers to the number of subordinates who report to a particular manager. A wide span of management occurs in flat organizations; a narrow one exists in tall organizations.

Compare and contrast some common forms of organizational structure.

Line structures have direct lines of authority that extend from the top manager to employees at the lowest level of the organization. The line-and-staff structure has a traditional line relationship between superiors and subordinates, and specialized staff managers are available to assist line managers. A multidivisional structure gathers departments into larger groups called divisions. A matrix, or project-management, structure sets up teams from different departments, thereby creating two or more intersecting lines of authority.

Distinguish between groups and teams.

A group is two or more persons who communicate, share a common identity, and have a common goal. A team is a small group whose members have complementary skills, a common purpose, goals, and approach and who hold themselves mutually accountable. The major distinction is that individual performance is most important in groups, while collective work group performance counts most in teams.

Identify the types of groups that exist in organizations.

Special kinds of groups include task forces, committees, project teams, product-development teams, quality-assurance teams, and self-directed work teams.

Describe how communication occurs in organizations.

Communication occurs both formally and informally in organizations. Formal communication may be downward, upward, horizontal, and even diagonal. Informal communication takes place through friendships and the grapevine.

Analyze a business's use of teams.

The "Solve the Dilemma" feature near the end of this chapter introduces Quest Star's attempt to restructure to a team environment. Based on the material presented in this chapter, you should be able to evaluate the Quest Star's efforts and make recommendations for resolving the problems that have developed.

Critical Thinking Questions

Enter the World of Business Questions

1. What types of communication does Keurig Green Mountain uses within its organization?

2. Why do you think strong communication is important in a decentralized organization with few layers of management?

3. What are some of the advantages of using functional departmentalization?

Learn the Terms

accountability 210
centralized organization 211
committee 218
customer departmentalization 210
decentralized organization 211
delegation of authority 210
departmentalization 208
functional departmentalization 209
geographical departmentalization 209
grapevine 221

group 217
line-and-staff structure 214
line structure 214
matrix structure 216
multidivisional structure 216
organizational chart 206
organizational culture 204
organizational layers 213
product departmentalization 209
product-development teams 219

project teams 219
quality-assurance teams (or quality circles) 219
responsibility 210
restructure 216
self-directed work team (SDWT) 220
span of management 212
specialization 207
structure 205
task force 218
team 217

Check Your Progress

1. Identify four types of departmentalization and give an example of each type.

2. Explain the difference between groups and teams.

3. What are self-managed work teams and what tasks might they perform that traditionally are performed by managers?

4. Explain how delegating authority, responsibility, and accountability are related.

5. Distinguish between centralization and decentralization. Under what circumstances is each appropriate?

6. Define *span of management*. Why do some organizations have narrow spans and others wide spans?

7. Discuss the different forms of organizational structure. What are the primary advantages and disadvantages of each form?

8. Discuss the role of the grapevine within organizations. How can managers use it to further the goals of the firm?

9. How have technological advances made electronic oversight a necessity in many companies?

10. Discuss how an organization's culture might influence its ability to achieve its objectives. Do you think that managers can "manage" the organization's culture?

Get Involved

1. Explain, using a specific example (perhaps your own future business), how an organizational structure might evolve. How would you handle the issues of specialization, delegation of authority, and centralization? Which structure would you use? Explain your answers.

2. Interview the department chairperson in charge of one of the academic departments in your college or university. Using Table 7.1 as a guideline, explore whether the professors function more like a group or a team. Contrast what you learned here with what you see on your school's basketball, football, or baseball team.

Build Your Skills

Teamwork

Background

Think about all the different kinds of groups and teams you have been a member of or been involved with. Here's a checklist to help you remember them—with "Other" spaces to fill in ones not listed. Check all that apply.

School Groups/Teams

☐ Sports teams
☐ Cheerleading squads
☐ Musical groups
☐ Hobby clubs
☐ Foreign language clubs
☐ Study groups
☐ Other ____

Community Groups/Teams

☐ Fund-raising groups
☐ Religious groups
☐ Sports teams
☐ Political groups
☐ Boy/Girl Scout troops
☐ Volunteer organizations
☐ Other ____

Employment Groups/Teams

☐ Problem-solving teams
☐ Work committees
☐ Project teams
☐ Labor union groups
☐ Work crews
☐ Other ____

Task

I. Of those you checked, circle those that you would categorize as a "really great team."

2. Examine the following table and circle those characteristics from columns 2 and 3 that were represented in your "really great" team experiences.[44]

Indicator	Good Team Experience	Not-So-Good Team Experience
Members arrive on time?	Members are prompt because they know others will be.	Members drift in sporadically, and some leave early.
Members prepared?	Members are prepared and know what to expect.	Members are unclear what the agenda is.
Meeting organized?	Members follow a planned agenda.	The agenda is tossed aside, and freewheeling discussion ensues.
Members contribute equally?	Members give each other a chance to speak; quiet members are encouraged.	Some members always dominate the discussion; some are reluctant to speak their minds.
Discussions help members make decisions?	Members learn from others' points of view, new facts are discussed, creative ideas evolve, and alternatives emerge.	Members reinforce their belief in their own points of view, or their decisions were made long before the meeting.
Any disagreement?	Members follow a conflict-resolution process established as part of the team's policies.	Conflict turns to argument, angry words, emotion, blaming.
More cooperation or more conflict?	Cooperation is clearly an important ingredient.	Conflict flares openly, as well as simmering below the surface.
Commitment to decisions?	Members reach consensus before leaving.	Compromise is the best outcome possible; some members don't care about the result.
Member feelings after team decision?	Members are satisfied and are valued for their ideas.	Members are glad it's over, not sure of results or outcome.
Members support decision afterward?	Members are committed to implementation.	Some members second-guess or undermine the team's decision.

3. What can you take with you from your positive team experiences and apply to a work-related group or team situation in which you might be involved?

_____

_____

_____

Solve the Dilemma

Quest Star in Transition

Quest Star (QS), which manufactures quality stereo loudspeakers, wants to improve its ability to compete against Japanese Arms. Accordingly, the company has launched a comprehensive quality-improvement program for its Iowa plant. The QS Intracommunication Leadership Initiative (ILI) has flattened the layers of management. The program uses teams and peer pressure to accomplish the plant's goals instead of multiple management layers with their limited opportunities for communication. Under the initiative, employees make all decisions within the boundaries of their responsibilities, and they elect team representatives to coordinate with other teams. Teams are also assigned tasks ranging from establishing policies to evaluating on-the-job safety.

However, employees who are not self-motivated team players are having difficulty getting used to their peers' authority within this system. Upper-level managers face stress

Analyze a business's use of teams.

and frustration because they must train workers to supervise themselves.

Critical Thinking Questions

I.   What techniques or skills should an employee have to assume a leadership role within a work group?

2.   If each work group has a team representative, what problems will be faced in supervising these representatives?

3.   What are the pros and cons of the system developed by QS?

Build Your Business Plan

Organization, Teamwork, and Communication

Developing a business plan as a team is a deliberate move of your instructor to encourage you to familiarize yourself with the concept of teamwork. You need to realize that you are going to spend a large part of your professional life working with others. At this point in time, you are working on the business plan for a grade, but after graduation, you will be "teaming" with co-workers, and the success of your endeavor may determine whether you get a raise or a bonus. It is important that you be comfortable as soon as possible with working with others and holding them accountable for their contributions.

Some people are natural "leaders," and leaders often feel that if team members are not doing their work, they take it upon themselves to "do it all." This is not leadership, but rather micro-managing.

Leadership means holding members accountable for their responsibilities. Your instructor may provide ideas on how this could be implemented, possibly by utilizing peer reviews. Remember, you are not doing a team member a favor by doing their work for them.

If you are a "follower" (someone who takes directions well) rather than a leader, try to get into a team where others are hard workers and you will rise to their level. There is nothing wrong with being a follower; not everyone can be a leader!

Visit Connect to practice building your business plan with the Business Plan Prep Exercises.

See for Yourself Videocase

A Freshii Approach to Food and Business

Freshii is a restaurant franchise that focuses on fast and healthy food options. With more than 395 locations in 18 countries, the company has a variety of customizable food offerings from burritos and salads to frozen yogurt and smoothies. Freshii takes pride in its flat organizational structure, which it believes lets the company cut excess layers of management and allow for easier communication. Its flat organizational structure also empowers employees to have more ownership in their segment of the business.

Matthew Corrin, CEO of Freshii, founded the company in 2005 when he identified a need for fast and healthy food options. He created Freshii to offer healthy food that is convenient and affordable across the world. Surprisingly, the restaurants have no stoves, ovens, fryers, ventilation, or freezers. Since Freshii started, it has become the fastest growing franchise in the world and is leading the industry with 20 consecutive quarters of same store sales growth. The company took the top spot in Fast Casual's Top 100 Movers & Shakers list in recognition of the Freshii's efforts to make fundamental changes for the industry.

Freshii aims to have a corporate culture that allows free communication and ideas but isn't bogged down by extra layers of management, and its organizational structure is a big part of this. The company's flat structure allows employees to make critical decisions in a timely manner and empowers employees to be responsible for their own tasks. Freshii considers all employees "Partners" and awards each employee with stock in the company so that employees have a real vested interest.

"We really feel that we empower our employees by giving them full ownership of their role for their area. They own their business from project ideation, research, and then all the way through execution and launching it, and then measuring the results out in the field," says Chief People Officer Ashley Dalziel. "And because of this structure, they get to see a project through from ideation, brainstorming, research, all the way through to execution."

Even though employees are given the power to design and execute their own projects, Freshii keeps a strong emphasis on communication. The company has seven key departments that are led by executives who have complete ownership of their area. Freshii's executive team, including its CEO, works in the same room around one large open table. This layout allows the executive team to communicate openly with each other in a space where the entire executive team can stay informed and be involved in the discussion.

Employees at Freshii are given a lot of responsibility to own their segments of the business. Because this can be overwhelming, the company rewards employees and builds chemistry through team activities. Freshii puts on monthly fitness challenges for employees where employees are encouraged to run, take spin classes, and exercise with each other. Some of Freshii's core values are to live an energized life and to drive to be better, which the company embraces through these activities.

Each Freshii store leader is encouraged to talk to employees and find out what they are passionate about. Then, the store leader can assign team members to roles where they are most passionate and most likely to succeed. This personal attention from management allows employees to find roles they are best suited for and gives them an opportunity to be in a role they care about.

Franchising has helped Freshii expand internationally and spread its brand. However, Freshii is careful about choosing franchise partners that share its mission. Freshii's mission statement is "To help citizens of the world live better by making healthy food convenient & affordable." It is important that franchise partners embrace this mission in order for the company to maintain its unified culture. Still, Freshii empowers franchise owners to try new things and create new offerings to create innovation.

Freshii's flat organizational structure fits well with its culture of open communication and personal responsibility. Because of its culture, the company finds employees who are competitive and passionate about providing healthy, convenient, and affordable food. Freshii strives to continue its rapid expansion while maintaining what made it unique from the start.[45]

Critical Thinking Questions

1. How does Freshii's organizational structure shape its culture?

2. In what ways are employees at Freshii empowered to contribute to the company?

3. What effect does Freshii's franchise model have on its company culture?

You can find the related video in the Video Library in Connect. Ask your instructor how you can access Connect.

Team Exercise

Assign the responsibility of providing the organizational structure for a company one of your team members has worked for. Was your organization centralized or decentralized in terms of decision making? Would you consider the span of control to be wide or narrow? Were any types of teams, committees, or task forces utilized in the organization? Report your work to the class.

Ask your instructor about the role-play exercises available with this book to practice working with a business team.

Notes

1. Jena McGregor, "Online Extra: Learning on the Front Lines," *Bloomberg Businessweek Magazine*, July 10, 2006, http://www.businessweek.com/stories/2006-07-09/online-extra-learning-on-the-front-lines (accessed January 1, 2018); Keurig Green Mountain website, http://www.keuriggreenmountain.com/ (accessed January 1, 2018); Dun & Bradstreet Inc., "Keurig Green Mountain," *Hoovers Online*, http://www.hoovers.com/company-information/cs/company-profile.keurig_green_mountain_inc.09ca839579577b55.html (accessed January 1, 2018); "Green Mountain Coffee Roasters Brews Formula for Success," in O.C. Ferrell, Geoffrey Hirt, and Linda Ferrell, *Business: A Changing World*, 6th ed. (New York: McGraw-Hill Irwin, 2008), pp. 233–234; Leslie Patton and Nikolaj Gammeltoft, "Green Mountain Drops as David Einhorn Says Market 'Limited,'" *Bloomberg Businessweek*, October 17, 2011, www.businessweek.com/news/2011-10-17/green-mountain-drops-as-david-einhorn-says-market-limited.html (accessed November 6, 2013); Christopher Faille, "Green Mountain Coffee's Trouble with Bean Counting," *Forbes*, June 23, 2011, https://www.forbes.com/sites/greatspeculations/2011/06/23/green-mountain-coffees-trouble-with-bean-counting/#2578180296dc (accessed January 1, 2018); *GMCR Fiscal 2012*

Annual Report, http://files.shareholder.com/downloads/ GMCR/2774887737x0x630863/FDBC5F63-78E8-493C-9BB9-8F33000C0465/GMCR_2012_ANNUAL_REPORT. pdf (accessed November 6, 2013); "Green Mountain Coffee: Starbucks Bursts Its Bubble," *Seeking Alpha,* March 9, 2012, http://seekingalpha.com/article/422241-green-mountain-coffee-starbucks-bursts-its-bubble (accessed January 1, 2018); "2016 By the Numbers," Keurig Green Mountain, January 3, 2017, https://www.keuriggreenmountain.com/en/ OurStories/Business/2016ByTheNumbers.aspx (accessed January 1, 2018); "Sustainability," Keurig Green Mountain, http://www.keuriggreenmountain.com/en/Sustainability/ Overview.aspx (accessed January 1, 2018); Steven Bruce, "HRWorks Sits Down with Keurig Green Mountain Coffee," *HR Daily Advisor,* October 9, 2017, http://hrdailyadvisor.blr. com/2017/10/09/hrworks-sits-keurig-green-mountain-coffee/ (accessed January 1, 2018); "World's Most Innovative Companies: #43 Keurig Green Mountain," *Forbes,* May 2015, https://www.forbes.com/companies/keurig-green-mountain/ (accessed January 1, 2018).

2. Horace Dediu, "Understanding Apple's Organizational Structure," *Asymco,* July 3, 2013, www.asymco. com/2013/07/03/understanding-apples-organizational-structure/ (accessed May 4, 2107); Sam Grobart, "How Samsung Became the World's No. 1 Smartphone Maker," *Bloomberg Businessweek,* March 28, 2013, www. businessweek.com/articles/2013-03-28/how-samsung-became-the-worlds-no-dot-1-smartphone-maker#p1 (accessed May 4, 2017); Jay Yarow, "Apple's New Organizational Structure Could Help It Move Faster," *Business Insider,* May 1, 2013, www.businessinsider.com/ apples-new-organizational-structure-could-help-it-move-faster-2013-5 (accessed May 4, 2017).

3. John D. Still, "Ford Aims to Pivot in Raising CEO's Pay," *The Wall Street Journal,* April 1, 2017, p. B1.

4. Megan McArdle, "'Unlimited Vacation' Is Code for 'No Vacation,'" September 30, 2015, *Bloomberg,* September 30, 2015, http://www.bloombergview.com/articles/2015-09-30/ unlimited-vacation-is-code-for-no-vacation- (accessed May 4, 2017).

5. Micah Solomon, "Take These Two to Rival Nordstrom's Customer Service Experience," *Forbes,* March 15, 2014, http://www.forbes.com/sites/micahsolomon/2014/03/15/ the-nordstrom-two-part-customer-experience-formula-lessons-for-your-business/#2ebc60a92335 (accessed May 4, 2017).

6. Billy Selekane "20 Ways Zappos Reinforces Its Company Culture," https://www.linkedin.com/pulse/20-ways-zappos-reinforces-its-company-culture-billy-selekane-csp-hof/ (accessed April 17, 2018).

7. "YouTube Revamped Its Ad System. AT&T Still Hasn't Returned," *The New York Times,* February 12, 2018, https:// www.nytimes.com/2018/02/12/business/media/att-youtube-advertising.html (accessed April 17, 2018).

8. Lauren Thomas, "New Ford CEO Promises to be a 'Cultural Change Agent,' Bill Ford Says," *CNBC,* May 22, 2017, https://www.cnbc.com/2017/05/22/ new-ford-ceo-promises-to-be-a-cultural-change-agent-bill-ford-says.html (accessed August 4, 2018).

9. "Holacracy and Self-Organization," *Zappos Insights,* https:// www.zapposinsights.com/about/holacracy (accessed April 18, 2018).

10. Adam Smith, *Wealth of Nations* (New York: Modern Library, 1937; originally published in 1776).

11. Ben Dipietro, "Automakers Face 'Herculean' Task in Implementing Supply Chain Guidelines," *The Wall Street Journal,* May 28, 2014, http://blogs.wsj.com/ riskandcompliance/2014/05/28/automakers-face-herculean-task-in-implementing-supply-chain-guidelines/ (accessed May 4, 2017).

12. "Tesla's Organizational Structure & Its Characteristics (Analysis)," *Panmore,* January 28, 2018, http:// panmore.com/tesla-motors-inc-organizational-structure-characteristics-analysis (accessed April 18, 2018).

13. "Brands," Unilever, https://www.unilever.com/brands/ (accessed March 10, 2016).

14. Global Divisions, PepsiCo Inc., http://www.pepsico.com/ About/global-divisions (accessed April 18, 2018).

15. "Regions," Diageo, www.diageo.com/en-row/ourbusiness/ ourregions/Pages/default.aspx (accessed March 10, 2016).

16. Matt Scholz, "The Three-Step Process That's Kept 3M Innovative for Decades," *Fast Company,* July 10, 2017, https://www.fastcompany.com/40437745/the-three-step-process-thats-kept-3m-innovative-for-decades (accessed April 19, 2018).

17. Sarah Whitten, "Fresh Beef Is Coming to a McDonald's Near You," *CNBC,* March 6, 2018, https://www.cnbc. com/2018/03/06/fresh-beef-is-coming-to-a-mcdonalds-near-you.html (accessed April 19, 2018).

18. "Why Work Here?" Whole Foods Market, www .wholefoodsmarket.com/careers/workhere.php (accessed March 10, 2016).

19. "Why We're a Great Place to Work," Whole Foods Market, http://www.wholefoodsmarket.com/careers/why-were-great-place-work (accessed December 30, 2017); "Whole Foods Market's Core Values," Whole Foods Market, www .wholefoodsmarket.com/values/corevalues.php#supporting (accessed December 30, 2017); "100 Best Companies to Work For: Whole Foods Market," *CNN Money,* 2011, http:// money.cnn.com/magazines/fortune/bestcompanies/2011/ snapshots/24.html (accessed December 30, 2017); "100 Best Companies to Work For: Whole Foods Market," *CNN Money,* 2013, http://money.cnn.com/magazines/fortune/ best-companies/2013/snapshots/71.html?iid=bc_fl_list (accessed September 9, 2013); Kerry A. Dolan, "America's Greenest Companies 2011," *Forbes,* April 18, 2011, www .forbes.com/2011/04/18/americas-greenest-companies. html (accessed December 30, 2017); Joseph Brownstein, "Is Whole Foods' Get Healthy Plan Fair?" *ABC News,* January 29, 2010, http://abcnews.go.com/Health/w_ DietAndFitnessNews/foods-incentives-make-employees-healthier/story?id=9680047 (accessed December 30, 2017); Deborah Dunham, "At Whole Foods Thinner Employees

Get Fatter Discounts," *That's Fit,* January 27, 2010, www .thatsfit.com/2010/01/27/whole-foods-thin-employees-get-discounts/ (accessed September 9, 2013); David Burkus, "Why Whole Foods Builds Its Entire Business On Teams," *Forbes,* June 8, 2016, https://www.forbes.com/sites/davidburkus/2016/06/08/why-whole-foods-build-their-entire-business-on-teams/#47872abc3fa1 (accessed December 30, 2017); Kate Taylor, "Here Are All the Changes Amazon Is Making to Whole Foods," *Business Insider,* November 15, 2017, http://www.businessinsider.com/amazon-changes-whole-foods-2017-9/#whole-foods-immediately-slashed-prices-and-announced-another-round-of-price-cuts-in-november-1 (accessed December 30, 2017).

20. Dinah Eng, "The Sweet Taste of Success," *Fortune,* June 1, 2015, pp. 9–20; "Press," Sugar Bowl Bakery, https://sugarbowlbakery.com/blogs/press/the-costco-connection (accessed March 23, 2018); Ilana DeBare, "Sugar Bowl Bakery Is a Family Affair, *SF Gate,* July 6, 2008, https://www.sfgate.com/business/article/Sugar-Bowl-Bakery-is-a-family-affair-3205820.php (accessed March 24, 2018); "About Sugar Bowl Bakery," Sugar Bowl Bakery, https://sugarbowlbakery.com/pages/about (accessed March 23, 2018).

21. "PepsiCo Unveils New Organizational Structure, Names CEOs of Three Principle Operating Units," *PR Newswire,* November 5, 2007, www.prnewswire.com/news-releases/pepsico-unveils-new-organizational-structure-names-ceos-of-three-principal-operating-units-58668152.html (accessed May 4, 2017); "The PepsiCo Family," PepsiCo, www.pepsico.com/Company/The-Pepsico-Family/PepsiCo-Americas-Beverages.html (accessed April 30, 2014).

22. Jon R. Katzenbach and Douglas K. Smith, "The Discipline of Teams," *Harvard Business Review* 71 (March–April 1993), p. 19.

23. Katzenbach and Smith, "The Discipline of Teams."

24. John Baldoni, "The Secret to Team Collaboration: Individuality," *Inc.,* January 18, 2012, www.inc.com/john-baldoni/the-secret-to-team-collaboration-is-individuality.html (accessed May 4, 2017).

25. Gregory Ciotti, "Why Remote Teams Are the Future (and How to Make Them Work)," *Help Scout,* October 23, 2013, www.helpscout.net/blog/virtual-teams/ (accessed March 11, 2016).

26. Anneke Seley, "Outside In: The Rise of the Inside Sales Team," *Salesforce.com Blog,* February 3, 2015, http://blogs.salesforce.com/company/2015/02/outside-in-rise-inside-sales-team-gp.html (accessed May 4, 2017).

27. Patrick Kiger, "Task Force Training Develops New Leaders, Solves Real Business Issues and Helps Cut Costs," *Workforce,* September 7, 2011, www.workforce.com/article/20070521/NEWS02/305219996/task-force-training-develops-new-leaders-solves-real-business-issues-and-helps-cut-costs (accessed March 10, 2016); Duane D. Stanford, "Coca-Cola Woman Board Nominee Bucks Slowing Diversity Trend," *Bloomberg,* February 22, 2013, www.bloomberg.com/news/2013-02-22/

coca-cola-s-woman-director-nominee-bucks-slowing-diversity-trend.html (accessed May 4, 2017).

28. Jerry Useem, "What's That Spell? TEAMWORK," *Fortune,* June 12, 2006, p. 66.

29. Courtney Connley, "Jeff Bezos' 'Two Pizza Rule' Can Help You Hold More Productive Meetings," *CNBC,* April 20, 2018, https://www.cnbc.com/2018/04/30/jeff-bezos-2-pizza-rule-can-help-you-hold-more-productive-meetings.html (accessed August 4, 2018).

30. Natasha Singer, "Intel's Sharp-Eyed Social Scientist," *The New York Times,* February 15, 2014, www.nytimes.com/2014/02/16/technology/intels-sharp-eyed-social-scientist.html?_r=0 (accessed May 4, 2017).

31. Mindy S. Lubber, "How Timberland, Levi's Use Teamwork to Advance Sustainability," *Green Biz,* May 9, 2011, https://www.greenbiz.com/blog/2011/05/09/how-companies-court-stakeholders-accelerate-sustainability (accessed December 31, 2017); *Engaging Employees: Timberland's Global Stewards Program 2009 Report,* Timberland Company, http://responsibility.timberland.com/wp-content/uploads/2011/05/Stewards_Program_2009.pdf (accessed August 6, 2013); Betsy Blaisdell and Nina Kruschwitz, "New Ways to Engage Employees, Suppliers and Competitors in CSR," *MIT Sloan,* November 14, 2012, https://sloanreview.mit.edu/article/new-ways-to-engage-employees-suppliers-and-competitors-in-csr/ (accessed December 31, 2017); *Focus—Corporate Social Responsibility Report 2001,* Timberland Company, http://responsibility.timberland.com/wp-content/uploads/2011/05/2001-CSR-Report.pdf (accessed August 6, 2013); Chuck Scofield, "Sharing Strength: Lessons about Getting by Giving," *LiNE Zine,* http://www.linezine.com/7.2/articles/cssslagbg.htm (accessed December 31, 2017); David Hellqvist, "Timberland: 40 Years of the Yellow Boot," *The Guardian,* April 16, 2013, https://www.theguardian.com/fashion/2013/apr/16/timberland-40-years-yellow-boot (accessed December 31, 2017); "About Us," Timberland, https://www.timberland.com/about-us.html (accessed December 31, 2017); "Responsibility," Timberland Company, https://www.timberland.com/responsibility.html (accessed December 31, 2017); "Timberland Employees Around the Globe Take Part in 18th Annual Serv-a-Palooza Service Event," Timberland Company, September 17, 2015, https://www.timberland.com/newsroom/press-releases/timberland-18th-annual-serv-a-palooza-service-event.html (accessed December 31, 2017).

32. Richard S. Wellins, William C. Byham, and Jeanne M. Wilson, *Empowered Teams: Creating Self-Directed Work Groups That Improve Quality, Productivity, and Participation* (San Francisco: Jossey-Bass Publishers, 1991), p. 5.

33. "2017 Employee Job Satisfaction and Engagement: The Doors of Opportunity Are Open," Society for Human Resource Management, April 24, 2017, https://www.shrm.org/hr-today/trends-and-forecasting/research-and-surveys/pages/2017-job-satisfaction-and-engagement-doors-of-opportunity-are-open.aspx (accessed April 21, 2018).

34. Peter Mell and Timothy Grance, "The NIST Definition of Cloud Computing," National Institute of Standards and Technology, Special Publication 800-145, September 2011, http://csrc.nist.gov/publications/nistpubs/800-145/SP800-145.pdf (accessed April 30, 2014).

35. Michael Christian, "Top 10 Ideas: Making the Most of Your Corporate Intranet," April 2, 2009, www.claromentis.com/blog/top-10-ideas-making-the-most-of-your-corporate-intranet/ (accessed May 4, 2017).

36. Verne Harnish, "Five Ways to Liberate Your Team from Email Overload," *Fortune,* June 16, 2014, p. 52.

37. Sue Shellenbarger, "They're Gossiping About You," *The Wall Street Journal,* October 8, 2014, pp. D1–D2.

38. Kim Komando, "Why You Need a Company Policy on Internet Use," www.microsoft.com/business/en-us/resources/management/employee-relations/why-you-need-a-company-policy-on-internet-use.aspx?fbid=HEChiHWK7CU (accessed April 30, 2014).

39. "Our Values and Standards: The Basis of Our Success," Merck, https://www.merck.com/abo0ut/code_of_conduct.pdf (accessed March 16, 2015).

40. Imani Moise, "New Tools Tell Bosses How You're Feeling," *The Wall Street Journal,* March 29, 2018, p. B6.

41. PBSNewsHour, "Apple Supplier Foxconn Pledges Better Working Conditions, but Will It Deliver?" *YouTube,* www.youtube.com/watch?v=ZduorbCkSBQ (accessed May 4, 2017).

42. Susan M. Heathfield, "Top Ten Employee Complaints," *About.com ,* http://humanresources.about.com/od/retention/a/emplo_complaint.htm (accessed March 16, 2015).

43. Gloria Mark, Shamsi T. Iqbal, Mary Czerwinski, Paul Johns, Akane Sano, and Yuliya Lutchyn, "Email Duration, Batching and Self-interruption: Patterns of Email Use on Productivity and Stress," *Proceedings of the 2016 CHI Conference on Human Factors in Computing Systems (CHI '16),* ACM, New York, NY, USA, p. 1717-1728; Andrew Blackman, "The Smartest Way to Use Email at Work," *The Wall Street Journal,* March 12, 2018, p. R1.

44. Michael D. Maginn, *Effective Teamwork,* 1994, p. 10. 1994; Richard D. Irwin, a Times Mirror Higher Education Group Inc. Company.

45. "How Did Freshii Become One of the Fastest Growing Restaurant Chains in the World?" *QSR,* June 2018, p. 1; William G. Nickels, James M. McHugh, and Susan M. McHugh, *Understanding Business* (New York: McGraw-Hill, 2019), p. 214; "Freshii Named FastCasual's Top Brand of the Year," *FastCasual,* May 22, 2018, https://www.fastcasual.com/articles/freshii-named-top-brand-of-the-year/ (accessed June 3, 2018); S. A. Whitehead, "Freshii Founder Tells How Missteps Can Be Key Teacher on Path to Success," *QSR Web,* April 16, 2018, https://www.qsrweb.com/articles/freshii-founder-tells-how-missteps-can-be-key-teachers-on-path-to-success/ (accessed June 3, 2018).

Credits

8 Managing Operations and Supply Chains

©QualityHD/Shutterstock

Chapter Outline

- Introduction
- The Nature of Operations Management
  - The Transformation Process
  - Operations Management in Service Businesses
- Planning and Designing Operations Systems
  - Planning the Product
  - Designing the Operations Processes
  - Planning Capacity
  - Planning Facilities
  - Sustainability and Manufacturing
- Managing the Supply Chain
  - Procurement
  - Managing Inventory
  - Outsourcing
  - Routing and Scheduling
- Managing Quality
  - International Organization for Standardization (ISO)
  - Inspection
  - Sampling
  - Integrating Operations and Supply Chain Management

Learning Objectives

After reading this chapter, you will be able to:

LO 8-1 Define *operations management*.

LO 8-2 Differentiate between operations and manufacturing.

LO 8-3 Explain how operations management differs in manufacturing and service firms.

LO 8-4 Describe the elements involved in planning and designing an operations system.

LO 8-5 Specify some techniques managers may use to manage the logistics of transforming inputs into finished products.

LO 8-6 Assess the importance of quality in operations management.

LO 8-7 Propose a solution to a business's operations dilemma.

Enter the World of Business

Trading Up: Trader Joe's Operational Success

First founded in 1967, Trader Joe's is composed of more than 470 grocery stores in 41 states. Despite its large reach, Trader Joe's exudes the same neighborhood store atmosphere it did in 1967. Trader Joe's excellent operations and customer service has generated a loyal customer following.

Trader Joe's maintains smaller facilities and product lines than comparable stores—a deliberate operational move to create its specialty image. The grocery store stocks about 4,000 items compared to the 50,000 stocked by the typical grocery store. Inventory control has been a key advantage to reducing costs and focusing on a limited number of quality products. While many markets sell as many as 50 selections of one food item, Trader Joe's sells only a few. Products that have low demand or high production costs often do not last long. The combination of lower inventory and higher product turnover creates greater operational efficiencies.

Another attribute that sets Trader Joe's apart is its supply chain. At any given time, management aims to minimize the number of hands that touch a product. Trader Joe's will purchase directly from manufacturers, ship straight to distribution centers, and then send products to stores. Trader Joe's has crafted its distribution process to create efficiency and reduce costs. This efficiency increases productivity and allows customers to receive premium products consistently.

Additionally, Trader Joe's excels at quality and supplier relationships. It employs product developers who travel the world in search of best product/price combinations. Suppliers covet contracts with the grocery store chain, which charges less in fees and is known for on-time payments. Trader Joe's expands only into areas that can support its streamlined distribution system. With its popularity continuing to rise, customers seem impressed by the way Trader Joe's is redefining the grocery shopping experience.[1]

Introduction

All organizations create products—goods, services, or ideas—for customers. Thus, organizations as diverse as Tesla, Subway, UPS, and a public hospital share a number of similarities relating to how they transform resources into the products we enjoy. Most hospitals use similar admission procedures, while online social media companies, like Facebook and Twitter, use their technology and operating systems to create social networking opportunities and sell advertising. Such similarities are to be expected. But even organizations in unrelated industries take similar steps in creating goods or services. The check-in procedures of hotels and commercial airlines are comparable, for example. The way Subway assembles a sandwich and the way Tesla assembles a car are similar (both use multiple employees in an assembly line). These similarities are the result of operations management, the focus of this chapter.

Here, we discuss the role of production or operations management in acquiring and managing the resources necessary to create goods and services. Production and operations management involve planning and designing the processes that will transform those resources into finished products, managing the movement of those resources through the transformation process, and ensuring that the products are of the quality expected by customers.

The Nature of Operations Management

Operations management (OM), the development and administration of the activities involved in transforming resources into goods and services, is of critical importance. Operations managers oversee the transformation process and the planning and designing of operations systems, managing logistics, quality, and productivity. Quality and productivity have become fundamental aspects of operations management because a company that cannot make products of the quality desired by consumers, using resources efficiently and effectively, will not be able to remain in business. OM is the "core" of most organizations because it is responsible for the creation of the organization's goods and services. Some organizations like General Motors produce tangible products, but service is an important part of the total product for the customer.

Historically, operations management has been called "production" or "manufacturing" primarily because of the view that it was limited to the manufacture of physical goods. Its focus was on methods and techniques required to operate a factory efficiently. The change from "production" to "operations" recognizes the increasing importance of organizations that provide services and ideas. Additionally, the term *operations* represents an interest in viewing the operations function as a whole rather than simply as an analysis of inputs and outputs.

Today, OM includes a wide range of organizational activities and situations outside of manufacturing, such as health care, food service, banking, entertainment, education, transportation, and charity. Thus, we use the terms **manufacturing** and **production** interchangeably to represent the activities and processes used in making *tangible* products, whereas we use the broader term **operations** to describe those processes used in the making of *both tangible and intangible products.* Manufacturing provides tangible products such as the Apple Watch, and operations also provides intangibles such as a stay at Wyndham Hotels and Resorts.

The Transformation Process

At the heart of operations management is the transformation process through which **inputs** (resources such as labor, money, materials, and energy) are converted into **outputs** (goods, services, and ideas). The transformation process combines inputs in predetermined ways using different equipment, administrative procedures, and technology to create a product (Figure 8.1). To ensure that this process generates quality products efficiently, operations managers control the process by taking measurements (feedback) at various points in the transformation process and comparing them to previously established standards. If there is any deviation between the actual and desired outputs, the manager may take some sort of corrective action. For example, if an airline has a standard of 90 percent of its flights departing on time but only 80 percent depart on time, a 10 percentage point negative deviation exists. All adjustments made to create a satisfying product are a part of the transformation process.

Transformation may take place through one or more processes. In a business that manufactures oak furniture, for example, inputs pass through several processes before being turned into the final outputs—furniture that has been designed to meet the desires of customers (Figure 8.2). The furniture maker must first strip the oak trees of their bark and saw them into appropriate sizes—one step in the transformation process. Next, the firm dries the strips of oak lumber, a second form of transformation.

inputs
the resources—such as labor, money, materials, and energy—that are converted into outputs.

outputs
the goods, services, and ideas that result from the conversion of inputs.

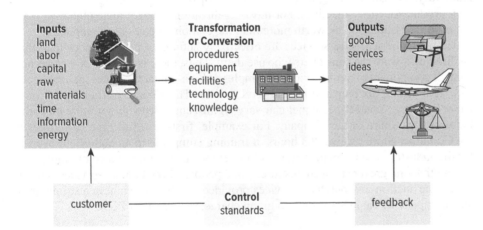

FIGURE 8.1
The Transformation Process of Operations Management

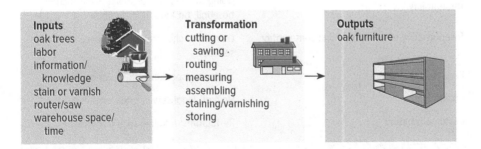

FIGURE 8.2
Inputs, Outputs, and Transformation Processes in the Manufacture of Oak Furniture

Third, the dried wood is routed into its appropriate shape and made smooth. Fourth, workers assemble and treat the wood pieces, then stain or varnish the piece of assembled furniture. Finally, the completed piece of furniture is stored until it can be shipped to customers at the appropriate time. Of course, many businesses choose to eliminate some of these stages by purchasing already processed materials—lumber, for example—or outsourcing some tasks to third-party firms with greater expertise.

LO 8-3

Explain how operations management differs in manufacturing and service firms.

Operations Management in Service Businesses

Different types of transformation processes take place in organizations that provide services, such as airlines, colleges, and most nonprofit organizations. An airline transforms inputs such as employees, time, money, and equipment through processes such as booking flights, flying airplanes, maintaining equipment, and training crews. The output of these processes is flying passengers and/or packages to their destinations. In a nonprofit organization like Habitat for Humanity, inputs such as money, materials, information, and volunteer time and labor are used to transform raw materials into homes for needy families. In this setting, transformation processes include fund-raising and promoting the cause in order to gain new volunteers and donations of supplies, as well as pouring concrete, raising walls, and setting roofs. Some companies such as Home Depot feel so strongly about a charitable cause that they donate materials and encourage their employees to volunteer for groups such as Habitat for Humanity. Transformation processes occur in all organizations, regardless of what they produce or their objectives. For most organizations, the ultimate objective is for the produced outputs to be worth more than the combined costs of the inputs.

Unlike tangible goods, services are effectively actions or performances that must be directed toward the consumers who use them. Thus, there is a significant customer-contact component to most services. Examples of high-contact services include health care, real estate, tax preparation, and food service. The amount of training for service or customer-contact personnel can vary significantly depending on the business, industry, or culture of the company. For example, first-year, full-time employees at the Container Store receive 263 hours of training compared to an eight-hour average for the industry. The Container Store culture embraces the importance of employees, stating that one great person equals three good people.[2] Low-contact services, such as the online auction-and-purchase services provided by eBay, often have a strong high-tech component. Table 8.1 shows common characteristics of services.

TABLE 8.1
Characteristics of Services

Service Characteristics	Examples
Intangibility	Going to a concert or sports event such as baseball, basketball, or football
Inseparability of production and consumption	Going to a chiropractor; air travel; veterinary services
Perishability	Seats at a speaker's presentation
Customization	Haircut; legal services; tax consultation
Customer contact	Restaurants; retailing such as Macy's

Sources: Adapted from Valerie A. Zeithaml, A. Parasuraman, and Leonard L. Berry, Delivering Quality Service: Balancing Customer Perceptions and Expectations (New York: Free Press, 1990); K. Douglas Hoffman and John E.G. Bateson, Essentials of Services Marketing (Mason, OH; Cengage Learning, 2001); Ian P. McCarthy, Leyland Pitt, and Pierre R. Berthon, "Service Customization Through Dramaturgy," Mass Customization, 2011, pp. 45–65.

Regardless of the level of customer contact, service businesses strive to provide a standardized process, and technology offers an interface that creates an automatic and structured response. The ideal service provider will be high tech and high touch. Amazon, for instance, has one of the highest customer service ratings. It provides a site that is easily navigable, and it has fast shipping times to deliver high-quality customer service. Amazon, through the commitment of CEO Jeff Bezos, encourages customers to e-mail him with any concerns. He routes the messages to the appropriate Amazon employee, asking for an explanation of why the problem occurred and how it can be prevented in the future.[3] Thus, service organizations must build their operations around good execution, which comes from hiring and training excellent employees, developing flexible systems, customizing services, and maintaining adjustable capacity to deal with fluctuating demand.[4]

Another challenge related to service operations is that the output is generally intangible and even perishable. Few services can be saved, stored, resold, or returned.[5] A seat on an airline or a table in a restaurant, for example, cannot be sold or used at a later date. Because of the perishability of services, it can be extremely difficult for service providers to accurately estimate the demand in order to match the right supply of a service. If an airline overestimates demand, for example, it will still have to fly each plane even with empty seats. The flight costs the same regardless of whether it is 50 percent full or 100 percent full, but the former will result in much higher costs per passenger. If the airline underestimates demand, the result can be long lines of annoyed customers or even the necessity of bumping some customers off of an overbooked flight.

Businesses that manufacture tangible goods and those that provide services or ideas are similar yet different. For example, both types of organizations must make design and operating decisions. Most goods are manufactured prior to purchase, but most services are performed after purchase. Flight attendants at Southwest Airlines, hotel service personnel at The Ritz Carlton, and even the Denver Broncos football team engage in performances that are a part of the total product. Though manufacturers and service providers often perform similar activities, they also differ in several respects. We can classify these differences in five basic ways.

Nature and Consumption of Output. First, manufacturers and service providers differ in the nature and consumption of their output. For example, the term *manufacturer* implies a firm that makes tangible products. A service provider, on the other hand, produces more intangible outputs such as U.S. Postal Service delivery of priority mail or a business stay in a Westin hotel. As mentioned earlier, the very nature of the service provider's product requires a higher degree of customer contact. Moreover, the actual performance of the service typically occurs at the point of consumption. At the Westin, business travelers may be very pleased with the Heavenly Bed and Shower System, both proprietary and luxury amenities. Or, customers may like getting points through their reward membership or upgrades based on their status with Starwood Preferred Guests. Automakers, on the other hand, can separate the production of a car from its actual use, but the service dimension requires closer contact with the consumer. Manufacturing, then, can occur in an isolated environment, away from the customer. However, service providers, because of their need for customer contact, are often more limited than manufacturers in selecting work methods, assigning jobs, scheduling work, and exercising control over operations. For this reason, Zappos adopted an online scheduling platform called Open Market to schedule its worker hours. Before, the process was complicated and involved workers having

Blaze Pizza's inputs are components such as pepperoni, mozzarella, mushrooms, onions, and sausage, while its outputs are customized pizzas.

©Sorbis/Shutterstock

to write their scheduling preferences on sheets of paper. This new scheduling platform uses surge-based pricing that provides higher compensation to workers who take shifts in periods of high demand.[6] The quality of the service experience is often controlled by a service-contact employee. However, some hospitals are studying the manufacturing processes and quality control mechanisms applied in the automotive industry in an effort to improve their service quality. By analyzing work processes to find unnecessary steps to eliminate and using teams to identify and address problems as soon as they occur, these hospitals are slashing patient waiting times, decreasing inventories of wheelchairs, readying operating rooms sooner, and generally moving patients through their hospital visit more quickly, with fewer errors, and at a lower cost.[7]

Uniformity of Inputs.   A second way to classify differences between manufacturers and service providers has to do with the uniformity of inputs. Manufacturers typically have more control over the amount of variability of the resources they use than do service providers. For example, each customer calling Fidelity Investments is likely to require different services due to differing needs, whereas many of the tasks required to manufacture a Ford Focus are the same across each unit of output. Consequently, the products of service organizations tend to be more "customized" than those of their manufacturing counterparts. Consider, for example, a haircut versus a bottle of shampoo. The haircut is much more likely to incorporate your specific desires (customization) than is the bottle of shampoo.

Uniformity of Output.   Manufacturers and service providers also differ in the uniformity of their output, the final product. Because of the human element inherent in providing services, each service tends to be performed differently. Not all grocery checkers, for example, wait on customers in the same way. If a barber or stylist performs 15 haircuts in a day, it is unlikely that any two of them will be exactly the same. Consequently, human and technological elements associated with a service can result in a different day-to-day or even hour-to-hour performance of that service. The service experience can even vary at McDonald's or Burger King despite the fact that the two chains employ very similar procedures and processes. Moreover, no two customers are exactly alike in their perception of the service experience. Health care offers another excellent example of this challenge. Every diagnosis, treatment, and surgery varies because every individual is different. In manufacturing, the high degree of automation available allows manufacturers to generate uniform outputs and, thus, the operations are more effective and efficient. For example, we expect luxury bicycles—such as the Giant TCR Advanced SL Disc, which sells for up to $10,500, and the Litespeed T1sl Disc Titanium Road Bike, which sells for up to $14,000—to have extremely high standards for quality and performance.

Labor Required.   A fourth point of difference is the amount of labor required to produce an output. Service providers are generally more labor-intensive (require more labor) because of the high level of customer contact, perishability of the output (must be consumed immediately), and high degree of variation of inputs and outputs (customization). For example, Adecco provides temporary support personnel. Each temporary worker's performance determines Adecco's product quality. A manufacturer, on the other hand, is likely to be more capital-intensive bècause of the machinery and technology used in the mass production of highly similar goods. For instance, it would take a considerable investment for Ford to make an electric car that has batteries with a longer life.

Measurement of Productivity.   The final distinction between service providers and manufacturers involves the measurement of productivity for each output produced. For manufacturers, measuring productivity is fairly straightforward because of the tangibility of the output and its high degree of uniformity. For the service provider, variations in demand (for example, higher demand for air travel in some seasons than in others), variations in service requirements from job to job, and the intangibility of the product make productivity measurement more difficult. Consider, for example, how much easier it is to measure the productivity of employees involved in the production of Intel computer processors as opposed to those serving the needs of Prudential Securities' clients.

It is convenient and simple to think of organizations as being either manufacturers or service providers as in the preceding discussion. In reality, however, most organizations are a combination of the two, with both tangible and intangible qualities embodied in what they produce. For example, Samsung provides customer services such as toll-free hotlines and warranty protection, while banks may sell checks and other tangible products that complement their primarily intangible product offering. Thus, we consider "products" to include both tangible physical goods and intangible service offerings. It is the level of tangibility of its principal product that tends to classify a company as either a manufacturer or a service provider. From an OM standpoint, this level of tangibility greatly influences the nature of the company's operational processes and procedures.

Planning and Designing Operations Systems

> **LO 8-4**
>
> Describe the elements involved in planning and designing an operations system.

Before a company can produce any product, it must first decide what it will produce and for what group of customers. It must then determine what processes it will use to make these products as well as the facilities it needs to produce them. These decisions comprise operations planning. Although planning was once the sole realm of the production and operations department, today's successful companies involve all departments within an organization, particularly marketing and research and development, in these decisions.

Planning the Product

Before making any product, a company first must determine what consumers want and then design a product to satisfy that want. Most companies use marketing research (discussed in the chapter titled "Customer-Driven Marketing") to determine the kinds of goods and services to provide and the features they must possess. Twitter and Facebook provide new opportunities for businesses to discover what consumers want, then design the product accordingly. Less than 50 percent of companies use social media in the new-product-development process.[8] Marketing research can also help gauge the demand for a product and how much consumers are willing to pay for it. But, when a market's environment changes, firms have to be flexible. Marketing research is advancing from wearables that measure skin response to geo-located survey delivery. Artificial intelligence can be used to measure people's expressions as they watch events.[9]

Developing a product can be a lengthy, expensive process. For example, Uber and Volvo are partnering together to bring a driverless car to consumers. The $300 million alliance brings Volvo's manufacturing and design expertise together with Uber's ride-sharing market and a staff that increasingly consists of former employees of Carnegie Mellon University's robotics department.[10] Most companies work to reduce development time and costs. By joining together, companies can pool their resources and reduce the time it takes to develop new products. Once a firm has developed an idea for a product that customers will buy, it must then plan how to produce the product.

Within a company, the engineering or research and development department is charged with turning a product idea into a workable design that can be produced economically. In smaller companies, a single individual (perhaps the owner) may be solely responsible for this crucial activity. Regardless of who is responsible for product design, planning does not stop with a blueprint for a product or a description of a service; it must also work out efficient production of the product to ensure that enough is available to satisfy consumer demand. How does a company like Snapper transform steel, aluminum, and other materials into a mower design that satisfies consumer and environmental requirements? Operations managers must plan for the types and quantities of materials needed to produce the product, the skills and quantity of people needed to make the product, and the actual processes through which the inputs must pass in their transformation to outputs.

Designing the Operations Processes

Before a firm can begin production, it must first determine the appropriate method of transforming resources into the desired product. Often, consumers' specific needs and desires dictate a process. A state's needs for toll booths for its highway systems may be very structured and consistent, and engineering and manufacturing would be standardized. On the other hand, a bridge often must be customized so that it is appropriate for the site and expected load; furthermore, the bridge must be constructed on site rather than in a factory. Typically, products are designed to be manufactured by one of three processes: standardization, modular design, or customization.

Standardization.   Most firms that manufacture products in large quantities for many customers have found that they can make them cheaper and faster by standardizing designs. **Standardization** is making identical, interchangeable components or even complete products. With standardization, a customer may not get exactly what he or she wants, but the product generally costs less than a custom-designed product. Television sets, ballpoint pens, and tortilla chips are standardized products; most are manufactured on an assembly line. Standardization speeds up production and quality control and reduces production costs. And, as in the example of the toll booths, standardization provides consistency so that customers who need certain products to function uniformly all the time will get a product that meets their expectations. Standardization becomes more complex on a global scale because different countries have different standards for quality. To help solve this problem, the International Organization for Standardization (ISO) has developed a list of global standards that companies can adopt to assure stakeholders that they are complying with the highest quality, environmental, and managerial guidelines. ISO standards are discussed later in the chapter.

standardization
the making of identical, interchangeable components or products.

Modular Design.   **Modular design** involves building an item in self-contained units, or modules, that can be combined or interchanged to create different products. The Container Store, for example, uses the modular design in many of its storage

modular design
the creation of an item in self-contained units, or modules, that can be combined or interchanged to create different products.

solutions. This allows for customers to mix and match components for customized design. Because many modular components are produced as integrated units, the failure of any portion of a modular component usually means replacing the entire component. Modular design allows products to be repaired quickly, thus reducing the cost of labor, but the component itself is expensive, raising the cost of repair materials. Many automobile manufacturers use modular design in the production process. Manufactured homes are built on a modular design and often cost about one-fourth the cost of a conventionally built house.

Lancôme sells Le Teint Particulier, a custom foundation with 72,000 possibilities. After customers have their skin tone scanned and select their skin type and desired coverage, the foundation is made on the spot. Because of the complexities of offering a customized product, the foundation is only available at select Nordstrom locations.

©SeongJoon Cho/Bloomberg via Getty Images

Customization. **Customization** is making products to meet a particular customer's needs or wants. Products produced in this way are generally unique. Such products include repair services, photocopy services, custom artwork, jewelry, and furniture, as well as large-scale products such as bridges, ships, and computer software. For instance, bicycles are popular products to customize. The company Villy Custom (named after the owner's dog)—supported by Mark Cuban and Barbara Corcoran on *Shark Tank*—has 10 "best seller" models of bicycles that can be customized for consumers. The company also has a customizable line of bikes for corporate use.[11] Mass customization relates to making products that meet the needs or wants of a large number of individual customers. The customer can select the model, size, color, style, or design of the product. Dell can customize a computer with the exact configuration that fits a customer's needs. Services such as fitness programs and travel packages can also be custom designed for a large number of individual customers. For both goods and services, customers get to make choices and have options to determine the final product.

customization
making products to meet a particular customer's needs or wants.

Blockchain. Blockchain information technology could alter processes throughout virtually every industry including supply chain, health care, and even online advertising. The blockchain is a secure, public database (or ledger) that records all transactions and is spread across multiple computers. The global blockchain is difficult to tamper with and growing rapidly. In fact, blockchain technology is projected to maintain a compound annual growth rate of over 48 percent between now and 2023.[12] In the age of tech-savvy consumers and continuous innovation, the adoption of blockchain technology is occurring quicker than ever. For example, Coda Coffee Company,

Bitcoin is a decentralized digital currency that works based on the blockchain technology without a central bank or single administrator. Bitcoins are generated at mining farms like this one.

©DR MANAGER/Shutterstock

Improving the Supply Chain with Blockchain

Although much of the technological hype revolves around cryptocurrency, perhaps a more important sector resides in the underlying technology: the blockchain. The blockchain could change many industries as we know them today. One of the most evident examples is the improvement throughout supply chains.

Blockchain is a secure database that records all transactions and is spread across multiple computers. Essentially, it is a ledger that, by its public nature, makes it difficult to tamper with. This makes it safer to make transfers (of any variety) without worry of security breaches. Bitcoin and other cryptocurrencies use the blockchain to make online transfers of currency; however, the blockchain is also used in global supply chains. Retailers such as Walmart use blockchain to improve logistics and transparency. Walmart uses blockchain for more than 1.1 million of its items, tracking their journey from suppliers to store shelves.

Companies are also using IBM's blockchain technology to improve their supply chains. For example, Maersk, the world's largest shipping container company, uses the technology to track shipping containers around the world, thus improving cargo tracking throughout its supply chain. IBM's blockchain more accurately displays all transactions in the ledger as it updates in real time and allows participants to see valuable information, including the location of the asset and who has it in their possession. Other companies employing IBM's technology are Kroger, Nestlé, and Tyson Foods.

Blockchain can be altered to fit different processes, participants, and types of networks. As it continues to rapidly grow, this technology could impact future transactions from online voting to goods that we buy and how we buy them.[13]

Critical Thinking Questions

1. What are some ways that blockchain increases transparency in the supply chain?
2. How can blockchain be used as a tool to manage inventory?
3. What are some ways that you think different industries could use blockchain to improve the quality of their operations and products?

---

located in Denver, Colorado, is allowing its customers to scan a QR code and get all of the information they could possibly want in regards to their coffee in what they call "the world's first blockchain-traced coffee." They are able to see the date and location of each transaction from collection at the farm through every process until they swipe their credit card at the register. Blockchain technology continues to be tested by companies across all industries and could change the way we do almost anything.[14]

Planning Capacity

Planning the operational processes for the organization involves two important areas: capacity planning and facilities planning. The term **capacity** basically refers to the maximum load that an organizational unit can carry or operate. The unit of measurement may be a worker or machine, a department, a branch, or even an entire plant. Maximum capacity can be stated in terms of the inputs or outputs provided. For example, an electric plant might state plant capacity in terms of the maximum number of kilowatt-hours that can be produced without causing a power outage, while a restaurant might state capacity in terms of the maximum number of customers who can be effectively—comfortably and courteously—served at any one particular time.

Efficiently planning the organization's capacity needs is an important process for the operations manager. Capacity levels that fall short can result in unmet demand, and consequently, lost customers. On the other hand, when there is more capacity available than needed, operating costs are driven up needlessly due to unused and often expensive resources. To avoid such situations, organizations must accurately

capacity
the maximum load that an organizational unit can carry or operate.

forecast demand and then plan capacity based on these forecasts. Another reason for the importance of efficient capacity planning has to do with long-term commitment of resources. Often, once a capacity decision—such as factory size—has been implemented, it is very difficult to change the decision without incurring substantial costs. Large companies have come to realize that although change can be expensive, not adjusting to future demand and stakeholder desires will be more expensive in the long run. Responding to consumers' concern for the environment, Toyota and its subsidiaries have acquired ISO 14001 certification for environmental management at many of its locations worldwide.[15] These systems help firms monitor their impact on the environment.

> **DID YOU KNOW?** Hershey's has the production capacity to make 390,000 Kit Kats per day.[16]

Planning Facilities

Once a company knows what process it will use to create its products, it then can design and build an appropriate facility in which to make them. Many products are manufactured in factories, but others are produced in stores, at home, or where the product ultimately will be used. Companies must decide where to locate their operations facilities, what layout is best for producing their particular product, and even what technology to apply to the transformation process.

Many firms are developing both a traditional organization for customer contact and a virtual organization. Charles Schwab Corporation, a securities brokerage and investment company, maintains traditional offices and has developed complete telephone and Internet services for customers. Through its website, investors can obtain personal investment information and trade securities over the Internet without leaving their home or office.

Facility Location.   Where to locate a firm's facilities is a significant question because, once the decision has been made and implemented, the firm must live with it due to the high costs involved. When a company decides to relocate or open a facility at a new location, it must pay careful attention to factors such as proximity to market, availability of raw materials, availability of transportation, availability of power, climatic influences, availability of labor, community characteristics (quality of life), and taxes and inducements. Inducements and tax reductions have become an increasingly important criterion in recent years. To increase production and to provide incentives for small startups, many states are offering tax inducements for solar companies. State governments are willing to forgo some tax revenue in exchange for job growth, getting in on a burgeoning industry, as well as the good publicity generated by the company. However, it is still less expensive for many firms to use overseas factories. Apple has followed the lead of other major companies by locating its manufacturing facilities in Asia to take advantage of lower labor and production costs. The facility-location decision is complex because it involves the evaluation of many factors, some of which cannot be measured with precision. Because of the long-term impact of the decision, however, it is one that cannot be taken lightly.

Facility Layout.   Arranging the physical layout of a facility is a complex, highly technical task. Some industrial architects specialize in the design and layout of certain types of businesses. There are three basic layouts: fixed-position, process, and product.

fixed-position layout
a layout that brings all resources required to create the product to a central location.

project organization
a company using a fixed-position layout because it is typically involved in large, complex projects such as construction or exploration.

process layout
a layout that organizes the transformation process into departments that group related processes.

intermittent organizations
organizations that deal with products of a lesser magnitude than do project organizations; their products are not necessarily unique but possess a significant number of differences.

product layout
a layout requiring that production be broken down into relatively simple tasks assigned to workers, who are usually positioned along an assembly line.

continuous manufacturing organizations
companies that use continuously running assembly lines, creating products with many similar characteristics.

computer-assisted design (CAD)
the design of components, products, and processes on computers instead of on paper.

computer-assisted manufacturing (CAM)
manufacturing that employs specialized computer systems to actually guide and control the transformation processes.

A company using a **fixed-position layout** brings all resources required to create the product to a central location. The product—perhaps an office building, house, hydro-electric plant, or bridge—does not move. A company using a fixed-position layout may be called a **project organization** because it is typically involved in large, complex projects such as construction or exploration. Project organizations generally make a unique product, rely on highly skilled labor, produce very few units, and have high production costs per unit.

Firms that use a **process layout** organize the transformation process into departments that group related processes. A metal fabrication plant, for example, may have a cutting department, a drilling department, and a polishing department. A hospital may have cardiology, urgent care, neurology, obstetrics and gynecology units, and so on. These types of organizations are sometimes called **intermittent organizations,** which deal with products of a lesser magnitude than do project organizations, and their products are not necessarily unique but possess a significant number of differences. Doctors, makers of custom-made cabinets, commercial printers, and advertising agencies are intermittent organizations because they tend to create products to customers' specifications and produce relatively few units of each product. Because of the low level of output, the cost per unit of product is generally high.

The **product layout** requires that production be broken down into relatively simple tasks assigned to workers, who are usually positioned along an assembly line. Workers remain in one location, and the product moves from one worker to another. Each person in turn performs his or her required tasks or activities. Companies that use assembly lines are usually known as **continuous manufacturing organizations,** so named because once they are set up, they run continuously, creating products with many similar characteristics. Examples of products produced on assembly lines are automobiles, television sets, vacuum cleaners, toothpaste, and meals from a cafeteria. Continuous manufacturing organizations using a product layout are characterized by the standardized product they produce, the large number of units produced, and the relatively low unit cost of production.

Many companies actually use a combination of layout designs. For example, an automobile manufacturer may rely on an assembly line (product layout) but may also use a process layout to manufacture parts.

Technology. Every industry has a basic, underlying technology that dictates the nature of its transformation process. Today business models are changing how the transformation process occurs. As mentioned earlier, blockchain Information systems, artificial intelligence, and computer integrated systems are driving changes throughout all industries.

Computers were introduced in the late 1950s by IBM. The operations function makes great use of computers in all phases of the transformation process. **Computer-assisted design (CAD),** for example, helps engineers design components, products, and processes on the computer instead of on paper. CAD software is used to develop a 3D image. Then, the CAD file is sent to the printer. The printer is able to use layers of liquid, powder, paper, or metal to construct a 3D model.[17]

Computer-assisted manufacturing (CAM) goes a step further, employing specialized computer systems to actually guide and control the transformation processes. Such systems can monitor the transformation process, gathering information about the equipment used to produce the products and about the product itself as it goes from one stage of the transformation process to the next. The computer provides information to an operator who may, if necessary, take corrective action. In some highly automated systems, the computer itself can take corrective action.

Using **flexible manufacturing,** computers can direct machinery to adapt to different versions of similar operations. For example, with instructions from a computer, one machine can be programmed to carry out its function for several different versions of an engine without shutting down the production line for refitting.

The use of drones in business operations would vastly change the technology landscape. Drones refer to unmanned aerial vehicles and have long been used in military operations. Amazon is pursuing methods to use drones for package delivery. Amazon has secured a patent to allow it to drop packages delivered by its fleet of drones by parachute. However, currently, the use of automated drones without direct supervision is illegal in the United States.[18] It is surprising that drones have not played a larger role thus far; the adoption of drones in organizations could increase and greatly impact everything from shipping to logistics and delivery in the near future. For example, one of the current common uses of this technology involves drones sending automatic signals for reorder when an order has been placed or when inventory for a product is low and thus greatly improves inventory management. Whether it is adapting drones to a specific company to increase efficiency and capability or expanding the available applications as a whole, it is evident that drones could potentially handle many activities associated with operations.[19]

Robots are also becoming increasingly useful in the transformation process. These "steel-collar" workers have become particularly important in industries such as nuclear power, hazardous-waste disposal, ocean research, and space construction and maintenance, in which human lives would otherwise be at risk. Robots are used in numerous applications by companies around the world. Many assembly operations—cars, electronics, metal products, plastics, chemicals, and numerous other products—depend on industrial robots. The economic impact of robots was quantified by two economists who found that for every robot per 1,000 employees, up to six employees lost their jobs and wages fell by three-fourths of a percent.[20] In the next 15 years, PwC projects that 38 percent of the jobs in the United States are at risk of being lost to robots and artificial intelligence.[21]

Researchers continue to make more sophisticated robots, extending their use beyond manufacturing and space programs to various industries, including laboratory research, education, medicine, and household activities. There are many advantages in using robotics, such as more successful surgeries, re-shoring manufacturing activities back to America, energy conservation, and safer work practices. The United States is the fourth largest market for industrial robots. The strongest market is China, with a 40 percent share of the market for industrial robots projected by the end of the decade.[22]

When all these technologies—CAD/CAM, flexible manufacturing, robotics, computer systems, and more—are integrated, the result is **computer-integrated manufacturing (CIM),** a complete system that designs products, manages machines and materials, and controls the operations function. Companies adopt CIM to boost productivity and quality and reduce costs. Such technology, and computers in particular, will continue to make strong inroads into operations on two fronts—one dealing with the technology involved in manufacturing and one dealing with the administrative functions and processes used by operations managers. The operations manager

flexible manufacturing
the direction of machinery by computers to adapt to different versions of similar operations.

computer-integrated manufacturing (CIM)
a complete system that designs products, manages machines and materials, and controls the operations function.

must be willing to work with computers and other forms of technology and to develop a high degree of computer literacy.

Sustainability and Manufacturing

Manufacturing and operations systems are moving quickly to establish environmental sustainability and minimize negative impact on the natural environment. Sustainability deals with conducting activities in such a way as to provide for the long-term well-being of the natural environment, including all biological entities. Sustainability issues are becoming increasingly important to stakeholders and consumers, as they pertain to the future health of the planet. The Hershey Company is committed to working toward a long-term, sustainable cocoa supply, protecting the natural environment as a part of its "Cocoa for Good" sustainability initiative.[23] Some sustainability issues include pollution of the land, air, and water, climate change, waste management, deforestation, urban sprawl, protection of biodiversity, and genetically modified foods. New Belgium Brewing is another company that illustrates green initiatives in operations and manufacturing. New Belgium was the first brewery to adopt 100 percent wind-powered electricity, reducing carbon emissions by 1,800 metric tons a year.

New Belgium Brewing demonstrates that reducing waste, recycling, conserving, and using renewable energy not only protect the environment, but can also gain the support of stakeholders. Green operations and manufacturing can improve a firm's reputation along with customer and employee loyalty, leading to improved profits.

Much of the movement to green manufacturing and operations is the belief that global warming and climate change must decline. In the United States, roughly 40 percent of carbon dioxide emissions are accounted for by buildings; however, Leadership in Energy and Environmental Design (LEED)-certified buildings maintain a 34 percent lower emissions rate and consume 25 percent less energy, 11 percent less water, and reduce more than 80 million tons of waste.[24] Companies like General Motors and Ford are adapting to stakeholder demands for greater sustainability by producing smaller and more fuel-efficient cars. Tesla has taken sustainability further by making a purely electric vehicle that also ranks at the top in safety. The company also makes sure that its manufacturing facilities operate sustainably by installing solar panels and other renewable sources of energy. Green products produced through green operations and manufacturing are our future. Cities around the United States are taking leadership roles, with Portland, Oregon, standing out as a leader in environmental initiatives. Portland boasts expansive parks, mass transit, sustainable eating, bicycle parking spots, and electricity-producing exercycles.[25] Government initiatives provide space for businesses to innovate their green operations and manufacturing.

More than 78 percent of Adobe employees work in LEED-certified buildings, emphasizing the company's focus on sustainability.[26]

©Katherine Welles/Shutterstock

Meet Your Meat: This Company Is Disrupting the Global Meat Market

De Vegetarische Slager (The Vegetarian Butcher) is a Dutch vegetarian food producer. At the end of the 20th century, founder Jaap Korteweg wanted to become a vegetarian. However, he missed meat's taste and texture. After spending years searching and developing meat-like vegetarian products, Jaap founded The Vegetarian Butcher to market his products. His success in managing the firm's operations involved transforming raw materials into plant-based food that taste like meat.

Since its launch, The Vegetarian Butcher has expanded to 3,500 stores in 15 countries. To earn such an achievement, The Vegetarian Butcher has focused extensively on product quality. From day one, the development team committed to producing plant-based products resembling meats without compromising taste or texture. With the right focus on quality, The Vegetarian Butcher convinced not only vegetarian customers to purchase its products, but also many meat eaters.

The meaty quality of its products allows The Vegetarian Butcher to take advantage of a market niche that targets both vegetarians and meat lovers. However, it also places greater pressure to maintain quality control standards. To ensure the unique taste and texture of each meat type, The Vegetarian Butcher has to set up production for each product line. Thus, expanding product lines while ensuring the meat-like experience is an operational challenge.

Managing products and the supply chain are necessary for The Vegetarian Butcher to maintain its competitive advantage for meat-like products. The firm has been an innovative leader in establishing standards and managing quality. If The Vegetarian Butcher can overcome operational challenges, global consumers can expect to have tastier vegetarian "meats" in the future.[27]

Critical Thinking Questions

1. Why do you think it took years for founder Jaap Korteweg to begin marketing his meat-like vegetarian products?
2. Why is quality control so important for The Vegetarian Butcher?
3. What are some of the operational challenges that The Vegetarian Butcher faces?

Managing the Supply Chain

A major function of operations is **supply chain management**, which refers to connecting and integrating all parties or members of the distribution system in order to satisfy customers.[28] Supply chain is a part of distribution that will be discussed in more detail in the chapter titled "Dimensions of Marketing Strategy," where we cover marketing channels, which are the groups of organizations that make decisions about moving products from producers to consumers. We discuss supply chains here because it is a major component of operations within a business. It may help to think of the firms involved in a total distribution system as existing along a conceptual line, the combined impact of which results in an effective supply chain. Firms that are "upstream" in the supply chain (for example, suppliers) and "downstream" (for example, wholesalers and retailers) work together to serve customers and generate competitive advantage. Supply chain management requires marketing managers to work with other managers in operations, logistics, and procurement.

Procurement involves the processes to obtain resources to create value through sourcing, purchasing, and recycling materials and information. Procurement for many is synonymous with "buying" or "purchasing," but this is only a small part of what goes into the procurement activities within a supply chain. Decisions about where the supplies (including services) come from are very important. They are not just about price but also relate to where they are sourced and the integrity of the supplier. Also, recycling impact on the environment after consumption is an important consideration. An important process is the creation of a digital platform to link everything from production to consumer and involves sensors, mobile devices, cameras, and other systems that capture information for procurement. We discuss purchasing more in the next section.

Logistical concerns involve physical distribution and the selection of transportation modes. In transportation, digital networks that integrate the movement of products provide insights to import service and reduce cost. Inbound logistics, outbound logistics, and third-party logistics are all important pieces of these transportation nodes. *Inbound logistics* involves the movement of the raw materials, packaging, information, and other goods and services from the suppliers to the producers. Similarly, *outbound logistics* follows the finished products and information from the business customers and then to the final consumer. In order to pull this transportation process together, some companies use *third-party logistics,* which involves employing outside firms to move goods because they can transport them more efficiently than the company can themselves. With predictive analytics and artificial intelligence orders, transportation decisions can be made based on customers' defined requirements, costs, and service options.[29] Manufacturers, distributors, and retailers need to communicate with their supply-chain partners to provide real-time information. This provides the opportunity to advance capabilities and efficiencies in the supply chain. For these same reasons, and with so many variables, logistical disruptions can be massively harmful to the ability to adequately satisfy customer expectations and can even be fatal to the firm. Logistics management is just as important in managing services, enabling and communicating with partners. For example, health care providers must rely on manufacturers and distributors to perform many supply-chain activities from various unrelated items; food, medicine, and supplies that meet required standard logistics involve deliveries from vendors with inventories that accommodate the ability to provide health care services.

Drone technology has its challenges but, as noted earlier, is nevertheless growing rapidly, with some estimating that drone services overall are worth more than $127 billion globally, $23 billion coming from transport. For example, Amazon is paving the way for new innovation with its ideas regarding consumer goods delivery. With skepticism and public safety concerns from consumers, Amazon is working more with lower-value, less fragile items to begin with before straying from traditional delivery methods for other goods. Amazon and drone technology as a whole could completely change the way logistics and the supply chain operates in the future.[30]

Operations managers are concerned with managing inventory to ensure that there is enough inventory in stock to meet demand.

©Amble Design/Shutterstock

Operations are often the most public and visible aspect of the supply chain. Consumers are increasingly concerned with an important question: "How are our products being made?" Fair trade, organic food products, working conditions, child labor, concerns with sending jobs overseas, and regulatory mandates create major factors in decision making about operations. Technology in operations is driving a more digital enterprise system. Robotics, predictive analytics, the Internet of Things (IoT), driverless cars, drones, automation in identification of inventory, and network inventory optimization tools are changing the landscape of operations. By making profitable and responsible use of the materials and products sourced to them, and utilizing the information and capabilities afforded them through logistics, operational personnel can create extensive financial and brand value.

In this section, we look at elements of supply chain, including purchasing, managing inventory, outsourcing, and scheduling, which are vital tasks in the transformation of raw materials into finished goods. To illustrate logistics, consider a hypothetical small business—we'll call it Rushing Water Canoes Inc.—that manufactures aluminum canoes, which it sells primarily to sporting goods stores and river-rafting expeditions. Our company also makes paddles and helmets, but the focus of the following discussion is the manufacture of the company's quality canoes as they proceed through the logistics process.

Procurement

Purchasing is a part in procurement involved in the buying of all the materials needed by the organization. The purchasing department aims to obtain items of the desired quality in the right quantities at the lowest possible cost. Rushing Water Canoes, for example, must procure not only aluminum and other raw materials, and various canoe parts and components, but also machines and equipment, manufacturing supplies (oil, electricity, and so on), and office supplies in order to make its canoes. People in the purchasing department locate and evaluate suppliers of these items. They must constantly be on the lookout for new materials or parts that will do a better job or cost less than those currently being used. The purchasing function can be quite complex and is one area made much easier and more efficient by technological advances. Advanced artificial intelligence can uncover the highest value opportunities and empower organizations to unlock savings and increase profit.[31]

Not all companies purchase all of the materials needed to create their products. Oftentimes, they can make some components more economically and efficiently than can an outside supplier. On the other hand, firms sometimes find that it is uneconomical to make or purchase an item, and instead arrange to lease it from another organization. Many organizations lease equipment such as copiers that are costly to own and maintain and where significant product improvements occur over time. Whether to purchase, make, or lease a needed item generally depends on cost, as well as on product availability and supplier reliability.

purchasing
the buying of all the materials needed by the organization; also called procurement.

Managing Inventory

Once the items needed to create a product have been procured, some provision has to be made for storing them until they are needed. Every raw material, component, completed or partially completed product, and piece of equipment a firm uses—its **inventory**—must be accounted for, or controlled. There are three basic types of inventory. *Finished-goods inventory* includes those products that are ready for sale, such as a fully assembled automobile ready to ship to a dealer. *Work-in-process inventory* consists of those products that are partly completed or are in some stage of the transformation process. At McDonald's, a cooking hamburger represents work-in-process inventory because it must go through several more stages before it can be sold to a customer. *Raw materials inventory* includes all the materials that have been purchased to be used as inputs for making other products. Nuts and bolts are raw materials for an automobile manufacturer, while hamburger patties, vegetables, and buns are raw materials for the fast-food restaurant. Our fictional Rushing Water Canoes has an inventory of materials for making canoes, paddles, and helmets, as well as its inventory of finished products for sale to consumers. **Inventory control** is the process of determining how many supplies and goods are needed and keeping track of quantities on hand, where each item is, and who is responsible for it.

inventory
all raw materials, components, completed or partially completed products, and pieces of equipment a firm uses.

inventory control
the process of determining how many supplies and goods are needed and keeping track of quantities on hand, where each item is, and who is responsible for it.

Operations management must be closely coordinated with inventory control. The production of televisions, for example, cannot be planned without some knowledge of the availability of all the necessary materials—the chassis, picture tubes, color guns, and so forth. Also, each item held in inventory—any type of inventory—carries with it a cost. For example, storing fully assembled televisions in a warehouse to sell to a dealer at a future date requires not only the use of space, but also the purchase of insurance to cover any losses that might occur due to fire or other unforeseen events.

Inventory managers spend a great deal of time trying to determine the proper inventory level for each item. The answer to the question of how many units to hold in inventory depends on variables such as the usage rate of the item, the cost of maintaining the item in inventory, future costs of inventory and other procedures associated with ordering or making the item, and the cost of the item itself. For example, radio-frequency identification (RFID) is a wireless system composed of tags and readers that use radio waves to communicate information (tags communicate to readers) through every phase of handling inventory. RFID has a broad range of applications, particularly throughout the supply chain, and has immensely improved shipment tracking and reduced cycle times. This technology can be used in everything from inventory control to helping to prevent the distribution of counterfeit drugs and medical devices.[32] Several approaches may be used to determine how many units of a given item should be procured at one time and when that procurement should take place.

The Economic Order Quantity Model. To control the number of items maintained in inventory, managers need to determine how much of any given item they should order. One popular approach is the **economic order quantity (EOQ) model**, which identifies the optimum number of items to order to minimize the costs of managing (ordering, storing, and using) them.

Just-in-Time Inventory Management. An increasingly popular technique is **just-in-time (JIT) inventory management**, which eliminates waste by using smaller quantities of materials that arrive "just in time" for use in the transformation process and, therefore, require less storage space and other inventory management expense. JIT minimizes inventory by providing an almost continuous flow of items from suppliers to the production facility. Many U.S. companies—including Hewlett-Packard, IBM, and Harley Davidson—have adopted JIT to reduce costs and boost efficiency.

Let's say that Rushing Water Canoes uses 20 units of aluminum from a supplier per day. Traditionally, its inventory manager might order enough for one month at a time: 440 units per order (20 units per day times 22 workdays per month). The expense of such a large inventory could be considerable because of the cost of insurance coverage, recordkeeping, rented storage space, and so on. The just-in-time approach would reduce these costs because aluminum would be purchased in smaller quantities, perhaps in lot sizes of 20, which the supplier would deliver once a day. Of course, for such an approach to be effective, the supplier must be extremely reliable and relatively close to the production facility.

On the other hand, there are some downsides to just-in-time inventory management that marketers must take into account. When an earthquake and tsunami hit Japan, resulting in a nuclear reactor crisis, several Japanese companies halted their operations. Some multinationals relied so much upon their Japanese suppliers that their supply chains were also affected. In the case of natural disasters, having only enough inventory to meet current needs could create delays in production and hurt the company's bottom line. For this reason, many economists suggest that businesses store components that are essential for production and diversify their supply chains.

economic order quantity (EOQ) model
a model that identifies the optimum number of items to order to minimize the costs of managing (ordering, storing, and using) them.

just-in-time (JIT) inventory management
a technique using smaller quantities of materials that arrive "just in time" for use in the transformation process and therefore require less storage space and other inventory management expense.

That way, if a natural disaster knocks out a major supplier, the company can continue to operate.[33]

Material-Requirements Planning.   Another inventory management technique is material-requirements planning (MRP), a planning system that schedules the precise quantity of materials needed to make the product. The basic components of MRP are a master production schedule, a bill of materials, and an inventory status file. At Rushing Water Canoes (RWC), for example, the inventory-control manager will look at the production schedule to determine how many canoes the company plans to make. He or she will then prepare a bill of materials—a list of all the materials needed to make that quantity of canoes. Next, the manager will determine the quantity of these items that RWC already holds in inventory (to avoid ordering excess materials) and then develop a schedule for ordering and accepting delivery of the right quantity of materials to satisfy the firm's needs. Because of the large number of parts and materials that go into a typical production process, MRP must be done on a computer. It can be, and often is, used in conjunction with just-in-time inventory management.

material-requirements planning (MRP)
a planning system that schedules the precise quantity of materials needed to make the product.

Outsourcing

Increasingly, outsourcing has become a component of supply chain management in operations. As we mentioned in the chapter titled "Business in a Borderless World," outsourcing refers to the contracting of manufacturing or other tasks to independent companies. The use of these outside firms, called third parties, is related to outsourcing logistical services. Many companies elect to outsource some aspects of their operations to companies that can provide these products more efficiently, at a lower cost, and with greater customer satisfaction. Globalization has put pressure on supply chain managers to improve speed and balance resources against competitive pressures. Companies outsourcing to China, in particular, face heavy regulation, high transportation costs, inadequate facilities, and unpredictable supply chain execution. Therefore, suppliers need to provide useful, timely, and accurate information about every aspect of the quality requirements, schedules, and solutions to dealing with problems. Companies that hire suppliers must also make certain that their suppliers are following company standards; failure to do so could lead to criticism of the parent company.

Many high-tech firms have outsourced the production of chips, computers, and telecom equipment to Asian companies. The hourly labor costs in countries such as China, India, and Vietnam are far less than in the United States, Europe, or even Mexico. These developing countries have improved their manufacturing capabilities, infrastructure, and technical and business skills, making them more attractive regions for global sourcing. For instance, Nike outsources almost all of its production to Asian countries such as China and Vietnam. On the other hand, the cost of outsourcing halfway around the world must be considered in decisions. While information technology is often outsourced today, transportation, human resources, services, and even marketing functions can be outsourced. Our hypothetical Rushing Water Canoes might contract with a local janitorial service to clean its offices and with a local accountant to handle routine bookkeeping and tax-preparation functions.

Outsourcing, once used primarily as a cost-cutting tactic, has increasingly been linked with the development of competitive advantage through improved product quality, speeding up the time it takes products to get to the customer, and increasing overall supply chain efficiencies. Table 8.2 describes five of the top 100 global outsourcing providers that assist mainly in information technology. Outsourcing allows

TABLE 8.2
Top Outsourcing Providers

Company	Services
ISS	Facility services
Accenture	Management consulting, technology, and outsourcing
Canon Business Process Services	Business process services, document management, and managed workforce services
CBRE	Commercial real estate services
Kelly Outsourcing and Consulting Group	Talent management solutions

Source: International Association of Outsourcing, The Global Outsourcing 100, 2016, https://www.iaop.org/FORTUNE (accessed April 12, 2017).

companies to free up time and resources to focus on what they do best and to create better opportunities to focus on customer satisfaction. Many executives view outsourcing as an innovative way to boost productivity and remain competitive against low-wage offshore factories. However, outsourcing may create conflict with labor and negative public opinion when it results in U.S. workers being replaced by lower-cost workers in other countries.

Routing and Scheduling

After all materials have been procured and their use determined, managers must then consider the **routing,** or sequence of operations through which the product must pass. Therefore, routing and scheduling is an important part of operations in the supply chain. For example, before employees at Rushing Water Canoes can form aluminum sheets into a canoe, the aluminum must be cut to size. Likewise, the canoe's flotation material must be installed before workers can secure the wood seats. The sequence depends on the product specifications developed by the engineering department of the company.

Once management knows the routing, the actual work can be scheduled. **Scheduling** assigns the tasks to be done to departments or even specific machines, workers, or teams. At Rushing Water, cutting aluminum for the company's canoes might be scheduled to be done by the "cutting and finishing" department on machines designed especially for that purpose.

Many approaches to scheduling have been developed, ranging from simple trial and error to highly sophisticated computer programs. One popular method is the *Program Evaluation and Review Technique (PERT),* which identifies all the major activities or events required to complete a project, arranges them in a sequence or path, determines the critical path, and estimates the time required for each event. Producing a McDonald's Big Mac, for example, involves removing meat, cheese, sauce, and vegetables from the refrigerator; grilling the hamburger patties; assembling the ingredients; placing the completed Big Mac in its package; and serving it to the customer (Figure 8.3). The cheese, pickles, onions, and sauce cannot be put on before the hamburger patty is completely grilled and placed on the bun. The path that requires the longest time from start to finish is called the *critical path* because it determines the minimum amount of time in which the process can be completed. If any of the activities on the critical path for production of the Big Mac fall behind schedule, the sandwich will not be completed on time, causing customers to wait longer than they usually would.

routing
the sequence of operations through which the product must pass.

scheduling
the assignment of required tasks to departments or even specific machines, workers, or teams.

FIGURE 8.3
A Hypothetical PERT Diagram for a McDonald's Big Mac

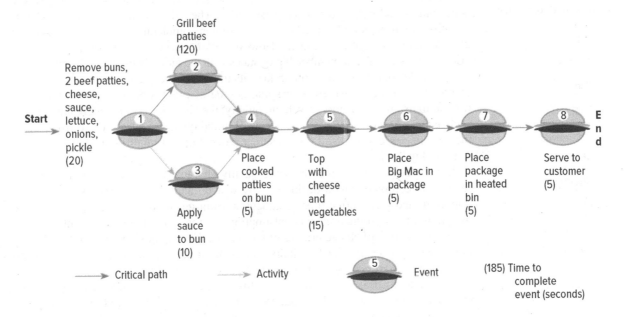

Grill beef patties (120)

Remove buns, 2 beef patties, cheese, sauce, lettuce, onions, pickle (20)

Start

Apply sauce to bun (10)

Place cooked patties on bun (5)

Top with cheese and vegetables (15)

Place Big Mac in package (5)

Place package in heated bin (5)

Serve to customer (5)

End

⟶ Critical path   ⟶ Activity   ◯ Event   (185) Time to complete event (seconds)

Managing Quality

LO 8-6

Assess the importance of quality in operations management.

Quality, like cost and efficiency, is a critical element of operations management, for defective products can quickly ruin a firm. Quality reflects the degree to which a good or service meets the demands and requirements of customers. Customers are increasingly dissatisfied with the quality of service provided by many airlines. Table 8.3 gives the

TABLE 8.3
Airline Scorecard (Best to Worst)

Rank	Overall Rank	On-Time Arrival	Cancelled Flights	Extreme Delays	2-Hour Tarmac Delays	Mishandled Baggage	Involuntary Bumping	Complaints
1	Delta	Delta	Frontier	Southwest	Southwest	Spirit	Delta	Southwest
2	Alaska/Virgin America	United	Delta	Alaska/Virgin America	Alaska/Virgin America	JetBlue	Alaska/Virgin America	Delta
3	Southwest	Alaska/Virgin America	Southwest	Delta	Spirit	Alaska/Virgin America	United	Alaska/Virgin America
4	United	American	United	American	Delta	Delta	Frontier	JetBlue
5	Frontier	Frontier	Alaska/Virgin America	United	Frontier	United	American	American
6	American	Southwest	American	Frontier	United	American	Southwest	United
7	Spirit	Spirit	JetBlue	Spirit	JetBlue	Southwest	JetBlue	Frontier
8	JetBlue	JetBlue	Spirit	JetBlue	American	Frontier	Spirit	Spirit

Sources: "2017 Airline Scorecard," The Wall Street Journal, https://graphics.wsj.com/table/Midseat_0III (accessed April 22, 2018).

TABLE 8.4
Top 10 J.D. Power
& Associates Initial
Automobile Quality Study

| 1. Kia |
| 2. Genesis |
| 3. Porsche |
| 4. Ford |
| 5. Ram |
| 6. BMW |
| 7. Chevrolet |
| 8. Hyundai |
| 9. Lincoln |
| 10. Nissan |

Source: J.D. Power & Associates,
"New-Vehicle Initial Quality is Best
Ever, J.D. Power Finds," June 21,
2017, http://www.jdpower.com/
press-releases/2017-us-initial-
quality-study-iqs (accessed April
21, 2018).

quality control
the processes an organization
uses to maintain its
established quality standards.

**total quality management
(TQM)**
a philosophy that uniform
commitment to quality in all
areas of an organization will
promote a culture that meets
customers' perceptions of
quality.

rankings of U.S. airlines in certain operational areas. Determining quality can be difficult because it depends on customers' perceptions of how well the product meets or exceeds their expectations. For example, customer satisfaction on airlines can vary wildly depending on individual customers' perspectives. The airline industry is notorious for its dissatisfied customers. Flight delays are a common complaint from airline passengers; 20 percent of all flights arrive late.[34] However, most consumers select airlines based on price, route, schedule, or membership or status with the airline's frequent-flyer program.

The fuel economy of an automobile or its reliability (defined in terms of frequency of repairs) can be measured with some degree of precision. Although automakers rely on their own measures of vehicle quality, they also look to independent sources such as the J.D. Power & Associates annual initial quality survey for confirmation of their quality assessment and for consumer perceptions of quality for the industry, as indicated in Table 8.4.

It is especially difficult to measure quality characteristics when the product is a service. A company has to decide exactly which quality characteristics it considers important and then define those characteristics in terms that can be measured. The inseparability of production and consumption and the level of customer contact influence the selection of characteristics of the service that are most important. Employees in high-contact services such as hairstyling, education, and legal services—and even the barista at Starbucks—are an important part of the product.

The Malcolm Baldrige National Quality Award is given each year to companies that meet rigorous standards of quality. The Baldrige criteria are (1) leadership, (2) information and analysis, (3) strategic planning, (4) human resource development and management, (5) process management, (6) business results, and (7) customer focus and satisfaction. The criteria have become a worldwide framework for driving business improvement. Four organizations won the award for 2016: Don Chalmers Ford (small business), Momentum Group (small business), Kindred Nursing and Rehabilitation Center (health care), and Memorial Herman Sugar Land Hospital (health care).[35]

Quality is so important that we need to examine it in the context of operations management. **Quality control** refers to the processes an organization uses to maintain its established quality standards. Quality has become a major concern in many organizations, particularly in light of intense foreign competition and increasingly demanding customers. To regain a competitive edge, a number of firms have adopted a total quality management approach. **Total quality management (TQM)** is a philosophy that uniform commitment to quality in all areas of the organization will promote a culture that meets customers' perceptions of quality. It involves coordinating efforts to improve customer satisfaction, increasing employee participation, forming and strengthening supplier partnerships, and facilitating an organizational culture of continuous quality improvement. TQM requires constant improvements in all areas of the company as well as employee empowerment.

Continuous improvement of an organization's goods and services is built around the notion that quality is free; by contrast, *not* having high-quality goods and services can be very expensive, especially in terms of dissatisfied customers.[36] A primary tool of the continuous improvement process is *benchmarking,* the measuring and evaluating of the quality of the organization's goods, services, or processes as compared with the quality produced by the best-performing companies in the industry.[37] Benchmarking lets the organization know where it stands competitively in its industry, thus giving it a goal to aim for over time. Now that online digital media are becoming more important in businesses, benchmarking tools are also becoming more popular. These tools allow companies to monitor and compare the success of their websites as they track traffic to their site versus competitors' sites.

Entrepreneurship in Action

El Pinto: When You Are Hot, You Are Hot

El Pinto
Founders: Jack and Connie Thomas
Founded: 1962, in Albuquerque, New Mexico
Success: El Pinto has become a successful restaurant and manufacturing powerhouse that sells its salsa across the nation.

El Pinto has come far from its humble origins as a one-room New Mexican restaurant. Today, El Pinto is owned by the original founders' twin sons, Jim and John Thomas, and consists of 12 acres of land, seats for more than 1,200 guests, and an 8,000-square-foot manufacturing facility. The restaurant consists of a cantina, five patios, three dining rooms, and a wall with photos of celebrities who have patronized the business, including President Obama, President Bush, Hillary Clinton, Mel Gibson, and Snoop Dogg.

Outside the Southwest, consumers are more likely to recognize El Pinto from its salsa. During the 1990s, El Pinto's customers began requesting products of the company's famous salsa and sauces. In 2000, it launched its salsa product line, manufacturing its salsas in an on-premises facility. Its products can be found at Walmart, HEB, Albertson's, Kroger, and Costco. Currently, the site is able to produce 4,000 cases of salsas and sauces per day. By operating both in the service and manufacturing industries, the entrepreneurs running El Pinto have taken advantage of multiple opportunities for expansion and growth.[38]

Critical Thinking Questions

1. How is El Pinto both a service organization *and* a manufacturing organization?
2. What are the inputs and outputs used in El Pinto's manufacturing of salsa?
3. Managing quality consistently is important to El Pinto's successful operations. What are some of the challenges of managing consistency of service in El Pinto's restaurant business that it may not have with its manufacturing business?

Companies employing TQM programs know that quality control should be incorporated throughout the transformation process, from the initial plans to the development of a specific product through the product and production-facility design processes to the actual manufacture of the product. In other words, they view quality control as an element of the product itself, rather than as simply a function of the operations process. When a company makes the product correctly from the outset, it eliminates the need to rework defective products, expedites the transformation process itself, and allows employees to make better use of their time and materials. One method through which many companies have tried to improve quality is **statistical process control,** a system in which management collects and analyzes information about the production process to pinpoint quality problems in the production system.

statistical process control a system in which management collects and analyzes information about the production process to pinpoint quality problems in the production system.

International Organization for Standardization (ISO)

Regardless of whether a company has a TQM program for quality control, it must first determine what standard of quality it desires and then assess whether its products meet that standard. Product specifications and quality standards must be set so the company can create a product that will compete in the marketplace. Rushing Water Canoes, for example, may specify that each of its canoes has aluminum walls of a specified uniform thickness, that the front and back be reinforced with a specified level of steel, and that each contain a specified amount of flotation material for safety. Production facilities must be designed that can produce products with the desired specifications.

Quality standards can be incorporated into service businesses as well. A hamburger chain, for example, may establish standards relating to how long it takes to cook an order and serve it to customers, how many fries are in each order, how thick the burgers are, or how many customer complaints might be acceptable. Once the

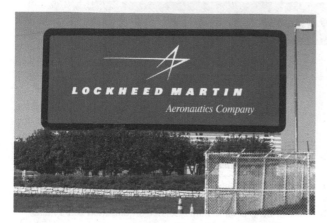

Lockheed Martin Aeronautics Company in Fort Worth, Texas, has achieved ISO 14001 certification.

©Ronald Martinez/Getty Images

ISO 9000
a series of quality assurance standards designed by the International Organization for Standardization (ISO) to ensure consistent product quality under many conditions.

ISO 14000
a comprehensive set of environmental standards that encourages a cleaner and safer world by promoting a more uniform approach to environmental management and helping companies attain and measure improvements in their environmental performance.

ISO 19600
a comprehensive set of guidelines for compliance management that address risks, legal requirements, and stakeholder needs.

desired quality characteristics, specifications, and standards have been stated in measurable terms, the next step is inspection.

The International Organization for Standardization (ISO) has created a series of quality management standards—**ISO 9000**—designed to ensure the customer's quality standards are met. The standards provide a framework for documenting how a certified business keeps records, trains employees, tests products, and fixes defects. To obtain ISO 9001 certification, an independent auditor must verify that a business's factory, laboratory, or office meets the quality standards spelled out by the International Organization for Standardization. The certification process can require significant investment, but for many companies, the process is essential to being able to compete. Thousands of companies have been certified, including General Electric Analytical Instruments, which has applied ISO standards to everything from the design to the manufacturing practices of its global facilities.[39] Certification has become a virtual necessity for doing business in Europe in some high-technology businesses. ISO 9002 certification was established for service providers.

ISO 14000 is a comprehensive set of environmental standards that encourages a cleaner and safer world. ISO 14000 is a valuable standard because currently considerable variation exists between the regulations in different nations, and even regions within a nation. These variations make it difficult for organizations committed to sustainability to find acceptable global solutions to problems. The goal of the ISO 14000 standards is to promote a more uniform approach to environmental management and to help companies attain and measure improvements in their environmental performance. **ISO 19600** provides guidance for establishing, developing, implementing, evaluating, maintaining, and improving an effective and responsive compliance management system within an organization. The guidelines are applicable to all types of organizations. The extent of the application of these guidelines depends on the size, structure, nature, and complexity of the organization. This guidance is based on the principles of good governance, transparency, and sustainability.[40]

Inspection

Inspection reveals whether a product meets quality standards. Some product characteristics may be discerned by fairly simple inspection techniques—weighing the contents of cereal boxes or measuring the time it takes for a customer to receive his or her hamburger. As part of the ongoing quality assurance program at Hershey Foods, all wrapped Hershey Kisses are checked, and all imperfectly wrapped kisses are rejected. Other inspection techniques are more elaborate. Automobile manufacturers use automated machines to open and close car doors to test the durability of latches and hinges. The food-processing and pharmaceutical industries use various chemical tests to determine the quality of their output. Rushing Water Canoes might use a special device that can precisely measure the thickness of each canoe wall to ensure that it meets the company's specifications.

Organizations normally inspect purchased items, work-in-process, and finished items. The inspection of purchased items and finished items takes place after the fact; the inspection of work-in-process is preventive. In other words, the purpose of inspection of purchased items and finished items is to determine what the quality level is. For items that are being worked on—an automobile moving down the assembly line or a canoe being assembled—the purpose of the inspection is to find defects before the product is completed so that necessary corrections can be made.

Sampling

An important question relating to inspection is how many items should be inspected. Should all canoes produced by Rushing Water be inspected or just some of them? Whether to inspect 100 percent of the output or only part of it is related to the cost of the inspection process, the destructiveness of the inspection process (some tests last until the product fails), and the potential cost of product flaws in terms of human lives and safety.

Some inspection procedures are quite expensive, use elaborate testing equipment, destroy products, and/or require a significant number of hours to complete. In such cases, it is usually desirable to test only a sample of the output. If the sample passes inspection, the inspector may assume that all the items in the lot from which the sample was drawn would also pass inspection. By using principles of statistical inference, management can employ sampling techniques that ensure a relatively high probability of reaching the right conclusion—that is, rejecting a lot that does not meet standards and accepting a lot that does. Nevertheless, there will always be a risk of making an incorrect conclusion—accepting a population that *does not* meet standards (because the sample was satisfactory) or rejecting a population that *does* meet standards (because the sample contained too many defective items).

Sampling is likely to be used when inspection tests are destructive. Determining the life expectancy of light bulbs by turning them on and recording how long they last would be foolish: There is no market for burned-out light bulbs. Instead, a generalization based on the quality of a sample would be applied to the entire population of light bulbs from which the sample was drawn. However, human lives and safety often depend on the proper functioning of specific items, such as the navigational systems installed in commercial airliners. For such items, even though the inspection process is costly, the potential cost of flawed systems is too great not to inspect 100 percent of the output.

Integrating Operations and Supply Chain Management

Managing operations and supply chains can be complex and challenging due to the number of independent organizations that must perform their responsibilities in creating product quality. Managing supply chains requires constant vigilance and the ability to make quick tactical changes. When allegations arose that seafood sold in the United States may have come from forced labor on ships around Thailand, companies had to immediately investigate these allegations. Therefore, managing the various partners involved in supply chains and operations is important because many stakeholders hold the firm responsible for appropriate conduct related to product quality. This requires that the company exercise oversight over all suppliers involved in producing a product. Encouraging suppliers to report problems, issues, or concerns requires excellent communication systems to obtain feedback. Ideally, suppliers will report potential problems before they reach the next level of the supply chain, which reduces damage.

Despite the challenges of monitoring global operations and supply chains, there are steps businesses can take to manage these risks. All companies who work with global suppliers should adopt a Global Supplier Code of Conduct and ensure that it is effectively communicated. Additionally, companies should encourage compliance and procurement employees to work together to find ethical suppliers at reasonable costs. Those in procurement are concerned with the costs of obtaining materials for the company. As a result, supply chain and procurement managers must work together to make operational decisions to ensure the selection of the best suppliers from an ethical and cost-effective standpoint. Businesses must also work to make certain that their supply chains are diverse. Having only a few suppliers in one area can disrupt operations should a disaster strike. Finally, companies must perform regular audits on its suppliers and take action against those found to be in violation of company standards.[41] Kellogg's offers a Global Supplier Code of Conduct on its website in 13 different languages, showing the importance of having access to the code throughout its supply chain.[42]

So You Want a Job in Operations Management

While you might not have been familiar with terms such as *supply chain* or *logistics* or *total quality management* before taking this course, careers abound in the operations management field. You will find these careers in a wide variety of organizations—manufacturers, retailers, transportation companies, third-party logistics firms, government agencies, and service firms. Approximately $1.3 trillion is spent on transportation, inventory, and related logistics activities, and logistics alone accounts for more than 9.5 percent of U.S. gross domestic product.[43] Closely managing how a company's inputs and outputs flow from raw materials to the end consumer is vital to a firm's success. Successful companies also need to ensure that quality is measured and actively managed at each step.

Supply chain managers have a tremendous impact on the success of an organization. These managers are engaged in every facet of the business process, including planning, purchasing, production, transportation, storage and distribution, customer service, and more. Their performance helps organizations control expenses, boost sales, and maximize profits.

Warehouse managers are a vital part of manufacturing operations. A typical warehouse manager's duties include overseeing and recording deliveries and pickups, maintaining inventory records and the product tracking system, and adjusting inventory levels to reflect receipts and disbursements. Warehouse managers also have to keep in mind customer service and employee issues. Warehouse managers can earn up to $60,000 in some cases.

Operations management is also required in service businesses. With more than 80 percent of the U.S. economy in services, jobs exist for services operations. Many service contact operations require standardized processes that often use technology to provide an interface that provides an automatic quality performance. Consider jobs in health care, the travel industry, fast food, and entertainment. Think of any job or task that is a part of the final product in these industries. Even an online retailer such as Amazon has a transformation process that includes information technology and human activities that facilitate a transaction. These services have a standardized process and can be evaluated based on their level of achieved service quality.

Total quality management is becoming a key attribute for companies to ensure that quality pervades all aspects of the organization. Quality assurance managers make a median salary of $72,000. These managers monitor and advise on how a company's quality management system is performing and publish data and reports regarding company performance in both manufacturing and service industries.[44]

Review Your Understanding

Define *operations management*.

Operations management (OM) is the development and administration of the activities involved in transforming resources into goods and services. Operations managers oversee the transformation process and the planning and designing of operations systems, managing logistics, quality, and productivity.

Differentiate between operations and manufacturing.

The terms *manufacturing* and *production* are used interchangeably to describe the activities and processes used in making tangible products, whereas *operations* is a broader term used to describe the process of making both tangible and intangible products.

Explain how operations management differs in manufacturing and service firms.

Manufacturers and service firms both transform inputs into outputs, but service providers differ from manufacturers in several ways: They have greater customer contact because the service typically occurs at the point of consumption; their inputs and outputs are more variable than manufacturers', largely because of the human element; service providers are generally more labor intensive; and their productivity measurement is more complex.

Describe the elements involved in planning and designing an operations system.

Operations planning relates to decisions about what product(s) to make, for whom, and what processes and facilities are needed to produce them. OM is often joined by marketing and research and development in these decisions. Common facility layouts include fixed-position layouts, process layouts, or product layouts. Where to locate operations facilities is a crucial decision that depends on proximity to the market, availability of raw materials, availability of transportation, availability of power, climatic influences, availability of labor, and community characteristics. Technology is also vital to operations, particularly computer-assisted design, computer-assisted manufacturing, flexible manufacturing, robotics, and computer-integrated manufacturing.

Specify some techniques managers may use to manage the logistics of transforming inputs into finished products.

Logistics, or supply chain management, includes all the activities involved in obtaining and managing raw materials and component parts, managing finished products, packaging them, and getting them to customers. The organization must first make or purchase (procure) all the materials it needs. Next, it must control its inventory by determining how many supplies and goods it needs and keeping track of every raw material, component, completed or partially completed product, and piece of equipment, how many of each are on hand, where they are, and who has responsibility for them. Common approaches to inventory control include the economic order quantity (EOQ) model, the just-in-time (JIT) inventory concept, and material-requirements planning (MRP). Logistics also includes routing and scheduling processes and activities to complete products.

Assess the importance of quality in operations management.

Quality is a critical element of OM because low-quality products can hurt people and harm the business. Quality control refers to the processes an organization uses to maintain its established quality standards. To control quality, a company must establish what standard of quality it desires and then determine whether its products meet that standard through inspection.

Propose a solution to a business's operations dilemma.

Based on this chapter and the facts presented in the "Solve the Dilemma" feature near the end of this chapter, you should be able to evaluate the business's problem and propose one or more solutions for resolving it.

Critical Thinking Questions

Enter the World of Business Questions

1. How does Trader Joe's use inventory control as a way to control operational efficiencies?

2. Why do you think it is important for Trader Joe's to maintain a simple supply chain?

3. How do Trader Joe's strong operational efficiencies allow it to compete against its larger rivals?

Learn the Terms

capacity   244

computer-assisted design (CAD)   246

computer-assisted manufacturing (CAM)   246

computer-integrated manufacturing (CIM)   247

continuous manufacturing organizations   246

customization   243

economic order quantity (EOQ) model   252

fixed-position layout   246

flexible manufacturing   247

inputs   237

intermittent organizations   246

inventory   251

inventory control   251

Check Your Progress

1. What is operations management?

2. Differentiate among the terms *operations, production,* and *manufacturing.*

3. Compare and contrast a manufacturer versus a service provider in terms of operations management.

4. Who is involved in planning products?

5. In what industry would the fixed-position layout be most efficient? The process layout? The product layout? Use real examples.

6. What criteria do businesses use when deciding where to locate a plant?

7. What is flexible manufacturing? How can it help firms improve quality?

8. Define supply chain management and summarize the activities it involves.

9. Describe some of the methods a firm may use to control inventory.

10. When might a firm decide to inspect a sample of its products rather than test every product for quality?

Get Involved

1. Compare and contrast OM at McDonald's with that of Honda of America. Compare and contrast OM at McDonald's with that of a bank in your neighborhood.

2. Find a real company that uses JIT, either in your local community or in a business journal. Why did the company decide to use JIT? What have been the advantages and disadvantages of using JIT for that particular company? What has been the overall effect on the quality of the company's goods or services? What has been the overall effect on the company's bottom line?

3. Interview someone from your local Chamber of Commerce and ask him or her what incentives the community offers to encourage organizations to locate there. (See if these incentives relate to the criteria firms use to make location decisions.)

Build Your Skills

Reducing Cycle Time

Background

An important goal of production and operations management is reducing cycle time—the time it takes to complete a task or process. The goal in cycle time reduction is to reduce costs and/or increase customer service.[45] Many experts believe that the rate of change in our society is so fast that a firm must master speed and connectivity.[46] Connectivity refers to a seamless integration of customers, suppliers, employees, and organizational, production, and operations management. The use of the Internet and other telecommunications systems helps many organizations connect and reduce cycle time.

Task

Break up into pairs throughout the class. Select two businesses (local restaurants, retail stores, etc.) that both of you frequent, are employed by, and/or are fairly well acquainted with. For the first business, one of you will role-play the "manager" and the other will role-play the "customer." Reverse roles for the second business you have selected. As managers at your respective businesses, you are to prepare a list of five questions you will ask the customer during the role-play. The questions you prepare should be designed to get the customer's viewpoint on how good the cycle time is at your business. If one of the responses leads to a problem area, you may need to ask a follow-up question to determine the nature of the dissatisfaction. Prepare one main question and a follow-up, if necessary, for each of the five dimensions of cycle time:

1. **Speed**—the delivery of goods and services in the minimum time; efficient communications; the elimination of wasted time.

2. **Connectivity**—all operations and systems in the business appear connected with the customer.

3. **Interactive relationships**—a continual dialog exists among operations units, service providers, and customers that permits the exchange of feedback on concerns or needs.

4. **Customization**—each product is tailored to the needs of the customer.

5. **Responsiveness**—the willingness to make adjustments and be flexible to help customers and to provide prompt service when a problem develops.

Begin the two role-plays. When it is your turn to be the manager, listen carefully when your partner answers your prepared questions. You need to elicit information on how to improve the cycle time at your business. You will achieve this by identifying the problem areas (weaknesses) that need attention.

After completing both role-play situations, fill out the accompanying form for the role-play when you were the manager. You may not have gathered enough information to fill in all the boxes. For example, for some categories, the customer may have had only good things to say; for others, the comments may all be negative. Be prepared to share the information you gain with the rest of the class. I role-played the manager at (business). After listening carefully to the customer's responses to my five questions, I determined the following strengths and weaknesses as they relate to the cycle time at my business:

Dimension	Strength	Weakness
Speed		
Connectivity		
Interactive relationships		
Customization		
Responsiveness		

Solve the Dilemma

Planning for Pizza

 McKing Corporation operates fast-food restaurants in 50 states, selling hamburgers, roast beef and chicken sandwiches, french fries, and salads. The company wants to diversify into the growing pizza business. Six months of tests revealed that the ideal pizza to sell was a 16-inch pie in three varieties: cheese, pepperoni, and deluxe (multiple toppings). Research found the size and toppings acceptable to families as well as to individuals (single buyers could freeze the leftovers), and the price was acceptable for a fast-food restaurant ($7.99 for cheese, $8.49 for pepperoni, and $9.99 for deluxe).

Marketing and human resources personnel prepared training manuals for employees, advertising materials, and the rationale to present to the restaurant managers (many stores are franchised). Store managers, franchisees, and employees are excited about the new plan. There is just one problem.

The drive-through windows in current restaurants are too small for a 16-inch pizza to pass through. The largest size the present windows can accommodate is a 12-inch pie. The managers and franchisees are concerned that if this aspect of operations has been overlooked, perhaps the product is not ready to be launched. Maybe there are other problems yet to be uncovered.

 LO 8-7

Propose a solution to a business's operations dilemma.

Critical Thinking Questions

1. What mistake did McKing make in approaching the introduction of pizza?

2. How could this product introduction have been coordinated to avoid the problems that were encountered?

3. If you were an executive at McKing, how would you proceed with the introduction of pizza into the restaurants?

Build Your Business Plan

Managing Service and Manufacturing Operations

For your business, you need to determine if you are providing raw materials that will be used in further production, or if you are a reseller of goods and services, known as a retailer. If you are the former, you need to determine what processes you go through in making your product.

The text provides ideas of breaking the process into inputs, transformation processes, and outputs. If you are a provider of a service or a link in the supply chain, you need to know exactly what your customer expectations are. Services are intangible, so it is all the more important to better understand what exactly the customer is looking for in resolving a problem or filling a need.

Visit Connect to practice building your business plan with the Business Plan Prep Exercises.

See for Yourself Videocase

How Sweet It Is: Creating Supply Chain Efficiencies at the Cocoa Exchange

Looking for a way to create incremental, non-cannibalizing growth, Mars, Incorporated, the world's largest chocolate company, launched The Cocoa Exchange, a stand-alone subsidiary. While Mars focuses on mass producing products like Snickers and M&M'S that are available in all distribution channels, The Cocoa Exchange sells exclusive and premium chocolate products directly to consumers through a commission based sales force under three product lines: Pod & Bean, Dove Signature, and Pure Dark. The Cocoa Exchange's mission is to create incremental, non-cannibalizing growth for Mars through niche products targeting individual consumers rather than the mass market. The company accomplishes this by creating supply-chain efficiencies, thanks to its parent company, and using the direct selling business model. Direct selling can be a more relationship-driven, customized form of selling and allows for even greater market segmentation.

The Cocoa Exchange is able to source its cocoa beans directly from farmers, just like Mars. Purchasing and sourcing can be challenging in the chocolate industry because cocoa only grows in a few places around the world. Complicating the matter is the fact that cocoa is produced by very small family farmers who often struggle to make ends meet. The majority of the cocoa sourced by The Cocoa Exchange is from developing countries like the Ivory Coast and Ghana in West Africa. The company develops relationships with the farmers to improve their yields and income as well as improve its own supply chain.

This close relationship between Mars, The Cocoa Exchange and the farmers allows for improvements in sustainability practices. Demand continues to grow for chocolate, however, sourcing cocoa in a sustainable manner is limited, since the crop can only grow in certain areas near the equator. Because of these challenges, The Cocoa Exchange aims to improve conditions for farmers so that they can create a larger supply

of cocoa. Mars is working with farmers as well to improve their productivity and economic viability through tripling farm yields in three to five years. By partnering with a humanitarian group called The CARE organization, The Cocoa Exchange helps farmers save money and secure loans. By investing in the suppliers from which The Cocoa Exchange sources and purchases its chocolate, the company has greater influence over the sustainability of its supply chain.

The concept of marketing and selling directly to the consumer, through an independent contractor salesforce, is called direct selling.. The Cocoa Exchange embraces the direct selling model and has created a commission based labor force called "curators" to sell its products through in-house parties. Party attendees can sample a range of products from The Cocoa Exchange and purchase the products they like online. Curators receive a 25 to 40 percent commission on products sold from their individual online store and gain access to exclusive discounts up to 50 percent depending on performance. Curators also have the opportunity to increase their earning power by building and training a team, receiving 3 to 5 percent commission on team sales. The Cocoa Exchange subsidizes the starter kits and curators host parties and provide free shipping on the items ordered. Additionally, as curators sales grow, they earn credits they can use to shop. The company also ships products directly to the consumer, so curators aren't burdened with inventory management and protection. Premium chocolates and other Cocoa Exchange products are carefully handled so that they arrive fresh and in 'mint' condition. Direct selling encourages an entrepreneurial spirit in their chocolate sellers, allows The Cocoa Exchange to find new customers, and helps the company maintain a robust and efficient supply chain.

As a subsidiary of Mars, Incorporated, The Cocoa Exchange benefits from an established and extensive supply chain where it can secure all the materials needed to create its chocolate products. The Cocoa Exchange is uniquely poised to sell premium and exclusive products directly to customers because of its premium food products and careful sourcing. Its initiatives to improve the operations of the rural cocoa farmers has benefited the supply chain as well as The Cocoa Exchange's curators, creating a strong, mutually beneficial relationship.[47]

Critical Thinking Questions

1. Describe how The Cocoa Exchange benefits from Mars, Incorporated's supply chain.

2. How does The Cocoa Exchange benefit from building relationships with cocoa farmers?

3. Why does the exclusive nature of these products increases engagement in the company's sales force?

You can find the related video in the Video Library in Connect. Ask your instructor how you can access Connect.

Team Exercise

Form groups and assign the responsibility of finding companies that outsource their production to other countries. What are the key advantages of this outsourcing decision? Do you see any drawbacks or weaknesses in this approach? Why would a company not outsource when such a tactic can be undertaken to cut manufacturing costs? Report your findings to the class.

Ask your instructor about the role play exercises available with this book to practice working with a business team.

Notes

1. Trader Joe's website, www.traderjoes.com (accessed March 24, 2018); Beth Kowitt, "Inside the Secret World of Trader Joe's—Full Version," *Fortune,* August 23, 2010, http://archive. fortune.com/2010/08/20/news/companies/inside_trader_ joes_full_version.fortune/index.htm (accessed January 1, 2018); "2010 World's Most Ethical Companies," *Ethisphere,* http://m1.ethisphere.com/wme2013/index.html (accessed October 1, 2013); Lisa Scherzer, "Trader Joe's Tops List of Best Grocery Store Chains," *Yahoo Finance!,* July 24, 2013, https://finance.yahoo.com/blogs/the-exchange/trader-joe-tops-list-best-grocery-store-chains-182739789.html (accessed January 1, 2018); Nancy Luna, "Trader Joe's Expanding to Texas," *Orange County Register,* May 5, 2011, https://www .ocregister.com/2011/05/05/trader-joes-expanding-to-texas-2/ (accessed January 1, 2018); Progressive Grocer, "Trader Joe's Opens Fewer-Than-Average Stores in 2017," October 16, 2017, https://progressivegrocer.com/trader-joes-opens-fewer-average-stores-2017 (accessed January 1, 2018); "Trader Joe's: Groceries for the 'Overeducated and Underpaid'," Technology and Operations Management—A Course at Harvard Business School, December 9, 2015, https://rctom.hbs.org/submission/ trader-joes-groceries-for-the-overeducated-and-underpaid/ (accessed January 1, 2018); John Boyle, "As Earth Fare Grows, Some Workers Feeling 'Squeezed,'" *Citizen Times,* September 3, 2016, http://www.citizen-times.com/story/news/ local/2016/09/03/earth-fare-grows-some-workers-feeling-squeezed/88412772/ (accessed January 1, 2018); Sheiresa Ngo, "15 Secrets Trader Joe's Shoppers Should Know," *Business Insider,* May 8, 2017, http://www.businessinsider. com/15-secrets-trader-joes-shoppers-should-know-2017-5/#3-shoppers-can-win-prizes-3 (accessed January 1, 2018); "Trader Joe's: Food for Thought," Technology and Operations Management—A Course at Harvard Business School, December 9, 2015, https://rctom.hbs.org/submission/trader-joes-food-for-thought/ (accessed January 1, 2018).

2. "Our Employee First Culture," *Stand For,* 2017, http:// standfor.containerstore.com/putting-our-employees-first/ (accessed April 11, 2017).

3. "Personified: 11 Things You Didn't Know about Amazon's CEO Jeff Bezos," *Gadgets Now,* January 12, 2017, http://www.gadgetsnow.com/checklist/personified-11-things-you-didnt-know-about-amazons-ceo-jeff-bezos/ checklistshow/56500597.cms (accessed April 11, 2017).

4. Leonard L. Berry, *Discovering the Soul of Service* (New York: The Free Press, 1999), pp. 86–96.

5. Valerie A. Zeithaml and Mary Jo Bitner, *Services Marketing,* 3rd ed. (Boston: McGraw-Hill Irwin, 2003), pp. 3, 22.

6. Claire Zillman, "Zappos Is Bringing Uber-like Surge Pay to the Workplace," *Fortune,* January 28, 2015, http://fortune. com/2015/01/28/zappos-employee-pay/ (accessed May 4, 2017).

7. Bernard Wysocki Jr., "To Fix Health Care, Hospitals Take Tips from the Factory Floor," *The Wall Street Journal,* April 9, 2004, via www.chcanys.org/clientuploads/downloads/ Clinical_resources/Leadership%20Articles/LeanThinking_ ACF28EB.pdf (accessed May 4, 2017).

8. Deborah L. Roberts and Frank T. Piller, "Finding the Right Role for Social Media in Innovation," *MIT Sloan Management Review,* March 15, 2016, http://sloanreview. mit.edu/article/finding-the-right-role-for-social-media-in-innovation/ (accessed April 11, 2017).

9. Debbie Qaqish, "So You Want to Do Experiential Marketing? Consider These 4 Things First," *Marketing News,* April 2018, pp. 4–6.

10. Danielle Muoio, "These 19 Companies Are Racing to Put Driverless Cars on the Road by 2020," *Business Insider,* August 18, 2016, http://www.businessinsider.com/

companies-making-driverless-cars-by-2020-2016-8/ (accessed April 11, 2017).

11. "Best Sellers," Villy Customs, 2017, https://villycustoms.com/collections/the-10 (accessed April 12, 2017).

12. Jonathan Dyble, "Global Blockchain in Supply Chain Market to Reach $424 Million by 2023," *Supply Chain Digital*, www.supplychaindigital.com/technology/global-blockchain-supply-chain-market-reach-424mn-2023 (accessed April 13, 2018).

13. Christopher Mims, "Why Blockchain Will Survive, Even if Bitcoin Doesn't," *The Wall Street Journal*, March 12, 2018, https://www.wsj.com/articles/why-blockchain-will-survive-even-if-bitcoin-doesnt-1520769600 (accessed April 5, 2018); Steve Banker, "The Growing Maturity of Blockchain for Supply Chain Management," February 22, 2018 https://www.forbes.com/sites/stevebanker/2018/02/22/the-growing-maturity-of-blockchain-for-supply-chain-management/#3048ec9811da (accessed April 5, 2018); Jon-Amerin Vorabutra, "Why Blockchain Is a Game Changer for Supply Chain Transparency," *Supply Chain 247*, October 3, 2016, http://www.supplychain247.com/article/why_blockchain_is_a_game_changer_for_the_supply_chain (accessed April 5, 2018); IBM, "Blockchain for Supply Chain," https://www.ibm.com/blockchain/supply-chain/ (accessed April 5, 2018); Deloitte, "Using Blockchain to Drive Supply Chain Transparency," https://www2.deloitte.com/us/en/pages/operations/articles/blockchain-supply-chain-innovation.html (accessed April 5, 2018).

14. Erica Philips, "Bringing Blockchain to the Coffee Cup," *The Wall Street Journal*, April 15, 2018, p. B2.

15. "Improvements to the Engineering Planning and Development Process," Toyota, 2017, http://www.toyota-global.com/company/history_of_toyota/75years/data/automotive_business/products_technology/research/engineering_planning/details_window.html (accessed April 12, 2017).

16. Michelle Maynard, "Kit Kat Lovers, Listen Up: Hershey Is Betting Big That You'll Break Off More," *Forbes*, March 11, 2018, https://www.forbes.com/sites/michelinemaynard/2018/03/11/kit-kat-lovers-listen-up-hershey-is-betting-big-that-youll-break-off-more/#7c61790d7975 (accessed April 22, 2018).

17. Ross Toro, "How 3D Printers Work (Infographic)," *Live Science*, June 18, 2013, www.livescience.com/37513-how-3d-printers-work-infographic.html (accessed May 4, 2017).

18. Matt McFarland, "Amazon's Delivery Drones May Drop Packages via Parachute," *CNN Tech*, February 14, 2017, http://money.cnn.com/2017/02/14/technology/amazon-drone-patent/ (accessed April 12, 2017).

19. Stuart Hodge. "SAP—Why Drones Have a Key Role to Play in the Future of Procurement," http://www.supplychaindigital.com/procurement/sap-why-drones-have-key-role-play-future-procurement (accessed April 16, 2018).

20. Claire Cain Miller, "Evidence That Robots Are Winning the Race for American Jobs," *The New York Times*, March 28, 2017, https://www.nytimes.com/2017/03/28/upshot/evidence-that-robots-are-winning-the-race-for-american-jobs.html?_r=1 (accessed April 12, 2017).

21. Alanna Petroff, "U.S. Workers Face Higher Risk of Being Replaced by Robots. Here's Why," *CNN Tech*, March 24, 2017, http://money.cnn.com/2017/03/24/technology/robots-jobs-us-workers-uk/ (accessed April 12, 2017).

22. "Are There Enough Robots?" *Robotics Tomorrow*, February 14, 2017, http://www.roboticstomorrow.com/article/2017/02/are-there-enough-robots/9507 (accessed April 12, 2017).

23. James Henderson, "Hershey Pledges $500mn to Improve Supply Chain Sustainability," *Supply Chain Digital*, http://www.supplychaindigital.com/technology/hershey-pledges-500mn-improve-supply-chain-sustainability (accessed April 5, 2018).

24. "Benefits of Green Building," *USGBE*, April 1, 2016, http://www.usgbc.org/articles/green-building-facts (accessed April 12, 2017).

25. Katrina Brown Hunt, "Counting Down the 20 Greenest Cities in America," *Travel Leisure*, 2017, http://www.travelandleisure.com/slideshows/americas-greenest-cities/20 (accessed April 12, 2017).

26. Adobe, "Sustainability," https://www.adobe.com/corporate-responsibility/sustainability.html?red=av (accessed April 22, 2018).

27. Caryn Ginsberg, "The Market for Vegetarian Foods," The Vegetarian Resource Group, http://www.vrg.org/nutshell/market.htm (accessed December 31, 2017); Jaag Korteweg, Founder of De Vegetarische Slager, November 2017 (J. Wienen, Interviewer); Nielsen, "Green Generation: Millennials Say Sustainability Is A Shopping Priority," November 5, 2015, http://www.nielsen.com/us/en/insights/news/2015/green-generation-millennials-say-sustainability-is-a-shopping-priority.html (accessed December 31, 2017); The Vegetarian Butcher, "About Us," 2017, https://www.thevegetarianbutcher.com/about-us/since-1962 (accessed December 31, 2017); Niamh Michail, "Vegetarian Butcher Slams Dutch Food Authority for Double Standards over 'Misleading' Meat Name Ban," *Food Navigator*, October 5, 2017, https://www.foodnavigator.com/Article/2017/10/05/Vegetarian-Butcher-slams-Dutch-food-authority-for-double-standards-over-misleading-meat-name-ban (accessed December 31, 2017); Bryan Walsh, "The Triple Whopper Environmental Impact of Global Meat Production," *Time*, December 16, 2013, http://science.time.com/2013/12/16/the-triple-whopper-environmental-impact-of-global-meat-production/ (accessed December 31, 2017).

28. O.C. Ferrell and Michael D. Hartline, *Marketing Strategy* (Mason, OH: South Western, 2011), p. 215.

29. "E-Commerce: Cultivating a New Logistics Landscape," *Inbound Logistics*, p. 13.

30. Mike Danby, "Innovation Beyond Drones—Transforming the Supply Chain," *Supply Chain Digital*, http://www.supplychaindigital.com/technology/comment-innovation-beyond-drones-transforming-supply-chain (accessed April 14, 2018).

31. James Henderson, "AI-Driven Procurement Platform, Suplari, Attracts $10.3 Million in Funding," *Supply Chain Digital*, http://www.supplychaindigital.com/technology/ai-driven-procurement-platform-suplari-attracts-103mn-funding (accessed April 6, 2018).

32. William Pride and O.C. Ferrell, *Marketing* (Boston: Cengage Learning, 2018), p. 443.

33. "Broken Links," *The Economist,* March 31, 2011, www
.economist.com/node/18486015 (accessed May 4, 2017).

34. Scott McCartney, "The Best and Worst Airlines of
2016," *The Wall Street Journal,* January 11, 2017,
https://www.wsj.com/articles/the-best-and-worst-
airlines-of-2016-1484149294 (accessed April 11, 2017).

35. "Four U.S. Organizations Receive Nation's Highest Honor
for Performance Excellence," NIST, November 17, 2016,
https://www.nist.gov/news-events/news/2016/11/four-us-
organizations-receive-nations-highest-honor-performance-
excellence (accessed April 12, 2017).

36. Philip B. Crosby, *Quality Is Free: The Art of Making Quality
Certain* (New York: McGraw-Hill, 1979), p. 9–10.

37. Nigel F. Piercy, *Market-Led Strategic Change* (Newton, MA:
Butterworth-Heinemann, 1992), pp. 374–385.

38. El Pinto, "The Salsa Twins Story," http://www.elpinto.
com/salsa-twins-story#.WksSdFQ-dAY (accessed
January 1, 2018); El Pinto, "12 Count Single Serving El
Pinto 4 Oz. Medium Green Chile," https://www.elpinto.
com/index.php?option=com_virtuemart&category_
id=29&flypage=flypage.tpl&manufacturer_
id=1&page=shop.product_details&product_id=97 (accessed
January 1, 2018); El Pinto, "Gov. Susanna Martinez
Announces El Pinto Foods Growing to Create up to 25
New Jobs, $7 Million Investment to Support 20,000 Sq.
Ft. of Manufacturing Facilities," January 31, 2017, https://
www.elpinto.com/el-pinto-blog (accessed January 1, 2018);
El Pinto, "History," http://www.elpinto.com/history#.
WksM0VQ-dAY (accessed January 1, 2018); El Pinto,
"History," https://www.elpinto.com/index.php?page=shop.
cart&option=com_content&Itemid=53&view=category&la
yout=blog&id=34&limitstart=5 (accessed January 1, 2018);
El Pinto, "Biggest Little Tequila Bar in Albuquerque," http://
www.elpinto.com/tequila-bar-albuquerque#.WksNj1Q-
dAY (accessed January 1, 2018); El Pinto, "Looking for
Scorpion Salsa?" https://www.elpinto.com/looking-for-
scorpion-salsa#.WksUYVQ-dAY (accessed January 1,
2018); Steve Ginsberg, "El Pinto's Thomas Twins Seek
Salsa's Superstar Status," *Albuquerque Business First,*
August 23, 2009, https://www.bizjournals.com/albuquerque/
stories/2009/08/24/story7.html (accessed January 1, 2018);
El Pinto, "Family," http://www.elpinto.com/family#.
UflCLPXLQ4k (accessed January 1, 2018); El Pinto, "El
Pinto Fires Up the History Channel," http://www.elpinto.
com/press-releases (accessed January 1, 2018).

39. "ISO Certification and Accreditations," GE Power, 2017,
http://www.geinstruments.com/company/iso-certification-
and-accreditations.html (accessed April 12, 2017).

40. "Compliance Management Systems—Guidelines,"
International Organization for Standardization, 2017, https://
www.iso.org/standard/62342.html (accessed April 12, 2017).

41. "Monitoring and Auditing Global Supply Chains Is a Must,"
Ethisphere, Q3 2011, pp. 38–45.

42. "Global Supplier Code of Conduct," Kellogg's, 2017, http://
www.kelloggcompany.com/en_US/supplier-relations/
transparency-in-supply-chain.html (accessed April 12, 2017).

43. "Employment Opportunities," *Careers in Supply Chain
Management,* www.careersinsupplychain.org/career-outlook/
empopp.asp (accessed March 16, 2016).

44. "Best Jobs in America," *CNN Money,* http://money.cnn.
com/magazines/moneymag/bestjobs/2009/snapshots/48.html
(accessed April 24, 2014); PayScale, "Quality Assurance
Manager Salary," http://www.payscale.com/research/US/
Job=Quality_Assurance_Manager/Salary (accessed March
16, 2016).

45. James Wetherbe, "Principles of Cycle Time Reduction,"
Cycle Time Research, 1995, p. iv.

46. Stan Davis and Christopher Meyer, *Blur: The Speed of
Change in the Connected Economy* (Reading, MA: Addison-
Wesley, 1998), p. 5.

47. International Cocoa Organization, "How Exactly is Cocoa
Harvested?" May 26, 1998, https://www.icco.org/faq/58-
cocoa-harvesting/130-how-exactly-is-cocoa-harvested.html
(accessed June 19, 2018); The Cocoa Exchange, "Sustainable
and Responsibly-Sourced Cocoa," http://www.mytcesite.
com/pws/homeoffice/tabs/sustainable-cocoa.aspx (accessed
June 19, 2018); Ashley Beyer, "The Cocoa Exchange Visits
Fresh Living," *KUTV,* May 31, 2018, http://kutv.com/
features/fresh-living/the-cocoa-exchange-visits-fresh-living
(accessed June 19, 2018); Oliver Nieburg, "'The Cocoa
Exchange': Mars Sets Up E-Commerce Party Program for
Chocolate," *Confectionary News,* May 16, 2017, https://www.
confectionerynews.com/Article/2017/05/15/Cocoa-Exchange-
Mars-sets-up-e-commerce-party-program-for-chocolate
(accessed June 19, 2018); Keith Loria, "Candy Maker Mars
Gets into the Party Business," *Food Dive,* May 10, 2017,
https://www.fooddive.com/news/candy-maker-mars-gets-
into-the-party-business/442160/ (accessed June 19, 2018);
Bernie Pacyniak, "One-on-One: Berta de Pablos-Barbier,
President of Mars Wrigley Confectionary U.S.," *Candy
Industry,* November 1, 2017, https://www.candyindustry.com/
articles/87940-one-on-one-berta-de-pablos-barbier-president-
of-mars-wrigley-confectionery-us (accessed June 19, 2018);
"Cocoa-Caring for the Future of Cocoa," https://www.mars.
com/global/sustainable-in-a-generation/our-approach-to-
sustainability/raw-materials/cocoa (accessed July 10, 2018).

Credits

Creating the Human Resource Advantage

PART **4**

Creating the Human Resource Advantage

9 Motivating the Workforce

©Tony Avelar/Bloomberg via Getty Images

Chapter Outline

Learning Objectives

After reading this chapter, you will be able to:

LO 9-1 Explain why the study of *human relations* is important.

LO 9-2 Summarize early studies that laid the groundwork for understanding employee motivation.

LO 9-3 Compare and contrast the human relations theories of Abraham Maslow and Frederick Herzberg.

LO 9-4 Investigate various theories of motivation, including Theories X, Y, and Z; equity theory; expectancy theory; and goal-setting theory.

LO 9-5 Describe some of the strategies that managers use to motivate employees.

LO 9-6 Critique a business's program for motivating its sales force.

Enter the World of Business

Facebook Knows How to Motivate Employees

Facebook has been ranked as the second-best company to work for in America. This success largely stems from Facebook's leadership, specifically CEO Mark Zuckerberg. Zuckerberg was awarded a 99.3 percent approval rating by nearly 19,000 Facebook employees.

At Facebook, employees are encouraged to take risks and stay innovative. Zuckerberg has been known to meet with entry-level employees to hear their ideas. Engineers are encouraged to consistently create new software builds and can test this software on 10,000 to 50,000 users. More importantly, engineers are not punished if their testing leads to mistakes. A former intern once crashed Facebook when testing a solution for a bug and was later hired by the company. Not only are engineers and employees given free rein on their work, they are also given the ability to choose the team with which they would like to work. A key success factor of Facebook is that the company focuses on an individual's strengths, rather than fixing his or her weaknesses. This encourages employees to find their "best fit" in the company.

Facebook has great employee benefits, including free lunches, laundry services, shuttle buses, flexible work hours, and the ability to work at home when needed. Research suggests that Facebook pulls employees from Apple 11 times more than Apple does from Facebook, and Facebook holds a 15:1 advantage over Google and 30:1 over Microsoft.

The mission of Facebook is to make the world more open and connected, which is demonstrated by its internal focus on an open and innovative work environment. Creativity and flexibility at Facebook lead to happier employees, allowing Facebook to earn its title as one of the best companies to work for in the United States.[1]

Introduction

Because employees do the actual work of the business and influence whether the firm achieves its objectives, most top managers agree that employees are an organization's most valuable resource. To achieve organizational objectives, employees must have the motivation, ability (appropriate knowledge and skills), and tools (proper training and equipment) to perform their jobs. The chapter titled "Managing Human Resources" covers topics such as those listed earlier. This chapter focuses on how to motivate employees.

We examine employees' needs and motivation, managers' views of workers, and several strategies for motivating employees. Managers who understand the needs of their employees can help them reach higher levels of productivity and thus better contribute to the achievement of organizational goals.

Nature of Human Relations

Explain why the study of *human relations* is important.

human relations
the study of the behavior of individuals and groups in organizational settings.

motivation
an inner drive that directs a person's behavior toward goals.

What motivates employees to perform on the job is the focus of **human relations,** the study of the behavior of individuals and groups in organizational settings. In business, human relations involves motivating employees to achieve organizational objectives efficiently and effectively. The field of human relations has become increasingly important over the years as businesses strive to understand how to boost workplace morale, maximize employees' productivity and creativity, and motivate their ever-more-diverse employees to be more effective.

Motivation is an inner drive that directs a person's behavior toward goals. A goal is the satisfaction of some need, and a need is the difference between an actual state and a desired state. Both needs and goals can be motivating. Motivation explains why people behave as they do; similarly, a lack of motivation explains, at times, why people avoid doing what they should do. Motivating employees to do the wrong things or for the wrong reasons can be problematic, however. At Wells Fargo, for instance, employees created at least 3.5 million fake customer accounts to meet unrealistic sales goals. The company paid millions in fines.[2] On the other hand, motivating employees to achieve realistic company objectives can greatly enhance an organization's productivity.

Motivation is important both in business and outside of it. For instance, coaches motivate athletes before major games to increase their chances they will play their best.

©Chris Szagola/CSM/REX/Shutterstock

A person who recognizes or feels a need is motivated to take action to satisfy the need and achieve a goal (Figure 9.1). Consider a person who takes a job as a salesperson. If that person's performance is far below other salespeople's, the person will likely recognize a need to increase sales. To satisfy that need and achieve success, the person may try to acquire new insights from successful salespeople or obtain additional training to improve sales skills. In addition, a sales manager might try different means to motivate the salesperson to work harder and to improve his or her skills. Human relations is concerned with the needs of employees, their goals and how they try to achieve them, and the impact of those needs and goals on job performance.

Effectively motivating employees helps keep them engaged in their work. Engagement involves emotional involvement and commitment. Being

FIGURE 9.1
The Motivation Process

engaged results in carrying out the expectations and obligations of employment. Many employees are actively engaged in their jobs, while others are not. Some employees do the minimum amount of work required to get by, and some employees are completely disengaged.

Motivating employees to stay engaged is a key responsibility of management. For example, to test if his onsite production managers were fully engaged in their jobs, former Van Halen frontman David Lee Roth placed a line in the band's rider asking for a bowl of M&M'S with the brown ones removed. It was a means for the band to test local stage production crews' attention to detail. Because their shows were highly technical, David Lee Roth would demand a complete recheck of everything if he found brown M&M'S in the bowl.[3]

One prominent aspect of human relations is **morale** on an employee's attitude toward his or her job, employer, and colleagues. High morale contributes to high levels of productivity, high returns to stakeholders, and employee loyalty. Conversely, low morale may cause high rates of absenteeism and turnover (when employees quit or are fired and must be replaced by new employees). Some companies go to great lengths to retain employees and value their contributions. Apple is a great example of how to motivate employees with its benefits and corporate culture—demonstrated in an average employee retention of five years. Besides traditional benefits, Apple employees receive 25 percent off Apple products; four weeks paid leave to expectant mothers with 14 weeks after delivery; and its AppleCare College Program, which helps student employees pay tuition. Employees on Glassdoor gave Apple's benefits 4.5 out of 5 stars.[4]

Employees are motivated by their perceptions of extrinsic and intrinsic rewards. An **intrinsic reward** is the personal satisfaction and enjoyment that you feel from attaining a goal. For example, in this class you may feel personal enjoyment in learning how business works and aspire to have a career in business or to operate your own business one day. **Extrinsic rewards** are benefits and/or recognition that you receive from someone else. In this class, your grade is extrinsic recognition of your efforts and success in the class. In business, praise and recognition, pay increases, and bonuses are extrinsic rewards. If you believe that your job provides an opportunity to contribute to society or the environment, then that aspect would represent an intrinsic reward. Both intrinsic and extrinsic rewards contribute to motivation that stimulates employees to do their best in contributing to business goals.

morale
an employee's attitude toward his or her job, employer, and colleagues.

intrinsic rewards
the personal satisfaction and enjoyment felt after attaining a goal.

extrinsic rewards
benefits and/or recognition received from someone else.

TABLE 9.1
How to Retain Good Employees

1. Offer training and mentoring
2. Create a positive organizational culture
3. Build credibility through communication
4. Blend compensation, benefits, and recognition
5. Encourage referrals and don't overlook internal recruiting
6. Give coaching and feedback
7. Provide growth opportunities
8. Create work/life balance and minimize stress
9. Foster trust, respect and confidence in senior leadership

Source: Sarah K. Yazinski, "Strategies for Retaining Employees and Minimizing Turnover," HR.BLR.com, August 3, 2009, https://hr.blr.com/whitepapers/Staffing-Training/Employee-Turnover/Strategies-for-Retaining-Employees-and-Minimizing- (accessed April 25, 2018).

Respect, involvement, appreciation, adequate compensation, promotions, a pleasant work environment, and a positive organizational culture are all morale boosters. Patagonia, for instance, has a positive organizational culture that encourages employees to act ethically and contribute their ideas. Ensuring that employee values are aligned with the company's values is extremely important for Patagonia. The company prides itself on charitable giving and doing what's right, and it hires individuals accordingly. For example, in 2016 employees were the ones to suggest that the company give away all Black Friday sales to environmental organizations. The idea was accepted by CEO Rose Marcario within 30 minutes via text message.[5] Table 9.1 lists some ways to retain and motivate good employees. Many companies offer a diverse array of benefits designed to improve the quality of employees' lives and increase their morale and satisfaction. Some of the "best companies to work for" offer onsite day care, concierge services (for example, dry cleaning, shoe repair, prescription renewal), domestic partner benefits to same-sex couples, and fully paid sabbaticals.

Historical Perspectives on Employee Motivation

LO 9-2

Summarize early studies that laid the groundwork for understanding employee motivation.

Throughout the 20th century, researchers have conducted numerous studies to try to identify ways to motivate workers and increase productivity. From these studies have come theories that have been applied to workers with varying degrees of success. A brief discussion of two of these theories—the classical theory of motivation and the Hawthorne studies—provides a background for understanding the present state of human relations.

Classical Theory of Motivation

The birth of the study of human relations can be traced to time and motion studies conducted at the turn of the century by Frederick W. Taylor and Frank and Lillian Gilbreth. Their studies analyzed how workers perform specific work tasks in an effort to improve the employees' productivity. These efforts led to the application of scientific principles to management.

DID YOU KNOW? Absenteeism costs U.S. employers about $225.8 billion annually.[6]

King Arthur: Baked to Perfection

You might think flour is boring, but at King Arthur Flour, employees love the stuff! Founded the year after George Washington was elected president, the company has grown slowly from five employees in 1790 to more than 300 today. What makes these employees so passionate about flour? They own the business. King Arthur Flour is 100 percent employee owned.

The company practices an open-book form of management in which employee-owners are provided access to financial information. Employees are openly encouraged to provide input. For instance, when King Arthur Flour experienced a difficult month due to rising flour prices and supplier cancellations, employees were informed about the situation and asked to collaborate on how to address it. Together, the employees came up with plans to help the company bounce back.

In 1990, King Arthur was essentially a mail order business. Today it sells to bakeries and grocery stores, runs a retail store in Vermont, and excels in public relations and marketing. Employees throughout the business are honored as valuable members of a team. All employees are offered baking/ cooking workshops, free products, and store discounts. The company has won awards based on its structure: number four on Best Places to Work in Vermont in the large employer category and part of *The Wall Street Journal's* Top Small Workplaces.

By owning the business, employees are motivated to produce premium products and provide excellent service. If an employee has a suggestion for an improved product or process, he or she is encouraged to share it with management. King Arthur Flour is a place where those who love cooking, baking, and sustainability can come together in partnership to run their own company.[7]

Critical Thinking Questions

1. What are some ways that King Arthur Flour motivates employees?
2. Why do you think making the company 100 percent employee owned encourages employees to ensure King Arthur Flour is successful?
3. How does King Arthur's participative culture allow the company to handle challenging situations and develop innovative new products or processes?

According to the **classical theory of motivation**, money is the sole motivator for workers. Taylor suggested that workers who were paid more would produce more, an idea that would benefit both companies and workers. To improve productivity, Taylor thought that managers should break down each job into its component tasks (specialization), determine the best way to perform each task, and specify the output to be achieved by a worker performing the task. Taylor also believed that incentives would motivate employees to be more productive. Thus, he suggested that managers link workers' pay directly to their output. He developed the piece-rate system, under which employees were paid a certain amount for each unit they produced; those who exceeded their quota were paid a higher rate per unit for all the units they produced.

We can still see Taylor's ideas in practice today in the use of financial incentives for productivity. Moreover, companies are increasingly striving to relate pay to performance at both the hourly and managerial level. Incentive planners choose an individual incentive to motivate and reward their employees. In contrast, team incentives are used to generate partnership and collaboration to accomplish organizational goals. Boeing develops sales teams for most of its products, including commercial airplanes. The team dedicated to each product shares in the sales incentive program.

More and more corporations are tying pay to performance in order to motivate—even up to the CEO level. The topic of executive pay has become controversial in recent years, and many corporate boards of directors have taken steps to link executive compensation more closely to corporate performance. Despite changes in linking pay to performance, there are many CEOs who receive extremely large compensation packages. Mark Hurd, co-CEO of Oracle, received an annual compensation package of $41.1 million and is one of the highest paid CEOs.[8]

classical theory of motivation
theory suggesting that money is the sole motivator for workers.

Some companies let people bring their pets to work as an added incentive to make the workplace feel more friendly.

©Image Source Plus/Alamy Stock Photo

Like most managers of the early 20th century, Taylor believed that satisfactory pay and job security would motivate employees to work hard. However, later studies showed that other factors are also important in motivating workers.

The Hawthorne Studies

Elton Mayo and a team of researchers from Harvard University wanted to determine what physical conditions in the workplace—such as light and noise levels—would stimulate employees to be most productive. From 1924 to 1932, they studied a group of workers at the Hawthorne Works Plant of the Western Electric Company and measured their productivity under various physical conditions.

What the researchers discovered was quite unexpected and very puzzling: Productivity increased regardless of the physical conditions. This phenomenon has been labeled the Hawthorne effect. When questioned about their behavior, the employees expressed satisfaction because their co-workers in the experiments were friendly and, more importantly, because their supervisors had asked for their help and cooperation in the study. In other words, they were responding to the attention they received, not the changing physical work conditions. The researchers concluded that social and psychological factors could significantly affect productivity and morale. The United Services Automobile Association (USAA) has a built-in psychological factor that influences employee morale. The work of the financial services company serves military and veteran families, which enlivens employees. This shows how important it is for employees to feel like their work matters.

The Hawthorne experiments marked the beginning of a concern for human relations in the workplace. They revealed that human factors do influence workers' behavior and that managers who understand the needs, beliefs, and expectations of people have the greatest success in motivating their workers.

Theories of Employee Motivation

LO 9-3

Compare and contrast the human relations theories of Abraham Maslow and Frederick Herzberg.

Maslow's hierarchy
a theory that arranges the five basic needs of people—physiological, security, social, esteem, and self-actualization—into the order in which people strive to satisfy them.

The research of Taylor, Mayo, and many others has led to the development of a number of theories that attempt to describe what motivates employees to perform. In this section, we discuss some of the most important of these theories. The successful implementation of ideas based on these theories will vary, of course, depending on the company, its management, and its employees. It should be noted, too, that what worked in the past may no longer work today. Good managers must have the ability to adapt their ideas to an ever-changing, diverse group of employees.

Maslow's Hierarchy of Needs

Psychologist Abraham Maslow theorized that people have five basic needs: physiological, security, social, esteem, and self-actualization. **Maslow's hierarchy** arranges these needs into the order in which people strive to satisfy them (Figure 9.2).

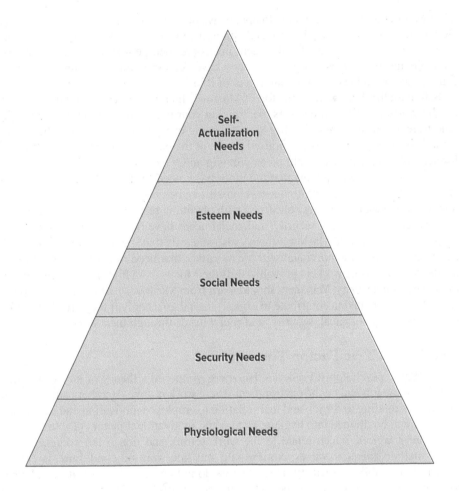

FIGURE 9.2
Maslow's Hierarchy of Needs

Source: Adapted from Abraham H. Maslow, "A Theory of Human Motivation," Psychology Review 50 (1943), pp. 370–396. American Psychology Association.

Physiological needs, the most basic and first needs to be satisfied, are the essentials for living—water, food, shelter, and clothing. According to Maslow, humans devote all their efforts to satisfying physiological needs until they are met. Only when these needs are met can people focus their attention on satisfying the next level of needs—security.

Security needs relate to protecting yourself from physical and economic harm. Actions that may be taken to achieve security include reporting a dangerous workplace condition to management, maintaining safety equipment, and purchasing insurance with income protection in the event you become unable to work. Once security needs have been satisfied, people may strive for social goals.

Social needs are the need for love, companionship, and friendship—the desire for acceptance by others. To fulfill social needs, a person may try many things: making friends with a co-worker, joining a group, volunteering at a hospital, throwing a party, and so on. Once their social needs have been satisfied, people attempt to satisfy their need for esteem.

Esteem needs relate to respect—both self-respect and respect from others. One aspect of esteem needs is competition—the need to feel that you can do something better than anyone else. Competition often motivates people to increase their productivity. Esteem needs are not as easily satisfied as the needs at lower levels in

physiological needs
the most basic human needs to be satisfied—water, food, shelter, and clothing.

security needs
the need to protect oneself from physical and economic harm.

social needs
the need for love, companionship, and friendship—the desire for acceptance by others.

esteem needs
the need for respect—both self-respect and respect from others.

Maslow's hierarchy because they do not always provide tangible evidence of success. However, these needs can be realized through rewards and increased involvement in organizational activities. Until esteem needs are met, people focus their attention on achieving respect. When they feel they have achieved some measure of respect, self-actualization becomes the major goal of life.

Self-actualization needs, at the top of Maslow's hierarchy, mean being the best you can be. Self-actualization involves maximizing your potential. Self-actualized people tend to feel that they are living life to its fullest in every way. For J.K. Rowling, self-actualization might mean being praised as one of the best fiction writers in the world; for actress Viola Davis, it might mean winning an Oscar.

Maslow's theory maintains that the more basic needs at the bottom of the hierarchy must be satisfied before higher-level goals can be pursued. Thus, people who are hungry and homeless are not concerned with obtaining respect from their colleagues. Only when physiological, security, and social needs have been more or less satisfied do people seek esteem. Maslow's theory also suggests that if a low-level need is suddenly reactivated, the individual will try to satisfy that need rather than higher-level needs. Many laid-off workers probably shift their focus from high-level esteem needs to the need for security. Managers should learn from Maslow's hierarchy that employees will be motivated to contribute to organizational goals only if they are able to first satisfy their physiological, security, and social needs through their work.

Herzberg's Two-Factor Theory

In the 1950s, psychologist Frederick Herzberg proposed a theory of motivation that focuses on the job and on the environment where work is done. Herzberg studied various factors relating to the job and their relation to employee motivation and concluded that they can be divided into hygiene factors and motivational factors (Table 9.2).[9]

Hygiene factors, which relate to the work setting and not to the content of the work, include adequate wages, comfortable and safe working conditions, fair company policies, and job security. These factors do not necessarily motivate employees to excel, but their absence may be a potential source of dissatisfaction and high turnover. Employee safety and comfort are clearly hygiene factors.

Many people feel that a good salary is one of the most important job factors, even more important than job security and the chance to use one's mind and abilities. Salary and security, two of the hygiene factors identified by Herzberg, make it possible for employees to satisfy the physiological and security needs identified by Maslow. However, the presence of hygiene factors is unlikely to motivate employees to work harder. For example, many people do not feel motivated to pursue a career as

self-actualization needs the need to be the best one can be; at the top of Maslow's hierarchy.

hygiene factors aspects of Herzberg's theory of motivation that focus on the work setting and not the content of the work; these aspects include adequate wages, comfortable and safe working conditions, fair company policies, and job security.

TABLE 9.2
Herzberg's Hygiene and Motivational Factors

Hygiene Factors	Motivational Factors
Company policies	Achievement
Supervision	Recognition
Working conditions	Work itself
Relationships with peers, supervisors, and subordinates	Responsibility
Salary	Advancement
Security	Personal growth

a gastroenterologist (doctors who specialize in the digestive system). The median annual salary of the job is more than $350,000; however, patients do not look forward to the visits and the process is fairly routine.[10]

Motivational factors, which relate to the content of the work itself, include achievement, recognition, involvement, responsibility, and advancement. The absence of motivational factors may not result in dissatisfaction, but their presence is likely to motivate employees to excel. Many companies are beginning to employ methods to give employees more responsibility and control and to involve them more in their work, which serves to motivate them to higher levels of productivity and quality. Hotels are adopting more employee-centric processes in order to better their offerings. Service businesses, such as hotels and airlines, recognize the benefit of happy employees—they are not only 12 percent more productive, but they work to generate happy customers. Many companies who value employee happiness consider the following benefits, where possible: implementing flexible hours and the ability to work from home (the average commute is over 25 minutes), making the office "pet friendly," maintaining a break room with recreational items (ping pong, video games, etc.), and celebrating successes and special occasions.[11]

Herzberg's motivational factors and Maslow's esteem and self-actualization needs are similar. Workers' low-level needs (physiological and security) have largely been satisfied by minimum-wage laws and occupational-safety standards set by various government agencies and are therefore not motivators. Consequently, to improve productivity, management should focus on satisfying workers' higher-level needs (motivational factors) by providing opportunities for achievement, involvement, and advancement and by recognizing good performance.

Google's employee-friendly offices feature elements like basketball courts, pinball machines, and photo booths to foster creativity and make work more enjoyable.

©Daniel Brenner/Bloomberg via Getty Images

motivational factors aspects of Herzberg's theory of motivation that focus on the content of the work itself; these aspects include achievement, recognition, involvement, responsibility, and advancement.

McGregor's Theory X and Theory Y

In *The Human Side of Enterprise,* Douglas McGregor related Maslow's ideas about personal needs to management. McGregor contrasted two views of management—the traditional view, which he called Theory X, and a humanistic view, which he called Theory Y.

According to McGregor, managers adopting **Theory X** assume that workers generally dislike work and must be forced to do their jobs. They believe that the following statements are true of workers:

1. The average person naturally dislikes work and will avoid it when possible.
2. Most workers must be coerced, controlled, directed, or threatened with punishment to get them to work toward the achievement of organizational objectives.
3. The average worker prefers to be directed and to avoid responsibility, has relatively little ambition, and wants security.[12]

Managers who subscribe to the Theory X view maintain tight control over workers, provide almost constant supervision, try to motivate through fear, and make decisions in an autocratic fashion, eliciting little or no input from their subordinates. The Theory X

LO 9-4

Investigate various theories of motivation, including Theories X, Y, and Z; equity theory; expectancy theory; and goal-setting theory.

Theory X
McGregor's traditional view of management whereby it is assumed that workers generally dislike work and must be forced to do their jobs.

style of management focuses on physiological and security needs and virtually ignores the higher needs discussed by Maslow. Computer Science Corporation seemed to adopt the Theory X perspective when it initiated an employee ranking system that ranked 40 percent of employees as below expectations. Employees felt that the system was unfair and the company did not have a good work/life balance. The outcry was so intense that Computer Science Corporation eventually relaxed some of its ratings criteria.[13]

The Theory X view of management does not take into account people's needs for companionship, esteem, and personal growth, whereas Theory Y, the contrasting view of management, does. Managers subscribing to the **Theory Y** view assume that workers like to work and that under proper conditions employees will seek out responsibility in an attempt to satisfy their social, esteem, and self-actualization needs. McGregor describes the assumptions behind Theory Y in the following way:

Theory Y
McGregor's humanistic view of management whereby it is assumed that workers like to work and that under proper conditions employees will seek out responsibility in an attempt to satisfy their social, esteem, and self-actualization needs.

1. The expenditure of physical and mental effort in work is as natural as play or rest.
2. People will exercise self-direction and self-control to achieve objectives to which they are committed.
3. People will commit to objectives when they realize that the achievement of those goals will bring them personal reward.
4. The average person will accept and seek responsibility.
5. Imagination, ingenuity, and creativity can help solve organizational problems, but most organizations do not make adequate use of these characteristics in their employees.
6. Organizations today do not make full use of workers' intellectual potential.[14]

Obviously, managers subscribing to the Theory Y philosophy have a management style very different from managers subscribing to the Theory X philosophy. Theory Y managers maintain less control and supervision; do not use fear as the primary motivator; and are more democratic in decision making, allowing subordinates to participate in the process. Theory Y managers address the high-level needs in Maslow's hierarchy as well as physiological and security needs. For example, H&R Block is empowering its employees to have "extra help" in interpreting the 74,000 pages of tax code. Employees can use IBM's Watson to assist them in analyzing returns and maximizing customer refunds. This empowerment gives employees the ability to offer more personalized service and improve customer satisfaction in return.[15] Today, Theory Y enjoys widespread support and may have displaced Theory X.

Theory Z

Theory Z
a management philosophy that stresses employee participation in all aspects of company decision making.

Theory Z is a management philosophy that stresses employee participation in all aspects of company decision making. It was first described by William Ouchi in his book, *Theory Z—How American Business Can Meet the Japanese Challenge.*[16] Theory Z incorporates many elements associated with the Japanese approach to management, such as trust and intimacy, but Japanese ideas have been adapted for use in the United States. In a Theory Z organization, managers and workers share responsibilities; the management style is participative; and employment is long term and, often, lifelong. Japan has faced a significant period of slowing economic progress and competition from China and other Asian nations. This has led to experts questioning Theory Z, particularly at firms such as Sony and Toyota. On the other hand, Theory Z results in employees feeling organizational ownership. Research has found that such feelings of ownership may produce positive attitudinal and behavioral effects for employees.[17]

Entrepreneurship in Action

Namasté Solar Embraces Employee Democracy Model of Management

Namasté Solar
Founders: Blake Jones, Wes Kennedy, and Ray Tuomey
Founded: 2005, in Boulder, Colorado
Success: Namasté Solar has motivated employees to help the company achieve more than $20 million in annual revenues and maintain a 50 percent compound annual growth rate.

At Namasté Solar, CEO Blake Jones's ideas can easily be shot down by employees. And this is the way he designed it. Jones became disillusioned with power inequality in business. So when he got together with his co-founders to construct a solar energy technology firm, Jones wanted to adopt a model in which employees and managers were equal.

To make this into a reality, Namasté Solar gives employees the right to purchase stock in the company, has an equal pay scale for all employees, allows six-week paid vacations, and embraces what it calls FOH—frank, open, honest communication—to encourage collaboration and discourage gossip. Each employee at Namasté Solar is eligible to purchase one share of voting stock after a 12-month candidacy period, with common stock only available for employees (nonvoting preferred stock has been offered to external investors). Its flat structure means that decisions are usually arrived at through employee consensus. Although board members tackle some major decisions, employees elect the board members for one- to two-year terms. At Namasté Solar, employees truly have a voice in all business operations.[18]

Critical Thinking Questions:

1. How have employee satisfaction and accountability contributed to the success of Namasté Solar?
2. According to Herzberg's theory, is giving employees the ability to purchase stock a hygiene factor or a motivation factor? Why?
3. How does Namasté Solar use employee empowerment to encourage motivation?

In a Theory Y organization, managers focus on assumptions about the nature of the worker. The two theories can be seen as complementary. Table 9.3 compares Theory X, Theory Y, and Theory Z.

Equity Theory

According to **equity theory,** how much people are willing to contribute to an organization depends on their assessment of the fairness, or equity, of the rewards they will receive in exchange. In a fair situation, workers receive rewards proportional

equity theory
an assumption that how much people are willing to contribute to an organization depends on their assessment of the fairness, or equity, of the rewards they will receive in exchange.

TABLE 9.3 Comparisons of Theories X, Y, and Z

	Theory X	Theory Y	Theory Z
Countries that use this style	China	United States	Japan
Philosophy	Tight control over workers	Assume workers will seek out responsibility and satisfy social needs	Employee participation in all aspects of company decision making
Job description	Considerable specialization	Less control and supervision; address higher levels of Maslow's hierarchy	Trust and intimacy with workers sharing responsibilities
Control	Tight control	Commitment to objectives with self-direction	Relaxed but required expectations
Worker welfare	Limited concern	Democratic	Commitment to worker's total lives
Responsibility	Managerial	Collaborative	Participative

to the contribution they make to the organization. However, in practice, equity is a subjective notion. Workers regularly develop personal input-output ratios by taking stock of their contributions (inputs) to the organization in time, effort, skills, and experience and assessing the rewards (outputs) offered by the organization in pay, benefits, recognition, and promotions. Workers compare their ratios to the input-output ratio of some other person's "comparison other," who may be a coworker, a friend working in another organization, or an "average" of several people working in the organization. If the two ratios are close, workers will feel that they are being treated equitably.

Let's say you have a high-school education and earn $25,000 a year. When you compare your input-output ratio with that of a co-worker who has a college degree and makes $35,000 a year, you will probably feel that you are being paid fairly. However, if you perceive that your personal input-output ratio is lower than that of your college-educated co-worker, you may feel that you are being treated unfairly and be motivated to seek change. Or if you learn that your co-worker who makes $35,000 has only a high-school diploma, you may feel cheated by your employer. To achieve equity, you could try to increase your outputs by asking for a raise or promotion. You could also try to have your co-worker's inputs increased or his or her outputs decreased. Failing to achieve equity, you may be motivated to look for a job at a different company.

Equity theory might explain why many consumers are upset about CEO compensation. Although the job of the CEO can be incredibly stressful, the fact that they take home millions in compensation, bonuses, and stock options has been questioned. The high unemployment rate coupled with the misconduct that occurred at some large corporations prior to the recession contributed largely to consumer frustration with executive compensation packages. To counter this perception of pay inequality, several corporations have now begun to tie CEO compensation with company performance. If the company performs poorly for the year, then firms such as Goldman Sachs will cut bonuses and other compensation.[19] While lower compensation rates might appease the general public, some companies are worried that lower pay might deter talented individuals from wanting to assume the position of CEO at their firms.

Managers should be transparent with employees about opportunities for advancement. According to expectancy theory, your motivation depends not only on how much you want something, but also on how likely you are to get it.

©Monkey Business Images/Shutterstock

Because almost all the issues involved in equity theory are subjective, they can be problematic. Author David Callahan has argued that feelings of inequity may underlie some unethical or illegal behavior in business.[20] The National Retail Federation reports that shoplifting and employee theft was approximately 1.44 percent of retail sales or nearly $49 billion a year. Some employees may take company resources to restore what they perceive to be an inequity (inadequate pay, working hours, or other deficient benefits).[21] The FBI notes that employee theft is one of the fastest-growing crimes in the United States.[22] Callahan believes that employees who do not feel they are being treated equitably may be motivated to equalize the situation by lying, cheating, or otherwise "improving" their pay, perhaps by stealing.[23] Managers should try to avoid equity problems by ensuring that rewards are distributed on the basis of performance and that all employees clearly understand the basis for their pay and benefits.

Expectancy Theory

Psychologist Victor Vroom described **expectancy theory,** which states that motivation depends not only on how much a person wants something, but also on the person's perception of how likely he or she is to get it. A person who wants something and has reason to be optimistic will be strongly motivated. For example, say you really want a promotion. And let's say because you have taken some night classes to improve your skills, and moreover, have just made a large, significant sale, you feel confident that you are qualified and able to handle the new position. Therefore, you are motivated to try to get the promotion. In contrast, if you do not believe you are likely to get what you want, you may not be motivated to try to get it, even though you really want it.

expectancy theory
the assumption that motivation depends not only on how much a person wants something but also on how likely he or she is to get it.

Goal-Setting Theory

Goal-setting theory refers to the impact that setting goals has on performance. According to this philosophy, goals act as motivators to focus employee efforts on achieving certain performance outcomes. Setting goals can positively affect performance because goals help employees direct their efforts and attention toward the outcome, mobilize their efforts, develop consistent behavior patterns, and create strategies to obtain desired outcomes.[24] When Cinnabon introduced two new hot chocolates, Ghirardelli and Cinnamon Roll, it had specific goals for sales. To support these sales goals and generate awareness of the new drinks, the company might have employees at the counter suggest these new options or offer other forms of promotional support.

In 1954, Peter Drucker introduced the term *management by objectives (MBO)* that has since become important to goal-setting theory. MBO refers to the need to develop goals that both managers and employees can understand and agree upon.[25] This requires managers to work with employees to set personal objectives that will be used to further organizational objectives. By linking managerial objectives with personal objectives, employees often feel a greater sense of commitment toward achieving organizational goals. Hewlett-Packard was an early adopter of MBO as a management style.[26]

goal-setting theory
refers to the impact that setting goals has on performance.

LO 9-5

Describe some of the strategies that managers use to motivate employees.

behavior modification
changing behavior and encouraging appropriate actions by relating the consequences of behavior to the behavior itself.

Strategies for Motivating Employees

Based on the various theories that attempt to explain what motivates employees, businesses have developed several strategies for motivating their employees and boosting morale and productivity. Some of these techniques include behavior modification and job design, as well as the already described employee involvement programs and work teams.

Behavior Modification

Behavior modification involves changing behavior and encouraging appropriate actions by relating the consequences of behavior to the behavior itself. Behavior modification is the most widely discussed

Setting goals can have a positive impact on employee performance.
©Zoonar GmbH/Alamy Stock Photo

PricewaterhouseCoopers promotes job rotation to allow employees to learn new skills.

©ricochet64/Shutterstock

reinforcement theory
the theory that behavior can be strengthened or weakened through the use of rewards and punishments.

job rotation
movement of employees from one job to another in an effort to relieve the boredom often associated with job specialization.

application of **reinforcement theory,** the theory that behavior can be strengthened or weakened through the use of rewards and punishments. The concept of behavior modification was developed by psychologist B. F. Skinner. Skinner found that behavior that is rewarded will tend to be repeated, while behavior that is punished will tend to be eliminated. For example, employees who know that they will receive a bonus such as an expensive restaurant meal for making a sale over $2,000 may be more motivated to make sales. Workers who know they will be punished for being tardy are likely to make a greater effort to get to work on time.

However, the two strategies may not be equally effective. Punishing unacceptable behavior may provide quick results but may lead to undesirable long-term side effects, such as employee dissatisfaction and increased turnover. In general, rewarding appropriate behavior is a more effective way to modify behavior.

Job Design

Herzberg identified the job itself as a motivational factor. Managers have several strategies that they can use to design jobs to help improve employee motivation. These include job rotation, job enlargement, job enrichment, and flexible scheduling strategies.

Job Rotation.   **Job rotation** allows employees to move from one job to another in an effort to relieve the boredom that is often associated with job specialization. Businesses often turn to specialization in hopes of increasing productivity, but there is a negative side effect to this type of job design: Employees become bored and dissatisfied, and productivity declines. Job rotation reduces this boredom by allowing workers to undertake a greater variety of tasks and by giving them the opportunity to learn new skills. With job rotation, an employee spends a specified amount of time performing one job and then moves on to another, different job. The worker eventually returns to the initial job and begins the cycle again. Table 9.4 offers additional benefits of job rotation.

Job rotation is a good idea, but it has one major drawback. Because employees may eventually become bored with all the jobs in the cycle, job rotation does not totally

TABLE 9.4
Benefits of Job Rotation

1. Exposure to a diversity of viewpoints
2. Motivating ongoing lifelong learning
3. Preparing for promotion and leadership roles
4. Building specific skills and abilities
5. Supporting recruitment efforts
6. Boosting overall productivity
7. Retaining employees

Source: Tim Hird, "The Lasting Benefits of Job Rotation," Treasury & Risk, January 24, 2017, https://www.treasuryandrisk.com/sites/treasuryandrisk/2017/01/24/the-lasting-benefits-of-job-rotation/ (accessed April 30, 2018).

eliminate the problem of boredom. Job rotation is extremely useful, however, in situations where a person is being trained for a position that requires an understanding of various units in an organization. Many businesses and departments understand the benefits of job rotation. Roughly 44 percent of CFOs said their companies promote job rotation. Finance employees can end up in accounting operations, finance, internal audit, compliance, and tax—and can benefit from the diverse exposure and experience.[27] Many executive training programs require trainees to spend time learning a variety of specialized jobs. Job rotation is also used to cross-train today's self-directed work teams.

Job Enlargement. Job enlargement adds more tasks to a job instead of treating each task as separate. Like job rotation, job enlargement was developed to overcome the boredom associated with specialization. The rationale behind this strategy is that jobs are more satisfying as the number of tasks performed by an individual increases. Employees sometimes enlarge, or craft, their jobs by noticing what needs to be done and then changing tasks and relationship boundaries to adjust. Individual orientation and motivation shape opportunities to craft new jobs and job relationships. Job enlargement strategies have been more successful in increasing job satisfaction than have job rotation strategies. IBM, AT&T, and Maytag are among the many companies that have used job enlargement to motivate employees.

job enlargement
the addition of more tasks to a job instead of treating each task as separate.

Job Enrichment. Job enrichment incorporates motivational factors such as opportunity for achievement, recognition, responsibility, and advancement into a job. It gives workers not only more tasks within the job, but more control and authority over the job. Job enrichment programs enhance a worker's feeling of responsibility and provide opportunities for growth and advancement when the worker is able to take on the more challenging tasks. Hyatt Hotels Corporation and Clif Bar use job enrichment to improve the quality of work life for their employees. The potential benefits of job enrichment are great, but it requires careful planning and execution.

job enrichment
the incorporation of motivational factors, such as opportunity for achievement, recognition, responsibility, and advancement, into a job.

Flexible Scheduling Strategies. Many U.S. workers work a traditional 40-hour workweek consisting of five 8-hour days with fixed starting and ending times. Facing problems of poor morale and high absenteeism as well as a diverse workforce with changing needs, many managers have turned to flexible scheduling strategies such as flextime, compressed workweeks, job sharing, part-time work, and telecommuting.

Flextime is a program that allows employees to choose their starting and ending times, as long as they are at work during a specified core period (Figure 9.3). FlexJobs defines flexible jobs as those that are professional, have a flexible schedule, include

flextime
a program that allows employees to choose their starting and ending times, provided that they are at work during a specified core period.

FIGURE 9.3
Flextime, Showing Core and Flexible Hours

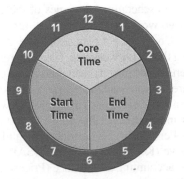

Patagonia Attracts and Empowers Passionate Employees

What type of organization lets employees take off during the day to go surfing? The answer is Patagonia, an outdoor clothing and gear company. When founder Yvon Chouinard first developed the company, he was not interested in pursuing profits as the firm's main goal. Instead, he wanted to improve the planet. New employees are hired based in large part on their passion for the firm's goals.

Chouinard decided that the company would produce only products of the highest quality manufactured in the most responsible way. He selected the following mission statement: "Build the best product, cause no unnecessary harm, use business to inspire and implement solutions to the environmental crisis." Patagonia ensures employees understand and support these values, using a mix of in-person training, video training, and online instruction to help employees learn about company expectations.

Patagonia creates excitement for the company's mission with a fun, informal work environment for employees. It instituted a flextime policy that allows employees to go surfing during the day. Solar panels, Tibetan prayer flags, and sheds full of rescued or recuperating owls and hawks are all a part of corporate headquarters. Patagonia also developed an internship program that enables employees to leave the company for two months to volunteer at the environmental organization of their choice. At Patagonia, employees are viewed as important partners toward advancing environmental preservation.[28]

Critical Thinking Questions

1. How does Patagonia use training and a fun, informal work environment to encourage employees to support Patagonia's mission?
2. Why do you think Yvon Chouinard decided to make his company so focused on employee satisfaction?
3. Do you think that an informal work environment that provides employees freedom to take off during the day will be harmful or helpful in the long run? Why?

---

a telecommuting component, and are part-time or freelance.[29] It does not reduce the total number of hours that employees work; instead, it gives employees more flexibility in choosing which hours they work. A firm may specify that employees must be present from 10:00 a.m. to 3:00 p.m. One employee may choose to come in at 7:00 a.m. and leave at the end of the core time, perhaps to attend classes at a nearby college after work. Another employee, a mother who lives in the suburbs, may come in at 9:00 a.m. in order to have time to drop off her children at a day care center and commute by public transportation to her job. Flextime provides many benefits, including improved ability to recruit and retain workers who wish to balance work and home life. Customers can be better served by allowing customer service over longer hours, workstations and facilities can be better utilized by staggering employee use, and rush hour traffic may be reduced. In addition, flexible schedules have been associated with an increase in job satisfaction on the part of employees. More flexible schedules are associated with higher job satisfaction, less burnout, and better work-to-family balance.[30]

Related to flextime are the scheduling strategies of the compressed workweek and job sharing. The **compressed workweek** is a four-day (or shorter) period in which an employee works 40 hours. Under such a plan, employees typically work 10 hours per day for four days and have a three-day weekend. The compressed workweek reduces the company's operating expenses because its actual hours of operation are reduced. It is also sometimes used by parents who want to have more days off to spend with their families. Millennials and Generation Z are groups of employees who value flexibility in their work schedules. Industries and companies that are the largest flextime employers include medical and health (Kaiser Permanente), education and training (Kaplan), computer and IT (VMWare), administrative (Kelly Services), sales (SAP, AT&T), customer services (Williams & Sonoma), and accounting and finance (PricewaterhouseCoopers).[31]

Job sharing occurs when two people do one job. One person may work from 8:00 a.m. to 12:30 p.m.; the second person comes in at 12:30 p.m. and works

compressed workweek
a four-day (or shorter) period during which an employee works 40 hours.

job sharing
performance of one full-time job by two people on part-time hours.

until 5:00 p.m. Job sharing gives both people the opportunity to work as well as time to fulfill other obligations, such as parenting or school. With job sharing, the company has the benefit of the skills of two people for one job, often at a lower total cost for salaries and benefits than one person working eight hours a day would be paid.

Two other flexible scheduling strategies attaining wider use include allowing full-time workers to work part-time for a certain period and allowing workers to work at home either full- or part-time. Employees at some firms may be permitted to work part-time for several months in order to care for a new baby or an elderly parent or just to slow down for a little while to "recharge their batteries." When the employees return to full-time work, they are usually given a position comparable to their original full-time position. Other firms

Working remotely is becoming increasingly common. Telecommuting, job sharing, and flextime can be beneficial for employees who cannot work normal work hours.

©nd3000/Shutterstock

are allowing employees to telecommute or telework (work at home a few days of the week), staying connected via computers and telephones. Most telecommuters tend to combine going into the office with working from home. At Dell, for instance, an estimated 25 percent of the workforce work from home either full-time or a few days a week.[32]

Although many employees ask for the option of working at home to ease the responsibilities of caring for family members, some have discovered that they are more productive at home without the distractions of the workplace. Workers like telecommuting because they can improve their overall productivity through a variety of means, besides just saving on driving time. They do not have office distractions such as people coming into their workspace to talk and office politics to detract from their productivity.[33] Other employees, however, have discovered that they are not suited for working at home. For telecommuting to work, it must be a feasible alternative and must not create significant costs for the company.[34] Work-at-home programs can also help reduce overhead costs for businesses. For example, some companies used to maintain a surplus of office space but have reduced the surplus through employee telecommuting, "hoteling" (being assigned to a desk through a reservation system), and "hot-desking" (several people using the same desk but at different times).

Companies are turning to flexible work schedules to provide more options to employees who are trying to juggle their work duties with other responsibilities and needs. Preliminary results indicate that flexible scheduling plans increase job satisfaction, which, in turn, leads to increases in productivity. Some recent research, however, has indicated there are potential problems with telecommuting. Some managers are reluctant to adopt the practice because the pace of change in today's workplace is faster than ever, and telecommuters may be left behind or actually cause managers more work in helping them stay abreast of changes. Some employers also worry that telecommuting workers create a security risk by creating more opportunities for computer hackers or equipment thieves. Some employees have found that working outside the office may hurt career advancement opportunities, and some report that instead of helping them balance work and family responsibilities, telecommuting increases the strain by blurring the barriers between the office and home. Co-workers call at all hours, and telecommuters are apt to continue to work when they are not supposed to (after regular business hours or during vacation time).

Businesses have come up with different ways to motivate employees, including rewards such as trophies and plaques to show the company's appreciation.

©Caiaimage/Paul Bradbury/Getty Images

Importance of Motivational Strategies

Motivation is more than a tool that managers can use to foster employee loyalty and boost productivity. It is a process that affects all the relationships within an organization and influences many areas such as pay, promotion, job design, training opportunities, and reporting relationships. Employees are motivated by the nature of the relationships they have with their supervisors, by the nature of their jobs, and by characteristics of the organization. Table 9.5 shows companies with excellent motivational strategies, along with the types of strategies they use to motivate employees. Even the economic environment can change an employee's motivation. In a slow growth or recession economy, sales can flatten or decrease and morale can drop because of the need to cut jobs. The firm may have to work harder to keep good employees and to motivate all employees to work to overcome obstacles. In good economic times, employees may be more demanding and be on the lookout for better opportunities. New rewards or incentives may help motivate workers in such economies. Motivation tools, then, must be varied as well. Managers can further nurture motivation by being honest, supportive, empathic, accessible, fair, and open. Motivating employees to increase satisfaction and productivity is an important concern for organizations seeking to remain competitive in the global marketplace.

TABLE 9.5
Companies with Excellent Motivational Strategies

Company	Motivational Strategies
3M	Gives employees 15–20 percent of their time to pursue own projects
Google	Perks include a massage every other week, free gourmet lunches, tuition reimbursement, a volleyball court, and time to work on own projects
Whole Foods	Employees receive 20 percent discounts on company products, the opportunity to gain stock options, and the ability to make major decisions in small teams
Patagonia	Provides areas for yoga and aerobics, in-house child care services, organic food in its café, and opportunities to go surfing during the day
The Container Store	Provides more than 260 hours of employee training and hosts "We Love Our Employees" Day
Southwest Airlines	Gives employees permission to interact with passengers as they see fit, provides free or discounted flights, and hosts the "Adopt-a-Pilot" program to connect pilots with students across the nation
Nike	Offers tuition assistance, product discounts, onsite fitness centers, and the ability for employees to give insights on how to improve the firm
Apple	Creates a fast paced, innovative work environment where employees are encouraged to debate ideas
Marriott International	Offers discounts at hotels across the world as well as free hotel stays and travel opportunities for employees with exceptional service
Zappos	Creates a fun, zany work environment for employees and empowers them to take as much time as needed to answer customer concerns

So You Think You May Be Good at Motivating a Workforce

If you are good at mediation, smoothing conflict, and have a good understanding of motivation and human relations theories, then you might be a good leader, human resource manager, or training expert. Most organizations, especially as they grow, will need to implement human relations programs. These are necessary to teach employees about sensitivity to other cultures, religions, and beliefs, as well as for teaching the workforce about the organization so that they understand how they fit in the larger picture. Employees need to appreciate the benefits of working together to make the firm run smoothly, and they also need to understand how their contributions help the firm. To stay motivated, most employees need to feel like what they do each day contributes something of value to the firm. Disclosing information and including employees in decision-making processes will also help employees feel valuable and wanted within the firm.

There are many different ways employers can reward and encourage employees. However, employers must be careful when considering what kinds of incentives to use. Different cultures value different kinds of incentives more highly than others. For example, a Japanese worker would probably not like it if she were singled out from the group and given a large cash bonus as a reward for her work. Japanese workers tend to be more group oriented, and therefore, anything that singles out individuals would not be an effective way of rewarding and motivating. American workers, on the other hand, are very individualistic, and a raise and public praise might be more effective. However, what might motivate a younger employee (bonuses, raises, and perks) may not be the same as what motivates a more seasoned, experienced, and financially successful employee (recognition, opportunity for greater influence, and increased training). Motivation is not an easy thing to understand, especially as firms become more global and more diverse.

Another important part of motivation is enjoying where you work and your career opportunities. Here is a list of the best places to do business and start careers in the United States, according to *Forbes* magazine. Chances are, workers who live in these places have encountered fewer frustrations than those placed at the bottom of the list and, therefore, would probably be more content with where they work.[35]

Best Places for Businesses and Careers

Rank	Metro Area	Job Growth Rank	Population
1.	Portland, Oregon	40	2,432,600
2.	Raleigh, North Carolina	20	1,307,600
3.	Seattle, Washington	32	2,944,900
4.	Denver, Colorado	19	2,857,700
5.	Des Moines, Iowa	69	636,000
6.	Provo, Utah	1	605,400
7.	Charlotte, North Carolina	21	2,479,600
8.	Austin, Texas	5	2,063,500
9.	Atlanta, Georgia	34	5,802,100
10.	Dallas, Texas	15	4,805,300

Source: "The Best Places for Business and Careers," Forbes, 2017, https://www.forbes.com/best-places-for-business/list/ (accessed April 22, 2018).

Review Your Understanding

Explain why the study of *human relations* is important.

Human relations is the study of the behavior of individuals and groups in organizational settings. Its focus is what motivates employees to perform on the job. Human relations is important because businesses need to understand how to motivate their increasingly diverse employees to be more effective, boost workplace morale, and maximize employees' productivity and creativity.

Summarize early studies that laid the groundwork for understanding employee motivation.

Time and motion studies by Frederick Taylor and others helped them analyze how employees perform specific work tasks in an effort to improve their productivity. Taylor and the early practitioners of the classical theory of motivation felt that money and job security were the primary motivators of employees. However, the Hawthorne studies revealed that human factors also influence workers' behavior.

Compare and contrast the human relations theories of Abraham Maslow and Frederick Herzberg.

Abraham Maslow defined five basic needs of all people and arranged them in the order in which they must be satisfied: physiological, security, social, esteem, and self-actualization. Frederick Herzberg divided characteristics of the job into hygiene factors and motivational factors. Hygiene factors relate to the work environment and must be present for employees to remain in a job. Motivational factors—recognition, responsibility, and advancement—relate to the work itself. They encourage employees to be productive. Herzberg's hygiene factors can be compared to Maslow's physiological and security needs; motivational factors may include Maslow's social, esteem, and self-actualization needs.

Investigate various theories of motivation, including Theories X, Y, and Z; equity theory; expectancy theory; and goal-setting theory.

Douglas McGregor contrasted two views of management: Theory X (traditional) suggests workers dislike work, while Theory Y (humanistic) suggests that workers not only like work but seek out responsibility to satisfy their higher order needs. Theory Z stresses employee participation in all aspects of company decision making, often through participative management programs and self-directed work teams. According to equity theory, how much people are willing to contribute to an organization depends on their assessment of the fairness, or equity, of the rewards they will receive in exchange. Expectancy theory states that motivation depends not only on how much a person wants something but also on the person's perception of how likely he or she is to get it.

Goal-setting theory refers to the impact that setting goals has on performance.

Describe some of the strategies that managers use to motivate employees.

Strategies for motivating workers include behavior modification (changing behavior and encouraging appropriate actions by relating the consequences of behavior to the behavior itself) and job design. Among the job design strategies businesses use are job rotation (allowing employees to move from one job to another to try to relieve the boredom associated with job specialization), job enlargement (adding tasks to a job instead of treating each task as a separate job), job enrichment (incorporating motivational factors into a job situation), and flexible scheduling strategies (flextime, compressed workweeks, job sharing, part-time work, and telecommuting).

Critique a business's program for motivating its sales force.

Using the information presented in the chapter, you should be able to analyze and defend Eagle Pharmaceutical's motivation program in the "Solve the Dilemma" feature near the end of this chapter, including the motivation theories the firm is applying to boost morale and productivity

Critical Thinking Questions

Enter the World of Business Questions

1. According to Maslow's hierarchy of needs, what type of needs do you think Facebook is meeting among its employees?

2. What motivational factors used at Facebook contribute to employee satisfaction?

3. Does Mark Zuckerberg take more of a Theory X or a Theory Y approach to his employees? Explain your reasoning.

Learn the Terms

Check Your Progress

1. Why do managers need to understand the needs of their employees?
2. Describe the motivation process.
3. What was the goal of the Hawthorne studies? What was the outcome of those studies?
4. Explain Maslow's hierarchy of needs. What does it tell us about employee motivation?
5. What are Herzberg's hygiene and motivational factors? How can managers use them to motivate workers?
6. Contrast the assumptions of Theory X and Theory Y. Why has Theory Y replaced Theory X in management today?
7. What is Theory Z? How can businesses apply Theory Z to the workplace?
8. Identify and describe four job design strategies.
9. Name and describe some flexible scheduling strategies. How can flexible schedules help motivate workers?
10. Why are motivational strategies important to both employees and employers?

Get Involved

1. Consider a person who is homeless: How would he or she be motivated and what actions would that person take? Use the motivation process to explain. Which of the needs in Maslow's hierarchy are likely to be most important? Least important?
2. View the video *Cheaper by the Dozen* (1950) and report on how the Gilbreths tried to incorporate their passion for efficiency into their family life.
3. What events and trends in society, technology, and economics do you think will shape human relations management theory in the future?

Build Your Skills

Motivating

Background

Do you think that, if employers could make work more like play, employees would be as enthusiastic about their jobs as they are about what they do in their leisure time? Let's see where this idea might take us.

Task

After reading the "Characteristics of PLAY," place a √ in column one for those characteristics you have experienced in your leisure time activities. Likewise, check column one on the next page for those "Characteristics of WORK" you have experienced in any of the jobs you've held.

All That Apply	Characteristics of PLAY
	1. New games can be played on different days.
	2. Flexible duration of play.
	3. Flexible time of when to play.
	4. Opportunity to express oneself.
	5. Opportunity to use one's talents.
	6. Skillful play brings applause, praise, and recognition from spectators.
	7. Healthy competition, rivalry, and challenge exist.
	8. Opportunity for social interaction.
	9. Mechanisms for scoring one's performance are available (feedback).
	10. Rules ensure basic fairness and justice.

All That Apply	Characteristics of WORK
	1. Job enrichment, job enlargement, or job rotation.
	2. Job sharing.
	3. Flextime, telecommuting.
	4. Encourage and implement employee suggestions.
	5. Assignment of challenging projects.
	6. Employee-of-the-month awards, press releases, employee newsletter announcements.
	7. Production goals with competition to see which team does best.
	8. Employee softball or bowling teams.
	9. Profit sharing; peer performance appraisals.
	10. Use tactful and consistent discipline.

Critical Thinking Questions

1. What prevents managers from making work more like play?

2. Are these forces real or imagined?

3. What would be the likely (positive and negative) results of making work more like play?

4. Could others in the organization accept such creative behaviors?

Solve the Dilemma

Motivating to Win

Eagle Pharmaceutical has long been recognized for its innovative techniques for motivating its salesforce. It features the salesperson who has been the most successful during the previous quarter in the company newsletter, "Touchdown." The salesperson also receives a football jersey, a plaque, and $1,000 worth of Eagle stock. Eagle's "Superbowl Club" is for employees who reach or exceed their sales goal, and a "Heisman Award," which includes a trip to the Caribbean, is given annually to the top 20 salespeople in terms of goal achievement.

Eagle employs a video conference hookup between the honored salesperson and four regional sales managers to capture some of the successful tactics and strategies the winning salesperson uses to succeed. The managers summarize these ideas and pass them along to the salespeople they manage. Sales managers feel strongly that programs such as this are important and that, by sharing strategies and tactics with one another, they can be a successful team.

LO 9-6
Critique a business's program for motivating its sales force.

Critical Thinking Questions

1. Which motivational theories are in use at Eagle?

2. What is the value of getting employees to compete against a goal instead of against one another?

3. Put yourself in the shoes of one of the four regional sales managers and argue against potential cutbacks to the motivational program.

Build Your Business Plan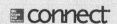

Motivating the Workforce

As you determine the size of your workforce, you are going to face the reality that you cannot provide the level of financial compensation that you would like to your employees, especially when you are starting your business.

Many employees are motivated by other things than money. Knowing that they are appreciated and doing a good job can bring great satisfaction to employees. Known as "stroking," it can provide employees with internal gratification that can be valued even more than financial incentives. Listening to your employees' suggestions, involving them in discussions about future growth, and valuing their input can go a long way toward building loyal employees and reducing employee turnover.

Think about what you could do in your business to motivate your employees without spending much money. Maybe you will have lunch brought in once a week or offer tickets to a local sporting event to the employee with the most sales. Whatever you elect to do, you must be consistent and fair with all your employees.

Visit Connect to practice building your business plan with the Business Plan Prep Exercises.

See for Yourself Videocase

The Container Store's Secret to Success: Employee Satisfaction

 Can you form a successful company by selling containers, boxes, and other storage products? The Container Store has shown that the answer is yes. With more than 10,000 products available, The Container Store sells items such as hanging bins, drawers, trash cans, and other items to help make a hectic life more organized. While there might have been skeptics in the beginning, particularly as the firm started out in a small 1,600-square-foot store, the idea quickly caught on with consumers.

"Word of mouth spread incredibly. It was just the oddest collection of merchandising anybody had ever seen to organize your home and to organize your life. A week into it, we knew we had something," co-founder and former CEO Kip Tindell said.

Yet in addition to its unusual product mix, The Container Store is also unique for its dedication to employees. The retail world can be difficult for employees because of high turnover, different hours every week, lower benefits, and constant interaction with people. The turnover rate in the retail industry is about 100 percent. This increases training costs, which can cause companies to decrease the amount of training offered.

This trend is reversed at The Container Store. First-year full-time employees receive 263 hours of training, much higher than the industry average of eight hours. Employees receive 50 percent more pay than at other retail establishments. Employee turnover at The Container Store is a low 10 percent. For more than a decade, The Container Store has been elected as one of the top 100 companies to work for by *Fortune* magazine.

Yet, the satisfaction that employees feel toward the company is not solely a result of higher pay. According to Frederick Herzberg's two-factor theory, good workplace conditions can prevent dissatisfaction but do not motivate the employee to go above and beyond what is required of them. As CEO of the company, Melissa Reiff realizes that the amount of time and effort an employee gives to the company will determine productivity.

"The first 25 percent for any employee is mandatory. If they don't do that, they're going to get fired. But the next 75 percent of an employee's productivity for any business in the world, I believe, is more or less voluntary. You do more or less of it depending upon how you feel about your boss and your product and your company."

Therefore, to enhance productivity, the firm has made employee satisfaction a priority. Employees come first, followed by customers and then shareholders. One of the ways that The Container Store motivates its employees is by creating an open communication culture. Employees are encouraged to approach their managers on any topic. This causes employees to feel as if the organization cares about them enough to take their concerns seriously. The Container Store also holds several events to show their appreciation of employee efforts. For instance, every February 14, The Container Store holds its "We Love Our Employees" day. At one of the events, the company announced the establishment of an emergency fund for employees. The company contributed $100,000 for unexpected costs that employees may find themselves having to pay for due to natural disasters, terrorist attacks, or significant medical issues.

Employees also receive many perks for working at The Container Store, and even part-time employees are eligible for health care benefits. Because employees take precedence at The Container Store, the company bases its decisions on what is best for employees even during hard times. During the recession, for instance, The Container Store refused to lay off employees. In addition to benefits, employees have access to all company data, including financial reports.

Does this mean that The Container Store ignores its customers for the sake of its employees? The high-quality customer service that The Container Store offers suggests just the opposite. In fact, the extensive employee training and the company's values demonstrate its high commitment to customer satisfaction. The company believes the key to great customer service, however, is highly motivated employees.

"We believe that if you take better care of your employees than anybody else, they'll take better care of the customer than anybody else, and if those two guys are ecstatic, ironically enough, your shareholder's going to be ecstatic too."[36]

Critical Thinking Questions

1. Name some of the hygiene factors at The Container Store.

2. Name some of the ways that The Container Store motivates its employees.

3. Do you believe that highly satisfied employees will lead to highly satisfied customers and shareholders?

You can find the related video in the Video Library in Connect. Ask your instructor how you can access Connect.

Team Exercise

Form groups and outline a compensation package that you would consider ideal in motivating an employee, recognizing performance, and assisting the company in attaining its cost-to-performance objectives. Think about the impact of intrinsic and extrinsic motivation and recognition. How can flexible scheduling strategies be used effectively to motivate employees? Report your compensation package to the class.

Ask your instructor about the role-play exercises available with this book to practice working with a business team.

Notes

1. Tanner Christensen, "How Facebook Keeps Employees Happy in the World's Largest Open Office," *Inc.,* March 9, 2016. https://www.inc.com/tanner-christensen/how-facebook-keeps-employees-happy-in-the-worlds-largest-open-office.html (accessed December 5, 2017); Catherine Clifford, "How Mark Zuckerberg keeps Facebook's 18,000+ Employees Innovating: 'Is This Going to Destroy the Company? If Not, Let Them Test It,'" *CNBC,* June 5, 2017. https://www.cnbc.com/2017/06/05/how-mark-zuckerberg-keeps-facebook-employees-innovating.html (accessed December 5, 2017); Mike Hoefflinger, "How Facebook Keeps Its Employees the Happiest, According to a Former Insider," *Business Insider,* April 11, 2017, http://www.businessinsider.com/how-facebook-keeps-employees-happy-2017-4 (accessed December 5, 2017); Steve Kux, "10 Reasons Why 99% of Facebook Employees Love Mark Zuckerberg," *Lifehack,* http://www.lifehack.org/articles/work/10-reasons-why-99-facebook-employees-love-mark-zuckerberg.html (accessed December 5, 2017); Marguerite Ward, "The 25 Best Companies to Work For in America," *CNBC,* December 7, 2016. https://www.cnbc.com/2016/12/07/the-25-best-companies-to-work-for-in-america.html (accessed December 5, 2017).

2. Jackie Wattles, Ben Geier, Matt Egan, and Danielle Wiener-Bronner, "Wells Fargo's 20-Month Nightmare," *CNN Money,* April 24, 2018, http://money.cnn.com/2018/04/24/news/companies/wells-fargo-timeline-shareholders/index.html (accessed April 25, 2018).

3. Dan Heath and Chip Heath, "Business Advice from Van Halen," *Fast Company,* March 1, 2010, www.fastcompany.com/1550881/business-advice-van-halen (accessed April 21, 2018).

4. Áine Cain, "6 Incredible Perks for Apple Employees," *Business Insider,* https://www.inc.com/business-insider/apple-employee-benefits-perks-glassdoor.html (accessed April 18, 2018).

5. Jeff Beer, "Patagonia: For Growing Its Business Every Time It Amplifies Its Social Mission," *Fast Company,* "The World's 50 Most Innovative Companies," pp. 33–34, March/April 2018.

6. "Worker Illness and Injury Costs U.S. Employers $225.8 Billion Annually," CDC Foundation, January 28, 2015, https://www.cdcfoundation.org/pr/2015/worker-illness-and-injury-costs-us-employers-225-billion-annually (accessed April 30, 2018).

7. Kelly K. Spors, "Top Small Workplaces 2008," *The Wall Street Journal,* February 22, 2009, https://www.wsj.com/articles/SB122347733961315417 (accessed December 30, 2017); King Arthur Flour website, www.kingarthurflour.com (accessed December 30, 2017); "King Arthur Flour Leads the Way with B-Corp Logo," *CSRwire,* February 15, 2008, www.csrwire.com/press/press_release/14672-King-Arthur-Flour (accessed December 30, 2017); "2013 'Best Places to Work in Vermont," http://bestplacestoworkinvt.com/index.php?option=com_content&task=view&id=50 (accessed December 30, 2017); "About the White House: Presidents," "1. George Washington," https://www.whitehouse.gov/about-the-white-house/presidents/george-washington/ (accessed December 30, 2017); Bruce Edwards, "VEDA aids King Arthur expansion," *Rutland Herald,* June 12, 2013, https://www.timesargus.com/articles/veda-aids-king-arthur-flour-growth/ (accessed December 30, 2017); King Arthur Flour, "Why Employee Ownership Matters," https://www.kingarthurflour.com/our-story/article/employee-owned.html (accessed December 30, 2017); King Arthur Flour, "Our Story – Community," https://www.kingarthurflour.com/our-story/community/ (accessed December 30, 2017); King Arthur Flour, "Bake for Good: Kids," https://www.kingarthurflour.com/our-story/article/bake-for-good-kids.html (accessed December 30, 2017).

8. Samuel Stebbins, "Highest Paid CEOs in 2017," *24/7 Wall St.,* December 28, 2017, https://247wallst.com/special-report/2017/12/28/highest-paid-ceos-in-2017/6/ (accessed April 22, 2018).

9. Frederick Herzberg, Bernard Mausner, and Barbara B. Snyderman, *The Motivation to Work,* 2nd ed. (New York: John Wiley, 1959).

10. "Physician—Gastroenterology Salaries," Salary.com, 2017, http://www1.salary.com/gastroenterologist-Salary.html (accessed April 22, 2018).

11. Shelia Eugenio, "4 Innovative Ways to Motivate Your Team," *Entrepreneur,* March 6, 2017, https://www.entrepreneur.com/article/289560 (access April 22, 2018).

12. Douglas McGregor, *The Human Side of Enterprise* (New York: McGraw-Hill, 1960), pp. 33–34.

13. Steven Pearlstein, "A Computer Specialist Reshapes Computer Sciences Corp.," *Washington Post,* August 1, 2015, https://www.washingtonpost.com/business/a-turnaround-specialist-reshapes-computer-sciences-corp/2015/07/30/c2be7c9c-32dd-11e5-97ae-30a30cca95d7_story.html?noredirect=on&utm_term=.814b0efe3d35 (accessed April 22, 2018).

14. McGregor, *The Human Side of Enterprise.*

15. Scott Koegler, "The Empowered Employee: How 6 Companies Are Arming Their Teams with Data," *IBM Watson,* March 24, 2017, https://www.ibm.com/blogs/watson/2017/03/empowered-employee-6-companies-arming-teams-data/ (access April 22, 2018).

16. William Ouchi, *Theory Z—How American Business Can Meet the Japanese Challenge* (New York: Perseus Group, 1981).

17. Jon L. Pierce, Tatiana Kostova, and Kurt T. Kirks, "Toward a Theory of Psychological Ownership in Organizations," *Academy of Management Review* 26, no. 2 (2001), p. 298.

18. Namaste Solar, *B Corporation,* https://www.bcorporation.net/community/namaste-solar (accessed April 7, 2018); Namasté Solar, "Employee Ownership," https://www.namastesolar.com/about-us/employee-ownership/ (accessed April 7, 2018); Namasté Solar, "Company Highlights," http://www.namastesolar.com/wp-content/uploads/2018/03/Namaste-Solar_Company-Model-Highlights.pdf (accessed April 7, 2018); Namasté Solar, "Namasté Solar Raises over $3.1 Million in Unconventional Stock Offering," June 6, 2017, https://www.namastesolar.com/namaste-solar-raises-3-1-million-unconventional-stock-offering/ (accessed April); Leigh Buchanan, "Where the CEO Is Just Another Guy with a Vote," *Inc.,* June 2011, pp. 64–66; "Unique Business Aims to Spread Solar Power," *CNN,* July 17, 2018, http://www.cnn.com/2008/TECH/07/17/solar.office/index.html (accessed April 7, 2018).

19. Matt Egan, "Lloyd Blankfein Takes Pay Cut at Goldman Sachs," *CNN Money,* March 17, 2017, http://money.cnn.com/2017/03/17/investing/lloyd-blankfein-goldman-sachs-pay-cut/index.html (accessed April 22, 2018).

20. David Callahan, *The Cheating Culture: Why More Americans Are Doing Wrong to Get Ahead* (Orlando, FL: Harcourt, 2004).

21. Katie Reilly, "Shoplifting and Other Fraud Cost Retailers Nearly $50 Billion Last Year," *Time,* June 22, 2017, http://time.com/money/4829684/shoplifting-fraud-retail-survey/ (accessed April 22, 2018).

22. Arthur Schwartz, "The 5 Most Common Unethical Behaviors in the Workplace," *Philadelphia Business Journal,* January 26, 2015, http://www.bizjournals.com/philadelphia/blog/guest-comment/2015/01/most-common-unethical-behaviors-in-the.html (accessed 22, 2018).

23. Archie Carroll, "Carroll: Do We Live in a Cheating Culture?" *Athens Banner-Herald,* February 21, 2004, www.onlineathens.com/stories/022204/bus_20040222028.shtml (accessed May 4, 2017).

24. Edwin A. Locke, K. M. Shaw, and Gary P. Latham, "Goal Setting and Task Performance: 1969–1980," *Psychological Bulletin* 90 (1981), pp. 125–152.

25. Peter Drucker, *The Practice of Management* (New York: Harper & Row, 1954).

26. "Management by Objectives," *The Economist,* October 21, 2009, http://www.economist.com/node/14299761 (accessed April 22, 2018).

27. Tim Hird, "The Lasting Benefits of Job Rotation," *Treasury & Risk,* January 24, 2017, https://www.treasuryandrisk.com/sites/treasuryandrisk/2017/01/24/the-lasting-benefits-of-job-rotation/ (accessed April 22, 2018).

28. Patagonia, "Environmental Internship Program," http://www.patagonia.com/us/patagonia.go?assetid=80524 (accessed on December 10, 2017); Giselle Abramovich, "Inside Patagonia's Content Machine," *Digiday,* January 31, 2013, http://digiday.com/brands/inside-patagonias-content-machine/ (accessed December 10, 2017); Leigh Buchanan, "How Patagonia's Roving CEO Stays in the Loop," *Inc.,* March 18, 2013, https://www.inc.com/leigh-buchanan/patagonia-founder-yvon-chouinard-15five.html (accessed December 10, 2017); "Patagonia: A Sustainable Outlook on Business," Daniels Fund Ethics Initiative, http://danielsethics.mgt.unm.edu/pdf/patagonia.pdf (accessed December 10, 2017); "Patagonia: Case Study," Lynda.com, 2013, http://cdn.lynda.com/cms/asset/text/patagonia-case-study--1931751689.pdf (accessed December 10, 2017); Patagonia, "Patagonia's Mission Statement," http://www.patagonia.com/company-info.html (accessed December 10, 2017).

29. Karsten Strauss, "The 250 Companies Offering the Most Flexible-Time Jobs," *Forbes,* October 11, 2016, www.forbes.com/sites/karstenstrauss/2016/10/11/the-250-companies-offering-the-most-flexible-time-jobs/#5827bead4424 (accessed April 22, 2018).

30. American Sociological Association, "Work Flexibility Benefits Employees, Study Says," *Phys.org,* January 13, 2016, http://abcnews.go.com/Health/Healthday/story?id=4509753 (accessed April 22, 2018).

31. Karsten Strauss, "Flextime Jobs: The 10 Best Career Fields For Making Your Own Schedule," *Forbes,* October 12, 2017, https://www.forbes.com/sites/karstenstrauss/2017/10/12/flex-time-jobs-the-10-best-career-fields-for-making-your-own-schedule/#f9bc0e376b9b (accessed April 22, 2018).

32. Jeanne Sahadi, "Dell *Really* Wants You to Work from Home. . .if You Want," *CNN,* June 9, 2016, http://money.cnn.com/2016/06/09/pf/dell-work-from-home/?iid=EL (accessed April 22, 2018).

33. Rebecca Wilson, "Work from Home to Increase Productivity," *Recruitment International,* March 31, 2017, https://www.recruitment-international.co.uk/blog/2017/03/work-from-home-to-increase-productivity (accessed April 22, 2018).

34. Dori Meinert, "Make Telecommuting Pay Off," Society for Human Resource Management, June 1, 2011, www.shrm.org/Publications/hrmagazine/EditorialContent/2011/0611/Pages/0611meinert.aspx (accessed April 22, 2018).

35. "The Best Places for Business and Careers," *Forbes*, 2017, https://www.forbes.com/best-places-for-business/list/ (accessed April 22, 2018).

36. "100 Best Companies to Work for 2009," *Fortune*, http://archive.fortune.com/magazines/fortune/bestcompanies/2009/snapshots/32.html (accessed August 5, 2016); "The Container Store: Employee Centric Retailer," UNM Daniels Fund Business Ethics Initiative, https://danielsethics.mgt.unm.edu/pdf/Container%20Store%20Case.pdf (accessed August 5, 2016); Fortune 100 Best Companies to Work For, "The Container Store," *CNN Money,* 2013, http://archive.fortune.com/magazines/fortune/best-companies/2013/snapshots/16.html (accessed August 5, 2016); The Container Store, "Employee First Culture," http://standfor.containerstore.com/putting-our-employees-first/ (accessed August 5, 2016); Bureau of Labor Statistics, "How to Become a Retail Sales Worker," January 8, 2014, http://www.bls.gov/ooh/Sales/Retail-sales-workers.htm (accessed August 5, 2016); The Container Store, "Communication IS Leadership," http://standfor.containerstore.com/our-foundation-principles/communication-is-leadership/ (accessed August 5, 2016); The Container Store, "Happy National We Love Our Employees Day," February 13, 2014, http://standfor.containerstore.com/happy-national-we-love-our-employees-day/ (accessed August 5, 2016); The Container Store, "Careers" http://www.containerstore.com/careers/index.html (accessed August 5, 2016); Maria Halkias, "The Container Store Set Up an Emergency Fund for Its Employee," *Dallas News,* February 14, 2014, http://bizbeatblog.dallasnews.com/2014/02/the-container-store-set-up-an-emergency-fund-for-its-employees.html/ (accessed August 5, 2016).

Credits

10 Managing Human Resources

©Krista Kennell/Shutterstock

Chapter Outline

Learning Objectives

After reading this chapter, you will be able to:

LO 10-1 Explain the significance of human resource management.

LO 10-2 Summarize the processes of recruiting and selecting human resources for a company.

LO 10-3 Describe how workers are trained and their performance appraised.

LO 10-4 Identify the types of turnover companies may experience.

LO 10-5 Explain why turnover is an important issue.

LO 10-6 Specify the various ways a worker may be compensated.

LO 10-7 Evaluate some of the issues associated with unionized employees, including collective bargaining and dispute resolution.

LO 10-8 Describe the importance of diversity in the workforce.

LO 10-9 Assess an organization's efforts to reduce its workforce size and to manage the resulting effects.

Enter the World of Business

Sheryl Sandberg: Wonder Woman

Sheryl Sandberg, chief operating officer (COO) of Facebook, is a role model for Facebook and women everywhere who want to assume greater managerial roles. Many experts credit Sandberg's leadership as one of the reasons Facebook is so successful. Experts believe that despite his innovativeness and leadership qualities, CEO Mark Zuckerberg lacks some of the human relations and marketing skills that Sandberg possesses.

When Sandberg became COO, she personally introduced herself to hundreds of employees throughout the company. She tries to solicit feedback from employees and encourage debate. Her strong communication skills, interpersonal interactions, and information sharing has led workers to identify with and admire Sandberg. *Fortune* magazine ranked Sandberg as the fifth most powerful woman in business in 2017.

Sandberg has noted the discrepancy that exists between the number of women in management positions compared to men. She wrote a book, *Lean In,* to address the empowerment of women in leadership positions. The book has created some 34,000 Lean In circles globally. Small groups of women meet to provide education, expert advice, and discussion about leadership. The idea is to gain new skills and expand the ability to influence others.

Sandberg also became concerned with helping others through hardship after her husband passed away in 2015. She coauthored a book, *Option B,* to help people deal with grief, illness, or adversity. Sandberg acknowledges that employees are expected to leave their emotions at the door when they enter the workplace. Sandberg believes leaving humanity out of the workplace is a mistake. She now has Option B groups on Facebook that provide resources to those dealing with grief. Sandberg's ability to challenge traditional employee roles and behaviors is helping her transform the workplace for the better.[1]

Today's organizations are more diverse, with a greater range of women, minorities, and older workers.

©Rawpixel.com/Shutterstock

Introduction

If a business is to achieve success, it must have sufficient numbers of employees who are qualified and motivated to perform the required duties. Thus, managing the quantity (from hiring to firing) and quality (through training, compensating, and so on) of employees is an important business function. Meeting the challenge of managing increasingly diverse human resources effectively can give a company a competitive edge in a global marketplace.

This chapter focuses on the quantity and quality of human resources. First, we look at how human resource managers plan for, recruit, and select qualified employees. Next, we look at training, appraising, and compensating employees—aspects of human resource management designed to retain valued employees. Along the way, we'll also consider the challenges of managing unionized employees and workplace diversity.

The Nature of Human Resource Management

In the "Dynamics of Business and Economics" chapter, *human resources* was defined as labor, the physical and mental abilities that people use to produce goods and services. **Human resource management (HRM)** refers to all the activities involved in determining an organization's human resource needs, as well as acquiring, training, and compensating people to fill those needs. Human resource managers are concerned with maximizing the satisfaction of employees and motivating them to meet organizational objectives productively. In some companies, this function is called personnel management.

HRM has increased in importance over the past few decades, in part because managers have developed a better understanding of human relations through the work of Maslow, Herzberg, and others. How employees are treated is also important to consumers. The conduct of an organization has an impact on the attitudes and behaviors of consumers. According to an Aflac survey, 92 percent of millennials are more likely to buy from an ethical company, and 82 percent would prefer to work for a company that has been publicly recognized for its ethical conduct.[2] Moreover, the human resources themselves are changing. Employees today are concerned not only about how much a job pays; they are concerned also with job satisfaction, personal performance, recreation, benefits, the work environment, and their opportunities for advancement. Today's workforce includes significantly more women, African Americans, Hispanics, and other minorities, as well as disabled and older workers, than in the past. Human resource managers must be aware of these changes and leverage them to increase the productivity of their employees. Every manager practices some of the functions of human resource management at all times.

Planning for Human Resource Needs

When planning and developing strategies for reaching the organization's overall objectives, a company must consider whether it will have the human resources necessary to carry out its plans. After determining how many employees and what skills are needed to satisfy the overall plans, the human resource department (which may range from the owner in a small business to hundreds of people in a large corporation) ascertains how many employees the company currently has and how many will be retiring or otherwise leaving the organization during the planning period. With this information, the human resource manager can then forecast how many more employees the company will need to hire and what qualifications they must have or determine if layoffs are required to meet demand more efficiently. HRM planning also requires forecasting the availability of people in the workforce who will have the necessary qualifications to meet the organization's future needs. The human resource manager then develops a strategy for satisfying the organization's human resource needs. As organizations strive to increase efficiency through outsourcing, automation, or learning to effectively use temporary workers, hiring needs can change dramatically.

Next, managers analyze the jobs within the organization so that they can match the human resources to the available assignments. **Job analysis** determines, through observation and study, pertinent information about a job—the specific tasks that comprise it; the knowledge, skills, and abilities necessary to perform it; and the environment in which it will be performed. Managers use the information obtained through a job analysis to develop job descriptions and job specifications.

A **job description** is a formal, written explanation of a specific job that usually includes job title, tasks to be performed (for instance, waiting on customers), relationship with other jobs, physical and mental skills required (such as lifting heavy boxes or calculating data), duties, responsibilities, and working conditions. Job seekers might turn to online websites or databases to help find job descriptions for specific occupations. For instance, the Occupational Information Network has an online database with hundreds of occupational descriptors. These descriptors list the skills, knowledge, and education needed to fulfill a particular occupation (for example, human resources).[3] A **job specification** describes the qualifications necessary for a specific job, in terms of education (some jobs require a college degree), experience, personal characteristics (ads frequently request outgoing, hardworking persons), and physical characteristics. Both the job description and job specification are used to develop recruiting materials such as newspapers, trade publications, and online advertisements.

job analysis
the determination, through observation and study, of pertinent information about a job—including specific tasks and necessary abilities, knowledge, and skills.

job description
a formal, written explanation of a specific job, usually including job title, tasks, relationship with other jobs, physical and mental skills required, duties, responsibilities, and working conditions.

job specification
a description of the qualifications necessary for a specific job, in terms of education, experience, and personal and physical characteristics.

Recruiting and Selecting New Employees

After forecasting the firm's human resource needs and comparing them to existing human resources, the human resource manager should have a general idea of how many new employees the firm needs to hire. With the aid of job analyses, management can then recruit and select employees who are qualified to fill specific job openings.

LO 10-2

Summarize the processes of recruiting and selecting human resources for a company.

Recruiting

Recruiting means forming a pool of qualified applicants from which management can select employees. There are two sources from which to develop this pool of applicants—internal and external.

Internal sources of applicants include the organization's current employees. Many firms have a policy of giving first consideration to their own employees—or promoting from within. The cost of hiring current employees to fill job openings is inexpensive when compared with the cost of hiring from external sources, and it is good for employee morale. However, hiring from within creates another job vacancy to be filled.

External sources of applicants consist of advertisements in newspapers and professional journals, employment agencies, colleges, vocational schools, recommendations from current employees, competing firms, unsolicited applications, online websites, and social networking sites such as LinkedIn.

recruiting
forming a pool of qualified applicants from which management can select employees.

Recruiters Embrace Nontraditional Recruitment Methods

Traditionally, recruiters have used résumés to gauge applicants' fit for a job. However, some organizations are realizing that initially judging applicants' suitability for jobs based on résumés—and immediately discarding those that do not fit the criteria—is a flawed system that can overlook talented candidates.

Part of the problem with using traditional recruiting methods is their inflexibility. Journalist George Anders claims that some of the best candidates are not the ones with great GPAs or job backgrounds, but those who possess analytical and conceptual skills to think outside of the box.

For this reason, some businesses are changing their recruitment methods. Employee referrals remain the top recruitment tool, with 30 percent of all hires recruited his way. Others are taking more unique approaches. Chipotle holds national hiring days when its locations hold open interviews for store positions.

Facebook sends out coding puzzles for programmers to solve; this enables candidates to test their abilities despite their previous work background. To recruit summer employees, McDonald's created a tool called "Snaplications." It placed 10-second videos on Snapchat about how great it is to work at McDonald's.

Although résumés will likely remain an important part of the recruitment process, employers are increasingly finding that résumés only show part of the picture. Using nontraditional recruitment tools gives companies the opportunity to test talents that may not be readily visible in a résumé, such as creativity or problem-solving skills.[4]

Critical Thinking Questions

1. What might be some of the limitations of résumés as the primary recruitment tool?
2. Why do you think referrals are valued so highly as a recruitment tool?
3. Do you believe that résumés are still important, or should they be discarded as a recruitment tool?

Internships are also a good way to solicit for potential employees. Many companies hire college students or recent graduates to low-paying internships that give them the opportunity to get hands-on experience on the job. If the intern proves to be a good fit, an organization may then hire the intern as a full-time worker.

There are also hundreds of websites where employers can post job openings and job seekers can post their résumés, including Monster.com, USAJobs, Simply Hired, and CareerBuilder. TheLadders.com is a website that focuses on career-driven professionals who make salaries of $100,000 or more. Employers looking for employees for specialized jobs can use more focused sites such as computerwork.com which lists jobs for people in technical careers. Increasingly, companies can turn to their own websites for potential candidates: Nearly all of the *Fortune* 500 firms provide career websites where they recruit, provide employment information, and take applications.

LinkedIn, a social network for professionals, has more than 500 million users.

©aradaphotography/Shutterstock

Using these sources of applicants is generally more expensive than hiring from within, but it may be necessary if there are no current employees who meet the job specifications or there are better-qualified people outside of the organization. Recruiting for entry-level managerial and professional positions is often carried out on college and university campuses. For managerial or professional positions above the entry level, companies sometimes depend on employment agencies or executive search firms, sometimes called *headhunters*, that specialize in luring qualified people away from other companies. Employers are also increasingly using professional social networking sites such LinkedIn (the most popular), Facebook, Instagram, Twitter, and blogs as recruitment tools.[5]

Selection

Selection is the process of collecting information about applicants and using that information to decide which ones to hire. It includes the application itself, as well as interviewing, testing, and reference checking. This process can be quite lengthy and expensive and is increasingly being completed online. Online job applications take from 1 minute to 52 minutes to complete, with a median completion time of 13 minutes. The longer the application process, the more applicants drop out and don't compete the application. *Business News Daily* found that 30 percent of candidates and 57 percent of those earning more than $100,000 will not spend more than 15 minutes on an application. With this in mind, AT&T revamped its online applications, cutting the number of screen shots by 50 percent. Its dropout rate fell 55 percent and generated 100,000 more desirable, quality applicants.[6] Companies are working to improve the online application process to attract the best pool of candidates for the job. Such rigorous scrutiny is necessary to find those applicants who can do the work expected and fit into the firm's structure and culture. If an organization finds the "right" employees through its recruiting and selection process, it will not have to spend as much money later in recruiting, selecting, and training replacement employees.

selection
the process of collecting information about applicants and using that information to make hiring decisions.

The Application.   In the first stage of the selection process, the individual fills out an application form and perhaps has a brief interview. The application form asks for the applicant's name, address, telephone number, education, and previous work experience. The goal of this stage of the selection process is to get acquainted with the applicants and to weed out those who are obviously not qualified for the job. For employees with work experience, most companies ask for the following information before contacting a potential candidate: current salary, reason for seeking a new job, years of experience, availability, and level of interest in the position. In addition to identifying obvious qualifications, the application can provide subtle clues about whether a person is appropriate for a particular job. For example, an applicant who gives unusually creative answers may be perfect for a position at an advertising agency; a person who turns in a sloppy, hurriedly scrawled application probably would not be appropriate for a technical job requiring precise adjustments. Most companies now accept online applications. To get a better view of the fit between the applicant and the company, the online application contains a questionnaire that asks applicants more specific questions, from how they might react in a certain situation to personality attributes like self-esteem or ability to interact with people.

The Interview.   The next phase of the selection process involves interviewing applicants. Table 10.1 lists some of the most common questions asked by interviewers. The interviewer can answer the applicant's questions about the requirements for the job, compensation, working conditions, company policies, organizational culture, and so on. A potential employee's questions may be just as revealing as his or her answers. Today's students might be surprised to have an interviewer ask them, "What's on your Facebook account?" or have them show the interviewer their Facebook accounts. Currently, these are legal questions for an interviewer to ask, although some states have passed laws banning employers from asking applicants to divulge their user names and passwords to their private Facebook accounts.[7] Approximately 70 percent of recruiters use social media sites to screen job candidates.[8] It is also legal and

TABLE 10.1   Most Common Questions Asked during the Interview

1. Tell me about yourself.
2. What are your biggest weaknesses?
3. What are your biggest strengths?
4. Where do you see yourself in 5 years?
5. Out of all of the other candidates, why should I hire you?
6. How did you learn about the opening?
7. Why do you want the job?
8. What do you consider your biggest professional achievement?
9. Tell me the last time a coworker or customer got angry with you. What happened?
10. Describe your dream job.

Source: Jeff Haden, "27 Most Common Job Interview Questions and Answers," Inc., 2017, https://www.inc.com/jeff-haden/27-most-common-job-interview-questions-and-answers.html (accessed April 27, 2018).

common for companies to monitor employee work habits and e-mails. While this can be important for monitoring outside threats such as hacking or information leaks, employees might view this as the company's way of saying it does not trust them.

Testing.   Another step in the selection process is testing. Ability and performance tests are used to determine whether an applicant has the skills necessary for the job. Aptitude, IQ, or personality tests may be used to assess an applicant's potential for a certain kind of work as well as the applicant's ability to fit into the organization's culture. One of the most commonly used tests is the Myers-Briggs Type Indicator. The Myers-Briggs Type Indicator Test is used worldwide by millions of people each year. Although polygraph ("lie detector") tests were once a common technique for evaluating the honesty of applicants, in 1988 their use was restricted to specific government jobs and those involving security or access to drugs. Applicants may also undergo physical examinations to determine their suitability for some jobs, and many companies require applicants to be screened for illegal drug use.

Drug and alcohol abuse can be particularly damaging to business. There is a growing opioid addiction in this country, with more than 70 percent of employers having experienced prescription drug misuse. Of those, 65 percent feel justified in terminating employees for this abuse.[9] The cost to companies for dealing with drug and alcohol abuse is staggering; abuse of prescription painkillers alone is estimated to cost companies $42 billion due to lost productivity or absenteeism.[10] Small businesses may have a higher percentage of these employees because they do not engage in systematic drug testing. E-cigarettes are another growing concern. Many organizations have restrictions on cigarette use, but these policies do not necessarily apply to e-cigarettes. As e-cigarette use grows among employees, companies are having to establish policies and requirements for their proper use.[11]

Reference Checking.   Before making a job offer, the company should always check an applicant's references. Reference checking usually involves verifying educational background and previous work experience. An Internet search is often done to determine social media activities or other public activities. Some of the employment trends related to background checks include the fact that criminal background checks are increasingly being delayed until after the interview or offer has been extended. With the growth in the gig and sharing economy, there will be more background checks for contingent workers. More companies will engage in continuous or ongoing background checks to keep the workplace safe. Social media screening will continue as nearly half of those using the technique find information that casts a negative light on the candidate.[12] Public companies are likely to do more extensive background searches to make sure applicants are not misrepresenting themselves.

Personality tests such as Myers-Briggs are used to assess an applicant's potential for a certain kind of job. For instance, extroversion and a love of people would be good qualities for a retail job.

©stockphoto mania/Shutterstock

Background checking is important because applicants may misrepresent themselves on their applications or résumés. Research has shown that those who are willing to exaggerate or lie on their résumés are more likely to engage in unethical behaviors.[13] As Table 10.2 illustrates, most of the common résumé lies relate to the applicant's college experience.

Reference checking is a vital, albeit often overlooked, stage in the selection process. Managers charged with hiring should be aware, however, that many organizations will confirm only that an applicant is a former employee, perhaps with beginning and ending work dates, and will not release details about the quality of the employee's work.

Legal Issues in Recruiting and Selecting

Legal constraints and regulations are present in almost every phase of the recruitment and selection process, and a violation of these regulations can result in lawsuits and

TABLE 10.2   Top Ten Most Common Résumé Lies

1. College you graduated from
2. Foreign language fluency
3. Academic degree
4. College major
5. GPA
6. Former employment or work history
7. Awards or accomplishments
8. College minor
9. Projects or portfolio
10. Job title

Source: Mike Timmermann, "The #1 Resume Lie That Could Cost You a New Job," Clark, http://clark.com/employment-military/worst-resume-lies/ (accessed April 26, 2018).

fines. Therefore, managers should be aware of these restrictions to avoid legal problems. Some of the laws affecting human resource management are discussed here.

Because one law pervades all areas of human resource management, we'll take a quick look at it now. **Title VII of the Civil Rights Act** of 1964 prohibits discrimination in employment. It also created the Equal Employment Opportunity Commission (EEOC), a federal agency dedicated to increasing job opportunities for women and minorities and eliminating job discrimination based on race, religion, color, gender identity, sexual orientation, national origin, or handicap. As a result of Title VII, employers must not impose sex distinctions in job specifications, job descriptions, or newspaper advertisements. In 2017, workplace discrimination charges filed with the EEOC were 84,254. The top five types of discrimination include retaliation, race, disability, sex, and age.[14] The Civil Rights Act of 1964 also outlaws the use of discriminatory tests for applicants. Aptitude tests and other indirect tests must be validated; in other words, employers must be able to demonstrate that scores on such tests are related to job performance so that no one race has an advantage in taking the tests or is alternatively discriminated against. Although many hope for improvements in diversity, according to Deloitte's "Missing Pieces Report," *Fortune* 500 firms have 10.6 percent minority men on their boards and 3.8 percent minority women.[15]

Other laws affecting HRM include the Americans with Disabilities Act (ADA), which prevents discrimination against disabled persons. It also classifies people with AIDS as handicapped and, consequently, prohibits using a positive AIDS test as reason to deny an applicant employment. The Age Discrimination in Employment Act specifically outlaws discrimination based on age. Its focus is banning hiring practices that discriminate against people 40 years and older. Generally, when companies need employees, recruiters head to college campuses, and when downsizing is necessary, many older workers are offered early retirement. Forced retirement based on age, however, is generally considered to be illegal in the United States, although claims of forced retirement still abound. Indeed, there are many benefits that companies are realizing in hiring older workers. Some of these benefits include the fact that they are more dedicated, punctual, honest, and detail-oriented; are good listeners; take pride in their work; exhibit good organizational skills; are efficient and confident; are mature; can be seen as role models; have good communication skills; and offer an opportunity for a reduced labor cost because of already having insurance plans.[16]

The Equal Pay Act mandates that men and women who do equal work must receive the same wage. Wage differences are acceptable only if they are attributed to seniority, performance, or qualifications. It is estimated that women make 80 cents on the dollar what their male counterparts make. The wage gap increases for African American and Hispanic women, at 62.5 percent and 54.4 percent what their male counterparts make, respectively.[17] While gender wage gaps vary by industry, a financial analytics company found that the wage gap tends to decrease among jobs that pay lower wages, at just under 5 percent. Professions with the lowest wage gaps include transportation, storage, and distribution managers; stock clerks and order fillers; and

Title VII of the Civil Rights Act prohibits discrimination in employment and created the Equal Employment Opportunity Commission.

The Department of Labor has oversight over workplace safety, wages and work hours, unemployment benefits, and more. It often files lawsuits against firms that it believes are treating workers unfairly and violating labor laws.

©B Christopher/Alamy Stock Photo

counselors (female counselors actually earn more on average than male counselors). The widest gap is among personal financial advisors, estimated to be at 55.6 percent.[18] However, despite the wage inequalities that still exist, women are becoming increasingly accepted in the workplace. The working mother is no longer a novelty; in fact, many working mothers seek the same amount of achievement as any other worker—with or without children.

Developing the Workforce

Once the most qualified applicants have been selected, have been offered positions, and have accepted their offers, they must be formally introduced to the organization and trained so they can begin to be productive members of the workforce. **Orientation** familiarizes the newly hired employees with fellow workers, company procedures, and the physical properties of the company. It generally includes a tour of the building; introductions to supervisors, co-workers, and subordinates; and the distribution of organizational manuals describing the organization's policy on vacations, absenteeism, lunch breaks, company benefits, and so on. Orientation also involves socializing the new employee into the ethics and culture of the new company. Many larger companies now show videos of procedures, facilities, and key personnel in the organization to help speed the adjustment process.

Training and Development

Although recruiting and selection are designed to find employees who have the knowledge, skills, and abilities the company needs, new employees still must undergo **training** to learn how to do their specific job tasks. *On-the-job training* allows workers to learn by actually performing the tasks of the job, while *classroom training* teaches employees with lectures, conferences, videos, case studies, and web-based training. McDonald's has always had a strong training presence at the Fred L. Turner Training Center, Hamburger University, in Oak Brook, Illinois. With major management transitions and plans for 4,000 company-owned restaurants to become franchises, the company is now investing in a virtual training platform through cloud-based technology to reach many of these new global operators.[19]

Some companies will go even further and ask a more experienced individual in the organization to mentor a new employee. **Mentoring** involves supporting, training, and guiding an employee in his or her professional development. Mentoring provides employees with more of a one-on-one interaction with somebody in the organization who not only teaches them, but also acts as their supporter as they progress in their jobs. Another benefit of mentoring is that companies can use this process to attract talent from underrepresented areas. For instance, mentoring has been suggested as a way to attract more women into male-dominated industries.

Development is training that augments the skills and knowledge of managers and professionals. Training and development are also used to improve

orientation
familiarizing newly hired employees with fellow workers, company procedures, and the physical properties of the company.

training
teaching employees to do specific job tasks through either classroom development or on-the-job experience.

mentoring
involves supporting, training, and guiding an employee in his or her professional development.

> ### LO 10-3
>
> Describe how workers are trained and their performance appraised.

development
training that augments the skills and knowledge of managers and professionals.

McDonald's Hamburger University, with locations in Chicago, Tokyo, London, Sydney, Munich, Shanghai, and São Paulo, provides learning and training for its employees and partners to build long-lasting careers.

©Qilai Shen/Bloomberg via Getty Images

the skills of employees in their present positions and to prepare them for increased responsibility and job promotions. Training is, therefore, a vital function of human resource management. At The Container Store, for example, first-year sales personnel receive more than 200 hours of formal training about the company's products versus an industry average of eight hours.[20] Companies are engaging in more experiential and involvement-oriented training exercises for employees. Use of role-plays, simulations, and online training methods are becoming increasingly popular in employee training.

Assessing Performance

Assessing employees' performance—their strengths and weaknesses on the job—is one of the most difficult tasks for managers. However, performance appraisal is crucial because it gives employees feedback on how they are doing and what they need to do to improve. It also provides a basis for determining how to compensate and reward employees, and it generates information about the quality of the firm's selection, training, and development activities.

Performance appraisals may be objective or subjective. An objective assessment is quantifiable. For example, a Westinghouse employee might be judged by how many circuit boards the employee typically produces in one day or by how many of the employee's boards have defects. A RE/MAX real estate agent might be judged by the number of listings the agent has shown or the number of sales the agent has closed. A company can also use tests as an objective method of assessment. Whatever method they use, managers must take into account the work environment when they appraise performance objectively.

When jobs do not lend themselves to objective appraisal, the manager must relate the employee's performance to some other standard. One popular tool used in subjective assessment is the ranking system, which lists various performance factors on which the manager ranks employees against each other. Although used by many large companies, ranking systems are unpopular with many employees. Qualitative criteria, such as teamwork and communication skills, used to evaluate employees are generally hard to gauge. Such grading systems have triggered employee lawsuits that allege discrimination in grade/ranking assignments. For example, one manager may grade a

Performance appraisals are important because they provide employees with feedback on how well they are doing as well as areas for improvement.

©bluedog studio/Shutterstock

company's employees one way, while another manager grades a group more harshly depending on the managers' grading style. If layoffs occur, then employees graded by the second manager may be more likely to lose their jobs. Other criticisms of grading systems include unclear wording or inappropriate words that a manager may unintentionally write in a performance evaluation, like *youthful* or *attractive* to describe an employee's appearance. These liabilities can all be fodder for lawsuits should employees allege that they were treated unfairly. Therefore, it is crucial that managers use clear language in performance evaluations and be consistent with all employees.

Another performance appraisal method used by many companies is the 360-degree feedback system, which provides feedback from a panel that typically includes superiors, peers, and subordinates. Because of the tensions it may cause, peer appraisal appears to be difficult for many. However, companies that have success with 360-degree feedback tend to be open to learning, willing to experiment, and are led by executives who are direct about the expected benefits as well as the challenges.[21] Managers and leaders with a high emotional intelligence (sensitivity to their own as well as others' emotions) assess and reflect upon their interactions with colleagues on a daily basis. In addition, they conduct follow-up analysis on their projects, asking the right questions and listening carefully to responses without getting defensive of their actions.[22]

Another trend occurring at some companies is the decrease of negative employee feedback. Executives have begun to recognize that hard tactics can harm employee confidence. Negative feedback tends to overshadow positive feedback, so employees may get discouraged if performance reviews are phrased too negatively. At the same time, it is important for managers to provide constructive criticism on employee weaknesses in addition to their strengths so workers know what to expect and how they are viewed.[23]

Whether the assessment is objective or subjective, it is vital that managers discuss the results with their employees, so that they know how well they are doing their jobs. The results of a performance appraisal become useful only when they are communicated, tactfully, to employees and presented as a tool to allow the employees to grow and improve in their positions and beyond. Performance appraisals are also used to determine whether an employee should be promoted, transferred, or terminated from the organization.

turnover
occurs when employees quit or are fired and must be replaced by new employees.

Turnover

LO 10-4

Identify the types of turnover companies may experience.

Turnover, which occurs when employees quit or are fired and must be replaced by new employees, results in lost productivity from the vacancy, costs to recruit replacement employees, management time devoted to interviewing, training, and socialization expenses for new employees. Gallup research shows that approximately 75 percent of the reasons for voluntary turnover involve areas that management can influence.[24] Companies can therefore significantly impact turnover rates. Of course, turnover is not always an unhappy occasion when it takes the form of a promotion or transfer.

A **promotion** is an advancement to a higher-level job with increased authority, responsibility, and pay. In some companies and most labor unions, seniority—the length of time a person has been with the company or at a particular job classification—is the key issue in determining who should be promoted. Most managers base promotions on seniority only when they have candidates with equal qualifications. Managers prefer to base promotions on merit.

A **transfer** is a move to another job within the company at essentially the same level and wage. Transfers allow workers to obtain new skills or to find a new

promotion
a persuasive form of communication that attempts to expedite a marketing exchange by influencing individuals, groups, and organizations to accept goods, services, and ideas.

transfer
a move to another job within the company at essentially the same level and wage.

Many companies in recent years are choosing to downsize by eliminating jobs. Reasons for downsizing might be due to financial constraints or the need to become more productive and competitive.

©belterz/iStockphoto

separations

employment changes involving resignation, retirement, termination, or layoff.

LO 10-5

Explain why turnover is an important issue.

position within an organization when their old position has been eliminated because of automation or downsizing.

Separations occur when employees resign, retire, are terminated, or are laid off. Employees may be terminated, or fired, for poor performance, violation of work rules, absenteeism, and so on. Businesses have traditionally been able to fire employees *at will*—that is, for any reason other than for race, religion, sex, or age, or because an employee is a union organizer. However, recent legislation and court decisions now require that companies fire employees fairly, for just cause only. Managers must take care, then, to warn employees when their performance is unacceptable and may lead to dismissal, elevating the importance of performance evaluations. They should also document all problems and warnings in employees' work records. To avoid the possibility of lawsuits from individuals who may feel they have been fired unfairly, employers should provide clear, business-related reasons for any firing, supported by written documentation if possible. Employee disciplinary procedures should be carefully explained to all employees and should be set forth in employee handbooks. Table 10.3 illustrates what to do and what *not* to do when you are terminated.

Many companies have downsized in recent years, laying off tens of thousands of employees in their effort to become more productive and competitive. Layoffs are sometimes temporary; employees may be brought back when business conditions improve. When layoffs are to be permanent, employers often help employees find other jobs and may extend benefits while the employees search for new employment. Such actions help lessen the trauma of the layoffs. Fortunately, there are several business areas that are choosing not to downsize.

A well-organized human resource department strives to minimize losses due to separations and transfers because recruiting and training new employees is very expensive. Note that a high turnover rate in a company may signal problems with the selection and training process, the compensation program, or even the type of company. To help reduce turnover, companies have tried a number of strategies, including giving employees more interesting job responsibilities (job enrichment), allowing for increased job flexibility, and providing more employee benefits. When employees do choose to leave the organization, the company will often ask them to participate in an *exit interview*. An exit interview is a survey used to determine why the employee is leaving the

TABLE 10.3   Actions You Should and Shouldn't Take When You Are Terminated

1. Do not criticize your boss who terminated you.
2. Do not take files or property that is not yours.
3. Do try to get a reference letter.
4. Do not criticize your former employer during job interviews.
5. Do look to the future and be positive about new job opportunities.

organization. The company hopes that this feedback will alert them to processes they can improve upon to dissuade valuable employees from leaving in the future.

Compensating the Workforce

LO 10-6

Specify the various ways a worker may be compensated.

People generally don't work for free, and how much they are paid for their work is a complicated issue. Also, designing a fair compensation plan is an important task because pay and benefits represent a substantial portion of an organization's expenses. Wages that are too high may result in the company's products being priced too high, making them uncompetitive in the market. Wages that are too low may damage employee morale and result in costly turnover. Remember that compensation is one of the hygiene factors identified by Herzberg.

Designing a fair compensation plan is a difficult task because it involves evaluating the relative worth of all jobs within the business while allowing for individual efforts. Compensation for a specific job is typically determined through a **wage/salary survey,** which tells the company how much compensation comparable firms are paying for specific jobs that the firms have in common. Compensation for individuals within a specific job category depends on both the compensation for that job and the individual's productivity. Therefore, two employees with identical jobs may not receive exactly the same pay because of individual differences in performance.

wage/salary survey
a study that tells a company how much compensation comparable firms are paying for specific jobs that the firms have in common.

Financial Compensation

Financial compensation falls into two general categories—wages and salaries. **Wages** are financial rewards based on the number of hours the employee works or the level of output achieved. Wages based on the number of hours worked are called time wages. The federal minimum wage in 2018 is $7.25 per hour for covered nonexempt workers.[25] Tipped wages may be $2.13 per hour as long as tips plus the wage of $2.13 per hour equal the minimum wage of $7.25 per hour, although some states require minimum wage and then tips added on top.[26] Many states also mandate minimum wages; in the case where this conflicts with the federal minimum wage, the higher of the two wages prevails. There may even be differences between city and state minimum wages. In New Mexico, the minimum wage is $7.50, whereas in the state capitol of Santa Fe, the minimum wage is $11.40, due to a higher cost of living.[27] Time wages are appropriate when employees are continually interrupted and when quality is more important than quantity. Assembly-line workers, clerks, and maintenance personnel are commonly paid on a time-wage basis. The advantage of time wages is the ease of computation. The disadvantage is that time wages provide no incentive to increase productivity. In fact, time wages may encourage employees to be less productive.

wages
financial rewards based on the number of hours the employee works or the level of output achieved.

To overcome these disadvantages, many companies pay on an incentive system, using piece wages or commissions. Piece wages are based on the level of output achieved. A major advantage of piece wages is that they motivate employees to supervise their own activities and to increase output. Skilled craftworkers are often paid on a piece-wage basis.

The other incentive system, **commission,** pays a fixed amount or a percentage of the employee's sales. Skincare direct seller Rodan + Fields uses independent contractors to sell its products. Consultants earn commissions on whatever products they sell. This method motivates employees to sell as much as they can. Some companies also combine payment based on commission with time wages or salaries.

commission
an incentive system that pays a fixed amount or a percentage of the employee's sales.

A **salary** is a financial reward calculated on a weekly, monthly, or annual basis. Salaries are associated with white-collar workers such as office personnel, executives,

salary
a financial reward calculated on a weekly, monthly, or annual basis.

and professional employees. Although a salary provides a stable stream of income, salaried workers may be required to work beyond usual hours without additional financial compensation.

In addition to the basic wages or salaries paid to employees, a company may offer **bonuses** for exceptional performance as an incentive to increase productivity further. Many workers receive a bonus as a "thank you" for good work and an incentive to continue working hard. Many owners and managers are recognizing that simple bonuses and perks foster happier employees and reduce turnover. Bonuses are especially popular among Wall Street firms. Wall Street executives' bonuses rise when the United States has an improving economy and strong performance in the financial sector.[28]

Another form of compensation is **profit sharing,** which distributes a percentage of company profits to the employees whose work helped to generate those profits. Some profit-sharing plans involve distributing shares of company stock to employees. Usually referred to as *ESOPs*—employee stock ownership plans—they have been gaining popularity in recent years. One reason for the popularity of ESOPs is the sense of partnership that they create between the organization and employees. Profit sharing can also motivate employees to work hard because increased productivity and sales mean that the profits or the stock dividends will increase. Many organizations offer employees a stake in the company through stock purchase plans, ESOPs, or stock investments through 401(k) plans. Companies are adopting broad-based stock option plans to build a stronger link between employees' interests and the organization's interests. Businesses have found employee stock options a great way to boost productivity and increase morale.

Benefits

Benefits are nonfinancial forms of compensation provided to employees, such as pension plans for retirement; health, disability, and life insurance; holidays and paid days off for vacation or illness; credit union membership; health programs; child care; elder care; assistance with adoption; and more. According to the Bureau of Labor Statistics, employer costs for employee compensation for civilian workers in the United States averaged $35.87 per hour worked. Wages and salaries accounted for 68.3 percent of these costs, while benefits account for 31.7 percent of the cost.[29] Such benefits increase employee security and, to a certain extent, their morale and motivation.

Although health insurance is a common benefit for full-time employees, rising health care costs have forced a growing number of employers to trim this benefit. Even government workers, whose wages and benefits used to be virtually guaranteed safe, have seen reductions in health care and other benefits. On the other hand, employee loyalty tends to increase when employees feel that the firm cares about them. Starbucks recognizes the importance of how benefits can significantly affect an employee's health and well-being. As a result, it offers

Onsite child care is just one of the benefits large companies have begun to offer employees.

©Ariel Skelley/Blend Images/Getty Images

bonuses
monetary rewards offered by companies for exceptional performance as incentives to further increase productivity.

profit sharing
a form of compensation whereby a percentage of company profits is distributed to the employees whose work helped to generate them.

benefits
nonfinancial forms of compensation provided to employees, such as pension plans, health insurance, paid vacation and holidays, and the like.

Employees Like Green Incentives

Some large businesses are providing incentives for employees to adopt green practices both at work and at home. For instance, Sony Pictures' Alternative Vehicle Incentive offers cash back for the purchase of fuel-efficient hybrid or electric vehicles. Its Greener World Grant Program allows employees to nominate environmental organizations in their communities for grants, and its Idea to Action Grant Program provides funding for eligible employees to pursue ways to make Sony's operations greener.

Intel has also been recognized for its green incentives. The company ties a portion of employee compensation to achieving sustainability goals. It also has an Environmental Excellence Award Program that recognizes employees for their contributions toward improving sustainability at the company and in the greater community.

Google is another company that has become well known for its green incentives. Google has placed Google Bikes, also known as GBikes, all around its campus to discourage the use of vehicles. The company encourages carpooling and alternative transportation methods for employees. It estimates that single-occupancy commuting to its headquarters has decreased by 45 percent. In addition, to promote more environmentally friendly vehicles, 10 percent of parking spaces have electric vehicle charging stations at its Bay Area headquarters. With companies like Sony, Intel, and Google encouraging green behaviors among employees, perhaps it is businesses that will make the greatest contribution toward sustainable environmental practices.[30]

Critical Thinking Questions

1. Why do you think businesses are providing employees with incentives to go green?
2. What are some of the incentives that Sony Pictures, Intel, and Google offer employees to engage in more sustainable practices?
3. Do you think these incentives are genuine efforts to encourage sustainability? Or are they more of a marketing initiative to make the businesses look good?

---

its part-time employees health insurance. Additionally, Starbucks began offering employees a benefit called College Achievement Plan in which it will pay full tuition for employees to finish a bachelor's degree at Arizona State University.[31]

A benefit increasingly offered is the employee assistance program (EAP). Each company's EAP is different, but most offer counseling for and assistance with those employees' personal problems that might hurt their job performance if not addressed. The most common counseling services offered include drug- and alcohol-abuse treatment programs, fitness programs, smoking-cessation clinics, stress-management clinics, financial counseling, family counseling, and career counseling. Home Depot offers an employee assistance program that is called CARE/Solutions for Life. The program focuses on three areas: free financial consultation by phone; free legal consultation; and three free face-to-face counseling sessions to assist with personal, family, or work life.[32] EAPs help reduce costs associated with poor productivity, absenteeism, and other workplace issues by helping employees deal with personal problems that contribute to these issues. For example, exercise and fitness programs reduce health insurance costs by helping employees stay healthy. Family counseling may help workers trying to cope with a divorce or other personal problems to better focus on their jobs.

Companies try to provide the benefits they believe their employees want, but diverse people may want different things. In recent years, some single workers have felt that co-workers with spouses and children seem to get "special breaks" and extra time off to deal with family issues. Some companies use flexible benefit programs to allow employees to choose the benefits they would like, up to a specified amount.

Fringe benefits include sick leave, vacation pay, pension plans, health plans, and any other extra compensation. Many states and cities are adopting new policies on sick leave that mandate a certain number of paid sick days a worker can take. It is often lower-wage employees who do not receive paid sick leave, yet they are the ones who usually cannot afford to take a day off if it is unpaid.[33] *Soft benefits* include perks that help balance life and work. They include onsite child care, spas, food service, and even laundry services and hair salons. These soft benefits motivate employees and give them more time to focus on their job. They also inspire loyalty to the company. Facebook is known for its positive work/life balance and for the numerous benefits it provides employees. The importance of management fueling a positive work/life balance of employees can be seen through Facebook's constant employee support after the Cambridge Analytica breach. Employees' attitudes toward the firm remained unwavering, with many saying the anger at Facebook by the public is misplaced. In fact, Facebook's rating on Glassdoor remained between 4.5 and 4.6 out of 5, even after the breach was announced.[34]

Cafeteria benefit plans provide a financial amount to employees so that they can select the specific benefits that fit their needs. The key is making benefits flexible, rather than giving employees identical benefits. As firms go global, the need for cafeteria or flexible benefit plans becomes even more important. For some employees, benefits are a greater motivator and differentiator in jobs than wages. For many Starbucks employees who receive health insurance when working part-time, this benefit could be the most important compensation. Over the past two decades, the list of fringe benefits offered by employers has grown dramatically, and new benefits are being added every year.

LO 10-7

Evaluate some of the issues associated with unionized employees, including collective bargaining and dispute resolution.

labor unions
employee organizations formed to deal with employers for achieving better pay, hours, and working conditions.

Managing Unionized Employees

Employees who are dissatisfied with their working conditions or compensation have to negotiate with management to bring about change. Dealing with management on an individual basis is not always effective, however, so employees may organize themselves into **labor unions** to deal with employers and to achieve better pay, hours, and working conditions. Organized employees are backed by the power of a large group that can hire specialists to represent the entire union in its dealings with management. Union workers make significantly more than nonunion employees. The United States has roughly 10.7 percent of wage and salary workers who are members of unions, unchanged from the previous year.[35] On average, the median usual weekly earnings of unionized full-time and salary workers are about $200 more than their nonunion counterparts.[36]

However, union growth has slowed in recent years, and prospects for growth do not look good. One reason is that most blue-collar workers, the traditional members of unions, have already been organized. Factories have become more automated and need fewer blue-collar workers. The United States has shifted from a manufacturing to a service economy, further reducing the demand for blue-collar workers. Moreover, in response to foreign competition, U.S. companies are scrambling to find ways to become more productive and cost-efficient. Job enrichment programs and participative management have blurred the line between management and workers. Because workers' say in the way plants are run is increasing, their need for union protection is decreasing. Many workers do not see the benefits of union membership if they do not have complaints or grievances against their employers.[37]

Nonetheless, labor unions have been successful in organizing blue-collar manufacturing, government, and health care workers, as well as smaller percentages of employees in other industries. Consequently, significant aspects of HRM, particularly compensation, are dictated to a large degree by union contracts at many companies. Therefore, we'll take a brief look at collective bargaining and dispute resolution in this section.

Collective Bargaining

Collective bargaining is the negotiation process through which management and unions reach an agreement about compensation, working hours, and working conditions for the bargaining unit (Figure 10.1). The objective of negotiations is to reach agreement about a **labor contract**, the formal, written document that spells out the relationship between the union and management for a specified period of time, usually two or three years.

In collective bargaining, each side tries to negotiate an agreement that meets its demands; compromise is frequently necessary. Management tries to negotiate a labor contract that permits the company to retain control over things like work schedules; the hiring and firing of workers; production standards; promotions, transfers, and separations; the span of management in each department; and discipline. Unions tend to focus on contract issues such as magnitude of wages; better pay rates for overtime,

collective bargaining
the negotiation process through which management and unions reach an agreement about compensation, working hours, and working conditions for the bargaining unit.

labor contract
the formal, written document that spells out the relationship between the union and management for a specified period of time—usually two or three years.

FIGURE 10.1
The Collective Bargaining Process

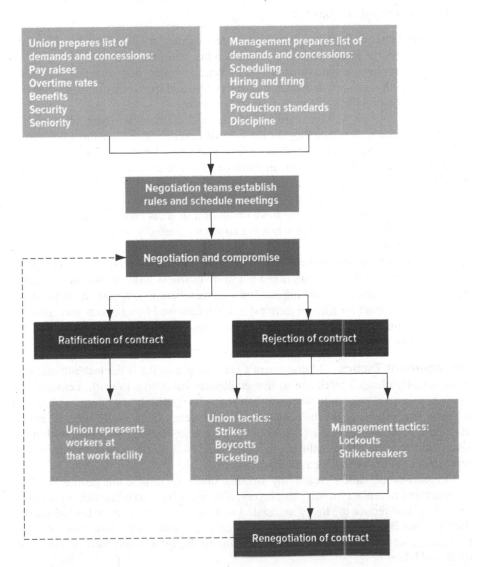

holidays, and undesirable shifts; scheduling of pay increases; and benefits. These issues will be spelled out in the labor contract, which union members will vote to either accept (and abide by) or reject.

Many labor contracts contain a *cost-of-living escalator* (or *adjustment*) *(COLA) clause,* which calls for automatic wage increases during periods of inflation to protect the "real" income of the employees. During tough economic times, unions may be forced to accept *givebacks*—wage and benefit concessions made to employers to allow them to remain competitive or, in some cases, to survive and continue to provide jobs for union workers.

Resolving Disputes

Sometimes, management and labor simply cannot agree on a contract. Most labor disputes are handled through collective bargaining or through grievance procedures. When these processes break down, however, either side may resort to more drastic measures to achieve its objectives.

picketing
a public protest against management practices that involves union members marching and carrying antimanagement signs at the employer's plant.

Labor Tactics. **Picketing** is a public protest against management practices and involves union members marching (often waving antimanagement signs and placards) at the employer's plant or work site. Picketing workers hope that their signs will arouse sympathy for their demands from the public and from other unions. Picketing may occur as a protest or in conjunction with a strike.

strikes
employee walkouts; one of the most effective weapons labor has.

Strikes (employee walkouts) are one of the most effective weapons labor has. By striking, a union makes carrying out the normal operations of a business difficult at best and impossible at worst. Strikes receive widespread publicity, but they remain a weapon of last resort. However, in extreme cases, workers may organize a strike with the help of unions and coalitions. Fast-food workers at McDonald's restaurants in Boston went on strike on Labor Day to demand $15/hour pay and union rights.[38] While it is mostly the case that the mere threat of a strike is enough to make management back down, there are times when the issues are heatedly debated and regulatory agencies become involved.[39]

boycott
an attempt to keep people from purchasing the products of a company.

A **boycott** is an attempt to keep people from purchasing the products of a company. In a boycott, union members are asked not to do business with the boycotted organization. Some unions may even impose fines on members who ignore the boycott. To gain further support for their objectives, a union involved in a boycott may also ask the public—through picketing and advertising—not to purchase the products of the picketed firm.

lockout
management's version of a strike, wherein a work site is closed so that employees cannot go to work.

Management Tactics. Management's version of a strike is the **lockout**; management actually closes a work site so that employees cannot go to work. Lockouts are used, as a general rule, only when a union strike has partially shut down a plant and it seems less expensive for the plant to close completely. Honeywell workers, after a 10-month lockout, returned to their jobs after agreeing to a new contract with management. More than 350 employees in Indiana and New York returned to work where replacement workers had been filling their jobs.[40]

strikebreakers
people hired by management to replace striking employees; called "scabs" by striking union members.

Strikebreakers, called "scabs" by striking union members, are people hired by management to replace striking employees. Managers hire strikebreakers to continue operations and reduce the losses associated with strikes—and to show the unions that they will not bow to their demands. Strikebreaking is generally a last-resort measure for management because it does great damage to the relationship between management and labor.

Outside Resolution.   Management and union members normally reach mutually agreeable decisions without outside assistance. Sometimes though, even after lengthy negotiations, strikes, lockouts, and other tactics, management and labor still cannot resolve a contract dispute. In such cases, they have three choices: conciliation, mediation, and arbitration. **Conciliation** brings in a neutral third party to keep labor and management talking. The conciliator has no formal power over union representatives or over management. The conciliator's goal is to get both parties to focus on the issues and to prevent negotiations from breaking down. Like conciliation, **mediation** involves bringing in a neutral third party, but the mediator's role is to suggest or propose a solution to the problem. With **arbitration,** a neutral third party is brought in to settle the dispute, but the arbitrator's solution is legally binding and enforceable. Generally, arbitration takes place on a voluntary basis—management and labor must agree to it, and they usually split the cost (the arbitrator's fee and expenses) between them. Occasionally, management and labor submit to *compulsory arbitration,* in which an outside party (usually the federal government) requests arbitration as a means of eliminating a prolonged strike that threatens to disrupt the economy.

The Importance of Workforce Diversity

Customers, employees, suppliers—all the participants in the world of business—come in different ages, genders, races, ethnicities, nationalities, and abilities, a truth that business has come to label **diversity.** Understanding this diversity means recognizing and accepting differences as well as valuing the unique perspectives such differences can bring to the workplace.

The Characteristics of Diversity

When managers speak of diverse workforces, they typically mean differences in gender and race. While gender and race are important characteristics of diversity, others are also important. We can divide these differences into primary and secondary characteristics of diversity. In the lower segment of Figure 10.2, sexual orientation, age, gender, race, ethnicity, and abilities represent *primary characteristics* of diversity. In the upper section of Figure 10.2 are eight *secondary characteristics* of diversity—education, work background, income, marital status, parental status, military experience, religious beliefs, geographic location. We acquire, change, and discard these characteristics as we progress through our lives.

Defining characteristics of diversity as either primary or secondary enhances our understanding, but we must remember that each person is defined by the interrelation of all characteristics. In dealing with diversity in the workforce, managers must consider the complete person—not one or a few of a person's differences.

Why Is Diversity Important?

The U.S. workforce is becoming increasingly diverse. Once dominated by white men, today's workforce includes significantly more women, African Americans, Hispanics, and other minorities, as well as disabled and older workers. The Census Bureau predicts that by 2044, minorities will be the majority in the U.S.[41] These groups have traditionally faced discrimination and higher unemployment rates and have been denied opportunities to assume leadership roles in corporate America. Consequently, more

---

conciliation
a method of outside resolution of labor and management differences in which a third party is brought in to keep the two sides talking.

mediation
a method of outside resolution of labor and management differences in which the third party's role is to suggest or propose a solution to the problem.

arbitration
settlement of a labor/management dispute by a third party whose solution is legally binding and enforceable.

 connect

▶ Need help understanding mediation vs. arbitration? Visit your Connect ebook video tab for a brief animated explanation.

LO 10-8

Describe the importance of diversity in the workforce.

diversity
the participation of different ages, genders, races, ethnicities, nationalities, and abilities in the workplace.

FIGURE 10.2
Characteristics of Diversity

Source: Marilyn Loden and Judy B. Rosener, *Workforce America! Managing Employee Diversity as a Vital Resource* (New York: McGraw-Hill, 1991), p. 20.

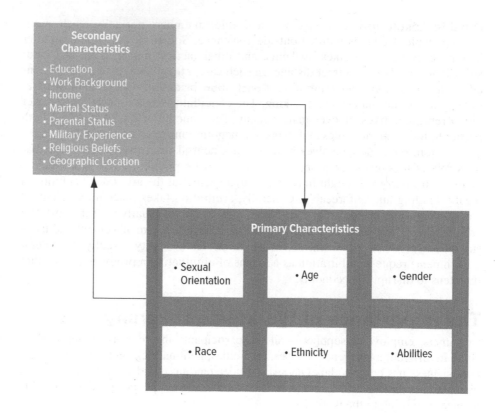

and more companies are trying to improve HRM programs to recruit, develop, and retain more diverse employees to better serve their diverse customers. Some firms are providing special programs such as sponsored affinity groups, mentoring programs, and special career development opportunities. Silicon Valley startup Social Capital and its CEO Chamath Palihapitiya have focused on investing in and nurturing more startups in some of the controversial industries like education and health care because they are imperative and typically handicapped by bias or perceived risk. Their program, Capital-as-a-Service (CAAS), automates early investment decisions and thus eliminates bias against unconventional startups. CAAS received 30,000 applications from startups and invested in 30, with half the funded CEOs being nonwhite and 40 percent female.[42] Table 10.4 shows the top companies for minorities according to a study by *DiversityInc*. Effectively managing diversity in the workforce involves cultivating and valuing its benefits and minimizing its problems.

The Benefits of Workforce Diversity

There are a number of benefits to fostering and valuing workforce diversity, including the following:

1. More productive use of a company's human resources.
2. Reduced conflict among employees of different ethnicities, races, religions, and sexual orientations as they learn to respect each other's differences.
3. More productive working relationships among diverse employees as they learn more about and accept each other.

TABLE 10.4   Top 10 Companies for Diversity

1. IBM
2. Accenture
3. EY
4. Dell
5. Sodexo
6. Wyndham Worldwide
7. Procter & Gamble
8. Deloitte
9. Caterpillar
10. MasterCard

Source: "The DiversityInc Top 10 Companies for Global Diversity," DiversityInc, http://di.diversityinc.com/top-10-companies-global-diversity/ (accessed April 26, 2018).

4. Increased commitment to and sharing of organizational goals among diverse employees at all organizational levels.

5. Increased innovation and creativity as diverse employees bring new, unique perspectives to decision-making and problem-solving tasks.

6. Increased ability to serve the needs of an increasingly diverse customer base.[43]

Companies that do not value their diverse employees are likely to experience greater conflict, as well as prejudice and discrimination. Among individual employees, for example, racial slurs and gestures, sexist comments, and other behaviors by co-workers harm the individuals at whom such behavior is directed. The victims of such behavior may feel hurt, depressed, or even threatened and may suffer from lowered self-esteem, all of which harm their productivity and morale. In such cases, women and minority employees may simply leave the firm, wasting the time, money, and other resources spent on hiring and training them. When discrimination comes from a supervisor, employees may also fear for their jobs. A discriminatory atmosphere not only can harm productivity and increase turnover, but it may also subject a firm to costly lawsuits and negative publicity.

Astute businesses recognize that they need to modify their human resource management programs to target the needs of *all* their diverse employees as well as the needs of the firm itself. They realize that the benefits of diversity are long term in nature and come only to those organizations willing to make the commitment. Most importantly, as workforce diversity becomes a valued organizational asset, companies spend less time managing

Some of the major benefits of diversity include a wider range of employee perspectives, greater innovation and creativity, and the ability to target a diverse customer base more effectively.

©Inti St Clair/Blend Images

conflict and more time accomplishing tasks and satisfying customers, which is, after all, the purpose of business.

Affirmative Action

Many companies strive to improve their working environment through **affirmative action programs,** legally mandated plans that try to increase job opportunities for minority groups by analyzing the current pool of workers; identifying areas where women and minorities are underrepresented; and establishing specific hiring and promotion goals, along with target dates, for meeting those goals to resolve the discrepancy. Affirmative action began in 1965 as Lyndon B. Johnson issued the first of a series of presidential directives. It was designed to make up for past hiring and promotion prejudices, to overcome workplace discrimination, and to provide equal employment opportunities for blacks and whites. Since then, minorities have made solid gains.

Legislation passed in 1991 reinforces affirmative action but prohibits organizations from setting hiring quotas that might result in reverse discrimination. Reverse discrimination occurs when a company' policies force it to consider only minorities or women instead of concentrating on hiring the person who is best qualified. More companies are arguing that affirmative action stifles their ability to hire the best employees, regardless of their minority status. Because of these problems, affirmative action became politically questionable.

Trends in Management of the Workforce

Advances in information technology as well as the last recession and financial crisis have had a major impact on employment. Employee benefits, especially health care, remain a significant and controversial issue. The nature of the workplace is changing as well. Microentrepreneurs, small-scale businesses with no more than five employees, are growing rapidly. The gig economy where individuals move from one project to another is growing quickly as well. The sharing economy (online gig economy) is providing opportunities for independent contractors to own their businesses and contract their time and resources as they see fit. The rise of Uber, Airbnb, and Lyft illustrates this new form of employment. Space is also increasingly shared among employees and small businesses, a new trend called *coworking.* Companies such as WeWork and Industrious Office are capitalizing on these coworking opportunities by offering flexible, agile workspaces to businesses ranging from freelancers to *Fortune* 500 companies.[44]

This new job market has been supported by smartphones, tablet computers, and other technologies. However, a major challenge with the increasing use of smartphones and tablet computers is the blurring between leisure and work time, with some employers calling employees after hours.[45] Employees themselves are mixing work and personal time by using social media in the office. This is requiring companies to come up with new policies that limit how employees can use social media in the workplace. On the other hand, it is estimated that 40 percent of full-time workers log more that 50 hours per week, while 20 percent log more than 60 hours. Many in this country are concerned about burnout and the negative impact that these work schedules can have on productivity.[46] Clearly, technology is changing the dynamics of the workplace in both positive and negative ways.

Off to WeWork We Go

WeWork

Founder: Adam Neumann and Miguel McKelvey

Founded: 2010, in New York, New York

Success: With more than 280 locations worldwide, WeWork is one of the highest valued private U.S. tech firms at an estimated worth of $20 billion.

WeWork sells more than just office space—it sells a community experience. WeWork, a frontrunner in the coworking industry, rents shared workspaces to clients and customizes the space with unique value-added services. The company has different packages based upon client needs, including a private office space starting at $450/month, a dedicated desk at $350/month, and a hot desk—or a workspace in a common area—starting at $220/month.

Many of WeWork's locations are rented and communally shared by different businesses or entrepreneurs. This common workspace is appealing to those who do not require a permanent office space. Clients can rent by the month, allowing them to use WeWork's services as needed. By co-sharing space, WeWork's clients have a place to work while being part of a larger community of workers. WeWork provides many amenities in these spaces, including craft beer, Internet, printing, daily cleaning, and mailing and packaging.

WeWork is also attracting the attention of larger firms. General Electric, for instance, uses WeWork to provide workspace for employees in six locations. WeWork is redesigning and consolidating workspace to save on costs while optimizing collaboration. If its popularity continues, the firm may help foster a revolution in office workspace design.[48]

Critical Thinking Questions

1. What services does WeWork offer that could help improve workspace?
2. What impact do you think WeWork's services have on the satisfaction of employees and entrepreneurs that use its workspaces? Why?
3. WeWork offers a flexible work environment for employers and entrepreneurs. How might this flexible work environment save on costs?

It is important for human resource managers to be aware of legal issues regarding worker rights. Strict criteria—such as having management responsibilities, having advanced degrees, or making more than $455 a week—determine whether an employee is exempt from overtime pay.[49] Interestingly, although it might currently be legal for certain employers to request an applicant's Facebook password, employees who "rant" about their employers on Facebook can receive some form of legal protection. Under the National Labor Relations Act of 1935, certain private-sector employees are allowed to complain about working conditions and pay—which seems to apply to social media sites as well. Threats, on the other hand, are not protected. Even then, however, courts sometimes differ from employers as to what constitutes a threat.[50] Additionally, the Affordable Care Act has extended coverage to millions of Americans. The new law has placed certain health care responsibilities on employers that employ full-time employees.[51] Hence, human resource managers should understand these issues to ensure that the company adheres to all applicable employment laws.

Despite the grim outlook of the past few years, hiring trends are on the rise. Food retailers are fighting for workers in the most competitive U.S. labor market in almost two decades. Kroger Co. is trying to stay competitive by hiring 11,000 workers to improve customer service and efficiency. Also, with the savings from the new federal tax law, Kroger says it will spend $500 million on boosting compensation.[52] This uptick in hiring will require firms not only to know about relevant employee laws, but also to understand how benefits and employee morale can contribute to overall productivity. Many of the most successful firms have discovered ways to balance costs with the well-being of their employees.

Managing human resources is a challenging and creative facet of a business. It is the department that handles the recruiting, hiring, training, and firing of employees. Because of the diligence and detail required in hiring and the sensitivity required in firing, human resource managers have a broad skill set. Human resources, therefore, is vital to the overall functioning of the business because, without the right staff, a firm will not be able to effectively carry out its plans. Like in basketball, a team is only as strong as its individual players, and those players must be able to work together to enhance strengths and downplay weaknesses. In addition, a good human resource manager can anticipate upcoming needs and changes in the business, hiring in line with the dynamics of the market and organization.

Once a good workforce is in place, human resource managers must ensure that employees are properly trained and oriented and that they clearly understand some elements of what the organization expects. Hiring new people is expensive, time consuming, and turbulent; thus, it is imperative that all employees are carefully selected, trained, and motivated so that they will remain committed and loyal to the company. This is not an easy task, but it is one of the responsibilities of the HR manager. Even with references, a résumé, background checks, and an interview, it can be hard to tell how a person will fit in the organization; therefore, the HR manager needs to have skills to be able to anticipate how every individual will "fit in." HR jobs include compensation, labor relations, benefits, training, ethics, and compliance managers. All of the tasks associated with the interface with hiring, developing, and maintaining employee motivation come into play in human resource management. Jobs are diverse and salaries will depend on responsibilities, education, and experience.

One of the major considerations for an HR manager is workforce diversity. A multicultural, multiethnic workforce consisting of men and women will help to bring in a variety of viewpoints and improve the quality and creativity of organizational decision making. Diversity is an asset and can help a company from having blind spots or harmony in thought, background, and perspective, which stifles good team decisions. However, a diverse workforce can present some management challenges. Human resource management is often responsible for managing diversity training and compliance to make sure employees do not violate the ethical culture of the organization or break the law. Different people have different goals, motivations, and ways of thinking about issues that are informed by their culture, religion, and the people closest to them. No one way of thinking is more right or more wrong than others, and they are all valuable. A human resource manager's job can become very complicated, however, because of diversity. To be good at human resources, you should be aware of the value of differences, strive to be culturally sensitive, and ideally should have a strong understanding and appreciation of different cultures and religions. Human resources managers' ability to manage diversity and those differences will affect their overall career success.

Review Your Understanding

Explain the significance of human resources management.

Human resource, or personnel, management refers to all the activities involved in determining an organization's human resource needs and acquiring, training, and compensating people to fill those needs. It is concerned with maximizing the satisfaction of employees and improving their efficiency to meet organizational objectives.

Summarize the processes of recruiting and selecting human resources for a company

First, the human resource manager must determine the firm's future human resource needs and develop a strategy to meet them. Recruiting is the formation of a pool of qualified applicants from which management will select employees; it takes place both internally and externally. Selection is the process of collecting information about applicants and using that information to decide which ones to hire; it includes the application, interviewing, testing, and reference checking.

Describe how workers are trained and their performance appraised.

Training teaches employees how to do their specific job tasks; development is training that augments the skills and knowledge of managers and professionals, as well as current employees. Appraising performance involves identifying an employee's strengths and weaknesses on the job. Performance appraisals may be subjective or objective.

Identify the types of turnover companies may experience.

A promotion is an advancement to a higher-level job with increased authority, responsibility, and pay. A transfer is a move to another job within the company at essentially the same level and wage. Separations occur when employees resign, retire, are terminated, or are laid off.

Explain why turnover is an important issue.

Turnovers due to separation are expensive because of the time, money, and effort required to select, train, and manage new employees.

Specify the various ways a worker may be compensated.

Wages are financial compensation based on the number of hours worked (time wages) or the number of units produced (piece wages). Commissions are a fixed amount or a percentage of a sale paid as compensation. Salaries are compensation calculated on a weekly, monthly, or annual basis, regardless of the number of hours worked or the number of items produced. Bonuses and profit sharing are types of financial incentives. Benefits are nonfinancial forms of compensation, such as vacation, insurance, and sick leave.

Evaluate some of the issues associated with unionized employees, including collective bargaining and dispute resolution.

Collective bargaining is the negotiation process through which management and unions reach an agreement on a labor

contract—the formal, written document that spells out the relationship written between the union and management. If labor and management cannot agree on a contract, labor union members may picket, strike, or boycott the firm, while management may lock out striking employees, hire strikebreakers, or form employers' associations. In a deadlock, labor disputes may be resolved by a third party—a conciliator, mediator, or arbitrator.

Describe the importance of diversity in the workforce.

When companies value and effectively manage their diverse workforces, they experience more productive use of human resources, reduced conflict, better work relationships among workers, increased commitment to and sharing of organizational goals, increased innovation and creativity, and enhanced ability to serve diverse customers.

Assess an organization's efforts to reduce its workforce size and to manage the resulting effects.

Based on the material in this chapter, you should be able to answer the questions posed in the "Solve the Dilemma" feature near the end of the chapter and evaluate the company's efforts to manage the human consequences of its downsizing.

Critical Thinking Questions

Enter the World of Business Questions

1. What are some ways that Sheryl Sandberg demonstrates strong human resource skills?

2. Why do you think communicating with employees is crucial to Sandberg's popularity at Facebook?

3. How is Sandberg attempting to change the workplace in ways that would benefit employees?

Learn the Terms

affirmative action programs 320
arbitration 317
benefits 312
bonuses 312
boycott 316
collective bargaining 315
commission 311
conciliation 317
development 307
diversity 317
human resource management
  (HRM) 300

job analysis 301
job description 301
job specification 301
labor contract 315
labor unions 314
lockout 316
mediation 317
mentoring 307
orientation 307
picketing 316
profit sharing 312
promotion 309

recruiting 301
salary 311
selection 303
separations 310
strikebreakers 316
strikes 316
Title VII of the Civil Rights Act 306
training 307
transfer 309
turnover 309
wage/salary survey 311
wages 311

Check Your Progress

1. Distinguish among job analysis, job descriptions, and job specifications. How do they relate to planning in human resources management?

2. What activities are involved in acquiring and maintaining the appropriate level of qualified human resources? Name the stages of the selection process.

3. What are the two types of training programs? Relate training to kinds of jobs.

4. What is the significance of a performance appraisal? How do managers appraise employees?

5. Why does turnover occur? List the types of turnover. Why do businesses want to reduce turnover due to separations?

6. Relate wages, salaries, bonuses, and benefits to Herzberg's distinction between hygiene and motivation factors. How does the form of compensation relate to the type of job?

7. What is the role of benefits? Name some examples of benefits.

8. Describe the negotiation process through which management and unions reach an agreement on a contract.

9. Besides collective bargaining and the grievance procedures, what other alternatives are available to labor and management to handle labor disputes?

10. What are the benefits associated with a diverse workforce?

Get Involved

1. Although many companies screen applicants and test employees for illegal drug use, such testing is somewhat controversial. Find some companies in your community that test applicants and/or employees for drugs. Why do they have such a policy? How do the employees feel about it? Using this information, debate the pros and cons of drug testing in the workplace.

2. If collective bargaining and the grievance procedures have not been able to settle a current labor dispute,

what tactics would you and other employees adopt? Which tactics would be best for which situations? Give examples.

3. Find some examples of companies that value their diverse workforces, perhaps some of the companies mentioned in the chapter. In what ways have these firms derived benefits from promoting cultural diversity? How have they dealt with the problems associated with cultural diversity?

Build Your Skills

Appreciating and Valuing Diversity

Background

Here's a quick self-assessment to get you to think about diversity issues and evaluate the behaviors you exhibit that reflect your level of appreciation of other cultures:

Do you . . .	Regularly	Sometimes	Never
1. Make a conscious effort not to think stereotypically?			
2. Listen with interest to the ideas of people who don't think like you do?			
3. Respect other people's opinions, even when you disagree?			
4. Spend time with friends who are not your age, race, gender, or the same economic status and education?			
5. Believe your way is not the only way?			
6. Adapt well to change and new situations?			

Do you . . .	Regularly	Sometimes	Never
7. Enjoy traveling, seeing new places, eating different foods, and experiencing other cultures?			
8. Try not to offend or hurt others?			
9. Allow extra time to communicate with someone whose first language is not yours?			
10. Consider the effect of cultural differences on the messages you send and adjust them accordingly?			

Scoring

Number of **Regularly** checks _____ multiplied by 5 = _____

Number of **Sometimes** checks _____ multiplied by 3 = _____

Number of **Never** checks _____ multiplied by 0 = _____

TOTAL _____

Indications from score

40–50   You appear to understand the importance of valuing diversity and exhibit behaviors that support your appreciation of diversity.

6–39   You appear to have a basic understanding of the importance of valuing diversity and exhibit some behaviors that support that understanding.

13–5   You appear to lack a thorough understanding of the importance of valuing diversity and exhibit only some behaviors related to valuing diversity.

0–1   You appear to lack an understanding of valuing diversity and exhibit few, if any, behaviors of an individual who appreciates and values diversity.

Task

In a small group or class discussion, share the results of your assessment. After reading the following list of ways you can increase your knowledge and understanding of other cultures, select one of the items that you have done and share how it helped you learn more about another culture. Finish your discussion by generating your own ideas on other ways you can learn about and understand other cultures and fill in those ideas on the blank lines at the end.

- Be alert to and take advantage of opportunities to talk to and get to know people from other races and ethnic groups. You can find them in your neighborhood, in your classes, at your fitness center, at a concert or sporting event—just about anywhere you go. Take the initiative to strike up a conversation and show a genuine interest in getting to know the other person.

- Select a culture you're interested in and immerse yourself in that culture. Read novels, look at art, take courses, see plays.

- College students often have unique opportunities to travel inexpensively to other countries—for example, as a member of a performing arts group, with a humanitarian mission group, or as part of a college course studying abroad. Actively seek out travel opportunities that will expose you to as many cultures as possible during your college education.

- Study a foreign language.

- Expand your taste buds. The next time you're going to go to a restaurant, instead of choosing that old familiar favorite, find a restaurant that serves ethnic food you've never tried before.

- Many large metropolitan cities sponsor ethnic festivals, particularly in the summertime, where you can go and take in the sights and sounds of other cultures. Take advantage of these opportunities to have a fun time learning about cultures that are different from yours.

- _____

- _____

Solve the Dilemma

Morale among the Survivors

Medallion Corporation manufactures quality carpeting and linoleum for homes throughout the United States. A recession and subsequent downturn in home sales has sharply cut the company's sales. Medallion found itself in the unenviable position of having to lay off hundreds of employees in the home office (the manufacturing facilities) as well as many salespeople. Employees were called in on Friday afternoon and told about their status in individual meetings with their supervisors. The laid-off employees were given one additional month of work and a month of severance pay, along with the opportunity to sign up for classes to help with the transition, including job search tactics and résumé writing.

Several months after the cutbacks, morale was at an all-time low for the company, although productivity had improved. Medallion brought in consultants, who suggested that the leaner, flatter organizational structure would be suitable for more team activities. Medallion, therefore, set up task forces and teams to deal with employee concerns, but the diversity of the workforce led to conflict and misunderstandings among team members. Medallion is evaluating how to proceed with this new team approach.

LO 10-9

Assess an organization's efforts to reduce its workforce size and to manage the resulting effects.

Critical Thinking Questions

1. What did Medallion's HRM department do right in dealing with the employees who were laid off?

2. What are some of the potential problems that must be dealt with after an organization experiences a major trauma such as massive layoffs?

3. What can Medallion do to make the team approach work more smoothly? What role do you think diversity training should play?

Build Your Business Plan   ▦ connect

Managing Human Resources

Now is the time to start thinking about the employees you will need to hire to implement your business plan. What kinds of background/skills are you going to look for in potential employees? Are you going to require a certain amount of work experience?

When you are starting a business, you are often only able to hire part-time employees because you cannot afford to pay the benefits for a full-time employee. Remember at the end of the last chapter we discussed how important it is to think of ways to motivate your employees when you cannot afford to pay them what you would like.

You need to consider how you are going to recruit your employees. When you are first starting your business, it is often a good idea to ask people you respect (and not necessarily members of your family) for any recommendations of potential employees they might have. You probably won't be able to afford to advertise in the classifieds, so announcements in sources such as church bulletins or community bulletin boards should be considered as an excellent way to attract potential candidates with little, if any, investment.

Finally, you need to think about hiring employees from diverse backgrounds, especially if you are considering targeting diverse segments. The more diverse your employees, the greater the chance you will be able to draw in diverse customers.

Visit Connect to practice building your business plan with the Business Plan Prep Exercises.

See for Yourself Videocase

The Importance of Hollywood Labor Unions

You might be familiar with unions for teachers or autoworkers. But what about unions for actors, radio artists, and screenwriters? Because we tend to view Hollywood as a glamorous place, we are tempted to view unions as unnecessary for these types of professions. Yet Hollywood unions were, and continue to be, important players in the careers of Hollywood artists.

When actors first became mainstream in the early 20th century, working conditions for the industry included long work weeks and low pay. Studios essentially "owned" their artists, which meant that rival studios would not hire actors or actresses once their contracts ended. Actors were forced to work for the same studio to advance their careers. Negotiations with studios often proved fruitless.

Because strikes can be so disruptive and risky, they are often used as a last resort. Yet, in 1919 the Actors' Equity Association, a union for theatrical performers, and the American Federation of Labor staged a Broadway strike to protest harsh working conditions. The strike resulted in a five-year contract and promises to improve labor conditions. Although the event happened off Broadway and not in Hollywood, it would inspire other artists to begin forming their own unions.

Working conditions might have improved somewhat for theatrical actors, but radio artists, film actors, and screenwriters still had to bear with hard conditions. For instance, radio artists might do an entire show and then receive only a dollar. Film actors would work around the clock with few (if any) breaks. Screenwriters experienced salary cuts. As individuals, these artists did not have much bargaining power with studios. Realizing that banding together could improve conditions, the Masquers Club (later the Screen Actors Guild of America) was created in 1925. It was followed by the Screenwriters Guild (later renamed the Writers Guild of America) in 1933 and the American Federation of Radio Artists in 1937.

It would take lawsuits, strikes, and hardline negotiations for Hollywood artists to receive more rights. This often required artists to take risks such as suspensions or firings in the hope of better treatment. In 1988, the Writers Guild of America organized the longest strike in Hollywood history after disagreements with producers over payments and creative rights. The strike, which lasted five months, was estimated to cost the industry $500 million.

Now that working conditions seem to have improved, it might be tempting to discard Hollywood unions as no longer useful. Yet, even today, conflicts often occur between artists and studios. For instance, the Screen Actors Guild of America (SAG) watches to make sure that low-budget actors know their rights and are not exploited by producers. Unequal treatment still happens in the entertainment industry. For instance, minorities and female actors/actresses still tend to be paid less than Caucasian and male actors. The introduction of new media venues, particularly the Internet, may also warrant additional negotiations.

To address these challenges, many unions are banding together to address mutual concerns in the industry. In 2012, SAG united with the American Federation of Television and Radio Artists (AFTRA). This increases the bargaining power of the combined union. Additionally, Hollywood unions work closely with the American Federation of Labor and Congress of Industrial Organizations (AFL-CIO), which represents a federation of labor unions. The merger between SAG and AFTRA creates solidarity in the industry's unions through the formation of an Industry Coordinating Committee. This allows for coordination of activity among 10 or 12 major unions within the industry.

Although these unions are for movie stars and other artists, the goal is much the same as for other unions across the nation. As the industry expands, Hollywood unions feel that they must work together to secure benefits for their members while striving to arrive at mutually beneficial agreements with studios, producers, and other stakeholders in the entertainment industry.

Critical Thinking Questions

1. Why are Hollywood labor unions considered necessary?

2. Why is striking often avoided if possible?

3. Why do you think unions in the entertainment industry are banding together?

You can find the related video in the Video Library in Connect. Ask your instructor how you can access Connect.

Team Exercise

Form groups and go to Monster.com and look up job descriptions for positions in business (account executive in advertising, marketing manager, human resource director, production supervisor, financial analyst, bank teller, etc.). What are the key requirements for the position that you have been assigned (education, work experience, language/ computer skills, etc.)? Does the position announcement provide a thorough understanding of the job? Was any key information that you would have expected omitted? Report your findings to the class.

Ask your instructor about the role play exercises available with this book to practice working with a business team.

Notes

1. Sarah Elbert, "Modern Super Hero," *Delta Sky*, December 2017, pp. 78–81, 178–179; Henna Inam, "Sheryl Sandberg on Being Human," *Transformational Leadership: Coaching & Leadership Development*, June 21, 2015, http://www .transformleaders.tv/sheryl-sandberg-on-being-human/ (accessed January 14, 2018); Shawn Doyle, "What Every Boss Can Learn about Leadership from Sheryl Sandberg from Facebook," *Inc.*, September 7, 2017, https://www

.inc.com/shawn-doyle/what-every-boss-can-learn-about-leadership-from-sh.html (accessed January 14, 2018); Alexandra Topping and Decca Aitkenhead, "Sheryl Sandberg Credits Mark Zuckerberg with Saving Her Life," *The Guardian,* April 15, 207, https://www.theguardian.com/technology/2017/apr/15/sheryl-sandberg-credits-mark-zuckerberg-with-saving-her-life (accessed January 14, 2018).

2. "Consumers Prefer Ethical Companies Survey Shows," *ECI Connector,* January 7, 2016. https://www.ethics.org/blogs/eci-connector/2016/01/07/consumers-prefer-ethical-companies-survey-shows (accessed April 27, 2018).

3. "Dictionary of Occupational Titles," http://www.occupationalinfo.org/ (accessed April 27, 2018).

4. George Anders, "The Rare Find," *Bloomberg Businessweek,* October 17–23, 2011, pp. 106–112; Philip Delves Broughton, "Spotting the Exceptionally Talented," *Financial Times,* October 13, 2011, https://www.ft.com/content/25dc1872-f4ba-11e0-a286-00144feab49a (accessed December 27, 2017); Joe Light, "Recruiters Rethink Online Playbook," *The Wall Street Journal,* January 18, 2011, http://online.wsj.com/article/SB10001424052748704307404576080492613858846.html (accessed December 27, 2017); Kris Maher, "A Tactical Recruiting Effort Pays Off," *The Wall Street Journal,* October 24, 2011, p. R6; Jeanne Meister, "The Death of the Resume: Five Ways to Re-Imagine Recruiting," *Forbes,* July 23, 2012, http://www.forbes.com/sites/jeannemeister/2012/07/23/the-death-of-the-resume-five-ways-to-re-imagine-recruiting/ (accessed December 27, 2017); Instagram, "Salesforce.com," https://www.instagram.com/salesforcejobs/?hl=en (accessed December 27, 2017); Erin Engstrom, "Recruiting on Instagram: 4 Companies That Are Crushing It," https://recruiterbox.com/blog/recruiting-on-instagram-4-companies-crushing-it/ (accessed December 27, 2017); J. T. O'Donnell, "A Powerful Way to Use Instagram to Recruit Employees," *Inc.,* June 25, 2015, https://www.inc.com/jt-odonnell/a-powerful-way-to-use-instagram-to-recruit-employees.html (accessed December 27, 2017); Zlati Meyer and Kellie Ell, "McDonald's Looks to Snapchat to Hire 250,000 for Summer," *USA Today,* June 14, 2017, https://www.usatoday.com/story/money/business/2017/06/12/mcdonalds-hires-teens-via-snapchat/102782168/ (accessed December 27, 2017); Vineeta Sawkar, "General Mills Recruits Millennials with a High-Tech Virtual Tour," *Star Tribune,* September 15, 2015, http://www.startribune.com/general-mills-recruits-millennials-with-gopro-goggle-tour/327721891/ (accessed December 27, 2017); John Sullivan, "12 Innovative Recruiting Strategies That Savvy Companies Are Using," July 10, 2017, https://business.linkedin.com/talent-solutions/blog/recruiting-strategy/2017/12-innovative-recruiting-strategies-that-savvy-companies-are-usi (accessed December 27, 2017); Roy Maurer, "Employee Referrals Remain Top Source for Hires," Society for Human Resource Management, June 23, 2017, https://www.shrm.org/resourcesandtools/hr-topics/talent-acquisition/pages/employee-referrals-remains-top-source-hires.aspx (accessed December 27, 2017).

5. Alison Doyle, "Best Social Media Sites for Job Searching," *The Balance,* June 7, 2016, https://www.thebalance.com/best-social-media-sites-for-job-searching-2062617 (accessed April 18, 2017); Susan P. Joyce, "Guide to Social Media and Job Search," *Job-Hunt,* https://www.job-hunt.org/social-networking/social-media.shtml (accessed April 27, 2018).

6. "Study: Which Companies Offer the Fastest Online Job Application Process?" *Indeed Blog,* August 15, 2016, http://blog.indeed.com/2016/08/15/fastest-job-application-process/ (accessed April 27, 2018).

7. Sachi Barreiro, "Can Potential Employers Check Out Your Facebook Page?" *Nolo.com,* https://www.nolo.com/legal-encyclopedia/can-potential-employers-check-your-facebook-page.html (accessed April 27, 2018); Diana Coker, "Should the Facebook Password Be Given in a Job Interview?" *HR Digest,* December 29, 2016, https://www.thehrdigest.com/facebook-password-given-job-interview/ (accessed April 27, 2018).

8. "Eight Ways to Leverage Social Media as a Hiring Tool," *Forbes,* February 19, 2018, https://www.forbes.com/sites/forbesagencycouncil/2018/02/19/eight-ways-to-leverage-social-media-as-a-hiring-tool/#785ba1e3750c (accessed April 27, 2018).

9. "70% of Employers Say Prescription Drug Abuse Affects Workplace," *Insurance Journal,* March 10, 2017, http://www.insurancejournal.com/news/national/2017/03/10/444117.htm (accessed April 27, 2018).

10. Alexia Elejalde-Ruiz, "Cost of Substance Abuse Hits Employers Hard, New Tool Shows," *Chicago Tribune,* April 6, 2017, http://www.chicagotribune.com/business/ct-workplace-substance-abuse-0407-biz-20170406-story.html (accessed April 27, 2018).

11. Lauren Weber and Mike Esterl, "E-Cigarette Rise Poses Quandary for Employers," *The Wall Street Journal,* January 16, 2014, p. A2.

12. Roy Maurer, "Know Before You Hire: 2017 Employment Screening Trends," Society for Human Resource Management, January 25, 2017, https://www.shrm.org/resourcesandtools/hr-topics/talent-acquisition/pages/2017-employment-screening-trends.aspx (accessed April 27, 2018).

13. Allison Linn, "Desperate Measures: Why Some People Fake Their Resumes," *CNBC,* February 7, 2014, www.cnbc.com/id/101397212 (accessed April 27, 2018).

14. U.S. Equal Employment Opportunity Commission, "EEOC Releases Fiscal Year 2017 Enforcement and Litigation Data," January 25, 2018, https://www.eeoc.gov/eeoc/newsroom/release/1-25-18.cfm (accessed April 27, 2018).

15. "Missing Pieces Report: The 2016 Board Diversity Census of Women and Minorities on Fortune 500 Boards," Deloitte, 2016, https://www2.deloitte.com/content/dam/Deloitte/us/Documents/center-for-corporate-governance/us-board-diversity-census-missing-pieces.pdf (accessed April 27, 2018).

16. Stephen Bastien, "12 Benefits of Hiring Older Workers," *Entrepreneur.com,* September 20, 2006, www.entrepreneur.com/article/167500 (accessed April 27, 2018).

17. Jennifer Calfas, "Women Have Pushed for Equal Pay for Decades. It's Sad How Little Progress We've Made," *Time,* April 10, 2018, http://time.com/money/5225986/equal-pay-day-2018-gender-wage-gap/ (accessed April 27, 2018).

18. Lydia Dishman, "These Jobs Have the Largest and Smallest Gender Pay Gaps," *Fast Company,* September 27, 2017, https://www.fastcompany.com/40474000/these-jobs-have-the-largest-and-smallest-gender-pay-gaps (accessed April 27, 2018).

19. Camaron Santos and Kellye Whitney, "McDonald's Virtual Strategy Aids Big Business Changes," *Chief Learning Officer,* February 8, 2017, http://www.clomedia.com/2017/02/08/restaurants-orders-up-virtual-training/ (accessed April 27, 2018).

20. "Our Employee First Culture," *Stand for,* 2017, http://standfor.containerstore.com/putting-our-employees-first/ (accessed April 27, 2018).

21. Maury A. Peiperl, "Getting 360-Degree Feedback Right," *Harvard Business Review,* January 2001, pp. 142–148.

22. Chris Musselwhite, "Self-Awareness and the Effective Leader," *Inc.com,* www.inc.com/resources/leadership/articles/20071001/musselwhite.html (accessed April 27, 2018).

23. Rachel Feintzeig, "You're Awesome! Firms Scrap Negative Feedback," *The Wall Street Journal,* February 11, 2015, pp. B1, B5.

24. Marcel Schwantes, "Study Says It Comes Down to Any of These 6 Reasons," *Inc.,* October 23, 2017, https://www.inc.com/marcel-schwantes/why-are-your-employees-quitting-a-study-says-it-comes-down-to-any-of-these-6-reasons.html (accessed April 27, 2018).

25. Alison Doyle, "2018 Federal and State Minimum Wage Rates," *The Balance,* March 26, 2018, https://www.thebalancecareers.com/2017-federal-state-minimum-wage-rates-2061043 (accessed April 27, 2018).

26. "Wage and Hour Division (WHD)," U.S. Department of Labor, January 1, 2018, https://www.dol.gov/whd/state/tipped.htm (accessed April 27, 2018).

27. Trip Stelnicki, "Santa Fe's Minimum Wage Set to Rise 2.82% March 1," *Santa Fe New Mexican,* February 2, 2018, http://www.santafenewmexican.com/news/local_news/santa-fe-s-minimum-wage-set-to-rise-march/article_79e6e4b4-5d4b-54f1-a13d-188520cd10d0.html (accessed April 27, 2018).

28. Kevin Dugan, "Wall Street Bonuses Rise for the First Time in Three Years," *New York Post,* March 15, 2017, http://nypost.com/2017/03/15/wall-street-bonuses-rise-for-the-first-time-in-three-years/ (accessed April 27, 2018).

29. Bureau of Labor Statistics, U.S. Department of Labor, "Employer Costs for Employee Compensation—December 2017," Bureau of Labor Statistics, March 20, 2018, https://www.bls.gov/news.release/pdf/ecec.pdf (accessed April 27, 2018).

30. Google, *Environmental Report: 2017 Progress Update,* https://storage.googleapis.com/gweb-environment.appspot.com/pdf/google-2017-environmental-report.pdf (accessed April 20, 2018); Sony Pictures, "Employee Eco-Incentives," *Sony Pictures: A Greener World,* http://www.sonypictures.com/green/act/employee-involvement/employee-incentives.php (accessed April 20, 2018); Sony Pictures, "Greening the Office—Idea to Action Grants and Certification," *Sony Pictures: A Greener World,* http://www.sonypictures.com/green/act/empowering-employees/green-office.php (accessed April 20, 2018); Sony Pictures, "Grant Programs," *Sony Pictures: A Greener World,* http://www.sonypicturesgreenerworld.com/employee/grant-programs (accessed April 20, 2018); Sony Pictures, "Greener World Grant," *Sony Pictures: A Greener World,* http://www.sonypictures.com/green/act/empowering-employees/greener-grant.php (accessed April 20, 2018); Sony Pictures, "Employee Involvement," *Sony Pictures: A Greener World* http://www.sonypictures.com/green/act/employee-involvement/employee-involvement.php (accessed April 20, 2018); Intel, "Intel and the Environment," https://www.intel.com/content/www/us/en/environment/intel-and-the-environment.html (accessed April 20, 2018); D G McCullough, "Putting Your Money Where Your Mouth Is: Companies Link Green Goals to Pay," *The Guardian,* June 26, 2014, https://www.theguardian.com/sustainable-business/2014/jun/26/green-executive-compensation-intel-alcoa-pay (accessed April 20, 2018).

31. Christine Birkner, "Taking Care of Their Own," *Marketing News,* February 2015, pp. 44–49.

32. "Employee Assistance Program," Home Depot, 2017, https://secure.livethehealthyorangelife.com/healthy_living/employee_assistance_program (accessed April 27, 2018).

33. Angus Loten and Sarah E. Needleman, "Laws on Paid Sick Leave Divide Businesses," *The Wall Street Journal,* February 6, 2014, p. B5.

34. Georgia Wells, Deepa Seetharaman, and Yoree Koh, "Employees Appear to Be Unfazed by Crisis," *The Wall Street Journal,* April 10, 2018, p. B4.

35. U.S. Bureau of Labor Statistics, "Union Members Summary," Bureau of Labor Statistics, January 19, 2018 https://www.bls.gov/news.release/union2.nr0.htm (accessed April 27, 2018).

36. U.S. Bureau of Labor Statistics, "Union Members Summary," Bureau of Labor Statistics, January 19, 2018 https://www.bls.gov/news.release/union2.nr0.htm (accessed April 27, 2018).

37. Tom Walsh, "UAW Needs Stronger Message," *USA Today,* February 17, 2014, p. 1B.

38. Boston 29 News, "Fast Food Workers Strike for $15 Minimum Wage," *Fox News,* September 24, 2017, http://www.foxnews.com/us/2017/09/04/fast-food-workers-strike-for-15-minimum-wage.html (accessed April 27, 2018).

39. "NLRB Says Walmart Retaliated against Workers," *CBS News,* January 15, 2014, https://www.cbsnews.com/news/nlrb-says-walmart-retaliated-against-workers/ (accessed April 27, 2018); Steven Greenhouse and Michael J. De La Merced, "At Judge's Urging, Hostess and Union Agree to Mediation," *The New York Times,* November 19, 2012, https://dealbook.nytimes.com/2012/11/19/hostess-and-bakers-union-agree-to-mediation/ (accessed April 27, 2018).

40. Ted Goodman, "Honeywell Union Workers Agree to New Contract after 10-Month Lockout," *Daily Caller,* February 27, 2017, http://dailycaller.com/2017/02/27/honeywell-union-workers-agree-to-new-contract-after-10-month-lockout/ (accessed April 27, 2018).

41. Evan Horowitz, "When Will Minorities Be the Majority?" *Boston Globe,* February 26, 2016, https://www.bostonglobe.com/news/politics/2016/02/26/when-will-minorities-majority/9v5m1Jj8hdGcXvpXtbQT5I/story.html (accessed April 27, 2018).

42. Ainsley Harris, "Social Capital: For Putting Values into Its Ventures," *Fast Company,* "The World's 50 Most Innovative Companies," March/April 2018, pp. 52–54.

43. Taylor H. Cox Jr., "The Multicultural Organization," *Academy of Management Executives* 5 (May 1991), pp. 34–47; Marilyn Loden and Judy B. Rosener, *Workforce America! Managing Employee Diversity as a Vital Resource* (Homewood, IL: Business One Irwin, 1991).

44. Industrious Office website, https://www.industriousoffice.com/ (accessed April 26, 2018); WeWork website, https://www.wework.com/ (accessed April 26, 2018); BS Interactive Inc., segment narrated by Tony Dokoupil, "Co-Working: The New Way to Work," April 15, 2018, https://www.cbsnews.com/news/co-working-the-new-way-to-work/ (accessed April 21, 2018).

45. Paul Davidson, "Overworked and Underpaid?" *USA Today,* April 16, 2012, pp. 1A–2A.

46. Maurie Backman, "Here's a New Reason to Work Fewer Hours," *CNN,* June 20, 2017, http://money.cnn.com/2017/06/20/pf/work-hours/index.html (accessed April 27, 2018).

47. "The World's 30 Biggest Employers Will Surprise You," *MSN Money,* September 18, 2017, https://www.msn.com/en-us/money/careersandeducation/the-worlds-30-biggest-employers-will-surprise-you/ss-AArVuGH#image=1 (accessed April 27, 2018).

48. WeWork website, https://www.wework.com/ (accessed April 21, 2018); CBS Interactive Inc., segment narrated by Tony Dokoupil, "Co-Working: The New Way to Work," April 15, 2018, https://www.cbsnews.com/news/co-working-the-new-way-to-work/ (accessed April 21, 2018); Jessi Hempel, "Why WeWork Thinks It's Worth $20 Billion," *Wired,* https://www.wired.com/story/this-is-why-wework-thinks-its-worth-20-billion/ (accessed April 21, 2018).

49. U.S. Department of Labor: Wage and Hour Division, "Fact Sheet #17A: Exemption for Executive, Administrative, Professional, Computer & Outside Sales Employees Under the Fair Labor Standards Act (FLSA)," July 2008, https://www.dol.gov/whd/overtime/fs17a_overview.pdf (accessed April 27, 2018).

50. Parker Poe Adams & Bernstein LLP, "Expletive-Laced Facebook Rant Protected under Federal Law," *Employment Law Alliance,* May 15, 2017, https://www.employmentlawalliance.com/firms/parkerpoe/articles/expletive-laced-facebook-rant-protected-under-federal-labor-law (accessed April 27, 2018); Melanie Trottman, "For Angry Employees, Legal Cover for Rants," *The Wall Street Journal,* December 2, 2011, http://online.wsj.com/article/SB10001424052970203710704577049822809710332.html (accessed April 27, 2018).

51. Bruce Japsen, "Insurer Profits Rise as Trump Waves White Flag on Obamacare Repeal," *Forbes,* February 4, 2018, https://www.forbes.com/sites/brucejapsen/2018/02/04/insurer-profits-rise-as-trump-waves-white-flag-on-obamacare-repeal/#33cdaa56e4a2 (accessed April 27, 2018); IRS, "Affordable Care Act Tax Provisions for Large Employers," https://www.irs.gov/Affordable-Care-Act/Employers/Affordable-Care-Act-Tax-Provisions-for-Large-Employers (accessed April 27, 2018).

52. Heather Haddon, "Kroger to Bulk Up Store Staffing," *The Wall Street Journal,* April 11, 2018, p. B3.

Credits

Marketing: Developing Relationships

PART 5

Marketing: Developing Relationships

11 Customer-Driven Marketing

©James Leynse/Corbis via Getty Images

Chapter Outline

Learning Objectives

After reading this chapter, you will be able to:

LO 11-1 Define *marketing*.

LO 11-2 Describe the exchange process.

LO 11-3 Specify the functions of marketing.

LO 11-4 Explain the marketing concept and its implications for developing marketing strategies.

LO 11-5 Examine the development of a marketing strategy, including market segmentation and marketing mix.

LO 11-6 Describe how marketers conduct marketing research and study buying behavior.

LO 11-7 Summarize the environmental forces that influence marketing decisions.

LO 11-8 Propose a solution for resolving a problem with a company's marketing plans.

Enter the World of Business

Dollar General Narrows in on Strategic Segmentation

Dollar General has overtaken the competition by recognizing that the large income gap in the United States leaves a significant market of locations untouched by rival stores. As such, Dollar General's business model focuses on selling small-ticket items to a target market of low-income consumers on tight budgets. In places like Decatur, Arkansas, a high poverty rate and a medium household income under $35,000 makes them an unattractive location for larger stores. However, understanding this market is where Dollar General found its niche.

Larger chain stores—such as Walmart or Whole Foods—find it difficult to sustain operations in these markets. Walmart faces startup costs of $15 million when creating a new Supercenter, in contrast with the $250,000 that Dollar General spends on a new store. Dollar General makes good use of shelving space, fitting a variety of different products within a space one-tenth the size of a Walmart store. Understanding that its leading target market shops paycheck-to-paycheck, the store avoids selling items in bulk like many competitors in the industry.

In 2016, Dollar General's CEO Todd Vasos discussed the company's vision to target all of the opportunities in the United States and add to their existing 12,483 stores. The idea is to seek out small towns consisting of low-income households and individuals that rely on government assistance. These are areas that are not looking to thrive economically, but primarily to survive in a growing economy. Dollar General announced it would add around 1,000 new stores as part of its $22 billion expansion plan.

Although it may not provide high-salary jobs, Dollar General offers an option for those that would not otherwise have one. It is committed to continuing its strategy of recognizing and catering to "Anytown, USA."[1]

Introduction

Marketing involves planning and executing the development, pricing, promotion, and distribution of ideas, goods, and services to create exchanges that satisfy individual and organizational goals. These activities ensure that the products consumers want to buy are available at a price they are willing to pay and that consumers are provided with information about product features and availability. Organizations of all sizes and objectives engage in these activities.

In this chapter, we focus on the basic principles of marketing. First we define and examine the nature of marketing. Then we look at how marketers develop marketing strategies to satisfy the needs and wants of their customers. Next we discuss buying behavior and how marketers use research to determine what consumers want to buy and why. Finally, we explore the impact of the environment on marketing activities.

Nature of Marketing

LO 11-1

Define *marketing*.

marketing
a group of activities designed to expedite transactions by creating, distributing, pricing, and promoting goods, services, and ideas.

exchange
the act of giving up one thing (money, credit, labor, goods) in return for something else (goods, services, or ideas).

A vital part of any business undertaking, **marketing** is a group of activities designed to expedite transactions by creating, distributing, pricing, and promoting goods, services, and ideas. These activities create value by allowing individuals and organizations to obtain what they need and want. A business cannot achieve its objectives unless it provides something that customers value. But just creating an innovative product that meets many users' needs isn't sufficient in today's volatile global marketplace. Products must be conveniently available, competitively priced, and uniquely promoted.

Marketing is an important part of a firm's overall strategy. Other functional areas of the business—such as operations, finance, and all areas of management—must be coordinated with marketing decisions. Marketing has the important function of providing revenue to sustain a firm. Only by creating trust and effective relationships with customers can a firm succeed in the long run. Businesses try to respond to consumer wants and needs and to anticipate changes in the environment. Unfortunately, it is difficult to understand and predict what consumers want: Motives are often unclear; few principles can be applied consistently; and markets tend to fragment, each desiring customized products, new value, or better service.

It is important to note what marketing is not: It is not manipulating consumers to get them to buy products they do not want. It is not just advertising and selling; it is a systematic approach to satisfying consumers. Marketing focuses on the many activities—planning, pricing, promoting, and distributing products—that foster exchanges. Unfortunately, the mass media and movies sometimes portray marketing as unethical or as not adding value to business. In this chapter, we point out that marketing is essential and provides important benefits in making products available to consumers.

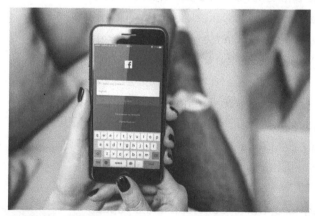

Companies find that communicating with customers through social media sites can enhance customer relationships and create value for their brands.

©Denis Rozhnovsky/Alamy Stock Photo

LO 11-2

Describe the exchange process.

The Exchange Relationship

At the heart of all business is the **exchange**, the act of giving up one thing (money, credit, labor, goods)

in return for something else (goods, services, or ideas). Businesses exchange their goods, services, or ideas for money or credit supplied by customers in a voluntary *exchange relationship,* illustrated in Figure 11.1. The buyer must feel good about the purchase, or the exchange will not continue. If your cell phone service works everywhere, you will probably feel good about using its services. But if you have a lot of dropped calls, you will probably use another phone service next time.

For an exchange to occur, certain conditions are required. As indicated by the arrows in Figure 11.1, buyers and sellers must be able to communicate about the "something of value" available to each. An exchange does not necessarily take place just because buyers and sellers have something of value to exchange. Each participant must be willing to give up his or her respective "something of value" to receive the "something" held by the other. You are willing to exchange your "something of value"—your money or credit—for soft drinks, football tickets, or new shoes because you consider those products more valuable or more important than holding on to your cash or credit potential.

When you think of marketing products, you may think of tangible things—cars, smartphones, or books, for example. What most consumers want, however, is a way to get a job done, solve a problem, or gain some enjoyment. You may purchase a Hoover vacuum cleaner not because you want a vacuum cleaner, but because you want clean carpets. Starbucks serves coffee drinks at a premium price, providing convenience, quality, and an inviting environment. It claims that it is not in the "coffee business serving people" but is in the "people business serving coffee." Therefore, the tangible product itself may not be as important as the image or the benefits associated with the product. This intangible "something of value" may be capability gained from using a product or the image evoked by it, or even the brand name. Good examples of items with brand names that are easy to remember include ColourPop Lippie Stix, Tide detergent, and the Ford Mustang. The label or brand name, such as Ravenswood or Smoking Loon wine, may also offer the added bonus of being a conversation piece in a social environment.

Specify the functions of marketing.

Functions of Marketing

Marketing focuses on a complex set of activities that must be performed to accomplish objectives and generate exchanges. These activities include buying, selling, transporting, storing, grading, financing, marketing research, and risk taking.

Buying.   Everyone who shops for products (consumers, stores, businesses, governments) decides whether and what to buy. A marketer must understand buyers' needs and desires to determine what products to make available.

Selling.   The exchange process is expedited through selling. Marketers usually view selling as a persuasive activity that is accomplished through promotion (advertising, personal selling, sales promotion, publicity, and packaging).

Transporting.   Transporting is the process of moving products from the seller to the buyer. Marketers focus on transportation costs and services.

Storing.   Like transporting, storing is part of the physical distribution of products and includes warehousing goods. Warehouses hold some products for lengthy periods in order to create time utility. Time utility has to do with being able to satisfy demand in a timely manner. This especially pertains to a seasonal good such as orange juice. Fresh oranges are only available for a few months annually, but consumers demand juice throughout the entire year. Sellers must arrange for cold storage of orange juice concentrate so that they can maintain a steady supply all of the time.

Grading.   Grading refers to standardizing products by dividing them into subgroups and displaying and labeling them so that consumers clearly understand their nature and quality. Many products, such as meat, steel, and fruit, are graded according to a set of standards that often are established by the state or federal government.

Financing.   For many products, especially large items such as automobiles, refrigerators, and new homes, the marketer arranges credit to expedite the purchase.

Marketing Research.   Through research, marketers ascertain the need for new goods and services. By gathering information regularly, marketers can detect new trends and changes in consumer tastes.

Risk Taking.   Risk is the chance of loss associated with marketing decisions. Developing a new product creates a chance of loss if consumers do not like it enough to buy it. Spending money to hire a sales force or to conduct marketing research also involves risk. The implication of risk is that most marketing decisions result in either success or failure.

Creating Value with Marketing

value
a customer's subjective assessment of benefits relative to costs in determining the worth of a product.

Value is an important element of managing long-term customer relationships and implementing the marking concept. We view **value** as a customer's subjective assessment of benefits relative to costs in determining the worth of a product (customer value = customer benefits − customer costs).

Customer benefits include anything a buyer receives in an exchange. Hotels and motels, for example, basically provide a room with a bed and bathroom, but each firm provides a different level of service, amenities, and atmosphere to satisfy its guests. Motel 6 offers the minimum services necessary to maintain a quality, efficient, low-price overnight accommodation. In contrast, the Ritz-Carlton provides

Responding to Business Challenges

Cool Beans: Starbucks Refines the Customer Experience

Starbucks is brewing up higher sales through new beverages and cafés in global markets. It also continues to refine the retail environment to increase customer value. To speed up purchases, it offers pay-by-cell phone and order-ahead options. To pay with the Starbucks app, consumers with iPhone, Android, or Windows cell phones download the app and let cashiers scan the Starbucks code on the screen during checkout. The app links to the customer's Starbucks Card, combining the rewards of a loyalty program with the convenience of a prepaid card, for making purchases. Nearly a third of Starbucks' sales volume now comes through its mobile wallet.

Starbucks aims to open 12,000 additional stores, refocus on its coffee beverages, and expand its focus on technology and its rewards program. With company food sales reaching 21 percent of total sales in the United States, Starbucks aims to elevate its food offerings. The company invested in high-end Italian bakery Princi and opened a bakery inside the company's flagship Reserve Roastery in Seattle. Starbucks' Roasteries are designed to be larger than their traditional café concept and will incorporate Princi bakeries, including those slated for Shanghai, Milan, New York, Tokyo, and Chicago.

When it comes to innovation, Starbucks proactively monitors the marketing environment to identify new trends. Its many successful initiatives demonstrate that the firm is ready and willing to make changes to maintain its competitive advantage and gain market share.[2]

Critical Thinking Questions

1. How did Starbucks introduce technological changes to speed up purchases?
2. Why must Starbucks remain proactive in monitoring the marketing environment?
3. Do you think a greater investment in food offerings will benefit Starbucks in the long run? Why or why not?

---

every imaginable service a guest might desire and strives to ensure that all service is of the highest quality. Customers judge which type of accommodation offers them the best value according to the benefits they desire and their willingness and ability to pay for the costs associated with the benefits.

Customer costs include anything a buyer must give up to obtain the benefits the product provides. The most obvious cost is the monetary price of the product, but nonmonetary costs can be equally important in a customer's determination of value. Two nonmonetary costs are the time and effort customers expend to find and purchase desired products. To reduce time and effort, a company can increase product availability, thereby making it more convenient for buyers to purchase the firm's products. Another nonmonetary cost is risk, which can be reduced by offering good basic warranties for an additional charge. Another risk-reduction strategy is increasingly popular in today's catalog/telephone/Internet shopping environment. L.L. Bean, for example, uses a guarantee to reduce the risk involved in ordering merchandise from its catalogs, and many online retailers like Nordstrom and Sephora offer free return shipping.

In developing marketing activities, it is important to recognize that customers receive benefits based on their experiences. For example, many computer buyers consider services such as fast delivery, ease of installation, technical advice, and training assistance to be important elements of the product. Customers also derive benefits from the act of shopping and selecting products. These benefits can be affected by the atmosphere or environment of a store, such as Red Lobster's nautical/seafood theme.

The Marketing Concept

A basic philosophy that guides all marketing activities is the **marketing concept,** the idea that an organization should try to satisfy customers' needs through coordinated activities that also allow it to achieve its own goals. According to the marketing

marketing concept
the idea that an organization should try to satisfy customers' needs through coordinated activities that also allow it to achieve its own goals.

LO 11-4

Explain the marketing concept and its implications for developing marketing strategies.

Drones provide an example of a product that is based on advanced technology but is finding many uses from being a toy or recreational product to commercial uses. The marketing concept is based on the philosophy that consumers purchase the satisfaction and value they derive from a product not the product itself.

©Stock Image/Shutterstock

concept, a business must find out what consumers desire and then develop the good, service, or idea that fulfills their needs or wants. The business must then get the product to the customer. In addition, the business must continually alter, adapt, and develop products to keep pace with changing consumer needs and wants. For instance, after years of criticism regarding her unrealistic body shape, Mattel released its iconic Barbie doll in three different shape sizes. Barbie will also have different skin tones and hairstyles. These new dolls are meant to reflect the multicultural diversity around the world.[3] To remain competitive, companies must be prepared to add to or adapt their product lines to satisfy customers' desires for new fads or changes in eating habits. General Mills is focused on removing artificial colors and flavors from its cereals and providing organic food products to adapt to changing consumer demand. With consumer preference rapidly evolving, General Mills is changing to satisfy a growing target market.[4] Each business must determine how best to implement the marketing concept, given its own goals and resources.

Trying to determine customers' true needs is increasingly difficult because no one fully understands what motivates people to buy things. However, Estée Lauder, founder of her namesake cosmetics company, had a pretty good idea. When a prestigious store in Paris rejected her perfume in the 1960s, she "accidentally" dropped a bottle on the floor where nearby customers could get a whiff of it. So many asked about the scent that Galeries Lafayette was obliged to place an order. Lauder ultimately built an empire using then-unheard-of tactics like free samples and gifts with purchase to market her "jars of hope."[5]

Although customer satisfaction is the goal of the marketing concept, a business must also achieve its own objectives, such as boosting productivity, reducing costs, or achieving a percentage of a specific market. If it does not, it will not survive. For example, Lenovo could sell computers for $50 and give customers a lifetime guarantee, which would be great for customers but not so great for Lenovo. Obviously, the company must strike a balance between achieving organizational objectives and satisfying customers.

To implement the marketing concept, a firm must have good information about what consumers want, adopt a consumer orientation, and coordinate its efforts throughout the entire organization; otherwise, it may be awash with goods, services, and ideas that consumers do not want or need. Successfully implementing the marketing concept requires that a business view the customer's perception of value as the ultimate measure of work performance and improving value, and the rate at which this is done, as the measure of success.[6] Everyone in the organization who interacts with customers—*all* customer-contact employees—must know what customers want. They are selling ideas, benefits, philosophies, and experiences—not just goods and services.

Someone once said that if you build a better mousetrap, the world will beat a path to your door. Suppose you do build a better mousetrap. What will happen? Actually, consumers are not likely to beat a path to your door because the market is so competitive. A coordinated effort by everyone involved with the mousetrap is needed to sell the product. Your company must reach out to customers and tell them about your mousetrap, especially how your mousetrap works better than those offered by competitors. If you do not make the benefits of your product widely known, in most cases, it will not be successful. One reason that Apple is so successful is because of its stores. Apple's more than 400 national and international retail stores market computers and electronics in a way unlike any other computer manufacturer or retail establishment. The upscale stores, located in high-rent shopping districts, show off Apple's products in modern, spacious settings to encourage consumers to try new things—like making a movie on a computer.[7] So for some companies, like Apple, you need to create stores to sell your product to consumers. You could also find stores that are willing to sell your product to consumers for you. In either situation, you must implement the marketing concept by making a product with satisfying benefits and making it available and visible.

Kellogg's knows its customers want a fast breakfast. Kellogg's Nutri-Grain Soft Baked Breakfast Bars are made with whole grains and real fruit and satisfy consumers' desires for a healthy breakfast option.

©Sheila Fitzgerald/Shutterstock

Orville Wright said that an airplane is "a group of separate parts flying in close formation." This is what most companies are trying to accomplish: They are striving for a team effort to deliver the right good or service to customers. A breakdown at any point in the organization—whether it be in production, purchasing, sales, distribution, or advertising—can result in lost sales, lost revenue, and dissatisfied customers.

Evolution of the Marketing Concept

The marketing concept may seem like the obvious approach to running a business and building relationships with customers. However, businesspeople are not always focused on customers when they create and operate businesses. Many companies fail to grasp the importance of customer relationships and fail to implement customer strategies. A firm's marketing department needs to share information about customers and their desires with the entire organization. Our society and economic system have changed over time, and marketing has become more important as markets have become more competitive. Although this is an oversimplification, these time periods help us to understand how marketing has evolved. There have always been some firms that have practiced the marketing concept.

The Production Orientation. During the second half of the 19th century, the Industrial Revolution was well under way in the United States. New technologies, such as electricity, railroads, internal combustion engines, and mass-production techniques, made it possible to manufacture goods with ever increasing efficiency. Together with new management ideas and ways of using labor, products poured into the marketplace, where demand for manufactured goods was strong.

The Sales Orientation. By the early part of the 20th century, supply caught up with and then exceeded demand, and businesspeople began to realize they would have to "sell" products to buyers. During the first half of the 20th century, businesspeople viewed sales as the primary means of increasing profits in what has become known as a sales orientation. Those who adopted the sales orientation perspective believed the most important marketing activities were personal selling and advertising. Today, some people still inaccurately equate marketing with a sales orientation.

The Market Orientation.   By the 1950s, some businesspeople began to recognize that even efficient production and extensive promotion did not guarantee sales. These businesses, and many others since, found that they must first determine what customers want and then produce it, rather than making the products first and then trying to persuade customers that they need them. Managers at General Electric first suggested that the marketing concept was a companywide philosophy of doing business. As more organizations realized the importance of satisfying customers' needs, U.S. businesses entered the marketing era, one of market orientation. A **market orientation** requires organizations to gather information about customer needs, share that information throughout the entire firm, and use it to help build long-term relationships with customers. Market orientation is linked to new product innovation by developing a strategic focus to explore and develop new products to serve target markets.[8] For example, the popular outdoor store REI allows consumers to return most products within one year for a full refund. Electronics have a shorter window for returns. A "user-friendly" return policy helps REI better serve its customers' desires to return merchandise without time pressures.[9] Top executives, marketing managers, nonmarketing managers (those in production, finance, human resources, and so on), and customers all become mutually dependent and cooperate in developing and carrying out a market orientation. Nonmarketing managers must communicate with marketing managers to share information important to understanding the customer. Consider the nearly 130-year history of Wrigley's gum. In 1891, the gum was given away to promote sales of baking powder (the company's original product). The gum was launched as its own product in 1893, and after four generations of Wrigley family CEOs, the company continues to reinvent itself and focus on consumers. Eventually, the family made the decision to sell the company to Mars. Wrigley now functions as a stand-alone subsidiary of Mars. The deal combined such popular brands as Wrigley's gums and Life Savers with Mars' M&M'S, Snickers, and Skittles to form the world's largest confectionary company.

Trying to assess what customers want, which is difficult to begin with, is further complicated by the rate at which trends, fashions, and tastes can change. Businesses today want to satisfy customers and build meaningful long-term relationships with them. It is more efficient and less expensive for the company to retain existing customers and even increase the amount of business each customer provides the organization than to find new customers. Most companies' success depends on increasing the amount of repeat business; therefore, relationship building between company and customer is key. Many companies are turning to technologies associated with customer relationship management to help build relationships and boost business with existing customers. A market orientation involves being responsive to ever-changing customer needs and wants. For example, Weight Watchers made its program easier for consumers to use by adapting its points system. The "Freestyle" program builds upon understanding a "healthy lifestyle" with more than 200 ingredients deemed healthy as opposed to penalizing members with higher points for less healthy options. The sustainability and results exceed those of their other programs.

Although it might be easy to dismiss customer relationship management as time-consuming and expensive, this mistake could destroy a company. Customer relationship

market orientation
an approach requiring organizations to gather information about customer needs, share that information throughout the firm, and use that information to help build long-term relationships with customers.

management (CRM) is important in a market orientation because it can result in loyal and profitable customers. Without loyal customers, businesses would not survive; therefore, achieving the full profit potential of each customer relationship should be the goal of every marketing strategy. At the most basic level, profits can be obtained through relationships by acquiring new customers, enhancing the profitability of existing customers, and extending the duration of customer relationships. The profitability of loyal customers throughout their relationship with the company (their lifetime customer value) should not be underestimated. Starbucks, for example, is estimated to have a customer lifetime value of more than $14,000.[10]

Communication remains a major element of any strategy to develop and manage long-term customer relationships. By providing multiple points of interactions with customers—that is, websites, telephone, fax, e-mail, and personal contact—companies can personalize customer relationships.[11] Like many online retailers, Amazon stores and analyzes purchase data in an attempt to understand each customer's interests. This information helps the online retailer improve its ability to satisfy individual customers and thereby increase sales of books, music, movies, and other products to each customer. The ability to identify individual customers allows marketers to shift their focus from targeting groups of similar customers to increasing their share of an individual customer's purchases. Regardless of the medium through which communication occurs, customers should ultimately be the drivers of marketing strategy because they understand what they want. Customer relationship management systems should ensure that marketers listen to customers in order to respond to their needs and concerns and build long-term relationships.

Developing a Marketing Strategy

To implement the marketing concept and customer relationship management, a business needs to develop and maintain a **marketing strategy**, a plan of action for developing, pricing, distributing, and promoting products that meet the needs of specific customers. This definition has two major components: selecting a target market and developing an appropriate marketing mix to satisfy that target market.

Selecting a Target Market

A **market** is a group of people who have a need, purchasing power, and the desire and authority to spend money on goods, services, and ideas. A **target market** is a more specific group of consumers on whose needs and wants a company focuses its marketing efforts. Target markets can be further segmented into business markets and consumer markets.

Business-to-business marketing (B2B) involves marketing products to customers who will use the product for resale, direct use in daily operations, or direct use in making other products. John Deere, for instance, sells earth-moving equipment to construction firms and tractors to farmers. Most people, however, tend to think of *business-to-consumer marketing (B2C)*, or marketing directly to the end consumer. Sometimes products are used by both types of markets. For example, Glo Skin Beauty sells its cosmetics and skin care products wholesale to salons and spas as well as to consumers directly via its website.

Marketing managers may define a target market as a relatively small number of people within a larger market, or they may define it as the total market (Figure 11.2). Rolls-Royce, for example, targets its products at a very exclusive, high-income market—people who want the ultimate in prestige in an automobile. On the other

LO 11-5

Examine the development of a marketing strategy, including market segmentation and marketing mix.

marketing strategy
a plan of action for developing, pricing, distributing, and promoting products that meet the needs of specific customers.

market
a group of people who have a need, purchasing power, and the desire and authority to spend money on goods, services, and ideas.

target market
a specific group of consumers on whose needs and wants a company focuses its marketing efforts.

FIGURE 11.2
Target Market Strategies

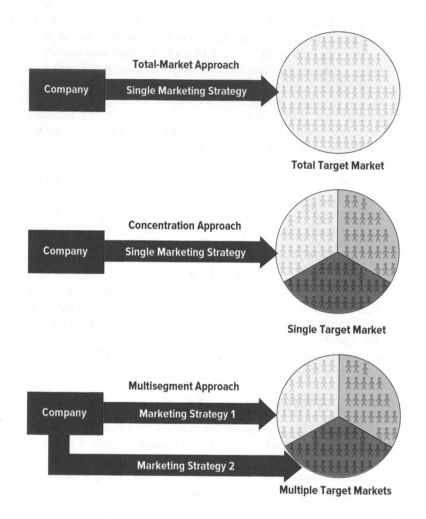

hand, Ford Motor Company manufactures a variety of vehicles including Lincolns, Mercurys, and Ford Trucks in order to appeal to varied tastes, needs, and desires.

Some firms use a **total-market approach,** in which they try to appeal to everyone and assume that all buyers have similar needs and wants. Sellers of salt, sugar, and many agricultural products use a total-market approach because everyone is a potential consumer of these products. This approach is also referred to as *mass marketing.* Most firms, though, use **market segmentation** and divide the total market into groups of people. A **market segment** is a collection of individuals, groups, or organizations who share one or more characteristics and thus have relatively similar product needs and desires. For example, women are a large market segment. At the household level, segmentation can identify each woman's social attributes, culture, and stages in life to determine preferences and needs.

Another market segment on which many marketers are focusing is the growing Hispanic population with a population of more than 59 million and buying power of more than $1.5 trillion.[12] For instance, online retail giant Amazon includes Spanish-language options to its site.[13] The companies hope to create relationships with Hispanic consumers in order to gain their loyalty. One of the challenges for marketers

total-market approach
an approach whereby a firm tries to appeal to everyone and assumes that all buyers have similar needs.

market segmentation
a strategy whereby a firm divides the total market into groups of people who have relatively similar product needs.

market segment
a collection of individuals, groups, or organizations who share one or more characteristics and thus have relatively similar product needs and desires.

	1990	2000	2010	2012	2015	2016
Total	$4,200	$7,300	$11,200	$12,200	$15,100	$13,900
Black	316	600	947	1,000	1,300	1,200
Native American	20	40	87	103	148	133
Asian	115	272	609	718	1,000	891
Hispanic	210	488	1,000	1,200	1,700	1,400

TABLE 11.1
Buying Power of U.S. Minorities by Race (billions)

Source: Jeffrey M. Humphreys, The Multicultural Economy 2012, Athens, GA: The University of Georgia Terry College of Business Selig Center for Economic Growth, 2013; UGA News Service, "Minority Buying Power Growing, UGA Study Says," OnlineAthens, March 11, 2017 http://onlineathens.com/local-news/business/2017-03-11/minority-buying-power-growing-uga-study-says (accessed April 21, 2018).

in the future will be to effectively address an increasingly racially diverse United States. In future decades, the purchasing power of minority market segments is set to grow by leaps and bounds. Table 11.1 shows the buying power of minority groups in the United States. Today, multicultural buying power exceeds $3.4 trillion.[14] Companies will have to learn how to most effectively reach these growing segments. Companies use market segmentation to focus their efforts and resources on specific target markets so that they can develop a productive marketing strategy. Two common approaches to segmenting markets are the concentration approach and the multisegment approach.

Market Segmentation Approaches. In the **concentration approach,** a company develops one marketing strategy for a single market segment. The concentration approach allows a firm to specialize, focusing all its efforts on the one market segment. Porsche, for example, directs all its marketing efforts toward high-income individuals who want to own high-performance vehicles. A firm can generate a large sales volume by penetrating a single market segment deeply. The concentration approach may be especially effective when a firm can identify and develop products for a segment ignored by other companies in the industry.

In the **multisegment approach,** the marketer aims its marketing efforts at two or more segments, developing a marketing strategy for each. Many firms use a multisegment approach that includes different advertising messages for different segments. Companies also develop product variations to appeal to different market segments. The U.S. Post Office, for example, offers personalized stamps, while Mars Inc. sells personalized M&M'S through mymms.com. Many other firms also attempt to use a multisegment approach to market segmentation, such as the manufacturer of Raleigh bicycles, which has designed separate marketing strategies for racers, tourers, commuters, and children.

Niche marketing is a narrow market segment focus when efforts are on one small, well-defined group that has a unique, specific set of needs. For example, Chrome Industries was founded in Boulder, Colorado, to make unique, durable bags for bike messengers. This target market showed such affinity for the product that the company has since expanded with a variety of bags as well as clothing and shoes.[15] Niche segments are usually very small compared to the total market for the products. Freshpet makes all-natural gourmet pet food that mimics the type of food humans like. This company is targeting a growing niche of pet owners who want the best food for their pets.[16]

concentration approach a market segmentation approach whereby a company develops one marketing strategy for a single market segment.

multisegment approach a market segmentation approach whereby the marketer aims its efforts at two or more segments, developing a marketing strategy for each.

For a firm to successfully use a concentration or multisegment approach to market segmentation, several requirements must be met:

1. Consumers' needs for the product must be heterogeneous.
2. The segments must be identifiable and divisible.
3. The total market must be divided in a way that allows estimated sales potential, cost, and profits of the segments to be compared.
4. At least one segment must have enough profit potential to justify developing and maintaining a special marketing strategy.
5. The firm must be able to reach the chosen market segment with a particular market strategy.

Bases for Segmenting Markets.   Companies segment markets on the basis of several variables:

1. *Demographic*—age, sex, race, ethnicity, income, education, occupation, family size, religion, social class. These characteristics are often closely related to customers' product needs and purchasing behavior, and they can be readily measured. For example, yogurt companies often segment by age: nonfat, high protein yogurt for adults and easy-to-eat tubes for children.
2. *Geographic*—climate, terrain, natural resources, population density, subcultural values. These influence consumer needs and product usage. Climate, for example, influences consumer purchases of clothing, automobiles, heating and air conditioning equipment, and leisure activity equipment.
3. *Psychographic*—personality characteristics, motives, lifestyles. Soft-drink marketers provide their products in several types of packaging, including two-liter bottles and cases of cans, to satisfy different lifestyles and motives.
4. *Behavioristic*—some characteristic of the consumer's behavior toward the product. These characteristics commonly involve some aspect of product use. Benefit segmentation is also a type of behavioristic segmentation. For instance, low-fat, low-carb food products would target those who desire the benefits of a healthier diet.

Developing a Marketing Mix

marketing mix
the four marketing activities—product, price, promotion, and distribution—that the firm can control to achieve specific goals within a dynamic marketing environment.

The second step in developing a marketing strategy is to create and maintain a satisfying marketing mix. The **marketing mix** refers to four marketing activities—product, price, distribution, and promotion—that the firm can control to achieve specific goals within a dynamic marketing environment (Figure 11.3). The buyer or the target market is the central focus of all marketing activities. Amazon is well known for its implementation of the marketing mix. It routinely engages in research and development to create new products like its digital assistant Echo. It promotes its products through advertising, social media, and media events. Best Buy and other retailers provide these products at a premium price to convey their quality and effectiveness.

Product.   A product—whether a good, a service, an idea, or some combination—is a complex mix of tangible and intangible attributes that provide satisfaction and benefits. A *good* is a physical entity you can touch. A Tesla Model 3, a Nerf Blaster, and Beats by Dre available for adoption at an animal shelter are examples of goods. A *service* is the application of human and mechanical efforts to people or objects to

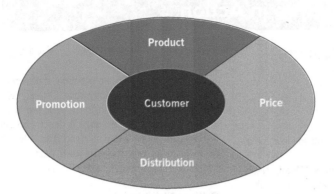

Marketing Environment

provide intangible benefits to customers. Air travel, dry cleaning, haircuts, facials, banking, insurance, medical care, and day care are examples of services. *Ideas* include concepts, philosophies, images, and issues. For instance, an attorney, for a fee, may advise you about what rights you have in the event that the IRS decides to audit your tax return. Other marketers of ideas include political parties, churches, and schools.

A product has emotional and psychological as well as physical characteristics that include everything that the buyer receives from an exchange. This definition includes supporting services such as installation, guarantees, product information, and promises of repair. For example, Icelandic Provisions introduced paper wrapped cups for its skyr, a yogurt-like dairy product, because they can be recycled easier than printed shrink-sleeved cups.[17] The company adapted its product's packaging to provide a healthier and "greener" offering. Products usually have both favorable and unfavorable attributes; therefore, almost every purchase or exchange involves trade-offs as consumers try to maximize their benefits and satisfaction and minimize unfavorable attributes.

Products are among a firm's most visible contacts with consumers. If they do not meet consumer needs and expectations, sales will be difficult, and product life spans will be brief. On the other hand, Silicon Valley car maker Tesla has overtaken market values of Ford and General Motors. Although its sales of cars are modest, Tesla's growth reflects the belief that electric motors are the future and Tesla is best positioned to bring advanced self-driving technology to the highways.[18] The product is an important variable—often the central focus—of the marketing mix; the other variables (price, promotion, and distribution) must be coordinated with product decisions.

price
a value placed on an object exchanged between a buyer and a seller.

Price. Almost anything can be assessed by a **price**, a value placed on an object exchanged between a buyer and a seller. Although the seller usually establishes the price, it may be negotiated between the buyer and the seller. The buyer usually exchanges purchasing power—income, credit, wealth—for the

AirPods, iPads and iMacs are all examples of products made by Apple.
©ROMSVETNIK/Shutterstock

Wanelo: $100 Million Valuation, Zero Products

Wanelo
Founder: Deena Varshavskaya
Founded: 2012, in San Francisco, California
Success: Wanelo's user base exploded from 1 million to 11 million users in a one-year period.

Wanelo, a shortened mix of the words "Want, Need, Love," is attempting to be for shopping what "Facebook is for friends" and "Instagram is for pictures," as founder Deena Varshavskaya puts it. In other words, it tries to "optimize" shopping in a deeply social manner. Founded in 2012, Wanelo has no products of its own; rather, it connects users to millions of products from hundreds of thousands of stores worldwide. This "digital mall" allows users to save, post, share, and tag products for others to view and share. They can instantly buy items they like with the click of a button. Wanelo also recommends products it thinks users will like based on their past viewing patterns.

Wanelo not only has no products, it also does no traditional marketing. Founder Varshavskaya instead focuses on further improving the Wanelo platform. Your product or service will market itself, she argues, once it directly and elegantly solves the problem your customers need it to solve. Other entrepreneurs could do well to learn from Wanelo's unusual product and marketing strategy choices.[19]

Critical Thinking Questions

1. Describe the marketing strategy that Wanelo uses. Which part of the marketing mix does Wanelo seem to particularly excel at?
2. How does Wanelo use the "social experience" of an online mall to appeal to both stores and consumers?
3. If Wanelo does not use traditional advertising, how do you think consumers find out about it?

DID YOU KNOW? During its first year of operation, sales of Coca-Cola averaged just nine drinks per day for total first-year sales of $50. Today, Coca-Cola products are consumed at the rate of 1.9 billion drinks per day.[20]

satisfaction or utility associated with a product. Because financial price is the measure of value commonly used in an exchange, it quantifies value and is the basis of most market exchanges.

Marketers view price as much more than a way of assessing value, however. It is a key element of the marketing mix because it relates directly to the generation of revenue and profits. Prices can also be changed quickly to stimulate demand or respond to competitors' actions. The sudden increase in the cost of commodities such as oil can create price increases or a drop in consumer demand for a product. When gas prices rise, consumers purchase more fuel-efficient cars; when prices fall, consumers return to larger vehicles.[21]

distribution
making products available to customers in the quantities desired.

Distribution. **Distribution** (sometimes referred to as "place" because it helps to remember the marketing mix as the "4 Ps") is making products available to customers in the quantities desired. For example, consumers can rent movies and videogames from a physical store, a vending machine, or an online service. Intermediaries, usually wholesalers and retailers, perform many of the activities required to move products efficiently from producers to consumers or industrial buyers. These activities involve transporting, warehousing, materials handling, and inventory control, as well as packaging and communication.

Critics who suggest that eliminating wholesalers and other intermediaries would result in lower prices for consumers do not recognize that eliminating intermediaries would not do away with the need for their services. Other institutions would have to perform those services, and consumers would still have to pay for them. In addition, in the absence of wholesalers, all producers would have to deal directly with retailers or customers, keeping voluminous records and hiring extra people to deal with customers. Supply chain management (SCM) involves maintaining a flow of

Going Green

Viridian Energy: Green Energy Gets Personal

We commonly associate personal selling with tangible goods such as jewelry. . .not green energy. Viridian Energy sells eco-responsible electricity, solar power, and natural gas. Founded in 2009, it has more than 20,000 associates to promote its products. With more than $250 million in revenues, Viridian has become the world's 59th largest direct selling firm.

Viridian's energy products are sourced from or offset by renewable energy sources. Its Pure Green electricity package is sourced from wind power. Viridian has also partnered with Sungevity to provide co-branded solar energy offerings. Viridian received publicity when it announced an initiative to donate one solar power system for every 15 new residential solar installations.

Associates market Viridian's green solutions to customers through personal selling, informing them about how their choice will benefit the environment. Although personal selling is the costliest form of promotion, it allows Viridian associates to tailor the marketing message to the individual consumer. Viridian associates receive payments for each customer they recruit for each month the customer remains with the company.

Despite its success, Viridian has faced challenges. In 2018, the firm reached a settlement with the Massachusetts Attorney General's Office for alleged deceptive marketing tactics and agreed to stop marketing its products door-to-door for a two-year period. As it continues to evolve as a socially responsible company, Viridian must ensure its promotional strategy remains transparent with customers.[22]

Critical Thinking Questions

1. Why is personal selling an effective type of promotion for Viridian Energy?
2. In addition to personal selling, what are some other ways that Viridian engages in promotion?
3. Why is it important that Viridian maintain transparent communications with customers, especially in the personal selling industry?

---

products through physical distribution activities. This includes acquiring resources, inventory, and the interlinked networks that make products available to customers through purchasing, logistics, and operations. SCM has become very important to the success of online marketers. Consider Amazon, the largest and most successful online retailer, which takes on typical wholesale and retail functions by storing inventory in warehouses and is developing its own shipping and delivery capacity. As an online retailer, it provides easy online access and customer service.[23] Companies now can make their products available throughout the world without maintaining facilities in each country. For instance, Pandora, Spotify and Apple Music have benefited from the ability to stream music over the Internet. Customers can listen to music for free with commercial interruptions, or they can pay to upgrade to listen without commercials. Pandora has 73.3 million active users, while Spotify has 140 million and Apple Music has 1 million.[24]

Promotion. Promotion is a persuasive form of communication that attempts to expedite a marketing exchange by influencing individuals, groups, and organizations to accept goods, services, and ideas. Promotion includes advertising, personal selling, publicity, and sales promotion, all of which we will look at more closely in the "Dimensions of Marketing Strategy" chapter.

The aim of promotion is to communicate directly or indirectly with individuals, groups, and organizations to facilitate exchanges. When marketers use advertising and other forms of promotion, they must effectively manage their promotional resources and understand product and target-market characteristics to ensure that these promotional activities contribute to the firm's objectives.

Most major companies have set up websites on the Internet to promote themselves and their products. For example, L'Oréal operates Makeup.com, a beauty website

promotion
a persuasive form of communication that attempts to expedite a marketing exchange by influencing individuals, groups, and organizations to accept goods, services, and ideas.

that discusses beauty trends and shares makeup tutorials using products from L'Oréal brands like Urban Decay, Maybelline, and NYX.[25] While traditional advertising media such as television, radio, newspapers, and magazines remain important, digital advertising on websites and social media sites is growing. Not only can digital advertising be less expensive, but advertising offerings such as Google AdWords allow companies to only pay when users click on the link or advertisement.[26] Additionally, social media sites offer advertising opportunities for both large and small companies. Firms can create a Facebook page and post corporate updates for free. Not to be outdone, Twitter also allows advertisers to purchase Promoted Tweets on the site. Promoted Tweets are just like regular tweets (except for the name), allowing users to respond or re-tweet them to their friends.[27]

LO 11-6

Describe how marketers conduct marketing research and study buying behavior.

marketing research
a systematic, objective process of getting information about potential customers to guide marketing decisions.

primary data
marketing information that is observed, recorded, or collected directly from respondents.

Marketing Research and Information Systems

Before marketers can develop a marketing mix, they must collect in-depth, up-to-date information about customer needs. **Marketing research** is a systematic, objective process of getting information about potential customers to guide marketing decisions. Such information might include data about the age, income, ethnicity, gender, and educational level of people in the target market, their preferences for product features, their attitudes toward competitors' products, and the frequency with which they use the product. For instance, marketing research has revealed that consumers often make in-store purchase decisions in three seconds or less.[28] Marketing research is vital because the marketing concept cannot be implemented without information about customers.

A marketing information system is a framework for accessing information about customers from sources both inside and outside the organization. Inside the organization, there is a continuous flow of information about prices, sales, and expenses. Outside the organization, data are readily available through private or public reports and census statistics, as well as from many other sources. Computer networking technology provides a framework for companies to connect to useful databases and customers with instantaneous information about product acceptance, sales performance, and buying behavior. This information is important to planning and marketing strategy development.

Two types of data are usually available to decision makers. **Primary data** are observed, recorded, or collected directly from respondents. If you've ever participated in a telephone survey about a product, recorded your TV viewing habits for ACNielsen or Arbitron, or even responded to a political opinion poll, you provided the researcher with primary data. Primary data must be gathered by researchers who develop a method to observe phenomena or research respondents. Many companies use "mystery shoppers" to visit their retail establishments and report on whether the stores were adhering to the companies' standards of service. These undercover customers document their observations of store appearance, employee effectiveness, and customer treatment. Mystery shoppers provide valuable information that helps companies improve their organizations and refine their marketing strategies.[29] Companies also use surveys and focus groups to gauge customer opinion. Table 11.2 shows the companies regarded as having the best customer service. A weakness of surveys is that respondents are sometimes untruthful in order to avoid seeming foolish or ignorant. Although focus groups can be more expensive than surveys, they allow marketers to understand how consumers express themselves as well as observe their behavior patterns.[30]

Rank	Companies	Experience Rating (%)
1	Publix	84
2	Chick-fil-A	83
2	H-E-B	83
4	Regions	82
4	Hardee's	82
4	Chipotle Mexican Grill	82
4	Hannaford	82
8	Subway	81
8	QVC	81
8	BJ's Wholesale Club	81
8	Ace Hardware	81
8	Food Lion	81
8	Trader Joe's	81

TABLE 11.2
Companies with the Best Customer Service

Source: Michael B. Sauter, Thomas C. Frohlich, and Samuel Stebbins, "Customer Service Hall of Fame," 24/7 Wall St., July 23, 2015, http://247wallst.com/special-report/2015/07/23/customer-service-hall-of-fame-2/ (accessed April 28, 2017).

Some methods for marketing research use passive observation of consumer behavior and open-ended questioning techniques. Called ethnographic or observational research, the approach can help marketers determine what consumers really think about their products and how different ethnic or demographic groups react to them.

Secondary data are compiled inside or outside the organization for some purpose other than changing the current situation. Marketers typically use information compiled by the U.S. census bureau and other government agencies, databases created by marketing research firms, as well as sales and other internal reports, to gain information about customers. For example, the average television viewership for NASCAR declined from 9 million in 2005 to 4.6 million in 2016. In addition, the audience is becoming older. This finding could indicate the need for additional research to determine approaches to increase viewership.[31]

secondary data information that is compiled inside or outside an organization for some purpose other than changing the current situation.

Online Marketing Research

The marketing of products and collecting of data about buying behavior—information on what people actually buy and how they buy it—represents marketing research of the future. New information technologies are changing the way businesses learn about their customers and market their products. Interactive multimedia research, or *virtual testing,* combines sight, sound, and animation to facilitate the testing of concepts as well as packaging and design features for consumer products. The evolving development of telecommunications and computer technologies is allowing marketing researchers quick and easy access to a growing number of online services and a vast database of potential respondents.

Coffee shops and restaurants attempt to influence consumers' buying behavior by offering free Wi-Fi and a comfortable retail environment. Businesses can also gather data about their customers with Wi-Fi services like Purple, Aislelabs, and Yelp WiFi.

©Image Source

Marketing research can use digital media and social networking sites to gather useful information for marketing decisions. Sites such as Twitter, Facebook, and LinkedIn can be good substitutes for focus groups. Online surveys can serve as an alternative to mail, telephone, or personal interviews. There are fewer landlines, in fact, 49 percent of households have a cell phone but no landline telephones.[32]

Social networks are a great way to obtain information from consumers who are willing to share their experiences about products and companies. In a way, this process identifies those consumers who develop an identity or passion for certain products, as well as those consumers who have concerns about quality or performance. It is possible for firms to tap into existing online social networks and simply "listen" to what consumers have on their mind. Firms can also encourage consumers to join a community or group so that they can share their opinions with the business.

A good outcome from using social networks is the opportunity to reach new voices and gain varied perspectives on the creative process of developing new products and promotions. For instance, Kickstarter gives aspiring entrepreneurs the ability to market their ideas online. Funders can then choose whether to fund those ideas in return for a finished product or a steep discount.[33] To some extent, social networking is democratizing design by welcoming consumers to join in the development process for new products.[34]

Online surveys are becoming an important part of marketing research. Traditionally, the process of conducting surveys online involved sending questionnaires to respondents either through e-mail or through a website. However, digital communication has increased the ability of marketers to conduct polls on blogs and social networking sites. The benefits of online market research include lower costs and quicker feedback. For instance, LEGO encourages fans to submit their own ideas for LEGO sets on their LEGO Ideas crowdsourcing platform. Other fans can vote on the ideas, and enough votes make the idea eligible for review and possible commercialization. This allows LEGO to solicit creative ideas from passionate fans at low cost.[35] By monitoring consumers' feedback, companies can understand customer needs and adapt their goods or services.

Finally, *marketing analytics* uses data that has been collected to measure, interpret, and evaluate marketing decisions. This emerging area uses advanced software that can track, store, and analyze data. Marketing analytics is becoming an increasingly important part of a company's marketing activities that are integrated into daily decision making. Dashboards are a data management tool that provide visual information related to sales, inventory, and product information as well as key performance indicators related to financial results. For example, Wave Analytics provides software that the marketer can see instantly on any device. The ability of marketing analytics to harness trillions of gigabytes of data provides marketing research findings on demand. In the future, marketing analytics will use artificial intelligence to turn data into new ways of making marketing decisions. IBM Watson is a leader in this area turning data into more efficient and accurate decisions to drive marketing strategies. Frequent-flyer programs enable airlines to track individual information about customers, using databases that can help airlines understand what different customers want, and treat customers differently depending on their flying habits and overall value to the company. Airlines, hotels, and other service providers are also increasingly gathering a greater 'share of customer,' as discussed below, by tying a company branded credit card to enhance overall value for the user and the customer. Many airlines require you fly certain levels of mileage and charge a minimum dollar value to their credit card to retain premium benefits. Relationship-building efforts like frequent-flyer and credit card programs have been shown to increase customer value.[36]

Buying Behavior

Carrying out the marketing concept is impossible unless marketers know what, where, when, and how consumers buy; conducting marketing research into the factors that influence buying behavior helps marketers develop effective marketing strategies. **Buying behavior** refers to the decision processes and actions of people who purchase and use products. It includes the behavior of both consumers purchasing products for personal or household use and organizations buying products for business use. Marketers analyze buying behavior because a firm's marketing strategy should be guided by an understanding of buyers. For instance, men's shopping habits are changing. Men's shopping habits used to be more targeted and fast. Today, they tend to buy more clothes on impulse, search websites or mobile phones for style ideas, and try out new brands. As a result, retailers such as Saks and Barney's have begun expanding the men's areas in their stores.[37] Both psychological and social variables are important to an understanding of buying behavior.

buying behavior
the decision processes and actions of people who purchase and use products.

Psychological Variables of Buying Behavior

Psychological factors include the following:

- **Perception** is the process by which a person selects, organizes, and interprets information received from that person's senses, as when experiencing an advertisement or touching a product to better understand it.

- **Motivation,** as we said in the "Motivating the Workforce" chapter, is an inner drive that directs a person's behavior toward goals. A customer's behavior is influenced by a set of motives rather than by a single motive. A buyer of a tablet computer, for example, may be motivated by ease of use, ability to communicate with the office, and price.

- **Learning** brings about changes in a person's behavior based on information and experience. For instance, a smartphone app that provides digital news or magazine content could eliminate the need for print copies. If a person's actions result in a reward, that person is likely to behave the same way in similar situations. If a person's actions bring about a negative result, however—such as feeling ill after eating at a certain restaurant—that person will probably not repeat that action.

- **Attitude** is knowledge and positive or negative feelings about something. For example, a person who feels strongly about protecting the environment may refuse to buy products that harm the earth and its inhabitants.

- **Personality** refers to the organization of an individual's distinguishing character traits, attitudes, or habits. Although market research on the relationship between personality and buying behavior has been inconclusive, some marketers believe that the type of car or clothing a person buys reflects that person's personality.

perception
the process by which a person selects, organizes, and interprets information received from his or her senses.

motivation
an inner drive that directs a person's behavior toward goals.

learning
changes in a person's behavior based on information and experience.

attitude
knowledge and positive or negative feelings about something.

personality
the organization of an individual's distinguishing character traits, attitudes, or habits.

Social Variables of Buying Behavior

social roles
a set of expectations for individuals based on some position they occupy.

Social factors include **social roles,** which are a set of expectations for individuals based on some position they occupy. A person may have many roles: parent, spouse, student, executive. Each of these roles can influence buying behavior. Consider a parent choosing an automobile. As a parent, that person might prefer to purchase a safe, gasoline-efficient car such as a Volvo. The person's environmentally supportive work colleagues might encourage the use of public transportation and Uber instead of buying a car. Because millennials (those born between 1981–1996) tend to prefer vehicles that represent how they see themselves, even that person's children might want different vehicles: one might want a Ford Explorer to take on camping trips, while the other might prefer a cool, classy car such as a Ford Mustang.[38] Thus, in choosing which car to buy, the parent's buying behavior may be affected by the opinions and experiences of family and friends and by that person's role as a parent and employee.

Other social factors include reference groups, social classes, and culture.

reference groups
groups with whom buyers identify and whose values or attitudes they adopt.

- **Reference groups** include families, professional groups, civic organizations, and other groups with whom buyers identify and whose values or attitudes they adopt. A person may use a reference group as a point of comparison or a source of information. A person new to a community may ask other group members to recommend a family doctor, for example.

social classes
a ranking of people into higher or lower positions of respect.

- **Social classes** are determined by ranking people into higher or lower positions of respect. Criteria vary from one society to another. People within a particular social class may develop common patterns of behavior. People in the upper-middle class, for example, might buy a Lexus or a BMW as a symbol of their social class.

culture
the integrated, accepted pattern of human behavior, including thought, speech, beliefs, actions, and artifacts.

- **Culture** is the integrated, accepted pattern of human behavior, including thought, speech, beliefs, actions, and artifacts. Culture determines what people wear and eat and where they live and travel. Many Hispanic Texans and New Mexicans, for example, buy *masa trigo,* the dough used to prepare flour tortillas, which are basic to Southwestern and Mexican cuisine.

Understanding Buying Behavior

Although marketers try to understand buying behavior, it is extremely difficult to explain exactly why a buyer purchases a particular product. The tools and techniques for analyzing consumers are not exact. Marketers may not be able to determine accurately what is highly satisfying to buyers, but they know that trying to understand consumer wants and needs is the best way to satisfy them. Marriott International's Innovation Lab, for instance, tests out new hotel designs for its brands to target millennials and other desirable demographics. Wi-Fi, lighting, and more soundproof rooms are among the top desirable traits travelers desire. Another trend is that travelers are not unpacking their suitcases as much. As a result, Marriott has begun reducing the size of closets and the number of hangers to save room.[39]

LO 11-7

Summarize the environmental forces that influence marketing decisions.

The Marketing Environment

A number of external forces directly or indirectly influence the development of marketing strategies; the following political, legal, regulatory, social, competitive, economic, and technological forces comprise the marketing environment.

- *Political, legal, and regulatory forces*—laws and regulators' interpretation of laws, law enforcement and regulatory activities, regulatory bodies, legislators and legislation, and political actions of interest groups. Specific laws, for example, require that advertisements be truthful and that all health claims be documented.
- *Social forces*—the public's opinions and attitudes toward issues such as living standards, ethics, the environment, lifestyles, and quality of life. For example, social concerns have led marketers to design and market safer toys for children.
- *Competitive and economic forces*—competitive relationships such as those in the technology industry, unemployment, purchasing power, and general economic conditions (prosperity, recession, depression, recovery, product shortages, and inflation).
- *Technological forces*—computers and other technological advances that improve distribution, promotion, and new-product development.

Marketing requires creativity and consumer focus because environmental forces can change quickly and dramatically. Changes can arise from social concerns and economic forces such as price increases, product shortages, and altering levels of demand for commodities. Recently, climate change, global warming, and the impact of carbon emissions on our environment have become social concerns and are causing businesses to rethink marketing strategies. These environmental issues have persuaded governments to institute stricter limits on greenhouse gas emissions. For instance, in the United States, the government has mandated that by 2025, vehicles must be able to reach 54.5 miles per gallon on average.[40] This is causing automobile companies like General Motors to investigate ways to make their cars more fuel-efficient without significantly raising the price. *Newsweek* magazine ranks the top "Green Companies" each year. Recent highly ranked companies include Hasboro Inc., Nike, Hershey Co., NVIDIA Corp., and Biogen Inc.[41]

Because such environmental forces are interconnected, changes in one may cause changes in others. Consider that because of evidence linking children's consumption of soft drinks and fast foods to health issues such as obesity, diabetes, and osteoporosis, marketers of such products have experienced negative publicity and calls for legislation regulating the sale of soft drinks in public schools.

Although the forces in the marketing environment are sometimes called uncontrollables, they are not totally so. A marketing manager can influence some environmental variables. For example, businesses can lobby legislators to dissuade them from passing unfavorable legislation. From a social responsibility perspective, firms like BMW try to make a positive contribution. The BMW Ultimate Drive program involves specially marked BMWs that drive across the United States to increase breast cancer awareness and raise funds for breast cancer research. BMW donates $1 to the Susan G. Komen Breast Cancer Foundation for every mile these cars are test driven by participants.[42] Figure 11.4 shows the variables in the marketing environment that affect the marketing mix and the buyer.

Importance of Marketing to Business and Society

As this chapter has shown, marketing is a necessary function to reaching consumers, establishing relationships, and creating revenue. While some critics might view marketing as a way to change what consumers want, marketing is essential in communicating the value of goods and services. For consumers, marketing is necessary to

FIGURE 11.4

The Marketing Mix and the Marketing Environment

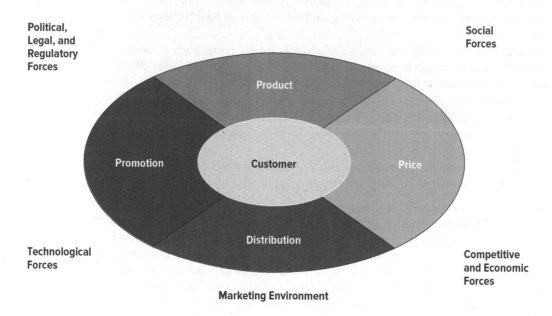

ensure that they get the products they desire at the right places in the right quantities at a reasonable price. From the perspective of businesses, marketing is necessary in order to form valuable relationships with customers to increase profitability and customer support.

It is not just for-profit businesses that engage in marketing activities. Nonprofit organizations, government institutions, and even people must market themselves to spread awareness and achieve desired outcomes. All organizations must reach their target markets, communicate their offerings, and establish high-quality services. For instance, nonprofit organization The Leukemia and Lymphoma Society uses print, radio, web, and other forms of media to market its Team in Training racing events to recruit participants and solicit support. Without marketing, it would be nearly impossible for organizations to connect with their target audiences. Marketing is, therefore, an important contributor to business and societal well-being.

You probably did not think as a child how great it would be to grow up and become a marketer. That's because, often, marketing is associated with sales jobs, but opportunities in marketing, public relations, product management, advertising, e-marketing, and customer relationship management and beyond represent almost one-third of all jobs in today's business world. To enter any job in the marketing field, you must balance an awareness of customer needs with business knowledge while mixing in creativity and the ability to obtain useful information to make smart business decisions.

Marketing starts with understanding the customer. Marketing research is a vital aspect in marketing decision making and presents many job opportunities. Market researchers survey customers to determine their habits, preferences, and aspirations. Activities include concept testing, product testing, package testing, test-market research, and new-product research. Salaries vary, depending on the nature and level of the position as well as the type, size, and location of the firm. A market analyst may make between $36,000 and $76,000, while a market research director earns a median salary of more than $107,000.

One of the most dynamic areas in marketing is direct marketing, where a seller solicits a response from a consumer using direct communications methods such as telephone, online communication, direct mail, or catalogs. Jobs in direct marketing include buyers, catalog managers, research/mail-list managers,

or order fulfillment managers. Most positions in direct marketing involve planning and market analysis. Some require the use of databases to sort and analyze customer information and sales history.

Use of the Internet for retail sales is growing, and the Internet continues to be very useful for business-to-business sales. E-marketing offers many career opportunities, including customer relationship management (CRM). CRM helps companies market to customers through relationships, maintaining customer loyalty. Information technology plays a huge role in such marketing jobs because you need to combine technical skills and marketing knowledge to effectively communicate with customers. Job titles include e-marketing manager, customer relationship manager, and e-services manager. A CRM manager earns a median salary of approximately $67,000, and experienced individuals may earn as much as $116,000.

A job in any of these marketing fields will require a strong sense of the current trends in business and marketing. Customer service is vital to many aspects of marketing, so the ability to work with customers and to communicate their needs and wants is important. Marketing is everywhere, from the corner grocery or local nonprofit organization to the largest multinational corporations, making it a shrewd choice for an ambitious and creative person. We will provide additional job opportunities in marketing in the "Dimensions of Marketing Strategy" chapter.[43]

Review Your Understanding

Define *marketing*.

Marketing is a group of activities designed to expedite transactions by creating, distributing, pricing, and promoting goods, services, and ideas.

Describe the exchange process.

Marketing facilitates the exchange, the act of giving up one thing in return for something else. The central focus of marketing is to satisfy needs.

Specify the functions of marketing.

Marketing includes many varied and interrelated activities: buying, selling, transporting, storing, grading, financing, marketing research, and risk taking.

Explain the marketing concept and its implications for developing marketing strategies.

The marketing concept is the idea that an organization should try to satisfy customers' needs through coordinated activities

that also allow it to achieve its goals. If a company does not implement the marketing concept by providing products that consumers need and want while achieving its own objectives, it will not survive.

Examine the development of a marketing strategy, including market segmentation and marketing mix.

A marketing strategy is a plan of action for creating a marketing mix (product, price, distribution, promotion) for a specific target market (a specific group of consumers on whose needs and wants a company focuses its marketing efforts). Some firms use a total-market approach, designating everyone as the target market. Most firms divide the total market into segments of people who have relatively similar product needs. A company using a concentration approach develops one marketing strategy for a single market segment, whereas a multisegment approach aims marketing efforts at two or more segments, developing a different marketing strategy for each.

Describe how marketers conduct marketing research and study buying behavior.

Carrying out the marketing concept is impossible unless marketers know what, where, when, and how consumers buy; marketing research into the factors that influence buying behavior helps marketers develop effective marketing strategies. Marketing research is a systematic, objective process of getting information about potential customers to guide marketing decisions. Buying behavior is the decision processes and actions of people who purchase and use products.

Summarize the environmental forces that influence marketing decisions.

There are several forces that influence marketing activities: political, legal, regulatory, social, competitive, economic, and technological.

Propose a solution for resolving a problem with a company's marketing plans.

Based on the material in this chapter, you should be able to answer the questions posed in the "Solve the Dilemma" feature near the end of the chapter and help the business understand what went wrong and how to correct it.

Critical Thinking Questions

Enter the World of Business Questions

1. How has Dollar General found success in catering to a unique market niche ignored by many of its rivals?

2. How does Dollar General keep costs low enough to maintain its low prices?

3. How do stores like Dollar General differ from other stores in a fluctuating economy?

Learn the Terms

attitude   351
buying behavior   351
concentration approach   343
culture   352
distribution   346
exchange   334
learning   351
market   341
market orientation   340
market segment   342

market segmentation   342
marketing   334
marketing concept   337
marketing mix   344
marketing research   348
marketing strategy   341
motivation   351
multisegment approach   343
perception   351
personality   351

price   345
primary data   348
promotion   347
reference groups   352
secondary data   349
social classes   352
social roles   352
target market   341
total-market approach   342
value   336

Check Your Progress

1. What is marketing? How does it facilitate exchanges?

2. Name the functions of marketing. How does an organization use marketing activities to achieve its objectives?

3. What is the marketing concept? Why is it so important?

4. What is a marketing strategy?

5. What is market segmentation? Describe three target market strategies.

6. List the variables in the marketing mix. How is each used in a marketing strategy?

7. Why are marketing research and information systems important to an organization's planning and development of strategy?

8. Briefly describe the factors that influence buying behavior. How does understanding buying behavior help marketers?

9. Discuss the impact of technological forces and political and legal forces on the market.

Get Involved

1. With some or all of your classmates, watch several hours of television, paying close attention to the commercials. Pick three commercials for products with which you are somewhat familiar. Based on the commercials, determine who the target market is. Can you surmise the marketing strategy for each of the three?

2. Discuss the decision process and influences involved in purchasing a personal computer.

Build Your Skills

The Marketing Mix

Background

You've learned the four variables—product, promotion, price, and distribution—that the marketer can select to achieve specific goals within a dynamic marketing environment. This exercise will give you an opportunity to analyze the marketing strategies of some well-known companies to determine which of the variables received the most emphasis to help the company achieve its goals.

Task

In groups of three to five students, discuss the examples below and decide which variable received the most emphasis.

A. Product

B. Distribution

C. Promotion

D. Price

_____ 1. Starbucks Coffee began selling bagged premium specialty coffee through an agreement with Kraft Foods to gain access to more than 30,000 supermarkets.

_____ 2. Skype is a software application that allows consumers to make telephone calls over the web. Calls to Skype subscribers are free, while calls to land line and mobile phones cost around 2 cents per minute.

_____ 3. Amid great anticipation, Apple released its iPad, selling more than 3 million within three months. The slim tablet computer is a major step forward in reading e-books, watching movies, and playing games.

_____ 4. After decades on the market, WD-40 is in about 80 percent of U.S. households—more than any other branded product. Although WD-40 is promoted as a product that can stop squeaks, protect metal, loosen rusted parts, and free sticky mechanisms, the WD-40 Company has received letters from customers who have sprayed the product on bait to attract fish, on pets to cure mange, and even on people to cure arthritis. Despite more than 200 proposals to expand the WD-40 product line and ideas to change the packaging and labeling, the company stands firmly behind its one highly successful and respected original product.

_____ 5. Southwest Airlines makes flying fun. Flight attendants try to entertain passengers, and the airline has an impeccable customer service record. Employees play a key role and take classes that emphasize that having fun translates into great customer service.

_____ 6. Hewlett-Packard offered a $100 rebate on a $799 HP LaserJet printer when purchased with an HP LaserJet toner cartridge. To receive the rebate, the buyer had to return a mail-in certificate to certify the purchase. A one-page ad with a coupon was used in *USA Today* stating, "We're taking $100 off the top."

_____ 7. Denny's, the largest full-service family restaurant chain in the United States, serves more than 1 million customers a day. The restaurants offer the Build Your Own Grand Slam Breakfast for about $7.15, lunch basket specials for $4 to $6, and a value menu with prices ranging from $2 to $8.

Solve the Dilemma

Will It Go?

Ventura Motors makes midsized and luxury automobiles in the United States. Best-selling models include its basic four-door sedans (priced from $20,000 to $25,000) and two-door and four-door luxury automobiles (priced from $40,000 to $55,000). The success of two-seat sports cars like the Mazda RX-8 started the company evaluating the market for a two-seat sports car priced midway between the moderate and luxury market. Research found that there was indeed significant demand and that Ventura needed to act quickly to take advantage of this market opportunity.

Ventura took the platform of the car from a popular model in its moderate line, borrowing the internal design from its luxury line. The car was designed, engineered, and produced in just over two years, but the coordination needed to bring the design together resulted in higher than anticipated costs. The price for this two-seat car, the Olympus, was set at $32,000. Dealers were anxious to take delivery on the car, and salespeople were well trained on techniques to sell this new model.

However, initial sales have been slow, and company executives are surprised and concerned. The Olympus was introduced relatively quickly, made available at all Ventura dealers, priced midway between luxury and moderate models, and advertised heavily since its introduction.

Critical Thinking Questions

LO 11-8

Propose a solution for resolving a problem with a company's marketing plans.

1. What do you think were the main concerns with the Olympus two-door sports coupe? Is there a market for a two-seat, $32,000 sports car when the RX-8 sells for significantly less?

2. What is the role of the marketing mix in the Olympus introduction?

3. What are some of the marketing strategies auto manufacturers use to stimulate sales of certain makes of automobiles?

Build Your Business Plan

connect

Customer-Driven Marketing

The first step is to develop a marketing strategy for your good or service. Who will be the target market you will specifically try to reach? What group(s) of people has the need, ability, and willingness to purchase this product? How will you segment customers within your target market? Segmenting by demographic and geographic variables are often the easiest segmentation strategies to attempt. Remember that you would like to have the customers in your segment be as homogeneous and accessible as possible. You might target several segments if you feel your good or service has broad appeal.

The second step in your marketing strategy is to develop the marketing mix for your good or service. Whether you are dealing with an established product or you are creating your own good or service, you need to think about what is the differential advantage your product offers. What makes it unique? How should it be priced? Should the product be priced below, above, or at the market? How will you distribute the product? And last but certainly not least, you need to think about the promotional strategy for your product.

What about the uncontrollable variables you need to be aware of? Is your product something that can constantly be technologically advanced? Is your product a luxury that will not be considered by consumers when the economy is in a downturn?

See for Yourself Videocase

Marriott: Your Home Away from Home

With 4,400 hotel properties in 87 countries, Marriott knows how to target different types of travelers. Its 19 brands use careful segmentation strategies based on demographic and psychographic variables to determine what their guests want and how best to meet their needs. Its extensive customer research allows the firm to identify its customers, understand cultural or generational changes, and adapt its hotels to target these different travelers more effectively. It also helps Marriott develop strategies to stay ahead of the competition and the emerging rental sharing sites such as Airbnb and VRBO (vacation rental by owner).

"Segmentation is really important because it really helps us to design the guest experience for a particular brand," says Tina Edmundson, global brand officer, Luxury & Lifestyle Brands for Marriott International.

Marriott uses price and service as a major form of differentiation among its hotels. For instance, it separates its brands using terms like Upper Upscale and Select Service. Upper Upscale has all the amenities a traveler is looking for in a luxury hotel, including room service, a bar, doormen, and more. Marriott's Ritz-Carlton brand, known for offering the highest in guest amenities, fits this category. Select Service still offers guests a great experience but lacks some features such as room service and bellhops. Marriott's Moxy Hotels is more of a do-it-yourself hotel without doormen or room service. Moxy is positioned as being a stylish, but affordable, hotel. Each one targets a specific type of customer: Upper Upscale is for those who are willing to pay high prices for a luxurious experience, while Select Service targets customers who are looking for quality experiences at lower prices.

Marriott also segments its customers based on lifestyle characteristics. The Discoverer category is interested in exploration and experiencing local culture. Marriott's luxurious Renaissance

Hotels targets this market with "local" themes. Renaissance Hotels come equipped with a focused concierge service called Navigator to provide suggestions for local experiences.

Although marketers at Marriott have become experts at customer segmentation, they constantly face new challenges. Millennials, for instance, differ significantly from baby boomers in their hotel preferences. Baby boomers enjoy familiarity and comfort, something Marriott's high-quality hotels have been able to achieve. For many consumers, consistency of the service and atmosphere is important—they want the J.W. Marriott they stay at in Milan, Italy, to be similar in service and design as the one in New York City.

Millennials go in the opposite direction. They desire an unpredictable adventure and are less likely to want to stay at large hotel chains. They are also more likely to travel globally; millennials are 23 percent more interested in traveling abroad than non-millennials. These wide-scale differences among two large demographic groups represents a challenge for Marriott—one it is tackling head-on.

For this reason, Marriott partnered with consulting firm Fahrenheit 212, which specializes in marketing to millennials. Preliminary research shows that millennials need to be convinced of the value of a hotel, requiring the Marriott to "tell a story" about its hotels. Marriott also tapped into the insights of local entrepreneurs and hotel employees. They were challenged to come up with creative ideas that would attract both travelers and locals to the hotel. Teams whose ideas were adopted were awarded $50,000. In this way, Marriott is able to utilize local talent who are familiar with the city and know the culture.

One hotel that has been developed to target millennials is Moxy Hotels, discussed earlier. Marriott partnered with IKEA to develop a hotel that is *specifically targeted* to the millennial traveler. Moxy Hotels combine a stylish atmosphere, functional guest rooms, affordable prices, and the types of amenities that tech-savvy millennials care about the most (for example, Wi-Fi, televisions, and public areas with computers). Unlike some of the other generations, millennials tend to be more self-sufficient, so Moxy does not have room service or bellhops. This keeps prices down for millennials, who tend to have higher unemployment or lower-paying jobs than other age groups.

No matter who the target market is, they all have one thing in common: They seek a pleasurable experience. As Tina Edmundson explains, "It's no longer about filling rooms anymore. It is really about providing experiences that resonate with guests personally."[44]

Critical Thinking Questions

1. How does Marriott use psychographic and demographic variables to segment the market?

2. Why is the millennial traveler market posing challenges for Marriott and other hotels?

3. What are some ways that Marriott uses market research to discover what its customers want?

You can find the related video in the Video Library in Connect. Ask your instructor how you can access Connect.

Team Exercise

Form groups and assign the responsibility of finding examples of companies that excel in one dimension of the marketing mix (price, product, promotion, and distribution). Provide several company and product examples, and defend why this would be an exemplary case. Present your research to the class.

Notes

1. Hayley Peterson, "Dollar General Is Defying the Retail Apocalypse and Opening 1,000 Stores," *Aol.com,* April 8, 2017, https://www.aol.com/article/finance/2017/04/08/dollar-general-is-defying-the-retail-apocalypse-and-opening-1-00/22031477/ (accessed November 22, 2017); Krystina Gustafson, "Dollar General Is Starting to Look a Lot Like Walmart," *CNBC,* March 16, 2017, https://www.cnbc.com/2017/03/16/dollar-general-is-starting-to-look-a-lot-like-wal-mart.html (accessed November 22, 2017); "Dollar General (DG)," *Yahoo! Finance,* https://finance.yahoo.com/quote/DG/financials?p=DG (accessed November 22, 2017); Mya Frazier, "Dollar General Hits a Goldmine in America," *Bloomberg Businessweek,* October 11, 2017, https://www.bloomberg.com/news/features/2017-10-11/dollar-general-hits-a-gold-mine-in-rural-america (accessed November 22, 2017).

2. Jackie Wattles, "Starbucks: Nearly a Third of Sales were Made Digitally Last Quarter," *CNN,* April 27, 2017, http://money.cnn.com/2017/04/27/news/companies/starbucks-digital-sales/index.html (accessed November 28, 2017); Sarah Whitten, "Starbucks Opens First Princi Location, Teases More to Come in 2018," *CNBC,* November 7, 2017, https://www.cnbc.com/2017/11/07/starbucks-opens-first-princi-location-teases-more-to-come-in-2018.html (accessed November 28, 2017); Marguerite Ward, "3 Ways CEO Kevin Johnson's Leadership Style Could Shape Starbucks," *CNBC,* April 3, 2017, https://www.cnbc.com/2017/04/03/3-things-you-need-to-know-about-new-starbucks-ceo-kevin-johnson.html (accessed November

28, 2017); Sarah Whitten, "Starbucks Shares Up as CEO Says He's 'Optimistic' He Can Exceed New Growth Targets," *CNBC,* November 3, 2017, https://www.cnbc.com/2017/11/03/starbucks-ceo-says-hes-optimistic-he-can-exceed-new-growth-targets.html (accessed November 28, 2017); Jennifer Van Grove, "Starbucks Apps Account for 42M Payments," *VentureBeat,* April 9, 2012, http://venturebeat.com/2012/04/09/starbucks-42m-mobile-pay/ (accessed July 6, 2015); Roemmele, "Why Is the Starbucks Mobile Payments App So Successful?" *Forbes,* June 13, 2014, http://www.forbes.com/sites/quora/2014/06/13/why-is-the-starbucks-mobile-payments-app-so-successful/ (accessed July 6, 2015); Trefis Team, "Here's How Starbucks Will Be Impacted by a Change in Management," *Forbes,* December 5, 2016, https://www.forbes.com/sites/greatspeculations/2016/12/05/here-how-starbucks-will-be-impacted-by-a-change-in-management/#639879401f56 (accessed December 22, 2017).

3. Mary Bowerman and Hadley Malcolm, "New Barbies Are Tall, Petite, Curvy," *USA Today,* January 29, 2016, p. 3B.

4. Sarah Elbert, "Food for Thought: Interview with CEO Ken Powell," *Delta Sky Magazine,* December 2016, p. 66.

5. "Beauty Queen," *People,* May 10, 2004, p. 187.

6. Michael Treacy and Fred Wiersema, *The Discipline of Market Leaders* (Reading, MA: Addison Wesley, 1995), p. 176.

7. Jefferson Graham, "At Apple Stores, iPads at Your Service," *USA Today,* May 23, 2011, p. 1B; Ana Swanson, "How the Apple Store Took over the World," *Washington Post,* July 21, 2015, https://www.washingtonpost.com/news/wonk/wp/2015/07/21/the-unlikely-success-story-of-the-apple-retail-store/ (accessed May 5, 2017).

8. Kwaku Atuahene-Gima, "Resolving the Capability-Rigidity Paradox in New Product Innovation," *Journal of Marketing* 69, 4 (October 2005), pp. 61–83.

9. Maryalene LaPonsie, "15 Stores with the Best Return Policies," *CBS,* November 22, 2017, https://www.cbsnews.com/news/15-stores-with-the-best-return-policies/ (accessed January 6, 2018).

10. Jeff Desjardins, "This Is the Lifetime Value of a Starbucks Customer," *Business Insider,* January 28, 2016, www.businessinsider.com/lifetime-value-of-a-starbucks-customer-2016-1 (accessed December 12, 2017).

11. Venky Shankar, "Multiple Touch Point Marketing," American Marketing Association, Faculty Consortium on Electronic Commerce, Texas A&M University, July 14–17, 2001.

12. Isaac Mizrahi, "The Hispanic Market 'Long Tail': Five Hidden Growth Opportunities for U.S. CMOs to Win in 2017," *Forbes,* January 18, 2017, https://www.forbes.com/sites/onmarketing/2017/01/18/the-hispanic-market-long-tail-five-hidden-growth-opportunities-for-u-s-cmos-to-win-in-2017/#4e14aa5e22ca (accessed December 10, 2017)

13. Tessa Berenson, "Amazon Is Adding Spanish to Its U.S. Website," *Fortune,* March 10, 2017, http://fortune.com/2017/03/10/amazon-spanish-website/ (accessed December 10, 2017).

14. Nielsen, "The Making of a Multicultural Super Consumer," March 18, 2015, http://www.nielsen.com/us/en/insights/news/2015/the-making-of-a-multicultural-super-consumer-.html (accessed April 4, 2016).

15. Chrome Industries website, http://www.chromeindustries.com/our-story.html (accessed December 26, 2016).

16. Craig Giammona, "Hey Mom, Set Another Place at Dinner for Fido," *Bloomberg Businessweek,* November 12, 2105, pp. 26–27.

17. Rick Lingle, "Skyr Retells Its Story via Innovative Packaging," *Packaging Digest,* December 5, 2017, www.packagingdigest.com/food-packaging/skyr-retells-story-innovative-pkg1712 (accessed December 10, 2017).

18. Tim Higgins, "Tesla Steps on the Gas Overtakes Ford in Value," *The Wall Street Journal,* April 4, 2017, p. A1.

19. Kurt Soller, "With Enough Customers, You Don't Need a Product," *Bloomberg Businessweek,* April 28–May 4, 2014, p. 70; Tomio Geron, "Inside Wanelo, The Hot Social Shopping. Experience," *Forbes,* March 27, 2013, http://www.forbes.com/sites/tomiogeron/2013/03/27/inside-wanelo-the-hot-social-shopping-service/ (accessed April 21, 2018); Kate Brodock, "Wanelo Founder: I Got 10M Members with No Marketing," *Women 2.0,* August 21, 2013, http://women2.com/2013/08/21/wanelo-founder-how-i-got-10m-members-with-no-marketing/ (accessed April 21, 2018); Bryan Pearson, "Retail Social Communities: How OpenSky, Wanelo and Others Engage by the Crowd," *Forbes,* May 9, 2017, https://www.forbes.com/sites/bryanpearson/2017/05/09/retail-social-communities-how-opensky-wanelo-and-others-engage-by-the-crowd/#5ce0ebc477ea (accessed April 21, 2018).

20. "FAQs," The Coca-Cola Company, https://www.coca-colacompany.com/contact-us/faqs (accessed April 21, 2018)

21. William Fierman, "This Is Why the SUV Is Here to Stay," *Business Insider,* March 9, 2016, http://www.businessinsider.com/suv-sales-continue-to-grow (accessed May 5, 2017).

22. Barbara Seale, "Viridian Energy: Energized by Global Vision," *Direct Selling News,* June 29, 2014, http://directsellingnews.com/index.php/view/viridian_energy_energized_by_global_vision#.VFe7dWddVc9 (accessed April 21, 2018); Viridian website, http://www.viridian.com/index.asp?CO_LA=US_EN (accessed April 21, 2018); Pam Zekman, "2 Investigators: Alternative Power Suppliers May Not Be Cheaper," *CBS,* May 9, 2014, http://chicago.cbslocal.com/2014/05/09/2-investigators-alternative-power-suppliers-may-not-be-cheaper/ (accessed April 21, 2018); Better Business Bureau, "Viridian Energy LLC," https://www.bbb.org/connecticut/business-reviews/electric-companies/viridian-energy-llc-in-norwalk-ct-87068321/Alerts-and-Actions (accessed April 21, 2018); "Viridian Energy Selects Sungevity to Offer Solar Energy with the Power to Make a Difference," *PR Newswire,* September 4, 2015, https://www.prnewswire.com/news-releases/viridian-energy-selects-sungevity-to-offer-solar-energy-with-the-power-to-make-a-difference-300142062.html (accessed April 21, 2018); "Viridian," *Direct Selling News,* 2016, https://www.directsellingnews.com/company-profiles/viridian/ (accessed April 21, 2018).

23. Laura Stevens, "Amazon's Profit Exceeds Target," *The Wall Street Journal,* February 8, 2017, p. B1.

24. Hugh McIntyre, "Pandora Has Lost Nearly 8 Million Listeners in 9 Months," *Forbes,* November 3, 2017, https://www.forbes.com/sites/hughmcintyre/2017/11/03/pandora-is-losing-850000-listeners-every-month/#25ef89955a93 (accessed December 10, 2017); Micah Singleton, "Spotify Now Has 140 Million Active Users," *The Verge,* June 15, 2017, https://www.theverge.com/2017/6/15/15807826/spotify-140-million-active-users (accessed December 10, 2017).

25. "About Us," Makeup.com, https://www.makeup.com/about (accessed December 10, 2017).

26. "Google AdWords," Google, https://www.google.com/adwords/ (accessed April 4, 2016).

27. Twitter, "What Are Promoted Tweets?" Twitter Help Center, https://support.twitter.com/articles/142101-what-are-promoted-tweets# (accessed April 4, 2016).

28. Christine Birkner, "10 Minutes with . . . Raul Murguia Villegas," *Marketing News,* July 30, 2011, pp. 26–27.

29. "MSPA North America," Mystery Shopping Providers Association, http://mysteryshop.org/ (accessed April 4, 2016).

30. Piet Levy, "10 Minutes with . . . Robert J. Morais," *Marketing News,* May 30, 2011, pp. 22–23.

31. Tripp Mickle and Valerie Bauerlein, "Nascar, Once a Cultural Icon, Hits the Skids," *The Wall Street Journal,* February 21, 2017, https://www.wsj.com/articles/long-in-victory-lane-nascar-hits-the-skids-1487686349?tesla=y (accessed March 12, 2017).

32. Brittany Wallman, "Rollout of Better Cellphone, Wireless Service Coming," *Sun Sentinel,* March 15, 2017, http://www.sun-sentinel.com/local/broward/fl-reg-cell-reception-microtowers-20170315-story.html (accessed March 18, 2017).

33. Steven Kurutz, "On Kickstarter, Designers' Dreams Materialize," *The New York Times,* September 21, 2011, www.nytimes.com/2011/09/22/garden/on-kickstarter-designers-dreams-materialize.html (accessed May 5, 2017).

34. Mya Frazier, "CrowdSourcing," *Delta Sky Mag,* February 2010, p. 73.

35. "LEGO Ideas—How It Works," LEGO, https://ideas.lego.com/howitworks (accessed April 4, 2016).

36. Robert W. Palmatier, Lisa K. Scheer, and Jan-Benedict E. M. Steenkamp, "Customer Loyalty to Whom? Managing the Benefits and Risks of Salesperson-Owned Loyalty," *Journal of Marketing Research* XLIV (May 2007), pp. 185–199.

37. Ray A. Smith, "Men Shop More Like Women," *The Wall Street Journal,* February 17, 2016, pp. D1–D2.

38. Mike Floyd, "Editor's Letter: What Drives Millennials?" *Automobile Magazine,* May 2015, p. 12.

39. Andrea Petersen, "Secrets of a Hotel Test Lab," *The Wall Street Journal,* October 1, 2015, pp. D1–D2.

40. Environmental Protection Agency, "EPA and NHTSA Set Standards to Reduce Greenhouse Gases and Improve Fuel Economy for Model Years 2017-2025 Cars and Light Trucks," https://www3.epa.gov/otaq/climate/documents/420f12051.pdf (accessed May 20, 2016).

41. "Top Green Companies in the U.S. 2016," *Newsweek,* 2016, http://www.newsweek.com/green-2016/top-green-companies-us-2016 (accessed December 27, 2016).

42. "BMW Ultimate Drive for the Cure," National Capital Chapter BMW Car Club of America, http://old.nccbmwcca.org/index.php?cure (accessed December 27, 2016).

43. "Customer Relationship Manager," PayScale Inc., http://www.payscale.com/research/US/Job=Customer_Relationship_Management_(CRM)_Manager/Salary (accessed April 5, 2016); "Marketing Analyst Salary," PayScale Inc., http://www.payscale.com/research/US/Job=Marketing_Analyst/Salary (accessed April 5, 2016); "Marketing Research Director," PayScale Inc., http://www.payscale.com/research/US/Job=Marketing_Research_Director/Salary (accessed April 5, 2016).

44. McGraw-Hill video, http://www.viddler.com/embed/16eb2415/?f=1&autoplay=0&player=full&disablebranding=0 (accessed April 11, 2016); Elizabeth Segran, "Inside Marriot's Attempt to Win oOver Millennials," *Fast Company,* June 26, 2015, http://www.fastcompany.com/3047872/innovation-agents/inside-marriotts-attempt-to-win-over-millennials (accessed April 8, 2016); Brad Tuttle, "Marriott & IKEA Launch a Hotel Brand for Millennials: What Does That Even Mean?" *Time,* March 8, 2013, http://business.time.com/2013/03/08/marriott-ikea-launch-a-hotel-brand-for-millennials-what-does-that-even-mean/ (accessed April 8, 2016); Larry Olmstead, "Luxury Hotels: Marriott Bets Big with 8 Brands, Explosive Growth," *Forbes,* June 6, 2014, http://www.forbes.com/sites/larryolmsted/2014/06/06/luxury-hotels-marriott-bets-big-with-8-brands-explosive-growth/#3414c894886d (accessed April 8, 2016); Marriott website, https://www.marriott.com/marriott/aboutmarriott.mi (accessed April 8, 2016).

Credits

12 Dimensions of Marketing Strategy

©McGraw-Hill Education

Chapter Outline

Learning Objectives

After reading this chapter, you will be able to:

LO 12-1 Describe the role of product in the marketing mix, including how products are developed, classified, and identified.

LO 12-2 Explain the importance of price in the marketing mix, including various pricing strategies a firm might employ.

LO 12-3 Identify factors affecting distribution decisions, such as marketing channels and intensity of market coverage.

LO 12-4 Specify the activities involved in promotion, as well as promotional strategies and promotional positioning.

LO 12-5 Evaluate an organization's marketing strategy plans.

Enter the World of Business

Harley-Davidson Rolls in the New Products

Although baby boomers consist of much of Harley-Davidson's market, Harley-Davidson recognizes it must also acquire younger customers. Doing this requires it to reinvent itself so that young people do not view it as a brand for their parents. However, as the average Harley costs over $30,000, traditional Harley products remain unaffordable for many of the younger generation.

Harley-Davidson addressed this problem by creating a stripped-down bike that is affordable for younger people. Its Street bike is designed for younger riders with a price point of about $7,500. To keep costs low, the Street model does not have a fuel gauge, tachometer, or a clock but does have warning lights that will flash for low fuel or oil pressure. The product was a hit, particularly for price-conscious consumers in Brazil, South Africa, and India.

Like many car companies, Harley-Davidson recognized that electric vehicles could one day replace more traditional vehicles—particularly with consumers' and governments' growing concern over sustainability. In 2016, Harley-Davidson showed off its prototype electric bike called Livewire. The motorcycle manufacturer announced its plans to make 100 new motorcycles over the next 10 years, including an entire range of electric vehicles. Vice President Bill Davidson believes that electric bikes are key to Harley Davidson's success in the future.

Despite its recent success in courting younger consumers, many young people are choosing to purchase used Harleys over new ones from the dealer. Harley-Davidson has therefore begun encouraging its authorized dealers to sell more used bikes in their inventory. While Harley-Davidson will face many challenges in the coming years, its expertise in marketing and its willingness to adapt its goals provides it with a greater chance at a comeback.[1]

Introduction

The key to developing a marketing strategy is selecting a target market and maintaining a marketing mix that creates long-term relationships with customers. Getting just the right mix of product, price, promotion, and distribution is critical if a business is to satisfy its target customers and achieve its own objectives (implement the marketing concept).

In the "Customer-Driven Marketing" chapter, we introduced the marketing concept and the various activities important in developing a marketing strategy. In this chapter, we'll take a closer look at the four dimensions of the marketing mix—product, price, distribution, and promotion—used to develop the marketing strategy. As we mentioned in the "Customer-Driven Marketing" chapter, sometimes the marketing mix is called the 4Ps with distribution referred to as place. The focus of these marketing mix elements is a marketing strategy that builds customer relationships and satisfaction.

The Marketing Mix

The marketing mix is the part of marketing strategy that involves decisions regarding controllable variables. After selecting a target market, marketers have to develop and manage the dimensions of the marketing mix to give their firm an advantage over competitors. Successful companies offer at least one dimension of value usually associated with a marketing mix element that surpasses all competitors in the marketplace in meeting customer expectations. However, this does not mean that a company can ignore the other dimensions of the marketing mix; it must maintain acceptable, and if possible distinguishable, differences in the other dimensions as well.

Walmart, for example, emphasizes price ("Save money, live better"). Procter & Gamble is well known for its products with top consumer brands such as Tide, Cheer, Crest, Ivory, and Head & Shoulders. Many successful marketers have worked with advertising agencies to create catchy jingles to help consumers recall ads. Campbell Soup's "Mmm Mmm Good," McDonald's "I'm lovin' it," and Coca-Cola's "I'd like to Teach the World to Sing" are all iconic jingles.[2] Finally, Amazon has become so successful by having a highly efficient distribution system.

LO 12-1

Describe the role of product in the marketing mix, including how products are developed, classified, and identified.

Product Strategy

As mentioned previously, the term *product* refers to goods, services, and ideas. Because the product is often the most visible of the marketing mix dimensions, managing product decisions is crucial. In this section, we'll consider product development, classification, mix, life cycle, and identification.

Developing New Products

Each year, thousands of products are introduced, but few of them succeed. Even established firms launch unsuccessful products. For example, Mini Cooper discontinued its Coupe and Roadster models after overexpanding its product lines.[3] Figure 12.1 shows the different steps in the product development process. Before introducing a new product, a business must follow a multistep process: idea development, the screening of new ideas, business analysis, product development, test marketing, and commercialization. A firm can take considerable time to get a product ready for the market: It took more than 20 years for the first photocopier, for example. Additionally, sometimes an idea or product prototype might be shelved only to be returned to later. Former Apple

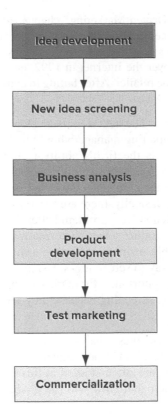

FIGURE 12.1
Product Development
Process

CEO Steve Jobs admitted that the iPad actually came before the iPhone in the product development process. Once it was realized that the scrolling mechanism he was thinking of using could be used to develop a phone, the iPad idea was placed on a shelf for the time being. Apple later returned to develop the product and released the iPad in 2010.[4]

Idea Development. New ideas can come from marketing research, engineers, and outside sources such as advertising agencies and management consultants. Nike has a separate division—Nike Sport Research Lab—where scientists, athletes, engineers, and designers work together to develop technology of the future. The teams research ideas in biomechanics, perception, athletic performance, and physiology to create unique, relevant, and innovative products. These final products are tested in environmental chambers with real athletes to ensure functionality and quality before being introduced into the market.[6] As we

Mayfield Robotics developed Kuri, a home assistant robot. The idea for Kuri came from Kaijen Hsiao and Sarah Osentoski, who originally set out to create a home security robot.[5]

©ZCHE/Kuri (Supplied by WENN)/Newscom

said in the "Customer-Driven Marketing" chapter, ideas sometimes come from customers, too. Other sources are brainstorming and intracompany incentives or rewards for good ideas. New ideas can even create a company. When Jeff Bezos came up with the idea to sell books over the Internet in 1992, he had no idea it would evolve into the world's largest online retailer. After failing to convince his boss of the value of his idea, Bezos left to start Amazon.[7]

New Idea Screening.   The next step in developing a new product is idea screening. In this phase, a marketing manager should look at the organization's resources and objectives and assess the firm's ability to produce and market the product. Important aspects to be considered at this stage are consumer desires; the competition; technological changes; social trends; and political, economic, and environmental considerations. Basically, there are two reasons new products succeed: They are able to meet a need or solve a problem better than products already available, or they add variety to the product selection currently on the market. Bringing together a team of knowledgeable people—including designers, engineers, marketers, and customers—is a great way to screen ideas. Using the Internet to encourage collaboration represents a rich opportunity for marketers to screen ideas. Most new-product ideas are rejected during screening because they seem inappropriate or impractical for the organization.

Business Analysis.   Business analysis is a basic assessment of a product's compatibility in the marketplace and its potential profitability. Both the size of the market and competing products are often studied at this point. The most important question relates to market demand: How will the product affect the firm's sales, costs, and profits?

Product Development.   If a product survives the first three steps, it is developed into a prototype that should reveal the intangible attributes it possesses as perceived by the consumer. Product development is often expensive, and few product ideas make it to this stage. New product research and development costs vary. Adding a new color to an existing item may cost $100,000 to $200,000, but launching a completely new product can cost millions of dollars. During product development, various elements of the marketing mix must be developed for testing. Copyrights, tentative advertising copy, packaging, labeling, and descriptions of a target market are integrated to develop an overall marketing strategy.

test marketing
a trial minilaunch of a product in limited areas that represent the potential market.

Test Marketing.   Test marketing is a trial minilaunch of a product in limited areas that represent the potential market. It allows a complete test of the marketing strategy in a natural environment, giving the organization an opportunity to discover weaknesses and eliminate them before the product is fully launched. Mamma Chia test marketed one of its products in Portland through a company known as SamplingLab. SamplingLab provides free samples for consumers in a retail environment, acting as a type of focus group. In exchange for free samples, consumers fill out surveys about their perceptions of the products.[8] Because test marketing requires significant resources and expertise, market research companies like ACNielsen can assist firms in test marketing their products. Figure 12.2 shows a sample of test markets that marketing research firms often use to test products to predict how successful they might be on a nationwide scale.

DID YOU KNOW? Less than 20 percent of new products succeed in the marketplace. The success rate for consumer goods is about 51 percent.[10]

FIGURE 12.2
Common Test Market Cities

Source: The Nielsen Company

Market Decisions
Test Market Locations

Portland
Albany
Boise
Columbus
Peoria
Provo   Evansville
Colorado   St. Louis
Springs   Lexington
Little Rock   Charlotte
                       Charleston
Tucson
                       Jacksonville
Baton Rouge

■ Permanent Test Markets
  ("Data Markets")

★ Additional "Custom" Test Markets

Commercialization. Commercialization is the full introduction of a complete marketing strategy and the launch of the product for commercial success. During commercialization, the firm gears up for full-scale production, distribution, and promotion. Firms such as AquAdvantage Salmon are getting ready to release genetically modified salmon into the market. The Food and Drug Administration has approved the salmon as fit for consumption. Federal approval is one major step for AquAdvantage in its plans for large-scale commercialization. However, even with federal regulatory approval, AquAdvantage may face hurdles because of consumer and environmental groups.[9]

commercialization
the full introduction of a complete marketing strategy and the launch of the product for commercial success.

consumer products
products intended for household or family use.

Classifying Products

Products are usually classified as either consumer products or industrial products. Consumer products are for household or family use; they are not intended for any purpose other than daily living. They can be further classified as convenience products, shopping products, and specialty products on the basis of consumers' buying behavior and intentions.

- *Convenience products,* such as beverages, granola bars, gasoline, and batteries, are bought frequently, without a lengthy search, and often for immediate consumption.

The Museum of Failure in Los Angeles is the largest collection of failed products and services such as Colgate Beef Lasagna, which was originally launched in the 1980s.

©ROBYN BECK/AFP/Getty Images

Consumers spend virtually no time planning where to purchase these products and usually accept whatever brand is available.

- *Shopping products,* such as computers, smartphones, clothing, and sporting goods, are purchased after the consumer has compared competitive products and "shopped around." Price, product features, quality, style, service, and image all influence the decision to buy.

- *Specialty products,* such as motorcycles, designer clothing, art, and rock concerts, require even greater research and shopping effort. Consumers know what they want and go out of their way to find it; they are not willing to accept a substitute.

business products products that are used directly or indirectly in the operation or manufacturing processes of businesses.

Business products are used directly or indirectly in the operation or manufacturing processes of businesses. They are usually purchased for the operation of an organization or the production of other products; thus, their purchase is tied to specific goals and objectives. They too can be further classified:

- *Raw materials* are natural products taken from the earth, oceans, and recycled solid waste. Iron ore, bauxite, lumber, cotton, and fruits and vegetables are examples.

- *Major equipment* covers large, expensive items used in production. Examples include earth-moving equipment, stamping machines, and robotic equipment used on auto assembly lines.

- *Accessory equipment* includes items used for production, office, or management purposes, which usually do not become part of the final product. Computers, calculators, and hand tools are examples.

- *Component parts* are finished items, ready to be assembled into the company's final products. Tires, window glass, batteries, and spark plugs are component parts of automobiles.

- *Processed materials* are things used directly in production or management operations but are not readily identifiable as component parts. Varnish, for example, is a processed material for a furniture manufacturer.

- *Supplies* include materials that make production, management, and other operations possible, such as paper, pencils, paint, cleaning supplies, and so on.

- *Industrial services* include financial, legal, marketing research, security, janitorial, and exterminating services. Purchasers decide whether to provide these services internally or to acquire them from an outside supplier.

Product Line and Product Mix

product line a group of closely related products that are treated as a unit because of similar marketing strategy, production, or end-use considerations.

Product relationships within an organization are of key importance. A **product line** is a group of closely related products that are treated as a unit because of a similar marketing strategy. At Colgate-Palmolive, for example, the personal-care product line includes deodorant, body wash, bar soap, liquid soap, and toiletries for men. A **product mix** is all the products offered by an organization. Figure 12.3 displays a sampling of the product mix and product lines of the Colgate-Palmolive Company.

product mix all the products offered by an organization.

Product Life Cycle

Like people, products are born, grow, mature, and eventually die. Some products have very long lives. Ivory Soap was introduced in 1879 and still exists (although

FIGURE 12.3
Colgate-Palmolive's Product Mix and Product Lines

	Oral Care	Personal Care	Home Care	Pet Nutrition
	Toothpaste	*Deodorant*	*Dishwashing*	Hill's Prescription Diet
	Colgate Total	Speed Stick	Palmolive	Hill's Science Diet
	Colgate Optic White	Lady Speed Stick	AJAX	Hill's Ideal Balance
	Colgate Enamel Health		Dermassage	
	Colgate Sensitive	*Body Wash*		
	MaxFresh	Softsoap	*Fabric Conditioner*	
		Irish Spring	Suavitel	
	Colgate Kids			
	Dora the Explorer	*Bar Soap*	*Household cleaner*	
	SpongeBob SquarePants	Irish Spring	Murphy Oil Soap	
	Teenage Mutant Ninja Turtles	Softsoap	Fabuloso	
	Monster High		AJAX	
	Transformers	*Liquid Soap*		
	Minion	Softsoap		
	Toothbrushes	*Toiletries for Men*		
	Colgate 360°	Afta		
	Colgate MaxFresh	Skin Bracer		
	Colgate Total			
	Colgate Optic White			

(Product Mix across the top; Product Lines down the left side)

Source: Colgate Palmolive, "Colgate World of Care," www.colgatepalmolive.com/app/Colgate/US/CompanyHomePage.cvsp (accessed April 5, 2016).

competition leading to decreased sales may soon put the future of Ivory Soap in question). In contrast, a new computer chip is usually outdated within a year because of technological breakthroughs and rapid changes in the computer industry. There are four stages in the life cycle of a product: introduction, growth, maturity, and decline (Figure 12.4). The stage a product is in helps determine marketing strategy. In the personal computer industry, desktop computers are in the decline stage, laptop computers have reached the maturity stage, and tablet computers are currently in the growth stage of the product life cycle (although growth has slowed in recent years). Manufacturers of these products are adopting different advertising and pricing strategies to maintain or increase demand for these types of computers.

In the *introductory stage,* consumer awareness and acceptance of the product are limited, sales are zero, and profits are negative. Profits are negative because the firm has spent money on research, development, and marketing to launch the product. During the introductory stage, marketers focus on making consumers aware of the product and its benefits. The smartwatch is still in the early stages of the product development cycle. However, it may quickly jump into the growth stage, with analysts predicting it achieving 12 percent global market growth by 2020.[11] It is not unusual for technology products to go quickly through the life cycle as the rate of new technology innovations continues to increase. Consider the work that Energizer has done to provide the first battery made from recycled batteries and branded as "EcoAdvanced." Innovation of this type is welcomed by consumers and the public at large. Table 12.1 shows some familiar products at different stages of

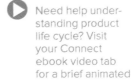

Need help understanding product life cycle? Visit your Connect ebook video tab for a brief animated explanation.

FIGURE 12.4
The Life Cycle of a Product

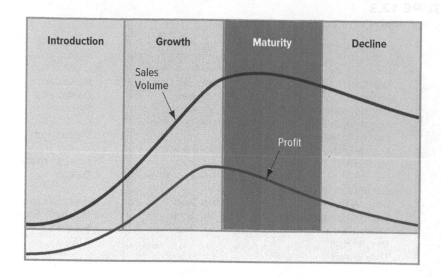

the product life cycle. Sales accelerate as the product enters the growth stage of the life cycle.

In the *growth stage,* sales increase rapidly and profits peak, then start to decline. One reason profits start to decline during the growth stage is that new companies enter the market, driving prices down and increasing marketing expenses. Drones for both recreational and business uses are growing rapidly and are a good example of a product in the growth stage. During the growth stage, the firm tries to strengthen its position in the market by emphasizing the product's benefits and identifying market segments that want these benefits.

Sales continue to increase at the beginning of the *maturity stage,* but then the sales curve peaks and starts to decline while profits continue to decline. This stage is characterized by severe competition and heavy expenditures. In the United States, soft drinks have hit the maturity stage. For example, the seven best-selling soft drinks are Coca-Cola, Diet Coke, Pepsi-Cola, Mountain Dew, Dr. Pepper, Sprite, and Diet Pepsi. In 2016, Diet Coke passed Pepsi-Cola to move into second place in U.S. soft-drink popularity. This represents a move to less sugary drinks as consumers have also turned to flavored water, bottled water, and tea. The soft-drink market is highly competitive, with many alternative products for consumers.[12]

During the *decline stage,* sales continue to fall rapidly. Profits also decline and may even become losses as prices are cut and necessary marketing expenditures are made. As profits drop, firms may eliminate certain models or items. To reduce expenses

TABLE 12.1
Products at Different Stages of the Product Life Cycle

Introduction	Growth	Maturity	Decline
Ultra HD 4K television	3D printers	Laptop computer	Desktop computers
Home assistants	Airbnb lodging sharing	Disney theme parks	Landline phones
Hydrogen fuel automobiles	Paleo products	Soft drinks	Print newspaper

and squeeze out any remaining profits, marketing expenditures may be cut back, even though such cutbacks accelerate the sales decline. Finally, plans must be made for phasing out the product and introducing new ones to take its place. Consoles for video games are often phased out and a new one takes its place. Nintendo's launch of the Switch that has a 6.2-inch touchscreen allows taking games on the go but required developing a strong lineup of game titles.[13]

At the same time, it should be noted that product stages do not always go one way. Some products that have moved to the maturity stage or to the decline stage can still rebound through redesign or new uses for the product. For example, United Record Pressing LLC is enjoying the new interest in vinyl records. The company produces 30 to 40 percent of all vinyl records available in stores. This new interest in vinyl signals a resurgence in the sales of vinyl records to consumers and has created a growth mode for this privately held company—as well as a commitment to the company's rich musical history, which dates back to 1949.[14]

Identifying Products

Branding, packaging, and labeling can be used to identify or distinguish one product from others. As a result, they are key marketing activities that help position a product appropriately for its target market.

Branding.   **Branding** is the process of naming and identifying products. A *brand* is a name, term, symbol, design, or combination that identifies a product and distinguishes it from other products. Consider that Google, iTunes, and TiVo are brand names that are used to identify entire product categories, much like Xerox has become synonymous with photocopying and Kleenex with tissues. Protecting a brand name is important in maintaining a brand identity. The world's 10 most valuable brands are shown in Table 12.2. The brand name is the part of the brand that can be spoken and consists of letters, words, and numbers—such as WD-40 lubricant. A *brand mark* is

branding
the process of naming and identifying products.

Rank	Brand	Brand Value ($ billion)	Brand Value (% change from prior year)
1	Amazon	150.8	42
2	Apple	146.3	37
3	Google	120.9	10
4	Samsung	92.3	39
5	Facebook	89.7	45
6	AT&T	82.4	-5
7	Microsoft	81.2	6
8	Verizon	62.8	7
9	Walmart	61.5	-1
10	ICBC	59.2	24

TABLE 12.2
The 10 Most Valuable Brands in the World

Source: Jessica Tyler, "The 10 Most Valuable Brands in the World," *Business Insider, February 5, 2018, http://www.businessinsider. com/most-valuable-brands-in-the-world-for-2018-brand-finance-2018-2 (accessed April 29, 2018).*

trademark
a brand that is registered with the U.S. Patent and Trademark Office and is thus legally protected from use by any other firm.

manufacturer brands
brands initiated and owned by the manufacturer to identify products from the point of production to the point of purchase.

private distributor brands
brands, which may cost less than manufacturer brands, that are owned and controlled by a wholesaler or retailer.

generic products
products with no brand name that often come in simple packages and carry only their generic name.

packaging
the external container that holds and describes the product.

the part of the brand that is a distinctive design, such as the silver star on the hood of a Mercedes or McDonald's golden arches logo. A **trademark** is a brand that is registered with the U.S. Patent and Trademark Office and is thus legally protected from use by any other firm.

Two major categories of brands are manufacturer brands and private distributor brands. **Manufacturer brands** are brands initiated and owned by the manufacturer to identify products from the point of production to the point of purchase. Kellogg's, Sony, and Chevron are examples. **Private distributor brands,** which may be less expensive than manufacturer brands, are owned and controlled by a wholesaler or retailer, such as Nice! (Walgreens), Pantry Essentials (Safeway), 365 Everyday Value (Whole Foods), Great Value (Walmart), and Trader Joe's. The names of private brands do not usually identify their manufacturer. While private-label brands were once considered cheaper and of poor quality, such as Walmart's Ol'Roy dog food, many private-label brands are increasing in quality and image and are competing with national brands. Even Amazon has noticed the lucrative opportunities of private-label brands. The firm has begun hiring employees with fashion experience to help develop its own private-label clothing brand.[15] Today, there are private-label brands in nearly every food and beverage category, and the private-label food and beverage market is expected to grow by nearly 5 percent annually.[16] Manufacturer brands are fighting hard against private distributor brands to retain their market share.

Another type of brand that has developed is **generic products**—products with no brand name at all. They often come in plain simple packages that carry only the generic name of the product—peanut butter, tomato juice, aspirin, dog food, and so on. They appeal to consumers who may be willing to sacrifice quality or product consistency to get a lower price. Sales of generic brands have significantly decreased in recent years, although generic pharmaceuticals are commonly purchased due to their lower prices.

Companies use two basic approaches to branding multiple products. In one, a company gives each product within its complete product mix its own brand name. Unilever sells many well-known consumer products—Dove, Axe, Knorr, Hellman's, Dermalogica—each individually branded. This branding policy ensures that the name of one product does not affect the names of others, and different brands can be targeted at different segments of the same market, increasing the company's market share (its percentage of the sales for the total market for a product). Another approach to branding is to develop a family of brands with each of the firm's products carrying the same name or at least part of the name. Gillette, Sara Lee, and IBM use this approach. Finally, consumers may react differently to domestic versus foreign brands. The quality of Chinese brands may be questioned compared to German or U.S. brands.

Packaging. The **packaging,** or external container that holds and describes the product, influences consumers' attitudes and their buying decisions. Surveys have shown that consumers are willing to pay more for certain packaging attributes. For example, after facing criticism for its cardboard packaging, Kylie Cosmetics released higher-quality plastic eyeshadow palettes.[17] One of the attributes includes clearly stated nutrition and ingredient labeling, especially those characteristics indicating whether a product is organic, gluten free, or environmentally friendly. Recyclable and biodegradable packaging is also popular.[18] It is estimated that consumers' eyes linger

Product characteristics and attributes, such as gluten-free or whole grain, can be included on product packaging to help differentiate items to customers.

©Tyler McKay/Shutterstock

only 2.5 seconds on each product on an average shopping trip; therefore, product packaging should be designed to attract and hold consumers' attention.

A package can perform several functions, including protection, economy, convenience, and promotion. IKEA is constantly trying to investigate new ways for more efficient packaging to save on shipping costs.[19] Packaging can also be used to appeal to emotions. For example, pet food packaging appeals to the emotions of pet owners with illustrations of animals happily running, eating, or looking serene.[20] On the other hand, organizations must also exert caution before changing the designs of highly popular products. For example, Coca-Cola released a new global design for its products that features the traditional red as the main color. The "one brand" strategy is to create a unified global presence for its flagship soda.

Labeling.   Labeling, the presentation of important information on the package, is closely associated with packaging. The content of a label, often required by law, may include ingredients or content, nutrition facts (calories, fat, etc.), care instructions, suggestions for use (such as recipes), the manufacturer's address and toll-free number, website, and other useful information. This information can have a strong impact on sales. The labels of many products, particularly food and drugs, must carry warnings, instructions, certifications, or manufacturers' identifications.

labeling
the presentation of important information on a package.

Product Quality.   Quality reflects the degree to which a good, service, or idea meets the demands and requirements of customers. Quality products are often referred to as reliable, durable, easily maintained, easily used, a good value, or a trusted brand name. The level of quality is the amount of quality that a product possesses, and the consistency of quality depends on the product maintaining the same level of quality over time.

Quality of service is difficult to gauge because it depends on customers' perceptions of how well the service meets or exceeds their expectations. In other words, service quality is judged by consumers, not the service providers. For this reason, it is quite common for perceptions of quality to fluctuate from year to year. For instance, Volkswagen suffered significant reputational and financial impact from its emissions scandal with attempting to cover up the actual emissions impact of

quality
the degree to which a good, service, or idea meets the demands and requirements of customers.

TABLE 12.3
Personal Care and
Cleaning Products
Customer Satisfaction
Ratings

Company	Score
Clorox	84
Dial	83
Procter & Gamble	82
Colgate-Palmolive	81
Unilever	81
Johnson & Johnson	81

Source: American Customer Satisfaction Index, "Benchmarks by Industry: Personal Care and Cleaning Products," 2017, https://theacsi.org/index.php?option=com_content&view=article&id=147&catid=&Itemid=212&i=Personal+Care+and+Cleaning+Products (accessed April 29, 2018).

its diesel vehicles. VW faced buybacks, repairs, and financial settlements with defrauded customers.

Consumers expect quality projects and truthful and transparent information.[21] A bank may define service quality as employing friendly and knowledgeable employees, but the bank's customers may be more concerned with waiting time, ATM access, security, and statement accuracy. Similarly, an airline traveler considers on-time arrival, on-board Internet or TV connections, and satisfaction with the ticketing and boarding process. The American Customer Satisfaction Index produces customer satisfaction scores for 10 economic sectors, 44 industries, and more than 300 companies. The latest results show that overall customer satisfaction was 76.7 (out of a possible 100), with increases in some industries balancing out drops in others.[22] Table 12.3 shows the customer satisfaction rankings of some of the most popular personal care and cleaning product companies.

The quality of services provided by businesses on the Internet can be gauged by consumers on such sites as ConsumerReports.org and BBBOnline. The subscription service offered by ConsumerReports.org provides consumers with a view of digital marketing sites' business, security, and privacy policies, while BBBOnline is dedicated to promoting responsibility online. As consumers join in by posting business and product reviews on the Internet on sites such as Yelp, the public can often get a much better idea of the quality of certain goods and services. Quality can also be associated with where the product is made. For example, "Made in USA" labeling can be perceived as having a different value and quality. This includes strict laws on how much of a product can be made outside of the United States to still qualify for the "Made in USA" label. There are differences in the perception of quality and value between U.S. consumers and Europeans when comparing products made in the United States, Japan, Korea, and China.[23] Chinese brands are usually perceived as lower quality, while Japanese and Korean products are perceived as being of higher quality. However, China is trying to change consumer perceptions of its low brand quality. The increase in middle and upper classes in China has stimulated a rise in Chinese-branded luxury goods.[24]

According to the American Customer Satisfaction Index, consumers are most satisfied with Clorox cleaning products.

©McGraw-Hill Education, Jill Braaten photographer

FarmLinks: The Ultimate Alabama Resort

FarmLinks

Founders: Jimmy and David Pursell

Founded: 2003, in Sylacauga, Alabama

Success: Ranked by *Golfweek* as the number-one public golf course in Alabama, FarmLinks is a unique demonstration/education golf course that attracts accomplished golfers from across the nation.

FarmLinks started out as a marketing strategy for fertilizer. It was launched when Jimmy Pursell and his son decided to create a demonstration/education golf course on their family farm to promote their fertilizer, a coated time-release formula used for golf courses, agriculture, and consumer home lawn and garden care.

What started as a marketing strategy grew into a much larger business. Today, Purcell Farms is a family resort targeting everyone from golfers to businesspeople to engaged couples. Its location one hour from Birmingham and a few hours from Atlanta gives it a strategic advantage, and the Pursells price their resort services competitively against other hotels offering fewer services. Among the resort's many features is top-class dining; a 40-room inn; a shooting range; tours of the farm; and FarmLinks, its 7,444-yard championship golf course. FarmLinks draws in accomplished golfers with its challenging courses, such as its 615-yard 18th hole. The resort is a popular venue for conferences, company retreats, and weddings.

While FarmLinks engages in some advertising, its promotion mainly comes from its website, word-of-mouth marketing, social networking, and personal contacts. Its promotional strategy and stellar reputation seem to be working. The golf course alone was ranked as the number-one public golf course in Alabama by *Golfweek* and one of the top courses in the nation by *Golf Digest.*[25]

Critical Thinking Questions

1. How does FarmLinks use the marketing mix?
2. How does FarmLinks' location provide it with a strategic advantage?
3. Why might word-of-mouth marketing be more effective for FarmLinks as a promotional tool over advertising?

Pricing Strategy

LO 12-2

Previously, we defined price as the value placed on an object exchanged between a buyer and a seller. Buyers' interest in price stems from their expectations about the usefulness of a product or the satisfaction they may derive from it. Because buyers have limited resources, they must allocate those resources to obtain the products they most desire. They must decide whether the benefits gained in an exchange are worth the buying power sacrificed. Almost anything of value can be assessed by a price. Many factors may influence the evaluation of value, including time constraints, price levels, perceived quality, and motivations to use available information about prices.[26] Figure 12.5 illustrates a method for calculating the value of a product. Indeed, consumers vary in their response to price: Some focus solely on the lowest price, while others consider quality or the prestige associated with a product and its price. Some types of consumers are increasingly "trading up" to more status-conscious products, such as automobiles, home appliances, restaurants, and even pet food, yet remain price-conscious for other products such as cleaning and grocery goods. In setting prices, marketers must consider not just a company's cost to produce a good or service, but the perceived value of that item in the marketplace. Products' perceived value has benefited marketers at Starbucks, Sub-Zero, BMW, and Petco—which can charge premium prices for high-quality, prestige products—as well as Sam's Clubs and Costco—which offer basic household products at everyday low prices.

Explain the importance of price in the marketing mix, including various pricing strategies a firm might employ.

FIGURE 12.5

Calculating the Value of a Product

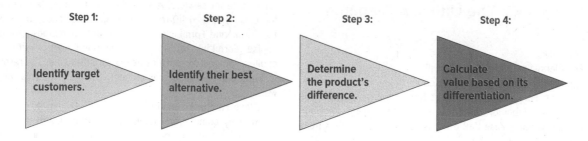

Step 1:
Identify target customers.

Step 2:
Identify their best alternative.

Step 3:
Determine the product's difference.

Step 4:
Calculate value based on its differentiation.

Price is a key element in the marketing mix because it relates directly to the generation of revenue and profits. In large part, the ability to set a price depends on the supply of and demand for a product. For most products, the quantity demanded goes up as the price goes down, and as the price goes up, the quantity demanded goes down. Changes in buyers' needs, variations in the effectiveness of other marketing mix variables, the presence of substitutes, and competition can influence demand. Faced with competition from online razor startups such as Harry's and Dollar Shave Club, Gillette saw its market share drop 16 percent in six years. This resulted in cutting prices by 20 percent in 2017.[27]

Price is probably the most flexible variable in the marketing mix. Although it may take years to develop a product, establish channels of distribution, and design and implement promotion, a product's price may be set and changed in a few minutes. Under certain circumstances, of course, the price may not be so flexible, especially if government regulations prevent dealers from controlling prices. Of course, price also depends on the cost to manufacture a good or provide a service or idea. A firm may temporarily sell products below cost to match competition, to generate cash flow, or even to increase market share, but in the long run, it cannot survive by selling its products below cost.

Pricing Objectives

Pricing objectives specify the role of price in an organization's marketing mix and strategy. They usually are influenced not only by marketing mix decisions, but also by finance, accounting, and production factors. Maximizing profits and sales, boosting market share, maintaining the status quo, and survival are four common pricing objectives.

Specific Pricing Strategies

Pricing strategies provide guidelines for achieving the company's pricing objectives and overall marketing strategy. They specify how price will be used as a variable in the marketing mix. Significant pricing strategies relate to the pricing of new products, psychological pricing, reference pricing, and price discounting.

Pricing New Products.   Setting the price for a new product is critical: The right price leads to profitability; the wrong price may kill the product. In general, there are two basic strategies to setting the base price for a new product. **Price skimming** is charging the highest possible price that buyers who want the product will pay. Price skimming is used with luxury items. Ultra-D (3D television without the use of glasses) often run into the thousands of dollars. Price skimming is often used to allow the company to generate much-needed revenue to help offset the costs of research and development. Conversely, a **penetration price** is a low price designed to help a product enter the market and gain market share rapidly. When Netflix entered the market, it offered its rentals at prices much lower than the average rental stores and did not charge late fees. Netflix quickly gained market share and eventually drove many rental stores out of business. Penetration pricing is less flexible than price skimming; it is more difficult to raise a penetration price than to lower a skimming price. Penetration pricing is used most often when marketers suspect that competitors will enter the market shortly after the product has been introduced.

Psychological Pricing. **Psychological pricing** encourages purchases based on emotional rather than rational responses to the price. For example, the assumption behind *even/odd pricing* is that people will buy more of a product for $9.99 than $10 because it seems to be a bargain at the odd price. The assumption behind *symbolic/prestige pricing* is that high prices connote high quality. Thus the prices of certain fragrances and cosmetics are set artificially high to give the impression of superior quality. Some over-the-counter drugs are priced high because consumers associate a drug's price with potency.

Reference Pricing. **Reference pricing** is a type of psychological pricing in which a lower-priced item is compared to a more expensive brand in hopes that the consumer will use the higher price as a comparison price. The main idea is to make the item appear less expensive compared with other alternatives. For example, Walmart might place its Great Value brand next to a manufacturer's brand such as Hefty or Heinz so that the Great Value brand will look like a better deal.

Price Discounting. Temporary price reductions, or **discounts,** are often employed to boost sales. Although there are many types, quantity, seasonal, and promotional discounts are among the most widely used. Quantity discounts reflect the economies of purchasing in large volumes. Seasonal discounts to buyers who purchase goods or services out of season help even out production capacity. Promotional discounts attempt to improve sales by advertising price reductions on selected products to increase customer interest. Often, promotional pricing is geared toward increased profits. For instance, bare-bones German grocery chain Aldi is attempting to compete against Trader Joe's as it expands in the United States through the offering of higher-end food and price discounts.[28] Aldi has about 1,700 stores in the U.S. and plans to open approximately 800 more by 2022.[29]

Distribution Strategy

In the "Managing Operations and Supply Chains" chapter, we discussed supply chain management that involves connecting and integrating all members of the supply chain. While supply chain management involves operations, procurement, and logistics, we take a closer look at the role of marketing channels and a part of logistics related to physical distribution.

The best products in the world will not be successful unless companies make them available where and when customers want to buy them. In this section, we will

price skimming
charging the highest possible price that buyers who want the product will pay.

penetration price
a low price designed to help a product enter the market and gain market share rapidly.

psychological pricing
encouraging purchases based on emotional rather than rational responses to the price.

reference pricing
a type of psychological pricing in which a lower-priced item is compared to a more expensive brand in hopes that the consumer will use the higher price as a comparison price.

discounts
temporary price reductions, often employed to boost sales.

LO 12-3

Identify factors affecting distribution decisions, such as marketing channels and intensity of market coverage.

How Archer Daniels Midlands Connects Farm, Factory, and Table

Archer Daniels Midland (ADM), one of the world's largest agricultural processors, is a vital link among farm, factory, and table. The Chicago-based multinational rings up almost $62 billion in annual net sales of soybeans, corn gluten, and other products provided by farmers worldwide. ADM's clients include several well-known companies. Unilever, for instance, purchases oil from soybeans procured and processed by ADM. Unilever uses this oil to produce its mayonnaise products.

To stay ahead of demand for sustainable agricultural processes, ADM has partnered with Unilever to enroll its farmers in the Iowa Sustainable Soy Fieldprint Project to monitor factors like soil conservation and land use. In the project's first year, 43 farms with land totaling more than 44,000 acres were enrolled. ADM also introduced a cost-share initiative to soybean farmers in Iowa to increase the pool of farms with sustainably grown soybeans and provide financial assistance to farmers. The initiative works with Unilever's Sustainable Living Plan, which includes a commitment to source all agricultural raw materials sustainably in the coming years. These factors make ADM an ideal soybean oil supplier for Unilever's Hellmann's mayonnaise.

In response to changing external forces, ADM is also making changes to its product lines. The company recently introduced Nutriance, a new range of wheat protein concentrates that are vegetarian and vegan-friendly. By recognizing potential new areas of profitability, ADM demonstrates that it is carefully monitoring the external environment and changing accordingly.[30]

Critical Thinking Questions

1. How is Archer Daniels Midland making changes to its products to appeal to changing consumer preferences?
2. Archer Daniels Midland is expending resources to engage in more sustainable agriculture. Do you think these costs will be beneficial for Archer Daniels Midland? Why or why not?
3. Why is it necessary for Archer Daniels Midland to maintain strong relationships with the companies that source its products?

---

explore dimensions of distribution strategy, including the channels through which products are distributed, the intensity of market coverage, and the physical handling of products during distribution.

Marketing Channels

marketing channel
a group of organizations that moves products from their producer to customers; also called a channel of distribution.

A **marketing channel**, or channel of distribution, is a group of organizations that moves products from their producer to customers. Marketing channels make products available to buyers when and where they desire to purchase them. Organizations that bridge the gap between a product's manufacturer and the ultimate consumer are called *middlemen*, or intermediaries. They create time, place, and ownership utility. Two intermediary organizations are retailers and wholesalers.

retailers
intermediaries who buy products from manufacturers (or other intermediaries) and sell them to consumers for home and household use rather than for resale or for use in producing other products.

Retailers buy products from manufacturers (or other intermediaries) and sell them to consumers for home and household use rather than for resale or for use in producing other products. Dick's Sporting Goods, for example, buys products from Nike and other manufacturers and resells them to consumers. By bringing together an assortment of products from competing producers, retailers create utility. Retailers arrange for products to be moved from producers to a convenient retail establishment (place utility). They maintain hours of operation for their retail stores to make merchandise available when consumers want it (time utility). They also assume the risk of ownership of inventories (ownership utility). Table 12.4 describes various types of general merchandise retailers.

Amazon represents an Internet retailer business model that is disrupting the competitive structure of retail markets. Traditional retailers are developing their own

TABLE 12.4
General Merchandise Retailers

Type of Retailer	Description	Examples
Department store	Large, full-service stores organized by departments	Nordstrom, Macy's, Neiman Marcus
Internet retailer	A direct marketer providing most products over the Internet.	Amazon, Alibaba
Discount store	Offers less services than department stores; store atmosphere reflects value pricing	Walmart, Stein Mart, Target
Convenience store	Small, self-service stores carrying many items for immediate consumption	Circle K, 7-Eleven, Allsup's
Supermarket	Large stores carrying most food items as well as nonfood items for daily family use	Trader Joe's, Albertsons, Wegmans
Superstore	Very large stores that carry most food and nonfood products that are routinely purchased	Super Walmart, Meijer
Hypermarket	The largest retail stores that take the foundation of the discount store and provide even more food and nonfood products	Carrefour, Tesco Extra
Warehouse club	Large membership establishments with food and nonfood products and deep discounts	Costco, BJ's Wholesale Club, Sam's Club
Warehouse showroom	Large facilities with products displayed that are often retrieved from a less expensive adjacent warehouse	IKEA, Cost Plus

online operations to compete with Amazon. The company accounts for 4 percent of retail spending and 44 percent of e-commerce sales in America and is growing rapidly.[31] While currently only half the size of the largest retailer, Walmart, it is the largest online retailer. Amazon is changing the nature of competition in the retail environment. The company sells almost every retail item and is challenging department and other retail stores. Also, Amazon allows other retailers to use its e-commerce platform, warehouses, and other services.[32] In many ways, Amazon has remade retailing. The company is even expanding into brick-and-mortar stores to push into the grocery business. With this, Amazon is rolling out Amazon Fresh markets that mix online and in-store shopping features.[33]

General merchandise retailers, especially department stores, are feeling the competitive threat, with Macy's, Sears, and others closing stores. Sears is even doubtful as to whether it can keep operating after many years of losses.[34] Online retailers, such as Amazon, pose a competitive threat to traditional retailers, undercutting store-based sellers on prices and options of many products. Many sporting goods store chains have gone out of business, but Dick's Sporting Goods is one that remains very successful today.[35]

Another type of retail is **direct marketing,** which is the use of nonpersonal media to communicate products, information, and the opportunity to purchase via media such as mail, telephone, or the Internet. For example, Duluth Trading has stores but specializes in catalog marketing, especially with products such as jeans, work boots, and hats. Another form of nonstore retailing is **direct selling,** which involves the marketing of products to ultimate consumers through face-to-face sales presentations at

direct marketing
the use of nonpersonal media to communicate products, information, and the opportunity to purchase via media such as mail, telephone, or the Internet.

direct selling
the marketing of products to ultimate consumers through face-to-face sales presentations at home or in the workplace.

home or in the workplace. The top three global direct selling companies are Amway, Avon, and Herbalife. Most individuals who engage in direct selling work on a part-time basis because they like the product and often sell to their own social networks.

wholesalers
intermediaries who buy from producers or from other wholesalers and sell to retailers.

Wholesalers are intermediaries who buy from producers or from other wholesalers and sell to retailers. They usually do not sell in significant quantities to ultimate consumers. Wholesalers perform the functions listed in Table 12.5.

Wholesalers are extremely important because of the marketing activities they perform, particularly for consumer products. Although it is true that wholesalers can be eliminated, their functions must be passed on to some other entity, such as the producer, another intermediary, or even the customer. Wholesalers help consumers and retailers by buying in large quantities, then selling to retailers in smaller quantities. By stocking an assortment of products, wholesalers match products to demand. Sysco is a food wholesaler for the food services industry. The company provides food, preparation, and serving products to restaurants, hospitals, and other institutions that provide meals outside of the home.[36] *Merchant wholesalers* like Sysco take title to the goods, assume risks, and sell to other wholesalers, business customers, or retailers. *Agents* negotiate sales, do not own products, and perform a limited number of functions in exchange for a commission.

Supply Chain Management.   In an effort to improve distribution channel relationships among manufacturers and other channel intermediaries, supply chain management creates alliances between channel members. In the "Managing Operations and Supply Chains" chapter, we defined supply chain management as connecting and integrating all parties or members of the distribution system in order to satisfy customers. It involves long-term partnerships among marketing channel members working together to reduce costs, waste, and unnecessary movement in the entire

TABLE 12.5
Major Wholesaling
Functions

Physical distribution	• Inventory management • Transportation • Warehousing • Materials handling
Promotion	• Personal selling • Publicity • Sales promotion • Advertising
Inventory control and data processing	• Management information systems • Inventory control • Transaction monitoring • Financial and accounting data analysis
Risk-taking	• Inventory decisions • Product deterioration • Theft control
Financing and budgeting	• Investment capital • Credit management • Managing cash flow and receivables
Marketing research and information systems	• Conducting primary market research • Analyzing big data • Utilizing marketing analytics

marketing channel in order to satisfy customers. It goes beyond traditional channel members (producers, wholesalers, retailers, customers) to include *all* organizations involved in moving products from the producer to the ultimate customer. In a survey of business managers, a disruption in the supply chain was viewed as the number-one crisis that could decrease revenue.[37]

The focus shifts from one of selling to the next level in the channel to one of selling products *through* the channel to a satisfied ultimate customer. Information, once provided on a guarded, "as-needed" basis, is now open, honest, and ongoing. Perhaps most importantly, the points of contact in the relationship expand from one-on-one at the salesperson-buyer level to multiple interfaces at all levels and in all functional areas of the various organizations. Predictive analytics are being used for forecasting and coordinating the integration of supply chain members. For example, Amazon ships products before it receives a customer order based upon predictive models that relate to customer purchasing history.[38]

Channels for Consumer Products.   Typical marketing channels for consumer products are shown in Figure 12.6. In channel A, the product moves from the producer directly to the consumer. Farmers who sell their fruit and vegetables to consumers at roadside stands or farmers' markets use a direct-from-producer-to-consumer marketing channel.

In channel B, the product goes from producer to retailer to consumer. This type of channel is used for products such as college textbooks, automobiles, and appliances. In channel C, the product is handled by a wholesaler and a retailer before it reaches the consumer. Producer-to-wholesaler-to-retailer-to-consumer marketing channels

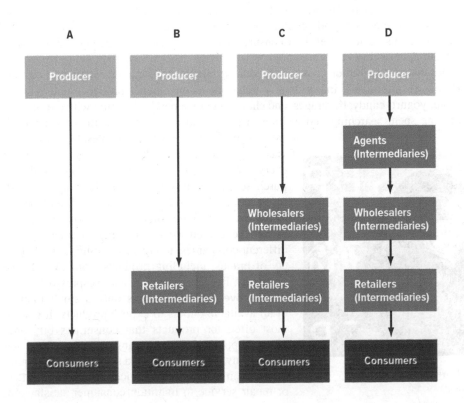

FIGURE 12.6
Marketing Channels for Consumer Products

distribute a wide range of products including refrigerators, televisions, soft drinks, cigarettes, clocks, watches, and office products. In channel D, the product goes to an agent, a wholesaler, and a retailer before going to the consumer. This long channel of distribution is especially useful for convenience products. Candy and some produce are often sold by agents who bring buyers and sellers together.

Services are usually distributed through direct marketing channels because they are generally produced *and* consumed simultaneously. For example, you cannot take a haircut home for later use. Many services require the customer's presence and participation: The sick patient must visit the physician to receive treatment; the child must be at the day care center to receive care; the tourist must be present to sightsee and consume tourism services.

Channels for Business Products. In contrast to consumer goods, more than half of all business products, especially expensive equipment or technically complex products, are sold through direct marketing channels. Business customers like to communicate directly with producers of such products to gain the technical assistance and personal assurances that only the producer can offer. For this reason, business buyers prefer to purchase expensive and highly complex mainframe computers directly from IBM, Unisys, and other mainframe producers. Other business products may be distributed through channels employing wholesaling intermediaries such as industrial distributors and/or manufacturer's agents.

Intensity of Market Coverage

intensive distribution
a form of market coverage whereby a product is made available in as many outlets as possible.

A major distribution decision is how widely to distribute a product—that is, how many and what type of outlets should carry it. The intensity of market coverage depends on buyer behavior, as well as the nature of the target market and the competition. Wholesalers and retailers provide various intensities of market coverage and must be selected carefully to ensure success. Market coverage may be intensive, selective, or exclusive.

selective distribution
a form of market coverage whereby only a small number of all available outlets are used to expose products.

Intensive distribution makes a product available in as many outlets as possible. Because availability is important to purchasers of convenience products such as toothpaste, yogurt, candy, beverages, and chewing gum, a nearby location with a minimum of time spent searching and waiting in line is most important to the consumer. To saturate markets intensively, wholesalers and many varied retailers try to make the product available at every location where a consumer might desire to purchase it. ZoomSystems provides robotic vending machines for products beyond candy and drinks. Zoom has 1,500 machines in airports and hotels across the United States. Through partnering with different companies, today's ZoomShops sell a variety of brands, including products from The Honest Company, Best Buy, Macy's, and Nespresso.[39]

Selective distribution uses only a small number of all available outlets to expose products. It is used most often for products that consumers buy only after shopping and comparing price, quality, and style. Many products sold on a selective basis require salesperson assistance, technical advice, warranties, or repair service to maintain consumer satisfaction.

Kat Von D makeup is available exclusively through Sephora.

©Oscar Gonzalez/NurPhoto/REX/Shutterstock

Typical products include automobiles, major appliances, clothes, and furniture. Ralph Lauren is a brand that uses selective distribution.

Exclusive distribution exists when a manufacturer gives an intermediary the sole right to sell a product in a defined geographic territory. Such exclusivity provides an incentive for a dealer to handle a product that has a limited market. Exclusive distribution is the opposite of intensive distribution in that products are purchased and consumed over a long period of time, and service or information is required to develop a satisfactory sales relationship. Products distributed on an exclusive basis include high-quality musical instruments, yachts, airplanes, and high-fashion leather goods. For example, Singapore's Triple Pte is the exclusive distributor for Under Armour in Southeast Asia.[40]

exclusive distribution
the awarding by a manufacturer to an intermediary of the sole right to sell a product in a defined geographic territory.

Physical Distribution

Logistics includes the planning and coordination of inbound and outbound as well as third-party services. **Physical distribution** is the part of logistics that focuses on transportation modes, warehousing, and materials handling. Physical distribution creates time and place utility by making products available when they are wanted, with adequate service and at minimum cost. Both goods and services require physical distribution. Many physical distribution activities are part of supply chain management, which we discussed in the "Managing Operations and Supply Chains" chapter; we'll take a brief look at a few more now.

logistics
the planning and coordination of inbound and outbound as well as third party services.

physical distribution
the part of logistics that focuses on transportation modes, warehousing, and materials handling.

Transportation. **Transportation,** the shipment of products to buyers, creates time and place utility for products, and thus is a key element in the flow of goods and services from producer to consumer. The five major modes of transportation used to move products between cities in the United States are railways, motor vehicles, inland waterways, pipelines, and airways.

transportation
the shipment of products to buyers.

Railroads are a cost-effective method of transportation for many products. Heavy commodities, foodstuffs, raw materials, and coal are examples of products carried by railroads. Trucks have greater flexibility than railroads because they can reach more locations. Trucks handle freight quickly and economically, offer door-to-door service, and are more flexible in their packaging requirements than are ships or airplanes. Air transport offers speed and a high degree of dependability but is the most expensive means of transportation; transport by ship is less expensive and is the slowest form. Pipelines are used to transport petroleum, natural gas, semiliquid coal, wood chips, and certain chemicals. Pipelines have the lowest costs for products that can be transported via this method. Many products can be moved most efficiently by using more than one mode of transportation.

warehousing
the design and operation of facilities to receive, store, and ship products.

Factors affecting the selection of a mode of transportation include cost, capability to handle the product, reliability, and availability; and, as suggested, selecting transportation modes requires trade-offs. Unique characteristics of the product and consumer desires often determine the mode selected.

Warehousing. **Warehousing** is the design and operation of facilities to receive, store, and ship products. A warehouse facility receives, identifies,

Amazon has fulfillment centers across the country from Lakeland, Florida, to San Bernardino, California.
©SWNS/Alamy Stock Photo

sorts, and dispatches goods to storage; stores them; recalls, selects, or picks goods; assembles the shipment; and finally, dispatches the shipment.

Companies often own and operate their own private warehouses that store, handle, and move their own products. Firms might want to own or lease a private warehouse when their goods require special handling and storage or when it has large warehousing needs in a specific geographic area. Private warehouses are beneficial because they provide customers with more control over their goods. However, fixed costs for maintaining these warehouses can be quite high.[41] They can also rent storage and related physical distribution services from public warehouses. While public warehouses store goods for more than one company, providing firms with less control over distribution, they are often less expensive than private warehouses and are useful for seasonal production or low-volume storage.[42] For example, Next Level Resource Partners offers warehousing services and fulfillment including picking, packing, and shipping products for clients like Pure Barre, Mineral Fusion, and Flywheel.[43] Regardless of whether a private or a public warehouse is used, warehousing is important because it makes products available for shipment to match demand at different geographic locations.

materials handling
the physical handling and movement of products in warehousing and transportation.

Materials Handling.   Materials handling is the physical handling and movement of products in warehousing and transportation. Handling processes may vary significantly due to product characteristics. Efficient materials-handling procedures increase a warehouse's useful capacity and improve customer service. Well-coordinated loading and movement systems increase efficiency and reduce costs.

Importance of Distribution in a Marketing Strategy

Distribution decisions are among the least flexible marketing mix decisions. Products can be changed over time, prices can be changed quickly, and promotion is usually changed regularly. But distribution decisions often commit resources and establish contractual relationships that are slow to change. As a company attempts to expand into new markets, it may require a complete change in distribution. Moreover, if a firm does not manage its marketing channel in the most efficient manner and provide the best service, then a new competitor will evolve to create a more effective distribution system. The growth of online retailing is increasing consumers' expectations for more choices, faster delivery, order status updates, and easy returns at no extra cost. Amazon is transforming the supply chain, and retailers such as Target are revamping their distribution systems to use digital technology to develop a smart network.[44] Target plans to redesign about 600 stores as part of its goal of creating a smart network where its stores, digital channels, and supply chain work together to support the guest.[45]

LO 12-4

Specify the activities involved in promotion, as well as promotional strategies and promotional positioning.

Promotion Strategy

The role of promotion is to communicate with individuals, groups, and organizations to facilitate an exchange directly or indirectly. It encourages marketing exchanges by attempting to persuade individuals, groups, and organizations to accept goods, services, and ideas. Promotion is used not only to sell products, but also to influence opinions and attitudes toward an organization, person, or cause. The state of Michigan, for example, has successfully used its "Pure Michigan" campaign to influence tourists to visit Michigan. The economic impact of the campaign was estimated at $1.2 billion.[46] Most people probably equate promotion with advertising, but it also includes personal selling, publicity, and sales promotion. The role that these elements play in a marketing strategy is extremely important.

Amazon Sells Everything from A-to-Z

Before Google or eBay came on the scene, Jeff Bezos quit his job in finance and founded Amazon in 1994 as an online bookstore. When books proved to be a successful product, he asked some of his customers what else he should sell on the site. The response was overwhelming. Bezos realized Amazon met a customer need that was greater than books: convenience.

Today, Amazon sells everything from toys and clothing to eBooks and groceries. Amazon makes up approximately 5 percent of the retail sales in the United States and is approaching half of all online retail sales in the United States. Bezos attributes Amazon's success to its focus on the customer instead of the competition. For example, in order to maintain and build a relationship with its customers, Amazon offers an A-to-z Guarantee that protects against damaged goods and late deliveries.

In another move to become an indispensable part of every household, Amazon has continued to evolve the Amazon Echo,

a voice-controlled digital assistant. Through the Echo, users can check the weather, get news alerts, play games, control connected smart-home devices, and more. Prime users, roughly two-thirds of U.S. households, can also order prime-eligible items. Additionally, Amazon is transforming grocery shopping with its recent purchase of Whole Foods. The company plans to integrate various aspects of Whole Foods, Prime Now two-hour delivery, and Amazon Fresh grocery delivery to explore new ways to meet customer needs.[47]

Critical Thinking Questions

1. What marketing channel does Amazon use to reach consumers?
2. What are some of the ways in which Amazon tries to guarantee quality in its services?
3. How is Amazon modifying its product and service offerings to appeal to consumers?

The Promotion Mix

Advertising, personal selling, publicity, and sales promotion are collectively known as the promotion mix because a strong promotion program results from the careful selection and blending of these elements. The process of coordinating the promotion mix elements and synchronizing promotion as a unified effort is called **integrated marketing communications.** When planning promotional activities, an integrated marketing communications approach results in the desired message for customers. Different elements of the promotion mix are coordinated to play their appropriate roles in delivery of the message on a consistent basis. Integrated communication creates a reason for purchase. Tide developed an integrated marketing campaign for the Super Bowl to highlight Tide's dominance in the detergent market as well as demonstrate its high performance. Tide's series of ads that ran during each quarter featured David Harbour (*Stranger Things*) and implied that any commercial with clean clothes is a Tide ad. The company coordinated social media posts with Old Spice spokesman Isaiah Mustafa as well as Betty White, Danica Patrick, Antonio Brown, and Drew Brees as part of the campaign. The result was more than 52,000 tweets during the game.[48]

Advertising. Perhaps the best-known form of promotion, **advertising** is a paid form of nonpersonal communication transmitted through a mass medium, such as television commercials, magazine advertisements, or online ads. Pharmaceutical firms have long used advertisements to promote medications for lifestyle conditions. However, more recently it has begun releasing advertisements promoting life-saving, often expensive, medications that specialist doctors prescribe.[49] Commercials featuring celebrities, customers, or unique creations serve to grab viewers' attention and pique their interest in a product. Peyton Manning, former quarterback for the Denver

integrated marketing communications
coordinating the promotion mix elements and synchronizing promotion as a unified effort.

 connect

Need help understanding integrated marketing communications? Visit your Connect ebook video tab for a brief animated explanation.

advertising
a paid form of nonpersonal communication transmitted through a mass medium, such as television commercials or magazine advertisements.

Former ABC *Bachelor* contestant and *Bachelor in Paradise* star Jade Roper Tolbert was tapped for a brand endorsement for Ava, an ovulation tracking bracelet.

©Broadimage/REX/Shutterstock

advertising campaign designing a series of advertisements and placing them in various media to reach a particular target market.

Broncos, has done celebrity endorsements for Oreo, Gatorade, Direct TV, Nationwide Insurance, and Buick. On the other hand, there are downsides to using celebrity endorsers when they act inappropriately. Nike has terminated or suspended contracts with celebrity endorsers for domestic violence allegations and doping allegations.

An **advertising campaign** involves designing a series of advertisements and placing them in various media to reach a particular target audience. The basic content and form of an advertising campaign are a function of several factors. A product's features, uses, and benefits affect the content of the campaign message and individual ads. Characteristics of the people in the target audience–gender, age, education, race, income, occupation, lifestyle, and other attributes–influence both content and form. When Procter & Gamble promotes Crest toothpaste to children, the company emphasizes daily brushing and cavity control, whereas it promotes tartar control and whiter teeth when marketing to adults. To communicate effectively, advertisers use words, symbols, and illustrations that are meaningful, familiar, and attractive to people in the target audience.

An advertising campaign's objectives and platform also affect the content and form of its messages. If a firm's advertising objectives involve large sales increases, the message may include hard-hitting, high-impact language and symbols. When campaign objectives aim at increasing brand awareness, the message may use much repetition of the brand name and words and illustrations associated with it. Thus, the advertising platform is the foundation on which campaign messages are built.

Advertising media are the vehicles or forms of communication used to reach a desired audience. Print media include newspapers, magazines, direct mail, and billboards, while electronic media include television, radio, and Internet advertising. Choice of media obviously influences the content and form of the message. Effective outdoor displays and short broadcast spot announcements require concise, simple messages. Magazine and newspaper advertisements can include considerable detail and long explanations. Because several kinds of media offer geographic selectivity, a precise message can be tailored to a particular geographic section of the target audience. For example, a company advertising in *Time* might decide to use one message in the New England region and another in the rest of the nation. A company may also choose to advertise in only one region. Such geographic selectivity lets a firm use the same message in different regions at different times. On the other hand, some companies are willing to pay extensive amounts of money to reach national audiences. Marketers spent approximately $5 million for one 30-second advertising slot during the Super Bowl due to its national reach and popularity.[50]

The use of online advertising is increasing. However, advertisers are demanding more for their ad dollars and proof that they are working, which is why Google AdWords only charges companies when users click on the ad. Certain types of ads are more popular than pop-up ads and banner ads that consumers find annoying. One technique is to blur the lines between television and online advertising. TV commercials may point viewers to a website for more information, where short "advertainment" films continue the marketing message. Marketers might also use the Internet to show advertisements or videos that were not accepted by mainstream television. People for the Ethical Treatment of Animals (PETA) often develop racy commercials that are denied Super Bowl spots. However, these ads can be viewed online through YouTube and other sites.[51]

Infomercials—typically 30-minute blocks of radio or television air time featuring a celebrity or upbeat host talking about and demonstrating a product—have evolved as an advertising method. Under Armour teamed with famed Olympic athlete Michael Phelps to create a day in his life called "The Water Diviner." The writing, visuals, and lighting make this a compelling piece and won it recognition by *Advertising Age* as one of the top 10 best branded content partnerships. As the most decorated Olympian in history, there remains significant fascination with Michael Phelps, and Under Armour has found an innovative way to create a partnership that enhances its branding.[52] Toll-free numbers and website addresses are usually provided so consumers can conveniently purchase the product or obtain additional information. Although many consumers and companies have negative feelings about infomercials, apparently they get results.

Personal Selling.   **Personal selling** is direct, two-way communication with buyers and potential buyers. For many products—especially large, expensive ones with specialized uses, such as cars, appliances, and houses–interaction between a salesperson and the customer is probably the most important promotional tool.

personal selling
direct, two-way communication with buyers and potential buyers.

Personal selling is the most flexible of the promotional methods because it gives marketers the greatest opportunity to communicate specific information that might trigger a purchase. Only personal selling can zero in on a prospect and attempt to persuade that person to make a purchase. Although personal selling has a lot of advantages, it is one of the costliest forms of promotion. A sales call on an industrial customer can cost more than $400.

There are three distinct categories of salespersons: order takers (for example, retail sales clerks and route salespeople), creative salespersons (for example, automobile, furniture, and insurance salespeople), and support salespersons (for example, customer educators and goodwill builders who usually do not take orders). For most of these salespeople, personal selling is a six-step process:

publicity
nonpersonal communication transmitted through the mass media but not paid for directly by the firm.

1. *Prospecting:* Identifying potential buyers of the product.
2. *Approaching:* Using a referral or calling on a customer without prior notice to determine interest in the product.
3. *Presenting:* Getting the prospect's attention with a product demonstration.
4. *Handling objections:* Countering reasons for not buying the product.
5. *Closing:* Asking the prospect to buy the product.
6. *Following up:* Checking customer satisfaction with the purchased product.

Personal selling is important with high-risk items such as medical tools and devices. Sales representatives can assist customers in discussing the benefits of a product, financing arrangements, and any warranties or quarantees.

©Blend Images/ERproductions Ltd/Getty Images

Publicity.   **Publicity** is nonpersonal communication transmitted through the mass media but not paid for directly by the firm. A firm does not pay the media cost for publicity and is not identified as the originator of the message; instead, the message is presented in news story form. Obviously, a company can benefit from publicity by releasing to news sources newsworthy messages about the firm and its involvement with the public. Many companies have *public relations* departments to try to gain favorable publicity and minimize negative publicity for the firm.

Although advertising and publicity are both carried by the mass media, they differ in several major ways. Advertising messages tend to be informative, persuasive, or both; publicity is mainly informative. Advertising is often designed to have an immediate impact or to provide specific information to persuade a person to act; publicity describes what a firm is doing, what products it is launching, or other newsworthy information, but seldom calls for action. When advertising is used, the organization must pay for media time and select the media that will best reach target audiences. The mass media willingly carry publicity because they believe it has general public interest. Advertising can be repeated a number of times; most publicity appears in the mass media once and is not repeated.

Advertising, personal selling, and sales promotion are especially useful for influencing an exchange directly. Publicity is extremely important when communication focuses on a company's activities and products and is directed at interest groups, current and potential investors, regulatory agencies, and society in general.

A variation of traditional advertising is buzz marketing, in which marketers attempt to create a trend or acceptance of a product. Companies seek out trendsetters in communities and get them to "talk up" a brand to their friends, family, co-workers, and others. One unusual method that Nagoya, Japan, used to attract more job applicants to the city was adopting a spokes-ape. The city used an ape at the local zoo that not only became a recruitment tool, with his face on posters and T-shirts, but also a national celebrity.[53] Other marketers using the buzz technique include Hebrew National ("mom squads" grilled the company's hot dogs), and Red Bull (its sponsorship of the stratosphere space diving project). The idea behind buzz marketing is that an accepted member of a particular social group will be more credible than any form of paid communication.[54] The concept works best as part of an integrated marketing communication program that also includes traditional advertising, personal selling, sales promotion, and publicity.

A related concept is viral marketing, which describes the concept of getting Internet users to pass on ads and promotions to others. For example, Canadian Tire, an automotive company based in Toronto, uploaded two 60-second ads during the summer Olympics that attracted viral attention. In one, a group of kids invite a boy in a wheelchair to play basketball. The video attracted more than 200 million views and almost 4 million shares. Positive opinion of Canadian Tire increased by 13 percent.[55]

Sales Promotion. Sales promotion involves direct inducements offering added value or some other incentive for buyers to enter into an exchange. Sales promotions are generally easier to measure and less expensive than advertising. The major tools of

sales promotion
direct inducements offering added value or some other incentive for buyers to enter into an exchange.

Ibotta is a digital coupon app that allows users to select deals and receive cash back from retailers like Kroger and Target after scanning their receipts.

©Joe Amon/The Denver Post via Getty Images

sales promotion are store displays, premiums, samples and demonstrations, coupons, contests and sweepstakes, refunds, and trade shows. Coupon-clipping in particular became common due to the recent recession. While coupons in the past decade traditionally had a fairly low redemption rate, with about 2 percent being redeemed, the recent recession caused an upsurge in coupon usage. There has also been a major upsurge in the use of mobile coupons, or coupons sent to consumers over mobile devices. The redemption rate for mobile coupons is eight times higher than that of traditional coupons.[56] While coupons can be a valuable tool in sales promotion, they cannot be relied upon to stand by themselves, but should be part of an overall promotion mix. Sales promotion stimulates customer purchasing and increases dealer effectiveness in selling products. It is used to enhance and supplement other forms of promotion. Sampling a product may also encourage consumers to buy. This is why many grocery stores provide free samples in the hopes of influencing consumers' purchasing decisions. In a given year, almost three-fourths of consumer product companies may use sampling.

Promotion Strategies: To Push or to Pull

In developing a promotion mix, organizations must decide whether to fashion a mix that pushes or pulls the product (Figure 12.7). A **push strategy** attempts to motivate intermediaries to push the product down to their customers. When a push strategy is used, the company attempts to motivate wholesalers and retailers to make the product available to their customers. Sales personnel may be used to persuade intermediaries to offer the product, distribute promotional materials, and offer special promotional incentives for those who agree to carry the product. For example, salespeople from

push strategy
an attempt to motivate intermediaries to push the product down to their customers.

FIGURE 12.7
Push and Pull Strategies

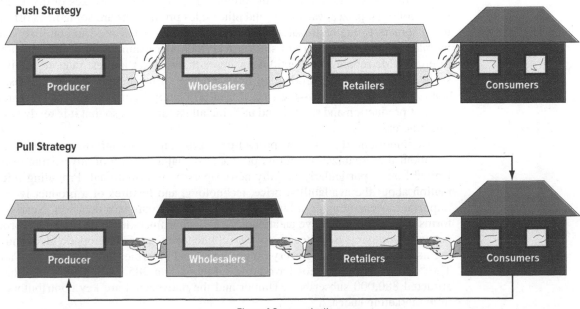

Flow of Communications

pull strategy
the use of promotion to create consumer demand for a product so that consumers exert pressure on marketing channel members to make it available.

pharmaceutical companies will often market new products to doctors in the hope that the doctors will recommend their products to their clients. A **pull strategy** uses promotion to create consumer demand for a product so that consumers exert pressure on marketing channel members to make it available. For Example, Mountain Dew regularly does a limited release of its Baja Blast flavor every few years to stoke demand. Most recently, the soda company brought back the fan-favorite flavor for a limited time only after its follower on Twitter championed the hashtag #bringbajablastback.[57] Many brands like Parmer Water Company, Cinnabon, and Campbell's have used Twitter to encourage followers to demand the brand's products at their local grocery store. Additionally, offering free samples prior to a product rollout encourages consumers to request the product from their favorite retailer.

A company can use either strategy, or it can use a variation or combination of the two. The exclusive use of advertising indicates a pull strategy. Personal selling to marketing channel members indicates a push strategy. The allocation of promotional resources to various marketing mix elements probably determines which strategy a marketer uses.

Objectives of Promotion

The marketing mix a company uses depends on its objectives. It is important to recognize that promotion is only one element of the marketing strategy and must be tied carefully to the goals of the firm, its overall marketing objectives, and the other elements of the marketing strategy. Firms use promotion for many reasons, but typical objectives are to stimulate demand, to stabilize sales, and to inform, remind, and reinforce customers.

Increasing demand for a product is probably the most typical promotional objective. Stimulating demand, often through advertising, publicity, and sales promotion, is particularly important when a firm is using a pull strategy.

Another goal of promotion is to stabilize sales by maintaining the status quo—that is, the current sales level of the product. During periods of slack or decreasing sales, contests, prizes, vacations, and other sales promotions are sometimes offered to customers to maintain sales goals. Advertising is often used to stabilize sales by making customers aware of slack use periods. For example, auto manufacturers often provide rebates, free options, or lower-than-market interest rates to stabilize sales and thereby keep production lines moving during temporary slowdowns. A stable sales pattern allows the firm to run efficiently by maintaining a consistent level of production and storage and utilizing all its functions so that it is ready when sales increase.

An important role of any promotional program is to inform potential buyers about the organization and its products. A major portion of advertising in the United States, particularly in daily newspapers, is informational. Providing information about the availability, price, technology, and features of a product is very important in encouraging a buyer to move toward a purchase decision. Nearly all forms of promotion involve an attempt to help consumers learn more about a product and a company. Blendtec, the powerful kitchen blender, developed videos showing its blender pulverizing everything from rakes to marbles and even Apple iPads. The "Will it Blend" videos have attracted more than 285 million views and have attracted 880,000 subscribers. Humor and the unexpected are key contributors to viral marketing success.[58]

Promotion is also used to remind consumers that an established organization is still around and sells certain products that have uses and benefits. Often advertising reminds customers that they may need to use a product more frequently or in certain situations. Pennzoil, for example, has run television commercials reminding car owners that they need to change their oil every 3,000 miles to ensure proper performance of their cars.

Reinforcement promotion attempts to assure current users of the product that they have made the right choice and tells them how to get the most satisfaction from the product. Also, a company could release publicity statements through the news media about a new use for a product. Additionally, firms can have salespeople communicate with current and potential customers about the proper use and maintenance of a product–all in the hope of developing a repeat customer.

Promotional Positioning

Promotional positioning uses promotion to create and maintain an image of a product in buyers' minds. It is a natural result of market segmentation. In both promotional positioning and market segmentation, the firm targets a given product or brand at a portion of the total market. A promotional strategy helps differentiate the product and makes it appeal to a particular market segment. For example, to appeal to safety-conscious consumers, Volvo heavily promotes the safety and crashworthiness of Volvo automobiles in its advertising. Promotion can be used to change or reinforce an image. Effective promotion influences customers and persuades them to buy.

promotional positioning the use of promotion to create and maintain an image of a product in buyers' minds.

Importance of Marketing Strategy

Marketing creates value through the marketing mix. For customers, value means receiving a product in which the benefit of the product outweighs the cost, or price paid for it. For marketers, value means that the benefits (usually monetary) received from selling the product outweigh the costs it takes to develop and sell it. This requires carefully integrating the marketing mix into an effective marketing strategy. One misstep could mean a loss in profits, whether it be from a failed product idea, shortages or oversupply of a product, a failure to effectively promote the product, or prices that are too high or too low. And while some of these marketing mix elements can be easily fixed, other marketing mix elements such as distribution can be harder to adapt.

On the other hand, firms that develop an effective marketing mix to meet customer needs will gain competitive advantages over those that do not. Often, these advantages occur when the firm excels at one or more elements of the marketing mix. Aldi has a reputation for low prices, while Christian Louboutin is known for its high-quality, luxury shoes. However, exceling at one element of the marketing mix does not mean that a company can neglect the others. The best product cannot succeed if consumers do not know about it or if they cannot find it in stores. Additionally, firms must constantly monitor the market environment to understand how demand is changing and whether adaptations in the marketing mix are needed. It is therefore essential that every element of the marketing mix be carefully evaluated and synchronized with the marketing strategy. Only then will firms be able to achieve the marketing concept of providing products that satisfy customers' needs while allowing the organization to achieve its goals.

So You Want to Be a Marketing Manager

Many jobs in marketing are closely tied to the marketing mix functions: distribution, product, promotion, and price. Often the job titles could be sales manager, distribution or supply chain manager, advertising account executive, or store manager.

A distribution manager arranges for transportation of goods within firms and through marketing channels. Transportation can be costly, and time is always an important factor, so minimizing their effects is vital to the success of a firm. Distribution managers must choose one or a combination of transportation modes from a vast array of options, taking into account local, federal, and international regulations for different freight classifications; the weight, size, and fragility of products to be shipped; time schedules; and loss and damage ratios. Manufacturing firms are the largest employers of distribution managers.

A product manager is responsible for the success or failure of a product line. This requires a general knowledge of advertising, transportation modes, inventory control, selling and sales management, promotion, marketing research, packaging, and pricing. Frequently, several years of selling and sales management experience are prerequisites for such a position as well as college training in business administration. Being a product manager can be rewarding both financially and psychologically.

Some of the most creative roles in the business world are in the area of advertising. Advertising pervades our daily lives, as businesses and other organizations try to grab our attention and tell us about what they have to offer. Copywriters, artists, and account executives in advertising must have creativity, imagination, artistic talent, and expertise in expression and persuasion. Advertising is an area of business in which a wide variety of educational backgrounds may be useful, from degrees in advertising itself, to journalism or liberal arts degrees. Common entry-level positions in an advertising agency are found in the traffic department, account service (account coordinator), or the media department (media assistant). Advertising jobs are also available in many manufacturing or retail firms, nonprofit organizations, banks, professional associations, utility companies, and other arenas outside of an advertising agency.

Although a career in retailing may begin in sales, there is much more to retailing than simply selling. Many retail personnel occupy management positions, focusing on selecting and ordering merchandise, promotional activities, inventory control, customer credit operations, accounting, personnel, and store security. Many specific examples of retailing jobs can be found in large department stores. A section manager coordinates inventory and promotions and interacts with buyers, salespeople, and consumers. The buyer's job is fast-paced, often involving much travel and pressure. Buyers must be open-minded and foresighted in their hunt for new, potentially successful items. Regional managers coordinate the activities of several retail stores within a specific geographic area, usually monitoring and supporting sales, promotions, and general procedures. Retail management can be exciting and challenging. Growth in retailing is expected to accompany the growth in population and is likely to create substantial opportunities in the coming years.

While a career in marketing can be very rewarding, marketers today agree that the job is getting tougher. Many advertising and marketing executives say the job has gotten much more demanding in the past 10 years, viewing their number one challenge as balancing work and personal obligations. Other challenges include staying current on industry trends or technologies, keeping motivated/inspired on the job, and measuring success. If you are up to the challenge, you may find that a career in marketing is just right for you to utilize your business knowledge while exercising your creative side as well.

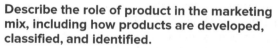

Review Your Understanding

Describe the role of product in the marketing mix, including how products are developed, classified, and identified.

Products (goods, services, ideas) are among a firm's most visible contacts with consumers and must meet consumers' needs and expectations to ensure success. New-product development is a multistep process: idea development, the screening of new ideas, business analysis, product development, test marketing, and commercialization. Products are usually classified as either consumer or business products. Consumer products can be further classified as convenience, shopping, or specialty products. The business product classifications are raw materials, major equipment, accessory equipment, component parts, processed materials, supplies, and industrial services. Products also can be classified by the stage of the product life cycle (introduction, growth, maturity, and decline). Identifying products includes branding (the process of naming and identifying products), packaging (the product's container), and labeling (of information, such as content and warnings, on the package).

Explain the importance of price in the marketing mix, including various pricing strategies a firm might employ.

Price is the value placed on an object exchanged between a buyer and a seller. It is probably the most flexible variable of the marketing mix. Pricing objectives include survival, maximization of profits and sales volume, and maintaining the status quo. When a firm introduces a new product, it may use price skimming or penetration pricing. Psychological pricing and price discounting are other strategies.

Identify factors affecting distribution decisions, such as marketing channels and intensity of market coverage.

Making products available to customers is facilitated by middlemen, or intermediaries, who bridge the gap between the producer of the product and its ultimate user. A marketing channel is a group of marketing organizations that directs the flow of products from producers to consumers. Market coverage relates to the number and variety of outlets that make products available to customers; it may be intensive, selective, or exclusive. Physical distribution is all the activities necessary to move products from producers to consumers, including inventory planning and control, transportation, warehousing, and materials handling.

Specify the activities involved in promotion, as well as promotional strategies and promotional positioning.

Promotion encourages marketing exchanges by persuading individuals, groups, and organizations to accept goods, services,

and ideas. The promotion mix includes advertising (a paid form of nonpersonal communication transmitted through a mass medium), personal selling (direct, two-way communication with buyers and potential buyers), publicity (nonpersonal communication transmitted through the mass media but not paid for directly by the firm), and sales promotion (direct inducements offering added value or some other incentive for buyers to enter into an exchange). A push strategy attempts to motivate intermediaries to push the product down to their customers, whereas a pull strategy tries to create consumer demand for a product so that consumers exert pressure on marketing channel members to make the product available. Typical promotion objectives are to stimulate demand; stabilize sales; and inform, remind, and reinforce customers. Promotional positioning is the use of promotion to create and maintain in the buyer's mind an image of a product.

Evaluate an organization's marketing strategy plans.

Based on the material in this chapter, you should be able to answer the questions posed in the "Solve the Dilemma" feature near the end of the chapter and evaluate the company's marketing strategy plans, including its target market and marketing mix.

Critical Thinking Questions

Enter the World of Business Questions

1. How did Harley-Davidson modify its marketing strategy to appeal to younger consumers?

2. What are some of the challenges Harley-Davidson faces as it adapts to customer preferences?

3. Do you think Harley-Davidson's marketing initiatives to younger consumers has the potential to alienate its baby boomer customers? Why or why not?

Learn the Terms

Check Your Progress

1. What steps do companies generally take to develop and introduce a new product?

2. What is the product life cycle? How does a product's life cycle stage affect its marketing strategy?

3. Which marketing mix variable is probably the most flexible? Why?

4. Distinguish between the two ways to set the base price for a new product.

5. What is probably the least flexible marketing mix variable? Why?

6. Describe the typical marketing channels for consumer products.

7. What activities are involved in physical distribution? What functions does a warehouse perform?

8. How do publicity and advertising differ? How are they related?

9. What does the personal selling process involve? Briefly discuss the process.

10. List the circumstances in which the push and pull promotional strategies are used.

Get Involved

1. Pick three products you use every day (in school, at work, or for pleasure—perhaps one of each). Determine what phase of the product life cycle each is in. Evaluate the marketer's strategy (product, price, promotion, and distribution) for the product and whether it is appropriate for the life-cycle stage.

2. Design a distribution channel for a manufacturer of stuffed toys.

3. Pick a nearby store, and briefly describe the kinds of sales promotion used and their effectiveness.

Build Your Skills

Analyzing Motel 6's Marketing Strategy

Background

Made famous through the well-known radio and TV commercials spoken in the distinctive "down-home" voice of Tom Bodett, the Dallas-based Motel 6 chain of budget motels is probably familiar to you. Based on the information provided here and any personal knowledge you may have about the company, you will analyze the marketing strategy of Motel 6.

Task

Read the following paragraphs; then complete the questions that follow.

Motel 6 was established in 1962 with the original name emphasizing its low-cost, no-frills approach. Rooms at that time were $6 per night. Today, Motel 6 has more than 760 units, and the average nightly cost is $49.99. Motel 6 is the largest company-owned and -operated lodging chain in the United States. Customers receive HBO, ESPN, free morning coffee, and free local phone calls, and most units have pools and some business services. Motel 6 has made a name for itself by offering clean, comfortable rooms at the lowest prices of any national motel chain and by standardizing both its product offering and its operating policies and procedures. The company's national spokesperson, Tom Bodett, is featured in radio and television commercials that use humorous stories to show why it makes sense to stay at Motel 6 rather than a pricey hotel.

In appealing to pleasure travelers on a budget as well as business travelers looking to get the most for their dollar, one commercial makes the point that all hotel and motel rooms look the same at night when the lights are out—when customers are getting what they came for, a good night's sleep. Motel 6 location sites are selected based on whether they provide convenient access to the highway system and whether they are close to areas such as shopping centers, tourist attractions, or business districts.

1. In SELECTING A TARGET MARKET, which approach is Motel 6 using to segment markets?

   a. concentration approach

   b. multisegment approach

2. In DEVELOPING A MARKETING MIX, identify in the second column of the table what the current strategy is and then identify any changes you think Motel 6 should consider for carrying it successfully through the next five years.

Marketing Mix Variables	Current Strategy	5-Year Strategy
a. Product		
b. Price		
c. Distribution		
d. Promotion		

Solve the Dilemma

Better Health with Snacks

Deluxe Chips is one of the leading companies in the salty-snack industry, with almost one-fourth of the $10 billion market. Its Deluxos tortilla chips are the number-one selling brand in North America, and its Ridgerunner potato chip is also a market share leader. Deluxe Chips wants to stay on top of the market by changing marketing strategies to match changing consumer needs and preferences. Promoting specific brands to market segments with the appropriate price and distribution channel is helping Deluxe Chips succeed.

As many middle-aged consumers modify their snacking habits, Deluxe Chips is considering a new product line of light snack foods with less fat and cholesterol and targeted at the 35- to 50-year-old consumer who enjoys snacking but wants to be more health conscious. Marketing research suggests that the product will succeed as long as it tastes good and that

consumers may be willing to pay more for it. Large expenditures on advertising may be necessary to overcome the competition. However, it may be possible to analyze customer profiles and retail store characteristics and then match the right product with the right neighborhood. Store-specific micromarketing would allow Deluxe Chips to spend its promotional dollars more efficiently.

LO 12-5

Evaluate an organization's marketing strategy plans.

Critical Thinking Questions

1. Design a marketing strategy for the new product line.
2. Critique your marketing strategy in terms of its strengths and weaknesses.
3. What are your suggestions for implementation of the marketing strategy?

Build Your Business Plan   ■ connect

Dimensions of Marketing Strategy

If you think your product/business is truly new to or unique to the market, you need to substantiate your claim. After a thorough exploration on the web, you want to make sure there has not been a similar business/product recently launched in your community. Check with your Chamber of Commerce or Economic Development Office that might be able to provide you with a history of recent business failures. If you are not confident about the ability or willingness of customers to try your new good or service, collecting your own primary data to ascertain demand is highly advisable.

The decision of where to initially set your prices is a critical one. If there are currently similar products in the market, you need to be aware of the competitors' prices before you

determine yours. If your product is new to the market, you can price it high (market skimming strategy) as long as you realize that the high price will probably attract competitors to the market more quickly (they will think they can make the same product for less), which will force you to drop your prices sooner than you would like. Another strategy to consider is market penetration pricing, a strategy that sets price lower and discourages competition from entering the market as quickly. Whatever strategy you decide to use, don't forget to examine your product elasticity.

At this time, you need to start thinking about how to promote your product. Why do you feel your product is different or new to the market? How do you want to position your product so customers view it favorably? Remember this is all occurring *within the consumer's mind.*

See for Yourself Videocase

Cutting Edge Quality: Cutco "Knives for Life"

What if you could have a high quality kitchen knife that would last forever? Look no further than Cutco. Cutco began manufacturing knives in Olean, New York in 1949 when Alcoa Corp and Case Cutlery formed a joint venture to begin creating high quality cutlery. Since then, the company has grown to become a nationally recognized brand known

for its high quality, American made products. All Cutco knives come with a "Forever Guarantee" that allows customers to have their knives serviced for free, regardless of when the knives were purchased, so the knives can be passed from generation to generation.

Since 1949, Cutco has used a direct sales channel. In 1981, Vector was founded in Philadelphia as the sales division. It has grown to over 250 district sales offices. The Vector management team at Vector started as sales representatives. Cutco's direct sales force, largely college students looking to make extra money during the summer, learn how to manage their time, how to interface with others, and how to express unique product attributes which set Cutco apart from the competition. There are two models of direct selling compensation: multilevel and single level marketing. Multilevel compensation means direct sellers earn commission from their own sales as well as commissions from those they have recruited and trained. Cutco uses a single level method of compensation, which means the sales representatives only earn a commission on the sales they make, and they do not recruit other sales representatives.

While the driver of the Cutco distribution model is direct selling, it has begun to experiment with different sales methods. For example, Cutco has opened 16 retail locations since 2005 in order to give existing customers a place to service their knives while allowing potential customers a place to view and handle its products. These stores are focused on giving a "Cutco experience" where customers can try out knives or take classes on cooking, floral arrangements, gardening, and more. The stores also offer more than 100 kinds of kitchen cutlery products in addition to its knives such as utensils, kitchen shears, flatware, cookware, and a full line of sporting knives.

Cutco is proud of its history and heritage as an American knife manufacturer. Quality and fair treatment of employees and the community are first and foremost the priorities for the company. Cutco's knife blades are made from a high carbon stainless steel, which goes through a 3-stage heat treatment process to ensure that blades can take a sharp edge, maximize their resistance to corrosion, and won't snap under pressure. All knives have "full-tang construction," which means that the blade's metal extends all the way through the handle of the knife to create stability. The handles are made from an acetal copolymer thermo-resin, which gives the handles high strength, toughness, and resistance to abrasion. Rivets that hold the handle together are made of a nickel-silver alloy that doesn't expand or contract from heat. This process creates a high quality knife that is built to last. However, wear and tear is inevitable, so customers can get their knives serviced at any time.

Cutco, which has manufactured its product in Olean, New York since it was created in 1949, is now one of the few American cutlery companies to keep manufacturing in the U.S. Cutco's American made commitment has not always been easy to abide by. After the company entered the flatware business, its American supplier went bankrupt and closed down its operation. All other flatware manufacturing was outside of the United States. In an unparalleled commitment to American jobs and American-made knives, Cutco purchased the fabricating equipment for its stainless table knives and insourced the product into its own factory in Olean. The company takes pride in the town of Olean where it has around 650 full-time employees. James Stitt, Executive Chairman, came to Olean in 1975 to work for Cutco. His son, James Stitt, Jr., joined the company in 1997 and is currently the President and CEO. Before letting his son take over the company, Stitt Sr. ensured that his son remained committed to the town which relies heavily on Cutco since it employs many of Olean's citizens. Corey Wiktor, The Executive Director of the Cattaraugus County Industrial Development Agency, described Cutco as ". . . the lifeblood of Cattaraugus County and the Southern Tier."

Thanks to Cutco's high quality products and unique business model, it is a nationally recognized brand for cutlery that is synonymous with quality. Cutco continues to expand its product offerings and has begun complimenting its direct sales distribution with retail and Internet sales. Even with its evolving marketing strategy, the high quality of its knives and its commitment to direct sales has remained constant since Cutco began in 1949.[59]

Critical Thinking Questions

1. How has Cutco used quality to differentiate its products?

2. How does Cutco add value to its products through the "Forever Guarantee"?

3. Does the direct selling channel help position Cutco's high quality knives in the marketplace?

You can find the related video in the Video Library in Connect. Ask your instructor how you can access Connect.

Team Exercise

Form groups and search for examples of convenience products, shopping products, specialty products, and business products. How are these products marketed? Provide examples of any ads that you can find to show examples of the promotional strategies for these products. Report your findings to the class.

Notes

1. James R. Hagerty, "Can Harley Spark a Movement?" *The Wall Street Journal,* June 20–21, 2015; Bill Saporito, "This Harley Is Electric," *Time,* June 2014, pp. 48–52; Charles Fleming, "First Times Ride: 2015 Harley-Davidson Street 750," *Los Angeles Times,* July 4, 2014, http://www.latimes.com/business/autos/la-fi-hy-first-times-ride-2015-harley-davidson-street-750-20140616-story.html (accessed November 30, 2017); Kyle Stock, "Is Harley-Davidson Losing Its Diehards?" *Bloomberg,* April 21, 2015, https://www.bloomberg.com/news/articles/2015-04-21/is-harley-davidson-losing-its-diehards- (accessed November 30, 2017); Harley-Davidson, "Our Company," http://www.harley-davidson.com/content/h-d/en_US/company.html (accessed December 5, 2017); Fred Lambert, "Harley-Davidson Will Bring to Market Its First All-Electric Bike 'within 5 Years,'" *electrek,* June 14, 2016, https://electrek.co/2016/06/14/harley-davidson-electric-bike-within-5-years/ (accessed December 22, 2017); Ankit Ajmera and Rachit Vats, "Harley Sales Recovery Far Off as Young Bikers Turn to Used Bikes," *Reuters,* August 30, 2017, https://www.reuters.com/article/us-harleydavidson-sales/harley-sales-recovery-far-off-as-young-buyers-turn-to-used-bikes-idUSKCN1BA26D (accessed December 22, 2017); Rob LeFebvre, "Harley-Davidson Embraces the Potential of Electric Motorcycles," *Endgadget,* May 11, 2017, https://www.engadget.com/2017/05/11/harley-davidson-electric-motorcycles/ (accessed December 22, 2017).

2. "The Power of the Jingle," *Advertising Age,* February 8, 2017, http://adage.com/article/iheart-media/power-jingle/307801/ (accessed April 5, 2017).

3. Robert Duffer, "Discontinued Car Models for 2016," *Chicago Tribune,* December 10, 2015, http://www.chicagotribune.com/classified/automotive/sc-2016-discontinued-cars-1022-20151016-story.html (accessed May 5, 2017).

4. Associated Press, "Jobs Says iPad Idea Came before iPhone," June 2, 2010, www.foxnews.com/tech/2010/06/02/jobs-says-ipad-idea-came-iphone/ (accessed May 5, 2017).

5. Matt Simon, "The Genesis of Kuri, the Friendly Home Robot," *Wired,* November 30, 2017, https://www.wired.com/story/the-genesis-of-kuri/ (accessed April 29, 2018).

6. Nike, "A Look Inside Nike's Sport Research Lab," September 8, 2014, http://news.nike.com/news/a-look-inside-nike-s-sport-research-lab (accessed April 11, 2016).

7. John A. Byrne, "Greatest Entrepreneurs of Our Time," *Fortune,* April 9, 2012, pp. 68–86; Google Finance, "Amazon.com, Inc.," April 30, 2014, www.google.com/finance?cid=660463 (accessed April 11, 2016).

8. Katy Muldoon, "Marketers Wonder: How Will It Play in Portland?" *The Wall Street Journal,* July 15, 2015, http://www.wsj.com/articles/marketers-wonder-how-will-it-play-in-portland-1436983102 (accessed May 5, 2017).

9. Andrew Pollack, "Genetically Engineered Salmon Approved for Consumption," *The New York Times,* November 19, 2015, http://www.nytimes.com/2015/11/20/business/genetically-engineered-salmon-approved-for-consumption.html (accessed May 5, 2017).

10. Kurt Schroeder, "Why So Many New Products Fail (and It's Not the Product)," *Business Journals,* March 14, 2017, https://www.bizjournals.com/bizjournals/how-to/marketing/2017/03/why-so-many-new-products-fail-and-it-s-not-the.html (accessed April 24, 2018).

11. Business Wire, "Global Smartwatch Market Growth of 12% by 2020—Analysis, Technologies & Forecast Report 2016–2020: Key Vendors: Apple, Fitbit, Pebble—Research and Markets," March 8, 2016, http://www.businesswire.com/news/home/20160308005941/en/Global-Smartwatch-Market-Growth-12-2020— (accessed April 11, 2016).

12. Peter Hartlaub, "Sweet! America's Top 10 Brands of Soda," *NBC News,* http://www.nbcnews.com/id/42255151/ns/business-us_business/t/sweet-americas-top-brands-soda/#.WHbiQrYrKV4 (accessed December 27, 2016).

13. Nathan Olivarez-Giles, "Nintendo's Switch: Elegant, but Unready," *The Wall Street Journal,* March 2, 2017, p. B1.

14. Burleson, "Vinyl Record Manufacturer in Nashville Is Said to Be Expanding."

15. Laura Lorenzetti, "Get Ready for Amazon's New Fashion Line," *Fortune,* February 17, 2016, http://fortune.com/2016/02/17/amazon-fashion-brand/ (accessed May 5, 2017).

16. Christopher Durham, "Private Label Market in the US 2015–2019," *My Private Brand,* March 20, 2016, http://mypbrand.com/2016/03/20/private-label-market-in-the-us-2015-2019/ (accessed April 11, 2016); Lisa Fickenscher, "Retailers Eyeing Private Brand Ownership as Key to Growth," *New York Post,* October 13, 2017, https://nypost.com/2017/10/13/retailers-eyeing-private-brand-ownership-as-key-to-growth/ (accessed April 29, 2018).

17. Nicola Dall'Asen, "Kyle Finally Upgraded Her Makeup Packaging to Feel Less Cheap," *Revelist,* April 19, 2018, http://www.revelist.com/beauty-news-/kylie-cosmetics-packaging/12476 (accessed April 29, 2018).

18. Mintel, "Beverage Packaging Trends—US—February 2014," February 2014, http://oxygen.mintel.com/sinatra/oxygen/list/id=680559&type=RCItem#0_1___page_RCItem=0 (accessed May 5, 2014).

19. Saabira Chaudhuri, "IKEA Can't Stop Obsessing about Its Packaging," *The Wall Street Journal,* June 17, 2015, http://www.wsj.com/articles/ikea-cant-stop-obsessing-about-its-packaging-1434533401 (accessed May 5, 2017).

20. Lindsey Beaton, "Packaging as Branding in Pet Food Marketing," Petfoodindustry.com, May 21, 2015, http://www.petfoodindustry.com/articles/5193-packaging-as-branding-in-pet-food-marketing (accessed May 1, 2015).

21. Nathan Bomey, "Volkswagen Will Buy Back 20k More Polluting Diesel Cars," *USA Today,* December 20, 2016, http://www.usatoday.com/story/money/cars/2016/12/20/volkswagen-3-liter-diesel-settlement/95661794/ (accessed April 18, 2016).

22. "U.S. Overall Customer Satisfaction," American Customer Satisfaction Index, http://www.theacsi.org/

national-economic-indicator/us-overall-customer-satisfaction (accessed April 29, 2018).

23. "American Demographics 2006 Consumer Perception Survey," *Advertising Age,* January 2, 2006, p. 9. Data by Synovate.

24. "The Top 10 Most Innovative Companies in China," *Fast Company,* 2014, http://www.fastcompany.com/most-innovative-companies/2014/industry/china (accessed May 5, 2017).

25. Personal interview of Jimmy Pursell by Diane Kroncke; Pursell Farms website, https://pursellfarms.com/ (accessed April 6, 2018).

26. Rajneesh Suri and Kent B. Monroe, "The Effects of Time Constraints on Consumers' Judgments of Prices and Products," *Journal of Consumer Research* 30 (June 2003), p. 92.

27. Sharon Terlep, "Gillette, in Change Shaves Prices," *The Wall Street Journal,* April 5, 2017, p. B1.

28. Leslie Patton, "Aldi Tries High-End Food and Discounts, Too," *Bloomberg,* August 6, 2015, http://www.bloomberg.com/news/articles/2015-08-06/aldi-grocery-chain-tries-high-end-food-and-discounts-too (accessed May 5, 2017).

29. Patt Johnson, "Discount Grocer Aldi Headed to Ankeny This Fall," *Des Moines Register,* April 25, 2018, https://www.desmoinesregister.com/story/money/business/new-business/2018/04/26/aldi-discount-grocer-ankeny-fall-north-ankeny-boulevard-produce-dairy-bakery/554046002/ (accessed April 29, 2018).

30. "ADM Expands Portfolio with Range of Wheat Protein Concentrates," Archer Daniels Midland Company, November 28, 2017, https://www.adm.com/news/news-releases/adm-expands-portfolio-with-range-of-wheat-protein-concentrates (accessed December 2, 2017); Tony Reid, "ADM Planning Layoffs," *Herald & Review,* July 20, 2017, http://herald-review.com/business/agriculture/adm-planning-layoffs/article_26e525e5-f10d-5a5f-a4ae-fab3958340d8.html (accessed December 2, 2017); Javier Blas, "Commodity Trading Arm in Sleepy Swiss Village Trips Up ADM Again," *Chicago Tribune,* February 8, 2017, http://www.chicagotribune.com/business/ct-archer-daniels-midland-commodity-trading-20170208-story.html (accessed December 5, 2017); "Partnerships across the Supply Chain Drive Continuous Improvement, Helping Meet Corporate Commitments," *Field to Market,* December 19, 2016, https://fieldtomarket.org/case-studies-series/partnerships-across-supply-chain/ (accessed December 5, 2017); Leah Guffey, "ADM Helps Unilever Bring Out the Best," *AgWired,* September 18, 2014, http://agwired.com/2014/09/18/adm-helps-unilever-bring-out-the-best/ (accessed July 6, 2015); Paul Demery, "Archer Daniels Midland Cultivates Its Handling of Purchase Orders and Invoices," *InternetRetailer,* September 9, 2014, https://www.internetretailer.com/2014/09/09/archer-daniels-midland-cleans-its-handling-purchase-orders (accessed December 23, 2017); Gregory Meyer, "Archer Daniels Midland Buys Wild Flavors for €2.3 bn," *Financial Times,* July 7, 2014, http://www.ft.com/intl/cms/s/0/636b8732-05b4-11e4-9baa-00144feab7de.html#axzz3LYqkltWg (accessed July 6, 2015); Archer Daniels Midland website, http://www.adm.com (accessed December 23, 2017).

31. Lauren Thomas, "Amazon Grabbed 4 Percent of All U.S. Retail Sales in 2017, New Study Says," *CNBC,* January 3, 2018, https://www.cnbc.com/2018/01/03/

amazon-grabbed-4-percent-of-all-us-retail-sales-in-2017-new-study.html (accessed April 29, 2018).

32. "Primed," *The Economist,* March 25, 2017, pp. 18–19.

33. Spencer Soper and Olivia Zaleski, "Jeff Bezos Goes Grocery Shopping," *Bloomberg Businessweek,* April 2–9, 2017, pp. 21–22.

34. Ann Steele, "Sears Creates Stir as It Casts Doubts about Its Future," *The Wall Street Journal,* March 23, 2017, p. B1.

35. Nathan Bomery, "For Sports Retailers Clock Has Run Out," *USA Today,* March 6, 2017, pp. B1–B2.

36. "The Sysco Story," Sysco, www.sysco.com/about-sysco.html# (accessed April 11, 2016).

37. "Top Threats to Revenue," *USA Today,* February 1, 2006, p. A1.

38. Paul Amyerson, "Predictive Analytics Takes Forecasting to a New Level," *Inbound Logistics,* March 2017, p. 46.

39. ZoomSystems, www.zoomsystems.com/ (accessed April 11, 2016).

40. Pitsinee Jitpleecheep, "Under Armour Set to Cut Back Prices," *Bangkok Post,* February 13, 2018, https://www.bangkokpost.com/business/news/1411475/under-armour-set-to-cut-back-prices (accessed April 29, 2018).

41. William Pride and O. C. Ferrell, *Marketing Foundations,* 5th ed. (Mason, OH: Cengage South-Western Learning, 2013), pp. 415–416.

42. Ibid.

43. Next Level Resource Partners, Homepage, https://www.nlrp.com/ (accessed April 29, 2018).

44. Karen M. Kroll, "Retail Logistics Bets on E-Commerce," *Inbound Logistics,* March 2017, p. 73.

45. "Target Reveals Design Elements of Next Generation of Stores," Target, March 20, 2017, https://corporate.target.com/press/releases/2017/03/target-reveals-design-elements-of-next-generation (accessed April 29, 2018).

46. Michelle Grinnell, "Pure Michigan Campaign Drives $1.2 Billion in Customer Spending," March 11, 2014, http://www.michigan.org/pressreleases/pure-michigan-campaign-drives-$1-2-billion-in-visitor-spending/ (accessed April 11, 2016).

47. Catherine Clifford, "How Amazon Founder Jeff Bezos Went from the Son of a Teen Mom to the World's Richest Person," *CNBC,* October 27, 2017, https://www.cnbc.com/2017/10/27/how-amazon-founder-jeff-bezos-went-from-the-son-of-a-teen-mom-to-the-worlds-richest-person.html (accessed December 18, 2017); Caroline Cakebread, "Amazon Launched 22 Years Ago This Week—Here's What Shopping on Amazon was Like Back in 1995," *Business Insider,* July 20, 2017, http://www.businessinsider.com/amazon-opened-22-years-ago-see-the-business-evolve-2017-7 (accessed December 18, 2017); Charisse Jones, "Amazon Wants Your Teen to Spend Your Money But Parents Have the Final Say," *USA Today,* October 11, 2017, https://www.usatoday.com/story/money/2017/10/11/amazon-let-parents-hand-over-shopping-keys-kids/751594001/ (accessed December 18, 2017); Daniel Keyes, "Amazon Prime May Be Reaching Saturation in the U.S.." *Business Insider,* December 18, 2017, http://www.businessinsider.com/amazon-prime-may-be-reaching-saturation-in-the-us-2017-12

(accessed December 18, 2017); "About A-to-z Guarantee," Amazon.com, https://www.amazon.com/gp/help/customer/display.html?nodeId=201889410 (accessed December 18, 2017); Alison Griswold, "Amazon Just Explained How Whole Foods Fits into Its Plan for World Domination," *Quartz,* October 27, 2017, https://qz.com/1113795/amazon-amzn-just-explained-how-whole-foods-fits-into-its-plan-for-world-domination/ (accessed December 23, 2017); Ry Crist, "After a Busy 2017, Alexa is Still on Top—and Still Evolving," *CNet,* December 20, 2017, https://www.cnet.com/news/after-a-busy-2017-alexa-is-still-on-top-and-still-evolving/ (accessed December 24, 2017).

48. Ilyse Liffreing, "Brand Winners and Losers of Super Bowl LII," *Digiday,* February 4, 2018, https://digiday.com/marketing/brand-winners-losers-super-bowl-lii/ (accessed February 21, 2018); Marty Swant, "Every Ad Is a Tide Ad: Inside Saatchi and P&G's Clever Super Bowl Takeover Starring David Harbour," *Adweek,* Feburary 4, 2018, www.adweek.com/brand-marketing/every-ad-is-a-tide-ad-inside-saatchi-and-pgs-clever-super-bowl-takeover-starring-david-harbour/ (accessed February 21, 2018); Jack Neff, "Tide Is Everywhere with Campaign to Own All Super Bowl Ads," *AdAge,* Feburary 4, 2018, http://adage.com/article/special-report-super-bowl/t/312249/ (accessed February 21, 2018).

49. Peter Loftus, "Ads for Costly Drugs Get Airtime," *The Wall Street Journal,* February 17, 2016, p. B1.

50. Andrew Gould, "Super Bowl Ads 2018: Latest Info on Cost of 2018 Super Bowl Commercials," *Bleacher Report,* February 4, 2018, http://bleacherreport.com/articles/2757119-super-bowl-ads-2018-latest-info-on-cost-of-2018-super-bowl-commercials (accessed April 29, 2018).

51. Natalie Evans, "See PETA 'Vegan Sex' Advert That's So Steamy It's Been Banned from the Super Bowl," *The Mirror,* January 26, 2016, http://www.mirror.co.uk/tv/tv-news/see-peta-vegan-sex-advert-7246273 (accessed May 5, 2017).

52. Jerrid Grimm, "The 10 Best Branded Content Partnerships of 2016," *Advertising Age,* http://adage.com/article/agency-viewpoint/10-branded-content-partnerships-2016/307284/ (accessed January 26, 2017).

53. Jun Hongo and Miho Inada, "To Lure Recruits, Japanese City Tries Gorilla Marketing," *The Wall Street Journal,* February 23, 2016, p. A1, A14.

54. Gerry Khermouch and Jeff Green, "Buzz Marketing," *BusinessWeek,* July 30, 2001, pp. 50–56.

55. Harmeet Signh, "Behind Canadian Tire's (Old) Viral Spot," *Strategy,* February 28, 2017, http://strategyonline.ca/2017/02/28/behind-canadian-tires-old-viral-spot/ (accessed April 29, 2018); "Wheels—The Most Viral Ad of 2017," *YouTube,* August 24, 2017, https://www.youtube.com/watch?v=2GjVG_DguQY&t=89s (accessed April 29, 2018).

56. "70% of Consumers Still Look to Traditional Paper-Based Coupons for Savings," *PR Newswire,* April 16, 2015, http://www.prnewswire.com/news-releases/70-of-consumers-still-look-to-traditional-paper-based-coupons-for-savings-300067097.html (accessed May 5, 2017).

57. Erica Sweeney, "Mountain Dew Shakes Up Baja Blast's Return with Bungalow Stunt," *Marketing Dive,* April 3, 2018, https://www.marketingdive.com/news/mountain-dew-shakes-up-baja-blasts-return-with-bungalow-stunt/520423/ (accessed April 29, 2018).

58. "Blendtec Celebrates 10 Years of Viral Marketing Success," *Nasdaq GlobeNewswire,* November 7, 2016, globenewswire.com/news-release/2016/11/07/887174/10165944/en/Blendtec-Celebrates-10-Years-of-Viral-Marketing-Success.html (accessed February 17, 2017); Blendtec, *YouTube,* https://www.youtube.com/user/Blendtec/about (accessed April 29, 2018).

59. Cutco, "The Cutco Story," https://www.cutco.com/ (accessed August 1, 2018); Dan Miner, "Long an Olean Anchor, Cutco Corp. Keeps Growing," *Buffalo Business First,* May 4, 2018, https://www.bizjournals.com/buffalo/news/2018/05/04/long-an-olean-anchor-cutco-corp-keeps-growing.html (accessed August 1, 2018); Bob Clark, "New, Former Cutco Heads Look to the Future," *Olean Times Herald,* December 3, 2017, http://www.oleantimesherald.com/news/new-former-cutco-heads-look-to-the-future/article_a7675356-d7d6-11e7-82da-3b9402086723.html (accessed August 1, 2018); John S. McClenahen, "The Cutting Edge: Cutco Cutlery Corp. Manufactures Knives in the United States to Maintain Quality and Flexibility," *Industry Week,* 255.4 p. 11, April 2006, http://link.galegroup.com/apps/doc/A145161140/ITOF?u=naal_aub&sid=ITOF&xid=6eb7ea52 (accessed August 1, 2018); Barbara E. Stefano, "Spotlight: Cutco Cutlery," *Feast,* December 4, 2013, http://www.feastmagazine.com/dine-out/spot-light/article_7814337e-5d39-11e3-909b-0019bb30f31a.html (accessed August 1, 2018); Kavita Kumar, "Cutco Sharpens Its Business by Adding Retail Stores," St. Louis Post-Dispatch, April 27, 2012, https://www.stltoday.com/business/local/cutco-sharpens-its-business-by-adding-retail-stores/article_707e846c-8fbc-11e1-a44a-0019bb30f31a.html (accessed August 1, 2018); Karen Dybis, "Knife Seller Aims to Keep an Edge," *The Detroit News,* April 12, 2013, https://advance.lexis.com/document/index?crid=fd685b54-4284-47df-bb22-421d70cc90a1&pdpermalink=45f47ae0-2d07-4192-8256-e49ea25f4f1b&pdmfid=1516831&pdisurlap i=true (accessed August 1, 2018).

Credits

Design elements: Part opener, ©Steve Allen/Getty Images; Consider Ethics and Social Responsibility icon, ©Design Pics/PunchStock; Think Globally icon, ©Sheff/Shutterstock; Responding to Business Challenges icon, ©Olivier LeMoal/Shutterstock; Going Green icon, ©Beboy/Shutterstock; Entrepreneurship in Action icon, ©Ruslan Grechka/Shutterstock; Test Prep icon, ©McGraw-Hill Education; Build Your Skills icon, ©Ilya Terentyev/Getty Images; Solve the Dilemma icon, ©Beautyimage/Shutterstock; Build Your Business Plan icon, ©ALMAGAMI/ Shutterstock; See for Yourself Videocase icon, ©MIKHAIL GRACHIKOV/Shutterstock.

13 Digital Marketing and Social Media

©catwalker/Shutterstock

Chapter Outline

Learning Objectives

After reading this chapter, you will be able to:

LO 13-1 Recognize the increasing value of digital media and digital marketing in strategic planning.

LO 13-2 Demonstrate the role of digital marketing in today's business environment.

LO 13-3 Show how digital media affect the marketing mix.

LO 13-4 Illustrate how businesses can use different types of social networking media.

LO 13-5 Explain online monitoring and analytics for social media.

LO 13-6 Identify legal and ethical considerations in digital media.

LO 13-7 Propose recommendations to a marketer's dilemma.

Enter the World of Business

Facebook in the Face of Data Privacy Controversy

Privacy concerns are catching up to Facebook. Facebook CEO Mark Zuckerberg was called to testify before Congress on Facebook's handling of user data after it was revealed that Cambridge Analytica, a U.K. affiliate of a U.S. consulting firm, had obtained data on 87 million Facebook users. Cambridge Analytica claims that it "uses data to change audience behavior." The firm was a supporter of the Republican campaigns prior to the election of President Donald Trump, leading critics to claim these data may have been used to further political initiatives.

To collect data on users, Cambridge Analytica hired a developer who developed a personality quiz app called This Is Your Digital Life. The app targeted 300,000 Facebook users. It collected information not only on these users, but also from their friends. Facebook allowed developers to collect information from users' friends if their Facebook privacy settings allowed it.

The reaction toward Facebook demonstrates that governments are beginning to take the mishandling of data more seriously. Regulators and privacy advocates are concerned with how much companies can infer about consumers based on their public profiles. At one point, before Facebook put a stop to it, lenders were even using user profiles to assess the creditworthiness of individuals.

Many are demanding stricter rules for data tracking and collection. Suggestions include simplifying permission structures so consumers can understand what they are allowing websites to gather, making privacy practices clear and at the forefront of websites' Terms of Use agreements, and prohibiting the collection of certain types of information such as medical data. In the meantime, some high-profile users are deleting their Facebook pages, including Elon Musk who deleted Facebook pages for Tesla and SpaceX.[1]

Introduction

The Internet and information technology have dramatically changed the environment for business. Marketers' new ability to convert all types of communications into digital media has created efficient, inexpensive ways of connecting businesses and consumers and has improved the flow and the usefulness of information. Businesses have the information they need to make more informed decisions, and consumers have access to a greater variety of products and more information about choices and quality. This has resulted in a shift in the balance of power between consumer and marketer.[2]

The defining characteristic of information technology in the 21st century is accelerating change. New systems and applications advance so rapidly that it is almost impossible to keep up with the latest developments. Startup companies emerge with systems that quickly overtake existing approaches to digital media. When Google first arrived on the scene, a number of search engines were fighting for dominance. With its fast, easy-to-use search engine, Google became number one and is now challenging many industries, including advertising, newspapers, mobile phones, and book publishing. Social media continues to advance as the channel most observers believe will dominate digital communication in the near future. Today, people spend more time on social networking sites, such as Facebook, than they spend on e-mail.

In this chapter, we first provide some key definitions related to digital marketing and social media. Next, we discuss using digital media in business and marketing. We look at marketing mix considerations when using digital media and pay special attention to social networking. Then we focus on digital marketing strategies—particularly new communication channels like social networks. We take a close look at media sharing, mobile marketing, applications, and widgets. Next the importance of online monitoring and analytics is discussed. Then we examine using digital media to learn about consumers. Finally, we examine the legal and social issues associated with information technology, digital media, and e-business.

e-business
carrying out the goals of business through utilization of the Internet.

digital media
electronic media that function using digital codes via computers, cellular phones, smartphones, and other digital devices that have been released in recent years.

digital marketing
uses all digital media, including the Internet and mobile and interactive channels, to develop communication and exchanges with customers.

LO 13-1

Recognize the increasing value of digital media and digital marketing in strategic planning.

Growth and Benefits of Digital Communication

Let's start with a clear understanding of our focus in this chapter. First, we can distinguish e-business from traditional business by noting that conducting **e-business** means carrying out the goals of business through the use of the Internet. **Digital media** are electronic media that function using digital codes—when we refer to digital media, we mean media available via computers and other digital devices, including mobile and wireless ones like smartphones.

Digital marketing uses all digital media, including the Internet and mobile and interactive channels, to develop communication and exchanges with customers. Digital marketing is a term we will use often because we are interested in all types of digital communications, regardless of the electronic channel that transmits the data. Digital marketing goes beyond the Internet and includes mobile phones, banner ads, digital outdoor marketing, and social networks.

The Internet has created tremendous opportunities for businesses to forge relationships with

Consumers are increasingly turning to mobile apps to access company information, earn loyalty rewards, and purchase products.

©Monika Wisniewska/Shutterstock

TABLE 13.1 Characteristics of Digital Marketing

Characteristic	Definition	Example
Addressability	The ability of the marketer to identify customers before they make a purchase	Amazon installs cookies on a user's computer that allows it to identify the owner when he or she returns to the website.
Interactivity	The ability of customers to express their needs and wants directly to the firm in response to its marketing communications	Texas Instruments interacts with its customers on its Facebook page by answering concerns and posting updates.
Accessibility	The ability for marketers to obtain digital information	Google can use web searches done through its search engine to learn about customer interests.
Connectivity	The ability for consumers to be connected with marketers along with other consumers	Volition Beauty's website encourages customers to submit their makeup and skin care product ideas, which can then be voted on by other users for the chance to be created.
Control	The customer's ability to regulate the information they view as well as the rate and exposure to that information	Consumers use Kayak to discover the best travel deals.

consumers and business customers, target markets more precisely, and even reach previously inaccessible markets at home and around the world. The Internet also facilitates business transactions, allowing companies to network with manufacturers, wholesalers, retailers, suppliers, and outsource firms to serve customers more quickly and more efficiently. The telecommunication opportunities created by the Internet have set the stage for digital marketing's development and growth.

Digital communication offers a completely new dimension in connecting with others. Some of the characteristics that distinguish digital from traditional communication are addressability, interactivity, accessibility, connectivity, and control. These terms are discussed in Table 13.1.

Using Digital Media in Business

The phenomenal growth of digital media has provided new ways of conducting business. Given almost instant communication with precisely defined consumer groups, firms can use real-time exchanges to create and stimulate interactive communication, forge closer relationships, and learn more accurately about consumer and supplier needs. Consider how Amazon is taking on department stores and big box stores such as Walmart and Home Depot.[3] In fact, store closings doubled in 2017 with the rise of online shopping.[4] Many of you may not remember a world before Amazon because it has completely transformed how many people shop.

Because it is fast and inexpensive, digital communication is making it easier for businesses to conduct marketing research, provide and obtain price and product information, and advertise, as well as to fulfill their business goals by selling goods and services online. Even the U.S. government engages in digital marketing activities—marketing everything from Treasury bonds and other financial instruments to oil-drilling leases and wild horses. Procter & Gamble uses the Internet as a fast, cost-effective means for marketing research, judging consumer demand for potential new

LO 13-2

Demonstrate the role of digital marketing in today's business environment.

Consumers use apps like Lyft and Uber to find local drivers to take them to their destinations.

©Kaspars Grinvalds/Shutterstock

products by inviting online consumers to sample new-product prototypes and provide feedback. If a product gets rave reviews from the samplers, the company might decide to introduce it.

New businesses and even industries are evolving that would not exist without digital media. Vimeo is a video website founded by filmmakers to share creative videos. The site lets users post or view videos from around the world. It has become the third most popular video website after YouTube and Netflix.[5]

The reality, however, is that Internet markets are more similar to traditional markets than they are different. Thus, successful digital marketing strategies, like traditional business strategies, focus on creating products that customers need or want, not merely developing a brand name or reducing the costs associated with online transactions. Instead of changing all industries, digital technology has had much more impact in certain industries where the cost of business and customer transactions has been very high. For example, investment trading is less expensive online because customers can buy and sell investments, such as stocks and mutual funds, on their own. Firms such as Charles Schwab Corp., the biggest online brokerage firm, have been innovators in promoting online trading. Traditional brokers such as Merrill Lynch have had to follow with online trading for their customers.

Digital media can also improve communication within and between businesses. In the future, most significant gains will come from productivity improvements within businesses. Communication is a key business function, and improving the speed and clarity of communication can help businesses save time and improve employee problem-solving abilities. Digital media can be a communications backbone that helps to store knowledge, information, and records in management information systems so co-workers can access it when faced with a problem to solve. A well-designed management information system that utilizes digital technology can, therefore, help reduce confusion, improve organization and efficiency, and facilitate clear communications. Given the crucial role of communication and information in business, the long-term impact of digital media on economic growth is substantial, and it will inevitably grow over time.

The dynamic nature of digital marketing can quickly change opportunities and create challenges. For example, digital assistants now function as a personal information manager and are being used to assist professionals in medicine, engineering, and other business areas. Smartphones, social networking, drones, and driverless cars are shaping a new marketing environment. While digital marketing has many benefits, challenges exist, especially in giving up privacy to use digital media.[6]

LO 13-3

Show how digital media affect the marketing mix.

Digital Media and the Marketing Mix

While digital marketing shares some similarities with conventional marketing techniques, a few valuable differences stand out. First, digital media make customer communications faster and interactive. Second, digital media help companies reach new target markets more easily, affordably, and quickly than ever before. Finally, digital media help marketers utilize new resources in seeking out and communicating with customers. One of the most important benefits of digital marketing is the ability of marketers and customers to easily share information. Through websites, social networks, and other digital media, consumers can learn about everything they consume and use in their lives, ask questions, voice complaints, indicate preferences, and otherwise communicate about their needs and desires. For example, IBM's digital assistant allows IBM customers to identify and digitally interact with key experts through a variety of platforms. IBM's Watson is an assistant in cognitive computing impacting fields as diverse as finance, medicine, and education. Many marketers use e-mail, mobile phones, social media, wikis, media sharing, blogs, videoconferencing, and other technologies to coordinate activities and communicate with employees, customers, and suppliers. Twitter, considered both a social network and a micro-blog, illustrates how these digital technologies can combine to create new communication opportunities.

Nielsen Marketing Research revealed that consumers now spend more time on social networking sites than they do on e-mail, and social network use is still growing. Figure 13.1 shows that while the majority of social network users are between the ages of 18 and 29, other age groups are not that far behind. With digital media, even small businesses can reach new markets through these inexpensive communication channels. Brick-and-mortar companies like Walmart utilize online catalogs and company websites and blogs to supplement their retail stores. Internet companies like Amazon and Zappos that lack physical stores let customers post reviews of their purchases on their websites, creating company-sponsored communities. Amazon is taking on department stores and big box stores such as Walmart and Home Depot.

FIGURE 13.1   Social Media Use by Platform

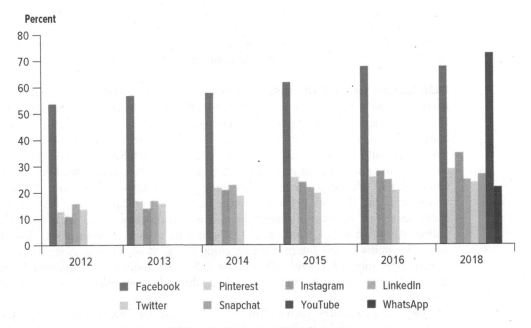

Note: U.S. adults. Pre-2018 telephone poll data are not available for YouTube, Snapchat, or WhatsApp.

Source: Pew Research Center survey conducted January 3–10, 2018. Trend data from previous Pew Research Center surveys.

Department stores such as Macy's and Sears have had to close hundreds of stores and some shopping centers have gone out of business.[7]

One aspect of marketing that has not changed with digital media is the importance of achieving the right marketing mix. Product, distribution, promotion, and pricing are as important as ever for successful online marketing strategies. More than 48 percent of the world's population now uses the Internet.[8] That means it is essential for businesses large and small to use digital media effectively, not only to grab or maintain market share, but also to streamline their organizations and offer customers entirely new benefits and convenience. Let's look at how businesses are using digital media to create effective marketing strategies on the web.

Product Considerations. Like traditional marketers, digital marketers must anticipate consumer needs and preferences, tailor their goods and services to meet these needs, and continually upgrade them to remain competitive. The connectivity created by digital media provides the opportunity for adding services and can enhance product benefits. Some products, such as online games, applications, and virtual worlds, are only available via digital media. The more than 1.5 million applications available on the iPad, for instance, provide examples of products that are only available in the digital world.[9] Businesses can often offer more items online than they could in a retail store.

The ability to access information for any product can have a major impact on buyer decision making. However, with larger companies now launching their own extensive marketing campaigns, and with the constant sophistication of digital technology, many businesses are finding it necessary to upgrade their product offerings to meet consumer needs. For example, Volition, a skin care and cosmetics company, crowdsources its new product ideas from its customers. If an idea makes it past the Volition team, tens of thousands of people in the Volition community will vote online whether the product should be produced and then receive a discount if the product is voted in. Using their fan base for new ideas has led to unique and innovative products, and as consumers share their product ideas to their social networks for support from friends and family, awareness increases for this beauty community.[10] The Internet provides a major resource for learning more about consumer wants and needs.

Distribution Considerations. The Internet is a new distribution channel for making products available at the right time, at the right place, and in the right quantities. Marketers' ability to process orders electronically and increase the speed of communications via the Internet reduces inefficiencies, costs, and redundancies while increasing speed throughout the marketing channel. Shipping times and costs have become an important consideration in attracting customers, prompting many companies to offer consumers low shipping costs or next-day delivery. Although consumers still flock to brick-and-mortar stores to purchase items, they tend to spend less time shopping because they have already determined what they want online. Approximately 88 percent of U.S. consumers research shoes, toys, clothing, and other items on the Internet before going to the store. Online shopping is also significantly increasing, with 209 million U.S. consumers finding and purchasing items online. Convenience and constant availability are two major reasons consumers prefer to shop online.[11]

Many online retailers, such as Birchbox, Blue Nile and Warby Parker, have established a presence in the traditional brick-and-mortar realm to create a physical presence and increase awareness. Unlike most, Blue Nile's shops, called a "Webroom," are showrooms only, meaning customers can touch and feel the products, but all orders are placed online, saving the company money in distribution costs and real estate costs associated with large storefronts. This trend is a result of increased online competition, as well as a trend toward **omni-channel retailing,** where retailers offer a seamless experience on mobile, desktop, or traditional retail spaces. For example, many retailers aim to offer consistent product assortments and pricing on all channels as well as streamline the return process. A customer may research a purchase online, shop in-store, browse an in-store digital catalog, and then use a coupon from the retailer's app at checkout. A survey revealed 73 percent of shoppers use multiple channels while shopping, making a seamless shopping experience a way to differentiate a retailer from its competitors. The survey revealed customers that research online before in-store shopping led to 13 percent more in sales among omni-channel shopper.[12]

omni-channel retailing
a type of retail that integrates the different methods of shopping available to consumers (for example, online, in a physical store, or by phone).

Promotion Considerations. Perhaps one of the best ways businesses can utilize digital media is for promotion purposes—whether they are increasing brand awareness, connecting with consumers, or taking advantage of social networks or virtual worlds (discussed later) to form relationships and generate positive publicity or "buzz" about their products. Thanks to online promotion, consumers can be more informed than ever, including reading customer-generated content before making purchasing decisions. Consumer consumption patterns are radically changing, and marketers must adapt their promotional efforts to meet them.

With the rise of bloggers and social media stars like Michelle Phan and Eva Gutowski, brands are turning to influencers to promote their products. Brands identify influencers who align with their brand image and often pay them for an endorsement or send complimentary product in exchange for a review. Roughly 22 percent of businesses are seeing higher customer purchases through influencer marketing than traditional channels like e-mail and web search marketing.[13] Brands can contact influencers directly or use paid platforms like TapInfluence and BrandBacker to identify ideal partners and manage campaigns. Influenster is a product sampling program that puts products in the hands of influencers and micro-influencers in exchange for authentic, user-generated content on social media. With a community of more than 4 million, brands are able to identify users who fall into their target demographic through data collected on the platform along with pre-qualification surveys. Once a user receives a set of products, Influenster drives social posts, in-store actions, and product reviews by incentivizing the influencers with a "Brand Badge." To receive the digital badge and retain membership privileges, users must participate in activities dictated by the brand, such a writing a review on a blog, posting a photo on Instagram, or uploading video using the products on YouTube.[14]

Pricing Considerations. Price is the most flexible element of the marketing mix. Digital marketing can enhance the value of products by providing extra benefits such as service, information, and convenience. Through digital media, discounts and other promotions can be quickly communicated. As consumers have become better informed about their options, the demand for low-priced products has grown, leading to the creation of deal sites where consumers can directly compare prices. Expedia, for instance, provides consumers with a wealth of travel information about everything from flights to hotels that lets them compare benefits and prices. Many marketers offer buying incentives like online coupons or free samples to generate consumer demand for their products. For the business that wants to compete on price, digital marketing provides unlimited opportunities.

Subscription-based social media network Vero is ad-free and has attracted 4 million users.

©ThomasDeco/Shutterstock

DID YOU KNOW? Facebook is the most popular social media network for businesses.[18]

Social Media Marketing

Social media marketing involves obtaining communications with consumers through social media sites. Social media marketing enables firms to promote a message and create online conversations through multiple platforms. Large markets can be targeted and reached through paid media, owned media, and earned media.[15] Traditional paid media includes traditional print and broadcast but is now joined by paid advertising on social networks such as Facebook and Twitter. Marketers can place ads on Google just like they place an ad on television. On Facebook, which has more than 2.5 million advertisers, brands can pay to boost posts, create compelling photo carousel ads, promote their page, and more.[16] In addition to placing ads, marketers can own their own media outlets and create messages on social networks. Most firms have owned websites but can also develop websites such as Facebook and LinkedIn.

Finally, markets can have earned media when consumers are communicating on social media sites. These digital word-of-mouth posts or interactions can promote a product or firm. Although it is not controllable like advertising, if the communication is positive, it increases sales.[17]

User-generated content relates to consumers who create, converse, rate, collect, join, or simply read online materials. Marketers can't always access the creative efforts of consumers who post or publish on publicly accessible websites, such as blogs, like A Beautiful Mess and Modern Martha, or on social networking sites such as LinkedIn. These user-generated sites often involve self-disclosure, where consumers share their knowledge, interests, and desire to join or associate with others. Participating in discussions to connect and network with others is a major motivating factor to influence others or to promote an interest or cause. There are many critics involved in user-generated content. These consumers post evaluations on blogs or post ratings and reviews. If you have ever posted a product review or rated a movie, you have engaged in this activity. Evaluating what critics post should be an important part in a company's digital marketing strategy. Of course, consumers read ratings to aid their shopping purchases. Yelp is one of the most comprehensive reviews sites on products and businesses. With more than 77 million reviews, Yelp continues to expand its platform, adding Questions and Answers for users to ask venue-specific questions for other users to answer.[19] Therefore, these rating sites can be helpful to collect information used in marketing research and to monitor firm reputation.

Marketers need to analyze their target markets and determine the best social media approach to support marketing objectives. Social media should be included in both the corporate and marketing strategy. It should be a part of the firm's marketing plan and implementation efforts. Social media can be used to monitor target market competitors and understand the social and economic environment as a whole. Social media has the potential of building campaigns that produce advocates and enthusiasts of a firm's products. For example, Dodge uses social media to release product teasers and news to its engaged fans. The brand rewarded its most engaged social media fans by inviting a limited number to the unveiling of the 2018 Dodge Challenger SRT Demon.[20] Marketing should be focused on relationship building and social media can influence consumer behavior and deliver value to the firm.

Consumer-Generated Marketing and Digital Media

Illustrate how businesses can use different types of social networking media.

While digital marketing has generated exciting opportunities for companies to interact with their customers, digital media are also more consumer-driven than traditional media. Internet users are creating and reading consumer-generated content as never before and are having a profound effect on marketing in the process.

Two factors have sparked the rise of consumer-generated information:

1. The increased tendency of consumers to post their own thoughts, opinions, reviews, and product discussions through blogs or digital media.
2. Consumers' tendencies to trust other consumers over corporations. Consumers often rely on the recommendations of friends, family, and fellow consumers when making purchasing decisions.

Marketers who know where online users are likely to express their thoughts and opinions can use these forums to interact with them, address problems, and promote their companies. Types of digital media in which Internet users are likely to participate include social networks, blogs, wikis, video-sharing sites, podcasts, virtual reality sites, and mobile applications. Let's look a little more closely at each.

Social Networks

The increase in social networking across the world is exponential. It is estimated that today's adults spend approximately 42.1 minutes per day on Facebook alone.[21] As social networks evolve, both marketers and the owners of social networking sites are realizing the opportunities such networks offer—an influx of advertising dollars for site owners and a large reach for the advertiser. As a result, marketers have begun investigating and experimenting with promotion on social networks. Two popular sites are Facebook and Twitter.

Facebook.   Facebook is the most popular social networking site in the world. Facebook users create profiles, which they can make public or private, and then search the network for people with whom to connect. The social networking giant has surpassed 1.5 billion users and is still growing. It has also acquired a number of companies as it expands into other services, including Instagram, WhatsApp, and Oculus.[22] Facebook also has a video feature that enables the sharing and tagging of videos.[23]

For this reason, many marketers are turning to Facebook to market products, interact with consumers, and gain free publicity. It is possible for a consumer to become a "fan" of a major company like Starbucks by clicking on the "Like" icon on the coffee retailer's Facebook page. Boosted posts, one

GoPro won a Shorty Award for its use of Facebook and Instagram to inspire and leverage user-generated content.

©Faiz Zaki/Shutterstock

of the features Facebook has to offer businesses, allows companies to develop an advertisement quickly from a post on their timelines, select the people they would like the advertisement to target, and select the budget they want to spend. Boosted posts appear higher up in the News Feeds of the advertisement's target market.[24] Facebook gained competitive advantage with its introduction of Facebook Messenger. The product allows up to six people to see each other live while chatting. There can be up to 50 connected, but only one person can be seen at a time. Facebook Messenger is a distinct improvement over its WhatsApp and competitor's products including Snap Inc.'s Snapchat, Apple's FaceTime, and Google Duo, which allow one-to-one communication.[25]

Additionally, social networking sites are useful for relationship marketing, or the creation of relationships that mutually benefit the business and customer. Approximately 30 percent of consumers (and 47 percent of millennials) claim social media has some influence on their purchasing decisions.[26] As a result, firms are spending more time on the quality of their Facebook interactions. Ritz-Carlton, for instance, spends a significant amount of time analyzing its social media conversations and reaching out to noncustomers. Businesses are shifting their emphasis from selling a product or promoting a brand to developing beneficial relationships in which brands are used to generate a positive outcome for the consumer.[27]

Twitter. Twitter is a hybrid of a social networking site and a micro-blogging site that asks users one simple question: "What's happening?" Members can post answers of up to 140 characters, which are then available for their registered "followers" to read. It sounds simple enough, but Twitter's effect on digital media has been immense. The site quickly progressed from a novelty to a social networking staple, attracting millions of viewers each month.[28] About 82 percent access the site from their mobile devices.[29]

Although 280 characters may not seem like enough for companies to send an effective message, shorter social media messages appear to be more effective. Tweets shorter than 100 characters are found to have a 17 percent higher engagement rate with users, and Facebook has shown similar data.[30] These efforts are having an impact; more than half of Twitter's active and monthly users follow companies or brands.[31]

Like other social networking tools, Twitter is also being used to build, or in some cases rebuild, customer relationships. For example, MoonPie uses Twitter to interact with consumers. MoonPie has been recognized for its humorous tweets and one-liners that have attracted viral attention.[32] On the other hand, approximately 70 percent of companies ignore complaints on Twitter. This failure acts as a missed opportunity to address customer concerns and maintain strong relationships.[33]

Snapchat. While Snap Inc. admits it may never achieve profitability, posting a net loss of approximately $3.4 billion in recent years, investors see value in Snapchat, which has more than 187 million daily active users.[34] The mobile app, launched in 2011, allows users to send messages and disappearing photos and videos to friends. The parent company prefers to think of itself as a camera company rather than a social media company and plans to release more lifestyle products like Spectacles, camera glasses sold at Snapchat pop-up shops, outside of its social media platform.[35] Marketers are looking at Snapchat as an opportunity to reach their young, highly engaged audience. Brands like Taco Bell, Sour Patch Kids, and Birchbox have taken to Snapchat to engage with their audiences.

Snapchat, which features skippable, vertical video ads and custom photo filters, is used mostly by users under the age of 34. In fact, 79 percent of daily users are

Consider Ethics and Social Responsibility

Recruiters Turn to Social Media to Find Network of Talent

Marketing is not just about promoting to external consumers. It is also about generating internal customers—in other words, employees. While traditionally firms have used newspaper classifieds and "Help Wanted" signs, today's firms are increasingly using digital media sites to target potential candidates.

Approximately 92 percent of recruiters polled in a survey claim they use social media to identify quality candidates. Social media sites like Facebook are also a great way to market job opportunities due to its global reach. While LinkedIn remains the most popular social media platform among recruiters, 35 to 55 percent of recruiters say they use Facebook as a recruitment tool.

Social media also offers companies the opportunity to interact with potential candidates directly. The use of mobile recruiting is increasing, with one-third of organizations claiming they use mobile recruiting to target candidates who own smartphones (77 percent of Americans). Social media is an effective tool for attracting younger job seekers because the majority have social media profiles. It is estimated that 86 percent of job seekers in their first decade of employment will turn to social media sites to look for jobs.

Companies across the world are able to use the Internet to fill their needs and locate the best talent. By opening opportunities for both businesses and job seekers, the Internet is revolutionizing the industry and keeping those "Help Wanted" signs in the closet.[36]

Critical Thinking Questions

1. What are some ways companies can use digital media to target potential job candidates that go beyond the traditional recruiting mechanisms?
2. Why do you think so many younger employees are turning toward social media sites to look for jobs?
3. Do you believe using social media to interact with job candidates will allow recruiters to hire more talented individuals? Why or why not?

---

between 18 and 34.[37] Sponsored Lenses are also shoppable, taking augmented reality to the next level. For example, Clairol took advantage of the feature for its Color Crave hair dye, allowing users to try on different hair colors. Users could tap the Shop Now button to go to Target's website to make a purchase.[38] One of Snapchat's biggest challenges will be scaling its advertising dollars to achieve its revenue goals.

YouTube.   Purchased by Google for $1.65 billion, YouTube allows users to upload and share videos worldwide. Users watch a billion hours of YouTube videos every day, making this popular video platform an important part of marketing strategy.[39] Though brands use the platform to release original video content, consumers far outnumber them on the platform. For example, beauty brands on YouTube are outnumbered by beauty vloggers in beauty searches by 14 to 1.[40] This makes it challenging for brands to control messaging about their products on the platform.

YouTube continues to diversify its video offering with YouTube Premium and YouTube TV. YouTube Premium expands upon the original platform, allowing users to pay for ad-free and offline video and original programming from top creators. As more homes cancel their cable packages, YouTube TV is an affordable alternative. For $40 per month, users can watch ABC, CBS, and NBC among other top networks, positioning the service as a competitor to Sling TV and DirecTV Now.[41]

LinkedIn.   LinkedIn is the top networking site for businesses and business professionals. This networking tool allows users to post a public profile, similar to a résumé, connect with colleagues, find job listings, and join private groups. Eighty percent of B2B marketers say LinkedIn is an effective business lead generator.[42] This platform can also be used to spread brand awareness and for corporate recruiting. HubSpot, an inbound marketing and sales platform with more than 215,000 followers, uses LinkedIn to spread its content, promote free webinars, and increase awareness around inbound marketing.[43]

Blogs and Wikis

Today's marketers must recognize that the impact of consumer-generated material like blogs and wikis and their significance to online consumers have increased a great deal. **Blogs** (short for web logs) are web-based journals in which writers can editorialize and interact with other Internet users. More than three-fourths of Internet users read blogs.[44] In fact, the blogging site Tumblr, which allows anyone to post text, hyperlinks, pictures, and other media for free, has been called "ground zero of the viral Internet." The site Buzzfeed, well known for its shareable and viral content, cites Tumblr as the top source from where it finds its stories. The site has 411 million blogs and more than 160 billion posts.[45] In 2013, Yahoo! purchased Tumblr for $1.1 billion.[46]

Blogs give consumers power, sometimes more than companies would like. Bloggers can post whatever they like about a company or its products, whether their opinions are positive or negative, true or false. For instance, although companies sometimes force bloggers to remove blogs, readers often create copies of the blog post and spread it across the Internet after the original's removal. In other cases, a positive review of a good or service posted on a popular blog can result in large increases in sales. Thus, blogs can represent a potent threat or opportunity to marketers.

Rather than trying to eliminate blogs that cast their companies in a negative light, some firms are using their own blogs, or employee blogs, to answer consumer concerns or defend their corporate reputations. Bill Marriott, son of the founder of Marriott International, maintains a blog called "Marriott on the Move" where he not only discusses the hotel business, but also posts on a number of insightful business and inspirational topics to engage his readers.[47] As blogging changes the face of media, smart companies are using it to build enthusiasm for their products and create relationships with consumers.

Wikis are websites where users can add to or edit the content of posted articles. One of the best known is Wikipedia, an online encyclopedia with more than 40 million entries in more than 299 languages on nearly every subject imaginable. For comparison, *Encyclopedia Britannica* only has 120,000 entries.[48] Wikipedia is one of the 10 most popular sites on the web, and because much of its content can be edited by anyone, it is easy for online consumers to add detail and supporting evidence and to correct inaccuracies in content. Wikipedia used to be completely open to editing, but in order to stop vandalism, the site had to make some topics off-limits that are now editable only by a small group of experts.

Like all digital media, wikis have advantages and disadvantages for companies. Wikis about controversial companies like Walmart and Nike often contain negative publicity about things such as workers' rights violations. However, monitoring relevant wikis can provide companies with a better idea of how consumers feel about the company or brand. Some companies also use wikis as internal tools for teams working on projects that require a great deal of documentation.[49]

There is too much at stake financially for marketers to ignore wikis and blogs. Despite this fact, statistics show that only about 36 percent of *Fortune* 500 companies have a corporate blog.[50] Marketers who want to form better customer relationships and promote their company's products must not underestimate the power of these two media outlets.

blogs
web-based journals in which writers can editorialize and interact with other Internet users.

wikis
websites where users can add to or edit the content of posted articles.

Media Sharing

Businesses can also share their corporate messages in more visual ways through media sharing sites. Media sharing sites allow marketers to share photos, videos, and podcasts. Media sharing sites are more limited in scope in how companies interact with consumers. They tend to be more promotional than reactive. This means that while firms can promote their products through videos or photos, they usually do not interact with consumers through personal messages or responses. At the same time, the popularity of these sites provides the potential to reach a global audience of consumers.

Video-sharing sites allow virtually anybody to upload videos, from professional marketers at *Fortune* 500 corporations to the average Internet user. Some of the most popular video-sharing sites include YouTube, Vimeo, and Dailymotion. Video-sharing sites give companies the opportunity to upload ads and informational videos about their products. A few videos become viral at any given time, and although many of these gain popularity because they embarrass the subject in some way, others reach viral status because people find them entertaining. **Viral marketing** occurs when a message gets sent from person to person to person. It can be an extremely effective tool for marketers—particularly on the Internet, where one click can send a message to dozens or hundreds of people simultaneously. Marketers are taking advantage of the viral nature of video-sharing sites like YouTube, either by creating their own unique videos or advertising on videos that have already reached viral status. Purple released a "Raw Egg Test" video on YouTube to demonstrate the support its mattress provides. To date the video has more than 166 million views.[51]

viral marketing a marketing tool that uses a networking effect to spread a message and create brand awareness. The purpose of this marketing technique is to encourage the consumer to share the message with friends, family, co-workers, and peers.

Posting videos on digital media sites also allows amateur entrepreneurs to showcase their talents for the chance to become successful. Michelle Phan started off posting makeup tutorials to YouTube in 2007. Her videos took off, catching the interest of women across the country who valued Phan's beauty advice.[52] She founded ipsy, a beauty subscription service that delivers its subscribers deluxe samples of popular cosmetics brands like Tarte and Ofra. Approximately 10,000 amateur beauty bloggers create videos on behalf of the subscription service monthly. The company is not as involved in selling products or advertising but has been valued at $800 million for driving subscriptions.[53]

Photo-sharing sites allow users to upload and share their photos and short videos with the world. Well-known photo-sharing sites include Instagram, Imgur, Shutterfly, and Photobucket. Instagram is the most popular mobile photo-sharing application. Instagram, owned by Facebook, allows users to be creative with their photos by using filters and tints and then sharing them with their friends. Chobani uses Instagram to build communities and suggest new uses for its yogurt products.[54] To compete against Snapchat, Instagram introduced Instagram Stories, a way for its users to send their friends messages that disappear in 24 hours.[55] Instagram is one of the fastest growing social networks.[56] With more and more people using mobile apps or accessing the Internet through their smartphones, the use of photo sharing through mobile devices is likely to increase.

Verizon sold Yahoo's photo sharing site Flickr to independent image-hosting company SmugMug. SmugMug plans to continue to operate Flickr separately without making changes to its plans or rates.[57]

©BigTunaOnline/Shutterstock

Other sites are emerging that take photo sharing to a new level. Pinterest is a photo-sharing bulletin board site that combines photo sharing with elements of bookmarking and social networking. Users can share photos and images among other Internet users, communicating mostly through images that they "pin" to their boards. Other users can "repin" these images to their boards, follow each other, "like" images, and make comments. Pinterest added a feature called Lens that allows users to take a picture of an object and find a list of pins with similar-looking objects, further establishing the platform as a discovery tool for shopping.[58] They also released functionality that allows users to filter hair and beauty search results by skin tone ranges.[59] Marketers have found that an effective way of marketing through Pinterest is to post images conveying a certain emotion that represents their brand.[60] Because Pinterest users create boards that deal with their interests, marketers also have a chance to develop marketing messages encouraging users to purchase the product or brand that interests them. Pinterest hopes to learn how to influence a customer to proceed from showing interest in a product to having an intent to purchase. This knowledge will be helpful to advertisers marketing through Pinterest's website.[61]

Photo sharing represents an opportunity for companies to market themselves visually by displaying snapshots of company events, company staff, and/or company products. Nike, Audi, and MTV have all used Instagram in digital marketing campaigns. Zales Jewelers has topic boards on Pinterest featuring rings as well as other themes of love, including songs, wedding cakes, and wedding dresses.[62] Digital marketing companies are also scanning photos and images on photo-sharing sites to gather insights about how brands are being displayed or used. They hope to offer these insights to big-name companies such as Kraft. The opportunities for marketers to use photo-sharing sites to gather information and promote brands appear limitless.[63]

podcast
audio or video file that can be downloaded from the Internet with a subscription that automatically delivers new content to listening devices or personal computers.

Podcasts are audio or video files that can be downloaded from the Internet via a subscription that automatically delivers new content to listening devices or personal computers. Podcasting offers the benefit of convenience, giving users the ability to listen to or view content when and where they choose. The markets podcasts reach are ideal for marketers, especially the 18–34 demographic, which includes the young and affluent.[64] They also affect consumer buying habits. For instance, listening to nutrition podcasts while in the grocery store increases the likelihood that shoppers will purchase healthier items.[65]

As podcasting continues to spread, radio stations and television networks like CBC Radio, NPR, MSNBC, and PBS are creating podcasts of their shows to profit from this growing trend. Many companies like GE, eBay, Basecamp, and Tinder hope to use podcasts to create brand awareness, promote their products, and encourage customer loyalty.

Mobile Marketing

As digital marketing becomes increasingly sophisticated, consumers are beginning to utilize mobile devices like smartphones as a highly functional communication method. The iPhone and iPad have changed the way consumers communicate, and a growing number of travelers are using their smartphones to find online maps, travel guides, and taxis. In industries such as hotels, airlines, and car rental agencies, mobile phones have become a primary method for booking reservations and communicating about services. Other marketing uses of mobile phones include sending shoppers timely messages related to discounts and shopping opportunities.[66] For these reasons, mobile marketing has exploded in recent years—mobile phones have become an important part of our everyday lives and can even affect how we shop. For instance, it is estimated that shoppers who are distracted by their phones in-store increased their unplanned purchases by 12 percent over those who are not.[67] To avoid being left behind, brands must recognize the importance of mobile marketing.

E-commerce sales on smartphones are also rapidly growing and are estimated to reach 50 percent of total online sales in the next couple of years.[68] This makes it essential for companies to understand how to use mobile tools to create effective campaigns. Figure 13.2 breaks down smartphone use in the United States. Some of the more common mobile marketing tools include the following:

- *SMS messages:* SMS messages are text messages of 160 characters or less. SMS messages have been an effective way to send coupons to prospective customers.[69]
- *Multimedia messages:* Multimedia messaging takes SMS messaging a step further by allowing companies to send video, audio, photos, and other types

FIGURE 13.2 Mobile App Activities Conducted by Smartphone Users

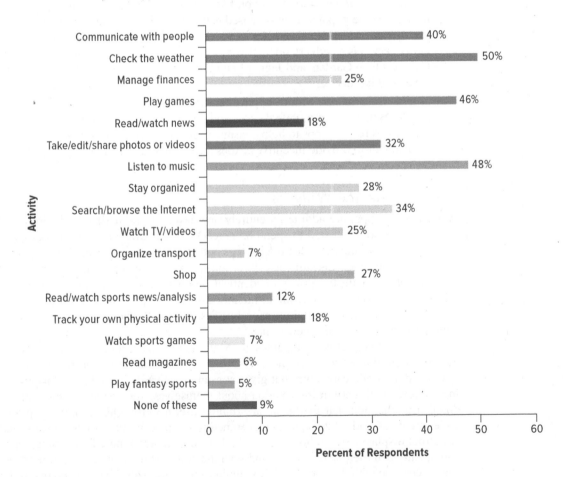

Note: ages 18–64; in the past 30 days

Source: Ipsos MORI, "Something for Everyone: Why the Growth of Mobile Apps is Good News for Brands," sponsored by Google, August 1, 2017.

Walgreens offers paperless coupons on its website and mobile app to allow customers to digitally "clip" coupons and use in-store through its Balance Rewards program.

©Andrew Resek/McGraw-Hill Education

of media over mobile devices. The global MMS market is estimated to become a $65 billion market by 2025. Approximately 98 percent of all U.S. cell phones can receive MMS.[70]

- *Mobile advertisements:* Mobile advertisements are visual advertisements that appear on mobile devices. Companies might choose to advertise through search engines, websites, or even games accessed on mobile devices. Mobile accounts for more than half of digital ad spending.[71]

- *Mobile websites:* Mobile websites are websites designed for mobile devices. More than 50 percent of e-commerce website traffic now comes through mobile devices.[72]

- *Location-based networks:* Location-based networks are built for mobile devices. Some popular location-based networks include Google Waze and Foursquare, which lets users check in and share their location with others. Foursquare has an advertising network called Pinpoint for marketers. Samsung Galaxy, Olive Garden, and Jaguar Land Rover are examples of companies that have used its service.[73]

- *Mobile applications:* Mobile applications (known as *apps*) are software programs that run on mobile devices and give users access to certain content.[74] Businesses release apps to help consumers access more information about their company or to provide incentives. Apps are discussed in further detail in the next section.

Applications and Widgets

Applications (apps) are adding an entirely new layer to the marketing environment as Americans are estimated to spend 85 percent of their time on smartphones using apps.[75] The most important feature of apps is the convenience and cost savings they offer to the consumer. Certain apps allow consumers to scan a product's barcode and then compare it with the prices of identical products in other stores. Mobile apps also enable customers to download in-store discounts. An estimated 68 percent of American adults have smartphones, so businesses cannot afford to miss out on the chance to profit from these new trends.[76]

To remain competitive, companies are beginning to use mobile marketing to offer additional incentives to consumers. As Unilever expands into Southeast Asia, it developed a mobile campaign that gives consumers rewards in exchange for providing Unilever with certain information about themselves, such as shopping habits.[77] Another application that marketers are finding useful is the QR scanning app. QR codes are black-and-white squares that sometimes appear in magazines, posters, and storefront displays. Smartphone users who have downloaded the QR scanning application can open their smartphones and scan the code, which contains a hidden message accessible with the app. The QR scanning app recognizes the code and opens the link, video, or image on the phone's screen. Marketers are using QR codes to promote their companies and offer consumer discounts.[78]

Going Green? There's an App for That

Although support for the green movement is growing, most consumers want to know how going green can save money. Green mobile apps offer a solution. They enable users to save money and/or locate green products. For example, EnergyElephant allows users to monitor their energy meters. By simply snapping a photo of their home energy meters, consumers worldwide can use the app to generate meter readings. EnergyElephant can submit these readings directly to utility companies, reducing the need to send people out to read the meter and reducing carbon emissions. Additionally, it provides consumers with helpful information including ways to reduce energy consumption and whether investing in renewable technologies will save on costs.

Mobile apps can also help combat greenwashing. Greenwashing occurs when marketers claim that a product is greener than it really is. One of the most popular green mobile apps, known as GoodGuide, is eliminating this problem. Consumers with the app can use their phones to photograph product barcodes. The app will then provide information on the sustainability and health impacts of the product while the consumer is still in the store.

Green apps have the potential to revolutionize the green movement as they meet consumer desires for cost savings and convenience. Perhaps best of all, many are relatively inexpensive, ranging from free to just a few dollars.[79]

Critical Thinking Questions

1. What are some of the barriers preventing consumers from buying green products?
2. How can mobile apps help consumers be more "green"?
3. What are some ways that mobile apps are combating greenwashing?

Mobile payments are also gaining traction, and companies like Google are working to capitalize on this opportunity.[80] Google Wallet and Apple Pay are mobile apps that store credit card information on the smartphone. When the shopper is ready to check out, he or she can tap the phone at the point of sale for the transaction to be registered.[81] Square is a company launched by Twitter co-founder Jack Dorsey. The company provides organizations with smartphone swiping devices for credit cards as well as tablets that can be used to tally purchases. Bitcoin is a virtual peer-to-peer currency that can be used to make a payment via smartphone. Smaller organizations have begun to accept Bitcoin at some of their stores. The success of mobile payments in revolutionizing the shopping experience will largely depend upon retailers adopting this payment system, but companies such as Starbucks are already jumping at the opportunity.

Widgets are small bits of software on a website, desktop, or mobile device that perform a simple purpose, such as providing stock quotes or blog updates. Marketers might use widgets to display news headlines, clocks, or games on their web pages.[82] For example, CNBC uses widgets to send alerts and financial news to subscribers. Widgets have been used by companies as a form of viral marketing—users can download the widget and send it to their friends with a click of a button.[83] Widgets downloaded to a user's desktop can update the user on the latest company or product information, enhancing relationship marketing between companies and their fans. Hotels, restaurants, and other tourist locations can download TripAdvisor widgets to their websites. These widgets display the latest company reviews, rewards, and other TripAdvisor content directly to the company's website.[84] Widgets are an innovative digital marketing tool to personalize web pages, alert users to the latest company information, and spread awareness of the company's products.

Explain online monitoring and analytics for social media.

Online Monitoring and Analytics

Without digital media monitoring and evaluation, it will not be possible to maximize resources and minimize costs in social media marketing. The strength of measurement relates to the ability to have online analytics and metrics. Social media monitoring involves activities to track, measure, and evaluate a firm's digital marketing initiatives.[85] An advantage of digital marketing evaluations is that there are methods to capture the metrics that indicate the outcomes of strategies. Therefore, establishing an expected level of performance against actual performance can be compared. Metrics develop from listening and tracking. For example, a firm could set up a hashtag and promote it. Metrics can be quantitative or qualitative. For example, click-through rate (CTR) determines the percentage of consumers who clicked on a link on a site as a quantitative measure. In addition, a qualitative metric could relate how consumers feel about a product.

Key performance indicators (KPIs) should be embedded at the onset of a social media strategy that can allow almost real-time measurement and evaluation. This provides a foundation for making iterative changes to implementation and tactical execution. Marketing analytics uses tools and methods to measure and interpret the effectiveness of marketing activities. Applying analytics to social media performance can help develop better targeted social media campaigns. Selecting valid metrics requires specific objectives that the social media strategy is to obtain. Objectives that are quantitative could include the number of likes on an Instagram post or the CTR of a Facebook post.

A comprehensive performance evaluation requires gathering all valid metrics and understanding the way the strategy meets performance standards or underperforms based on expectations. One way to approach this is to use Google Analytics, the largest analytics platform monitoring more than 30 million websites.[86] The Google Analytics dashboard is broken down into five sections: real time, audience, acquisition, behavior, and conversions. Table 13.2 explains the function of each section. Using this tool allows you to identify your website's strengths and weaknesses and uncover opportunities for growth. For example, you may find that organic search traffic is very high, but that your social media traffic is quite low, or you may see a spike in weekday traffic while weekends are slow. KPIs for your social media strategy can include likes, shares, reach, engagement rate, CTR, and conversions. In the conversions dashboard, marketers can set up custom conversion goals to see the impact social media has on their business.

By analyzing rich site traffic data, marketers can better understand their customers and measure the effectiveness of their marketing efforts. For example, PBS uses

TABLE 13.2
Google Analytics

Real time	Data updates are live so you can see pageviews, top social traffic, top referrals, top keywords, top active pages, and top locations in real time.
Audience	Audience reports provide insight into demographics, interests, geography, behavior, mobile use, and more.
Acquisition	In-bound traffic is monitored through acquisition reports, allowing you to compare traffic from search, referrals, e-mail, and social media.
Behavior	Use RSS feeds. Add tags to web pages or photos. "Vote" for websites online.
Joiners	Evaluate your site's content by seeing how visitors interact with your content Monitor landing pages, exit pages, site speed, bounce rate, and more.
Conversions	Google Analytics allows users to set goals and objectives to monitor web conversions, like signing up for an e-mail newsletter or completing a purchase.

Google Analytics to monitor the web performance for multiple properties and track key events like user registrations and video views. After analyzing search engine trends, PBS experienced 30 percent more site traffic in the first year after implementation.[87] Google Analytics is arguably the most robust web analytics tool available, and it is free to anyone with a Google account. A premium version, Google Analytics 360 Suite, designed to help companies target potential customers, is available for even more in-depth analytics. The tool identifies someone's habits from web and television to mobile, competing with companies like Salesforce and Oracle.[88]

M&M'S has used crowdsourcing to let consumers vote on new limited edition flavors.
©usersam2007/123RF

Using Digital Media to Learn about Consumers

Marketing research and information systems can use digital media and social networking sites to gather useful information about consumers and their preferences. Sites such as Twitter and Facebook can be good substitutes for focus groups. Online surveys can serve as an alternative to mail, telephone, or personal interviews.

Crowdsourcing describes how marketers use digital media to find out the opinions or needs of the crowd (or potential markets). Communities of interested consumers join sites like Threadless, which designs T-shirts, or Crowdspring, which creates logos and print and web designs. These companies give interested consumers opportunities to contribute and give feedback on product ideas. Crowdsourcing lets companies gather and utilize consumers' ideas in an interactive way when creating new products.

Consumer feedback is an important part of the digital media equation. Ratings and reviews have become exceptionally popular. Online reviews are estimated to influence the buying decisions of approximately 97 percent of U.S. consumers.[89] Retailers such as Amazon, Netflix, and Priceline allow consumers to post comments on their sites about the books, movies, and travel arrangements they sell. Today, most online shoppers search the Internet for ratings and reviews before making major purchase decisions.

While consumer-generated content about a firm can be either positive or negative, digital media forums do allow businesses to closely monitor what their customers are saying. In the case of negative feedback, businesses can communicate with consumers to address problems or complaints much more easily than through traditional communication channels. Yet despite the ease and obvious importance of online feedback, many companies do not yet take full advantage of the digital tools at their disposal.

Legal and Social Issues in Internet Marketing

LO 13-6

Propose recommendations to a marketer's dilemma.

The extraordinary growth of information technology, the Internet, and social networks has generated many legal and social issues for consumers and businesses. These issues include privacy concerns, the risk of identity theft and online fraud, and the need to protect intellectual property. The FTC rules for online marketing are the same as for any other form of communication or advertising. These rules help maintain the credibility of the Internet as an advertising medium. To avoid deception, all online communication must tell the truth and cannot mislead consumers. In addition, all claims must

Why Kayak.com Is a Travel Search Engine Worth Visiting

Kayak.com

Founders: Paul English and Steve Hafner

Founded: 2004, in Norwalk, Connecticut

Success: Kayak.com has won numerous awards and is ranked as one of the best flight search websites.

For the intrepid traveler looking for low-priced airfare, Kayak.com is there to meet your needs. Kayak.com is a travel search engine that compares hundreds of travels sites at once to determine the lowest travel prices. Using the power of the Internet, Kayak.com brings up travel information from hundreds of websites for users at the click of a button. After launching a beta site in 2004, Kayak.com has gained a reputation for offering the lowest prices in airfare, hotels, and car rentals when booking traveling.

In the digital world, Kayak.com has joined a number of companies that find the lowest travel prices available for consumers, facilitating both price and nonprice competition over the Internet. Its dynamic pricing model and a customer-driven focus are essential in allowing Kayak.com to compete against competitors like Orbitz and Expedia. Its success attracted the attention of Priceline.com, which purchased Kayak.com for $1.8 billion. More than a decade after its founding and with competition only increasing, Kayak.com is still ranked as one of the highest travel search engines for lowest overall prices.[90]

Critical Thinking Questions

1. What part of the marketing mix does Kayak.com seem to address particularly well?
2. How does Kayak.com provide users with more control over what they view?
3. Why do you think Kayak.com must also engage in nonprice competition?

be substantiated. If online communication is unfair and causes injury that is substantial and not reasonably avoidable and is not outweighed by other benefits, it is considered deceptive. The FTC identifies risk areas for online communication and issues warnings to consumers as misconduct is reported. Some of the areas include testimonials and endorsements, warranties and guarantees, free products, and mail and telephone orders. The FTC periodically joins with other law enforcement agencies to monitor the Internet for potentially false or deceptive online claims, including fraud, privacy, and intellectual property issues. We discuss these in this section, as well as steps that individuals, companies, and the government have taken to address them.

Privacy

Businesses have long tracked consumers' shopping habits with little controversy. However, observing the contents of a consumer's shopping cart or the process a consumer goes through when choosing a box of cereal generally does not result in the collection of specific, personally identifying data. Although using credit cards, shopping cards, and coupons forces consumers to give up a certain degree of anonymity in the traditional shopping process, they can still choose to remain anonymous by paying cash. Shopping on the Internet, however, allows businesses to track them on a far more personal level, from the contents of their online purchases to the websites they favor. Current technology has made it possible for marketers to amass vast quantities of personal information, often without consumers' knowledge, and to share and sell this information to interested third parties.

How is personal information collected on the web? Many sites follow users online by storing a "cookie," or an identifying string of text, on users' computers. Cookies permit website operators to track how often a user visits the site, what he or she looks at while there, and in what sequence. They also allow website visitors to customize services, such as virtual shopping carts, as well as the particular content they see when they log onto a web page. Users have the option of turning off cookies on their machines, but nevertheless, the potential for misuse has left many consumers uncomfortable with this technology.

The European Union passed a law requiring companies to get users' consent before using cookies to track their information. In the United States, one proposed solution for consumer Internet privacy is a "do not track" bill, similar to the "do not call" bill for telephones, to allow users to opt out of having their information tracked.[91] While consumers may welcome such added protections, web advertisers, who use consumer information to better target advertisements to online consumers, see it as a threat. In response to impending legislation, many web advertisers are attempting self-regulation in order to stay ahead of the game. For instance, the Digital Advertising Alliance (DAA) adopted privacy guidelines for online advertisers and created a "trusted mark" icon that websites adhering to their guidelines can display. However, because it is self-regulatory, not all digital advertisers will choose to participate in its programs.[92]

Influencer marketing, as we discussed earlier as a form of promotion, is relatively new compared with other forms of advertising, so it should be no surprise there have been road bumps for early adopters. Due to concerns about dishonest advertising, the Federal Trade Commission (FTC) requires influencers to clearly disclose any connection they have with brands they promote. Neglecting to make a disclosure is viewed as deceptive advertising. Cases have been filed against Warner Bros. Home Entertainment, which paid PewDiePie, YouTube's number one most subscribed channel, for an endorsement of its videogame Middle-Earth: Shadow of Mordor, and Lord & Taylor, which paid various influencers to promote their dresses, all without disclosures. According to the FTC, any level of compensation much be disclosed, whether a partnership is paid or an influencer strictly receives free product.[93]

Identity Theft and Online Fraud

Identity theft occurs when criminals obtain personal information that allows them to impersonate someone else in order to use the person's credit to access financial accounts and make purchases. This requires organizations to implement increased security measures to prevent database theft. As you can see in Figure 13.3, the most common complaints relate to employment or tax-related fraud, credit card fraud, phone/utilities fraud, bank fraud, loan/lease fraud, and government documents/benefits fraud. Sadly, cyberthieves have started targeting children's identities as they offer criminals "a clean slate" for them to commit fraud, such as applying for loans or credit cards.[94]

> **identity theft**
> when criminals obtain personal information that allows them to impersonate someone else in order to use their credit to access financial accounts and make purchases.

The Internet's relative anonymity and speed make possible both legal and illegal access to databases storing Social Security numbers, drivers' license numbers, dates of birth, mothers' maiden names, and other information that can be used to establish a credit card or bank account in another person's name in order to make fraudulent transactions. One growing scam used to initiate identity theft fraud is the practice of *phishing,* whereby con artists counterfeit a well-known website and send out e-mails directing victims to it. There, visitors find instructions to reveal sensitive information such as their credit card numbers. Phishing scams have faked websites for PayPal, AOL, and the Federal Deposit Insurance Corporation.

Some identity theft problems are resolved quickly, while other cases take weeks and hundreds of dollars before a victim's bank balances and credit standings are restored. To deter identity theft, the National Fraud Center wants financial institutions to implement new technologies such as digital certificates, digital signatures, and biometrics—the use of fingerprinting or retina scanning.

Online fraud includes any attempt to purposely deceive online. Many cybercriminals use hacking to commit online fraud. Hackers break into websites and steal users'

> **online fraud**
> any attempt to conduct fraudulent activities online.

FIGURE 13.3

Main Sources of Identity Theft

Source: "Consumer Sentinel Network Data Book: January–December 2016," Federal Trade Commission, March 2017, https://www.ftc.gov/system/files/documents/reports/consumer-sentinel-network-data-book-january-december-2016/csn_cy-2016_data_book.pdf (accessed April 27, 2017).

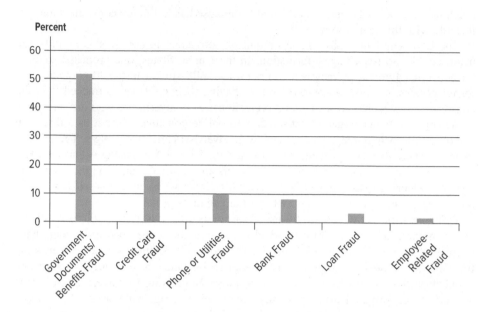

personal information. Home Depot, Target, and JPMorgan are some notable cases where cybercriminals hacked into these companies' systems and stole information. Sony experienced a devastating attack that shut down its entire computer network and resulted in the theft of 27 gigabytes of files.[95]

Using a different password for each website users visit is another important way to avoid becoming the victim of online fraud. Passwords should be complex enough that a cybercriminal cannot easily guess it. However, many consumers do not do this because of the hassle it takes in remembering complex passwords for multiple sites.[96]

Credit card fraud is a major type of fraud that occurs online. One way to tackle online fraud for credit cards is to use a pin number when doing online transactions. Banks are releasing credit cards with embedded chips rather than magnetic tape to make it harder for fraud to occur. In Europe, this type of credit card combined with the use of a pin number has deterred credit card fraud. This is because the consumers use their pin numbers as well as the embedded chip in their credit cards to make purchases. However, U.S. banks are not requiring Americans to input their pin numbers, which might limit their effectiveness in preventing online fraud within the United States.[97] Privacy advocates advise that the best way to stay out of trouble is to avoid giving out personal information, such as Social Security numbers or credit card information, unless the site is definitely legitimate.

Intellectual Property Theft and Other Illegal Activities

In addition to protecting personal privacy, Internet users and others want to protect their rights to property they may create, including songs, movies, books, and software. Such intellectual property consists of the ideas and creative materials developed to solve problems, carry out applications, and educate and entertain others.

Although intellectual property is generally protected by patents and copyrights, losses from the illegal copying of computer programs, music, movies, compact discs, and books reach billions of dollars each year in the United States alone. This has become a particular problem with digital media sites. YouTube has often faced lawsuits on intellectual property infringement. With millions of users uploading content to YouTube, it can be hard for Google to monitor and remove all the videos that may contain copyrighted materials.

Illegal sharing of content is another major intellectual property problem. Consumers rationalize the pirating of software, videogames, movies, and music for a number of reasons. First, many feel they just don't have the money to pay for what they want. Second, because their friends engage in piracy and swap digital content, some users feel influenced to engage in this activity. Others enjoy the thrill of getting away with something with a low risk of consequences. And finally, some people feel being tech-savvy allows them to take advantage of the opportunity to pirate content.[98]

Illicit online marketing is also becoming a serious issue for law enforcement across the globe. The ease of the Internet and the difficulty in pinpointing perpetrators are leading drug buyers to deal in illegal drugs over the Internet. Websites that deal in illegal drugs are looking increasingly legitimate, even employing marketing strategies and customer service.[99] Sales of counterfeit goods are another problem. Knockoffs of popular products seized by federal officials annually are valued at more than $1 billion. Counterfeit products, particularly from overseas, are thriving on the Internet because they can be shipped directly to customers without having to be examined by customs officials when shipped through ports. Some firms, including UGG Boots, are creating online services allowing users to type in the address to verify whether the electronic retailer is a legitimate seller.[100]

Elizabeth LaBau, the creator behind the food blog SugarHero! sued Food Network after it allegedly stole and recreated one of LaBau's DIY recipe videos.
©Owen Kolasinski/BFA/Shutterstock

Digital Media's Impact on Marketing

To be successful in business, you need to know much more than how to use a social media site to communicate with friends. Developing a strategic understanding of how digital marketing can make business more efficient and productive is increasingly necessary. If you are thinking of becoming an entrepreneur, then the digital world can open doors to new resources and customers. Smartphones, mobile broadband, and webcams are among the tools that can make the most of an online business world, creating greater efficiency at less cost. For example, rather than using traditional phone lines, Skype helps people make and receive calls via the Internet and provides free video calling and text messaging for about 10 percent of the cost of a landline.[101] It is up to businesses and entrepreneurs to develop strategies that achieve business success using existing and future technology, software, and networking opportunities.

Traditional businesses accustomed to using print media can find the transition to digital challenging. New media may require employees with new skills or additional training for current employees. There is often a gap between technical knowledge of how to develop sites and how to develop effective digital marketing strategies to enhance business success. Determining the correct blend of traditional and new media requires careful consideration; the mix will vary depending on the business, its size, and its target market. Future career opportunities will require skills in both traditional and digital media areas so that marketers properly understand and implement marketing strategies that help businesses achieve a competitive advantage.

So You Want to Be a Digital Marketer

The business world has grown increasingly dependent on digital marketing to maintain communication with stakeholders. Reaching customers is often a major concern, but digital marketing can also be used to communicate with suppliers, concerned community members, and special interest groups about issues related to sustainability, safety practices, and philanthropic activities. Many types of jobs exist: account executive directors of social media and director of marketing for digital products, as well as digital advertisers, online marketers, global digital marketers, and brand managers are prominently listed on career opportunity websites.

Entrepreneurs are taking advantage of the low cost of digital marketing, building social networking sites to help market their products. In fact, some small businesses such as specialty publishing, personal health and beauty, and other specialty products can use digital marketing as the primary channel for reaching consumers. Many small businesses are posting signs outside their stores with statements such as "Follow us on Twitter" or "Check out our Facebook page."

To utilize digital marketing, especially social networking, requires more than information technology skills related to constructing websites, graphics, videos, podcasts, etc. Most importantly, one must be able to determine how digital media can be used in implementing a marketing strategy.

All marketing starts with identifying a target market and developing a marketing mix to satisfy customers. Digital marketing is just another way to reach customers, provide information, and develop relationships. Therefore, your opportunity for a career in this field is greatly based on understanding the messages, desired level of interactivity, and connectivity that helps achieve marketing objectives.

As social media use skyrockets, digital marketing professionals will be in demand. The experience of many businesses and research indicate digital marketing is a powerful way to increase brand exposure and generate traffic. In fact, a study conducted on Social Media Examiner found that 85 percent of marketers surveyed believe generating exposure for their business is their number-one advantage in Internet marketing. As consumers use social networking for their personal communication, they will be more open to obtaining information about products through this channel. Digital marketing could be the fastest growing opportunity in business.

To prepare yourself for a digital marketing career, learn not only the technical aspects, but also how social media can be used to maximize marketing performance. A glance at careerbuilder. com indicates that management positions such as account manager, digital marketing manager, and digital product manager can pay from $60,000 to $170,000 or more per year.

Review Your Understanding

Recognize the increasing value of digital media and digital marketing in strategic planning.

Digital media are electronic media that function using digital codes and are available via computers, cellular phones, smartphones, and other digital devices. Digital marketing refers to the strategic process of distributing, promoting, pricing products, and discovering the desires of customers in the virtual environment of the Internet. Because they can enhance the exchange of information between the marketer and the customer, digital media have become an important component of firms' marketing strategies.

Demonstrate the role of digital marketing in today's business environment.

Digital communication facilitates marketing research and lowers the cost of communication and consumer service and support. Through websites, social networks, and other digital media, consumers can learn about everything they purchase and use in life and businesses can reach new markets through inexpensive and interactive communication channels.

Social networking is expanding so fast that no business can ignore its impact on customer relationships.

Show how digital media affect the marketing mix.

The ability to process orders electronically and increase the speed of communications via the Internet has reduced many distribution inefficiencies, costs, and redundancies while increasing speed throughout the marketing channel. Digital media help firms increase brand awareness, connect with consumers, form relationships, and spread positive publicity about their products. Because consumers are more informed than ever and consumer consumption patterns are changing, marketers must adapt their promotional efforts. The Internet gives consumers access to more information about costs and prices.

Illustrate how businesses can use different types of social networking media.

Social networking occurs when online consumers interact with other users on a web-based platform to discuss or view topics of interest. Types of social networking media include social networking sites, blogs, wikis, media sharing sites, virtual reality sites, mobile marketing, mobile applications, and widgets.

Blogs not only give consumers power, but also allow companies to answer consumer concerns and obtain free publicity. Wikis give marketers a better understanding of how consumers feel about their companies. Photo-sharing sites enable companies to share images of their businesses or products with consumers and often have links that connect users to company-sponsored blogs. Video sharing is allowing many businesses to engage in viral marketing. Amateur filmmakers are also becoming a potential low-cost, effective marketing venue for companies. Podcasts are audio or video files that can be downloaded from the Internet with a subscription that automatically delivers new content to listening devices or personal computers.

Marketers have begun joining and advertising on social networking sites like Facebook and Twitter due to their global reach. Virtual realities can be fun and creative ways to reach consumers, create brand loyalty, and use consumer knowledge to benefit companies. Mobile marketing includes advertising, text messages, and other types of digital marketing through mobile devices. Mobile apps can be anything from games, to news updates, to shopping assistance. They provide a way for marketers to reach consumers via their cell phones. Apps can help consumers to perform services and make purchases more easily, such as checking in at a hotel or comparing and contrasting the price of appliances or a new dress. Widgets are small bits of software on a website, desktop, or mobile device. They can be used to inform consumers about company updates and can easily go viral.

Identify legal and ethical considerations in digital media.

Increasing consumer concerns about privacy are prompting the FTC to look into regulating the types of information marketers can gather from Internet users, while many web advertisers and trade groups try to engage in self-regulation to prevent the passage of new Internet privacy laws. Online fraud includes any attempt to conduct fraudulent activities online. Intellectual property losses cost the United States billions of dollars and have become a particular problem for sites such as YouTube, which often finds it hard to monitor the millions of videos uploaded to its site for copyright infringement.

Propose recommendations to a marketer's dilemma.

Based on the material in this chapter, you should be able to answer the questions posed in the "Solve the Dilemma" feature near the end of this chapter and evaluate where the company's marketing strategy has failed. How could Paul utilize new digital media to help promote his product and gather data on how to improve it?

Critical Thinking Questions

Enter the World of Business Questions

1. What are the privacy concerns involved with allowing third parties access to user information?

2. What do you think might happen if the government passes stricter rules on data gathering and collection?

Do you think this might have some negative implications for users? Why or why not?

3. Facebook claims its data-gathering practices are common among similar firms such as Google. If this is true, why do you think users such as Elon Musk is reacting so negatively toward Facebook?

Learn the Terms

blogs 412
digital marketing 402
digital media 402
e-business 402

identity theft 421
omni-channel retailing 407
online fraud 421
podcast 414

viral marketing 413
wiki 412

Check Your Progress

1. What is digital marketing?

2. How can marketers utilize digital media to improve business?

3. Define *accessibility, addressability, connectivity, interactivity,* and *control.* What do these terms have to do with digital marketing?

4. What is e-business?

5. How is the Internet changing the practice of marketing?

6. What impact do digital media have on the marketing mix?

7. How can businesses utilize new digital and social networking channels in their marketing campaigns?

8. What are some of the privacy concerns associated with the Internet and e-business? How are these concerns being addressed in the United States?

9. What is identity theft? How can consumers protect themselves from this crime?

10. Why do creators want to protect their intellectual property? Provide an example on the Internet where intellectual property may not be protected or where a copyright has been infringed.

Get Involved

1. Amazon is one of the most recognized e-businesses. Visit the site (www.amazon.com) and identify the types of products the company sells. Explain its privacy policy.

2. Visit some of the social networking sites identified in this chapter. How do they differ in design, audience, and features? Why do you think some social networking sites like Facebook are more popular than others?

3. It has been stated that digital technology and the Internet are to business today what manufacturing was to business during the Industrial Revolution. The technology revolution requires a strategic understanding greater than learning the latest software and programs or determining which computer is the fastest. Leaders in business can no longer delegate digital media to specialists and must be the connectors and the strategists of how digital media will be used in the company. Outline a plan for how you will prepare yourself to function in a business world where digital marketing knowledge will be important to your success.

Build Your Skills

Planning a Digital Marketing and Social Networking Site

Background

Many companies today utilize digital media in a way that reflects their images and goals. They can also help to improve customer service, loyalty, and satisfaction while reaching out to new target markets. Companies use these sites in a variety of ways, sometimes setting up Facebook pages or Twitter accounts to gather customer feedback, to promote new products, or even to hold competitions.

The U.S. economy has experienced many ups and downs in recent decades, but e-commerce has been an area that has continued to grow throughout economic ups and downs. Many dot-com companies and social networking sites have risen and collapsed. Others such as Amazon, eBay, Facebook, and Twitter have not only survived, but thrived. Many that succeed are "niche players"; that is, they cater to a very specific market that a brick-and-mortar business (existing only in a physical marketplace) would find hard to reach. Others are able to compete with brick-and-mortar stores because they offer a wider variety of products, lower prices, or better customer service. Many new digital media outlets help companies compete on these fronts.

As a manager of Biodegradable Packaging Products Inc., a small business that produces packaging foam from recycled agricultural waste (mostly corn), you want to expand into e-business by using digital media to help market your product. Your major customers are other businesses and could include environmentally friendly companies like Tom's of Maine (natural toothpaste) and Celestial Seasonings (herbal tea). Your first need is to develop a social networking site or blog that will help you reach your potential customers. You must decide who your target market is and which medium will attract it the best.

Task

Plan a digital media marketing campaign using online social networking sites, blogs, or another digital media outlet using the following template.

Social networking/blog/other site: _____

Overall image and design of your site: _____

Strategy for attracting followers to your site: _____

Potential advertising partners to draw in more customers: _____

Solve the Dilemma

Developing Successful Freeware

Paul Easterwood, a recent graduate of Colorado State University with a degree in computer science, entered the job market during a slow point in the economy. Tech sector positions were hard to come by, and Paul felt he wouldn't be making anywhere near what he was worth. The only offer he received was from an entrepreneurial firm, Pentaverate Inc., that produced freeware. Freeware, or public domain software, is offered to consumers free of charge in exchange for revenues generated later. Makers of freeware (such as Adobe and Netscape) can earn high profits through advertisements their sites carry, from purchases made on the freeware site, or, for more specialized software, through fee-based tutorials and workshops offered to help end users. Paul did some research and found an article in *Worth* magazine documenting the enormous success of freeware.

Pentaverate Inc. offered compensation mainly in the form of stock options, which had the potential to be highly profitable if the company did well. Paul's job would be to develop freeware that people could download from the Internet and that would generate significant income for Pentaverate. With this in mind, he decided to accept the position, but he quickly realized he knew very little about business. With no real experience in marketing, Paul was at a loss to know what software he should produce that would make the company money. His first project,

IOWatch, was designed to take users on virtual tours of outer space, especially the moons of Jupiter (Paul's favorite subject), by continually searching the Internet for images and video clips associated with the cosmos and downloading them directly to a PC. The images would then appear as soon as the person logged on. Advertisements would accompany each download, generating income for Pentaverate.

Evaluate a marketer's dilemma and propose recommendations.

However, IOWatch experienced low end-user interest and drew little advertising income as a result. Historically at Pentaverate, employees were fired after two failed projects. Desperate to save his job, Paul decided to hire a consultant. He needed to figure out what customers might want so he could design some useful freeware for his second project. He also needed to know what went wrong with IOWatch because he loved the software and couldn't figure out why it had failed to find an audience. The job market has not improved, so Paul realizes how important it is for his second project to succeed.

Critical Thinking Questions

1. As a consultant, what would you do to help Paul figure out what went wrong with IOWatch?

2. What ideas for new freeware can you give Paul? What potential uses will the new software have?

3. How will it make money?

Build Your Business Plan

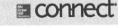

Digital Marketing and Social Networking

If you are considering developing a business plan for an established good or service, find out whether it is currently marketed digitally. If it is not, think about why that is the case. Can you think of how you might overcome any obstacles and engage in digital marketing on the Internet?

If you are thinking about introducing a new good or service, now is the time to think about whether you might want to market this product on the Internet. Remember, you do not have to have a brick-and-mortar store to open your own business anymore. Perhaps you might want to consider click instead of brick!

See for Yourself Videocase

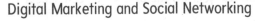

Should Employees Use Social Media Sites at Work?

As Facebook and other social media sites have gained popularity and expanded, managing their use at work has become an increasingly hot topic. Studies on the use of social media in the workplace conflict over how much it

inhibits productivity. Should employees be allowed to access social media at work? Approximately 38 percent of time wasters say they waste work time on social media sites. Many offices have banned access to Facebook. The results are as mixed as the research. A National Business Ethics Survey (NBES) revealed that 11 percent of employees who engage in social networking are "active" social networkers who spend 30 percent or more of the workday on social networking sites.

Many managers are conflicted as to whether this constitutes enough of a problem to be banned outright.

Another study conducted by Nucleus Research (an IT research company) found that 77 percent of Facebook users used the site during work for as much as two hours a day; 87 percent of those surveyed admitted they were using social media sites to waste time. NBES also found that active social networkers were more likely to find certain questionable behaviors to be acceptable, such as criticizing the company or its managers on social networking sites. Procter & Gamble realized that many of its employees were using social networking sites for non-work purposes. Its investigations revealed that employees across the company were watching an average of 50,000 five-minute YouTube videos and listening to 4,000 hours of music on Pandora daily.

However, an outright ban could cause problems. Some younger employees have expressed that they do not want to work for companies without social media access; they view restricting or eliminating access like removing a benefit. Employees at companies with an outright ban often resent the lack of trust associated with such a move and feel that management is censuring their activities. Other employees who use Facebook during their lunch hours or break times may feel that they are being punished because of others' actions. Additionally, Procter & Gamble uses YouTube and Facebook extensively for marketing purposes. Banning these sites would disrupt the firm's marketing efforts.

An Australian study indicates that employees taking time out to pursue Facebook and other social media were actually 9 percent more productive than those who did not. Brent Coker, the study's author and University of Melbourne faculty member, says people are more productive when they take time to "zone out" throughout the workday. Doing so can improve concentration. In the sales industry, a study of 100,000 employees revealed that social media "mavens" had 1.6 percent higher sales conversions.

Some companies actually encourage employees to use social networking as part of their integrated marketing strategy. In fact, not having a social media page such as Facebook or LinkedIn might be seen as a missed opportunity for marketing the firm. Even the law industry is starting to use social media on a more daily basis. One study of the top 50 highest ranked law firms in the country determine that 64 percent use Facebook and 90 percent are on Twitter. Approximately 80 percent post something every day or once a week. Although larger law firms tend not to use social media as effectively as smaller law firms, the use of social media to interact with clients is clearly gaining throughout the industry.

Despite the benefits that companies have received from allowing their employees to use social media, many companies have gone ahead with social media bans. Procter & Gamble has restricted the use of Netflix and Pandora, but not Facebook or YouTube. Companies all need to ask, "Can management use social media to benefit the company?" If so, it may be more advantageous to take the risks of employees using social media for personal use if they can also be encouraged to use social networks to publicize their organizations, connect with customers, and view consumer comments or complaints. By restricting social media use, companies may be forfeiting an effective marketing tool.[102]

Critical Thinking Questions

1. Why do you think results are so mixed on the use of social networking in the workplace?

2. What are some possible upsides to utilizing social media as part of an integrated marketing strategy, especially in digital marketing?

3. What are the downsides to restricting employee access to social networking sites?

You can find the related video in the Video Library in Connect. Ask your instructor how you can access Connect.

Team Exercise

Develop a digital marketing promotion for a local sports team. Use Twitter, Facebook, and other social networking media to promote ticket sales for next season's schedule. In your plan, provide specific details and ideas for the content you would use on the sites. Also, describe how you would encourage fans and potential fans to go to your site. How would you use digital media to motivate sports fans to purchase tickets and merchandise and attend games?

Notes

1. Jessica Rich, "Beyond Facebook: It's High Time for Stronger Privacy Laws," *Wired,* April 8, 2018, https://www.wired.com/story/beyond-facebook-its-high-time-for-stronger-privacy-laws/ (accessed April 14, 2018); Bloomberg,

"Facebook Cambridge Analytica Scandal: 10 Questions Answered," *Fortune,* April 10, 2018, http://fortune.com/2018/04/10/facebook-cambridge-analytica-what-happened/ (accessed April 14, 2018); Catherine Clifford,

"Elon Musk: Facebook 'Gives Me the Willies,'" *CNBC*, March 26, 2016, https://www.cnbc.com/2018/03/26/tesla-ceo-elon-musk-tweets-why-he-doesnt-like-facebook.html (accessed April 14, 2018); Daisuke Wakabayashi and Jack Nicas, "Facebook Takes the Punches while Rest of Silicon Valley Ducks," *The New York Times*, April 13, 2018, https://www.nytimes.com/2018/04/13/technology/facebook-silicon-valley.html (accessed April 14, 2018); Laura Lorenzetti, "Lenders Are Dropping Plans to Judge You by Your Facebook Friends," *Fortune*, February 24, 2016, http://fortune.com/2016/02/24/facebook-credit-score/ (accessed April 14, 2018); Kevin Granville, "Facebook and Cambridge Analytica: What You Need to Know as Fallout Widens," *The New York Times*, March 19, 2018, https://www.nytimes.com/2018/03/19/technology/facebook-cambridge-analytica-explained.html (accessed April 14, 2018).

2. Lauren I. Labrecque, Jonas vor dem Esche, Charla Mathwick, Thomas P. Novak, and Charles F. Hofacker, "Consumer Power: Evolution in the Digital Age," *Journal of Interactive Marketing* 27, no. 4 (November 2013), pp. 257–269.

3. Charisse Jones, "Department Stores Become Endangered," *USA Today*, January 68, 2017, p. 1A.

4. Suzanne Kapcr, "Store Closings Accelerating," *The Wall Street Journal*, April 22–23, 2017, p. B1.

5. "Top 15 Most Popular Video Websites," eBiz MBA, April 2016, www.ebizmba.com/articles/video-websites (accessed April 18, 2016); "About Vimeo," Vimeo, http://vimeo.com/about (accessed April 18, 2016).

6. O.C. Ferrell, "Broadening Marketing's Contribution to Data Privacy," *Journal of Academy of Marketing Science* 23, no. 2 (2017), p. 160-163.

7. Charisse Jones, "Department Stores Become Endangered," *USA Today*, January 6–8, 2017, p. 1A.

8. Ray Downs, "UN: Majority of World's Population Lacks Internet Access," *UPI*, September 18, 2017, https://www.upi.com/Top_News/World-News/2017/09/18/UN-Majority-of-worlds-population-lacks-internet-access/6571505782626/ (accessed April 28, 2018).

9. Daniel Nations, "How Many iPad Apps Are in the App Store?" *About.com*, September 22, 2015, http://ipad.about.com/od/iPad-FAQ/f/How-Many-iPad-Apps-Are-In-The-App-Store.htm (accessed April 18, 2016).

10. Emily Orofino, "Can This Genius Beauty Brand Crowdsource the Next Big Thing?" *POPSUGAR*, July 10, 2016, https://www.popsugar.com/beauty/Volition-Beauty-Custom-Made-Beauty-Products-41873521 (accessed March 25, 2017).

11. "Consumers Are Increasingly Researching Purchases Online," *PYMNTS*, January 9, 2018, https://www.pymnts.com/news/retail/2018/omichannel-ecommerce-consumer-habits/ (accessed April 28, 2018); "Number of Digital Shoppers in the United States from 2016 to 2021 (in millions)," *Statista*, 2018, https://www.statista.com/statistics/183755/number-of-us-internet-shoppers-since-2009/ (accessed April 28, 2018)

12. Emma Sopadjieva, Utpal M. Dholakia, and Beth Benjamin, "A Study of 46,000 Shoppers Shows That Omnichannel Retailing Works." *Harvard Business Review*, January 3, 2016.

13. Kate Rockwood, "Why Spending $1,000 on an Instagram Post Might Actually Be Worth it," *Inc.*, February 2017, http://www.inc.com/magazine/201702/kate-rockwood/tip-sheet-social-media-influencers.html (accessed March 5, 2017).

14. "About Influenster," *Influenster*, https://www.influenster.com/about (accessed April 28, 2018).

15. Melissa S. Barker, Donald I. Barker, Nicholas F. Bormann, Mary Lous Roberts, and Debra Zahay, *Social Media Marketing*, 2nd ed. (Mason, OH: Cengage Learning, 2016).

16. Ken Yeung, "Now with 2.5M Advertisers, Facebook Targets TV Budgets with New Ad-Buy Offering," *Venture Beat*, September 27, 2015, http://venturebeat.com/2015/09/27/now-with-2-5m-advertisers-facebook-targets-tv-budgets-with-new-ad-buying-offering/ (accessed March 2, 2017).

17. Barker et al., *Social Media Marketing*.

18. Kristen Herhold, "How Businesses Use Social Media: 2017 Survey," *Clutch*, September 14, 2017, https://clutch.co/agencies/social-media-marketing/resources/social-media-survey-2017 (accessed April 24, 2018).

19. Darius Fisher, "How to Handle Negative Yelp Reviews," *Huffington Post*, June 7, 2016, http://www.huffingtonpost.com/darius-fisher/how-to-handle-negative-ye_b_10324196.html (accessed March 2, 2017).

20. "Dodge Brand Connecting with Its Fans by Inviting a Limited Number to the Hotly Anticipated Reveal of the All-New 2018 Dodge Challenger SRT Demon in NYC," *Yahoo! Finance*, February 22, 2017, http://finance.yahoo.com/news/dodge-brand-connecting-fans-inviting-130000410.html (accessed March 2, 2017).

21. Shea Bennett, "This is How Much Time We Spend on Social Networks Every Day," *Ad Week*, November 18, 2014, http://www.adweek.com/socialtimes/social-media-minutes-day/503160 (accessed May 5, 2017).

22. Harry McCracken, "Inside Mark Zuckerberg's Bold Plan for the Future of Facebook," *Fast Company*, December 2015/January 2016, pp. 86–100, 136.

23. Wendy Boswell, "Video Websites: The Top Ten," *About Tech*, 2015, http://websearch.about.com/od/imagesearch/tp/popularvideosites.htm (accessed April 18, 2016).

24. "Boost a Post," Facebook for Business, https://www.facebook.com/business/a/boost-a-post (accessed April 18, 2016); Olsy Sorokina, "What Are Facebook Boost Posts and How Can They Help Your Business," *Hootsuite*, October 24, 2014, https://blog.hootsuite.com/how-does-facebook-boost-posts-work/ (accessed April 18, 2016).

25. Nathan Oliverez-Giles, "Facebook Messenger Brings Group Video Calling to Apps and Web," *The Wall Street Journal*, December 19, 2016, www.wsj.com/articles/facebook-messenger-brings-group-video-calling-to-apps-and-web-1482184260 (accessed December 28, 2016).

26. Peter Roesler, "How Social Media Influences Consumer Buying Decisions," *Business Journals*, May 29, 2015, https://www.bizjournals.com/bizjournals/how-to/marketing/2015/05/how-social-media-influences-consumer-buying.html (accessed April 28, 2018).

27. Jeff Elder, "Facing Reality, Companies Alter Social-Media Strategies," *The Wall Street Journal,* June 23, 2014, pp. B1–B2.

28. Jefferson Graham, "Cake Decorator Finds Twitter a Tweet Recipe for Success," *USA Today,* April 1, 2009, p. 5B.

29. Craig Smith, "Twitter Mobile Statistics and Facts," *DMR,* February 17, 2018, https://expandedramblings.com/index.php/twitter-mobile-statistics/ (accessed April 28, 2018).

30. Christine Birkner, "The Goldfish Conundrum," *Marketing News,* April 2015, pp. 18–19.

31. Stephanie Frasco, "100 Facts and Figures about Twitter, and Why They Matter for Your Business," *Social Media Today,* September 26, 2013, http://socialmediatoday.com/stephaniefrasco/1770161/100-facts-figures-about-twitter-business (accessed May 5, 2017).

32. Samantha Grossman, "The 13 Sassiest Brands on Twitter," *Time,* February 7, 2014, http://time.com/5151/sassiest-brands-on-twitter-ranked/ (accessed January 14, 2016); "Taco Bell," Twitter, https://twitter.com/tacobell (accessed January 14, 2016).

33. Belinda Parmar, "50 Companies That Get Twitter—and 50 That Don't," *Harvard Business Review,* April 27, 2015, https://hbr.org/2015/04/the-best-and-worst-corporate-tweeters (accessed May 5, 2017).

34. Nathan McAlone, "Investors Are Going Nuts for Snapchat—Here's How Snap Thinks It Can Turn a $500 Million Loss into Profit," *Business Insider,* March 5, 2017, http://www.businessinsider.com/how-will-snapchat-make-money-2017-3 (accessed March 5, 2017); "Global Annual Net Loss of Snap from 2015 to 2017 (in million U.S. dollars)," *Statista,* https://www.statista.com/statistics/668190/snapchat-annual-net-income-loss/ (accessed April 29, 2018); Garrett Sloane, "Oh Snap! Shares Soar as Snapchat Reports 9 Million More Daily Users and an Ad Surge," *AdAge,* February 6, 2018, http://adage.com/article/digital/snapchat-reports-9-million-daily-users-ad-surge/312286/ (accessed April 29, 2018).

35. Haley Tsukayama, "Snapchat Files for Its Initial Public Offering: Here Are the 10 Most Interesting Things We've Learned So Far," *Washington Post,* February 3, 2017, https://www.washingtonpost.com/news/the-switch/wp/2017/02/03/snapchat-files-for-its-initial-public-offering-here-are-the-10-most-interesting-things-weve-learned-so-far/?utm_term=.591d0ee20f98 (accessed March 5, 2017).

36. Elaine Pofeldt, "Freelancers Now Make Up 35% of U.S. Workforce," *Forbes,* October 6, 2016, https://www.forbes.com/sites/elainepofeldt/2016/10/06/new-survey-freelance-economy-shows-rapid-growth/#3ca2d55a7c3f (accessed December 26, 2017); "Upwork Releases First-Ever Quarterly Skills Index Revealing the Fastest-Growing Freelance Skills in the U.S.," Upwork, July 7, 2016, https://www.upwork.com/press/2016/07/07/upwork-releases-first-ever-quarterly-skills-index-revealing-fastest-growing-freelance-skills-u-s/ (accessed December 26, 2017); Susan M. Heathfield, "Use the Web for Recruiting Talent," *The Balance,* February 15, 2017, https://www.thebalance.com/use-the-web-for-recruiting-talent-1918951 (accessed December 26, 2017); "10 Simple Steps to a Great Company Career Page," ZipRecruiter blog, https://www.ziprecruiter.com/blog/10-simple-steps-to-a-great-company-career-page-on-your-website/ (accessed December 26, 2017); Jimmy Rohampton, "9 Tips for Recruiting Millennial Talent through Social Media," *Forbes,* January 10, 2017, https://www.forbes.com/sites/jimmyrohampton/2017/01/10/9-tips-for-recruiting-millennial-talent-through-social-media/#797707c0493d (accessed December 26, 2017); "How It Works," Upwork, https://www.upwork.com/i/how-it-works/client/ (accessed December 26, 2017); Kimberlee Morrison, "Survey: 92% of Recruiters Use Social Media to Find High-Quality Candidates," *Ad Week,* September 22, 2015, http://www.adweek.com/digital/survey-96-of-recruiters-use-social-media-to-find-high-quality-candidates/ (accessed December 26, 2017); "Using Social Media for Talent Acquisition," Society for Human Resource Management, September 20, 2017, https://www.shrm.org/hr-today/trends-and-forecasting/research-and-surveys/pages/social-media-recruiting-screening-2015.aspx (accessed December 26, 2017); Sarah K. White, "Recruiters Increasingly Rely on Social Media to Find Talent," *CIO,* May 23, 2016, https://www.cio.com/article/3073589/hiring/recruiters-increasingly-rely-on-social-media-to-find-talent.html (accessed December 26, 2017); "Mobile Fact Sheet," Pew Research Center, January 12, 2017, http://www.pewinternet.org/fact-sheet/mobile/ (accessed December 26, 2017); Laura Petrecca, "More College Grads Use Social Media to Find Jobs," *USA Today,* April 5, 2011, http://usatoday30.usatoday.com/money/workplace/2011-04-04-social-media-in-job-searches.htm (accessed December 26, 2017); Joseph Walker," LinkedIn Gets Closer to Job Seekers," *The Wall Street Journal,* September 13, 2011, http://online.wsj.com/article/SBB00014240531119042655045765689208904 53258.html (accessed December 26, 2017).

37. Artyom Dogtiev, "Snapchat Revenue and Usage Statistics 2017," *Business of Apps,* April 26, 2018, http://www.businessofapps.com/data/snapchat-statistics/ (accessed April 28, 2018).

38. Hillary Grigonis, "Snap, then Shop—Snapchat Rolls Out Shoppable AR Filters," *Digital Trends,* April 18, 2018, https://www.digitaltrends.com/social-media/snapchat-launches-shoppable-ar-filters/ (accessed April 28, 2018).

39. Douglas A. McIntyre, "YouTube Viewership Hits a Billion Hours of Content a Day," *24/7 Wall St.,* February 27, 2017, http://247wallst.com/apps-software/2017/02/27/youtube-viewership-hits-100-billion-hours-of-content-a-day/ (accessed March 5, 2017).

40. Rachel Strugatz, "Beauty's Battle for Views: Brands vs. Vloggers," *WWD,* February 23, 2017, http://wwd.com/beauty-industry-news/beauty-features/youtube-l2-beautys-battle-for-views-brands-vs-bloggers-10814639/ (access March 5, 2017).

41. Mike Snider, "How YouTube TV Compares to Rivals Sling, PlayStation, DirecTV," *USA Today,* March 5, 2017, http://www.usatoday.com/story/tech/talkingtech/2017/03/05/

how-youtube-tv-compares-rivals-sling-playstation-directv/98551276/ (accessed March 5, 2017).

42. Sujan Patel, "How LinkedIn Uses LinkedIn for Marketing," *Forbes,* March 4, 2017, https://www.forbes.com/sites/sujanpatel/2017/03/04/how-linkedin-uses-linkedin-for-marketing/2/#3b96f9d3657f (accessed March 5, 2017).

43. Amanda Walgrove, "5 B2B Brands That Rock LinkedIn Marketing," *Contently,* February 24, 2017, https://contently.com/strategist/2015/02/24/5-b2b-brands-that-rock-linkedin/ (accessed March 5, 2017).

44. Marcelina Hardy, "Statistics on Blogging," *ContentWriters.com,* August 19, 2014, https://contentwriters.com/blog/statistics-blogging/ (accessed April 18, 2016).

45. "About," Tumblr, https://www.tumblr.com/about (accessed April 29, 2018).

46. Caitlin Dewey, "2015 Is the Year That Tumblr Became the Front Page of the Internet," *Washington Post,* December 23, 2015, https://www.washingtonpost.com/news/the-intersect/wp/2015/03/11/move-over-reddit-tumblr-is-the-new-front-page-of-the-internet/ (accessed May 5, 2017).

47. Niall Harbison and Lauren Fisher, "40 of the Best Corporate Blogs to Inspire You," *Ragan's PR Daily,* September 13, 2012, http://www.prdaily.com/Main/Articles/40_of_the_best_corporate_blogs_to_inspire_you_12645.aspx (accessed May 5, 2017); "Wikipedia," *Wikipedia,* https://en.wikipedia.org/wiki/Wikipedia (accessed April 29, 2018).

48. Drake Bennett, "Ten Years of Inaccuracy and Remarkable Detail: Wikipedia," *Bloomberg Businessweek,* January 10–16, 2011, pp. 57–61; "Wikipedia: About," *Wikipedia,* http://en.wikipedia.org/wiki/Wikipedia:About (accessed May 5, 2017).

49. Charlene Li and Josh Bernoff, *Groundswell* (Boston: Harvard Business Press, 2008), pp. 25–26.

50. "Percentage of Fortune 500 Companies with Public Blogs from 2010 to 2016," *Statista,* https://www.statista.com/statistics/262466/share-of-fortune-500-companies-with-public-blogs/ (accessed April 29, 2018).

51. "How to Use a Raw Egg to Determine if Your Mattress Is Awful—Purple," *YouTube,* April 26, 2016, https://youtu.be/4BvwpjaGZCQ (accessed April 29, 2018).

52. Stephanie Hayes, "Michelle Phan, a YouTube Sensation for Her Makeup Tutorials, Has Transformed Her Life," *Tampa Bay Times,* August 22, 2009, http://www.tampabay.com/features/humaninterest/michelle-phan-a-youtube-sensation-for-her-makeup-tutorials-has-transformed/1029747 (accessed May 5, 2017).

53. Nicole LaPorte, "Serious Beauty," *Fast Company,* February 2016, pp. 27–28.

54. "Photoset," Instagram, http://blog.business.instagram.com/post/78694901404/how-yogurt-maker-chobani-uses-instagram-to-open (accessed January 15, 2016).

55. Bryan Clark, "Snapchat Growth Slowed Significantly after Launch of Instagram Stories," *The Next Web,* February 3, 3017, https://thenextweb.com/socialmedia/2017/02/03/snapchat-growth-slowed-significantly-after-the-launch-of-instagram-stories/#.tnw_360bhC9j (accessed March 5, 2017).

56. "4 Key Advantages for Video Marketing on Instagram vs. Vine," *Ad Week,* March 4, 2015, http://www.adweek.com/socialtimes/sumall-dane-atkinson-video-marketing-instagram-vs-vine/616331 (accessed May 5, 2017).

57. Samuel Gibbs, "Flickr Bought by SmugMug as Yahoo Breakup Begins," *The Guardian,* April 23, 2018, https://www.theguardian.com/technology/2018/apr/23/flickr-bought-by-smugmug-yahoo-breakup (accessed April 29, 2018).

58. Nicole Lee, "Pinterest Uses AI and Your Camera to Recommend Pins," *Engadget,* February 8, 2017, https://www.engadget.com/2017/02/08/pinterest-uses-ai-and-your-camera-to-recommend-pins/ (accessed March 5, 2017).

59. Shannon Liao, "Pinterest Now Lets You Filter Search Results by Skin Tone," *The Verge,* April 26, 2018, https://www.theverge.com/2018/4/26/17286898/pinterest-skin-tone-search-filter-results-feature-update (accessed April 29, 2018).

60. "Marketers' Interest in Pinterest," *Marketing News,* April 30, 2012, pp. 8–9; "PINTEREST INTEREST: Survey: 17 Percent of Marketers Currently Using or Planning to Join Pinterest," The Creative Group, August 22, 2012, http://creativegroup.mediaroom.com/pinterest-for-business (accessed April 18, 2016); Jason Falls, "How Pinterest Is Becoming the Next Big Thing in Social Media for Business," *Entrepreneur,* February 7, 2012, www.entrepreneur.com/article/222740 (accessed April 18, 2016); Pinterest website, http://pinterest.com/ (accessed April 18, 2016).

61. Jeff Bercovici, "Social Media's New Mad Men," *Forbes,* November 2014, pp. 71–82.

62. Zale Jewelers Pinterest page, www.pinterest.com/zalesjewelers (accessed April 18, 2016).

63. Douglas MacMillan and Elizabeth Dwoskin, "Smile! Marketers Are Mining Selfies," *The Wall Street Journal,* October 10, 2014, pp. B1–B2.

64. Natalie Wires, "The Rising Popularity of Podcasts: Why Listeners Are Rediscovering Podcasts," *Tunheim,* March 26, 2014, http://blog.tunheim.com/2014/03/26/rising-popularity-podcasts-listeners-rediscovering-podcasts/1438#.U2pMWYFdVc8 (accessed April 18, 2016).

65. Ann Lukits, "Podcasts Send Shoppers to Omega-3s," *The Wall Street Journal,* December 9, 2014, p. D2.

66. Roger Yu, "Smartphones Help Make Bon Voyages," *USA Today,* March 5, 2010, p. B1.

67. Sean Silverthorpe, "Should Retailers Worry about In-Store Mobile Use?" *Insights from Marketing Science Institute* 1 (2015), pp. 1–2.

68. April Berthene, "Mobile Accounts for Nearly 35% of 2017 E-Commerce Sales," *Digital Commerce 360,* February 19, 2018, https://www.digitalcommerce360.com/2018/02/19/mobile-accounts-for-nearly-35-of-2017-e8209commerce-sales/ (accessed April 29, 2018).

69. Mark Milian, "Why Text Messages Are Limited to 160 Characters," *Los Angeles Times,* May 3, 2009,

http://latimesblogs.latimes.com/technology/2009/05/invented-text-messaging.html (accessed May 5, 2017); "Eight Reasons Why Your Business Should Use SMS Marketing," *Mobile Marketing Ratings,* www.mobilemarketingratings.com/eight-reasons-sms-marketing.html (accessed April 18, 2016).

70. James Citron, "2014: The Year the MMS Upswing Arrives and How to Take Advantage of It," *Wired,* January 21, 2014, http://insights.wired.com/profiles/blogs/2014-the-year-the-mms-upswing-arrives-and-how-to-take-advantage#axzz3xLvg3dvO (accessed May 5, 2017); "Global $65 Billion Managed Mobility Services (MMS) Market Analysis & Trends Report 2017—Industry Forecast to 2025—Research and Markets," *Business Wire,* February 1, 2017, https://www.businesswire.com/news/home/20170201005533/en/Global-65-Billion-Managed-Mobility-Services-MMS (accessed April 29, 2018).

71. "Mobile to Account for More than Half of Digital Ad Spending in 2015," *eMarketer,* September 1, 2015, www.emarketer.com/Article/Mobile-Account-More-than-Half-of-Digital-Ad-Spending-2015/1012930 (accessed January 15, 2016).

72. Jake Jeffries, "10 Incredible Mobile Marketing Stats 2015 [INFOGRAPHIC]," *Social Media Today,* January 13, 2015, www.socialmediatoday.com/content/10-incredible-mobile-marketing-stats-2015-infographic (accessed January 15, 2016).

73. Christopher Heine, "Foursquare Unleashes Location Data for Cross-Mobile Ad Targeting," *AdWeek,* April 14, 2015, http://www.adweek.com/news/technology/foursquare-finally-unleashes-location-data-cross-mobile-ad-targeting-164069 (accessed May 5, 2017).

74. Anita Campbell, "What the Heck Is an App?" *Small Business Trends,* March 7, 2011, http://smallbiztrends.com/2011/03/what-is-an-app.html (accessed April 18, 2016).

75. Sarah Perez, "Consumers Spend 85% of Time on Smartphones in Apps, but Only 5 Apps See Heavy Use," *TechCrunch,* June 22, 2015, http://techcrunch.com/2015/06/22/consumers-spend-85-of-time-on-smartphones-in-apps-but-only-5-apps-see-heavy-use/ (accessed May 5, 2017).

76. Monica Anderson, "Technology Device Ownership: 2015," October 29, 2015, http://www.pewinternet.org/2015/10/29/technology-device-ownership-2015/ (accessed April 18, 2016).

77. Michelle Yeomans, "Unilever Opts for 'Mobile Marketing Platform' to Reach South-East Asia," *Cosmetics Design,* September 15, 2015, http://www.cosmeticsdesign-asia.com/Business-Financial/Unilever-opts-for-mobile-marketing-platform-to-reach-south-east-Asia (accessed January 19, 2016).

78. Umika Pidaparthy, "Marketers Embracing QR Codes, for Better or Worse,"*CNN Tech,* March 28, 2011, http://www.cnn.com/2011/TECH/mobile/03/28/qr.codes.marketing/ (accessed May 5, 2017).

79. GoodGuide website, https://www.goodguide.com/#/ (accessed April 7, 2018); EnergyElephant, "Benefits," https://energyelephant.com/benefits (accessed April 7, 2018); Google Play, "Energy Elephant," https://play.google.com/store/apps/details?id=com.energyelephant.meterReading&hl=en (accessed April 7, 2018); EnergyElephant, "App," https://energyelephant.com/app (accessed April 7, 2018); Jefferson Graham, "Mobile Apps Make It Easier to Go Green," *USA Today,* May 12, 2011, https://usatoday30.usatoday.com/tech/products/2011-05-12-green-tech_n.htm (accessed April 7, 2018); Lars Paronen, "Green Apps that Can Save You Money," *Reuters,* February 18, 2011, http://blogs.reuters.com/environment/2011/02/18/green-apps-that-can-save-you-money/ (accessed April 7, 2018); Jefferson Graham, "GoodGuide App Helps Navigate Green Products," *USA Today,* May 13, 2011, https://usatoday30.usatoday.com/tech/products/2011-05-12-GoodGuide-app_n.htm (accessed April 7, 2018).

80. Brad Stone and Olga Kharif, "Pay as You Go," *Bloomberg Businessweek,* July 18–24, 2011, pp. 66–71.

81. "Google Wallet," www.google.com/wallet/what-is-google-wallet.html (accessed April 18, 2016).

82. "All about Widgets," *Webopedia,* September 14, 2007, www.webopedia.com/DidYouKnow/Hardware_Software/widgets.asp (accessed April 18, 2016).

83. Rachael King, "Building a Brand with Widgets," *Bloomberg Businessweek,* March 3, 2008, www.businessweek.com/technology/content/feb2008/tc20080303_000743.htm (accessed May 5, 2017).

84. TripAdvisor, "Welcome to TripAdvisor's Widget Center," www.tripadvisor.com/Widgets (accessed April 18, 2016).

85. Barker et al., *Social Media Marketing.*

86. Matt McGee, "As Google Analytics Turns 10, We Ask: How Many Websites Use It?" *Marketing Land,* November 12, 2015, http://marketingland.com/as-google-analytics-turns-10-we-ask-how-many-websites-use-it-151892 (accessed March 5, 2017).

87. "By Tailoring the Features of Google Analytics, LunaMetrics Helps PBS Increase Both Conversions and Visits by 30%," *Google Analytics,* https://static.googleusercontent.com/media/www.google.com/en//intl/pt_ALL/analytics/customers/pdfs/pbs.pdf (accessed March 24, 2017).

88. Quentin Hardy, "Google Introduces Products That Will Sharpen Its Ad Focus," *The New York Times,* March 15, 2016, https://www.nytimes.com/2016/03/16/technology/google-introduces-products-that-will-sharpen-its-ad-focus.html (accessed March 24, 2017).

89. Nicole Lyn Pesce, "People Are Complete Suckers for Online Reviews," *New York Post,* August 23, 2017, https://nypost.com/2017/08/23/people-are-complete-suckers-for-online-reviews/ (accessed April 29, 2018).

90. Kayak.com website, www.kayak.com (accessed April 7, 2018); Craig Grannell, "The Best Flight Search Websites—Tried and Tested," *The Telegraph,* April 10, 2017, https://www.telegraph.co.uk/travel/advice/best-flight-booking-comparison-websites-apps/ (accessed April 7, 2018); Reid Bramblett, "The 10 Best (and Worst) Airfare Search Sites,"

Frommers, https://www.frommers.com/slideshows/848046-the-10-best-and-worst-airfare-search-sites (accessed April 7, 2018); Geoff Colvin, "Kayak Takes on the Big Dogs," *Fortune,* October 8, 2012, pp. 78–82; Associated Press, "Kayak Shares Soar after Travel Site Goes Public," *CBS,* July 20, 2012, https://www.cbsnews.com/news/kayak-shares-soar-after-travel-site-goes-public/ (accessed April 7, 2018); Cheryl Morris, "Boston Tech-Mafia Mondays: The Kayak Crew," *Bostinno,* October 25, 2010, https://www.americaninno.com/boston/boston-tech-mafia-mondays-the-kayak-crew/ (accessed April 7, 2018); "Kayak.com Leading the Travel Search Engine Market," *VC Gate,* March 14, 2009, https://www.vcgate.com/Kayak-com-Leading-The-Travel-Search-Engine-Market.asp (accessed April 7, 2018); "Company Overview of Kayak Software Corporation," *Bloomberg,* https://www.bloomberg.com/research/stocks/private/snapshot.asp?privcapId=12618272 (accessed April 7, 2018).

91. Jon Swartz, "Facebook Changes Its Status in Washington," *USA Today,* January 13, 2011, pp. 1B–2B; John W. Miller, "Yahoo Cookie Plan in Place," *The Wall Street Journal,* March 19, 2011, http://online.wsj.com/news/articles/SB10001424052748703512404576208700813815570 (accessed May 5, 2107).

92. Jesse Brody, "Terms and Conditions," *Marketing News,* November 2014, pp. 34–41.

93. "FTC Cracking Down on Social Influencers' Labeling of Paid Promotions," *AdAge,* August 5, 2016, http://adage.com/article/digital/ftc-cracking-social-influencers-labeling-promotions/305345/ (accessed March 5, 2017).

94. Priya Anand, "Cyberthieves Have a New Target: Children," *The Wall Street Journal,* February 1, 2016, p. R8.

95. Elizabeth Weise, "Sony Hack Leaves Intriguing Clues," *USA Today,* December 4, 2014, p. 1B.

96. Elizabeth Weise, "Consumers Have to Protect Themselves Online," *USA Today,* May 22, 2014, p. 1B.

97. Jim Zarroli, "U.S. Credit Cards Tackle Fraud with Embedded Chips, but No Pins," *NPR,* January 5, 2015, http://www.npr.org/blogs/alltechconsidered/2015/01/05/375164839/u-s-credit-cards-tackle-fraud-with-embedded-chips-but-no-pins (accessed May 5, 2017).

98. Kevin Shanahan and Mike Hyman, "Motivators and Enablers of SCOURing," *Journal of Business Research* 63 (September–October 2010), pp. 1095–1102.

99. "The Amazons of the Dark Net," *The Economist,* November 1, 2014, pp. 57–58.

100. Erica E. Phillips, "U.S. Officials Chase Counterfeit Goods Online," *The Wall Street Journal,* November 28, 2014, http://www.wsj.com/articles/u-s-officials-chase-counterfeit-goods-online-1417217763 (accessed May 5, 2017).

101. Max Chafkin, "The Case, and the Plan, for the Virtual Company," *Inc.,* April 2010, p. 68.

102. Anthony Balderrama, "Social Media at Work—Bane or Boon?" *CNN,* March 8, 2010, www.cnn.com/2010/LIVING/worklife/03/08/cb.social.media.banned/index.html (accessed April 25, 2016); Emily Glazer, "P&G Curbs Employees Internet Use," *The Wall Street Journal,* April 4, 2012, http://online.wsj.com/article/SB10001424052702304072004577324142847006340.html (accessed April 25, 2016); Ethics Resource Center, *2011 National Business Ethics Survey®: Ethics in Transition* (Arlington, VA: Ethics Resource Center, 2012); Miral Fahmy, "Facebook, YouTube at Work Make Better Employees: Study," *San Francisco Chronicle,* April 2, 2009, www.sanfranciscosentinel.com/?p=21639 (accessed April 25, 2016); Sharon Gaudin, "Study: Facebook Use Cuts Productivity at Work," *Computer World,* July 22, 2009, www.computerworld.com/s/article/9135795/Study_Facebook_use_cuts_productivity_at_work (accessed April 25, 2016); Sharon Gaudin, "Study: 54% of Companies Ban Facebook, Twitter at Work," *Computer World,* October 6, 2009, www.computerworld.com/s/article/9139020/Study_54_of_companies_ban_Facebook_Twitter_at_work (accessed April 25, 2016); Guy Alvarez, Brian Dalton, Joe Lamport, and Kristina Tsamis, "The Social Law Firm," *Above the Law,* http://good2bsocial.com/wp-content/uploads/2013/12/THE-SOCIAL-LAW-FIRM.pdf (accessed June 18, 2014); Cheryl Conner, "Wasting Time at Work: The Epidemic Continues," *Forbes,* July 31, 2015, http://www.forbes.com/sites/cherylsnappconner/2015/07/31/wasting-time-at-work-the-epidemic-continues/#72369bf83ac1 (accessed April 25, 2016); Barbara Siegel, "Social Media in the Workplace: Does It Impact Productivity?" Lake Forest Graduate School of Management, March 28, 2014, http://www.lakeforestmba.edu/blog/social-media-workplace-impact-productivity/ (accessed April 25, 2016).

Credits

Design elements: Part opener, ©Steve Allen/Getty Images; Consider Ethics and Social Responsibility icon, ©Design Pics/PunchStock; Think Globally icon, ©Sheff/Shutterstock; Responding to Business Challenges icon, ©Olivier LeMoal/Shutterstock; Going Green icon, ©Beboy/Shutterstock; Entrepreneurship in Action icon, ©Ruslan Grechka/Shutterstock; Test Prep icon, ©McGraw-Hill Education; Build Your Skills icon, ©Ilya Terentyev/Getty Images; Solve the Dilemma icon, ©Beautyimage/Shutterstock; Build Your Business Plan icon, ©ALMAGAMI/ Shutterstock; See for Yourself Videocase icon, ©MIKHAIL GRACHIKOV/Shutterstock.

Financing the Enterprise

PART **6**

Financing the Enterprise

14 Accounting and Financial Statements

©Amer Ghazzal/Shutterstock

Chapter Outline

Learning Objectives

After reading this chapter, you will be able to:

LO 14-1 Describe the different uses of accounting information.

LO 14-2 Demonstrate the accounting process.

LO 14-3 Examine the various components of an income statement in order to evaluate a firm's "bottom line."

LO 14-4 Interpret a company's balance sheet to determine its current financial position.

LO 14-5 Analyze financial statements, using ratio analysis, to evaluate a company's performance.

LO 14-6 Assess a company's financial position using its accounting statements and ratio analysis.

Enter the World of Business

By the Numbers: Deloitte Excels in the Big Four

When it comes to accounting, the Big Four dominate. What are the Big Four? The Big Four refer to the four largest accounting firms throughout the world: Ernst & Young, KPMG, PricewaterhouseCoopers, and Deloitte. These accounting firms provide a range of services to their clients, including external audit, taxation services, management and business consultancy, and risk assessment and control. The Big Four audit more than 80 percent of all U.S. public companies.

Deloitte was ranked number one in revenue, earning $36.8 billion globally with approximately 244,400 employees in more than 150 countries. Deloitte believes the best measure of success is the impact it makes on the world. It does not want to be known for the size or services that it offers; instead, it wants to make an impact in the world. Deloitte strives to have intelligent people work across more than 20 industry sectors with the purpose "to deliver measurable, lasting results . . . and help lead the way toward a stronger economy and a healthy society."

As part of the Big Four, it is important for Deloitte to continue its long-standing commitment to communities. It offers a pro bono and skills-based volunteering program that are key components to its commitment to youth education, inspiring future leaders, and creating opportunity for veterans. To focus on education, Deloitte's strategy has been to improve college readiness across America. It commits time, talent, and resources to help students persist through high school and college and transition into a great career.

Deloitte is not only highly profitable, but also has a reputation for innovating employees, processes, technologies, governance, and policy. This places it at a competitive advantage with the other Big Four leaders.[1]

Introduction

Although you may cover some of this material in your accounting course, reading this chapter will only strengthen your understanding of accounting. What professors find is that a little duplication and repetition goes a long way in helping the brain retain material. Accounting is the financial "language" that organizations use to record, measure, and interpret all of their financial transactions and records and is very important in business. All businesses—from a small family farm to a giant corporation—use the language of accounting to make sure they track their use of funds, measure profitability, and budget for future expenditures. Nonbusiness organizations such as charities and governments also use accounting to demonstrate to donors and taxpayers how well they use their funds to meet their stated objectives.

This chapter explores the role of accounting in business and its importance in making business decisions. First, we discuss the uses of accounting information and the accounting process. Then, we briefly look at some simple financial statements and accounting tools that are useful in analyzing organizations worldwide.

The Nature of Accounting

LO 14-1

Describe the different uses of accounting information.

accounting
the recording, measurement, and interpretation of financial information.

Simply stated, **accounting** is the recording, measurement, and interpretation of financial information. Large numbers of people and institutions, both within and outside businesses, use accounting tools to evaluate organizational operations. The Financial Accounting Standards Board has been setting the principles and standards of financial accounting and reporting in the private sector since 1973. Its mission is to establish and improve standards of financial accounting and reporting for the guidance and education of the public, including issuers, auditors, and users of financial information. However, accounting scandals at the turn of the last century resulted when many accounting firms and businesses failed to abide by generally accepted accounting principles, or GAAP. Consequently, the federal government has taken a greater role in making rules, requirements, and policies for accounting firms and businesses through the Securities and Exchange Commission's (SEC) Public Company Accounting Oversight Board (PCAOB). For example, the PCAOB has the ability to file a disciplinary order against a firm or individual that either temporarily or permanently prohibits that firm or individual from practicing accounting. To better understand the importance of accounting, we must first understand who prepares accounting information and how it is used.

Accountants

Many of the functions of accounting are carried out by public or private accountants.

certified public accountant (CPA)
an individual who has been state certified to provide accounting services ranging from the preparation of financial records and the filing of tax returns to complex audits of corporate financial records.

Public Accountants. Individuals and businesses can hire a **certified public accountant (CPA)**, an individual who has been certified by the state in which he or she practices to provide accounting services ranging from the preparation of financial records and the filing of tax returns to complex audits of corporate financial records. Certification gives a public accountant the right to express, officially, an unbiased opinion regarding the accuracy of the client's financial statements. Most public accountants are either self-employed or members of large public accounting firms such as Ernst & Young, KPMG, Deloitte, and PricewaterhouseCoopers, together referred to as "the Big Four." In addition, many CPAs work for one of the second-tier accounting firms that are much smaller than the Big Four. Table 14.1 shows the number of employees of the top 10 firms. Vault.com uses a weighted ranking system

Rank	Firm	Employees	Score	Location
I	PwC (PricewaterhouseCoopers) LLP	58,133	8.497	New York, NY
2	Deloitte LLP	80,000	8.327	New York, NY
3	Ernst & Young (EY) LLP	39,400	8.257	New York, NY
4	KPMG LLP	31,000	8.003	New York, NY
5	Grant Thornton LLP	7,530	7.715	Chicago, IL
6	BDO USA LLP	5,395	7.200	Chicago, IL
7	Plante Moran	2,200	7.131	Southfield, MI
8	Moss Adams LLP	2,500	7.072	Seattle, WA
9	Crowe LLP	4,100	6.971	Chicago, IL
IO	RSM US LLP	9,560	6.943	Chicago, IL

TABLE 14.1
Prestige Rankings of Accounting Firms in the U.S.

Note: U.S. only, does not include international operations.

Source: "Vault Accounting 50," Vault.com, http://www.vault.com/company-rankings/accounting/vault-accounting-50/ (accessed April 27, 2018).

based on survey results to create a score that represents work-life quality issues to reflect the prestige of the firm.

Although there will always be companies and individual money managers who can successfully hide illegal or misleading accounting practices for a while, eventually they are exposed. After the accounting scandals of Enron and Worldcom in the early 2000s, Congress passed the Sarbanes-Oxley Act, which required firms to be more rigorous in their accounting and reporting practices. Sarbanes-Oxley made accounting firms separate their consulting and auditing businesses and punished corporate executives with potential jail sentences for inaccurate, misleading, or illegal accounting statements. This seemed to reduce the accounting errors among nonfinancial companies, but declining housing prices exposed some of the questionable practices by banks and mortgage companies. Only five years after the passage of the Sarbanes-Oxley Act, the world experienced a financial crisis starting in 2008—part of which was due to excessive risk taking and inappropriate accounting practices. Many banks failed to understand the true state of their financial health. Banks also developed questionable lending practices and investments based on subprime mortgages made to individuals who had poor credit. When housing prices declined and people suddenly found that they owed more on their mortgages than their homes were worth, they began to default. To prevent a depression, the government intervened and bailed out some of the United States' largest banks. Congress passed the Dodd-Frank Act in 2010 to strengthen the oversight of financial institutions. This act gave the Federal Reserve Board the task of implementing the legislation. This legislation limits the types of assets commercial banks can buy; the amount of capital they must maintain; and the use of derivative instruments such as options, futures, and structured investment products. However, at the Economics Club of New York on February 16, 2017, Alan Greenspan, the former chair of the Federal Reserve Board, was quoted as saying the Dodd-Frank Act is one of the worst pieces of legislation he has ever seen.

ERNST & YOUNG

Ernst & Young is part of the Big Four, or the four largest international accounting firms. The other three are KPMG, PricewaterhouseCoopers, and Deloitte.

©Lars A. Niki

DID YOU KNOW? Corporate fraud costs are estimated at $4 trillion annually.[3]

---

private accountants
accountants employed by large corporations, government agencies, and other organizations to prepare and analyze their financial statements.

certified management accountants (CMAs)
private accountants who, after rigorous examination, are certified by the Institute of Management Accountants and who have some managerial responsibility.

A growing area for public accountants is *forensic accounting,* which is accounting that is fit for legal review. It involves analyzing financial documents in search of fraudulent entries or financial misconduct. Functioning as much like detectives as accountants, forensic accountants have been used since the 1930s. Many auditing firms are expanding their forensic or fraud-detection services. Additionally, many forensic accountants root out evidence of "cooked books" for federal agencies like the Federal Bureau of Investigation or the Internal Revenue Service. The Association of Certified Fraud Examiners, which certifies accounting professionals as *certified fraud examiners (CFEs),* has grown to more than 75,000 members.[2]

Private Accountants. Large corporations, government agencies, and other organizations may employ their own **private accountants** to prepare and analyze their financial statements. With titles such as controller, tax accountant, or internal auditor, private accountants are deeply involved in many of the most important financial decisions of the organizations for which they work. Private accountants can be CPAs and may become **certified management accountants (CMAs)** by passing a rigorous examination by the Institute of Management Accountants.

Accounting or Bookkeeping?

The terms *accounting* and *bookkeeping* are often mistakenly used interchangeably. Much narrower and far more mechanical than accounting, bookkeeping is typically limited to the routine, day-to-day recording of business transactions. Bookkeepers are responsible for obtaining and recording the information that accountants require to analyze a firm's financial position. They generally require less training than accountants. Accountants, on the other hand, usually complete course work beyond their basic four- or five-year college accounting degrees. This additional training allows accountants not only to record financial information, but to understand, interpret, and even develop the sophisticated accounting systems necessary to classify and analyze complex financial information.

The Uses of Accounting Information

Accountants summarize the information from a firm's business transactions in various financial statements (which we'll look at in a later section of this chapter) for a variety of stakeholders, including managers, investors, creditors, and government agencies. Many business failures may be directly linked to ignorance of the information "hidden" inside these financial statements. Likewise, most business successes can be traced to informed managers who understand the consequences of their decisions. While maintaining and even increasing short-run profits is desirable, the failure to plan sufficiently for the future can easily lead an otherwise successful company to insolvency and bankruptcy court.

Basically, managers and owners use financial statements (1) to aid in internal planning and control and (2) for external purposes such as reporting to the Internal Revenue Service, stockholders, creditors, customers, employees, and other interested parties. Figure 14.1 shows some of the users of the accounting information generated by organizations and other stakeholders.

Internal Uses.   **Managerial accounting** refers to the internal use of accounting statements by managers in planning and directing the organization's activities. Perhaps management's greatest single concern is **cash flow,** the movement of money through an organization over a daily, weekly, monthly, or yearly basis. Obviously, for any business to succeed, it needs to generate enough cash to pay its bills as they fall due. However, it is not at all unusual for highly successful and rapidly growing companies to struggle to make payments to employees, suppliers, and lenders because of an inadequate cash flow. One common reason for a so-called cash crunch, or shortfall, is poor managerial planning.

Managerial accountants also help prepare an organization's **budget,** an internal financial plan that forecasts expenses and income over a set period of time. It is not unusual for an organization to prepare separate daily, weekly, monthly, and yearly budgets. Think of a budget as a financial map, showing how the company expects to move from Point A to Point B over a specific period of time. While most companies prepare *master budgets* for the entire firm, many also prepare budgets for smaller segments of the organization such as divisions, departments, product lines, or projects. "Top-down" master budgets begin at the upper management level and filter down to the individual department level, while "bottom-up" budgets start at the department or project level and are combined at the chief executive's office. Generally, the larger and more rapidly growing an organization, the greater will be the likelihood that it will build its master budget from the ground up.

Regardless of focus, the principal value of a budget lies in its breakdown of cash inflows and outflows. Expected operating expenses (cash outflows such as wages, materials costs, and taxes) and operating revenues (cash inflows in the form of payments from

managerial accounting
the internal use of accounting statements by managers in planning and directing the organization's activities.

cash flow
the movement of money through an organization over a daily, weekly, monthly, or yearly basis.

budget
an internal financial plan that forecasts expenses and income over a set period of time.

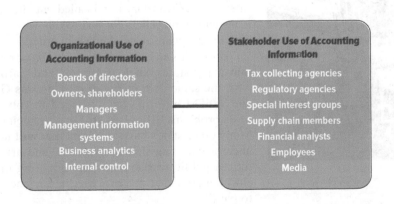

FIGURE 14.1
The Users of Accounting Information

Source: Adapted from *Principles of Accounting,* 4th edition. Houghton Mifflin Company, 1990. Authors: Belverd E. Needles, Henry R. Anderson, and James C. Caldwell.

UNITED STATES
SECURITIES AND EXCHANGE COMMISSION
Washington, D.C. 20549

FORM 10-K

☒ ANNUAL REPORT PURSUANT TO SECTION 13 OR 15(d) OF THE SECURITIES EXCHANGE ACT OF 1934

For the Fiscal Year Ended June 30, 2018

OR

☐ TRANSITION REPORT PURSUANT TO SECTION 13 OR 15(d) OF THE SECURITIES EXCHANGE ACT OF 1934

For the Transition Period From _____ to _____

Commission File Number 001-37845

MICROSOFT CORPORATION

WASHINGTON	91-1144442
(STATE OF INCORPORATION)	(I.R.S. ID)

ONE MICROSOFT WAY, REDMOND, WASHINGTON 98052-6399

(425) 882-8080

www.microsoft.com/investor

Securities registered pursuant to Section 12(b) of the Act:

COMMON STOCK, $0.00000625 par value per share   NASDAQ

Securities registered pursuant to Section 12(g) of the Act:

NONE

The annual report is a summary of the firm's financial information, products, and growth plans for owners and potential investors. Many investors look at a firm's annual report to determine how well the company is doing financially.

Source: U.S. Securities and Exchange Commission

annual report
summary of a firm's financial information, products, and growth plans for owners and potential investors.

Illinois has found itself in a financial crisis due to debt overload.

©Bilgin Sasmaz/Anadolu Agency/Getty Images

customers) over a set period of time are carefully forecast and subsequently compared with actual results. Deviations between the two serve as a "trip wire" or "feedback loop" to launch more detailed financial analyses in an effort to pinpoint trouble spots and opportunities.

External Uses. Managers also use accounting statements to report the business's financial performance to outsiders. Such statements are used for filing income taxes, obtaining credit from lenders, and reporting results to the firm's stockholders. They become the basis for the information provided in the official corporate **annual report,** a summary of the firm's financial information, products, and growth plans for owners and potential investors. While frequently presented between slick, glossy covers, the single most important component of an annual report is the signature of a certified public accountant attesting that the required financial statements are an accurate reflection of the underlying financial condition of the firm. Financial statements meeting these conditions are termed *audited.* The primary external users of audited accounting information are government agencies, stockholders and potential investors, lenders, suppliers, and employees.

During the global financial crisis, it was discovered that Greece had been engaging in deceptive accounting practices using financial techniques to hide massive amounts of debt from its public balance sheets. Eventually, the markets figured out the country might not be able to pay off its creditors. The European Union and the International Monetary Fund bailed out Greece with loans and credit relief, but tied to this was the message to "get your financial house in order."

To top this off, *The New York Times* reported that many states, such as Illinois and California, have the same debt overload problems as Greece, and while California has managed to solve most of its problems, Illinois is still in a financial mess. Many other states have "budgets that will not balance, accounting that masks debt, and armies of retired public workers who are counting on pension benefits that are proving harder and harder to pay."[4]

Entrepreneurship in Action

Perfectly Al Dente: How This Pasta Company Handles Its Finances

Al Dente® Pasta Company
Founder: Monique Deschaine
Founded: 1981, in Whitmore Lake, Michigan
Success: The company's Al Dente Perfect Pasta™ and Monique's Sumptuous Sauces™ are now sold nationwide.

Monique Deschaine dreamed of making the perfect pasta. She turned that dream into a reality in 1981 by launching the Al Dente® Pasta Company. Deschaine befriended celebrated Italian chef and cookbook author Marcella Hazan, using Hazan's restaurant kitchen after-hours as her pasta-making location. Dedication to the recipe and technique convinced Deschaine to market to niche consumers looking for perfect al dente pasta rather than competing with large pasta companies. Using word-of-mouth advertising, Al Dente Pasta managed to become a national player in the pasta market for its gourmet pastas that cook within three minutes.

In the beginning, Deschaine assumed the firm's accounting responsibilities. She was responsible for monitoring and responding to growth by keeping the company's financial statements up-to-date according to generally accepted accounting principles (GAAP). As the pasta's popularity has grown, so has the company. Al Dente now sells products nationwide and has been featured on The Food Network. Financial information is distributed throughout the company, allowing managers to examine strengths and weaknesses. Through Deschaine's institution of good accounting systems and proper management, Al Dente Pasta Company is achieving sustainable profits and growth.[5]

Critical Thinking Questions

1. Why did Monique Deschaine need to possess good accounting skills in her management of Al Dente Pasta?
2. Why is it important for Monique Deschaine to use GAAP? What could happen if she didn't?
3. As companies grow, do you think it is a good idea for the founders to continue assuming their firms' accounting responsibilities? Why or why not?

Financial statements evaluate the return on stockholders' investments and the overall quality of the firm's management team. As a result, poor performance, as documented in the financial statements, often results in changes in top management.

Potential investors study the financial statements in a firm's annual report to determine whether the company meets its investment requirements and whether the returns from a given firm are likely to compare favorably with other similar companies.

Banks and other lenders look at financial statements to determine a company's ability to meet current and future debt obligations if a loan or credit is granted. To determine this ability, a short-term lender examines a firm's cash flow to assess its ability to repay a loan quickly with cash generated from sales. A long-term lender is more interested in the company's profitability and indebtedness to other lenders.

Labor unions and employees use financial statements to establish reasonable expectations for salary and other benefit requests. Just as firms experiencing record profits are likely to face added pressure to increase employee wages, so too are employees unlikely to grant employers wage and benefit concessions without considerable evidence of financial distress.

As one of the biggest banks in the United States, Bank of America specializes in banking, mortgage, and financial services. The data provided can be used in financial statements.

©McGraw-Hill Education, Andrew Resek photographer

FIGURE 14.2

The Accounting Equation and Double-Entry Bookkeeping for Anna's Flowers

	Assets			= Liabilities	+	Owners' Equity
	Cash	Equipment	Inventory	Debts to suppliers	Loans	Equity
Cash invested by Anna	$2,500.00					$2,500.00
Loan from SBA	$5,000.00				$5,000.00	
Purchase of furnishings	−$3,000.00	$3,000.00				
Purchase of inventory	−$2,000.00		$2,000.00			
Purchase of roses			$325.00	$325.00		
First month sales	$2,000.00		−$1,500.00			$500.00
Totals	$4,500.00	$3,000.00	$825.00	$325.00	$5,000.00	$3,000.00

$8,325 = $5,325 + $3,000

$8,325 Assets = $8,325 (Liabilities + Owners' Equity)

The Accounting Process

Many view accounting as a primary business language. It is of little use, however, unless you know how to "speak" it. Fortunately, the fundamentals—the accounting equation and the double-entry bookkeeping system—are not difficult to learn. These two concepts serve as the starting point for all currently accepted accounting principles.

The Accounting Equation

Accountants are concerned with reporting an organization's assets, liabilities, and owners' equity. To help illustrate these concepts, consider a hypothetical floral shop called Anna's Flowers, owned by Anna Rodriguez. A firm's economic resources, or items of value that it owns, represent its **assets**—cash, inventory, land, equipment, buildings, and other tangible and intangible things. The assets of Anna's Flowers include counters, refrigerated display cases, flowers, decorations, vases, cards, and other gifts, as well as something known as "goodwill," which in this case is Anna's reputation for preparing and delivering beautiful floral arrangements on a timely basis. **Liabilities,** on the other hand, are debts the firm owes to others. Among the liabilities of Anna's Flowers are a loan from the Small Business Administration and money owed to flower suppliers and other creditors for items purchased. The **owners' equity** category contains all of the money that has ever been contributed to the company that never has to be paid back. The funds can come from investors who have given money or assets to the company, or it can come from past profitable operations. In the case of Anna's Flowers, if Anna were to sell off, or liquidate, her business, any money left over after selling all the shop's assets and paying off its

assets
a firm's economic resources, or items of value that it owns, such as cash, inventory, land, equipment, buildings, and other tangible and intangible things.

liabilities
debts that a firm owes to others.

owners' equity
equals assets minus liabilities and reflects historical values.

liabilities would comprise her owners' equity. The relationship among assets, liabilities, and owners' equity is a fundamental concept in accounting and is known as the **accounting equation:**

$$\text{Assets} = \text{Liabilities} + \text{Owners' equity}$$

Double-Entry Bookkeeping

Double-entry bookkeeping is a system of recording and classifying business transactions in separate accounts in order to maintain the balance of the accounting equation. Returning to Anna's Flowers, suppose Anna buys $325 worth of roses on credit from the Antique Rose Emporium to fill a wedding order. When she records this transaction, she will list the $325 as a liability or a debt to a supplier. At the same time, however, she will also record $325 worth of roses as an asset in an account

The owners' equity portion of this florist's balance sheet includes the money she has put into the firm.

©Jose Luis Pelaez Inc/Blend Images LLC

known as "inventory." Because the assets and liabilities are on different sides of the accounting equation, Anna's accounts increase in total size (by $325) but remain in balance:

$$\text{Assets} = \text{Liabilities} + \text{Owners' equity}$$
$$\$325 = \$325$$

Thus, to keep the accounting equation in balance, each business transaction must be recorded in two separate accounts.

In the final analysis, all business transactions are classified as assets, liabilities, or owners' equity. However, most organizations further break down these three accounts to provide more specific information about a transaction. For example, assets may be broken down into specific categories such as cash, inventory, and equipment, while liabilities may include bank loans, supplier credit, and other debts.

Figure 14.2 shows how Anna used the double-entry bookkeeping system to account for all of the transactions that took place in her first month of business. These transactions include her initial investment of $2,500, the loan from the Small Business Administration, purchases of equipment and inventory, and the purchase of roses on credit. In her first month of business, Anna generated revenues of $2,000 by selling $1,500 worth of inventory. Thus, she deducts, or (in accounting notation that is appropriate for assets) *credits,* $1,500 from inventory and adds, or *debits,* $2,000 to the cash account. The difference between Anna's $2,000 cash inflow and her $1,500 outflow is represented by a credit to owners' equity because it is money that belongs to her as the owner of the flower shop.

accounting equation
assets equal liabilities plus owners' equity.

double-entry bookkeeping
a system of recording and classifying business transactions that maintains the balance of the accounting equation.

accounting cycle
the four-step procedure of an accounting system: examining source documents, recording transactions in an accounting journal, posting recorded transactions, and preparing financial statements.

The Accounting Cycle

In any accounting system, financial data typically pass through a four-step procedure sometimes called the **accounting cycle.** The steps include examining source documents, recording transactions in an accounting journal, posting recorded transactions, and preparing financial statements. Figure 14.3 shows how Anna works through them. Traditionally, all of these steps were performed using paper, pencils, and erasers (lots of erasers!), but today, the process is often fully computerized.

FIGURE 14.3   The Accounting Process for Anna's Flowers

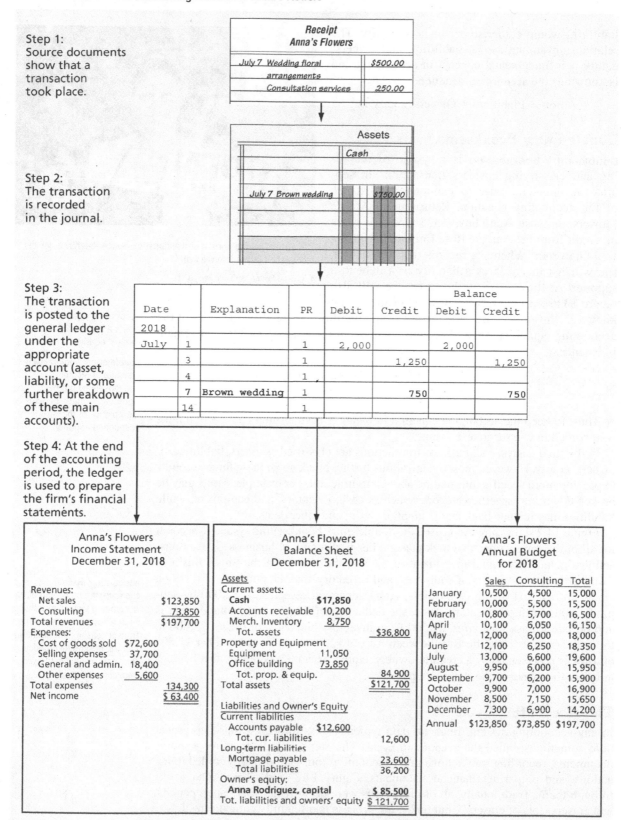

Step 1:
Source documents
show that a
transaction
took place.

Receipt Anna's Flowers	
July 7 Wedding floral	$500.00
arrangements	
Consultation services	250.00

Step 2:
The transaction
is recorded
in the journal.

Assets		
	Cash	
July 7 Brown wedding		$750.00

Step 3:
The transaction
is posted to the
general ledger
under the
appropriate
account (asset,
liability, or some
further breakdown
of these main
accounts).

						Balance	
Date		Explanation	PR	Debit	Credit	Debit	Credit
2018							
July	1		1	2,000		2,000	
	3		1		1,250		1,250
	4		1				
	7	Brown wedding	1		750		750
	14		1				

Step 4: At the end
of the accounting
period, the ledger
is used to prepare
the firm's financial
statements.

Anna's Flowers
Income Statement
December 31, 2018

Revenues:		
Net sales		$123,850
Consulting		73,850
Total revenues		$197,700
Expenses:		
Cost of goods sold	$72,600	
Selling expenses	37,700	
General and admin.	18,400	
Other expenses	5,600	
Total expenses		134,300
Net income		$ 63,400

Anna's Flowers
Balance Sheet
December 31, 2018

Assets		
Current assets:		
Cash	**$17,850**	
Accounts receivable	10,200	
Merch. Inventory	8,750	
Tot. assets		$36,800
Property and Equipment		
Equipment	11,050	
Office building	73,850	
Tot. prop. & equip.		84,900
Total assets		$121,700
Liabilities and Owner's Equity		
Current liabilities		
Accounts payable	$12,600	
Tot. cur. liabilities		12,600
Long-term liabilities		
Mortgage payable		23,600
Total liabilities		36,200
Owner's equity:		
Anna Rodriguez, capital		$ 85,500
Tot. liabilities and owners' equity		$ 121,700

Anna's Flowers
Annual Budget
for 2018

	Sales	Consulting	Total
January	10,500	4,500	15,000
February	10,000	5,500	15,500
March	10,800	5,700	16,500
April	10,100	6,050	16,150
May	12,000	6,000	18,000
June	12,100	6,250	18,350
July	13,000	6,600	19,600
August	9,950	6,000	15,950
September	9,700	6,200	15,900
October	9,900	7,000	16,900
November	8,500	7,150	15,650
December	7,300	6,900	14,200
Annual	$123,850	$73,850	$197,700

Step One: Examine Source Documents. Like all good managers, Anna Rodriguez begins the accounting cycle by gathering and examining source documents—checks, credit card receipts, sales slips, and other related evidence concerning specific transactions.

Step Two: Record Transactions. Next, Anna records each financial transaction in a journal, which is basically just a time-ordered list of account transactions. While most businesses keep a general journal in which all transactions are recorded, some classify transactions into specialized journals for specific types of transaction accounts.

Step Three: Post Transactions. Anna next transfers the information from her journal into a ledger, a book or computer program with separate files for each account. This process is known as *posting*. At the end of the accounting period (usually yearly, but occasionally quarterly or monthly), Anna prepares a *trial balance,* a summary of the balances of all the accounts in the general ledger. If, upon totaling, the trial balance doesn't balance (that is, the accounting equation is not in balance), Anna or her accountant must look for mistakes (typically an error in one or more of the ledger entries) and correct them. If the trial balance is correct, the accountant can then begin to prepare the financial statements.

Step Four: Prepare Financial Statements. The information from the trial balance is also used to prepare the company's financial statements. In the case of public corporations and certain other organizations, a CPA must *attest,* or certify, that the organization followed generally accepted accounting principles in preparing the financial statements. When these statements have been completed, the organization's books are "closed," and the accounting cycle begins anew for the next accounting period.

Financial Statements

The end result of the accounting process is a series of financial statements. The income statement, the balance sheet, and the statement of cash flows are the best-known examples of financial statements. They are provided to stockholders and potential investors in a firm's annual report as well as to other relevant outsiders such as creditors, government agencies, and the Internal Revenue Service.

It is important to recognize that not all financial statements follow precisely the same format. The fact that different organizations generate income in different ways suggests that when it comes to financial statements, one size definitely does not fit all. Manufacturing firms, service providers, and nonprofit organizations each use a different set of accounting principles or rules upon which the public accounting profession has agreed. As we have already mentioned, these are sometimes referred to as *generally accepted accounting principles (GAAP).* Each country has a different set of rules that the businesses within that country are required to use for their accounting process and financial statements. However, a number of countries have adopted a standard set of accounting principles known as International Financial Reporting Standards. The United States has discussed adopting these standards to create a more standardized system of reporting for global investors. Moreover, as is the case in many other disciplines, certain concepts have more than one name. For example, *sales* and *revenues* are often interchanged, as are *profits, income,* and *earnings.* Table 14.2 lists a few common equivalent terms that should help you decipher their meaning in accounting statements.

 connect

Need help understanding the accounting cycle? Visit your Connect ebook video tab for a brief animated explanation.

journal
a time-ordered list of account transactions.

ledger
a book or computer file with separate sections for each account.

TABLE 14.2
Equivalent Terms in
Accounting

Term	Equivalent Term
Revenues	Sales
	Goods or services sold
Gross profit	Gross income
	Gross earnings
Operating income	Operating profit
	Earnings before interest and taxes (EBIT)
	Income before interest and taxes (IBIT)
Income before taxes (IBT)	Earnings before taxes (EBT)
	Profit before taxes (PBT)
Net income (NI)	Earnings after taxes (EAT)
	Profit after taxes (PAT)
Income available to common stockholders	Earnings available to common stockholders

income statement
a financial report that shows an organization's profitability over a period of time—month, quarter, or year.

revenue
the total amount of money received from the sale of goods or services, as well as from related business activities.

cost of goods sold
the amount of money a firm spent to buy or produce the products it sold during the period to which the income statement applies.

The Income Statement

The question, "What's the bottom line?" derives from the income statement, where the bottom line shows the overall profit or loss of the company after taxes. Thus, the **income statement** is a financial report that shows an organization's profitability over a period of time, be that a month, quarter, or year. By its very design, the income statement offers one of the clearest possible pictures of the company's overall revenues and the costs incurred in generating those revenues. Other names for the income statement include profit and loss (P&L) statement or operating statement. A sample income statement in word form with line-by-line explanations is presented in Table 14.3, while Table 14.4 presents the income statement of Microsoft for the years 2016, 2017 and 2018. The income statement indicates the firm's profitability or income (the bottom line), which is derived by subtracting the firm's expenses from its revenues.

Revenue.   **Revenue** is the total amount of money received (or promised) from the sale of goods or services, as well as from other business activities such as the rental of property and investments. Nonbusiness entities typically obtain revenues through donations from individuals and/or grants from governments and private foundations. One of the controversies in accounting has been when a business should recognize revenue. For instance, should an organization book revenue during a project or after the project is completed? Differences in revenue recognition have caused similar organizations to book different accounting results. Generally accepted practice is that firms should book revenue when "it satisfies a performance obligation by transferring a promised good or service to a customer."[6]

For most manufacturing and retail concerns, the next major item included in the income statement is the **cost of goods sold,** the amount of money the firm spent (or promised to spend) to buy and/or produce the products it sold during the accounting period. This figure may be calculated as follows:

Cost of goods sold = Beginning inventory + Interim purchases − Ending inventory

TABLE 14.3   Sample Income Statement

Company Name for the Year Ended December 31	
Revenues (sales)	Total dollar amount of products sold (includes income from other business services such as rental-lease income and interest income).
Less: Cost of goods sold	The cost of producing the goods and services, including the cost of labor and raw materials as well as other expenses associated with production.
Gross profit	The income available after paying all expenses of production.
Less: Selling and administrative expense	The cost of promoting, advertising, and selling products as well as the overhead costs of managing the company. This includes the cost of management and corporate staff. One noncash expense included in this category is depreciation, which approximates the decline in the value of plant and equipment assets due to use over time. In most accounting statements, depreciation is not separated from selling and administrative expenses. However, financial analysts usually create statements that include this expense.
Income before interest and taxes (operating income or EBIT)	This line represents all income left over after operating expenses have been deducted. This is sometimes referred to as operating income because it represents all income after the expenses of operations have been accounted for. Occasionally, this is referred to as EBIT, or earnings before interest and taxes.
Less: Interest expense	Interest expense arises as a cost of borrowing money. This is a financial expense rather than an operating expense and is listed separately. As the amount of debt and the cost of debt increase, so will the interest expense. This covers the cost of both short-term and long-term borrowing.
Income before taxes (earnings before taxes—EBT)	The firm will pay a tax on this amount. This is what is left of revenues after subtracting all operating costs, depreciation costs, and interest costs.
Less: Taxes	The tax rate is specified in the federal tax code.
Net income	This is the amount of income left after taxes. The firm may decide to retain all or a portion of the income for reinvestment in new assets. Whatever it decides not to keep will usually be paid out in dividends to its stockholders.
Less: Preferred dividends	If the company has preferred stockholders, they are first in line for dividends. That is one reason their stock is called "preferred."
Income to common stockholders	This is the income left for the common stockholders. If the company has a good year, there may be a lot of income available for dividends. If the company has a bad year, income could be negative. The common stockholders are the ultimate owners and risk takers. They have the potential for very high or very poor returns because they get whatever is left after all other expenses.
Earnings per share	Earnings per share is found by taking the income available to the common stockholders and dividing by the number of shares of common stock outstanding. This is income generated by the company for each share of common stock.

TABLE 14.4
Microsoft Corporation
Consolidated Statement
of Earnings (in millions
except per share
amounts)

(In millions, except per share amounts)

Year Ended June 30,	2018	2017	2016
Revenue:			
Product	$ 64,497	$63,811	$67,336
Service and other	45,863	32,760	23,818
Total revenue	110,360	96,571	91,154
Cost of revenue:			
Product	15,420	15,175	17,880
Service and other	22,933	19,086	14,900
Total cost of revenue	38,353	34,261	32,780
Gross margin	72,007	62,310	58,374
Research and development	14,726	13,037	11,988
Sales and marketing	17,469	15,461	14,635
General and administrative	4,754	4,481	4,563
Impairment and restructuring	0	306	1,110
Operating income	35,058	29,025	26,078
Other income (expense), net	1,416	876	(439)
Income before income taxes	36,474	29,901	25,639
Provision for income taxes	19,903	4,412	5,100
Net income	$ 16,571	$25,489	$20,539
Earnings per share:			
Basic	$ 2.15	$ 3.29	$ 2.59
Diluted	$ 2.13	$ 3.25	$ 2.56
Weighted average shares outstanding:			
Basic	7,700	7,746	7,925
Diluted	7,794	7,832	8,013
Cash dividends declared per common share	$ 1.68	$ 1.56	$ 1.44

Let's say that Anna's Flowers began an accounting period with an inventory of goods for which it paid $5,000. During the period, Anna bought another $4,000 worth of goods, giving the shop a total inventory available for sale of $9,000. If, at the end of the accounting period, Anna's inventory was worth $5,500, the cost of goods sold during the period would have been $3,500 ($5,000 + $4,000 − $5,500 = $3,500). If Anna had total revenues of $10,000 over the same period of time, subtracting the cost of goods sold ($3,500) from the total revenues of $10,000 yields the store's **gross income** or **profit** (revenues minus the cost of goods sold required to generate the revenues): $6,500. The same process occurs at Microsoft. As indicated in Table 14.4, the cost of goods sold was more than $38 billion in 2018. Notice that Microsoft calls it cost of revenues, rather than cost of goods sold.

gross income
revenues minus the cost of goods sold required to generate the revenues.

profit
the difference between what it costs to make and sell a product and what a customer pays for it.

Expenses. Expenses are the costs incurred in the day-to-day operations of an organization. Three common expense accounts shown on income statements are (1) selling, general, and administrative expenses; (2) research, development, and engineering expenses; and (3) interest expenses (remember that the costs directly attributable to selling goods or services are included in the cost of goods sold). Selling expenses include advertising and sales salaries. General and administrative expenses include salaries of executives and their staff and the costs of owning and maintaining the general office. Research and development costs include scientific, engineering, and marketing personnel and the equipment and information used to design and build prototypes and samples. Interest expenses include the direct costs of borrowing money.

expenses
the costs incurred in the day-to-day operations of an organization.

The number and type of expense accounts vary from organization to organization. Included in the general and administrative category is a special type of expense known as **depreciation,** the process of spreading the costs of long-lived assets such as buildings and equipment over the total number of accounting periods in which they are expected to be used. Consider a manufacturer that purchases a $100,000 machine expected to last about 10 years. Rather than showing an expense of $100,000 in the first year and no expense for that equipment over the next nine years, the manufacturer is allowed to report depreciation expenses of $10,000 per year in each of the next 10 years because that better matches the cost of the machine to the years the machine is used. Each time this depreciation is "written off" as an expense, the book value of the machine is also reduced by $10,000. The fact that the equipment has a zero value on the firm's balance sheet when it is fully depreciated (in this case, after 10 years) does not necessarily mean that it can no longer be used or is economically worthless. Indeed, in some industries, machines used every day have been reported as having no book value whatsoever for more than 30 years.

depreciation
the process of spreading the costs of long-lived assets such as buildings and equipment over the total number of accounting periods in which they are expected to be used.

Net Income. Net income (or net earnings) is the total profit (or loss) after all expenses, including taxes, have been deducted from revenue. Generally, accountants divide profits into individual sections such as operating income and earnings before interest and taxes. Like most companies, Microsoft presents not only the current year's results, but also the previous two years' income statements to permit comparison of performance from one period to another.

net income
the total profit (or loss) after all expenses, including taxes, have been deducted from revenue; also called net earnings.

The Tax Cut and Jobs Act of 2017 became effective in 2018, causing Microsoft's taxes for 2018 to increase and net income to decrease. The U.S. switched to a territorial tax consistent with most developed countries where income is taxed where it is earned. The Tax Act of 2017 included a provision that all overseas income that had not been repatriated to the U.S. and taxed by the U.S. would now be taxed at 15.5 percent as a one-time transaction. The law also reduced the statutory rate on corporate

Walmart is the largest company in the United States, with more than $485 billion in revenue.

©Quality HD/Shutterstock

income from 35 percent to 21 percent. The one-time impact was to increase Microsoft's taxes by $13.7 billion. So, the decline in net income you observe is not an indication that Microsoft's business has suffered, just a one-time hit to cash flow. The impact of the tax changes will show up in the statement of cash flow and the ratio analysis later in the chapter.

Temporary Nature of the Income Statement Accounts. Companies record their operational activities in the revenue and expense accounts during an accounting period. Gross profit, earnings before interest and taxes, and net income are the results of calculations made from the revenues and expenses accounts; they are not actual accounts. At the end of each accounting period, the dollar amounts in all the revenue and expense accounts are moved into an account called "Retained Earnings," one of the owners' equity accounts. Revenues increase owners' equity, while expenses decrease it. The resulting change in the owners' equity account is exactly equal to the net income. This shifting of dollar values from the revenue and expense accounts allows the firm to begin the next accounting period with zero balances in those accounts. Zeroing out the balances enables a company to count how much it has sold and how many expenses have been incurred during a period of time. The basic accounting equation (Assets = Liabilities + Owners' equity) will not balance until the revenue and expense account balances have been moved or "closed out" to the owners' equity account.

One final note about income statements: You may remember that corporations may choose to make cash payments called dividends to shareholders out of their net earnings. When a corporation elects to pay dividends, it decreases the cash account (in the assets category of the balance sheet) as well as a capital account (in the owners' equity category of the balance sheet). During any period of time, the owners' equity account may change because of the sale of stock (or contributions/withdrawals by owners), the net income or loss, or the dividends paid.

The Balance Sheet

LO 14-4

Interpret a company's balance sheet to determine its current financial position.

balance sheet
a "snapshot" of an organization's financial position at a given moment.

The second basic financial statement is the **balance sheet,** which presents a "snapshot" of an organization's financial position at a given moment. As such, the balance sheet indicates what the organization owns or controls and the various sources of the funds used to pay for these assets, such as bank debt or owners' equity.

The balance sheet takes its name from its reliance on the accounting equation: Assets *must* equal liabilities plus owners' equity. Table 14.5 provides a sample balance sheet in word form with line-by-line explanations. Unlike the income statement, the balance sheet does not represent the result of transactions completed over a specified accounting period. Instead, the balance sheet is, by definition, an accumulation of all financial transactions conducted by an organization since its founding. Following long-established traditions, items on the balance sheet are listed on the basis of their original cost less accumulated depreciation, rather than their present values.

Balance sheets are often presented in two different formats. The traditional balance sheet format placed the organization's assets on the left side and its liabilities

Stranded Assets Pose a Financial and Environmental Risk

The term *stranded assets* is becoming more popular in financial reporting, especially for the oil and gas industry. Stranded assets are assets that are not recoverable. These assets come with a high environmental price and have the potential to be left unused if tougher environmental regulations are passed. For the oil and gas industry, this may mean more untapped oil and gas reserves. Stranded assets create significant risk for companies and their shareholders. According to the CDP—a group that gathers environmental data for shareholders—deforestation alone could result in $1 trillion of stranded assets for public companies, particularly those in the oil, lumber, and cattle industries.

As a result, more companies have begun to report environmental risks in their financial reports. One survey found that 27 percent of the risks identified in company sustainability reports were being reported in financial reports as well. Attracting (or appeasing) investors is one of the largest drivers of this trend. Additionally, if climate risk could potentially impact a firm's bottom line or the value of its assets, failing to report this information could be seen as misleading. For instance, the Securities and Exchange Commission investigated whether ExxonMobil had valued some of its assets correctly to account for increasing environmental regulations. This shift toward greener accountability has implications for the possible development of generally accepted accounting principles for sustainability concerns.[7]

Critical Thinking Questions

1. Should companies include environmental costs, or "stranded assets," into its accounting statements?
2. How might climate change and other environmental risks negatively impact a firm's assets or bottom line?
3. What might be some advantages to reporting and monitoring environmental costs?

TABLE 14.5 Sample Balance Sheet

Typical Company	December 31
Assets	This is the major category for all physical, monetary, or intangible goods that have some dollar value.
Current assets	Assets that are either cash or are expected to be turned into cash within the next 12 months.
Cash	Cash or checking accounts.
Marketable securities	Short-term investments in securities that can be converted to cash quickly (liquid assets).
Accounts receivable	Cash due from customers in payment for goods received. These arise from sales made on credit.
Inventory	Finished goods ready for sale, goods in the process of being finished, or raw materials used in the production of goods.
Prepaid expense	A future expense item that has already been paid, such as insurance premiums or rent.
Total current assets	The sum of the above accounts.
Fixed assets	Assets that are long term in nature and have a minimum life expectancy that exceeds one year.
Investments	Assets held as investments rather than assets owned for the production process. Most often, the assets include small ownership interests in other companies.
Gross property, plant, and equipment	Land, buildings, and other fixed assets listed at original cost.

(continued)

TABLE 14.5 Sample Balance Sheet *(continued)*

Typical Company	December 31
Less: Accumulated depreciation	The accumulated expense deductions applied to all plant and equipment over their life. Land may not be depreciated. The total amount represents, in general, the decline in value as equipment gets older and wears out. The maximum amount that can be deducted is set by the U.S. Federal Tax Code and varies by type of asset.
Net property, plant, and equipment	Gross property, plant, and equipment minus the accumulated depreciation. This amount reflects the book value of the fixed assets and not their value if sold.
Other assets	Any other asset that is long term and does not fit into the preceding categories. It could be patents or trademarks.
Total fixed assets	The sum of the above accounts.
Total assets	The sum of all the asset values.
Liabilities and stockholders' equity	This is the major category. Liabilities refer to all indebtedness and loans of both a long-term and short-term nature. Stockholders' equity refers to all money that has been contributed to the company over the life of the firm by the owners.
Current liabilities	Short-term debt expected to be paid off within the next 12 months.
Accounts payable	Money owed to suppliers for goods ordered. Firms usually have between 30 and 90 days to pay this account, depending on industry norms.
Wages payable	Money owed to employees for hours worked or salary. If workers receive checks every two weeks, the amount owed should be no more than two weeks' pay.
Taxes payable	Firms are required to pay corporate taxes quarterly. This refers to taxes owed based on earnings estimates for the quarter.
Notes payable	Short-term loans from banks or other lenders.
Other current liabilities	The other short-term debts that do not fit into the preceding categories.
Total current liabilities	The sum of the preceding accounts.
Long-term liabilities	All long-term debt that will not be paid off in the next 12 months.
Long-term debt	Loans of more than one year from banks, pension funds, insurance companies, or other lenders. These loans often take the form of bonds, which are securities that may be bought and sold in bond markets.
Deferred income taxes	This is a liability owed to the government but not due within one year.
Other long-term liabilities	Any other long-term debt that does not fit the preceding two categories.
Total long-term liabilities	The sum of the preceding accounts.
Total liabilities	The sum of all liability values.
Stockholders' equity	The following three categories are the owners' investment in the company.
Common stock	The tangible evidence of ownership is a security called common stock. The par value is stated value and does not indicate the company's worth.
Capital in excess of par (a.k.a. contributed capital)	When shares of stock were sold to the owners, they were recorded at the price at the time of the original sale. If the price paid was $10 per share, the extra $9 per share would show up in this account at 100,000 shares times $9 per share, or $900,000.

TABLE 14.5   Sample Balance Sheet *(continued)*

Typical Company	December 31
Retained earnings	The total amount of earnings the company has made during its life and not paid out to its stockholders as dividends. This account represents the owners' reinvestment of earnings into company assets rather than payments of cash dividends. This account does not represent cash.
Total stockholders' equity	This is the sum of the preceding equity accounts representing the owners' total investment in the company.
Total liabilities and stockholders' equity	The total short-term and long-term debt of the company plus the owners' total investment. This combined amount *must* equal total assets.

and owners' equity on the right. More recently, a vertical format, with assets on top followed by liabilities and owners' equity, has gained wide acceptance. Microsoft's balance sheet is presented in Table 14.6. In the sections that follow, we'll briefly describe the basic items found on the balance sheet; we'll take a closer look at a number of these in the "Financial Management and Securities Markets" chapter.

Assets.   All asset accounts are listed in descending order of *liquidity*—that is, how quickly each could be turned into cash. Current assets, also called short-term assets, are those that are used or converted into cash within the course of a calendar year. Cash is followed by temporary investments, accounts receivable, and inventory, in that order. Accounts receivable refers to money owed the company by its clients or customers who have promised to pay for the products at a later date. Accounts receivable usually includes an allowance for bad debts that management does not expect to collect. The bad-debts adjustment is normally based on historical collections experience and is deducted from the accounts receivable balance to present a more realistic view of the payments likely to be received in the future, called net receivables. Inventory may be held in the form of raw materials, work-in-progress, or finished goods ready for delivery.

Long-term or fixed assets represent a commitment of organizational funds of at least one year. Items classified as fixed include long-term investments, such as plants and equipment, and intangible assets, such as corporate "goodwill," or reputation, as well as patents and trademarks.

current assets
assets that are used or converted into cash within the course of a calendar year; also called short-term assets.

accounts receivable
money owed a company by its clients or customers who have promised to pay for the products at a later date.

Liabilities.   As seen in the accounting equation, total assets must be financed either through borrowing (liabilities) or through owner investments (owners' equity). Current liabilities include a firm's financial obligations to short-term creditors, which must be repaid within one year, while long-term liabilities have longer repayment terms. Accounts payable represents amounts owed to suppliers for goods and services purchased with credit. For example, if you buy gas with a BP credit card, the purchase represents an account payable for you (and an account receivable for BP). Other liabilities include wages earned by employees but not yet paid and taxes owed to the government. Occasionally, these accounts are consolidated into an accrued expenses account, representing all unpaid financial obligations incurred by the organization.

Owners' Equity.   Owners' equity includes the owners' contributions to the organization along with income earned by the organization and retained to finance

current liabilities
a firm's financial obligations to short-term creditors, which must be repaid within one year.

accounts payable
the amount a company owes to suppliers for goods and services purchased with credit.

accrued expenses
all unpaid financial obligations incurred by an organization.

TABLE 14.6   Microsoft Corporation Consolidated Balance Sheets (in millions, except per share data)

(In millions)		
June 30,	2018	2017
Assets		
Current assets:		
Cash and cash equivalents	$ 11,946	$ 7,663
Short-term investments	121,822	125,318
Total cash, cash equivalents, and short-term investments	133,768	132,981
Accounts receivable, net of allowance for doubtful accounts of $377 and $345	26,481	22,431
Inventories	2,662	2,181
Other	6,751	5,103
Total current assets	169,662	162,696
Property and equipment, net of accumulated depreciation of $29,223 and $24,179	29,460	23,734
Operating lease right-of-use assets	6,686	6,555
Equity and other investments	1,862	6,023
Goodwill	35,683	35,122
Intangible assets, net	8,053	10,106
Other long-term assets	7,442	6,076
Total assets	$258,848	$250,312
Liabilities and stockholders' equity		
Current liabilities:		
Accounts payable	$ 8,617	$ 7,390
Short-term debt	0	9,072
Current portion of long-term debt	3,998	1,049
Accrued compensation	6,103	5,819
Short-term income taxes	2,121	718
Short-term unearned revenue	28,905	24,013
Other	8,744	7,684
Total current liabilities	58,488	55,745
Long-term debt	72,242	76,073
Long-term income taxes	30,265	13,485
Long-term unearned revenue	3,815	2,643
Deferred income taxes	541	5,734
Operating lease liabilities	5,568	5,372
Other long-term liabilities	5,211	3,549
Total liabilities	176,130	162,601

TABLE 14.6   Microsoft Corporation Consolidated Balance Sheets (in millions, except per share data) *(continued)*

(In millions)		
June 30,	2018	2017
Commitments and contingencies		
Stockholders' equity:		
Common stock and paid-in capital – shares authorized 24,000; outstanding 7,677 and 7,708	71,223	69,315
Retained earnings	13,682	17,769
Accumulated other comprehensive income (loss)	(2,187)	627
Total stockholders' equity	82,718	87,711
Total liabilities and stockholders' equity	$258,848	$250,312

continued growth and development. If the organization were to sell off all of its assets and pay off all of its liabilities, any remaining funds would belong to the owners. Not surprisingly, the accounts listed as owners' equity on a balance sheet may differ dramatically from company to company. Corporations sell stock to investors, who then become the owners of the firm. Many corporations issue two, three, or even more different classes of common and preferred stock, each with different dividend payments and/or voting rights. Google has three classes of stock, with the class B stock having more voting rights than class A shares. These are sometimes called founder's shares and allow the founders to maintain control over the company even though they do not own the majority of the shares. Ford Motor has the same type of voting structure. Because each type of stock issued represents a different claim on the organization, each must be represented by a separate owners' equity account, called contributed capital.

The Statement of Cash Flows

The third primary financial statement is called the **statement of cash flows,** which explains how the company's cash changed from the beginning of the accounting period to the end. Cash, of course, is an asset shown on the balance sheet, which provides a snapshot of the firm's financial position at one point in time. However, many investors and other users of financial statements want more information about the cash flowing into and out of the firm than is provided on the balance sheet in order to better understand the company's financial health. The statement of cash flows takes the cash balance from one year's balance sheet and compares it with the next while providing detail about how the firm used the cash. Table 14.7 presents Microsoft's statement of cash flows.

The change in cash is explained through details in three categories: cash from (used for) operating activities, cash from (used for) financing activities, and cash from (used for) investing activities.

Cash from operating activities is calculated by combining the changes in the revenue accounts, expense accounts, current asset accounts, and current liability accounts. This category of cash flows includes all the accounts on the balance sheet that relate to computing revenues and expenses for the accounting period. If this amount is a

statement of cash flows explains how the company's cash changed from the beginning of the accounting period to the end.

TABLE 14.7   Microsoft Consolidated Statements of Cash Flows (in millions)

(In millions)			
Year Ended June 30,	2018	2017	2016
Operations			
Net income	$ 16,571	$ 25,489	$ 20,539
Adjustments to reconcile net income to net cash from operations:			
Asset impairments	0	0	630
Depreciation, amortization, and other	10,261	8,778	6,622
Stock-based compensation expense	3,940	3,266	2,668
Net recognized gains on investments and derivatives	(2,212)	(2,073)	(223)
Deferred income taxes	(5,143)	(829)	2,479
Changes in operating assets and liabilities:			
Accounts receivable	(3,862)	(1,216)	562
Inventories	(465)	50	600
Other current assets	(952)	1,028	(1,212)
Other long-term assets	(285)	(917)	(1,110)
Accounts payable	1,148	81	88
Unearned revenue	5,922	3,820	2,565
Income taxes	18,183	1,792	(298)
Other current liabilities	798	356	(179)
Other long-term liabilities	(20)	(118)	(406)
Net cash from operations	43,884	39,507	33,325
Financing			
Proceeds from issuance (repayments) of short-term debt, maturities of 90 days or less, net	(7,324)	(4,963)	7,195
Proceeds from issuance of debt	7,183	44,344	13,884
Repayments of debt	(10,060)	(7,922)	(2,796)
Common stock issued	1,002	772	668
Common stock repurchased	(10,721)	(11,788)	(15,969)
Common stock cash dividends paid	(12,699)	(11,845)	(11,006)
Other, net	(971)	(190)	(369)
Net cash from (used in) financing	(33,590)	8,408	(8,393)
Investing			

TABLE 14.7   Microsoft Consolidated Statements of Cash Flows (in millions) *(continued)*

(In millions)			
Year Ended June 30,	2018	2017	2016
Additions to property and equipment	(11,632)	(8,129)	(8,343)
Acquisition of companies, net of cash acquired, and purchases of intangible and other assets	(888)	(25,944)	(1,393)
Purchases of investments	(137,380)	(176,905)	(129,758)
Maturities of investments	26,360	28,044	22,054
Sales of investments	117,577	136,350	93,287
Securities lending payable	(98)	(197)	203
Net cash used in investing	(6,061)	(46,781)	(23,950)
Effect of foreign exchange rates on cash and cash equivalents	50	19	(67)
Net change in cash and cash equivalents	4,283	1,153	915
Cash and cash equivalents, beginning of period	7,663	6,510	5,595
Cash and cash equivalents, end of period	$ 11,946	$ 7,663	$ 6,510

positive number, as it is for Microsoft, then the business is making extra cash that it can use to invest in increased long-term capacity or to pay off debts such as loans or bonds.

Cash from financing activities is calculated from changes in the long-term liability accounts and the contributed capital accounts in owners' equity. If this amount is negative, the company is likely paying off long-term debt or returning contributed capital to investors. In the case of Microsoft, it sold debt and repurchased $10.721 billion of its own stock and paid dividends of $12.699 billion, which resulted in a negative $33.590 billion cash flow from financing.

Finally, *cash from investing activities* is calculated from changes in the long-term or fixed asset accounts. If this amount is negative, as is the case with Microsoft, we can see that the company bought $11.632 billion of property and equipment. It also purchased $137.380 billion of investments and sold $117.577 billion of investments and had $26.360 billion of investments mature for a total negative cash flow of $5.061 billion. A positive figure usually indicates a business that is selling off existing long-term assets and reducing its capacity for the future. So a negative number in this category is not a bad thing.

Ratio Analysis: Analyzing Financial Statements

The income statement shows a company's profit or loss, while the balance sheet itemizes the value of its assets, liabilities, and owners' equity. Together, the two statements provide the means to answer two critical questions: (1) How much did the firm make or lose? and (2) How much is the firm presently worth based on historical values

LO 14-5

Analyze financial statements, using ratio analysis, to evaluate a company's performance.

ratio analysis
calculations that measure an organization's financial health.

found on the balance sheet? **Ratio analysis,** calculations that measure an organization's financial health, brings the complex information from the income statement and balance sheet into sharper focus so that managers, lenders, owners, and other interested parties can measure and compare the organization's productivity, profitability, and financing mix with other similar entities.

As you know, a ratio is simply one number divided by another, with the result showing the relationship between the two numbers. For example, we measure fuel efficiency with miles per gallon. This is how we know that 55 mpg in a Toyota Prius is much better than the average car. We use ratios in all sports, such as earned run and batting averages in baseball, field goal percentage in basketball, and percentage of passes completed in football. But, to make sense out of ratios, you have to know what you want to measure. Financial ratios are used to weigh and evaluate a firm's performance. An absolute value such as earnings of $70,000 or accounts receivable of $200,000 almost never provides as much useful information as a well-constructed ratio. Whether those numbers are good or bad depends on their relation to other numbers. If a company earned $70,000 on $700,000 in sales (a 10 percent return), such an earnings level might be quite satisfactory. The president of a company earning this same $70,000 on sales of $7 million (a 1 percent return), however, should probably start looking for another job!

Ratios by themselves are not very useful. What matters is the relationship of the calculated ratios to the previous years performance, comparison to its competitors and the company's stated goals. Remember, while the profitability, asset utilization, liquidity, debt ratios, and per share data we'll look at here can be very useful, you will never see the forest by looking only at the trees.

Profitability Ratios

profitability ratios
ratios that measure the amount of operating income or net income an organization is able to generate relative to its assets, owners' equity, and sales.

Profitability ratios measure how much operating income or net income an organization is able to generate relative to its assets, owners' equity, and sales. The numerator (top number) used in these examples is always the net income after taxes. Common profitability ratios include profit margin, return on assets, and return on equity. The following examples are based on the 2018 income statement and balance sheet for Microsoft, as shown in Tables 14.4 and 14.6. Except where specified, all data are expressed in millions of dollars.

Microsoft's 2018 revenue grew $13.8 billion, driven by a $13 billion increase in service revenues that can mainly be attributed to Microsoft's successful commercial cloud storage business. Net income declined due to the $13.7 billion tax put in place by the Tax Cut and Jobs Act of 2017 mentioned previously.

profit margin
net income divided by sales.

The **profit margin,** computed by dividing net income by sales, shows the overall percentage of profits earned by the company. It is based solely upon data obtained from the income statement. The higher the profit margin, the better the cost controls within the company and the higher the return on every dollar of revenue. Microsoft's profit margin is calculated as follows:

$$\text{Profit margin} = \frac{\text{Net income}}{\text{Sales}}$$

$$= \frac{\$16,571}{\$110,360}$$

$$= 15.02\%$$

Thus, for every $1 in sales, Microsoft generated profits after taxes of almost 15 cents.

Return on assets, net income divided by assets, shows how much income the firm produces for every dollar invested in assets. A company with a low return on assets is probably not using its assets very productively—a key managerial failing. For its computation, the return on assets calculation requires data from both the income statement and the balance sheet.

return on assets
net income divided by assets.

$$\text{Return on assets} = \frac{\text{Net income}}{\text{Total assets}}$$
$$= \frac{\$16,571}{\$258,848}$$
$$= 6.40\%$$

In the case of Microsoft, every $1 of assets generated a return of 6.4 percent, or profits of 6.4 cents per dollar.

Stockholders are always concerned with how much money they will make on their investment, and they frequently use the return on equity ratio as one of their key performance yardsticks. **Return on equity** (also called return on investment [ROI]), calculated by dividing net income by owners' equity, shows how much income is generated by each $1 the owners have invested in the firm. Obviously, a low return on equity means low stockholder returns and may indicate a need for immediate managerial attention. Because some assets may have been financed with debt not contributed by the owners, the value of the owners' equity is usually considerably lower than the total value of the firm's assets. Microsoft's return on equity is calculated as follows:

return on equity
net income divided by owners' equity; also called return on investment (ROI).

$$\text{Return on equity} = \frac{\text{Net income}}{\text{Stockholders' equity}}$$
$$= \frac{\$16,571}{\$82,718}$$
$$= 20.03\%$$

For every dollar invested by Microsoft stockholders, the company earned a 20.03 percent return, or 20.03 cents per dollar invested.

Asset Utilization Ratios

Asset utilization ratios measure how well a firm uses its assets to generate each $1 of sales. Obviously, companies using their assets more productively will have higher returns on assets than their less efficient competitors. Similarly, managers can use asset utilization ratios to pinpoint areas of inefficiency in their operations. These ratios (receivables turnover, inventory turnover, and total asset turnover) relate balance sheet assets to sales, which are found on the income statement.

asset utilization ratios
ratios that measure how well a firm uses its assets to generate each $1 of sales.

The **receivables turnover,** sales divided by accounts receivable, indicates how many times a firm collects its accounts receivable in one year. It also demonstrates how quickly a firm is able to collect payments on its credit sales. Obviously, no payments means no profits. Microsoft collected its receivables 4.17 times per year, which translates to about 86 days that receivables are outstanding. These ratios are based

receivables turnover
sales divided by accounts receivable.

on a 360 day year. This is most likely due to the trade terms it gives its corporate customers.

$$\text{Receivables turnover} = \frac{\text{Sales}}{\text{Receivables}}$$

$$= \frac{\$110,360}{\$26,481}$$

$$= 4.17 \text{ times per year}$$

inventory turnover
sales divided by total inventory.

Inventory turnover, sales divided by total inventory, indicates how many times a firm sells and replaces its inventory over the course of a year. A high inventory turnover ratio may indicate great efficiency but may also suggest the possibility of lost sales due to insufficient stock levels. Microsoft's inventory turnover indicates that it replaced its inventory 41.46 times last year, or about every 8.7 days. This high inventory turnover is a reflection that Microsoft has very little physical inventory and instead downloads its Windows programs over the Internet.

$$\text{Inventory turnover} = \frac{\text{Sales}}{\text{Inventory}}$$

$$= \frac{\$110,360}{\$2,662}$$

$$= 41.46 \text{ times per year}$$

Accountants often like to calculate inventory turnover using cost of goods sold in the numerator because inventory is carried on the books at cost and so is cost of goods sold. Financial analysts prefer sales in the numerator to measure the efficiency of inventory in producing revenues. Microsoft has very little physical inventory as opposed to an industrial company so using cost of goods sold would not provide much meaningful information.

total asset turnover
sales divided by total assets.

Total asset turnover, sales divided by total assets, measures how well an organization uses all of its assets in creating sales. It indicates whether a company is using its assets productively. Microsoft generated $0.43 in sales for every $1 in total corporate assets. The cause of this low total asset turnover is the large cash balance that Microsoft has on its balance sheet. Cash does not produce sales dollars. At the end of 2018 more than 50 percent of its assets were held in cash or cash equivalents.

$$\text{Total asset turnover} = \frac{\text{Sales}}{\text{Total assets}}$$

$$= \frac{\$110,360}{\$258,848}$$

$$= 0.43 \text{ times per year}$$

Liquidity Ratios

liquidity ratios
ratios that measure the speed with which a company can turn its assets into cash to meet short-term debt.

Liquidity ratios compare current (short-term) assets to current liabilities to indicate the speed with which a company can turn its assets into cash to meet debts as they fall due. High liquidity ratios may satisfy a creditor's need for safety, but ratios that are too high may indicate that the organization is not using its current assets efficiently. Liquidity ratios are generally best examined in conjunction with asset utilization ratios because high turnover ratios imply that cash is flowing through an organization very quickly—a situation that dramatically reduces the need for the type of reserves measured by liquidity ratios.

The **current ratio** is calculated by dividing current assets by current liabilities. Microsoft's current ratio indicates that for every $1 of current liabilities, the firm had $2.90 of current assets on hand. The relatively high current ratio is also due to the $133,768 billion of cash, cash equivalents, and short-term investments on hand, which is part of the current asset total.

> **current ratio** current assets divided by current liabilities.

$$\text{Current ratio} = \frac{\text{Current assets}}{\text{Current liabilities}}$$

$$= \frac{\$169,662}{\$58,488}$$

$$= 2.90 \text{ times}$$

The **quick ratio** (also known as the **acid test**) is a far more stringent measure of liquidity because it eliminates inventory, the least liquid current asset. It measures how well an organization can meet its current obligations without resorting to the sale of its inventory. Because Microsoft has so little inventory, the quick ratio at 2.86 times is almost the same as the current ratio.

> **quick ratio (acid test)** a stringent measure of liquidity that eliminates inventory.

$$\text{Quick ratio} = \frac{\text{Current assets} - \text{Inventory}}{\text{Current liabilities}}$$

$$= \frac{\$167,000}{\$58,488}$$

$$= 2.86 \text{ times}$$

Debt Utilization Ratios

Debt utilization ratios provide information about how much debt an organization is using relative to other sources of capital, such as owners' equity. Because the use of debt carries an interest charge that must be paid regularly regardless of profitability, debt financing is much riskier than equity. Unforeseen negative events such as recessions affect heavily indebted firms to a far greater extent than those financed exclusively with owners' equity. Because of this and other factors, the managers of most firms tend to keep debt-to-asset levels below 50 percent. However, firms in very stable and/or regulated industries, such as electric utilities, often are able to carry debt ratios well in excess of 50 percent with no ill effects.

> **debt utilization ratios** ratios that measure how much debt an organization is using relative to other sources of capital, such as owners' equity.

The **debt to total assets ratio** indicates how much of the firm is financed by debt and how much by owners' equity. To find the value of Microsoft's total debt, you must add current liabilities to long-term debt and other liabilities.

> **debt to total assets ratio** a ratio indicating how much of the firm is financed by debt and how much by owners' equity.

$$\text{Debt to total assets} = \frac{\text{Debt (Assets} - \text{Equity)}}{\text{Total assets}}$$

$$= \frac{\$176,130}{\$258,848}$$

$$= 68\%$$

Thus, for every $1 of Microsoft's total assets, 68 percent is financed with debt. The remaining 32 percent is provided by owners' equity. Debt to total assets increased from 45 percent to 68 percent from 2013 to 2018 as Microsoft took advantage of low interest rates to sell debt and repurchase common stock.

times interest earned ratio
operating income divided by interest expense.

The **times interest earned ratio,** operating income divided by interest expense, is a measure of the safety margin a company has with respect to the interest payments it must make to its creditors. A low times interest earned ratio indicates that even a small decrease in earnings may lead the company into financial straits. Microsoft had so little interest expense that it did not list it as a separate item on the income statement. In this case, the analyst has to go searching through the footnotes to the financial statements. We find that interest expense was $2,733 billion. Putting this into the calculation, we find that interest expense is covered 5.06 times by operating income. Even though it is down from 10.47 times in 2018 because of the one time tax payment that reduced income, a lender would have no worries about receiving interest payments from Microsoft.

$$\text{Times interest earned} = \frac{\text{Income before interest and taxes}}{\text{Interest}}$$

$$= \frac{\$13,838}{\$2,733}$$

$$= 5.06 \text{ times}$$

Per Share Data

per share data
data used by investors to compare the performance of one company with another on an equal, per share, basis.

Investors may use **per share data** to compare the performance of one company with another on an equal, or per share, basis. Generally, the more shares of stock a company issues, the less income is available for each share.

earnings per share
net income or profit divided by the number of stock shares outstanding.

Earnings per share is calculated by dividing net income or profit by the number of shares of stock outstanding. This ratio is important because yearly changes in earnings per share, in combination with other economy-wide factors, determine a company's overall stock price. When earnings go up, so does a company's stock price—and so does the wealth of its stockholders.

$$\frac{\text{Diluted earnings}}{\text{per share}} = \frac{\text{Net income}}{\text{Number of shares outstanding (diluted)}}$$

$$= \frac{\$16,571}{7,794}$$

$$= \$2.13$$

We can see from the income statement that Microsoft's basic earnings per share increased from $2.59 per share in 2016 to $3.29 in 2017 and declined to $2.12 in 2018 because of the one-time tax on earnings held over seas. Financial analysts expect the company's earnings per share to improve. You can see from the income statement that diluted earnings per share include more shares than the basic calculation; this is because diluted shares include potential shares that could be issued due to the exercise of stock options or the conversion of certain types of debt into common stock. Investors generally pay more attention to diluted earnings per share than basic earnings per share.

dividends per share
the actual cash received for each share owned.

Dividends per share are paid by the corporation to the stockholders for each share owned. The payment is made from earnings after taxes by the corporation but is taxable income to the stockholder. Thus, dividends result in double taxation: The corporation pays tax once on its earnings, and the stockholder pays tax a second time on his or her dividend income. Since 2004, Microsoft has raised its dividend every year, from $0.16 per share to $1.68 per share. Note that the shares listed on the income statement versus shares listed as outstanding on the balance sheet are not the same. Share count for earnings per share are weighted average shares over the year. The

Responding to Business Challenges

How Data Analytics Is Changing the World of Accounting

Data analytics is the next frontier in accounting. While data analytics can be simple, today the term is most often used to describe the analysis of large volumes of data and/or high-velocity data, which presents unique computational and data-handling challenges.

Certified public accountants (CPAs) are constantly using data analytics in daily tasks. There are four types of data analytics: descriptive analytics ("What's happening?"), diagnostic analytics ("Why did it happen?"), predictive analytics ("What's going to happen?"), and prescriptive analytics ("What should happen?"). Descriptive analytics includes the categorization and classification of information. Accurate reporting with large quantities of data is a hallmark of solid accounting practices. Diagnostic analytics monitors changes in data. Accountants regularly analyze variances and calculate historical performance as an indicator of future performance. Predictive analytics assesses the likelihood of future outcomes. Accountants are instrumental in building forecasts and identifying patterns that shape those forecasts. Prescriptive analytics is used to create tangible actions and critical business decisions. Accountants use these forecasts to make recommendations for future growth opportunities.

Data analytics offers challenges and opportunities for CPAs. The challenges include obtaining the skills needed to initiate and support data analytics activities, as well as altering the present audit model to include appropriate audit analytics techniques. Opportunities include a technology-rich audit model and the ability to provide data analytics expertise to clients. With the mastery of data analytics, businesses can generate a higher profit margin and gain meaningful competitive advantages.[8]

Critical Thinking Questions

1. What types of opportunities are data analytics providing for CPAs?
2. Describe the different types of analytics and how they can be utilized in the accounting field.
3. What do you think will happen to CPAs that do not master data analytics techniques?

---

shares on the balance sheet are those outstanding at the year's end and do not necessarily represent the shares on which dividends were actually paid.

$$\text{Dividends per share} = \frac{\text{Dividends paid}}{\text{Number of shares outstanding (year end)}}$$
$$= \frac{\$12,917}{7,677}$$
$$= \$1.68$$

Importance of Integrity in Accounting

The financial crisis and the recession that followed provided another example of a failure in accounting reporting. Many firms attempted to exploit loopholes and manipulate accounting processes and statements. Banks and other financial institutions often held assets off their books by manipulating their accounts. If the accountants, the SEC, and the bank regulators had been more careful, these types of transactions would have been discovered and corrected.

On the other hand, strong compliance to accounting principles creates trust among stakeholders. Accounting and financial planning is important for all organizational entities, even cities. The City of Maricopa in Arizona received the Government Finance Officers Association of the United States and Canada (GFOA) Distinguished Budget Presentation Award for its governmental budgeting. The city scored proficient in its policy, financial plan, operations guide, and communications device. Integrity in accounting is crucial to creating trust, understanding the

So You Want to Be an Accountant

Do you like numbers and finances? Are you detail oriented, a perfectionist, and highly accountable for your decisions? If so, accounting may be a good field for you. If you are interested in accounting, there are always job opportunities available no matter the state of the economy. Accounting is one of the most secure job options in business. Of course, becoming an accountant is not easy. You will need at least a bachelor's degree in accounting to get a job, and many positions require additional training. Many states demand coursework beyond the 120 to 150 credit hours collegiate programs require for an accounting degree. If you are really serious about getting into the accounting field, you will probably want to consider getting your master's in accounting and taking the CPA exam. The field of accounting can be complicated, and the extra training provided through a master's in accounting program will prove invaluable when you go out looking for a good job. Accounting is a volatile discipline affected by changes in legislative initiatives.

With corporate accounting policies changing constantly and becoming more complex, accountants are needed to help keep a business running smoothly and within the bounds of the law. In fact, the number of jobs in the accounting and auditing field are expected to increase 10 percent between 2016 and 2026. Jobs in accounting tend to pay quite well, with the median salary standing at $69,350. If you go on to get your master's degree in accounting, expect to see an even higher starting wage. Of course, your earnings could be higher or lower than these averages, depending on where you work, your level of experience, the firm, and your particular position.

Accountants are needed in the public and the private sectors, in large and small firms, in for-profit and not-for-profit organizations. Accountants in firms are generally in charge of preparing and filing tax forms and financial reports. Public sector accountants are responsible for checking the veracity of corporate and personal records in order to prepare tax filings. Basically, any organization that has to deal with money and/or taxes in some way or another will be in need of an accountant, either for in-house service or occasional contract work. Requirements for audits under the Sarbanes-Oxley Act and rules from the Public Company Accounting Oversight Board are creating more jobs and increased responsibility to maintain internal controls and accounting ethics. The fact that accounting rules and tax filings tend to be complex virtually ensures that the demand for accountants will never decrease.[9]

financial position of an organization or entity, and making financial decisions that will benefit the organization.[10]

It is most important to remember that integrity in accounting processes requires ethical principles and compliance with both the spirit of the law and professional standards in the accounting profession. Most states require accountants preparing to take the CPA exam to take accounting ethics courses. Transparency and accuracy in reporting revenue, income, and assets develops trust from investors and other stakeholders.

Review Your Understanding

Describe the different uses of accounting information.

Accounting is the language businesses and other organizations use to record, measure, and interpret financial transactions. Financial statements are used internally to judge and control an organization's performance and to plan and direct its future activities and measure goal attainment. External organizations such as lenders, governments, customers, suppliers, and the Internal Revenue Service are major consumers of the information generated by the accounting process.

Demonstrate the accounting process.

Assets are an organization's economic resources; liabilities, debts the organization owes to others; and owners' equity, the difference between the value of an organization's assets and liabilities. This principle can be expressed as the accounting equation: Assets = Liabilities + Owners' equity. The double-entry bookkeeping system is a system of recording and classifying business transactions in accounts that maintain the balance of the accounting equation. The accounting cycle involves examining source documents, recording transactions in a journal, posting transactions, and preparing financial statements on a continuous basis throughout the life of the organization.

Examine the various components of an income statement in order to evaluate a firm's "bottom line."

The income statement indicates a company's profitability over a specific period of time. It shows the "bottom line," the total profit (or loss) after all expenses (the costs incurred in the day-to-day operations of the organization) have been deducted from revenue (the total amount of money received from the sale of goods or services and other business activities). The cash flow statement details how much cash is moving through the firm and thus adds insight to a firm's "bottom line."

Interpret a company's balance sheet to determine its current financial position.

The balance sheet, which summarizes the firm's assets, liabilities, and owners' equity since its inception, portrays its financial position as of a particular point in time. Major classifications included in the balance sheet are current assets (assets that can be converted to cash within one calendar year), fixed assets (assets of greater than one year's duration), current liabilities (bills owed by the organization within one calendar year), long-term liabilities (bills due more than one year hence), and owners' equity (the net value of the owners' investment).

Analyze financial statements, using ratio analysis, to evaluate a company's performance.

Ratio analysis is a series of calculations that brings the complex information from the income statement and balance sheet into sharper focus so that managers, lenders, owners, and other interested parties can measure and compare the organization's productivity, profitability, and financing mix with similar entities. Ratios may be classified in terms of profitability (measure dollars of return for each dollar of employed assets), asset utilization (measure how well the organization uses its assets to generate $1 in sales), liquidity (assess organizational risk by comparing current assets to current liabilities), debt utilization (measure how much debt the organization is using relative to other sources of capital), and per share data (compare the performance of one company with another on an equal basis).

Assess a company's financial position using its accounting statements and ratio analysis.

Based on the information presented in the chapter, you should be able to answer the questions posed in the "Solve the Dilemma" feature near the end of the chapter. Formulate a plan for determining BrainDrain's bottom line, current worth, and productivity.

Critical Thinking Questions

Enter the World of Business Questions

I. What is unique about Deloitte that would be important to clients?

2. What is the difference between a CPA at Deloitte and a bookkeeper?

3. Why is it important to have a high-integrity accounting firm provide services to clients?

Learn the Terms

accounting   438
accounting cycle   445
accounting equation   445
accounts payable   455
accounts receivable   455
accrued expenses   455
annual report   442
asset utilization ratios   461
assets   444
balance sheet   452
budget   441
cash flow   441
certified management accountants
  (CMAs)   440
certified public accountant (CPA)   438
cost of goods sold   448
current assets   455

current liabilities   455
current ratio   463
debt to total assets ratio   463
debt utilization ratios   463
depreciation   451
dividends per share   464
double-entry bookkeeping   445
earnings per share   464
expenses   451
gross income   451
income statement   448
inventory turnover   462
journal   447
ledger   447
liabilities   444
liquidity ratios   462
managerial accounting   441

net income   451
owners' equity   444
per share data   464
private accountants   440
profit   451
profit margin   460
profitability ratios   460
quick ratio (acid test)   463
ratio analysis   460
receivables turnover   461
return on assets   461
return on equity   461
revenue   448
statement of cash flows   457
times interest earned ratio   464
total asset turnover   462

Check Your Progress

1. Why are accountants so important to a corporation? What function do they perform?

2. Discuss the internal uses of accounting statements.

3. What is a budget?

4. Discuss the external uses of financial statements.

5. Describe the accounting process and cycle.

6. The income statements of all corporations are in the same format. True or false? Discuss.

7. Which accounts appear under "current liabilities"?

8. Together, the income statement and the balance sheet answer two basic questions. What are they?

9. What are the five basic ratio classifications? What ratios are found in each category?

10. Why are debt ratios important in assessing the risk of a firm?

Get Involved

1. Go to the library or the Internet and get the annual report of a company with which you are familiar. Read through the financial statements; then write up an analysis of the firm's performance using ratio analysis. Look at data over several years and analyze whether the firm's performance is changing through time.

2. Form a group of three or four students to perform an industry analysis. Each student should analyze a company in the same industry, and then all of you should compare your results. The following companies would make good group projects:

Automobiles: Fiat Chrysler, Ford, General Motors

Computers: Apple, Hewlett-Packard

Brewing: MillerCoors, Molson Coors, The Boston Beer Company

Chemicals: DuPont, Dow Chemical, Monsanto

Petroleum: Chevron, ExxonMobil, BP

Pharmaceuticals: Merck, Lilly, Amgen

Retail: Sears, JCPenney, Macy's, Express

Build Your Skills

Financial Analysis

Background

The income statement for Western Grain Company, a producer of agricultural products for industrial as well as consumer markets, is shown here. Western Grain's total assets are $4,237.1 million, and its equity is $1,713.4 million.

Consolidated Earnings and Retained Earnings Year Ended December 31

(Millions)	2017
Net sales	$6,295.4
Cost of goods sold	2,989.0
Selling and administrative expense	2,237.5
Operating profit	1,068.9
Interest expense	33.3
Other income (expense), net	(1.5)
Earnings before income taxes	1,034.1
Income taxes	353.4
Net earnings	680.7
(Net earnings per share)	$ 2.94
Retained earnings, beginning of year	3,033.9
Dividends paid	(305.2)
Retained earnings, end of year	$3,409.4

Task

Calculate the following profitability ratios: profit margin, return on assets, and return on equity. Assume that the industry averages for these ratios are as follows: profit margin, 12 percent; return on assets, 18 percent; and return on equity, 25 percent. Evaluate Western Grain's profitability relative to the industry averages. Why is this information useful?

Solve the Dilemma

Exploring the Secrets of Accounting

You have just been promoted from vice president of marketing of BrainDrain Corporation to president and CEO! That's the good news. Unfortunately, while you know marketing like the back of your hand, you know next to nothing about finance. Worse still, the "word on the street" is that BrainDrain is in danger of failure if steps to correct large and continuing financial losses are not taken immediately. Accordingly, you have asked the vice president of finance and accounting for a complete set of accounting statements detailing the financial operations of the company over the past several years.

Recovering from the dual shocks of your promotion and feeling the weight of the firm's complete accounting report for the very first time, you decide to attack the problem systematically and learn the "hidden secrets" of the company, statement by statement. With Mary Pruitt, the firm's trusted senior financial analyst, by your side, you delve into

the accounting statements as never before. You resolve to "get to the bottom" of the firm's financial problems and set a new course for the future—a course that will take the firm from insolvency and failure to financial recovery and perpetual prosperity.

Assess a company's financial position using its accounting statements and ratio analysis.

Critical Thinking Questions

1. Describe the three basic accounting statements. What types of information does each provide that can help you evaluate the situation?

2. Which of the financial ratios are likely to prove to be of greatest value in identifying problem areas in the company? Why? Which of your company's financial ratios might you expect to be especially poor?

3. What are the limitations of ratio analysis?

Build Your Business Plan

Accounting and Financial Statements

After you determine your initial *reasonable selling price,* you need to estimate your sales forecasts (in terms of units and dollars of sales) for the first year of operation. Remember to be conservative and set forecasts that are more modest.

While customers may initially try your business, many businesses have seasonal patterns. A good budgeting/planning system allows managers to anticipate problems, coordinate activities of the business (so that subunits within the organization are all working toward the common goal of the organization), and control operations (how we know whether spending is "in line").

The first financial statement you need to prepare is the income statement. Beginning with your estimated sales revenue, determine what expenses will be necessary to generate that level of sales revenue.

The second financial statement you need to create is your balance sheet. Your balance sheet is a snapshot of your financial position in a moment in time. Refer to Table 14-6 to assist you in listing your assets, liabilities, and owner's equity.

The last financial statement, the cash flow statement, is the most important one to a bank. It is a measure of your ability to get and repay the loan from the bank. Referring to Table 14.8, be as realistic as possible as you are completing it. Allow yourself enough cash on hand until the point in which the business starts to support itself.

See for Yourself Videocase

Goodwill Industries: Accounting in a Nonprofit

Goodwill Industries International Inc. consists of a network of 164 independent, community-based organizations located throughout the United States and Canada. The mission of this nonprofit is to enhance the lives of individuals, families, and communities "through learning and the power of work." Local Goodwill stores sell donated goods and then donate the proceeds to fund job training programs, placement services, education, and more. Despite its nonprofit status, Goodwill establishments are, in many ways, run similar to for-profit businesses. One of these similarities involves the accounting function.

Like for-profit firms, nonprofit organizations like Goodwill must provide detailed information about how they are using the donations that are provided to them. Indeed, fraud can occur just as easily at a nonprofit organization as for a for-profit company, making it necessary for nonprofits to reassure stakeholders that they are using their funds legitimately. Additionally, donors want to know how much of their donations are going toward activities such as job creation and how much is going toward operational and administrative expenses. It sometimes surprises people that nonprofits use part of the funds they receive for operational costs. Yet such a perspective fails to see that nonprofits must also pay for electricity, rent, wages, and other services.

"We have revenue and support for the revenue pieces, and then we have direct and indirect expenses for our program services, and then we have G and A, general administrative

services. And we have what's called the bottom line, or other people call net profit. We have what's called net change in assets. The concept is pretty much the same as far as accounting," says Jeff McCaw, CFO of Goodwill.

Goodwill creates the equivalent of a balance sheet and income statement. Yet because Goodwill is a nonprofit entity, its financials are known by the names *statement of financial position* and *statement of activities.* These financials have some differences compared to financial statements of for-profit companies. For instance, Goodwill's statement of financial position does not have shareholder's equity but, instead, has net assets. The organization's financials are audited, and stakeholders can find the firm's information in Form 990 through Goodwill's public website (Form 990 is the IRS form for nonprofits).

Because Goodwill sells goods at its stores, the company must also figure in costs of goods sold. In fact, most of the organization's revenue comes from its store activities. In one year, the retail division or sale of donated goods and contributed goods generated $3.94 billion. The contracts division generated $666 million, which provides custodial, janitorial, and lawn maintenance service contracts to government agencies. Grants from foundations, corporations, individuals, and government account for $185 million. The fact that Goodwill is able to generate much of its own funding through store activities and contracts is important. Many nonprofits that rely solely on donated funds find it hard to be sustainable in the long run, particularly during economic downturns.

Remember that even though nonprofits are different from for-profit companies, they must still make certain that their financial information is accurate. This requires nonprofit accountants to be meticulous and thorough in gathering and analyzing information. Like all accountants, accountants at Goodwill record transactions in journals and then carefully review the information before it is recorded in the general ledger. The organization uses trial balances to ensure that everything balances out, as well as advanced software to record transactions, reconcile any discrepancies, and provide an idea of how much cash the organization has on hand.

Finally, Goodwill uses ratio analysis to determine the financial health of the company. For instance, the common ratio allows Goodwill to determine how much revenue it brings in for every dollar it spends on costs. The organization also uses ratio analysis to compare its results to similar organizations. It is important for Goodwill to identify the best performers in its field so that it can generate ideas and even form partnerships with other organizations. By using accounting to identify how

best to use its resources, Goodwill is advancing its mission of helping others.[ll]

Critical Thinking Questions

1. What are some similarities between the type of accounting performed at Goodwill compared to accounting at for-profit companies?

2. What are some differences between the type of accounting performed at Goodwill compared to accounting at for-profit companies?

3. How can Goodwill use ratio analysis to improve its operations?

You can find the related video in the Video Library in Connect. Ask your instructor how you can access Connect.

Team Exercise

You can look at websites such as Yahoo! Finance (http://finance.yahoo.com/), under the company's "key statistics" link, to find many of its financial ratios, such as return on assets and return on equity. Have each member of your team look up a different company, and explain why you think there are differences in the ratio analysis for these two ratios among the selected companies.

Notes

1. "Big 4 Accounting Firms," *AccountingVerse,* http://www.accountingverse.com/articles/big-4-accounting-firms.html (accessed April 7, 2018); "The Big 4 Accounting Firms," The Big 4 Accounting Firms, http://big4accountingfirms.org/ (accessed April 7, 2018); "About Deloitte," Deloitte, https://www2.deloitte.com/us/en/pages/about-deloitte/articles/about-deloitte.html (accessed April 7, 2018); "Our Approach to Innovation—in Business and Beyond," Deloitte, https://www2.deloitte.com/us/en/pages/about-deloitte/articles/business-innovation-approach.html (accessed April 7, 2018); "Corporate Citizenship," Deloitte, https://www2.deloitte.com/us/en/pages/about-deloitte/articles/deloitte-corporate-citizenship.html?icid=top_deloitte-corporate-citizenship (accessed April 7, 2018)

2. "About the ACFE," Association of Certified Fraud Examiners, www.acfe.com/about-the-acfe.aspx (accessed April 20, 2016).

3. "Report to the Nations: 2018 Global Study On Occupational Fraud and Abuse," Association of Certified Fraud Examiners, 2018, https://s3-us-west-2.amazonaws.com/acfepublic/2018-report-to-the-nations.pdf (accessed April 24, 2018).

4. Mary Williams Walsh, "State Woes Grow Too Big to Camouflage," *CNBC,* March 30, 2010, www.cnbc.com/id/36096491/ (accessed April 17, 2017).

5. Stephanie Zonis, "Al Dente Pasta Co.," *Goodlifer,* September 17, 2009, http://www.goodlifer.com/2009/09/al-dente-pasta-co/ (accessed April 7, 2018); Al Dente Pasta website, http://www.aldentepasta.com (accessed April 7, 2018); "Marcella Hazan," HarperCollins, http://www.harpercollins.com/authors/4331/Marcella_Hazan/index.aspx (accessed April 7, 2018); Janet Miller, "After 30 Years, Al Dente Pasta Company Becomes a National Leader," *Ann Arbor News,* December 31, 2011, http://www.annarbor.com/business-review/al-dente-hand-rolls-out-30-years-of-making-pasta-to-become-a-national-leader-in-this-niche-market/ (accessed April 7, 2018); Susan Salasky, "Food Gifts from Michigan: Pasty, Coneys and Smoked Fish and More," *Detroit Free Press,* December 10, 2017, http://www.annarbor.com/business-review/al-dente-hand-rolls-out-30-years-of-making-pasta-to-become-a-national-leader-in-this-niche-market/ (accessed April 7, 2018).

6. Sarah Johnson, "Averting Revenue-Recognition Angst," *CFO,* April 2012, p. 21.

7. Terry Slavin, "Green Finance: Deforestation Brings Risk of $1trn in Stranded Assets, Warns CDP," *Ethical Corporation,* November 23, 2017, http://www.ethicalcorp

.com/green-finance-deforestation-brings-risk-1trn-stranded-assets-warns-cdp (accessed April 12, 2018); Kevin McCoy, "New York Attorney General Probing ExxonMobil's Accounting Amid Oil Slide," *USA Today,* September 16, 2016, https://www.usatoday.com/story/money/2016/09/16/exxonmobil-accounting-disclosures-examined/90476826/ (accessed April 12, 2018); Bradley Olson and Aruna Viswanatha, "SEC Probes Exxon over Accounting for Climate Change," *The Wall Street Journal,* September 20, 2016, https://www.wsj.com/articles/sec-investigating-exxon-on-valuing-of-assets-accounting-practices-1474393593 (accessed April 12, 2018); "How to Deal with Worries about Stranded Assets," *The Economist,* November 24, 2016, https://www.economist.com/news/special-report/21710632-oil-companies-need-heed-investors-concerns-how-deal-worries-about-stranded (accessed April 12, 2018); Felicia Jackson, "Disclosure Is about Risks and Opportunities, Not Politics," *Forbes,* April 12, 2018, https://www.forbes.com/sites/feliciajackson/2018/04/12/disclosure-is-about-risks-and-opportunities-not-politics/#50a8b1cf63be (accessed April 12, 2018).

8. Norbert Tschakert, Julia Kokina, Stephen Kozlowski, and Miklos Vasarhelyi, "The Next Frontier in Data Analytics," *Journal of Accountancy,* August 1, 2016, https://www.journalofaccountancy.com/issues/2016/aug/data-analytics-skills.html (accessed April 14, 2018); "What Is Data Analytics?" *Informatica,* (accessed April 14, 2018).

9. "Accountants and Auditors: Occupational Outlook Handbook," Bureau of Labor Statistics, April 13, 2018, www.bls.gov/ooh/Business-and-Financial/Accountants-and-auditors.htm (accessed September 2, 2018).

10. "City Finance Department Receives Distinguished Presentation Award for Its Budget," City of Maricopa, March 25, 2014, www.maricopa-az.gov/web/finance-administrativeservice-home/1029-city-s-finance-department-recieves-distinguished-budget-presentation-award-for-its-budget (accessed May 5, 2017).

11. Goodwill website, www.goodwill.org/ (accessed April 21, 2016); "About Us— Revenue Sources," Goodwill Industries International Inc., www.goodwill.org/about-us/ (accessed April 21, 2016).

Credits

15 Money and the Financial System

©panuwat phimpha/Shutterstock

Chapter Outline

Learning Objectives

After reading this chapter, you will be able to:

LO 15-1 Define *money,* its functions, and its characteristics.

LO 15-2 Describe various types of money.

LO 15-3 Specify how the Federal Reserve Board manages the money supply and regulates the American banking system.

LO 15-4 Compare and contrast commercial banks, savings and loan associations, credit unions, and mutual savings banks.

LO 15-5 Distinguish among nonbanking institutions such as insurance companies, pension funds, mutual funds, and finance companies.

LO 15-6 Analyze the challenges ahead for the banking industry.

LO 15-7 Recommend the most appropriate financial institution for a hypothetical small business.

Enter the World of Business

How Mobile Banking Is Breaking the Bank

Mobile banking has created a world of convenience and ease of use for consumers. Basic tasks like depositing checks and monitoring accounts can easily be done whenever and wherever the customer pleases. However, this increase in convenience can come at a price.

Transferring tasks to mobile devices escalates risks of cyberattacks and identity fraud. Most experts agree that mobile banking is generally secure and that many security issues do not stem from the technology itself, but because consumers do not understand how to properly use it.

According to reports, the average cost that financial services companies pay to address cyberattacks grew from $13 million in 2014 to $18.3 million in 2017. Financial services companies averaged 40 breaches in 2012, which increased to 125 in 2017. Cybersecurity breaches in mobile and online banking have skyrocketed. Consumers fail to consider that their phones work similarly to their computers. Antivirus software should be installed and regular security updates exercised. Incorrect perceptions around the safety of banking apps and Wi-Fi usage increase consumer vulnerability. According to a survey of senior IT executives, around 47 percent say their organizations are adopting mobile applications without assessing associated risks.

While there will always be some risk in regards to cyberspace, there are ways to minimize the potential for information theft. Consumers should educate themselves in how this technology works and its vulnerabilities. Just like a laptop, mobile phones offer security software that can monitor for attacks. Making sure all mobile banking and other apps are up-to-date is crucial. Older and outdated apps are more vulnerable. Consumers should closely monitor all activity and report anything suspicious. Doing so will allow them to reap the rewards of online banking without making themselves vulnerable to cyberattacks.[1]

Introduction

From Wall Street to Main Street, both overseas and at home, money is the one tool used to measure personal and business income and wealth. **Finance** is the study of how money is managed by individuals, companies, and governments. This chapter introduces you to the role of money and the financial system in the economy. Of course, if you have a checking account, automobile insurance, a college loan, or a credit card, you already have personal experience with some key players in the financial world.

We begin our discussion with a definition of money and then explore some of the many forms money may take. Next, we examine the roles of the Federal Reserve Board and other major institutions in the financial system. Finally, we explore the future of the finance industry and some of the changes likely to occur over the course of the next several years.

Money in the Financial System

Strictly defined, **money**, or *currency*, is anything generally accepted in exchange for goods and services. Materials as diverse as salt, cattle, fish, rocks, shells, and cloth, as well as precious metals such as gold, silver, and copper, have long been used by various cultures as money. Most of these materials were limited-supply commodities that had their own value to society (for example, salt can be used as a preservative and shells and metals as jewelry). The supply of these commodities therefore determined the supply of "money" in that society. The next step was the development of "IOUs," or slips of paper that could be exchanged for a specified supply of the underlying commodity. "Gold" notes, for instance, could be exchanged for gold, and the money supply was tied to the amount of gold available. While paper money was first used in North America in 1685 (and even earlier in Europe), the concept of *fiat money*—a paper money not readily convertible to a precious metal such as gold—did not gain full acceptance until the Great Depression in the 1930s. The United States abandoned its gold-backed currency standard largely in response to the Great Depression and converted to a fiduciary, or fiat, monetary system. In the United States, paper money is really a government "note" or promise, worth the value specified on the note.

Functions of Money

No matter what a particular society uses for money, its primary purpose is to enable a person or organization to trade money for a good or a service. These desires may be for entertainment actions like funding party expenses; operating actions, such as paying for rent, utilities, or employees; investing actions, such as buying property or equipment; or financing actions, such as starting or growing a business. Money serves three important functions: as a medium of exchange, a measure of value, and a store of value.

Medium of Exchange.   Before fiat money, the trade of goods and services was accomplished through *bartering*—trading one good or service for another of similar value. There had to be a simpler way, and that was to decide on a single item—money—that can be freely converted to any other good upon agreement between parties.

Measure of Value.   As a measure of value, money serves as a common standard or yardstick of the value of goods and services. For example, $2 will buy a dozen large eggs and $25,000 will buy a nice car in the United States. In Japan, where the currency is known as the yen, these same transactions would cost about 210 yen and

2.75 million yen, respectively. Money, then, is a common denominator that allows people to compare the different goods and services that can be consumed on a particular income level. While a star athlete and a minimum-wage earner are paid vastly different wages, each uses money as a measure of the value of their yearly earnings and purchases.

Store of Value.   As a store of value, money serves as a way to accumulate wealth (buying power) until it is needed. For example, a person making $1,000 per week who wants to buy a $500 computer could save $50 per week for each of the next 10 weeks. Unfortunately, the value of stored money is directly dependent on the health of the economy. If, due to rapid inflation, all prices double in one year, then the purchasing power value of the money "stuffed in the mattress" would fall by half. On the other hand, deflation occurs when prices of goods fall. Deflation might seem like a good thing for consumers, but in many ways it can be just as problematic as inflation. Periods of major deflation often lead to decreases in wages and increases in debt burdens.[2] Deflation also tends to be an indicator of problems in the economy. Deflation usually indicates slow economic growth and falling prices. Over the past 25 years, we have seen deflation in Japan, and Europe has continued to struggle with deflation off and on since the financial crisis. Given a choice, central banks like the Federal Reserve would rather have a small amount of inflation than deflation.

Characteristics of Money

To be used as a medium of exchange, money must be acceptable, divisible, portable, stable in value, durable, and difficult to counterfeit.

Acceptability.   To be effective, money must be readily acceptable for the purchase of goods and services and for the settlement of debts. Acceptability is probably the most important characteristic of money: If people do not trust the value of money, businesses will not accept it as a payment for goods and services, and consumers will have to find some other means of paying for their purchases.

Divisibility.   Given the widespread use of quarters, dimes, nickels, and pennies in the United States, it is no surprise that the principle of divisibility is an important one. With barter, the lack of divisibility often makes otherwise preferable trades impossible, as would be an attempt to trade a steer for a loaf of bread. For money to serve effectively as a measure of value, all items must be valued in terms of comparable units—dimes for a piece of bubble gum, quarters for laundry machines, and dollars (or dollars and coins) for everything else.

Portability.   Clearly, for money to function as a medium of exchange, it must be easily moved from one location to the next. Large colored rocks could be used as money, but you couldn't carry them around in your wallet. Paper currency and metal coins, on the other hand, are capable of transferring vast purchasing power into small, easily carried (and hidden!) bundles. Few Americans realize it, but more U.S. currency is in circulation outside the United

Zimbabwe adopted a multi-currency system in 2009 using up to eight official currencies. The country is now paving the way to establish a local currency.

©Feije Riemersma/Alamy Stock Photo

States than within. As of 2018, $1.69 trillion of U.S. currency was in circulation.[3] Some countries, such as Panama, even use the U.S. dollar as their currency. Retailers in other countries often state prices in dollars and in their local currency.

Stability.   Money must be stable and maintain its declared face value. A $10 bill should purchase the same amount of goods or services from one day to the next. The principle of stability allows people who wish to postpone purchases and save their money to do so without fear that it will decline in value. As mentioned earlier, money declines in value during periods of inflation, when economic conditions cause prices to rise. Thus, the same amount of money buys fewer and fewer goods and services. In some countries, people spend their money as fast as they can in order to keep it from losing any more of its value. Instability destroys confidence in a nation's money and its ability to store value and serve as an effective medium of exchange. It also has an impact on other countries. When Switzerland decided to no longer hold its Swiss franc at a fixed exchange rate with the euro, other countries were concerned because they tended to view the Swiss franc as relatively "safe" for investments. This change will make Swiss exports more expensive and imports less expensive. The investment community is wary of changes in the stability of a currency, and the change caused massive losses for investors and the plunging of the Swiss stock market.[4] Ultimately, people faced with spiraling price increases avoid the increasingly worthless paper money at all costs, storing all of their savings in the form of real assets such as gold and land.

Durability.   Money must be durable. The crisp new dollar bills you trade for products at the mall will make their way all around town for about six years before being replaced (see Table 15.1). Were the value of an old, faded bill to fall in line with the deterioration of its appearance, the principles of stability and universal acceptability would fail (but, no doubt, fewer bills would pass through the washer!). Although metal coins, due to their much longer useful life, would appear to be an ideal form of money, paper currency is far more portable than metal because of its light weight. Today, coins are used primarily to provide divisibility.

Difficulty to Counterfeit.   Finally, to remain stable and enjoy universal acceptance, it almost goes without saying that money must be very difficult to counterfeit—that is, to duplicate illegally. Every country takes steps to make counterfeiting difficult. Most use multicolored money, and many use specially watermarked papers that are virtually impossible to duplicate. It is becoming increasingly easy for counterfeiters to print money.[5] This illegal printing of money is fueled by hundreds of people who often circulate only small amounts of counterfeit bills. However, even rogue governments such as North Korea are known to make counterfeit U.S. currency. To thwart the problem of counterfeiting, the U.S. Treasury Department redesigned the U.S. currency, starting with the $20 bill in 2003, the $50 bill in 2004, the $10 bill in 2006, the $5 bill in 2008, and the $100 bill in 2013.[6] U.S. money includes subtle colors in addition to the traditional green, as well as enhanced security features, such as a watermark, security thread, and color-shifting ink.[7] Many countries are discontinuing large-denominated bills that are used in illegal trade such as drugs or terrorism. The idea is that it is more difficult to transport or hide €100 notes than €500 notes. Although counterfeiting is not as much of an issue with coins, U.S. metal coins are usually worth more for the metal than their face value. It has begun to cost more to manufacture coins than what they are worth monetarily.

TABLE 15.1   Life Expectancy of Money
How long is the life span of U.S. paper money?

When currency is deposited with a Federal Reserve Bank, the quality of each note is evaluated by sophisticated processing equipment. Notes that meet the strict quality criteria—that is, they are still in good condition—continue to circulate, while those that do not are taken out of circulation and destroyed. This process determines the life span of a Federal Reserve note.

Life span varies by denomination. One factor that influences the life span of each denomination is how the denomination is used by the public. For example, $100 notes are often used as a store of value. This means that they pass between users less frequently than lower denominations that are more often used for transactions, such as $5 notes. Thus, $100 notes typically last longer than $5 notes.

Denomination	Estimated Life Span*
$ 1	5.8 years
$ 5	5.5 years
$ 10	4.5 years
$ 20	7.9 years
$ 50	8.5 years
$ 100	15.0 years

*Estimated life spans as of December 2013. Because the $2 note does not widely circulate, we do not publish its estimated life span. The $100 bill estimate is from December 2012.

Source: Board of Governors of the Federal Reserve System, "How Long Is the Life Span of U.S. Paper Money?" www.federalreserve.gov/faqs/how-long-is-the-life-span-of-us-paper-money.htm (accessed April 28, 2018).

As Table 15.2 indicates, it costs more to produce pennies and nickels than their face value. For example, we can see that in 2017 it cost $0.0182 to produce a one-cent piece, or 82 percent more than it was worth. However, what the U.S. Mint loses on pennies and nickels it makes up for with profits on dimes, quarters, and dollars. The U.S. $1 coin proved to be so unpopular that the U.S. Mint discontinued producing it after 2013. Profits fluctuate over time because of the rising and falling costs of copper, zinc, and nickel, but dimes and quarters have always been profitable for the U.S. Mint.

checking account
money stored in an account at a bank or other financial institution that can be withdrawn without advance notice; also called a demand deposit.

Types of Money

While paper money and coins are the most visible types of money, the combined value of all of the printed bills and all of the minted coins is actually rather insignificant when compared with the value of money kept in checking accounts, savings accounts, and other monetary forms.

You probably have a **checking account** (also called a *demand deposit*), money stored in an account at a bank or other financial institution that can be withdrawn without advance notice. One way to withdraw funds from your account is by writing a *check*, a

The U.S. government redesigns currency to stay ahead of counterfeiters and protect the public.

©zefart/Shutterstock

Entrepreneurship in Action

Square Rounds Out Its Product Offering

Square Inc.

Founders: Jack Dorsey and Jim McKelvey

Founded: 2010, in San Francisco, California

Success: Square Inc. has released a number of products that help small businesses process credit cards, print receipts, manage inventory, and more—all from a mobile device.

Square Inc. is perhaps best known for the Square Reader, a small device that acts as a credit card reader when plugged into a mobile device's audio jack. However, Square's innovative products did not stop there. Three years after release of the Square Reader, Square introduced the Square Stand, a piece of hardware that easily converts an iPad into a point-of-sale system. The Square Stand enables the built-in card reader to connect with other accessories, allowing businesses to print receipts and scan bar codes. The swiveling stand, including a contactless card payment system and chip reader, is priced at $169.

To provide more innovation for retail businesses, in 2017 the company launched Square for Retail, a set of tools optimized for retailers that included a new point-of-sale iPad app. Key features include employee management, advanced inventory management and reporting, barcode label printing, and more. This newest product is priced at $999. Square's innovative mobile payment and point-of-sale products give businesses an easy and affordable way to grow and manage their operations.[9]

Critical Thinking Questions

1. How do you think Square is affecting the credit card industry?
2. Why is Square particularly popular among small businesses?
3. Will Square's products replace more traditional card readers and point-of-purchase systems? Why or why not?

LO 15-2

Describe various types of money.

written order to a bank to pay the indicated individual or business the amount specified on the check from money already on deposit. Figure 15.1 explains the significance of the numbers found on a typical U.S. check. As legal instruments, checks serve as a substitute for currency and coins and are preferred for many transactions due to their lower risk of loss. If you lose a $100 bill, anyone who finds or steals it can spend it. If you lose a blank check, however, the risk of catastrophic loss is quite low. Not only does your bank have a sample of your signature on file to compare with a suspected forged signature, but you can render the check immediately worthless by means of a stop-payment order at your bank.

There are several types of checking accounts, with different features available for different monthly fee levels or specific minimum account balances. Some checking accounts earn interest (a small percentage of the amount deposited in the account that the bank pays to the depositor). One such interest-bearing checking account is

TABLE 15.2
Costs to Produce U.S. Coins

Fiscal Year	Penny	Nickel	Dime	Quarter	Total Profit from Coins (Millions)
2017	$ 0.0182	$0.0666	$0.0333	$0.0082	$391.50
2016	$ 0.0150	$0.0632	$0.0308	$0.0763	$ 578.7
2015	$ 0.0143	$0.0744	$0.0354	$0.0844	$ 540.9
2014	$ 0.0166	$0.0809	$ 0.0391	$0.0895	$ 289.1
2013	$ 0.0183	$ 0.0941	$0.0456	$ 0.0105	$ 137.4
2012	$0.0200	$ 0.1009	$0.0499	$ 0.1130	$ 105.9
2011	$ 0.0241	$ 0.1118	$0.0565	$ 0.1114	$ 348.8

Source: Various annual reports of the U.S. Mint.

FIGURE 15.1   A Check

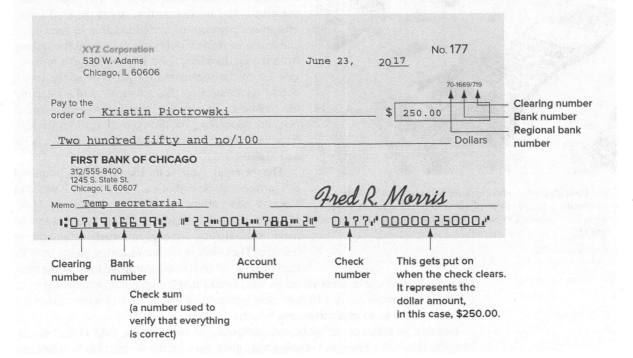

the *NOW (Negotiable Order of Withdrawal) account* offered by most financial institutions. The interest rate paid on such accounts varies with the interest rates available in the economy but is typically quite low (more recently less than 1 percent but in the past between 2 and 5 percent).

Savings accounts (also known as *time deposits*) are accounts with funds that usually cannot be withdrawn without advance notice and/or have limits on the number of withdrawals per period. While seldom enforced, the "fine print" governing most savings accounts prohibits withdrawals without two or three days' notice. Savings accounts are not generally used for transactions or as a medium of exchange, but their funds can be moved to a checking account or turned into cash.

Money market accounts are similar to interest-bearing checking accounts, but with more restrictions. Generally, in exchange for slightly higher interest rates, the owner of a money market account can write only a limited number of checks each month, and there may be a restriction on the minimum amount of each check.

Certificates of deposit (CDs) are savings accounts that guarantee a depositor a set interest rate over a specified interval of time as long as the funds are not withdrawn before the end of the interval—six months, one year, or seven years, for example. Money may be withdrawn from these accounts prematurely only after paying a substantial penalty. In general, the longer the term of the CD, the higher is the interest rate it earns. As with all interest rates, the rate offered and fixed at the time the account is opened fluctuates according to economic conditions.

Credit cards allow you to promise to pay at a later date by using preapproved lines of credit granted by a bank or finance company. They are a popular substitute for

savings accounts
accounts with funds that usually cannot be withdrawn without advance notice; also known as time deposits.

money market accounts
accounts that offer higher interest rates than standard bank rates but with greater restrictions.

certificates of deposit (CDs)
savings accounts that guarantee a depositor a set interest rate over a specified interval as long as the funds are not withdrawn before the end of the period—six months or one year, for example.

credit cards
means of access to preapproved lines of credit granted by a bank or finance company.

Credit cards have many advantages, including being able to buy expensive items and pay them off a little at a time. However, this can easily lead an individual to incur spiraling credit card debt that is hard to pay off.

©Piotr Adamowicz/123RF

reward cards
credit cards made available by stores that carry a benefit to the user.

debit card
a card that looks like a credit card but works like a check; using it results in a direct, immediate, electronic payment from the cardholder's checking account to a merchant or third party.

cash payments because of their convenience, easy access to credit, and acceptance by merchants around the world. The institution that issues the credit card guarantees payment of a credit charge to merchants and assumes responsibility for collecting the money from the cardholders. Card issuers charge a transaction fee to the merchants for performing the credit check, guaranteeing the payment, and collecting the payment. The fee is typically between 2 and 5 percent, depending on the type of card. American Express fees are usually higher than those for Visa and MasterCard.

The original American Express cards required full payment at the end of each month, but American Express now offers credit cards similar to Visa, MasterCard, and Discover that allow cardholders to make installment payments and carry a maximum balance. There is a minimum monthly payment with interest charged on the remaining balance. Some people pay off their credit cards monthly, while other make monthly payments. Charges for unpaid balances can run 18 percent or higher at an annual rate, making credit card debt one of the most expensive ways to borrow money.

Besides the major credit card companies, many stores—Target, Saks Fifth Avenue, Macy's, Bloomingdales, and others—have their own branded credit cards. They use credit rating agencies to check the credit of the cardholders and they generally make money on the finance charges. **Reward cards** are credit cards that carry a benefit to the user. For example, gas stations such as Mobil and Shell have branded credit cards so that when you use the card you save five or six cents per gallon. Others—such as airline cards for American, Delta, and United—reward you with miles that you can use for flights. And there are cash-back credit cards that give you 1 percent or more cash back on everything you spend.

The Credit CARD (Card Accountability Responsibility and Disclosure) Act of 2009 was passed to regulate the practices of credit card companies. The law limited the ability of card issuers to raise interest rates, limited credit to young adults, gave people more time to pay bills, and made clearer due dates on billing cycles, along with several other provisions. For college students, the most important part of the law is that young adults under the age of 21 have to have an adult co-signer or show proof that they have enough income to handle the debt limit on the card.

This act is important to all companies and cardholders. Research indicates that approximately 40 percent of lower- and middle-income households use credit cards to pay for basic necessities. Yet there is also good news. The average credit card debt for lower- and middle-income households has decreased in recent years. On the other hand, studies also show that college students tend to lack the financial literacy needed to understand credit cards and their requirements. Therefore, vulnerable segments of the population, such as college students, should be careful about which credit cards to choose and how often they use them.[10]

A **debit card** looks like a credit card but works like a check. The use of a debit card results in a direct, immediate, electronic payment from the cardholder's checking account to a merchant or other party. While they are convenient to carry and profitable for banks, they lack credit features, offer no purchase "grace period," and provide no hard "paper trail." Debit cards are gaining more acceptance with merchants, and

Decrypting Cryptocurrencies

The expansion of the Internet has helped develop the phenomenon of cryptocurrency. *Cryptocurrency* is a digital currency that uses cryptography (writing or solving codes) for security and is not issued by any central authority. The first cryptocurrency was Bitcoin. Launched in 2009, Bitcoin is backed by a decentralized network of computers that use computing power to keep up with and verify transactions. All transactions are validated, coded, and lumped together on a chronological, public ledger called a *blockchain*. It uses a public and private key (a string of numbers and letters strung together in an algorithm) as dual security to authorize and keep secret Bitcoin transactions.

Cryptocurrency is appealing for a variety of reasons, including easier transfer of funds between parties and fewer processing fees. On the other hand, the lack of a centralized entity makes it vulnerable to illicit activities, such as money laundering and tax evasion. Government-issued currency (fiat money) has a stable value, while Bitcoin and the other cryptocurrencies do not have a stable value and are subject to speculation.

Bitcoin has ranged in price from $19,500 per coin to $6,000 per coin.

Some are also concerned that cryptocurrency is being used in countries like North Korea and Venezuela to circumvent economic sanctions because the transfers of cryptocurrency going from buyer to seller are almost impossible to detect. Many consider cryptocurrencies to be securities, and the government wants to be able to track taxes from any gains or losses.

Although there is controversy about the sustainability of cryptocurrency as legitimate legal tender, many believe that the current stage is similar to that of the Internet in the 1990s. Given its unique qualities, there is potential for these digital currencies to make transactions simpler and less expensive.[11]

Critical Thinking Questions

1. What are some of the risks involved with cryptocurrencies?
2. What are some of the benefits involved with trading in cryptocurrencies?
3. Many governments are wary of cryptocurrencies. Why do you think this is?

---

consumers like debit cards because of the ease of getting cash from an increasing number of ATM machines. Financial institutions also want consumers to use debit cards because they reduce the number of teller transactions and check processing costs. Some cash management accounts at retail brokers like Merrill Lynch offer deferred debit cards. These act like a credit card but debit to the cash management account once a month. During that time, the cash earns a money market return.

Traveler's checks, money orders, and cashier's checks are other forms of "near money." Although each is slightly different from the others, they all share a common characteristic: A financial institution, bank, credit company, or neighborhood currency exchange issues them in exchange for cash and guarantees that the purchased note will be honored and exchanged for cash when it is presented to the institution making the guarantee.

A new type of money called cryptocurrency has become popular over the last several years with technology-oriented people (see cryptocurrency box above). Bitcoin is the most popular, but it has a fluctuating price, which makes it less desirable than fiat money like dollars, yens, and euros that are backed by governments. It also does not have a stable value or portability. But it does work in a digital economy with instantaneous payments where the transaction size is large and the value of the cryptocurrency is known by both the buyer and the seller.

Credit Card Fraud. More and more computer hackers have managed to steal credit card information and either use the information for Internet purchases or actually make a card exactly the same as the stolen card. Losses on credit card theft run into the billions, but consumers are usually not liable for the losses. However, consumers should be careful with debit cards because once the money is out of the account, the bank and credit card companies cannot get it back. Debit cards do not have the same level of protection as credit cards.

LO 15-3

Specify how the Federal Reserve Board manages the money supply and regulates the American banking system.

The American Financial System

The U.S. financial system fuels our economy by storing money, fostering investment opportunities, and making loans for new businesses and business expansion as well as for homes, cars, and college educations. This amazingly complex system includes banking institutions, nonbanking financial institutions such as finance companies, and systems that provide for the electronic transfer of funds throughout the world. Over the past 20 years, the rate at which money turns over, or changes hands, has increased exponentially. Different cultures place unique values on saving, spending, borrowing, and investing. The combination of this increased turnover rate and increasing interactions with people and organizations from other countries has created a complex money system. First, we need to meet the guardian of this complex system.

Federal Reserve Board an independent agency of the federal government established in 1913 to regulate the nation's banking and financial industry; also called "the Fed."

The Federal Reserve System

The guardian of the American financial system is the **Federal Reserve Board**, or "the Fed," as it is commonly called, an independent agency of the federal government established in 1913 to regulate the nation's banking and financial industry. The Federal Reserve System is organized into 12 regions, each with a Federal Reserve Bank that serves its defined area (Figure 15.2). All the Federal Reserve banks except those in

FIGURE 15.2   Federal Reserve System

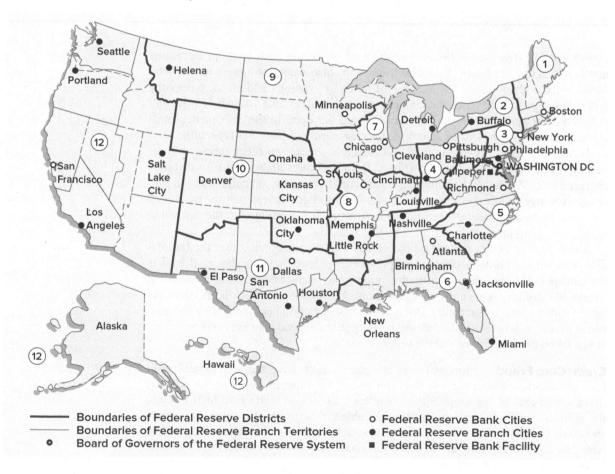

Boundaries of Federal Reserve Districts
Boundaries of Federal Reserve Branch Territories
○ Board of Governors of the Federal Reserve System

○ Federal Reserve Bank Cities
● Federal Reserve Branch Cities
■ Federal Reserve Bank Facility

Boston and Philadelphia have regional branches. The Cleveland Federal Reserve Bank, for example, is responsible for branch offices in Pittsburgh and Cincinnati.

The Federal Reserve Board is the chief economic policy arm of the United States. Working with Congress and the president, the Fed tries to create a positive economic environment capable of sustaining low inflation, high levels of employment, a balance in international payments, and long-term economic growth. To this end, the Federal Reserve Board has four major responsibilities: (1) to control the supply of money, or monetary policy; (2) to regulate banks and other financial institutions; (3) to manage regional and national checking account procedures, or check clearing; and (4) to supervise the federal deposit insurance programs of banks belonging to the Federal Reserve System.

Monetary Policy.   The Fed controls the amount of money available in the economy through **monetary policy**. Without this intervention, the supply of and demand for money might not balance. This could result in either rapid price increases (inflation) because of too little money or economic recession and a slowdown of price increases (disinflation) because of too little growth in the money supply. In very rare cases (the depression of the 1930s), the United States has suffered from deflation, when the actual purchasing power of the dollar has increased as prices declined. To effectively control the supply of money in the economy, the Fed must have a good idea of how much money is in circulation at any given time. This has become increasingly challenging because the global nature of our economy means that more and more U.S. dollars are circulating overseas. Using several different measures of the money supply, the Fed establishes specific growth targets that, presumably, ensure a close balance between money supply and money demand. The Fed fine-tunes money growth by using four basic tools: open market operations, reserve requirements, the discount rate, and credit controls (see Table 15.3). There is generally a lag of 6 to 18 months before the effect of these charges shows up in economic activity.

monetary policy
means by which the Fed controls the amount of money available in the economy.

Activity	Effect on the Money Supply and the Economy
Buy government securities	The money supply increases; economic activity increases.
Sell government securities	The money supply decreases; economic activity slows down.
Raise discount rate	Interest rates increase; the money supply decreases; economic activity slows down.
Lower discount rate	Interest rates decrease; the money supply increases; economic activity increases.
Increase reserve requirements	Banks make fewer loans; the money supply declines; economic activity slows down.
Decrease reserve requirements	Banks make more loans; the money supply increases; economic activity increases.
Relax credit controls	More people are encouraged to make major purchases, increasing economic activity.
Restrict credit controls	People are discouraged from making major purchases, decreasing economic activity.

TABLE 15.3
Fed Tools for Regulating the Money Supply

open market operations decisions to buy or sell U.S. Treasury bills (short-term debt issued by the U.S. government) and other investments in the open market.

Open market operations refer to decisions to buy or sell U.S. Treasury bills (short-term debt issued by the U.S. government; also called T-bills) and other investments in the open market. The actual purchase or sale of the investments is performed by the New York Federal Reserve Bank. This monetary tool, the most commonly employed of all Fed operations, is performed almost daily in an effort to control the money supply.

When the Fed buys securities, it writes a check on its own account to the seller of the investments. When the seller of the investments (usually a large bank) deposits the check, the Fed transfers the balance from the Federal Reserve account into the seller's account, thus increasing the supply of money in the economy and, hopefully, fueling economic growth. The opposite occurs when the Fed sells investments. The buyer writes a check to the Federal Reserve, and when the funds are transferred out of the purchaser's account, the amount of money in circulation falls, slowing economic growth to a desired level.

The second major monetary policy tool is the **reserve requirement**, the percentage of deposits that banking institutions must hold in reserve ("in the vault," as it were). Funds so held are not available for lending to businesses and consumers. For example, a bank holding $10 million in deposits, with a 10 percent reserve requirement, must have reserves of $1 million. If the Fed were to reduce the reserve requirement to, say, 5 percent, the bank would need to keep only $500,000 in reserves. The bank could then lend to customers the $500,000 difference between the old reserve level and the new lower reserve level, thus increasing the supply of money. Because the reserve requirement has such a powerful effect on the money supply, the Fed does not change it very often, relying instead on open market operations most of the time.

The third monetary policy tool, the **discount rate**, is the rate of interest the Fed charges to loan money to any banking institution to meet reserve requirements. The Fed is the lender of last resort for these banks. When a bank borrows from the Fed, it is said to have borrowed at the "discount window," and the interest rates charged there are often higher than those charged on loans of comparable risk elsewhere in the economy. This added interest expense, when it exists, serves to discourage banks from borrowing from the Fed.

When the Fed wants to expand the money supply, it lowers the discount rate to encourage borrowing. Conversely, when the Fed wants to decrease the money supply, it raises the discount rate. The increases in interest rates that occurred in the United States from 2003 through 2006 were the result of more than 16 quarter-point (0.25 percent) increases in the Fed discount rate. The purpose was to keep inflation under control and to raise rates to a more normal level as the economy recovered from the recession of 2001. The Fed lowered interest rates to combat the 2007 recession, and they have remained historically low for the past 10 years as the Fed tried to stimulate a slow-growing economy. As the economy showed improvement, the Fed began to raise rates in 2017. Not surprisingly, economists watch changes in this sensitive interest rate as an indicator of the Fed's monetary policy.

The final tool in the Fed's arsenal of weapons is **credit controls**—the authority to establish and enforce credit rules for financial institutions and some private investors. For example, the Fed can determine how large a down payment individuals and businesses must make on credit purchases of expensive items such as automobiles, and how much time they have to finish paying for the purchases. By raising and lowering minimum down payment amounts and payment periods, the Fed can stimulate or discourage credit purchases of "big ticket" items. The Fed also has the authority to

connect

Need help understanding how the Federal Reserve tries to stabilize the economy? Visit your Connect ebook video tab for a brief animated explanation.

reserve requirement the percentage of deposits that banking institutions must hold in reserve.

discount rate the rate of interest the Fed charges to loan money to any banking institution to meet reserve requirements.

credit controls the authority to establish and enforce credit rules for financial institutions and some private investors.

set the minimum down payment investors must use for the credit purchases of stock. Buying stock with credit—"buying on margin"—is a popular investment strategy among individual speculators. By altering the margin requirement (currently set at 50 percent of the price of the purchased stocks), the Fed can effectively control the total amount of credit borrowing in the stock market.

Regulatory Functions.   The second major responsibility of the Fed is to regulate banking institutions that are members of the Federal Reserve System. Accordingly, the Fed establishes and enforces banking rules that affect monetary policy and the overall level of the competition between different banks. It determines which non-banking activities, such as brokerage services, leasing, and insurance, are appropriate for banks and which should be prohibited. The Fed also has the authority to approve or disapprove mergers between banks and the formation of bank holding companies. In an effort to ensure that all rules are enforced and that correct accounting proce-dures are being followed at member banks, surprise bank examinations are conducted by bank examiners each year.

Check Clearing.   The Federal Reserve provides national check processing on a huge scale. Divisions of the Fed known as check clearinghouses handle almost all the checks written against a bank in one city and presented for deposit to a bank in a sec-ond city. Any banking institution can present the checks it has received from others around the country to its regional Federal Reserve Bank. The Fed passes the checks to the appropriate regional Federal Reserve Bank, which then sends the checks to the issuing bank for payment. With the advance of electronic payment systems and the passage of the Check Clearing for the 21st Century Act (Check 21 Act), checks can now be processed in a day. The Check 21 Act allows banks to clear checks electroni-cally by presenting an electronic image of the check. This eliminates mail delays and time-consuming paper processing.

Depository Insurance.   The Fed is also responsible for supervising the federal insurance funds that protect the deposits of member institutions. These insurance funds will be discussed in greater detail in the following section.

Banking Institutions

LO 15-4

Compare and contrast commercial banks, savings and loan associations, credit unions, and mutual savings banks.

Banking institutions accept money deposits from and make loans to individual con-sumers and businesses. Some of the most important banking institutions include com-mercial banks, savings and loan associations, credit unions, and mutual savings banks. Historically, these have all been separate institutions. However, new hybrid forms of banking institutions that perform two or more of these functions have emerged over the past two decades. They all have one thing in common: They are businesses whose objective is to earn money by managing, safeguarding, and lending money to others. Their sales revenues come from the fees and interest that they charge for providing these financial services.

Since the financial crisis, and during the 2016 political campaign, Wall Street and banks have been the target of politicians as they continue to take out their anger at the banking world for all the economic problems that exist in the United States. However, during the first two years of President Trump's administration, many of the regulations were relaxed, particularly for regional and community banks. You will see as you go through this section that the financial network is very complex.

Citibank is the consumer division of Citigroup, one of the four largest banks in the United States.

©Andriy Blokhin/Shutterstock

commercial banks
the largest and oldest of all financial institutions, relying mainly on checking and savings accounts as sources of funds for loans to businesses and individuals.

Commercial Banks. The largest and oldest of all financial institutions are **commercial banks**, which perform a variety of financial services. They rely mainly on checking and savings accounts as their major source of funds and use only a portion of these deposits to make loans to businesses and individuals. Because it is unlikely that all the depositors of any one bank will want to withdraw all of their funds at the same time, a bank can safely loan out a large percentage of its deposits.

Today, banks are quite diversified and offer a number of services. Commercial banks make loans for virtually any conceivable legal purpose, from vacations to cars, from homes to college educations. Banks in many states offer *home equity loans,* by which home owners can borrow against the appraised value of their already purchased homes. Banks also issue Visa and MasterCard credit cards and offer CDs and trusts (legal entities set up to hold and manage assets for a beneficiary). Many banks rent safe deposit boxes in bank vaults to customers who want to store jewelry, legal documents, artwork, and other valuables. In 1999, Congress passed the Financial Services Modernization Act, also known as the Gramm-Leach-Bliley Bill. This act repealed the Glass Steagall Act, which was enacted in 1929 after the stock market crash and prohibited commercial banks from being in the insurance and investment banking business. This puts U.S. commercial banks on the same competitive footing as European banks and provides a more level playing field for global banking competition. As commercial banks and investment banks have merged, the financial landscape has changed. Consolidation remains the norm in the U.S. banking industry. The financial crisis and the economic recession that began in 2007 and lasted into 2012 only accelerated the consolidation as large, healthy banks ended up buying weak banks that were in trouble. Most of these purchases were made with financial help from the U.S. Treasury and Federal Reserve. By 2012, the banks had paid back their loans, but the financial meltdown exposed some high-risk activities in the banking industry Congress wanted to curtail. The result was the passage of the Dodd-Frank Act. This act added many new regulations, but the two most important changes raised the required capital banks had to hold on their balance sheet and limited certain types of high-risk trading activities.

savings and loan associations (S&Ls)
financial institutions that primarily offer savings accounts and make long-term loans for residential mortgages; also called "thrifts."

Savings and Loan Associations. **Savings and loan associations (S&Ls)**, often called "thrifts," are financial institutions that primarily offer savings accounts and make long-term loans for residential mortgages. A mortgage is a loan made so that a business or individual can purchase real estate, typically a home; the real estate itself is pledged as a guarantee (called *collateral*) that the buyer will repay the loan. If the loan is not repaid, the savings and loan has the right to repossess the property. Prior to the 1970s, S&Ls focused almost exclusively on real estate lending and accepted only savings accounts. Today, following years of regulatory changes, S&Ls compete directly with commercial banks by offering many types of services.

Savings and loans have gone through a metamorphosis since the early 1990s, after having almost collapsed in the 1980s. Today, many of the largest savings and loans have merged with commercial banks. This segment of the financial services industry plays a diminished role in the mortgage lending market.

Credit Unions.   A credit union is a financial institution owned and controlled by its depositors, who usually have a common employer, profession, trade group, or religion. The Aggieland Credit Union in College Station, Texas, for example, provides banking services for faculty, employees, and current and former students of Texas A&M University. A savings account at a credit union is commonly referred to as a share account, while a checking account is termed a share draft account. Because the credit union is tied to a common organization, the members (depositors) are allowed to vote for directors and share in the credit union's profits in the form of higher interest rates on accounts and/or lower loan rates.

While credit unions were originally created to provide depositors with a short-term source of funds for low-interest consumer loans for items such as cars, home appliances, vacations, and college, today they offer a wide range of financial services. Generally, the larger the credit union, the more sophisticated its financial service offerings will be.

credit union
a financial institution owned and controlled by its depositors, who usually have a common employer, profession, trade group, or religion.

Mutual Savings Banks.   Mutual savings banks are similar to savings and loan associations, but, like credit unions, they are owned by their depositors. Among the oldest financial institutions in the United States, they were originally established to provide a safe place for savings of the working classes. Found mostly in New England, they are becoming more popular in the rest of the country as some S&Ls have converted to mutual savings banks to escape the stigma created by the widespread S&L failures in the 1980s.

mutual savings banks
financial institutions that are similar to savings and loan associations but, like credit unions, are owned by their depositors.

Insurance for Banking Institutions.   The Federal Deposit Insurance Corporation (FDIC), which insures individual bank accounts, was established in 1933 to help stop bank failures throughout the country during the Great Depression. Today, the FDIC insures personal accounts up to a maximum of $250,000 at nearly 6,000 FDIC member institutions.[12] While most major banks are insured by the FDIC, small institutions in some states may be insured by state insurance funds or private insurance companies. Should a member bank fail, its depositors can recover all of their funds, up to $250,000. Amounts over $250,000, while not legally covered by the insurance, are, in fact, usually covered because the Fed understands very well the enormous damage that would result to the financial system should these large depositors withdraw their money. When the financial crisis occurred, the FDIC increased the deposit insurance amount from $100,000 to $250,000 on a temporary basis to increase consumer confidence in the banking system. The Dodd-Frank Act passed on July 21, 2010, made the $250,000 insurance per account permanent. The *Federal Savings and Loan Insurance Corporation (FSLIC)* insured thrift deposits prior to its insolvency and failure during the S&L crisis of the 1980s. Now, the insurance functions once overseen by the FSLIC are handled directly by the FDIC through its Savings Association Insurance Fund. The National Credit Union Administration (NCUA) regulates and charters credit unions and insures their deposits through its National Credit Union Insurance Fund.

When they were originally established, Congress hoped that these insurance funds would make people feel secure about their

Federal Deposit Insurance Corporation (FDIC)
an insurance fund established in 1933 that insures individual bank accounts.

National Credit Union Administration (NCUA)
an agency that regulates and charters credit unions and insures their deposits through its National Credit Union Insurance Fund.

The National Credit Union Administration has the important job of regulating and chartering credit unions and insuring their deposits through its National Credit Union Insurance Fund.

Source: National Credit Union Administration
https://www.ncua.gov/newsroom/Pages/downloadable-graphics.aspx

Bank on It: Widespread Wells Fargo Fraud Uncovered

In 2016, Wells Fargo lost its place as the world's biggest commercial bank in the wake of a large-scale, cross-selling scandal. Wells Fargo employees faked 3.5 million customer accounts to meet short-term sales goals. Approximately 5,300 employees were fired, and the firm was slapped with a $185 million fine by the Consumer Financial Protection Bureau.

Investigations revealed that controversial sales goals encouraged employees to open accounts without customers' permission. Employees engaged in fraudulent activities such as opening up fake bank accounts and falsifying signatures to satisfy unrealistic sales goals.

Although the accusations claimed the misconduct reached back to 2011, managers at Wells Fargo claim these practices had been occurring long before. Several employees came forward to claim that they reported the misconduct and were fired as a result. If true, this would violate laws that protect whistleblowers from retaliation.

To reach its lofty sales goals, Wells Fargo set up increased commissions around the product being emphasized. These products were cross-sold to customers with an aggressive sales incentive program tied to employee compensation. This incentive program suggests that Wells Fargo executives, managers, and employees forgot that a bank's reputation is built on a basic cultural value of trust. Managers at many branches played a large role in the establishment of unauthorized accounts.

As a result of the scandal, bank customers felt deceived. The bank reported that checking account openings had fallen 43 percent and credit card applications 55 percent from the year before. Today, the bank finds its reputation in ruins thanks to unrealistic sales quotas and a coercive corporate environment.[13]

Critical Thinking Questions

1. How did Wells Fargo's focus on short-term gains cause it to violate the duty it owed to its customers?
2. Describe how the Wells Fargo scandal affected the opening of new accounts in the immediate aftermath. Do you think these business areas will recover in the near future?
3. How is the Wells Fargo scandal likely to affect trust in the financial and banking industry?

---

savings so they would not panic and withdraw their money when news of a bank failure was announced. The "bank run" scene in the perennial Christmas movie *It's a Wonderful Life*, when dozens of Bailey Building and Loan depositors attempted to withdraw their money (only to have the reassuring figure of Jimmy Stewart calm their fears), was not based on mere fiction. During the Great Depression, hundreds of banks failed and their depositors lost everything. The fact that large numbers of major financial institutions failed in the 1980s and 1990s—without a single major banking panic—underscores the effectiveness of the current insurance system. Large bank failures occurred once again during the most recent recession. According to the FDIC, 52 banks failed between January 2000 and December 31, 2008. Because of the financial crisis and long-lasting recession, 488 failed between January 1, 2009, and December 31, 2014. Only 21 insured banks failed from 2015 to the end of 2017, so it is clear the crisis is behind us. This reflects an improving economy and a healthier financial system. It is safe to say that most depositors go to sleep every night without worrying about the safety of their savings.

Nonbanking Institutions

LO 15-5

Distinguish among nonbanking institutions such as insurance companies, pension funds, mutual funds, and finance companies.

Nonbank financial institutions offer some financial services, such as short-term loans or investment products, but do not accept deposits. These include insurance companies, pension funds, mutual funds, brokerage firms, nonfinancial firms, and finance companies. Table 15.4 lists the assets of some diversified financial services firms.

Diversified Firms.   There are many nonfinancial firms that help finance their customers' purchases of expensive equipment. For example, Caterpillar (construction

	2017 Assets (in millions)	2016 Assets (in millions)	2015 Assets (in millions)	
JPMorgan Chase	$2,533,600	$2,490,972	$2,351,698	**TABLE 15.4**
Citigroup Inc.	1,842,465	1,792,077	1,731,210	Leading Diversified
BlackRock Group	220,217	220,177	225,261	Financial Services Firms
American Express	181,159	158,893	161,184	
Ameriprise Financial	147,470	139,821	145,342	
Visa	67,977	64,035	40,236	
Invesco	31,669	25,734	25,073	
The Blackstone Group	26,403	22,526	31,497	
Aon	26,088	26,615	27,164	
Marsh & McLennan Companies	20,429	18,190	18,216	
Apollo Investment Corp.	6,991	3,093	3,561	

Source: CFRA Stock Reports by S&P Global

equipment), Boeing (airplanes), and General Electric (jet engines and locomotives) help their customers finance these large-scale expensive purchases through their finance subsidiaries. At one time, General Electric's credit subsidiary accounted for 40 percent of the company's revenues, but this is slowly falling as the company divests itself of consumer credit operations. Automobile companies such as Ford have also traditionally had credit subsidiaries to help customers finance their cars.

Insurance Companies. **Insurance companies** are businesses that protect their clients against financial losses from certain specified risks (death, injury, disability, accident, fire, theft, and natural disasters, for example) in exchange for a fee, called a premium. Because insurance premiums flow into the companies regularly, but major insurance losses cannot be timed with great accuracy (though expected risks can be assessed with considerable precision), insurance companies generally have large amounts of excess funds. They typically invest these or make long-term loans, particularly to businesses in the form of commercial real estate loans.

insurance companies
businesses that protect their clients against financial losses from certain specified risks (death, accident, and theft, for example).

pension funds
managed investment pools set aside by individuals, corporations, unions, and some nonprofit organizations to provide retirement income for members.

Pension Funds. **Pension funds** are managed investment pools set aside by individuals, corporations, unions, and some nonprofit organizations to provide retirement income for members. One type of pension fund is the *individual retirement account (IRA)*, which is established by individuals to provide for their personal retirement needs. IRAs can be invested in a variety of financial assets, from risky

State Farm is the largest auto and home insurer. Its website allows users to input their information and receive an auto insurance quote quickly and conveniently.

©James R. Martin/Shutterstock

commodities such as oil or cocoa to low-risk financial "staples" such as U.S. Treasury securities. The choice is up to each person and is dictated solely by individual objectives and tolerance for risk. The interest earned by all of these investments may be deferred tax-free until retirement.

In 1997, Congress revised the IRA laws and created a Roth IRA. Although similar to a traditional IRA in that investors may contribute $5,500 per year, the money in a Roth IRA is considered an after-tax contribution. Workers over 50 can add an extra $1,000, but in all cases, if you make too much money, you cannot fund a Roth. When the money is withdrawn at retirement, no tax is paid on the distribution. The Roth IRA is beneficial to young people who can allow a long time for their money to compound and who may be able to have their parents or grandparents fund the Roth IRA with gift money.

Most major corporations provide some kind of pension plan for their employees. Many of these are established with bank trust departments or life insurance companies. Money is deposited in a separate account in the name of each individual employee, and when the employee retires, the total amount in the account can be either withdrawn in one lump sum or taken as monthly cash payments over some defined time period (usually for the remaining life of the retiree).

Social Security, the largest pension fund, is publicly financed. The federal government collects Social Security funds from payroll taxes paid by both employers and employees. The Social Security Administration then takes these monies and makes payments to those eligible to receive Social Security benefits—the retired, the disabled, and the young children of deceased parents.

Mutual Funds.   A **mutual fund** pools individual investor dollars and invests them in large numbers of well-diversified securities. Individual investors buy shares in a mutual fund in the hope of earning a high rate of return and in much the same way as people buy shares of stock. Because of the large numbers of people investing in any one mutual fund, the funds can afford to invest in hundreds (if not thousands) of securities at any one time, minimizing the risks of any single security that does not do well. Mutual funds provide professional financial management for people who lack the time and/or expertise to invest in particular securities, such as government bonds. While there are no hard-and-fast rules, investments in one or more mutual funds are one way for people to plan for financial independence at the time of retirement.

A special type of mutual fund called a *money market fund* invests specifically in short-term debt securities issued by governments and large corporations. Although they offer services such as check-writing privileges and reinvestment of interest income, money market funds differ from the money market accounts offered by banks primarily in that the former represent a pool of funds, while the latter are basically specialized, individual checking accounts. Money market funds usually offer slightly higher rates of interest than bank money market accounts.

Brokerage Firms and Investment Banks.   **Brokerage firms** buy and sell stocks, bonds, and other securities for their customers and provide other financial services. Larger brokerage firms like Merrill Lynch, Charles Schwab, and Edward Jones offer financial services unavailable at their smaller competitors. Merrill Lynch, for example, offers the Merrill Lynch Cash Management Account (CMA), which pays interest on deposits and allows clients to write checks, borrow money, and withdraw cash much like a commercial bank. The largest of the brokerage firms (including Merrill Lynch) have developed so many specialized services that they may be considered financial networks—organizations capable of offering virtually all of the services

mutual fund
an investment company that pools individual investor dollars and invests them in large numbers of well-diversified securities.

brokerage firms
firms that buy and sell stocks, bonds, and other securities for their customers and provide other financial services.

traditionally associated with commercial banks. The rise of online brokerage firms has helped investors who want to do it themselves at low costs. Firms like E-Trade and TD Ameritrade offer investors the ability to buy and sell securities for $5 to $7 per trade, while the same trade at Morgan Stanley might cost $125. E-Trade offers banking services, debit cards, wire transfers, and many of the same services that the traditional brokerage firms offer.

Most brokerage firms are really part financial conglomerates that provide many different kinds of services besides buying and selling securities for clients. For example, Merrill Lynch also is an investment banker, as are Morgan Stanley and Goldman Sachs. The **investment banker** underwrites new issues of securities for corporations, states, and municipalities needed to raise money in the capital markets. The new issue market is called a *primary market* because the sale of the securities is for the first time. After the first sale, the securities trade in the *secondary markets* by brokers. The investment banker advises on the price of the new securities and generally guarantees the sale while overseeing the distribution of the securities through the selling brokerage houses. Investment bankers also act as dealers who make markets in securities. They do this by offering to sell the securities at an asked price (which is a higher rate) and buy the securities at a bid price (which is a lower rate)—the difference in the two prices represents the profit for the dealer.

investment banker
underwrites new issues of securities for corporations, states, and municipalities.

Finance Companies.   **Finance companies** are businesses that offer short-term loans at substantially higher rates of interest than banks. Commercial finance companies make loans to businesses, requiring their borrowers to pledge assets such as equipment, inventories, or unpaid accounts as collateral for the loans. Consumer finance companies make loans to individuals. Like commercial finance companies, these firms require some sort of personal collateral as security against the borrower's possible inability to repay their loans. Because of the high interest rates they charge and other factors, finance companies typically are the lender of last resort for individuals and businesses whose credit limits have been exhausted and/or those with poor credit ratings.

finance companies
businesses that offer short-term loans at substantially higher rates of interest than banks.

Electronic Banking

Since the advent of the computer age, a wide range of technological innovations has made it possible to move money all across the world electronically. Such "paperless" transactions have allowed financial institutions to reduce costs in what has been, and continues to be, a virtual competitive battlefield. **Electronic funds transfer (EFT)** is any movement of funds by means of an electronic terminal, telephone, computer, or magnetic tape. Such transactions order a particular financial institution to subtract money from one account and add it to another. The most commonly used forms of EFT are automated teller machines, automated clearinghouses, and home banking systems.

electronic funds transfer (EFT)
any movement of funds by means of an electronic terminal, telephone, computer, or magnetic tape.

Automated Teller Machines.   Probably the most familiar form of electronic banking is the **automated teller machine (ATM)**, which dispenses cash, accepts deposits, and allows balance inquiries and cash transfers from one account to another. ATMs provide 24-hour banking services—both at home (through a local bank) and far away (via worldwide ATM networks such as Cirrus and Plus). Rapid growth, driven by both strong consumer acceptance and lower transaction costs for banks (about half the cost of teller transactions), has led to the installation of hundreds of thousands of ATMs worldwide. Table 15.5 presents some interesting statistics about ATMs.

automated teller machine (ATM)
the most familiar form of electronic banking, which dispenses cash, accepts deposits, and allows balance inquiries and cash transfers from one account to another.

TABLE 15.5
Facts about ATM Use

There are 425,000 ATM machines in use in the United States.
The average cash withdrawal from ATMs is $60.
The typical ATM consumer will visit an ATM 7.4 times per month.
The total ratio of people per ATM machine is 144:1.
The top ATM owners are Cardtronics, Payment Alliance International, Bank of America, JPMorgan Chase, and Wells Fargo.

Source: "ATM Machines Statistics," March 29, 2017, www.statisticbrain.com/atm-machine-statistics/ (accessed April 28, 2018).

automated clearinghouses (ACHs)
a system that permits payments such as deposits or withdrawals to be made to and from a bank account by magnetic computer tape.

Automated Clearinghouses. **Automated clearinghouses (ACHs)** permit payments such as deposits or withdrawals to be made to and from a bank account by magnetic computer tape. Most large U.S. employers, and many others worldwide, use ACHs to deposit their employees' paychecks directly to the employees' bank accounts. While direct deposit is used by only 50 percent of U.S. workers, nearly 100 percent of Japanese workers and more than 90 percent of European workers utilize it. The largest user of automated clearinghouses in the United States is the federal government, with 99 percent of federal government employees and 65 percent of the private workforce receiving their pay via direct deposit. More than 82 percent of all Social Security payments are made through an ACH system. The Social Security Administration is trying to reduce costs and reduce theft and fraud, so if you applied for Social Security benefits on or after May 1, 2011, you must receive your payments electronically.

The advantages of direct deposits to consumers include convenience, safety, and potential interest earnings. It is estimated that more than 4 million paychecks are lost or stolen annually, and FBI studies show that 2,000 fraudulent checks are cashed every day in the United States. Checks can never be lost or stolen with direct deposit. The benefits to businesses include decreased check-processing expenses and increased employee productivity. Research shows that businesses that use direct deposit can save more than $1.25 on each payroll check processed. Productivity could increase by $3 to $5 billion annually if all employees were to use direct deposit rather than taking time away from work to deposit their payroll checks.

Some companies also use ACHs for dividend and interest payments. Consumers can also use ACHs to make periodic (usually monthly) fixed payments to specific creditors without ever having to write a check or buy stamps. The estimated number of bills paid annually by consumers is 20 billion, and the total number paid through ACHs is estimated at only 8.5 billion. The average consumer who writes 10 to 15 checks each month would save $41 to $62 annually in postage alone.[14]

Online Banking. Many banking activities are now conducted on a computer at home or at work or through wireless devices such as cell phones and tablets anywhere there is a wireless "hot spot." Consumers and small businesses can now make a bewildering array of financial transactions at home or on the go 24 hours a day. Functioning much like a vast network of personal ATMs, companies like Google and Apple provide online banking services through mobile phones, allowing subscribers to make sophisticated banking transactions, buy and sell stocks and bonds, and purchase products and airline tickets without ever leaving home or speaking to another human being. Many banks allow customers to log directly into their accounts to check balances, transfer money between accounts, view their account statements, and pay bills via home computer or other Internet-enabled devices. Computer and advanced telecommunications technology have revolutionized world commerce; 62 percent of adults list Internet banking as their preferred banking method, making it the most popular banking method in the United States.[15]

Future of Banking

Rapid advances and innovations in technology are challenging the banking industry and requiring it to change. As we said earlier, more and more banks, both large and small, are offering electronic access to their financial services. ATM technology is rapidly changing, with machines now dispensing more than just cash. Online financial services, ATM technology, and bill presentation are just a few of the areas where rapidly changing technology is causing the banking industry to change as well.

Computers and handheld devices have made online banking extremely convenient. However, hackers have stolen millions from banking customers by tricking them into visiting websites and downloading malicious software that gives hackers access to their passwords.

©Kite_rin/Shutterstock

Impact of Financial Crisis.   The premise that banks will get bigger over the next 10 years is uncertain. During 2007–2008, the financial markets collapsed under the weight of declining housing prices, subprime mortgages (mortgages with low-qualifying borrowers), and risky securities backed by these subprime mortgages. Because the value of bank assets declined dramatically, most large banks like Citibank, Bank of America, and Wells Fargo had a shrinking capital base. That is, the amount of debt in relation to their equity was so high that they were below the minimum required capital requirements.

> **LO 15-6**
>
> Analyze the challenges ahead for the banking industry.

During this period, the Federal Reserve took unprecedented actions that included buying up troubled assets from the banks and lending money at the discount window to nonbanks such as investment banks and brokers. The Fed also entered into the financial markets by making markets in commercial paper and other securities where the markets had ceased to function in an orderly fashion. Additionally, the Fed began to pay interest on reserves banks kept at the Fed, and finally, it kept interest rates low to stimulate the economy and to help the banks regain their health. Because banks make money by the spread between their borrowing and lending rates, the Fed managed the spread between long- and short-term rates to generate a fairly large spread for the banks.

Additionally, to keep interest rates low and stimulate the economy, the Fed bought billions of dollars of mortgages and other financial assets on a monthly basis. By mid-2017, it had accumulated $4.5 trillion of securities on its balance sheet even though it stopped its asset purchases in 2015 as the economy improved. By the end of 2017, the Fed was slowly letting the securities in its portfolio mature without reinvesting the proceeds. As the economy picked up steam in 2018, the process of unwinding the Fed's portfolio helped push interest rates higher.

Lastly, the future of the structure of the banking system is in the hands of the U.S. Congress. In reaction to the financial meltdown and severe recession, Congress passed the Dodd-Frank Wall Street Reform and Consumer Protection Act. The full name implies that the intent of the act is to eliminate the ability of banks to create this type of problem in the future. The question that remains is how much of this legislation will be revised or reversed by the Trump administration.

Shadow Banking.   In broad general terms, *shadow banking* refers to companies performing banking functions of some sort that are not regulated by banking regulators. Shadow banking activities are increasing. In a letter to shareholders in its annual report, James Dimon, CEO and chair of JPMorgan Chase, was quoted as saying to his shareholders that the bank will face tough competitors, including shadow banking. He may have said it best in the following quote:

> Many of these institutions are smart and sophisticated and will benefit as banks move out of certain products and services. Non-bank financial competitors will look at every product we price, and if they can do it cheaper with their set of capital providers, they will. There is nothing inherently wrong with this—it is a natural state of affairs and, in some cases, may benefit the clients who get the better price. But regulators should—and will—be looking at how all financial companies (including non-bank competitors) need to be regulated and will be evaluating what is better to be done by banks vs. non-banks and vice versa.[16]

In addition to shadow banks mentioned by Mr. Dimon, there are the peer-to-peer lenders like Prosper, a company that matches investors and borrowers with loans of between $2,000 and $35,000. There are other sources of funding by Internet websites such as GoFundMe, which helps people enhance their life skills, raise money for health care issues, and more. Another similar website is Kickstarter, which funds creative projects in the worlds of art, film, games, music, publishing, and so on. In many cases, funds provided for these projects replace loans that might have been used to develop the project. These forms of funding are growing rapidly. Kickstarter was formed in October 2009 and has already received a total of $3 billion to fund more than 122,000 projects.[17] There is also the budding use of virtual money and other futuristic ideas, so only time will tell how the world of banking changes over time and how bank regulators will deal with these nonbank institutions. ■

So You're Interested in Financial Systems or Banking

You think you might be interested in going into finance or banking, but it is so hard to tell when you are a full-time student. Classes that seem interesting when you take them might not translate into an interesting work experience after graduation. A great way to see if you would excel at a career in finance is to get some experience in the industry. Internships, whether they are paid or unpaid, not only help you figure out what you might really want to do after you graduate, but they are also a great way to build up your résumé, put your learning to use, and start generating connections within the field.

For example, Pennsylvania's Delaware County District Attorney's Office has been accepting business students from Villanova University for a six-month internship. The student works in the economic crime division, analyzing documents of people under investigation for financial crimes ranging from fraud to money laundering. The students get actual experience in forensic accounting and have the chance to see whether this is the right career path. On top of that, the program has saved the county an average of $20,000 annually on consulting and accounting fees, not to mention that detectives now have more time to take on larger caseloads. One student who completed the program spent his six months investigating a case in which the owner of a sewage treatment company had embezzled a total of $1 million over the course of nine years. The student noted that the experience helped him gain an understanding about how different companies handle their financial statements, as well as how accounting can be applied in forensics and law enforcement.

Internship opportunities are plentiful all over the country, although you may need to do some research to find them. To start, talk to your program advisor and your professors about opportunities. Also, you can check company websites where you think you might like to work to see if they have any opportunities available. City, state, or federal government offices often provide student internships as well. No matter where you end up interning, the real-life skills you pick up, as well as the résumé boost you get, will be helpful in finding a job after you graduate. When you graduate, commercial banks and other financial institutions offer major employment opportunities. In 2008–2009, a major downturn in the financial industry resulted in mergers, acquisitions, and financial restructuring for many companies. While the immediate result was a decrease in job opportunities, as the industry recovers, there will be many challenging job opportunities available.[18]

Review Your Understanding

Define *money*, its functions, and its characteristics.

Money is anything generally accepted as a means of payment for goods and services. Money serves as a medium of exchange, a measure of value, and a store of wealth. To serve effectively in these functions, money must be acceptable, divisible, portable, durable, stable in value, and difficult to counterfeit.

Describe various types of money.

Money may take the form of currency, checking accounts, or other accounts. Checking accounts are funds left in an account in a financial institution that can be withdrawn (usually by writing a check) without advance notice. Other types of accounts include savings accounts (funds left in an interest-earning account that usually cannot be withdrawn without advance notice), money market accounts (an interest-bearing checking account that is invested in short-term debt instruments), certificates of deposit (deposits left in an institution for a specified period of time at a specified interest rate), credit cards (access to a preapproved line of credit granted by a bank or company), and debit cards (means of instant cash transfers between customer and merchant accounts), as well as traveler's checks, money orders, and cashier's checks.

Specify how the Federal Reserve Board manages the money supply and regulates the American banking system.

The Federal Reserve Board regulates the U.S. financial system. The Fed manages the money supply by buying and selling government securities, raising or lowering the discount rate (the rate of interest at which banks may borrow cash reserves from the Fed), raising or lowering bank reserve requirements (the percentage of funds on deposit at a bank that must be held to cover expected depositor withdrawals), and adjusting down payment and repayment terms for credit purchases. It also regulates banking practices, processes checks, and oversees federal depository insurance for institutions.

Compare and contrast commercial banks, savings and loan associations, credit unions, and mutual savings banks.

Commercial banks are financial institutions that take and hold deposits in accounts for and make loans to individuals and businesses. Savings and loan associations are financial institutions that primarily specialize in offering savings accounts and mortgage loans. Credit unions are financial institutions owned and controlled by their depositors. Mutual savings banks are similar to S&Ls except that they are owned by their depositors.

Distinguish among nonbanking institutions such as insurance companies, pension funds, mutual funds, and finance companies.

Insurance companies are businesses that protect their clients against financial losses due to certain circumstances, in exchange for a fee. Pension funds are investments set aside by organizations or individuals to meet retirement needs. Mutual funds pool investors' money and invest in large numbers of different types of securities. Brokerage firms buy and sell stocks and bonds for investors. Finance companies make short-term loans at higher interest rates than do banks.

Analyze the challenges ahead for the banking industry.

Future changes in financial regulations are likely to result in fewer but larger banks and other financial institutions.

Recommend the most appropriate financial institution for a hypothetical small business.

Using the information presented in this chapter, you should be able to answer the questions in the "Solve the Dilemma" feature near the end of this chapter and find the best institution for Hill Optometrics.

Critical Thinking Questions

Enter the World of Business Questions

1. What are some of the benefits of mobile banking? The risks?

2. Why do you think consumers fail to understand the risk factors of using online banking apps without taking appropriate security precautions?

3. What are some ways consumers can significantly reduce the risk of cyberattacks when engaging in mobile banking activities?

Learn the Terms

automated clearinghouses (ACHs) 494
automated teller machine (ATM) 493
brokerage firms 492
certificates of deposit (CDs) 481
checking account 479
commercial banks 488
credit cards 481
credit controls 486
credit union 489
debit card 482
discount rate 486

electronic funds transfer (EFT) 493
Federal Deposit Insurance Corporation
 (FDIC) 489
Federal Reserve Board 484
finance 476
finance companies 493
insurance companies 491
investment banker 493
monetary policy 485
money 476
money market accounts 481

mutual fund 492
mutual savings banks 489
National Credit Union Administration
 (NCUA) 489
open market operations 486
pension funds 491
reserve requirement 486
reward cards 482
savings accounts 481
savings and loan associations
 (S&Ls) 488

Check Your Progress

1. What are the six characteristics of money? Explain how the U.S. dollar has those six characteristics.

2. What is the difference between a credit card and a debit card? Why are credit cards considerably more popular with U.S. consumers?

3. Discuss the four economic goals the Federal Reserve must try to achieve with its monetary policy.

4. Explain how the Federal Reserve uses open market operations to expand and contract the money supply.

5. What are the basic differences between commercial banks and savings and loans?

6. Why do credit unions charge lower rates than commercial banks?

7. Why do finance companies charge higher interest rates than commercial banks?

8. How are mutual funds, money market funds, and pension funds similar? How are they different?

9. What are some of the advantages of electronic funds transfer systems?

Get Involved

1. Survey the banks, savings and loans, and credit unions in your area, and put together a list of interest rates paid on the various types of checking accounts. Find out what, if any, restrictions are in effect for NOW accounts and regular checking accounts. In which type of account and in what institution would you deposit your money? Why?

2. Survey the same institutions as in question 1, this time inquiring as to the rates asked for each of their various loans. Where would you prefer to obtain a car loan? A home loan? Why?

Build Your Skills

Managing Money

Background

You have just graduated from college and have received an offer for your dream job (annual salary: $35,000). This premium salary is a reward

for your hard work, perseverance, and good grades. It is also a reward for the social skills you developed in college doing service work as a tutor for high school students and interacting with the business community as the program chairman of the college business fraternity, Delta Sigma Pi. You are engaged and plan to be married this summer. You and your spouse will have a joint income of $60,000, and the two of you are trying to decide the best way to manage your money.

Task

Research available financial service institutions in your area, and answer the following questions.

1. What kinds of institutions and services can you use to help manage your money?

2. Do you want a full-service financial organization that can take care of your banking, insurance, and investing needs, or do you want to spread your business among individual specialists? Why have you made this choice?

3. What retirement alternatives do you have?

Solve the Dilemma

Seeing the Financial Side of Business

Dr. Stephen Hill, a successful optometrist in Indianapolis, Indiana, has tinkered with various inventions for years. Having finally developed what he believes is his first saleable product (a truly scratch-resistant and lightweight lens), Hill has decided to invest his life savings and open Hill Optometrics to manufacture and market his invention.

Unfortunately, despite possessing true genius in many areas, Hill is uncertain about the "finance side" of business and the various functions of different types of financial institutions in the economy. He is, however, fully aware that he will need financial services such as checking and savings accounts, various short-term investments that can easily and quickly be converted to cash as needs dictate, and sources of borrowing capacity—should the need for either short- or

long-term loans arise. Despite having read mounds of brochures from various local and national financial institutions, Hill is still somewhat unclear about the merits and capabilities of each type of financial institution. He has turned to you, his 11th patient of the day, for help.

LO 15-7

Recommend the most appropriate financial institution for a hypothetical small business.

Critical Thinking Questions

1. List the various types of U.S. financial institutions and the primary function of each.

2. What services of each financial institution is Hill's new company likely to need?

3. Which single financial institution is likely to be best able to meet Hill's small company's needs now? Why?

Build Your Business Plan

 connect

Money and the Financial System

This chapter provides you with the opportunity to think about money and the financial system and just how many new businesses fail every year. In some industries, the failure rate is as high as 80 percent. One reason for such a high failure rate is the inability to manage the finances of the

organization. From the start of the business, financial planning plays a key role. Try getting a loan without an accompanying budget/forecast of earnings and cash flow.

While obtaining a loan from a family member may be the easiest way to fund your business, it may cause more problems for you later on if you are unable to pay the money back as scheduled. Before heading to a lending officer at a bank, contact your local SBA center to see what assistance it might provide.

See for Yourself Videocase

Crowdfunding: Loans You Can Count On

Kiva is a nonprofit organization that accepts donations to crowdfund loans around the world. It was founded in 2005 when Premal Shah recognized that small business owners in

developing countries didn't have access to funding. Kiva takes donations on its site from people around the world who can donate a minimum of $25 for specific projects. Money is then given to recipients who are expected to pay back the loan over a set period of time. As the loan is paid back, the original lenders can then either withdraw the money they donated or use it to fund another loan. Kiva now backs thousands of loans each week at a variety of price points and has surpassed $1 billion in loans. Most Kiva loans go to help recipients build inventory or buy assets that will improve their productivity.

Technology is constantly changing the world around us. This is true in the financial world, too. Traditional banks and other financial entities are competing with new structures, like Kiva, that can provide similar services in an innovative way. By using crowdfunding on the Internet to award loans to broader audiences, Kiva has found a way to innovate in the financial loan industry. This is a form of shadow banking, where a company performs banking functions, like awarding loans, when the company is not a traditional banking institution and is not encumbered by traditional banking regulations. Lenders like Kiva, known as peer-to-peer lenders, create a human connection.

Even though Kiva does not have to follow many traditional banking regulations, the company must comply with regulations imposed by countries around the world that they are operating in. Developing countries often have poor regulations that increase costs for Kiva and complicate the loan process. For example, money invested in India must stay in India for at least three years, which can create major cash flow problems. Another challenge for Kiva is dealing with infrastructure. These countries have very poor financial infrastructure systems, which is another challenge that Kiva has to overcome when trying to fund those in need.

While Kiva faces many challenges to operate in developing countries, it also faces challenges in the United States. Kiva started operating in the U.S. in 2010 when it recognized a need for a nontraditional loan financing system domestically. While the U.S. is one of the best countries in the world to start a small businesses and secure capital, young people without a credit history and immigrants often have trouble securing loans for their businesses. Kiva now aims to aid those who can't secure a traditional loan in the United States through its existing crowdfunded loan system.

One of the most fascinating aspects of the Kiva system is Kiva's choice to crowdfund loans instead of using donations.

While donations are very helpful for aiding crisis efforts and natural disaster relief, Kiva believes that lending is a more effective way to combat poverty. The company believes that by lending, recipients will have a stable and ongoing source of capital for their business. Loan recipients can also use their Kiva loans as a credit history in order to try and secure loans with traditional financial institutions. Because of the payback structure of Kiva's loans, Kiva is able to reinvest money that is paid back to fund other projects.

"Access to capital isn't enough to end global poverty, but it is clearly necessary," says Elliot Collins, a research and evaluations manager for Kiva. "Kiva's network is reducing the cost of lending to entrepreneurs around the world, and our goal is to ensure that everybody can benefit from financial markets and help build their local economy."

Thanks to its innovative take on crowdfunding loans, Kiva is able to reach those in need all around the world. Donors can give money and then reinvest paid back loans into other projects. Because of Kiva's nontraditional structure, it is able to avoid many banking regulations. However, Kiva continues faces many challenges due to different government regulations and infrastructures. To stay competitive across the world, Kiva must continue to monitor regulations in the countries in which it operates and pay close attention to how governments will treat shadow banking entities in the future.[19]

Critical Thinking Questions

1. Explain the benefit of awarding loans instead of donations.

2. What are the differences between Kiva and more traditional financial institutions?

3. What challenges does Kiva currently face, and what future challenges could it face?

You can find the related video in the Video Library in Connect. Ask your instructor how you can access Connect.

Team Exercise

Mutual funds pool individual investor dollars and invest them in a number of different securities. Go to http://finance.yahoo.com/ and select some top-performing funds using criteria such as sector, style, or strategy. Assume that your group has $100,000 to invest in mutual funds. Select five funds in which to invest, representing a balanced (varied industries, risk, etc.) portfolio, and defend your selections.

Notes

1. Dan Weil, "How Secure Is Mobile Banking?" *The Wall Street Journal,* March 18, 2018, https://www.wsj.com/articles/how-secure-is-mobile-banking-1521424920 (accessed April 7, 2018); Haider Pasha, "5 Ways to Improve Your Digital Banking Security," *Gulf News—Personal Finance,* March 20, 2018, http://gulfnews.com/gn-focus/personal-finance/banking/5-ways-to-improve-your-digital-banking-security-1.2191358 (accessed April 7, 2018); Dan DiPietro, "Survey Roundup: Cost, Frequency of Cyberattacks Rise for Banks," *The Wall Street Journal,* February 14, 2018, https://blogs.wsj.com/riskandcompliance/2018/02/14/survey-roundup-cost-frequency-of-cyberattacks-rises-for-banks/ (accessed April 7, 2018).

2. Paul Krugman, "Why Is Deflation Bad?" *The New York Times,* August 2, 2010, http://krugman.blogs.nytimes.com/2010/08/02/why-is-deflation-bad/ (accessed May 5, 2017).

3. Board of Governors of the Federal Reserve System, "How Much U.S. Currency Is in Circulation?," September 26, 2018, https://www.federalreserve.gov/faqs/currency_12773.htm (accessed November 8, 2018).

4. "Why the Swiss Unpegged the Franc," January 18, 2015, *The Economist,* http://www.economist.com/blogs/economist-explains/2015/01/economist-explains-13 (accessed May 5, 2107); "Swiss-Made Products Become More Expensive as Franc Rises," *Sydney Morning Herald,* January 16, 2015, http://www.smh.com.au/national/swissmade-products-become-more-expensive-asfranc-rises-20150116-12riei.html (accessed May 5, 2017).

5. "Weird and Wonderful Money Facts and Trivia," *Happy Worker,* www.happyworker.com/magazine/facts/weird-and-wonderful-money-facts (accessed April 25, 2016).

6. "The History of American Currency," *U.S. Currency Education Program,* https://www.uscurrency.gov/history (accessed September 2, 2018).

7. "About the Redesigned Currency," Department of the Treasury, Bureau of Engraving and Printing, www.newmoney.gov/newmoney/currency/aboutnotes.htm (accessed April 2, 2010).

8. "Like Magic: The Tech That Goes into Making Money Harder to Fake," *NPR,* October 23, 2017, https://www.npr.org/sections/alltechconsidered/2017/10/23/559092168/like-magic-the-tech-that-goes-into-making-money-harder-to-fake (accessed April 24, 2017).

9. "About Us," Square Inc., https://squareup.com/about (accessed April 19, 2018); "Square Register," Square Inc., https://squareup.com/hardware/register (accessed April 19, 2018); Kaya Yurieff, "Square Unveils Cash Register of the Future," *CNN,* October 30, 2017, http://money.cnn.com/2017/10/30/technology/square-register/indcx.html (accessed April 19, 2018); Tomio Geron, "Square Launches Stand, a New Point of Sale Device," *Forbes,* May 14, 2013, https://www.forbes.com/sites/tomiogeron/2013/05/14/square-launches-stand-a-new-point-of-sale-device/#631b7f587c2e (accessed April 19, 2018).

10. Jessica Dickler, "Americans Still Relying on Credit Cards to Get By," *CNN Money,* May 23, 2012, http://money.cnn.com/2012/05/22/pf/credit-card/index.htm (accessed May 5, 2017); Martin Merzer, "Survey: Students Fail the Credit Card Test," *Fox Business,* April 16, 2012, www.creditcards.com/credit-card-news/survey-students-fail-credit-card-test-1279.php (accessed May 5, 2017).

11. Paul Vigna and Dave Michaels, "Has the Cryptocoin Market Met Its Match in the SEC?" *The Wall Street Journal,* March 20, 2018, https://www.wsj.com/articles/hot-cryptocoin-market-chilled-by-sec-scrutiny-1521557569 (accessed April 7, 2018); David Goodboy, "The 3 Best Cryptocurrency Exchanges," January 8, 2018, https://www.nasdaq.com/article/the-3-best-cryptocurrency-exchanges-cm902049 (accessed April 7, 2018); Laura Shin, "How to Explain Cryptocurrencics and Blockchains to the Average Person," *Forbes,* October 3, 2017, https://www.forbes.com/sites/laurashin/2017/10/03/how-to-explain-cryptocurrencies-and-blockchains-to-the-average-person/#1fc27ef9324d (accessed April 7, 2018); Paul Ford, "Bitcoin Is Ridiculous. Blockchain Is Dangerous," *Bloomberg,* https://www.bloomberg.com/news/features/2018-03-09/bitcoin-is-ridiculous-blockchain-is-dangerous-paul-ford (accessed April 7, 2018); "Cryptocurrency," *Investopedia,* https://www.investopedia.com/terms/c/cryptocurrency.asp (accessed April 7, 2018).

12. "Deposit Insurance Simplification Fact Sheet," FDIC website, www.unitedamericanbank.com/pdfs/FDIC-Insurance-Coverage-Fact-Sheet.pdf (accessed June 16, 2014).

13. Paul Blake, "Timeline of the Wells Fargo Accounts Scandal," *ABC News,* November 3, 2016, http://abcnews.go.com/Business/timeline-wells-fargo-accounts-scandal/story?id=42231128 (accessed April 14, 2017); Matt Egan, "I Called the Wells Fargo Ethics Line and Was Fired," *CNN Money,* September 21, 2016, http://money.cnn.com/2016/09/21/investing/wells-fargo-fired-workers-retaliation-fake-accounts/ (accessed April 14, 2017); Matt Egan, "Wells Fargo Workers: Fake Accounts Began Years Ago," *CNN Money,* September 26, 2016, http://money.cnn.com/2016/09/26/investing/wells-fargo-fake-accounts-before-2011/ (accessed April 14, 2017); James Venable, "Wells Fargo: Where Did They Go Wrong?" Working Paper, Harvard University, February 9, 2017, http://scholar.harvard.edu/files/jtv/files/wells_fargo_where_did_they_go_wrong_by_james_venable_pdf_02.pdf (accessed April 14, 2017); Curtis C. Verschoor, "Lessons from the Wells Fargo Scandal," *Strategic Finance,* November 1, 2016, http://sfmagazine.com/post-entry/november-2016-lessons-from-the-wells-fargo-scandal/ (accessed April 14, 2017); Lisa Cook, "The Wells Fargo Scandal:

Is the Profit Model to Blame?" *Knowledge@Wharton,* University of Pennsylvania, September 13, 2016, http://knowledge.wharton.upenn.edu/article/how-the-wells-fargo-scandal-will-reverberate/ (accessed April 14, 2017); Geoff Colvin, "The Wells Fargo Scandal Is Now Reaching VW Proportions," *Fortune,* January 26, 2017, http://fortune.com/2017/01/25/the-wells-fargo-scandal-is-now-reaching-vw-proportions/ (accessed April 14, 2017); Brian Tayan, "The Wells Fargo Cross-Selling Scandal," Harvard Law School Forum on Corporate Governance and Financial Regulations, December 19, 2016, https://corpgov.law.harvard.edu/2016/12/19/the-wells-fargo-cross-selling-scandal/ (accessed April 14, 2017); Michael Corkery, "Wells Fargo Struggling in the Aftermath of Fraud Scandal," *The New York Times,* January 13, 2017, https://www.nytimes.com/2017/01/13/business/dealbook/wells-fargo-earnings-report.html (accessed April 14, 2017); Lucinda Shen, "Wells Fargo Sales Scandal Could Hurt Growth Permanently," *Fortune,* April 13, 2017, http://fortune.com/2017/04/13/wells-fargo-report-earnings/ (accessed April 14, 2017); Matt Egan, "5,300 Wells Fargo Employees Fired Over 2 Million Phony Accounts," *CNN,* September 9, 2016, http://money.cnn.com/2016/09/08/investing/wells-fargo-created-phony-accounts-bank-fees/ (accessed April 14, 2017); Laura J. Keller and Katherine Chiglinsky, "Wells Fargo Eclipsed by JPMorgan as World's Most Valuable Bank," *Bloomberg,* September 13, 2016, https://www.bloomberg.com/news/articles/2016-09-13/wells-fargo-eclipsed-by-jpmorgan-as-world-s-most-valuable-bank (accessed April 14, 2017); Laura Lorenzetti, "This Is the Most Valuable Bank in the World," *Fortune,* July 23, 2015, http://fortune.com/2015/07/23/wells-fargo-worlds-most-valuable-bank/ (accessed April 14, 2017); Matt Egan, "Letter Warned Wells Fargo of 'Widespread' Fraud in 2007—Exclusive," *CNN Money,* October 18, 2016, http://money.cnn.com/2016/10/18/investing/wells-fargo-warned-fake-accounts-2007/ (accessed January 6, 2017); Stacy Cowley, "At Wells Fargo, Complaints about Fraudulent Accounts since 2005," *The New York Times,* October 11, 2016, http://www.nytimes.com/2016/10/12/business/dealbook/at-wells-fargo-complaints-about-fraudulent-accounts-since-2005.html (accessed January 6, 2017); Mark Snider, "Ex-Wells Fargo Bankers Sue over Firing amid Fraud," *USA Today,* September 25, 2016, http://www.usatoday.com/story/money/2016/09/25/ex-wells-fargo-employees-sue-over-scam/91079158/

(accessed January 6, 2017); Ian Mount, "Wells Fargo's Fake Accounts May Go Back More Than 10 Years," *Fortune,* October 12, 2016, http://fortune.com/2016/10/12/wells-fargo-fake-accounts-scandal/ (accessed April 14, 2017); Stacy Cowley and Jennifer A. Kingson, "Wells Fargo to Claw Back $75 Million from 2 Former Executives," *The New York Times,* April 10, 2017, https://www.nytimes.com/2017/04/10/business/wells-fargo-pay-executives-accounts-scandal.html (accessed April 14, 2017); Matt Egan, "Wells Fargo Uncovers Up to 1.4 Million More Fake Accounts," *CNN Money,* August 31, 2017, http://money.cnn.com/2017/08/31/investing/wells-fargo-fake-accounts/index.html (accessed December 28, 2017).

14. "NACHA Reports More Than 18 Billion ACH Payments in 2007," *The Free Library,* May 19, 2008, www.thefreelibrary.com/NACHA+Reports+More+Than+18+Billion+ACH+Payments+in+2007.-a0179156311 (accessed May 2, 2016).

15. "Online Banking Far and Away the Preferred Retail Channel, Survey Finds," September 13, 2011, http://www.bankersweb.com/headlines/article/online-banking-far-and-away-preferred-retail-channel (accessed November 19, 2017).

16. JP Morgan 2013 Annual Report, 10.

17. "About," Kickstarter, https://www.kickstarter.com/about (accessed April 2, 2017).

18. "CSI Pennsylvania," *CFO Magazine,* March 2008, p. 92.

19. "Leadership," Kiva, https://www.kiva.org/about/leadership (accessed May 22, 2018); Devishobha Chandramouli, "Study Details Why Women Entrepreneurs Have Greater Crowdfunding Success," *Entrepreneur,* May 17, 2018, https://www.entrepreneur.com/article/312964 (accessed June 3, 2018); Elizabeth MacBride, "Kiva Hits $1B in Loans, $25 at a Time. Here's One of the Hidden Keys to its Success," *Forbes,* July 31, 2017, https://www.forbes.com/sites/elizabethmacbride/2017/07/31/can-online-lenders-assess-your-character-to-a-certain-extent-yes/#6c984c5c1b2f (accessed June 3, 2018); Connie Loizos, "This Young Lending Startup Just Secured $70 Million to Lend $2 at a Time," *TechCrunch,* March 28, 2018, https://techcrunch.com/2018/03/28/this-young-lending-startup-just-secured-70-million-to-lend-2-at-a-time/ (accessed June 3, 2018).

Credits

Design elements: Part opener, ©Steve Allen/Getty Images; Consider Ethics and Social Responsibility icon, ©Design Pics/PunchStock; Think Globally icon, ©Sheff/Shutterstock; Responding to Business Challenges icon, ©Olivier LeMoal/Shutterstock; Going Green icon, ©Beboy/Shutterstock; Entrepreneurship in Action icon, ©Ruslan Grechka/Shutterstock;

Test Prep icon, ©McGraw-Hill Education; Build Your Skills icon, ©Ilya Terentyev/Getty Images; Solve the Dilemma icon, ©Beautyimage/Shutterstock; Build Your Business Plan icon, ©ALMAGAMI/ Shutterstock; See for Yourself Videocase icon, ©MIKHAIL GRACHIKOV/Shutterstock.

16 Financial Management and Securities Markets

©Carsten Reisinger/Shutterstock

Chapter Outline

Learning Objectives

After reading this chapter, you will be able to:

LO 16-1 Describe some common methods of managing current assets.

LO 16-2 Identify some sources of short-term financing (current liabilities).

LO 16-3 Summarize the importance of long-term assets and capital budgeting.

LO 16-4 Specify how companies finance their operations and manage fixed assets with long-term liabilities, particularly bonds.

LO 16-5 Explain how corporations can use equity financing by issuing stock through an investment banker.

LO 16-6 Describe the various securities markets in the United States.

LO 16-7 Critique the position of short-term assets and liabilities of a small manufacturer and recommend corrective action.

Enter the World of Business

No Shock: General Electric Struggles after Poor Financial Management

You would be hard-pressed to find a more influential firm in American history than General Electric (GE). Starting out as a light bulb company founded by Thomas Edison, in 2017 it listed the following divisions: Power, Renewable Energy, Oil & Gas, Aviation, Healthcare, Transportation, Lighting, and GE Capital.

Perhaps the biggest mistake was expanding GE Capital into consumer loans. Originally, GE Capital helped customers finance their purchases of its locomotives and airplane engines. However, GE expanded into consumer products such as credit cards, insurance, and mortgage lending through GE Capital. When the financial crisis hit, GE was forced to be regulated by the U.S. Federal Reserve as a bank. The decision was made to shrink GE Capital back to its original focus of lending to industrial customers. It sold off Synchrony Financial, its consumer finance area, and liquidated most of its consumer-oriented subsidiaries. GE Capital got out from under government regulation but lost almost $7 billion with more losses predicted.

The shareholder outrage at declining earnings peaked in 2017 when GE cut its dividend in half because of cash flow problems. After becoming the worst-performing stock in the Dow Jones Industrial Average, GE was booted from the Dow in 2018 causing the stock to plunge. The common stock continued to decline to under $10 per share as the company once again cut the dividend to $0.01 per quarter in October of 2018.

The problem GE faced was a mix of bad decisions and market conditions. It sold off divisions it bought years earlier for losses, bought companies that did not perform, and were in the wrong industries at the wrong time. To recover, GE plans to sell several divisions to create a simpler company. Time will tell whether GE can regain its reputation.[1]

Introduction

While it's certainly true that money makes the world go around, financial management is the discipline that makes the world turn more smoothly. Indeed, without effective management of assets, liabilities, and owners' equity, all business organizations are doomed to fail—regardless of the quality and innovativeness of their products. Financial management is the field that addresses the issues of obtaining and managing the funds and resources necessary to run a business successfully. It is not limited to business organizations: All organizations, from the corner store to the local nonprofit art museum, from giant corporations to county governments, must manage their resources effectively and efficiently if they are to achieve their objectives.

In this chapter, we look at both short- and long-term financial management. First, we discuss the management of short-term assets, which companies use to generate sales and conduct ordinary day-to-day business operations. Next, we turn our attention to the management of short-term liabilities, the sources of short-term funds used to finance the business. Then, we discuss the management of long-term assets such as plants, equipment, and the use of common stock (equity) and bonds (long-term liability) to finance these long-term corporate assets. Finally, we look at the securities markets, where stocks and bonds are traded.

LO 16-1

Describe some common methods of managing current assets.

Managing Current Assets and Liabilities

Managing short-term assets and liabilities involves managing the current assets and liabilities on the balance sheet (discussed in the "Accounting and Financial Statements" chapter). Current assets are short-term resources such as cash, investments, accounts receivable, and inventory. Current liabilities are short-term debts such as accounts payable, accrued salaries, accrued taxes, and short-term bank loans. We use the terms *current* and *short term* interchangeably because short-term assets and liabilities are usually replaced by new assets and liabilities within three or four months, and always within a year. Managing short-term assets and liabilities is sometimes called **working capital management** because short-term assets and liabilities continually flow through an organization and are thus said to be "working."

working capital management
the managing of short-term assets and liabilities.

Managing Current Assets

The chief goal of financial managers who focus on current assets and liabilities is to maximize the return to the business on cash, temporary investments of idle cash, accounts receivable, and inventory.

Managing Cash. A crucial element facing any financial manager is effectively managing the firm's cash flow. Remember that cash flow is the movement of money through an organization on a daily, weekly, monthly, or yearly basis. Ensuring that sufficient (but not excessive) funds are on hand to meet the company's obligations is one of the single most important facets of financial management.

Idle cash does not make money, and corporate checking accounts typically do not earn interest. As a result, astute money managers try to keep just enough cash on hand, called **transaction balances**, to pay bills—such as employee wages, supplies, and utilities—as they fall due. To manage the firm's cash and ensure that enough cash flows through the organization quickly and efficiently, companies try to speed up cash collections from customers.

transaction balances
cash kept on hand by a firm to pay normal daily expenses, such as employee wages and bills for supplies and utilities.

To facilitate collection, some companies have customers send their payments to a **lockbox,** which is simply an address for receiving payments, instead of directly to the company's main address. The manager of the lockbox, usually a commercial bank, collects payments directly from the lockbox several times a day and deposits them into the company's bank account. The bank can then start clearing the checks and get the money into the company's checking account much more quickly than if the payments had been submitted directly to the company. However, there is no free lunch: The costs associated with lockbox systems make them worthwhile only for those companies that receive thousands of checks from customers each business day.

lockbox
an address, usually a commercial bank, at which a company receives payments in order to speed collections from customers.

Large firms with many stores or offices around the country, such as HSBC Finance Corporation, frequently use electronic funds transfer to speed up collections. HSBC Finance Corporation's local offices deposit checks received each business day into their local banks and, at the end of the day, HSBC Finance Corporation's corporate office initiates the transfer of all collected funds to its central bank for overnight investment. This technique is especially attractive for major international companies, which face slow and sometimes uncertain physical delivery of payments and/or less-than-efficient check-clearing procedures.

More and more companies are now using electronic funds transfer systems to pay and collect bills online. Companies generally want to collect cash quickly but pay out cash slowly. When companies use electronic funds transfers between buyers and suppliers, the speed of collections and disbursements increases to one day. Only with the use of checks can companies delay the payment of cash by three or four days until the check is presented to their bank and the cash leaves their account.

Investing Idle Cash.   As companies sell products, they generate cash on a daily basis, and sometimes cash comes in faster than it is needed to pay bills. Organizations often invest this "extra" cash, for periods as short as one day (overnight) or for as long as one year, until it is needed. Such temporary investments of cash are known as **marketable securities.** Examples include U.S. Treasury bills, certificates of deposit, commercial paper, and eurodollar deposits. Table 16.1 summarizes a number of different marketable securities used by businesses and some sample interest rates on

marketable securities
temporary investment of "extra" cash by organizations for up to one year in U.S. Treasury bills, certificates of deposit, commercial paper, or eurodollar loans.

TABLE 16.1   Short-Term Investment Possibilities for Idle Cash

Type of Security	Maturity	Seller of Security	6/23/2006	4/18/2016	11/01/2018	Safety Level
U.S. Treasury bills	90 days	U.S. government	4.80%	0.22%	2.29%	Excellent
U.S. Treasury bills	180 days	U.S. government	5.05	0.35	2.43	Excellent
Commercial paper	30 days	Major corporations	5.14	0.46	2.25	Very good
Certificates of deposit	90 days	U.S. commercial banks	5.40	0.40	2.25	Very good
Certificates of deposit	180 days	U.S. commercial banks	5.43	0.45	2.35	Very good
Eurodollars	90 days	European commercial banks	5.48	0.65	2.58	Very good

(Interest Rate column spans 6/23/2006, 4/18/2016, 11/01/2018)

* Rate is as of April 19, 2018.

Sources: Board of Governors of the Federal Reserve System, "Selected Interest Rates (Weekly)—H.15," November 2, 2018, www.federalreserve.gov/releases/H15/current/default.htm (accessed November 7, 2018); Fidelity, "Certificates of Deposit," www.fidelity.com/fixed-income-bonds/cds (accessed November 7, 2018); Eurodollars, Bank of England, bankofengland.co.uk (accessed November 7, 2018).

these investments as of June 23, 2006, April 18, 2016, and November 1, 2018. The safety rankings are relative. While all of the listed securities are very low risk, the U.S. government securities are the safest. You can see from the table that interest rates declined between 2006 and 2016 but are on their way back up in 2018.

The Fed used monetary policy to lower interest rates to stimulate borrowing and investment during the recession of 2007–2009 and kept rates low into 2016 to stimulate employment and economic growth. The Fed raised interest rates 25 basis points (1/4 of a percent) in December 2015, again in 2016, three times in 2017, and three times in 2018. Most economists agree that rates will continue to rise as the economy continues its recovery.

Treasury bills (T-bills)
short-term debt obligations the U.S. government sells to raise money.

Many large companies invest idle cash in U.S. **Treasury bills (T-bills),** which are short-term debt obligations the U.S. government sells to raise money. Issued weekly by the U.S. Treasury, T-bills carry maturities of between one week and one year. U.S. T-bills are generally considered to be the safest of all investments and are called risk free because the U.S. government will not default on its debt.

commercial certificates of deposit (CDs)
certificates of deposit issued by commercial banks and brokerage companies, available in minimum amounts of $100,000, which may be traded prior to maturity.

Commercial certificates of deposit (CDs) are issued by commercial banks and brokerage companies. They are available in minimum amounts of $100,000 but are typically in units of $1 million for large corporations investing excess cash. Unlike consumer CDs (discussed in the "Money and the Financial System" chapter), which must be held until maturity, commercial CDs may be traded prior to maturity. Should a cash shortage occur, the organization can simply sell the CD on the open market and obtain needed funds.

commercial paper
a written promise from one company to another to pay a specific amount of money.

One of the most popular short-term investments for the largest business organizations is **commercial paper**—a written promise from one company to another to pay a specific amount of money. Because commercial paper is backed only by the name and reputation of the issuing company, sales of commercial paper are restricted to only the largest and most financially stable companies. As commercial paper is frequently bought and sold for durations as short as one business day, many "players" in the market find themselves buying commercial paper with excess cash on one day and selling it to gain extra money the following day.

eurodollar market
a market for trading U.S. dollars in foreign countries.

During 2007 and 2008, the commercial paper market simply stopped functioning. Investors no longer trusted the IOUs of even the best companies. Companies that had relied on commercial paper to fund short-term cash needs had to turn to the banks for borrowing. Those companies who had existing lines of credit at their bank were able to draw on their line of credit. Others were in a tight spot. Eventually, the Federal Reserve entered the market to buy and sell commercial paper for its own portfolio. This is something the Fed was not in the habit of doing. But it rescued the market, and the commercial paper market is now standing on its own two feet without the Fed's help.

Companies can invest their idle cash in marketable securities such as U.S. Treasury bills, commercial paper, and eurodollar deposits.
©RomanR/Shutterstock

Some companies invest idle cash in international markets such as the **eurodollar market,** a market for U.S. dollars held in foreign countries. Because the eurodollar market was originally developed by London banks, any dollar-denominated deposit in a non-U.S. bank is called a eurodollar deposit. For example, if you travel overseas and deposit $1,000

Making Every Woman Count: Improving Gender Diversity in Finance

Although more than half of financial managers in the U.S. financial services industry consist of women, only 13.8 percent of CFOs at *Fortune* 500 companies are female. This discrepancy is a concern to companies that are trying to emphasize diversity in senior management. Although sexual discrimination is still a problem, studies suggest that other factors contribute to this low percentage.

One theory is that women do not have as many connections with higher-level finance executives as men. According to a former portfolio consultant, because female employees do not often connect as well with male supervisors, they may inadvertently be passed up for future management opportunities. Another problem could be attitudinal bias by gender. An assertive personality in a male employee may be perceived as being "pushy" in a female employee. The support of senior executives is often essential for career advancement because these executives can more successfully advocate on behalf of the employee.

A solution that some companies have implemented is to encourage female executives to form long-lasting relationships with senior executives. Goldman Sachs offers leadership training for women, as well as mentorship and sponsorship opportunities where female employees are paired with senior leaders. Although the percentage of women in senior leadership finance positions is still low, its steady increase shows that more companies are taking gender diversity in finance seriously.[2]

Critical Thinking Questions

1. Why do you think there are so few female CFOs?
2. What are some ways that companies are trying to promote senior management positions to female finance employees?
3. Do you feel that a mentorship program will help close the diversity gap? Why or why not?

---

in a London bank, you will have "created" a eurodollar deposit in the amount of $1,000. Because the U.S. dollar is an international currency, these dollar deposits can be used by international companies to settle their accounts. The market created for trading such investments offers firms with extra dollars a chance to earn a slightly higher rate of return with just a little more risk than they would face by investing in U.S. Treasury bills.

Managing Accounts Receivable. After cash and marketable securities, the balance sheet lists accounts receivable and inventory. Remember that accounts receivable is money owed to a business by credit customers. Many businesses make the vast majority of their sales on credit, so managing accounts receivable is an important task.

Each credit sale represents an account receivable for the company, the terms of which typically require customers to pay the full amount due within 30, 60, or even 90 days from the date of the sale. To encourage quick payment, some businesses offer some of their customers discounts of between 1 and 2 percent if they pay off their balance within a specified period of time (usually between 10 and 30 days). On the other hand, late payment charges of between 1 and 1.5 percent serve to discourage slow payers from sitting on their bills forever. The larger the discount for early payment, the faster customers will tend to pay their accounts. Unfortunately, while discounts increase cash flow, they also reduce profitability. Finding the right balance between the added advantages of early cash receipt and the disadvantages of reduced profits is no simple matter. Similarly, determining the optimal balance between the higher sales likely to result from extending credit to customers with less than sterling credit ratings and the higher bad-debt losses likely to result from a more lenient credit policy is also challenging. Information on company credit ratings is provided by local credit bureaus, national credit-rating agencies such as Dun & Bradstreet, and industry trade groups.

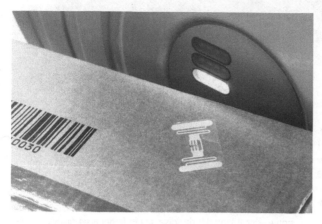

From materials management, race timing, or attendee tracking, RFID technology has many practical applications.

©nullplus/Getty Images

Optimizing Inventory. While the inventory that a firm holds is controlled by both production needs and marketing considerations, the financial manager has to coordinate inventory purchases to manage cash flows. The object is to minimize the firm's investment in inventory without experiencing production cutbacks as a result of critical materials shortfalls or lost sales due to insufficient finished goods inventories. Every dollar invested in inventory is a dollar unavailable for investment in some other area of the organization. Optimal inventory levels are determined in large part by the method of production. If a firm attempts to produce its goods just in time to meet sales demand, the level of inventory will be relatively low. If, on the other hand, the firm produces materials in a constant, level pattern, inventory increases when sales decrease and decreases when sales increase. One way companies manage inventory is through the use of radio frequency identification (RFID) technology. For example, Walmart manages its inventories by using RFID tags. An RFID tag, which contains a silicon chip and an antenna, allows a company to use radio waves to track and identify the products to which the tags are attached. These tags are primarily used to track inventory shipments from the manufacturer to the buyer's warehouses and then to the individual stores and also cut down on trucking theft because the delivery truck and its contents can be tracked in real time.

The automobile industry is an excellent example of an industry driven almost solely by inventory levels. Because it is inefficient to continually lay off workers in slow times and call them back in better times, Ford, General Motors, and Toyota try to set and stick to quarterly production quotas. Automakers typically try to keep a 60-day supply of unsold cars. During particularly slow periods, however, it is not unusual for inventories to exceed 100 days of sales.

Although less publicized, inventory shortages can be as much of a drag on potential profits as too much inventory. Not having an item on hand may send the customer to a competitor—forever. Complex computer inventory models are frequently employed to determine the optimum level of inventory a firm should hold to support a given level of sales. Such models can indicate how and when parts inventories should be ordered so that they are available exactly when required—and not a day before. Developing and maintaining such an intricate production and inventory system is difficult, but it can often prove to be the difference between experiencing average profits and spectacular ones.

LO 16-2

Identify some sources of short-term financing (current liabilities).

Managing Current Liabilities

While having extra cash on hand is a delightful surprise, the opposite situation—a temporary cash shortfall—can be a crisis. The good news is that there are several potential sources of short-term funds. Suppliers often serve as an important source through credit sales practices. Also, banks, finance companies, and other organizations offer short-term funds through loans and other business operations.

Accounts Payable. Remember from the "Accounting and Financial Statements" chapter that accounts payable is money an organization owes to suppliers for goods and services. Just as accounts receivable must be actively managed to ensure proper cash collections, so too must accounts payable be managed.

The most widely used source of short-term financing, and therefore the most important account payable, is **trade credit**—credit extended by suppliers for the purchase of their goods and services. Most trade credit agreements offer discounts to organizations that pay their bills early. A supplier, for example, may offer trade terms of "1/10 net 30," meaning that the purchasing organization may take a 1 percent discount from the invoice amount if it makes payment by the 10th day after receiving the bill. Otherwise, the entire amount is due within 30 days. For example, pretend that you are the financial manager in charge of payables. You owe Ajax Company $10,000, and it offers trade terms of 2/10 net 30. By paying the amount due within 10 days, you can save 2 percent of $10,000, or $200. Assume you place orders with Ajax once per month and have 12 bills of $10,000 each per year. By taking the discount every time, you will save 12 times $200, or $2,400, per year. Now assume you are the financial manager of Gigantic Corp., and it has monthly payables of $100 million per month. Two percent of $100 million is $2 million per month. Failure to take advantage of such trade discounts can add up to lost cash savings over the span of a year.

Bank Loans.   Virtually all organizations—large and small—obtain short-term funds for operations from banks. In most instances, the credit services granted by these firms take the form of a line of credit or fixed dollar loan. A **line of credit** is an arrangement by which a bank agrees to lend a specified amount of money to the organization upon request—provided that the bank has the required funds to make the loan. In general, a business line of credit is very similar to a consumer credit card, with the exception that the preset credit limit can amount to millions of dollars.

In addition to credit lines, banks also make **secured loans**—loans backed by collateral that the bank can claim if the borrowers do not repay the loans—and **unsecured loans**—loans backed only by the borrower's good reputation and previous credit rating. Both individuals and businesses build their credit rating from their history of borrowing and repaying borrowed funds on time and in full. The three national credit-rating services are Equifax, TransUnion, and Experian. A lack of credit history or a poor credit history can make it difficult to get loans from financial institutions. The *principal* is the amount of money borrowed; *interest* is a percentage of the principal that the bank charges for use of its money. As we mentioned in the "Money and the Financial System" chapter, banks also pay depositors interest on savings accounts and some checking accounts. Thus, banks charge borrowers interest for loans and pay interest to depositors for the use of their money. In addition, these loans may include origination fees.

One of the complaints from borrowers during the financial meltdown and recession was that banks weren't willing to lend. There were several causes. Banks were trying to rebuild their capital, and they didn't want to take the extra risk of making loans in an economic recession. They were drowning in bad debts and were not sure how future loan losses would affect their capital. Smaller regional banks did a better job of maintaining small business loans than the major money center banks who suffered most in the recession.

The **prime rate** is the interest rate commercial banks charge their best customers for short-term loans. For many years, loans at the prime rate represented funds at the lowest possible cost. For some companies, other alternatives may be cheaper, such as borrowing at the London Interbank Offer Rate (LIBOR) or using commercial paper.

The interest rates on commercial loans may be either fixed or variable. A variable- or floating-rate loan offers an advantage when interest rates are falling but represents

trade credit
credit extended by suppliers for the purchase of their goods and services.

line of credit
an arrangement by which a bank agrees to lend a specified amount of money to an organization upon request.

secured loans
loans backed by collateral that the bank can claim if the borrowers do not repay them.

unsecured loans
loans backed only by the borrower's good reputation and previous credit rating.

prime rate
the interest rate that commercial banks charge their best customers (usually large corporations) for short-term loans.

a distinct disadvantage when interest rates are rising. Between 1999 and 2004, interest rates plummeted, and borrowers refinanced their loans with low-cost, fixed-rate loans. Nowhere was this more visible than in the U.S. mortgage markets, where homeowners lined up to refinance their high-percentage home mortgages with lower-cost loans, in some cases as low as 5 percent on a 30-year loan. Mortgage rates rose to 6.5 percent by mid-2006 but fell again after 2012, and in late 2018, homeowners could still get a fixed-rate mortgage for slightly more than 4 percent. Individuals and corporations have the same motivation: to minimize their borrowing costs. During this period of historically low interest rates, companies ramped up their borrowing, bought back stock, and locked in large amounts of debt at low rates.

Nonbank Liabilities.   Banks are not the only source of short-term funds for businesses. Indeed, virtually all financial institutions—from insurance companies to pension funds, from money market funds to finance companies—make short-term loans to many organizations. The largest U.S. companies also actively engage in borrowing money from the eurodollar and commercial paper markets. As noted earlier, both of these funds' sources are typically slightly less expensive than bank loans.

factor
a finance company to which businesses sell their accounts receivable—usually for a percentage of the total face value.

In some instances, businesses actually sell their accounts receivable to a finance company known as a **factor**, which gives the selling organizations cash and assumes responsibility for collecting the accounts. For example, a factor might pay $80,000 for receivables with a total face value of $100,000 (80 percent of the total). The factor profits if it can collect more than what it paid for the accounts. Because the selling organization's customers send their payments to a lockbox, they may have no idea that a factor has bought their receivables.

Additional nonbank liabilities that must be efficiently managed to ensure maximum profitability are taxes owed to the government and wages owed to employees. Clearly, businesses are responsible for many different types of taxes, including federal, state, and local income taxes, property taxes, mineral rights taxes, unemployment taxes, Social Security taxes, workers' compensation taxes, excise taxes, and more. While the public tends to think that the only relevant taxes are on income and sales, many industries must pay other taxes that far exceed those levied against their income. Taxes and employees' wages represent debt obligations of the firm, which the financial manager must plan to meet as they fall due.

LO 16-3

Summarize the importance of long-term assets and capital budgeting.

Managing Fixed Assets

Up to this point, we have focused on the short-term aspects of financial management. While most business failures are the result of poor short-term planning, successful ventures must also consider the long-term financial consequences of their actions. Managing the long-term assets and liabilities and the owners' equity portion of the balance sheet is important for the long-term health of the business.

long-term (fixed) assets
production facilities (plants), offices, and equipment—all of which are expected to last for many years.

Long-term (fixed) assets are expected to last for many years—production facilities (plants), offices, equipment, heavy machinery, furniture, automobiles, and so on. In today's fast-paced world, companies need the most technologically advanced, modern facilities and equipment they can afford. Automobile, oil refining, and transportation companies are dependent on fixed assets.

Modern and high-tech equipment carry high price tags, and the financial arrangements required to support these investments are by no means trivial. Leasing is just one approach to financing. Obtaining major long-term financing can be challenging for even the most profitable organizations. For less successful firms, such challenges can prove nearly impossible. One approach is leasing assets such as equipment, machines, and buildings. Leasing involves paying a fee for usage rather than owning the asset. We'll take a closer look at long-term financing in a moment, but first let's address some issues associated with fixed assets, including capital budgeting, risk assessment, and the costs of financing fixed assets.

Capital Budgeting and Project Selection

One of the most important jobs performed by the financial manager is to decide what fixed assets, projects, and investments will earn profits for the firm beyond the costs necessary to fund them. The process of analyzing the needs of the business and selecting the fixed assets that will maximize its value is called **capital budgeting,** and the capital budget is the amount of money budgeted for investment in such long-term assets. But capital budgeting does not end with the selection and purchase of a particular piece of land, equipment, or major investment. All assets and projects must be continually reevaluated to ensure their compatibility with the organization's needs. If a particular asset does not live up to expectations, then management must determine why and take necessary corrective action. Budgeting is not an exact process, and managers must be flexible when new information is available.

capital budgeting
the process of analyzing the needs of the business and selecting the assets that will maximize its value.

Assessing Risk

Every investment carries some risk. Figure 16.1 ranks potential investment projects according to estimated risk. When considering investments overseas, risk assessments must include the political climate and economic stability of a region. For example, the decision to introduce a product or build a manufacturing facility in England would be much less risky than a decision to build one in the Middle East.

The longer a project or asset is expected to last, the greater its potential risk because it is hard to predict whether a piece of equipment will wear out or become obsolete in 5 or 10 years. Predicting cash flows one year down the road is difficult, but projecting them over the span of a 10-year project is a gamble.

The level of a project's risk is also affected by the stability and competitive nature of the marketplace and the world economy as a whole. The latest high-technology computer product is far more likely to become obsolete overnight than an electric utility plant. Dramatic changes in world markets are not uncommon. Indeed, uncertainty created by the rapid devaluation of Asian currencies in the late 1990s laid waste to the financial forecasts that hundreds of projects had relied on for their economic feasibility. Financial managers have to consider the probability of changing conditions that could affect their forecast when making long-term decisions about the purchase of fixed assets.

Pricing Long-Term Money

The ultimate success of any project depends not only on accurate assumptions of return on investment, but also on its cost of capital (equity and debt). Because a business must pay interest on money it borrows and generate returns for stockholders, the returns from any project must cover not only the costs of operating the project, but also the cost of capital used to finance the project. Unless an organization can effectively cover all of its costs—both financial and operating—it will eventually fail.

FIGURE 16.1
Qualitative Assessment of
Capital Budgeting Risk

Highest Risk

Introduce a New Product in
Foreign Markets (risk depends
on stability of country)

Expand into a New Market

Introduce a New Product in
a Familiar Area

Add to a Product Line

Buy New Equipment for
an Established Market

Repair Old Machinery

Lowest Risk

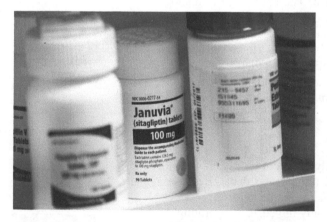

Pharmaceutical companies spend millions of dollars developing drugs
such as Januvia without knowing if the drug will pass FDA approval
and have a significant profit margin.

©George Frey/Bloomberg via Getty Images

Clearly, only a limited supply of funds is available for investment in any given enterprise. The most efficient and profitable companies can attract the lowest-cost funds because they typically offer reasonable financial returns at very low relative risks. Newer and less prosperous firms must pay higher costs to attract capital because these companies tend to be quite risky. One of the strongest motivations for companies to manage their financial resources wisely is that they will, over time, be able to reduce the costs of their funds and in so doing increase their overall profitability.

In our free-enterprise economy, new firms tend to enter industries that offer the greatest potential rewards for success. However, as more and more companies enter an industry, competition intensifies, eventually driving profits down to average levels.

Going Green

How Making Sustainability a Top Priority Helps the Bottom Line

Can sustainability improve a firm's bottom line? American chemical company DowDuPont proves the answer is yes. At DowDuPont, sustainability is seen as a market-driven process that enhances the bottom line and creates social value. Sustainability assists in the effective management of assets, liabilities, and owners' equity.

DowDuPont's quest for sustainability began two decades ago when the firm began analyzing how its plants were affecting the environment. It next turned its attention to energy and found that adopting more energy-efficient processes decreased energy costs. Since then, sustainability has morphed into a strategic tool. The company has reduced global greenhouse gas emissions by over 30 percent and water usage by over 19 percent since 2004. Its 2020 Sustainability Goals include reducing greenhouse gas emissions intensity by 7 percent (2015 baseline) and energy intensity by 10 percent (2010 baseline), developing

business-specific waste goals, and establishing water risk mitigation plans for select sites.

Finally, DowDuPont believes it can share what it has learned from best practices in sustainability with other companies. Its DowDuPont Sustainable Solutions business helps organizations adopt a triple-bottom-line (people-planet-profits) approach. This approach goes beyond the bottom line to incorporate human and environmental concerns. The company shares many of these practices with its suppliers to increase the sustainability of the supply chain. DowDuPont is an example of a successful company that has turned sustainability into a competitive advantage.[3]

Critical Thinking Questions

1. How is DowDuPont using sustainability as a competitive advantage?
2. Why is it important for DowDuPont to develop market-driven sustainability goals?
3. How can DowDuPont's sustainability initiatives improve financial management?

The digital music player market of the early 2000s provides an excellent example of the changes in profitability that typically accompany increasing competition. The sign of a successful capital budgeting program is that the new products create higher than normal profits and drive sales, profits, and the stock price up. This has certainly been true for Apple when it made the decision to enter the consumer electronics industry. In 2001, Apple introduced the first iPod, and as the iPod became more popular, it made the iTunes Store possible. The iPhone, introduced in 2007, has now gone through many updates, and over time, iPhones took the place of iPods as music players and have become sophisticated cameras and mini televisions. The iPad was introduced in 2010 and is now the third best product after second-place Mac computers. In 2015, Apple introduced the Apple Watch, which is now in its fourth upgrade. Financial analysts always talk about Apple's ecosystem, which allows all Apple products to be synchronized and be updated on whatever products the user owns.

Even with a well-planned capital budgeting program, it may be difficult for Apple to stay ahead of the competition, and many people think Apple has lost its innovation lead to Samsung and Amazon. However, Apple is now the most valuable company in the world, valued at over $1 trillion on October 3, 2018. On June 9, 2014, Apple split its stock seven for one, meaning that for every share you owned, you would get six more, for a total of seven shares. There is no real gain involved because the stock price is divided by 7, so stockholders

Apple stock trades at approximately 100 times what it did nearly 10 years ago.

©TuiPhotoEngineer/Shutterstock

still have the same value, just more shares at a lower price. An investor who bought $1,000 of Apple stock in 2003 for $0.91 (price adjusted for stock splits) would have Apple stock worth $229,109 on November 8, 2018. The problem is having the patience to continue to hold such a winner without taking some profits along the way.[4]

Maintaining market dominance is also difficult in the personal computer industry, particularly because tablet computers are taking away market share. With increasing competition, prices have fallen dramatically. Weaker companies have failed, leaving the most efficient producers/marketers scrambling for market share. The expanded market for personal computers dramatically reduced the financial returns generated by each dollar invested in productive assets. The "glory days" of the personal computer industry have long since passed into history. Personal computers have essentially become commodity items, and profit margins for companies in this industry have shrunk as the market matures and sales decline.

LO 16-4

Specify how companies finance their operations and manage fixed assets with long-term liabilities, particularly bonds.

Financing with Long-Term Liabilities

As we said earlier, long-term assets do not come cheaply, and few companies have the cash on hand to open a new store across town, build a new manufacturing facility, research and develop a new life-saving drug, or launch a new product worldwide. To develop such fixed assets, companies need to raise low-cost long-term funds to finance them. Two common choices for raising these funds are attracting new owners (*equity financing*), which we'll look at in a moment, and taking on long-term liabilities (*debt financing*), which we'll look at now.

long-term liabilities
debts that will be repaid over a number of years, such as long-term loans and bond issues.

Long-term liabilities are debts that will be repaid over a number of years, such as long-term bank loans and bond issues. These take many different forms, but in the end, the key word is *debt*. Companies may raise money by borrowing it from commercial banks in the form of lines of credit, short-term loans, or long-term loans. Many corporations acquire debt by borrowing from financial institutions such as pension funds, mutual funds, or life insurance funds.

Companies that rely too heavily on debt can get into serious trouble should the economy falter. During recessions, they may not earn enough operating income to make the required interest payments (remember the times interest earned ratio in the "Accounting and Financial Statements" chapter). In severe cases when the problem persists too long, creditors will not restructure loans but will instead sue for the interest and principal owed and force the company into bankruptcy or reorganization.

Bonds: Corporate IOUs

bonds
debt instruments that larger companies sell to raise long-term funds.

Much long-term debt takes the form of **bonds**, which are debt instruments that larger companies sell to raise long-term funds. In essence, the buyers of bonds (bondholders) loan the issuer of the bonds cash in exchange for regular interest payments until the loan is repaid on or before the specified maturity date. The bond itself is a certificate, much like an IOU, that represents the company's debt to the bondholder. Bonds are issued by a wide variety of entities, including corporations; national, state, and local governments; public utilities; and nonprofit corporations. Most bondholders need not hold their bonds until maturity; rather, the existence of active secondary markets of brokers and dealers allows for the transfer of bonds from the owner (seller) to the buyer at an agreed-upon price.

The bond contract, or *indenture,* specifies all of the terms of the agreement between the bondholders and the issuing organization. The indenture, which can

TABLE 16.2   U.S. Corporate Bond Quotes

·For the week ending April 20, 2018

Company (Ticker)	Coupon	Maturity	Last Price	Last Yield	Est. Spread*	UST**	Est $ Vol (000s)
Abbott Laboratories (ABT)	4.9	30-Nov-46	108.439	4.375	124	30	290,079
AT&T (T)	4.3	15-Feb-30	97.418	4.585	164	10	250,720
Citigroup (C)	4.075	23-Apr-29	98.611	4.247	129	10	654,204
General Mills (GIS)	4.2	17-Apr-28	99.186	4.301	134	10	228,466

*Estimated spreads, in basis points (100 basis points is one percentage point), over the 2-, 5-, 10-, or 30-year hot run Treasury note/bond.

**Comparable U.S. Treasury issue.

Coupon—the percentage in interest payment that the bond pays based on a $1,000 bond

Maturity—the day on which the issuer will reissue the principal

Last Price—last price at which the security is traded

Last Yield—yield-to-maturity for the investor that buys the bond today and holds it until it matures

Est. Spread—amount of additional yield the investor will earn each year compared to a U.S. Treasury bond. For example, the General Mills bond would earn 1.34% more than a U.S. Treasury bond.

UST—U.S. Treasury bond

Est $ Vol (000s)—number of individual bonds that were bought and sold on the date indicated

Sources: MarketAxess Corporate BondTicker, www.bondticker.com; Barron's, "Corporate Bonds," March 31, 2017, http://online.barrons.com/public/page/9_0210-corpbonds.html (accessed April 21, 2018).

run more than 100 pages, specifies the basic terms of the bond, such as its face value, maturity date, and the annual interest rate. Table 16.2 briefly explains how to determine these and more things about a bond from a bond quote, as it might appear in *Barron's* magazine. The face value of the bond, its initial sales price, is typically $1,000. After this, however, the price of the bond on the open market will fluctuate along with changes in the economy (particularly, changes in interest rates) and in the creditworthiness of the issuer. Bondholders receive the face value of the bond along with the final interest payment on the maturity date. The annual interest rate (often called the *coupon rate*) is the guaranteed percentage of face value that the company will pay to the bond owner every year. For example, a $1,000 bond with a coupon rate of 7 percent would pay $70 per year in interest. In most cases, bond indentures specify that interest payments be made every six months. In the preceding example, the $70 annual payment would be divided into two semiannual payments of $35.

In addition to the terms of interest payments and maturity date, the bond indenture typically covers other important topics, such as repayment methods, interest payment dates, procedures to be followed in case the organization fails to make the interest payments, conditions for the early repayment of the bonds, and any conditions requiring the pledging of assets as collateral.

Types of Bonds

Not surprisingly, there are a great many different types of bonds. Most are **unsecured bonds,** meaning that they are not backed by collateral; such bonds are termed *debentures.* **Secured bonds,** on the other hand, are backed by specific collateral that must be forfeited in the event that the issuing firm defaults. Whether secured or unsecured, bonds may be repaid in one lump sum or with many payments spread out over a period of time. **Serial bonds,** which are different from secured bonds, are actually a

unsecured bonds
debentures or bonds that are not backed by specific collateral.

secured bonds
bonds that are backed by specific collateral that must be forfeited in the event that the issuing firm defaults.

serial bonds
a sequence of small bond issues of progressively longer maturity.

floating-rate bonds
bonds with interest rates that change with current interest rates otherwise available in the economy.

junk bonds
a special type of high interest rate bond that carries higher inherent risks.

sequence of small bond issues of progressively longer maturity. The firm pays off each of the serial bonds as they mature. **Floating-rate bonds** do not have fixed interest payments; instead, the interest rate changes with current interest rates otherwise available in the economy and the rate is usually reset every six months.

High-yield bonds, or **junk bonds** as they are popularly known, offer relatively high rates of interest because they have higher inherent risks. Historically, junk bonds have been associated with companies in poor financial health and/or startup firms with limited track records. In the mid-1980s, however, junk bonds became a very attractive method of financing corporate mergers; they remain popular today with many investors as a result of their very high relative interest rates. But higher risks are associated with those higher returns and the average investor would be well-advised to heed those famous words: Look before you leap! The best strategy is to buy a mutual fund specializing in high yield bonds. This provides diversification across lots of risky bonds.

LO 16-5

Explain how corporations can use equity financing by issuing stock through an investment banker.

connect

 Need help understanding equity financing and debt financing? Visit your Connect ebook video tab for a brief animated explanation.

Financing with Owners' Equity

A second means of long-term financing is through equity. Remember from the "Accounting and Financial Statements" chapter that owners' equity refers to the owners' investment in an organization. Sole proprietors and partners own all or a part of their businesses outright, and their equity includes the money and assets they have brought into their ventures. Corporate owners, on the other hand, own stock or shares of their companies, which they hope will provide them with a return on their investment. Stockholders' equity includes common stock, preferred stock, and retained earnings.

Common stock (introduced in the "Options for Organizing Business" chapter) is the single most important source of capital for most new companies. On the balance sheet, the common stock account is separated into two basic parts—common stock at par and capital in excess of par. The *par value* of a stock is simply the stated value of one share of stock and has no relation to actual *market value*—the price at which the common stock is currently trading. The difference between a stock's par value and its offering price is called *capital in excess of par.* Except in the case of some very low-priced stocks, the capital in excess of par account is significantly larger than the par value account. Table 16.3 briefly explains how to gather important information from a stock quote, as it appears on Yahoo's website. You should be familiar with EPS from the "Accounting and Financial Statements" chapter. However, *beta* is a new term, and Nike's beta of 0.58 indicates that its stock price is 58 percent as volatile as the Standard & Poor's 500 Index. The market cap represents the total value of Nike's common stock. The target price of $71.27 is the analysts' consensus of the potential stock price in 12 months.

Preferred stock was defined in the "Accounting and Financial Statements" chapter as corporate ownership that gives the stockholder preference in the distribution of the company's profits but not the voting and control rights accorded to common stockholders. Thus, the primary advantage of owning preferred stock is that it is a safer investment than common stock.

retained earnings
earnings after expenses and taxes that are reinvested in the assets of the firm and belong to the owners in the form of equity.

All businesses exist to earn profits for their owners. Without the possibility of profit, there can be no incentive to risk investors' capital and succeed. When a corporation has profits left over after paying all of its expenses and taxes, it has the choice of retaining all or a portion of its earnings and/or paying them out to its shareholders in the form of dividends. **Retained earnings** are reinvested in the assets of the firm

TABLE 16.3 A Basic Stock Quote

Nike Inc. (NKE)—NYSE

66.09 ^0.36 (0.55%)

Previous close	65.73	Market cap	106.52IB
Open	65.75	Beta	0.58
Bid	0.00 × 0	PE ratio (TTM)	28.6
Ask	0.00 × 0	EPS (TTM)	2.31
Day's range	65.45-66.11	Earnings date	Jun 27, 2018-Jul 2, 2018
52-week range	50.35-70.25	Forward dividend and yield	0.80 (1.19%)
Volume	10,402,847	Ex-dividend date	3/2/18
Average volume	8,106,114	1-year target estimate	71.27

1. The **52-week high and low**—the highest and lowest prices, respectively, paid for the stock in the last year; for Nike stock, the highest was $70.25 and the lowest price, $50.35.

2. **Stock**—the name of the issuing company. When followed by the letters "pf," the stock is a preferred stock.

3. **Symbol**—the ticker tape symbol for the stock; NKE.

4. **Dividend**—the annual cash dividend paid to stockholders; Nike paid a dividend of $0.80 per share of stock outstanding.

5. **Dividend yield**—the dividend return on one share of common stock; 1.19 percent.

6. **Volume**—the number of shares traded on this day; for Nike, 10,402,847 while the average volume over the last three months was 8,106,114 shares.

7. **Net change**—the difference between the previous day's close and the close on the day being reported; Nike was up $0.36.

Source: Yahoo! Finance, http://finance.yahoo.com/q?s (accessed April 20, 2018); bid/ask from E-Trade after hours.

and belong to the owners in the form of equity. Retained earnings are an important source of funds and are, in fact, the only long-term funds that the company can generate internally.

When the board of directors distributes some of a corporation's profits to the owners, it issues them as cash dividend payments. But not all firms pay dividends. Many fast-growing firms like Facebook retain all of their earnings because they can earn high rates of return on the earnings they reinvest. Companies with fewer growth opportunities like Century Link or Verizon typically pay out large proportions of their earnings in the form of dividends, thereby allowing their stockholders to reinvest their dividend payments in other companies or spend their dividends. Retirees often prefer stocks that pay dividends. Table 16.4 presents a sample of companies and the dividend each paid on a single share of stock. As shown in the table, when the dividend is divided by the price the result is the **dividend yield.** The dividend yield is the cash return as a percentage of the price but does not reflect the total return an investor earns on the individual stock. If the dividend yield is 4.87 percent on Verizon and the stock price increases by 4 percent from $48.43 to $50.37, then the total return would be 8.87 percent. Most large U.S. companies pay their stockholders dividends on a quarterly basis. High-growth companies tend to pay no dividends.

dividend yield
the dividend per share divided by the stock price.

The last column in Table 16.4 is the payout ratio, which indicates what percentage of earnings is paid to the stockholders in dividends. Companies like Verizon and Procter & Gamble with high payout ratios have low expected growth rates. Companies like Amazon, Facebook, and Alphabet have zero payout ratios and are expected to grow fast because they are reinvesting cash flow into investment opportunities.

TABLE 16.4
Common Stock Prices, Earnings, Dividends and Ratios for Selected Companies

Ticker Symbol	Company Name	Price per Share	Dividend per Share	Dividend Yield	Earnings per Share*	Price Earnings Ratio	Payout Ratio
AEO	American Eagle	$ 21.29	$0.55	2.58%	$ 1.13	18.84	48.67%
AMZN	Amazon	$1,556.91	$0.00	0.00%	$ 6.15	253.16	0.00%
AXP	American Express	$ 102.37	1.40	1.37%	$ 2.97	34.47	47.14%
AAPL	Apple	$ 172.80	2.52	1.46%	$ 9.70	17.81	25.98%
CPB	Campbell Soup	$ 41.27	1.40	3.39%	$ 3.47	11.89	40.35%
DIS	Disney	$ 100.89	1.68	1.67%	$ 7.01	14.39	23.97%
F	Ford	$ 10.96	0.60	5.47%	$ 1.90	5.77	31.58%
FB	Facebook	$ 168.10	0.00	0.00%	$ 5.39	31.19	0.00%
GOOGL	Alphabet	$1,089.45	0.00	0.00%	$18.00	60.53	0.00%
HOG	Harley-Davidson	$ 40.96	1.48	3.61%	$ 3.02	13.56	49.01%
HD	Home Depot	$ 177.08	4.12	2.33%	$ 7.29	24.29	56.52%
MCD	McDonald's	$ 159.53	4.04	2.53%	$ 6.37	25.04	63.42%
PG	Procter & Gamble	$ 74.95	2.87	3.83%	$ 3.75	19.99	76.53%
LUV	Southwest Airlines	$ 54.80	0.50	0.91%	$ 5.79	9.46	8.64%
VZ	Verizon	$ 48.43	2.36	4.87%	$ 7.36	6.58	32.07%

* Earnings per share are for the latest 12-month period and do not necessarily match year-end numbers.

Source: Yahoo Finance, http://finance.yahoo.com/ (April 19, 2018).

Investment Banking

A company that needs money to expand may be able to obtain financing by issuing stock. The first-time sale of stocks and bonds directly to the public is called a *new issue* or an initial public offering (IPO) and creates a stock that can be traded in the secondary market. Companies that already have stocks or bonds outstanding may offer more stock or a new issue of bonds to raise additional funds for specific projects.

New issues of stocks and bonds are sold directly to the public and to institutions in what is known as the **primary market**—the market where firms raise financial capital. The primary market differs from **secondary markets,** which are stock exchanges and over-the-counter markets where investors can trade their securities with other investors rather than the company that issued the stock or bonds. Primary market transactions actually raise cash for the issuing corporations, while secondary market transactions do not. For example, when Facebook went public on May 18, 2012, its IPO raised $16 billion for the company and stockholders, who were cashing in on their success. Once the investment bankers distributed the stock to retail brokers, the brokers sold it to clients in the secondary market for $38 per share. The stock got off to a rocky start and hit a low of $17.73 in September 2012. However, by March 2014, it was at $71.97, and as you can see from Table 16.4, it was $168.10 on April 19, 2018. You might want to check out its current price for fun.

primary market
the market where firms raise financial capital.

secondary markets
stock exchanges and over-the-counter markets where investors can trade their securities with others.

DID YOU KNOW? A single share of Coca-Cola stock purchased during its original 1919 IPO would be worth more than $5 million today.[5]

Sounds Like Spotify's Non-IPO Was a Hit

Spotify

Founders: Daniel Ek and Martin Lorentzon

Founded: 2008 (launch date) in Stockholm, Sweden

Success: Spotify ended the first week as a public company with a market capitalization (total market value) of $27 billion.

Just like any entrepreneurial journey, Spotify started with an idea. Entrepreneurs Daniel Ek and Martin Lorentzon were looking up music on Ek's home-theater computer and thought the process could be improved. They began working on the Spotify application in 2005 and officially launched the application in 2008.

Spotify offers a free service where users can create, edit, and share playlists and tracks on social media, as well as make playlists with other users. Spotify provides access to more than 30 million songs and has more than 140 million monthly active users. Spotify pays royalties based on the number of artists' streams as a proportion of total songs streamed, instead of the traditional way of paying a fixed price per song or album sold.

On April 3, 2018, Spotify made its debut on the New York Stock Exchange with an opening price of $165.90 per share. The founders decided to do a direct listing, foregoing investment-banking underwriters and opting not to raise any money for itself. This saved the company tens of millions of dollars. About $940 million worth of Spotify's shares exchanged hands on the first day, making it the fourth largest opening trade in a company going public since 2010.[6]

Critical Thinking Questions

1. Why do you think Spotify was so popular among investors when it went public?
2. What were the benefits of doing a direct listing when going public? What does this indicate about how Spotify sees itself from a financial perspective?
3. How does Spotify's business royalty model differ from more traditional models, and why do you think this is important?

Investment banking, the sale of stocks and bonds for corporations, helps such companies raise funds by matching people and institutions who have money to invest with corporations in need of resources to exploit new opportunities. Corporations usually employ an investment banking firm to help sell their securities in the primary market. An investment banker helps firms establish appropriate offering prices for their securities. In addition, the investment banker takes care of the myriad details and securities regulations involved in any sale of securities to the public.

Just as large corporations such as IBM and Microsoft have a client relationship with a law firm and an accounting firm, they also have a client relationship with an investment banking firm. An investment banking firm such as Merrill Lynch, Goldman Sachs, or Morgan Stanley can provide advice about financing plans, dividend policy, or stock repurchases, as well as advice on mergers and acquisitions. Many now offer additional banking services, making them "one-stop shopping" banking centers. When Pixar merged with Disney, both companies used investment bankers to help them value the transaction. Each firm wanted an outside opinion about what it was worth to the other. Sometimes mergers fall apart because the companies cannot agree on the price each company is worth or the structure of management after the merger. The advising investment banker, working with management, often irons out these details. Of course, investment bankers do not provide these services for free. They usually charge a fee of between 1 and 1.5 percent of the transaction. A $20 billion merger can generate between $200 and $300 million in investment banking fees. The merger mania of the late 1990s allowed top investment bankers to earn huge sums. Unfortunately, this type of fee income is dependent on healthy stock markets, which seem to stimulate the merger fever among corporate executives.

investment banking
the sale of stocks and bonds for corporations.

LO 16-6

Describe the various securities markets in the United States.

securities markets
the mechanism for buying and selling securities.

The Securities Markets

Securities markets provide a mechanism for buying and selling securities. They make it possible for owners to sell their stocks and bonds to other investors. Thus, in the broadest sense, stocks and bonds markets may be thought of as providers of liquidity—the ability to turn security holdings into cash quickly and at minimal expense and effort. Without liquid securities markets, many potential investors would sit on the sidelines rather than invest their hard-earned savings in securities. Indeed, the ability to sell securities at well-established market prices is one of the very pillars of the capitalistic society that has developed over the years in the United States.

Unlike the primary market, in which corporations sell stocks directly to the public, secondary markets permit the trading of previously issued securities. There are many different secondary markets for both stocks and bonds. If you want to purchase 100 shares of Alphabet (formerly Google) common stock, for example, you must purchase this stock from another investor or institution. It is the active buying and selling by many thousands of investors that establishes the prices of all financial securities. Secondary market trades may take place on organized exchanges or in what is known as the over-the-counter market. Many brokerage houses exist to help investors with financial decisions, and many offer their services through the Internet. One such broker is Charles Schwab. Its site offers a wealth of information and provides educational material to individual investors.

Stock Markets

Stock markets exist around the world in New York, Tokyo, London, Frankfort, Paris, and other world locations. The two biggest stock markets in the United States are the New York Stock Exchange (NYSE) and the Nasdaq market.

Exchanges used to be divided into organized exchanges and over-the-counter markets, but during the past several years, dramatic changes have occurred in the markets. Both the NYSE and Nasdaq became publicly traded companies. They were previously not-for-profit organizations but are now for-profit companies. Additionally, both exchanges bought or merged with electronic exchanges. In an attempt to expand their markets, Nasdaq acquired the OMX, a Nordic stock exchange headquartered in Sweden, and the New York Stock Exchange merged with Euronext, a large European electronic exchange that trades options and futures contracts as well as common stock.

Traditionally, the Nasdaq market has been an electronic market, and many of the large technology companies such as Microsoft, Alphabet Inc., Apple, and Facebook trade on the Nasdaq market. The Nasdaq operates through dealers who buy and sell common stock (inventory) for their own accounts. The NYSE used to be primarily a floor-traded market, where brokers meet at trading posts on the floor of the New York Stock Exchange to buy and sell common stock, but now more than 80 percent of NYSE trading is electronic. The brokers act as agents for their clients and do not own their own inventory. This traditional division between the two markets is becoming less significant as the exchanges become electronic.

The New York Stock Exchange is the world's largest stock exchange by market capitalization.

©EdStock/iStockphoto.com

Electronic markets have grown quickly because of the speed, low cost, and efficiency of trading that they offer over floor trading. One of the fastest-growing electronic markets has been the Intercontinental Exchange (referred to as ICE). ICE, based in Atlanta, Georgia, primarily trades financial and commodity futures products. It started out as an energy futures exchange and has broadened its futures contracts into an array of commodities and derivative products. In December 2012, ICE made an offer to buy the New York Stock Exchange. When the NYSE became a public company and had common stock trading in the secondary market, rather than the hunter, it became the prey. On November 13, 2013, ICE completed its takeover of the NYSE. One condition of the takeover was that ICE had to divest itself of Euronext because international regulators thought the company would have a monopoly on European derivative markets. Also acquired as part of the NYSE family of exchanges was LIFFE, the London International Financial Futures Exchange. Many analysts thought that LIFFE was the major reason ICE bought the NYSE—not for its equity markets trading common stocks. What we are seeing is the globalization of securities markets and the increasing reliance on electronic trading.

The rise of electronic markets has led to the rise of robotic trading, sometimes referred to as "bots" or "algos" because they use algorithmic trading formulas to buy and sell based on market trends. Robotic trading looks for patterns in the market that will trigger a trade to buy or sell. These trading systems rely on programmed instructions that have the advantage of eliminating psychological trading errors. These systems also react to market trends faster than human traders. Using artificial intelligence to enhance trading systems is in its infancy but is expected to have an influence on markets over time.

The Over-the-Counter Market

Unlike the organized exchanges, the **over-the-counter (OTC) market** is a network of dealers all over the country linked by computers, telephones, and Teletype machines. It has no central location. Today, the OTC market for common stock consists of small stocks, illiquid bank stocks, penny stocks, and companies whose stocks trade on the "pink sheets." Because most corporate bonds and all U.S. government debt securities are traded over the counter, the OTC market regularly accounts for the largest total dollar value of all of the secondary markets.

over-the-counter (OTC) market
a network of dealers all over the country linked by computers, telephones, and Teletype machines.

Measuring Market Performance

Investors, especially professional money managers, want to know how well their investments are performing relative to the market as a whole. Financial managers also need to know how their companies' securities are performing when compared with their competitors'. Thus, performance measures—averages and indexes—are very important to many different people. They not only indicate the performance of a particular securities market, but also provide a measure of the overall health of the economy.

Indexes and averages are used to measure stock prices. An *index* compares current stock prices with those in a specified base period, such as 1944, 1967, or 1977. An *average* is the average of certain stock prices. The averages used are usually not simple calculations, however. Some stock market averages (such as the Standard & Poor's Composite Index) are weighted averages, where the weights employed are the total market values of each stock in the index (in this case 500). The Dow Jones Industrial Average (DJIA) is a price-weighted average. Regardless of how they are constructed, all market averages of stocks move together closely over time. See Figure 16.2, which

FIGURE 16.2

Recent Performance of Stock Market and Dow Jones Industrial Average (DJIA)

Source: "Dow Jones Industrial Average," Yahoo! Finance, https://finance.yahoo.com/quote/%5EDJI/chart?p=%5EDJI (accessed April 20, 2018).

graphs the Dow Jones Industrial Average. Notice the sharp downturn in the market during the 2008–2009 time period and the recovery that started in 2010. Investors perform better by keeping an eye on the long-term trend line and not the short-term fluctuations. Contrarian investors buy when everyone else is panicked and prices are low because they play the long-term trends. However, for many, this is psychologically a tough way to play the market.

Many investors follow the activity of the Dow Jones Industrial Average to see whether the stock market has gone up or down. Table 16.5 lists the 30 companies that currently make up the Dow. Although these companies are only a small fraction of the total number of companies listed on the New York Stock Exchange, because of their size, they account for about 25 percent of the total value of the NYSE.

The numbers listed in an index or average that tracks the performance of a stock market are expressed not as dollars, but as a number on a fixed scale. If you know, for example, that the Dow Jones Industrial Average climbed from 860 in August 1982 to a high of 11,497 at the beginning of 2000, you can see clearly that the value of the Dow

TABLE 16.5

The 30 Stocks in the Dow Jones Industrial Average

3M Co	Goldman Sachs	Pfizer
American Express Co	Home Depot	Procter & Gamble
Apple	Intel	Travelers Companies Inc.
Boeing	IBM	United Health Group
Caterpillar	Johnson & Johnson	United Technologies
Chevron	JPMorgan Chase	Verizon
Cisco Systems	McDonald's	Visa
Coca-Cola	Merck	Walgreens Boots Alliance Inc.
DowDuPont Inc.	Microsoft	Walmart
ExxonMobil	Nike	Walt Disney

Source: Barron's Market Laboratory, p. M35, November 5, 2018.

Jones Average increased more than 10 times in this 19-year period, making it one of the highest rate of return periods in the history of the stock market.

Unfortunately, prosperity did not last long once the Internet bubble burst. Technology stocks and new Internet companies were responsible for the huge increase in stock prices. Even companies with few sales and no earnings were selling at prices that were totally unreasonable. It is always easier to realize that a bubble existed after it has popped. By September 2002, the Dow Jones Industrial Average hit 7,461. The markets stabilized and the economy kept growing; investors were euphoric when the Dow Jones Industrial Average hit an all-time high of 14,198 in October 2007. However, once the housing bubble burst, the economy and the stock market went into a free fall. The Dow Jones Industrial Average bottomed out at 6,470 in March 2009. The market entered a period of recovery, and by April 2010, it hit a new high for the year of 10,975. By January 26, 2018, the Dow Jones Industrial Average hit an all-time record high of 26,616.71. The good news is that even when the market has been rather flat, an investor would have collected dividends, which are not reflected in the index. Perhaps this roller coaster ride indicates why some people are afraid to buy common stocks. If you look at the long-term trend and long-term returns in common stocks, they far outdistance bonds and government securities.

Recognizing financial bubbles can be difficult. It is too easy to get caught up in the enthusiasm that accompanies rising markets. Knowing what something is worth in economic terms is the test of true value. During the housing bubble, banks made loans to subprime borrowers to buy houses. (Remember that the prime rate is the rate for the highest quality borrowers and subprime loans are made to those who have low credit ratings.) As more money poured into the housing market, the obvious supply and demand relationship from economics would indicate that housing prices would rise. As prices rose, speculators entered the real estate market trying to make a fast buck. States such as Florida, Arizona, Nevada, and California were the favorite speculative spots and the states with the largest decline in house prices. To make matters worse, banks had created the home equity loan years ago so that borrowers could take out a second mortgage against their house and deduct the interest payment for tax purposes. Many homeowners no longer thought about paying off their mortgages but, instead, used the increase in the price of their houses to borrow more money. This behavior was unsustainable and created a real estate bubble that burst with dire financial consequences for the whole economy, workers, and investors.

People defaulted on loans when they could no longer afford to pay the mortgage. Many of these subprime borrowers shouldn't have been able to borrow in the first place. The defaults caused housing prices to fall, and some people who had home equity loans no longer had any equity left in their house. Some homeowners owed the bank more than the house was worth, and they walked away from their mortgage. At the same time, investors realized that the mortgage-backed securities they owned were not worth their face value, and prices of these assets plummeted. Banks and other financial service firms that had these assets on their books suffered a double whammy. They had loan losses and losses on

During the housing bubble, banks provided loans to riskier subprime borrowers. Although these loans were highly profitable, it was only a matter of time before the bubble burst.

©moodboard/Getty Images

mortgage-backed securities that another division of the bank had bought for investment purposes. Soon, many banks were close to violating their capital requirement, and the U.S. Treasury and Federal Reserve stepped in—with the help of funding from Congress—to make bank loans, buy securities that were illiquid, and invest in the capital of the banks by buying preferred stock.

The consensus of most economists is that through the actions of the U.S. Treasury and the Federal Reserve, the U.S. economy escaped what might have been another depression equal to or worse than the depression of the 1930s. The recession of 2007–2009 lasted 18 months and was the longest recession since the 1930s. Some worry that as the Federal Reserve lets interest rates rise, the rising interest rates will have a negative effect on stock prices. This is always possible if corporate earnings do not increase enough to outweigh the impact of higher required returns motivated by higher rates and higher inflation.

For investors to make sound financial decisions, it is important that they stay in touch with business news, markets, and indexes. Of course, business and investment magazines, such as *Bloomberg Businessweek, Fortune,* and *Money,* offer this type of information. Many Internet sites, including CNN/*Money, Business Wire, USA Today,* and others offer this information, as well. Many sites offer searchable databases of information by topic, company, or keyword. However, investors choose to receive and review business news, doing so is a necessity in today's market.

So You Want to Work in Financial Management or Securities

Taking classes in financial and securities management can provide many career options, from managing a small firm's accounts receivable to handling charitable giving for a multinational to investment banking to stock brokerage. We have entered into a less certain period for finance and securities jobs, however. In the world of investment banking, the past few years have been especially challenging. Tens of thousands of employees from Wall Street firms have lost their jobs. This phenomenon is not confined to New York City either, leaving the industry with a lot fewer jobs around the country. This type of phenomenon is not isolated to the finance sector. In the early 2000s, the tech sector experienced a similar downturn, from which it has subsequently largely recovered. Undoubtedly, markets will bounce back and job creation in finance and securities will increase again—but until that happens, the atmosphere across finance and securities will be more competitive than it has been in the past. However, this does not mean that there are no jobs. All firms need financial analysts to determine whether a project should be implemented, when to issue stocks or bonds, or when to initiate loans. These and other forward-looking questions, such as how to invest excess cash, must be addressed by financial managers. Economic uncertainty in the financial and securities market has made for more difficulty in finding the most desirable jobs.

Why this sudden downturn in financial industry prospects? A lot of these job cuts came in response to the subprime lending fallout and subsequent bank failures such as Bear Stearns, which alone lost around 7,000 employees. All of these people had to look for new jobs in new organizations, increasing the competitive level in a lot of different employment areas. For young jobseekers with relatively little experience, this may result in a great deal of frustration. Uncertainty results in hiring freezes and layoffs but leaves firms lean and ready to grow when the cycle turns around, resulting in hiring from the bottom up.

Many different industries require people with finance skills. So do not despair if you have a difficult time finding a job in exactly the right firm. Most people switch companies a number of times over the course of their careers. Many organizations require individuals trained in forecasting, statistics, economics, and finance. Even unlikely places like museums, aquariums, and zoos need people who are good at numbers. It may require some creativity, but if you are committed to a career in finance, look to less obvious sources—not just the large financial firms.[7]

Review Your Understanding

Describe some common methods of managing current assets.

Current assets are short-term resources such as cash, investments, accounts receivable, and inventory, which can be converted to cash within a year. Financial managers focus on minimizing the amount of cash kept on hand and increasing the speed of collections through lockboxes and electronic funds transfer and investing in marketable securities. Marketable securities include U.S. Treasury bills, certificates of deposit, commercial paper, and money market funds. Managing accounts receivable requires judging customer creditworthiness and creating credit terms that encourage prompt payment. Inventory management focuses on determining optimum inventory levels that minimize the cost of storing and ordering inventory without sacrificing too many lost sales due to inventory shortages.

Identify some sources of short-term financing (current liabilities).

Current liabilities are short-term debt obligations that must be repaid within one year, such as accounts payable, taxes payable, and notes (loans) payable. Trade credit is extended by suppliers for the purchase of their goods and services. A line of credit is an arrangement by which a bank agrees to lend a specified amount of money to a business whenever the business needs it. Secured loans are backed by collateral; unsecured loans are backed only by the borrower's good reputation.

Summarize the importance of long-term assets and capital budgeting.

Long-term, or fixed, assets are expected to last for many years, such as production facilities (plants), offices, and equipment. Businesses need modern, up-to-date equipment to succeed in today's competitive environment. Capital budgeting is the process of analyzing company needs and selecting the assets that will maximize its value; a capital budget is the amount of money budgeted for the purchase of fixed assets. Every investment in fixed assets carries some risk.

Specify how companies finance their operations and manage fixed assets with long-term liabilities, particularly bonds.

Two common choices for financing are equity financing (attracting new owners) and debt financing (taking on long-term liabilities). Long-term liabilities are debts that will be repaid over a number of years, such as long-term bank loans and bond issues. A bond is a long-term debt security that an organization sells to raise money. The bond indenture specifies the provisions of the bond contract—maturity date, coupon rate, repayment methods, and others.

Explain how corporations can use equity financing by issuing stock through an investment banker.

Owners' equity represents what owners have contributed to the company and includes common stock, preferred stock, and retained earnings (profits that have been reinvested in the assets of the firm). To finance operations, companies can issue new common and preferred stock through an investment banker that sells stocks and bonds for corporations.

Describe the various securities markets in the United States.

Securities markets provide the mechanism for buying and selling stocks and bonds. Primary markets allow companies to raise capital by selling new stock directly to investors through investment bankers. Secondary markets allow the buyers of previously issued shares of stock to sell them to other owners. The two biggest stock markets in the United States are the New York Stock Exchange (NYSE) and the Nasdaq market. Investors measure stock market performance by watching stock market averages and indexes such as the Dow Jones Industrial Average and the Standard & Poor's (S&P) Composite Index.

Critique the position of short-term assets and liabilities of a small manufacturer and recommend corrective action.

Using the information presented in this chapter, you should be able to answer the questions in the "Solve the Dilemma" feature near the end of the chapter presented by the current bleak working capital situation of Glasspray Corporation.

Critical Thinking Questions

Enter the World of Business Questions

1. Describe the financial mistakes GE made with GE Capital.
2. Do you think GE Capital's decision to expand into consumer financial products was a decision of bad timing (right before the financial crisis), or was it just a bad business move? Support your argument with reasoning.
3. Is a breakup of GE the best option for the company? Why or why not?

Learn the Terms

bonds   514

capital budgeting   511

commercial certificates of deposit
   (CDs)   506

commercial paper   506

dividend yield   517

eurodollar market   507

factor   510

floating-rate bonds   516

investment banking   519

junk bonds   516

line of credit   509

lockbox   505

long-term (fixed) assets   510

long-term liabilities   514

marketable securities   505

over-the-counter (OTC)
   market   521

primary market   518

prime rate   509

retained earnings   516

secondary markets   518

secured bonds   515

secured loans   509

securities markets   520

serial bonds   515

trade credit   509

transaction balances   504

Treasury bills (T-bills)   506

unsecured bonds   515

unsecured loans   509

working capital management   504

Check Your Progress

1. Define *working capital management.*
2. How can a company speed up cash flow? Why should it?
3. Describe the various types of marketable securities.
4. What does it mean to have a line of credit at a bank?
5. What are fixed assets? Why is assessing risk important in capital budgeting?
6. How can a company finance fixed assets?
7. What are bonds and what do companies do with them?
8. How can companies use equity to finance their operations and long-term growth?
9. What are the functions of securities markets?
10. What were some of the principal causes of the most recent recession?

Get Involved

1. Using your local newspaper or *The Wall Street Journal,* find the current rates of interest on the following marketable securities. If you were a financial manager for a large corporation, which would you invest extra cash in? Which would you invest in if you worked for a small business?
   a. Three-month T-bills
   b. Six-month T-bills
   c. Commercial certificates of deposit
   d. Commercial paper
   e. Eurodollar deposits
   f. Money market deposits

2. Select five of the Dow Jones Industrials from Table 16-5. Look up their earnings, dividends, and prices for the past five years. What kind of picture is presented by this information? Which stocks would you like to have owned over this past period? Do you think the next five years will present a similar picture?

Build Your Skills

Choosing among Projects

Background

As the senior executive in charge of exploration for High Octane Oil Co., you are constantly looking for projects that will add to the company's profitability—without increasing the company's risk. High Octane Oil is an international oil company with operations in Latin America, the Middle East, Africa, the United States, and Mexico. The company is one of the world's leading experts in deep-water exploration and drilling. High Octane currently produces 50 percent of its oil in the United States, 25 percent in the Middle East, 5 percent in Africa, 10 percent in Latin America, and 10 percent in Mexico. You are considering six projects from around the world.

Project 1—Your deep-water drilling platform in the Gulf of Mexico is producing at maximum capacity from the Valdez oil field, and High Octane's geological engineers think there is a high probability that there is oil in the Sanchez field, which is adjacent to Valdez. They recommend drilling a new series of wells. Once commercial quantities of oil have been discovered, it will take two more years to build the collection platform and pipelines. It will be four years before the discovered oil gets to the refineries.

Project 2—The Brazilian government has invited you to drill on some unexplored tracts in the middle of the central jungle region. There are roads to within 50 miles of the tract, and BP has found oil 500 miles away from this tract. It would take about three years to develop this property and several more years to build pipelines and pumping stations to carry the oil to the refineries. The Brazilian government wants 20 percent of all production as its fee for giving High Octane Oil Co. the drilling rights or a $500 million up-front fee and 5 percent of the output.

Project 3—Your fields in Saudi Arabia have been producing oil for 50 years. Several wells are old, and the pressure has diminished. Your engineers are sure that if you were to initiate high-pressure secondary recovery procedures, you would increase the output of these existing wells by 20 percent. High-pressure recovery methods pump water at high pressure into the underground limestone formations to enhance the movement of petroleum toward the surface.

Project 4—Your largest oil fields in Alaska have been producing from only 50 percent of the known deposits. Your geological engineers estimate that you could open up 10 percent of the remaining fields every two years and offset your current declining production from existing wells. The pipeline capacity is available and, while you can only drill during six months of the year, the fields could be producing oil in three years.

Project 5—Some of High Octane's west Texas oil fields produce in shallow stripper wells of 2,000- to 4,000-foot depths. Stripper wells produce anywhere from 10 to 2,000 barrels per day and can last for six months or 40 years. Generally, once you find a shallow deposit, there is an 80 percent chance that offset wells will find more oil. Because these wells are shallow, they can be drilled quickly at a low cost. High Octane's engineers estimate that in your largest tract, which is closest to the company's Houston refinery, you could increase production by 30 percent for the next 10 years by increasing the density of the wells per square mile.

Project 6—The government of a republic in Russia has invited you to drill for oil in Siberia. Russian geologists think that this oil field might be the largest in the world, but there have been no wells drilled and no infrastructure exists to carry oil if it should be found. The republic has no money to help you build the infrastructure but if you find oil, it will let you keep the first five years' production before taking its 25 percent share. Knowing that oil fields do not start producing at full capacity for many years after initial production, your engineers are not sure that your portion of the first five years of production will pay for the infrastructure they must build to get the oil to market. The republic also has been known to have a rather unstable government, and the last international oil company that began this project left the country when a new government demanded a higher than originally agreed-upon percentage of the expected output. If this field is, in fact, the largest in the world, High Octane's supply of oil would be ensured well into the 21st century.

Task

1. Working in groups, rank the six projects from lowest risk to highest risk.

2. Given the information provided, do the best you can to rank the projects from lowest cost to highest cost.

3. What political considerations might affect your project choice?

4. If you could choose one project, which would it be and why?

5. If you could choose three projects, which ones would you choose? In making this decision, consider which projects might be highly correlated to High Octane Oil's existing production and which ones might diversify the company's production on a geographical basis.

Solve the Dilemma

Surviving Rapid Growth

Glasspray Corporation is a small firm that makes industrial fiberglass spray equipment. Despite its size, the company supplies a range of firms from small mom-and-pop boatmakers to major industrial giants, both overseas and here at home. Indeed, just about every molded fiberglass resin product, from bathroom sinks and counters to portable spas and racing yachts, is constructed with the help of one or more of the company's machines.

Despite global acceptance of its products, Glasspray has repeatedly run into trouble with regard to the management of its current assets and liabilities as a result of extremely rapid and consistent increases in year-to-year sales. The firm's president and founder, Stephen T. Rose, recently lamented the sad state of his firm's working capital position: "Our current assets aren't, and our current liabilities are!" Rose shouted in a recent meeting of the firm's top officers. "We can't afford any more increases in sales! We're selling our way into bankruptcy! Frankly, our *working* capital doesn't!"

Critical Thinking Questions

1. Normally, rapidly increasing sales are a good thing. What seems to be the problem here?

2. What are the important components of a firm's working capital? Include both current assets and current liabilities.

3. What are some management techniques applied to current liabilities that Glasspray might use to improve its working capital position?

Build Your Business Plan

Financial Management and Securities Market

This chapter helps you realize that once you are making money, you need to be careful in determining how to invest it. Meanwhile, your team should consider the pros and cons of establishing a line of credit at the bank.

Remember the key to building your business plan is to be realistic!

See for Yourself Videocase

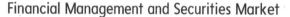

Tom & Eddie's: Fed through Nontraditional Channels

With the motto "We Put Good Taste in Everything We Do," Tom & Eddie's—a former line of gourmet burger restaurants in the Chicago area—was launched in 2009. It was a difficult time to launch a new business because it was at the height of the recession and banks were not doing much lending to small startup businesses. The owners of Tom & Eddie's, former McDonald's executives Tom Dentice and Ed Rensi, both agreed that access to cash was a major issue. However, they refused to be deterred.

It began when Tom and Ed decided they wanted to create their own upscale burger restaurant. Both men had retired from McDonald's during the 1990s, but retirement was not for them. They conceived of a burger restaurant totally different from McDonald's, with unique menu items, great customer service, and comfortable seating. The men had researched the market and knew that there was a strong demand for gourmet burgers. To meet this demand, the men settled on fresh burgers of 100 percent beef with high-quality cheese such as smoked gouda and other toppings. For those who dislike hamburger, the restaurant also has Ahi tuna, edamame, and turkey options. Tom & Eddie's used the nearby College of DuPage as a test facility for its menu items.

Although the men were certain their idea would be a success, they knew banks at the time were reluctant to lend to small businesses. They ended up using their own money to finance the business and took on another partner. They also hired Brian Gordon, whom they knew from McDonald's, to be the restaurant's CFO. They knew the restaurant needed someone skilled in financial management on board for things like operating expenditures, daily operational costs, and inventory management.

"In a small business, financial management is obviously very important. It's really the linchpin that holds everything together," said CFO Brian Gordon.

Immediate expenses for the restaurant included the cost for long-term assets, including the building, equipment, and furnishings and a sophisticated inventory management system. The restaurant uses trade credit to purchase its products in which it orders the products and is later invoiced for them.

One early decision they had to make was whether to build their own restaurant. Building a restaurant is tempting because the owners can customize it to meet their vision. However, it is also highly expensive. Tom & Eddie's made the decision to use buildings that already existed. As Brian Gordon explains, purchasing existing locations saves the company up-front costs. It also buys rather than leases its own equipment.

It was difficult to earn a profit initially. Tom & Eddie's struggled with the high cost of commodities, so it struggled to keep profit and loss under control. However, the restaurant proved popular among businesspeople, who tended to eat there during the day, as well as couples who came in the evening. The longer a restaurant stays in business, the more access it will have to lines of credit because it will be able to prove to the banks that it can earn money and pay back loans.

Tom & Eddie's had a fourth location in Deerfield, Illinois. After three years, however, the restaurant was not performing up to par, and the owners made the difficult decision to close it. The closure occurred before the lease expired, which meant the restaurant was still responsible for honoring its financial agreement. The restaurant did so and was able to maintain a good relationship with the landlords, which is crucial toward developing profitable business relationships in the future.

Tom & Eddie's overcame a number of obstacles, including a recession when banks were not lending to small businesses. While its strong financial management helped the business survive a period of major economic contraction, the restaurant quietly closed all locations in late 2017.[8]

Critical Thinking Questions

1. Why was it difficult to open a new business in 2009? How did the owners of Tom & Eddie's overcome this obstacle?

2. Describe some ways the restaurant exhibited strong financial management in the early days of its operations.

3. What are some of the costs involved with closing a business location that does not earn enough money?

You can find the related video in the Video Library in Connect. Ask your instructor how you can access Connect.

Team Exercise

Compare and contrast financing with long-term liabilities, such as bonds versus financing with owner's equity, typically retained earnings, common stock, and preferred stock. Form groups and suggest a good mix of long-term liabilities and owner's equity for a new firm that makes wind turbines for generating alternative energy and would like to grow quickly.

Notes

1. Matt Egan, "General Electric Gets Booted from the Dow," *CNN*, June 19, 2018, https://money.cnn.com/2018/06/19/investing/ge-dow-jones-walgreens/index.html (accessed September 6, 2018); Drake Bennett, "What the Hell Is Wrong with General Electric," *Bloomberg Businessweek*, February 5, 2018, pp. 42–49; Jeff Desjardins, "Chart: The Largest Companies by Market Cap over 15 Years," *Visual Capitalist*, August 12, 2016, http://www.visualcapitalist.com/chart-largest-companies-market-cap-15-years/ (accessed April 26, 2018); Tomas Gryta, Dana Mattiolo, and David Benoit, "GE Explores Further Deals," *The Wall Street Journal*, April 13, 2018, p. A1; Matt Egan, "What's Wrong with GE? An American Icon Is in 'Crisis' Mode," *CNN Money*, October 11, 2017, http://money.cnn.com/2017/10/11/investing/general-electric-stock-crisis-mode/index.html (accessed April 26, 2018).

2. Ed Zwirn, "Leaning In," *CFO*, May 4, 2016, http://ww2.cfo.com/people/2016/05/leaning-in-women-in-finance/ (accessed April 14, 2018); "Women in Financial Services," *Catalyst*, January 5, 2018, http://www.catalyst.org/knowledge/women-financial-services (accessed April 14, 2018); Emma Jacobs, "How Goldman Changed Its Ways to Boost Women at the Top," *Financial Times*, June 4, 2017, https://www.ft.com/content/51971b96-2f3e-11e7-9555-23ef563ecf9a (accessed April 14, 2018); Marielle Segarra, "Taking the Next Step," *CFO*, July 15, 2011, http://www.cfo.com/printable/article.cfm/14586563 (accessed April 14, 2018); Dan Fitzpatrick and Lisa Rappaport, "Financial Firms' Ceiling," *The Wall Street Journal*, September 8, 2011, https://www.wsj.com/articles/SB10001424053111904103404576557100384026220 (accessed April 14, 2018); Kyle Stock, "Ranks of Women on Wall Street Thin,"

The Wall Street Journal, September 20, 2010, https://www
.wsj.com/articles/SB1000142405274870485830457549807
1732136704 (accessed April 14, 2018).

3. DuPont, *DuPont 2017 Global Reporting Initiative
Report,* 2017, http://www.dupont.com/content/dam/
dupont/corporate/our-approach/sustainability/2017-
Documents/2017%20DuPont%20Sustainability%20
Report.pdf (accessed April 19, 2018); DuPont, *CDP 2017
Climate Change 2017 Information Request,* 2017, http://
www.dupont.com/content/dam/dupont/corporate/our-
approach/sustainability/documents/2016-sustainability-
documents/ProgrammeResponseClimate%20
Change%202017.pdf (accessed April 19, 2018); DuPont,
"Performance and Reporting," 2018, http://www.dupont.
com/corporate-functions/sustainability/sustainability-
commitments/performance-reporting.html (accessed
April 19, 2018).

4. Calculated by Geoff Hirt from Apple's annual reports and
website on June 16, 2014.

5. Kennon, Joshua, "Should You Invest in an IPO?" About.com,
http://beginnersinvest.about.com/od/investmentbanking/a/
aa073106a.htm (accessed May 4, 2016).

6. Maureen Farrell and Alexander Osipovich, "Spotify's
Splashy Debut Pressures Banks," *The Wall Street Journal,*
April 3, 2018, https://www.wsj.com/articles/spotify-shares-
jump-in-market-debut-1522773951 (accessed April 14,
2018); Ben Sisario, "As Spotify Goes Public, Sony Cashes
In," *The New York Times,* April 4, 2018, https://www.
nytimes.com/2018/04/04/business/media/as-spotify-goes-
public-sony-cashes-in.html (accessed April 14, 2018); Adam
Holownia, "How Spotify Started," *Medium,* February 1,
2017, https://medium.com/@obtaineudaimonia/how-spotify-
started-257e713fcd8f (accessed April 14, 2018).

7. Vincent Ryan, "From Wall Street to Main Street," *CFO
Magazine,* June 2008, pp. 85–86.

8. Tom & Eddie's website, http://www.tomandeddies.com/
index.php (accessed May 10, 2016); Steve Sadin, "Burger
Shop Tom & Eddie's Closes Its Doors in Deerfield,"
Chicago Tribune, September 12, 2014, http://www.
chicagotribune.com/dining/chi-sp-burger-shop-tom-eddies-
closes-its-doors-in-deerfield-20141210-story.html (accessed
May 10, 2016); "Tom & Eddie Talk Burgers & Fries,"
BurgerBusiness, November 3, 2010, http://burgerbusiness.
com/tom-eddie-talk-burgers-fries/ (accessed May 10, 2016).

Credits

Bonus Chapters

A The Legal and Regulatory Environment

©Robert Daly/Getty Images

Chapter Outline

Learning Objectives

After reading this chapter, you will be able to:

LO A-1 Describe how the legal system resolves disputes.

LO A-2 Outline regulatory agencies that oversee business conduct.

LO A-3 Describe the laws that address business practices.

LO A-4 Summarize laws that determine parameters and requirements for business decision making.

LO A-5 Evaluate legal issues related to the Internet.

LO A-6 Explain the Sarbanes-Oxley Act and its impact on business decisions.

LO A-7 Demonstrate the impact of the Dodd-Frank Wall Street Reform and Consumer Protection Act on business decisions.

Business law refers to the rules and regulations that govern the conduct of business. Problems in this area come from the failure to keep promises, misunderstandings, disagreements about expectations, or, in some cases, attempts to take advantage of others. The regulatory environment offers a framework and enforcement system in order to provide a fair playing field for all businesses. The regulatory environment is created based on inputs from competitors, customers, employees, special interest groups, and the public's elected representatives. Lobbying by pressure groups who try to influence legislation often shapes the legal and regulatory environment.

business law
the rules and regulations that govern the conduct of business.

Sources of Law

Laws are classified as either criminal or civil. *Criminal law* not only prohibits a specific kind of action, such as unfair competition or mail fraud, but also imposes a fine or imprisonment as punishment for violating the law. A violation of a criminal law is thus called a crime. *Civil law* defines all the laws not classified as criminal, and it specifies the rights and duties of individuals and organizations (including businesses). Violations of civil law may result in fines but not imprisonment. The primary difference between criminal and civil law is that criminal laws are enforced by the state or nation, whereas civil laws are enforced through the court system by individuals or organizations.

Criminal and civil laws are derived from four sources: the Constitution (constitutional law), precedents established by judges (common law), federal and state statutes (statutory law), and federal and state administrative agencies (administrative law). Federal administrative agencies established by Congress control and influence business by enforcing laws and regulations to encourage competition and protect consumers, workers, and the environment. The Supreme Court is the ultimate authority on legal and regulatory decisions for appropriate conduct in business.

Courts and the Resolution of Disputes

Describe how the legal system resolves disputes.

The primary method of resolving conflicts and business disputes is through **lawsuits,** where one individual or organization takes another to court using civil laws. The legal system, therefore, provides a forum for businesspeople to resolve disputes based on our legal foundations. The courts may decide when harm or damage results from the actions of others.

lawsuit
where one individual or organization takes another to court using civil laws.

Because lawsuits are so frequent in the world of business, it is important to understand more about the court system where such disputes are resolved. Both financial restitution and specific actions to undo wrongdoing can result from going before a court to resolve a conflict. All decisions made in the courts are based on criminal and civil laws derived from the legal and regulatory system.

A businessperson may win a lawsuit in court and receive a judgment, or court order, requiring the loser of the suit to pay monetary damages. However, this does not guarantee the victor will be able to collect those damages. If the loser of the suit lacks the financial resources to pay the judgment—for example, if the loser is a bankrupt business—the winner of the suit may not be able to collect the award. Most business lawsuits involve a request for a sum of money, but some lawsuits request that a court specifically order a person or organization to do or to refrain from doing a certain act, such as slamming telephone customers.

The Court System

Jurisdiction is the legal power of a court, through a judge, to interpret and apply the law and make a binding decision in a particular case. In some instances, other courts will not

jurisdiction
the legal power of a court, through a judge, to interpret and apply the law and make a binding decision in a particular case.

enforce the decision of a prior court because it lacked jurisdiction. Federal courts are granted jurisdiction by the Constitution or by Congress. State legislatures and constitutions determine which state courts hear certain types of cases. Courts of general jurisdiction hear all types of cases; those of limited jurisdiction hear only specific types of cases. The Federal Bankruptcy Court, for example, hears only cases involving bankruptcy. There is some combination of limited and general jurisdiction courts in every state.

In a **trial court** (whether in a court of general or limited jurisdiction and whether in the state or the federal system), two tasks must be completed. First, the court (acting through the judge or a jury) must determine the facts of the case. In other words, if there is conflicting evidence, the judge or jury must decide who to believe. Second, the judge must decide which law or set of laws is pertinent to the case and must then apply those laws to resolve the dispute.

An **appellate court,** on the other hand, deals solely with appeals relating to the interpretation of law. Thus, when you hear about a case being appealed, it is not retried, but rather reevaluated. Appellate judges do not hear witnesses but instead base their decisions on a written transcript of the original trial. Moreover, appellate courts do not draw factual conclusions; the appellate judge is limited to deciding whether the trial judge made a mistake in interpreting the law that probably affected the outcome of the trial. If the trial judge made no mistake (or if mistakes would not have changed the result of the trial), the appellate court will let the trial court's decision stand. If the appellate court finds a mistake, it usually sends the case back to the trial court so that the mistake can be corrected. Correction may involve the granting of a new trial. On occasion, appellate courts modify the verdict of the trial court without sending the case back to the trial court.

Alternative Dispute Resolution Methods

Although the main remedy for business disputes is a lawsuit, other dispute resolution methods are becoming popular. The schedules of state and federal trial courts are often crowded; long delays between the filing of a case and the trial date are common. Further, complex cases can become quite expensive to pursue. As a result, many businesspeople are turning to alternative methods of resolving business arguments: mediation and arbitration, the mini-trial, and litigation in a private court.

Mediation is a form of negotiation to resolve a dispute by bringing in one or more third-party mediators, usually chosen by the disputing parties, to help reach a settlement. The mediator suggests different ways to resolve a dispute between the parties. The mediator's resolution is nonbinding–that is, the parties do not have to accept the mediator's suggestions; they are strictly voluntary.

Arbitration involves submission of a dispute to one or more third-party arbitrators, usually chosen by the disputing parties, whose decision is final. Arbitration differs from mediation in that an arbitrator's decision must be followed, whereas a mediator merely offers suggestions and facilitates negotiations. Cases may be submitted to arbitration because a contract—such as a labor contract—requires it or because the parties agree to do so. Some consumers are barred from taking claims to court by agreements drafted by banks, brokers, health plans, and others. Instead, they are

trial court
when a court (acting through the judge or jury) must determine the facts of the case, decide which law or set of laws is pertinent to the case, and apply those laws to resolve the dispute.

appellate court
a court that deals solely with appeals relating to the interpretation of law.

mediation
a method of outside resolution of labor and management differences in which the third party's role is to suggest or propose a solution to the problem.

arbitration
settlement of a labor/management dispute by a third party whose solution is legally binding and enforceable.

When workers and management cannot come to an agreement, workers may choose to picket and go on strike. Going on strike is usually reserved as a last resort if mediation and arbitration fail.

©Tannen Maury/Epa/REX/Shutterstock

required to take complaints to mandatory arbitration. Arbitration can be an attractive alternative to a lawsuit because it is often cheaper and quicker, and the parties frequently can choose arbitrators who are knowledgeable about the particular area of business at issue.

A method of dispute resolution that may become increasingly important in settling complex disputes is the **mini-trial,** in which both parties agree to present a summarized version of their case to an independent third party. That person then advises them of his or her impression of the probable outcome if the case were to be tried. Representatives of both sides then attempt to negotiate a settlement based on the advisor's recommendations. For example, employees in a large corporation who believe they have muscular or skeletal stress injuries caused by the strain of repetitive motion in using a computer could agree to a mini-trial to address a dispute related to damages. Although the mini-trial itself does not resolve the dispute, it can help the parties resolve the case before going to court. Because the mini-trial is not subject to formal court rules, it can save companies a great deal of money, allowing them to recognize the weaknesses in a particular case.

In some areas of the country, disputes can be submitted to a private nongovernmental court for resolution. In a sense, a **private court system** is similar to arbitration in that an independent third party resolves the case after hearing both sides of the story. Trials in private courts may be either informal or highly formal, depending on the people involved. Businesses typically agree to have their disputes decided in private courts to save time and money.

mini-trial
a situation in which both parties agree to present a summarized version of their case to an independent third party; the third party advises them of his or her impression of the probable outcome if the case were to be tried.

private court system
similar to arbitration in that an independent third party resolves the case after hearing both sides of the story.

Federal Trade Commission (FTC)
the federal regulatory unit that most influences business activities related to questionable practices that create disputes between businesses and their customers.

Regulatory Administrative Agencies

LO A-2

Outline regulatory agencies that oversee business conduct.

Federal and state administrative agencies (listed in Table A.1) also have some judicial powers. Many administrative agencies, such as the Federal Trade Commission, decide disputes that involve their regulations. In such disputes, the resolution process is usually called a "hearing" rather than a trial. In these cases, an administrative law judge decides all issues.

Federal regulatory agencies influence many business activities and cover product liability, safety, and the regulation or deregulation of public utilities. Usually, these bodies have the power to enforce specific laws, such as the Federal Trade Commission Act, and have some discretion in establishing operating rules and regulations to guide certain types of industry practices. Because of this discretion and overlapping areas of responsibility, confusion or conflict regarding which agencies have jurisdiction over which activities is common.

Of all the federal regulatory units, the **Federal Trade Commission (FTC)** most influences business activities related to questionable practices that create disputes between businesses and their customers. Although the FTC regulates a variety of business practices, it allocates a large portion of resources to curbing false advertising, misleading pricing, and deceptive packaging and labeling. When it receives a complaint or otherwise has reason to believe that a firm is violating a law, the FTC issues a complaint stating that the business is in violation.

The Federal Trade Commission regulates business activities related to questionable practices in order to protect consumers.
©Carol M. Highsmith Archive/Library of Congress, Prints and Photographs Division

TABLE A.1   The Major Regulatory Agencies

Agency	Major Areas of Responsibility
Federal Trade Commission (FTC)	Enforces laws and guidelines regarding business practices; takes action to stop false and deceptive advertising and labeling.
Food and Drug Administration (FDA)	Enforces laws and regulations to prevent distribution of adulterated or misbranded foods, drugs, medical devices, cosmetics, veterinary products, and particularly hazardous consumer products.
Consumer Product Safety Commission (CPSC)	Ensures compliance with the Consumer Product Safety Act; protects the public from unreasonable risk of injury from any consumer product not covered by other regulatory agencies.
Interstate Commerce Commission (ICC)	Regulates franchises, rates, and finances of interstate rail, bus, truck, and water carriers.
Federal Communications Commission (FCC)	Regulates communication by wire, radio, and television in interstate and foreign commerce.
Environmental Protection Agency (EPA)	Develops and enforces environmental protection standards and conducts research into the adverse effects of pollution.
Federal Energy Regulatory Commission (FERC)	Regulates rates and sales of natural gas products, thereby affecting the supply and price of gas available to consumers; also regulates wholesale rates for electricity and gas, pipeline construction, and U.S. imports and exports of natural gas and electricity.
Equal Employment Opportunity Commission (EEOC)	Investigates and resolves discrimination in employment practices.
Federal Aviation Administration (FAA)	Oversees the policies and regulations of the airline industry.
Federal Highway Administration (FHA)	Regulates vehicle safety requirements.
Occupational Safety and Health Administration (OSHA)	Develops policy to promote worker safety and health and investigates infractions.
Securities and Exchange Commission (SEC)	Regulates corporate securities trading and develops protection from fraud and other abuses; provides an accounting oversight board.
Consumer Financial Protection Bureau	Regulates financial products and institutions to ensure consumer protection.

If a company continues the questionable practice, the FTC can issue a cease-and-desist order, which is an order for the business to stop doing whatever has caused the complaint. In such cases, the charged firm can appeal to the federal courts to have the order rescinded. However, the FTC can seek civil penalties in court—up to a maximum penalty of $10,000 a day for each infraction—if a cease-and-desist order is violated. In its battle against unfair pricing, the FTC has issued consent decrees alleging that corporate attempts to engage in price fixing or invitations to competitors to collude are violations even when the competitors in question refuse the invitations. The commission can also require companies to run corrective advertising in response to previous ads considered misleading.

The FTC also assists businesses in complying with laws. New marketing methods are evaluated every year. When general sets of guidelines are needed to improve business practices in a particular industry, the FTC sometimes encourages firms within that industry to establish a set of trade practices voluntarily. The FTC may even sponsor a conference bringing together industry leaders and consumers for the purpose of establishing acceptable trade practices.

Unlike the FTC, other regulatory units are limited to dealing with specific goods, services, or business activities. The Food and Drug Administration (FDA) enforces regulations prohibiting the sale and distribution of adulterated, misbranded, or hazardous food and drug products. For example, the FDA outlawed the sale and distribution of most over-the-counter hair-loss remedies after research indicated that few of the products were effective in restoring hair growth.

The Environmental Protection Agency (EPA) develops and enforces environmental protection standards and conducts research into the adverse effects of pollution. The Consumer Product Safety Commission recalls about 300 products a year, ranging from small, inexpensive toys to major appliances. The Consumer Product Safety Commission's website provides details regarding current recalls. The Consumer Product Safety Commission has fallen under increasing scrutiny in the wake of a number of product safety scandals involving children's toys. The most notable of these issues was lead paint discovered in toys produced in China. Some items are not even targeted to children but can be dangerous because children think they are food. Magnetic desk toys and Tide Pods have both been mistaken as candy by children.

Allstar Marketing Group, the business behind the Snuggie, faced claims from the FTC that its ads were misleading. The company was required to pay its customers $7.2 million in refunds.

©Cyrus McCrimmon/The Denver Post via Getty Images

Important Elements of Business Law

LO A-3

Describe the laws that address business practices.

To avoid violating criminal and civil laws, as well as discouraging lawsuits from consumers, employees, suppliers, and others, businesspeople need to be familiar with laws that address business practices.

The Uniform Commercial Code

At one time, states had their own specific laws governing various business practices, and transacting business across state lines was difficult because of the variation in the laws from state to state. To simplify commerce, every state—except Louisiana—has enacted the Uniform Commercial Code (Louisiana has enacted portions of the code). The **Uniform Commercial Code (UCC)** is a set of statutory laws covering several business law topics. Article II of the Uniform Commercial Code, which is discussed in the following paragraphs, has a significant impact on business.

Uniform Commercial Code (UCC)
set of statutory laws covering several business law topics.

Sales Agreements.   Article II of the Uniform Commercial Code covers sales agreements for goods and services such as installation but does not cover the sale of stocks and bonds, personal services, or real estate. Among its many provisions,

Article II stipulates that a sales agreement can be enforced even though it does not specify the selling price or the time or place of delivery. It also requires that a buyer pay a reasonable price for goods at the time of delivery if the buyer and seller have not reached an agreement on price. Specifically, Article II addresses the rights of buyers and sellers, transfers of ownership, warranties, and the legal placement of risk during manufacture and delivery.

Article II also deals with express and implied warranties. An **express warranty** stipulates the specific terms the seller will honor. Many automobile manufacturers, for example, provide three-year or 36,000-mile warranties on their vehicles, during which period they will fix any and all defects specified in the warranty. An **implied warranty** is imposed on the producer or seller by law, although it may not be a written document provided at the time of sale. Under Article II, a consumer may assume that the product for sale has a clear title (in other words, that it is not stolen) and that the product will serve the purpose for which it was made and sold as well as function as advertised.

The Law of Torts and Fraud

A **tort** is a private or civil wrong other than a breach of contract. For example, a tort can result if the driver of a Domino's Pizza delivery car loses control of the vehicle and damages property or injures a person. In the case of the delivery car accident, the injured persons might sue the driver and the owner of the company—Domino's in this case—for damages resulting from the accident.

Fraud is a purposefully unlawful act to deceive or manipulate in order to damage others. Thus, in some cases, fraud may also represent a violation of criminal law. Health care fraud has become a major issue in the courts.

An important aspect of tort law involves **product liability**—businesses' legal responsibility for any negligence in the design, production, sale, and consumption of products. Product liability laws have evolved from both common and statutory law. Some states have expanded the concept of product liability to include injuries by products whether or not the producer is proven negligent. Under this strict product liability, a consumer who files suit because of an injury has to prove only that the product was defective, that the defect caused the injury, and that the defect made the product unreasonably dangerous. For example, a carving knife is expected to be sharp and is not considered defective if you cut your finger using it. But an electric knife could be considered defective and unreasonably dangerous if it continued to operate after being switched off.

Reforming tort law, particularly in regard to product liability, has become a hot political issue as businesses look for relief from huge judgments in lawsuits. Although many lawsuits are warranted—few would disagree that a wrong has occurred when a patient dies because of negligence during a medical procedure or when a child is seriously injured by a defective toy and that the families deserve some compensation—many suits are not. Because of multimillion-dollar judgments, companies are trying to minimize their liability, and sometimes they pass on the costs of the damage awards to their customers in the form of higher prices. Some states have passed laws limiting damage awards and some tort reform is occurring at the federal level. Table A.2 lists the state courts systems the U.S. Chamber of Commerce's Institute for Legal Reform has identified as being "friendliest" and "least friendly" to business in terms of juries' fairness, judges' competence and impartiality, and other factors.

---

express warranty stipulates the specific terms the seller will honor.

implied warranty imposed on the producer or seller by law, although it may not be a written document provided at the time of sale.

tort a private or civil wrong other than breach of contract.

fraud a purposefully unlawful act to deceive or manipulate in order to damage others.

product liability businesses' legal responsibility for any negligence in the design, production, sale, and consumption of products.

TABLE A.2   State Court Systems' Reputations for Supporting Business

Most Friendly to Business	Least Friendly to Business
South Dakota	New Jersey
Vermont	Kentucky
Idaho	Alabama
Minnesota	Mississippi
New Hampshire	West Virginia
Alaska	Florida
Nebraska	California
Wyoming	Illinois
Maine	Missouri
Virginia	Louisiana

Source: U.S. Chamber Institute for Legal Reform, "States," www.instituteforlegalreform.com/states (accessed March 30, 2018).

The Law of Contracts

Virtually every business transaction is carried out by means of a **contract,** a mutual agreement between two or more parties that can be enforced in a court if one party chooses not to comply with the terms of the contract. If you rent an apartment or house, for example, your lease is a contract. If you have borrowed money under a student loan program, you have a contractual agreement to repay the money. Many aspects of contract law are covered under the Uniform Commercial Code.

A "handshake deal" is, in most cases, as fully and completely binding as a written, signed contract agreement. Indeed, many oil-drilling and construction contractors have for years agreed to take on projects on the basis of such handshake deals. However, individual states require that some contracts be in writing to be enforceable. Most states require that at least some of the following contracts be in writing:

- Contracts involving the sale of land or an interest in land.
- Contracts to pay somebody else's debt.
- Contracts that cannot be fulfilled within one year.
- Contracts for the sale of goods that cost more than $500 (required by the Uniform Commercial Code).

Only those contracts that meet certain requirements—called *elements*—are enforceable by the courts. A person or business seeking to enforce a contract must show that it contains the following elements: voluntary agreement, consideration, contractual capacity of the parties, and legality.

For any agreement to be considered a legal contract, all persons involved must agree to be bound by the terms of the contract. *Voluntary agreement* typically comes about when one party makes an offer and the other accepts. If both the offer and the acceptance are freely, voluntarily, and knowingly made, the acceptance forms the basis for the contract. If, however, either the offer or the acceptance is the result of

contract
a mutual agreement between two or more parties that can be enforced in a court if one party chooses not to comply with the terms of the contract.

fraud or force, the individual or organization subject to the fraud or force can void, or invalidate, the resulting agreement or receive compensation for damages.

The second requirement for enforcement of a contract is that it must be supported by *consideration*—that is, money or something of value must be given in return for fulfilling a contract. As a general rule, a person cannot be forced to abide by the terms of a promise unless that person receives a consideration. The "something of value" could be money, goods, services, or even a promise to do or not to do something.

Contractual capacity is the legal ability to enter into a contract. As a general rule, a court cannot enforce a contract if either party to the agreement lacks contractual capacity. A person's contractual capacity may be limited or nonexistent if he or she is a minor (under the age of 18), mentally unstable, intellectually disabled, insane, or intoxicated.

Legality is the state or condition of being lawful. For an otherwise binding contract to be enforceable, both the purpose of and the consideration for the contract must be legal. A contract in which a bank loans money at a rate of interest prohibited by law, a practice known as usury, would be an illegal contract, for example. The fact that one of the parties may commit an illegal act while performing a contract does not render the contract itself illegal, however.

Breach of contract is the failure or refusal of a party to a contract to live up to his or her promises. In the case of an apartment lease, failure to pay rent would be considered breach of contract. The breaching party—the one who fails to comply—may be liable for monetary damages that he or she causes the other person.

The Law of Agency

An **agency** is a common business relationship created when one person acts on behalf of another and under that person's control. Two parties are involved in an agency relationship: The **principal** is the one who wishes to have a specific task accomplished; the **agent** is the one who acts on behalf of the principal to accomplish the task. Authors, movie stars, and athletes often employ agents to help them obtain the best contract terms.

An agency relationship is created by the mutual agreement of the principal and the agent. It is usually not necessary that such an agreement be in writing, although putting it in writing is certainly advisable. An agency relationship continues as long as both the principal and the agent so desire. It can be terminated by mutual agreement, by fulfillment of the purpose of the agency, by the refusal of either party to continue in the relationship, or by the death of either the principal or the agent. In most cases, a principal grants authority to the agent through a formal *power of attorney,* which is a legal document authorizing a person to act as someone else's agent. The power of attorney can be used for any agency relationship, and its use is not limited to lawyers. For instance, in real estate transactions, often a lawyer or real estate agent is given power of attorney with the authority to purchase real estate for the buyer. Accounting firms often give employees agency relationships in making financial transactions.

Both officers and directors of corporations are fiduciaries, or people of trust, who use due care and loyalty as an agent in making decisions on behalf of the organization. This relationship creates a duty of care, also called duty of diligence, to make informed decisions. These agents of the corporation are not held responsible for negative outcomes if they are informed and diligent in their decisions. The duty of loyalty means that all decisions should be in the interests of the corporation and its

breach of contract
the failure or refusal of a party to a contract to live up to his or her promises.

agency
a common business relationship created when one person acts on behalf of another and under that person's control.

principal
the one in an agency relationship who wishes to have a specific task accomplished.

agent
the one in an agency relationship who acts on behalf of the principal to accomplish the task.

stakeholders. After Wells Fargo was found to be engaging in widespread misconduct, a new board chair and several new board members were brought in to improve risk management and oversight.[1]

The Law of Property

Property law is extremely broad in scope because it covers the ownership and transfer of all kinds of real, personal, and intellectual property. **Real property** consists of real estate and everything permanently attached to it; **personal property** basically is everything else. Personal property can be further subdivided into tangible and intangible property. *Tangible property* refers to items that have a physical existence, such as automobiles, business inventory, and clothing. *Intangible property* consists of rights and duties; its existence may be represented by a document or by some other tangible item. For example, accounts receivable, stock in a corporation, goodwill, and trademarks are all examples of intangible personal property. **Intellectual property** refers to property, such as musical works, artwork, books, and computer software, that is generated by a person's creative activities.

Copyrights, patents, and trademarks provide protection to the owners of property by giving them the exclusive right to use it. *Copyrights* protect the ownership rights on material (often intellectual property) such as books, music, videos, photos, and computer software. The creators of such works, or their heirs, generally have exclusive rights to the published or unpublished works for the creator's lifetime, plus 50 years. *Patents* give inventors exclusive rights to their invention for 20 years. The most intense competition for patents is in the pharmaceutical industry. Most patents take a minimum of 18 months to secure.

A *trademark* is a brand (name, mark, or symbol) that is registered with the U.S. Patent and Trademark Office and is thus legally protected from use by any other firm. Among the symbols that have been so protected are McDonald's golden arches. It is estimated that large multinational firms may have as many as 15,000 conflicts related to trademarks. Companies are diligent about protecting their trademarks both to avoid confusion in consumers' minds and because a term that becomes part of everyday language can no longer be trademarked. The names *aspirin* and *nylon,* for example, were once the exclusive property of their creators but became so widely used as product names (rather than brand names) that now anyone can use them. A related term is *trade dress,* which refers to the visual appearance of a product or its packaging. Coca-Cola's contoured bottle and Hershey's 12 rectangular panels for its chocolate bars are examples of visual characteristics protected as intellectual property. In order for these visual characteristics to receive protection, consumers must strongly associate the shape or design with the product itself.[2]

As the trend toward globalization of trade continues, and more and more businesses trade across national boundaries, protecting property rights, particularly intellectual property such as computer software, has become an increasing challenge. While a company may be able to register a trademark, a

real property
consists of real estate and everything permanently attached to it.

personal property
all other property that is not real property.

intellectual property
refers to property, such as musical works, artwork, books, and computer software, that is generated by a person's creative activities.

Christian Louboutin has received intellectual property protection for its signature red sole shoes. This type of protection is known as trade dress.
©denisfilm/123RF

Claire's, a teen jewelry and accessories chain, filed for Chapter II bankruptcy in the hopes of lowering its nearly $2 billion debt.

©James W Copeland/Shutterstock

brand name, or a symbol in its home country, it may not be able to secure that protection abroad. Some countries have copyright and patent laws that are less strict than those of the United States; some countries will not enforce U.S. laws. China, for example, has often been criticized for permitting U.S. goods to be counterfeited there. Pacific Mall in Markham in Ontario, Canada, has been labeled as one of several notorious markets selling counterfeit products at the same level as the Silk Market in Beijing and open air markets in Mexico City.[3] Such counterfeiting harms not only the sales of U.S. companies, but also their reputations if the knockoffs are of poor quality. Thus, businesses engaging in foreign trade may have to take extra steps to protect their property because local laws may be insufficient to protect them.

The Law of Bankruptcy

Although few businesses and individuals intentionally fail to repay (or default on) their debts, sometimes they cannot fulfill their financial obligations. Individuals may charge goods and services beyond their ability to pay for them. Businesses may take on too much debt in order to finance growth, or business events such as an increase in the cost of commodities can bankrupt a company. An option of last resort in these cases is bankruptcy, or legal insolvency. Some well-known companies that have declared bankruptcy include Hostess, American Apparel, and RadioShack.

Individuals or companies may ask a bankruptcy court to declare them unable to pay their debts and thus release them from the obligation of repaying those debts. The debtor's assets may then be sold to pay off as much of the debt as possible. In the case of a personal bankruptcy, although the individual is released from repaying debts and can start over with a clean slate, obtaining credit after bankruptcy proceedings is very difficult. However, a restrictive allows fewer consumers to use bankruptcy to eliminate their debts. The law makes it harder for consumers to prove that they should be allowed to clear their debts for what is called a "fresh start" or Chapter 7 bankruptcy. Although the person or company in debt usually initiates bankruptcy proceedings, creditors may also initiate them. The subprime mortgage crisis caused a string of bankruptcies among individuals, and Chapter 7 and Chapter 11 bankruptcies among banks and other businesses as well. Table A.3 describes the various levels of bankruptcy protection a business or individual may seek.

LO A-4

Summarize laws that determine parameters and requirements for business decision making.

Sherman Antitrust Act passed in I890 to prevent businesses from restraining trade and monopolizing markets.

Laws Affecting Business Practices

One of the government's many roles is to act as a watchdog to ensure that businesses behave in accordance with the wishes of society. Congress has enacted a number of laws that affect business practices; some of the most important of these are summarized in Table A.4. Many state legislatures have enacted similar laws governing business within specific states.

The **Sherman Antitrust Act,** passed in 1890 to prevent businesses from restraining trade and monopolizing markets, condemns "every contract, combination, or conspiracy in restraint of trade." For example, a request that a competitor agree to

TABLE A.3   Types of Bankruptcy

Chapter 7	Requires that the business be dissolved and its assets liquidated, or sold, to pay off the debts. Individuals declaring Chapter 7 retain a limited amount of exempt assets, the amount of which may be determined by state or federal law, at the debtor's option. Although the type and value of exempt assets vary from state to state, most states' laws allow a bankrupt individual to keep an automobile, some household goods, clothing, furnishings, and at least some of the value of the debtor's residence. All nonexempt assets must be sold to pay debts.
Chapter 11	Temporarily frees a business from its financial obligations while it reorganizes and works out a payment plan with its creditors. The indebted company continues to operate its business during bankruptcy proceedings. Often, the business sells off assets and less-profitable subsidiaries to raise cash to pay off its immediate obligations.
Chapter 13	Similar to Chapter 11 but limited to individuals. This proceeding allows an individual to establish a three- to five-year plan for repaying his or her debt. Under this plan, an individual ultimately may repay as little as 10 percent of his or her debt.

TABLE A.4   Major Federal Laws Affecting Business Practices

Act (Date Enacted)	Purpose
Sherman Antitrust Act (1890)	Prohibits contracts, combinations, or conspiracies to restrain trade; establishes as a misdemeanor monopolizing or attempting to monopolize.
Clayton Act (1914)	Prohibits specific practices such as price discrimination, exclusive dealer arrangements, and stock acquisitions in which the effect may notably lessen competition or tend to create a monopoly.
Federal Trade Commission Act (1914)	Created the Federal Trade Commission; also gives the FTC investigatory powers to be used in preventing unfair methods of competition.
Robinson-Patman Act (1936)	Prohibits price discrimination that lessens competition among wholesalers or retailers; prohibits producers from giving disproportionate services of facilities to large buyers.
Wheeler-Lea Act (1938)	Prohibits unfair and deceptive acts and practices regardless of whether competition is injured; places advertising of foods and drugs under the jurisdiction of the FTC.
Lanham Act (1946)	Provides protections and regulation of brand names, brand marks, trade names, and trademarks.
Celler-Kefauver Act (1950)	Prohibits any corporation engaged in commerce from acquiring the whole or any part of the stock or other share of the capital assets of another corporation when the effect substantially lessens competition or tends to create a monopoly.
Fair Packaging and Labeling Act (1966)	Makes illegal the unfair or deceptive packaging or labeling of consumer products.
Magnuson-Moss Warranty (FTC) Act (1975)	Provides for minimum disclosure standards for written consumer product warranties; defines minimum consent standards for written warranties; allows the FTC to prescribe interpretive rules in policy statements regarding unfair or deceptive practices.
Consumer Goods Pricing Act (1975)	Prohibits the use of price maintenance agreements among manufacturers and resellers in interstate commerce.
Antitrust Improvements Act (1976)	Requires large corporations to inform federal regulators of prospective mergers or acquisitions so that they can be studied for any possible violations of the law.

(continued)

TABLE A.4   (Continued)

Act (Date Enacted)	Purpose
Trademark Counterfeiting Act (1980)	Provides civil and criminal penalties against those who deal in counterfeit consumer goods or any counterfeit goods that can threaten health or safety.
Trademark Law Revision Act (1988)	Amends the Lanham Act to allow brands not yet introduced to be protected through registration with the Patent and Trademark Office.
Nutrition Labeling and Education Act (1990)	Prohibits exaggerated health claims and requires all processed foods to contain labels with nutritional information.
Telephone Consumer Protection Act (1991)	Establishes procedures to avoid unwanted telephone solicitations; prohibits marketers from using automated telephone dialing system or an artificial or prerecorded voice to certain telephone lines.
Federal Trademark Dilution Act (1995)	Provides trademark owners the right to protect trademarks and requires relinquishment of names that match or parallel existing trademarks.
Digital Millennium Copyright Act (1998)	Refined copyright laws to protect digital versions of copyrighted materials, including music and movies.
Children's Online Privacy Protection Act (2000)	Regulates the collection of personally identifiable information (name, address, e-mail address, hobbies, interests, or information collected through cookies) online from children under age 13.
Sarbanes-Oxley Act (2002)	Made securities fraud a criminal offense; stiffened penalties for corporate fraud; created an accounting oversight board; instituted numerous other provisions designed to increase corporate transparency and compliance.
Do Not Call Implementation Act (2003)	Directs FCC and FTC to coordinate so their rules are consistent regarding telemarketing call practices, including the Do Not Call Registry.
Dodd-Frank Wall Street Reform and Consumer Protection Act (2010)	Increases accountability and transparency in the financial industry; protects consumers from deceptive financial practices; establishes the Consumer Financial Protection Bureau.

fix prices or divide markets would, if accepted, result in a violation of the Sherman Antitrust Act. The FTC challenged the proposed merger between Staples and Office Depot because it believed the merger could significantly reduce the competition in the "consumable" office supply market.[4] The Sherman Antitrust Act, still highly relevant 100 years after its passage, is being copied throughout the world as the basis for regulating fair competition.

Because the provisions of the Sherman Antitrust Act are rather vague, courts have not always interpreted it as its creators intended. The Clayton Act was passed in 1914 to limit specific activities that can reduce competition. The **Clayton Act** prohibits price discrimination, tying and exclusive agreements, and the acquisition of stock in another corporation where the effect may be to substantially lessen competition or tend to create a monopoly. In addition, the Clayton Act prohibits members of one company's board of directors from holding seats on the boards of competing corporations. The act also exempts farm cooperatives and labor organizations from antitrust laws.

In spite of these laws regulating business practices, there are still many questions about the regulation of business. For instance, it is difficult to determine what constitutes an acceptable degree of competition and whether a monopoly is harmful to

Clayton Act
prohibits price discrimination, tying and exclusive agreements, and the acquisition of stock in another corporation where the effect may be to substantially lessen competition or tend to create a monopoly.

a particular market. Many mergers were permitted that resulted in less competition in the banking, publishing, and automobile industries. In some industries, such as utilities, it is not cost effective to have too many competitors. For this reason, the government permits utility monopolies, although recently, the telephone, electricity, and communications industries have been deregulated. Furthermore, the antitrust laws are often rather vague and require interpretation, which may vary from judge to judge and court to court. Thus, what one judge defines as a monopoly or trust today may be permitted by another judge a few years from now. Businesspeople need to understand what the law says on these issues and try to conduct their affairs within the bounds of these laws.

The Internet: Legal and Regulatory Issues

> **LO A-5**
>
> Evaluate legal issues related to the Internet.

Our use and dependence on the Internet is increasingly creating a potential legal problem for businesses. With this growing use come questions of maintaining an acceptable level of privacy for consumers and proper competitive use of the medium. Some might consider that tracking individuals who visit or "hit" their website by attaching a "cookie" (identifying you as a website visitor for potential recontact and tracking your movement throughout the site) is an improper use of the Internet for business purposes. Others may find such practices acceptable and similar to the practices of non-Internet retailers who copy information from checks or ask customers for their name, address, or phone number before they will process a transaction. There are few specific laws that regulate business on the Internet, but the standards for acceptable behavior that are reflected in the basic laws and regulations designed for traditional businesses can be applied to business on the Internet as well. One law aimed specifically at advertising on the Internet is the CAN-SPAM Act of 2004. The law restricts unsolicited e-mail advertisements by requiring the consent of the recipient. Furthermore, the CAN-SPAM Act follows the "opt-out" model wherein recipients can elect not to receive further e-mails from a sender simply by clicking on a link.[5] Individual e-mail violations are subject to penalties of up to $41,484.[6]

The central focus for future legislation of business conducted on the Internet is the protection of personal privacy. The present basis of personal privacy protection is the U.S. Constitution, various Supreme Court rulings, and laws such as the 1971 Fair Credit Reporting Act; the 1978 Right to Financial Privacy Act; and the 1974 Privacy Act, which deals with the release of government records. With few regulations on the use of information by businesses, companies legally buy and sell information on customers to gain competitive advantage. Sometimes existing laws are not enough to protect people, and the ease with which information on customers can be obtained becomes a problem. For example, identity theft has increased due to the proliferation of the use of the Internet. A disturbing trend is how many children have had their identities stolen. One study of 40,000 children revealed that more than 10 percent have had their Social Security numbers stolen.[7] It has been suggested that the treatment of personal data as property will ensure privacy rights by recognizing that customers have a right to control the use of their personal data.

Internet use is different from traditional interaction with businesses in that it is readily accessible, and most online businesses are able to develop databases of information on customers. Congress has restricted the development of databases on children using the Internet. The Children's Online Privacy Protection Act (COPPA) of 2000 prohibits website and Internet providers from seeking personal information from children under age 13 without parental consent. Companies are still running afoul of COPPA. Two app developers paid $360,000 to settle charges from the FTC that they had allowed third-party advertisers to collect personal information from children under the age of 13.[8]

The FTC rules for online advertising and marketing are the same as any form of communication or advertising. These rules help maintain the credibility of the Internet as an advertising medium. To avoid deception all online communication must tell the truth and cannot mislead consumers. In addition, all claims must be substantiated. Influencer marketing is relatively new compared with other forms of digital marketing, so it should be no surprise there have been road bumps for early adopters. Due to concerns about dishonest advertising, the FTC requires influencers to clearly disclose any connection they have with brands they promote. Neglecting to make a disclosure is viewed as deceptive advertising. Cases have been filed against Warner Bros. Home Entertainment, which paid PewDiePie (YouTube's number one most subscribed channel) for an endorsement of its video game Middle-Earth: Shadow of Mordor, and Lord & Taylor, which paid various influencers to promote their dresses, all without disclosures. According to the FTC, any level of compensation much be disclosed, whether a partnership is paid or an influencer strictly receives free product.[9]

Legal Pressure for Responsible Business Conduct

To ensure greater compliance with society's desires, both federal and state governments are moving toward increased organizational accountability for misconduct. Before 1991, laws mainly punished those employees directly responsible for an offense. Under new guidelines established by the Federal Sentencing Guidelines for Organizations (FSGO), however, both the responsible employees and the firms that employ them are held accountable for violations of federal law. Thus, the government now places responsibility for controlling and preventing misconduct squarely on the shoulders of top management. The main objectives of the federal guidelines are to train employees, self-monitor and supervise employee conduct, deter unethical acts, and punish those organizational members who engage in illegal acts.

A 2010 amendment to the FSGO directs ethics officers to report directly to the board of directors rather than simply the general counsel or top officers. This places the responsibility on the shoulders of the firm's leadership, usually the board of directors. The board must ensure that there is a high-ranking manager accountable for the day-to-day operational oversight of the ethics program. The board must provide for adequate authority, resources, and access to the board or an appropriate subcommittee of the board. The board must ensure that there are confidential mechanisms available so that the organization's employees and agents may report or seek guidance about potential or actual misconduct without fear of retaliation. Finally, the board is required to oversee the discovery of risks and to design, implement, and modify approaches to deal with those risks.

If an organization's culture and policies reward or provide opportunities to engage in misconduct through lack of managerial concern or failure to comply with the seven minimum requirements of the FSGO (provided in Table A.5), then the organization may incur not only penalties but also the loss of customer trust, public confidence, and other intangible assets. For this reason, organizations cannot succeed solely through a legalistic approach to compliance with the sentencing guidelines; top management must cultivate high ethical standards that will serve as barriers to illegal conduct.

Responding to Business Challenges

SEC Puts the Whistle Back in Whistleblowing

Whistleblowers may soon get new protections, thanks to a Securities and Exchange Commission (SEC) ruling limiting companies' abilities to censor and intimidate employees who report violations. The SEC offers financial incentives to some whistleblowers reporting large-scale misconduct. If a report results in a company receiving fines of more than $1 million, the SEC may award the whistleblower 10 to 30 percent of the fines or penalties levied against the company. The SEC's whistleblower program was created in 2010 as part of the Dodd-Frank Act. The intent of the program is to curb legal and ethical misconduct before it leads to disastrous consequences, such as the global financial crisis.

Some companies are suspected of forcing workers to sign nondisclosure agreements and harassing employees who voice concerns over misconduct. This includes clauses in employment contracts that limit the ability of whistleblowers to receive financial rewards for reporting misconduct, effectively neutralizing the incentive of employees to expose company violations. The Ethics Resource Center has found that 21 percent of employees experience retaliation after reporting misconduct.

The SEC is fighting back and taking steps to protect the rights of whistleblowers. The government agency is requesting years of corporate training procedures, confidentiality agreements, and even lists of terminated employees from certain companies to see whether they engaged in illegal discrimination against whistleblowers. These steps demonstrate the SEC's commitment to whistleblower protection and could lead to fewer companies suppressing future reporting of misconduct.[10]

Critical Thinking Questions

1. How is the SEC trying to prevent employers from silencing whistleblowers?
2. Why does the SEC feel that it is unfair for an employment contract to limit the ability of employees to receive financial rewards for blowing the whistle?
3. What is the SEC hoping to achieve by offering whistleblowers rewards for large misconduct convictions against firms?

The organization must want to be a good citizen and recognize the importance of compliance to successful workplace activities and relationships. In fact, the top concern of corporate lawyers is ethics and compliance. Implementing ethics and compliance ranks higher than any other concern, possibly due to the pressures placed on companies by the passage of Sarbanes-Oxley, the Dodd-Frank Act, and the Federal Sentencing Guidelines.[11]

The federal guidelines also require businesses to develop programs that can detect—and that will deter employees from engaging in—misconduct. To be considered effective, such compliance programs must include disclosure of any wrongdoing,

TABLE A.5 Seven Steps to Compliance

1. Develop standards and procedures to reduce the propensity for criminal conduct.
2. Designate a high-level compliance manager or ethics officer to oversee the compliance program.
3. Avoid delegating authority to people known to have a propensity to engage in misconduct.
4. Communicate standards and procedures to employees, other agents, and independent contractors through training programs and publications.
5. Establish systems to monitor and audit misconduct and to allow employees and agents to report criminal activity.
6. Enforce standards and punishments consistently across all employees in the organization.
7. Respond immediately to misconduct and take reasonable steps to prevent further criminal conduct.

cooperation with the government, and acceptance of responsibility for the misconduct. Codes of ethics, employee ethics training, hotlines (direct 800 phone numbers), compliance directors, newsletters, brochures, and other communication methods are typical components of a compliance program. The ethics component, discussed in the "Business Ethics and Social Responsibility" chapter, acts as a buffer, keeping firms away from the thin line that separates unethical and illegal conduct.

Despite the existing legislation, a number of ethics scandals in the early 2000s led Congress to pass—almost unanimously—the **Sarbanes-Oxley Act,** which criminalized securities fraud and strengthened penalties for corporate fraud. It also created an accounting oversight board that requires corporations to establish codes of ethics for financial reporting and to develop greater transparency in financial reports to investors and other interested parties. Additionally, the law requires top corporate executives to sign off on their firms' financial reports, and they risk fines and jail sentences if they misrepresent their companies' financial position. Table A.6 summarizes the major provisions of the Sarbanes-Oxley Act.

The Sarbanes-Oxley Act has created a number of concerns and is considered burdensome and expensive to corporations. Large corporations report spending more than $4 million each year to comply with the act, according to Financial Executives International. The act has caused more than 500 public companies a year to report problems in their accounting systems. Additionally, Sarbanes-Oxley failed to prevent and detect the widespread misconduct of financial institutions that led to the financial crisis.

Sarbanes-Oxley Act
a law that criminalized securities fraud and strengthened penalties for corporate fraud.

Explain the Sarbanes-Oxley Act and its impact on business decisions.

TABLE A.6   Major Provisions of the Sarbanes-Oxley Act

1. Requires the establishment of a Public Company Accounting Oversight Board in charge of regulations administered by the Securities and Exchange Commission.

2. Requires CEOs and CFOs to certify that their companies' financial statements are true and without misleading statements.

3. Requires that corporate boards of directors' audit committees consist of independent members who have no material interests in the company.

4. Prohibits corporations from making or offering loans to officers and board members.

5. Requires codes of ethics for senior financial officers; code must be registered with the SEC.

6. Prohibits accounting firms from providing both auditing and consulting services to the same client without the approval of the client firm's audit committee.

7. Requires company attorneys to report wrongdoing to top managers and, if necessary, to the board of directors; if managers and directors fail to respond to reports of wrongdoing, the attorney should stop representing the company.

8. Mandates "whistleblower protection" for persons who disclose wrongdoing to authorities.

9. Requires financial securities analysts to certify that their recommendations are based on objective reports.

10. Requires mutual fund managers to disclose how they vote shareholder proxies, giving investors information about how their shares influence decisions.

11. Establishes a 10-year penalty for mail/wire fraud.

12. Prohibits the two senior auditors from working on a corporation's account for more than five years; other auditors are prohibited from working on an account for more than seven years. In other words, accounting firms must rotate individual auditors from one account to another from time to time.

Source: Pub. L. 107-204, 116 Stat. 745 (2002).

On the other hand, there are many benefits, including greater accountability of top managers and boards of directors, that improve investor confidence and protect employees, especially their retirement plans. It is believed that the law has more benefits than drawbacks—with the greatest benefit being that boards of directors and top managers are better informed. Some companies such as Cisco and Pitney Bowes report improved efficiency and cost savings from better financial information.

In spite of the benefits Sarbanes-Oxley offers, it did not prevent widespread corporate corruption from leading to the most recent recession. The resulting financial crisis prompted the Obama administration to create new regulation to reform Wall Street and the financial industry. In 2010, the Dodd-Frank Wall Street Reform and Consumer Protection Act was passed. In addition to new regulations for financial institutions, the legislation created a Consumer Financial Protection Bureau (CFPB) to protect consumers from complex or deceptive financial products.

The Dodd-Frank Act contains 16 titles meant to increase consumer protection, enhance transparency and accountability in the financial sector, and create new financial agencies. In some ways, Dodd-Frank is attempting to improve upon provisions laid out in the Sarbanes-Oxley Act. For instance, Dodd-Frank takes whistleblower protection a step further by offering additional incentives to whistleblowers for reporting misconduct. If whistleblowers report misconduct that results in penalties of more than $1 million, the whistleblower will be entitled to a percentage of the settlement.[12] Additionally, complex financial instruments must now be made more transparent so that consumers will have a better understanding of what these instruments involve.

The act also created three new agencies: The Consumer Financial Protection Bureau (CFPB), the Office of Financial Research, and the Financial Stability Oversight Council. While the CFPB was created to protect consumers, the other two agencies work to maintain stability in the financial industry so such a crisis will not recur in the future.[13] Although it is too early to tell whether these regulations will serve to create wide-scale positive financial reform, the Dodd-Frank Act is certainly leading to major changes on Wall Street and in the financial sector.

LO A-7

Demonstrate the impact of the Dodd-Frank Wall Street Reform and Consumer Protection Act on business decisions.

Review Your Understanding

Describe how the legal system resolves disputes.

The primary method of resolving conflicts and business disputes is through lawsuits, where one individual or organization takes another to court using civil laws. All decisions made in the courts are based on criminal and civil laws derived from the legal and regulatory system. Courts must have jurisdiction, or the legal power, through a judge, to interpret and apply the law and make a binding decision in a particular case. In a trial court, the court must determine the facts of the case and the judge must decide which law or set of laws is pertinent to the case and must then apply those laws to resolve the dispute. An appellate court deals solely with appeals relating to the interpretation of law.

There are alternative methods of resolving disputes. Mediation is a form of negotiation to resolve a dispute by bringing in one or more third-party mediators, usually chosen by the disputing parties, to help reach a settlement. Arbitration

involves submission of a dispute to one or more third-party arbitrators, usually chosen by the disputing parties, whose decision usually is final. A mini-trial is when both parties agree to present a summarized version of their case to an independent third party. A private court system is similar to arbitration in that an independent third party resolves the case after hearing both sides of the story.

Outline regulatory agencies that oversee business conduct.

The Federal Trade Commission (FTC) most influences business activities related to questionable practices that create disputes between businesses and their customers. If a company continues the questionable practice, the FTC can issue a cease-and-desist order, which is an order for the business to stop doing whatever has caused the complaint. The Food and Drug Administration (FDA) enforces regulations prohibiting the sale and distribution of adulterated, misbranded, or hazardous food and drug products. The Environmental Protection Agency (EPA) develops and enforces environmental protection standards and conducts research into the adverse effects of pollution. The Consumer Product Safety Commission recalls about 300 products a year, ranging from small, inexpensive toys to major appliances.

Describe the laws that address business practices.

Important elements of business law include the Uniform Commercial Code (UCC), the law of torts and fraud, the law of contracts, the law of agency, the law of property, and the law of bankruptcy. The UCC is a set of statutory laws covering several business law topics. Article II of the UCC deals with express and implied warranties. An express warranty stipulates the specific terms the seller will honor. An implied warranty is imposed on the producer or seller by law, although it may not be a written document provided at the time of sale.

A tort is a private or civil wrong other than a breach of contract. An important part of tort law is product liability, businesses' legal responsibility for any negligence in the design, production, sale, and consumption of products. Fraud is a purposefully unlawful act to deceive or manipulate in order to damage others. A contract is a mutual agreement between two or more parties that can be enforced in a court if one party chooses not to comply with the terms of the contract.

An agency is a common business relationship created when one person acts on behalf of another and under that person's control. Property law is extremely broad in scope because it covers the ownership and transfer of all kinds of real, personal, and intellectual property. There are also laws dealing with bankruptcy, which occurs when individuals or companies ask a bankruptcy court to declare them unable to pay their debts and thus release them from the obligation of repaying those debts.

Summarize laws that determine parameters and requirements for business decision making.

Laws that determine parameters and requirements for business decision making include the Sherman Antitrust Act, the Clayton Act, and a number of other laws. The Sherman Antitrust Act, passed in 1890, is meant to prevent businesses from restraining trade and monopolizing markets. The Clayton Act prohibits price discrimination, tying and exclusive

agreements, and the acquisition of stock in another corporation where the effect may be to substantially lessen competition or tend to create a monopoly.

Evaluate legal issues related to the Internet.

With the growing use of the Internet, privacy is becoming a major concern. Some consider that tracking individuals who visit or "hit" their website by attaching a cookie is an invasion of privacy. The CAN-SPAM Act of 2004 restricts unsolicited email advertisements. The Children's Online Privacy Protection Act (COPPA) prohibits website and Internet providers from seeking personal information from children under age 13 without parental consent.

Explain the Sarbanes-Oxley Act and its impact on business decisions.

The Sarbanes-Oxley Act criminalized securities fraud and strengthened penalties for corporate fraud. It created an accounting oversight board that requires corporations to establish codes of ethics for financial reporting and to develop greater transparency in financial reports to investors and other interested parties. Additionally, the law requires top corporate executives to sign off on their firms' financial reports, and they risk fines and jail sentences if they misrepresent their companies' financial position.

Demonstrate the impact of the Dodd-Frank Wall Street Reform and Consumer Protection Act on business decisions.

In 2010, the Dodd-Frank Wall Street Reform and Consumer Protection Act was passed. In addition to new regulations for financial institutions, the legislation created a Consumer Financial Protection Bureau (CFPB) to protect consumers from complex or deceptive financial products. The Dodd-Frank Act contains 16 titles meant to increase consumer protection, enhance transparency and accountability in the financial sector, and create new financial agencies. Dodd-Frank takes whistleblower protection a step further by offering additional incentives to whistleblowers for reporting misconduct. If whistleblowers report misconduct that results in penalties of more than $1 million, the whistleblower will be entitled to a percentage of the settlement. The act also created two other agencies: the Office of Financial Research and the Financial Stability Oversight Council.

Learn the Terms

agency 542
agent 542
appellate court 536
arbitration 536
breach of contract 542
business law 535

Clayton Act 546
contract 541
express warranty 540
Federal Trade Commission (FTC) 537
fraud 540
implied warranty 540

intellectual property 543
jurisdiction 535
lawsuit 535
mediation 536
mini-trial 537
personal property 543

Check Your Progress

1. What are the sources of law relating to both criminal and civil actions?

2. How does the court system work to resolve business disputes?

3. Why is it important to understand how regulatory administrative agencies resolve disputes and protect the public?

4. What are the major elements of business law?

5. What are some of the most important laws that affect business practices?

6. Why is there the need for additional regulation to resolve issues related to the Internet?

7. What are the important steps to developing an ethics and compliance program?

8. What was the purpose of the Sarbanes-Oxley Act?

Build Your Skills

Developing an Ethics and Compliance Program

 The Federal Sentencing Guidelines for Organizations require a business's governing authority to develop an ethics and compliance program. This places responsibility for implementation on the firm's officers and board of directors. Organizations of all sizes can develop an ethics program. Below are the seven steps in developing an ethics and compliance program. After each step provide some activities that would help accomplish these steps.

1. Develop standards and procedures to reduce the propensity for criminal conduct.

    _____

2. Designate a high-level compliance manager or ethics officer to oversee the compliance program.

    _____

3. Avoid delegating authority to people known to have a propensity to engage in misconduct.

    _____

4. Communicate standards and procedures to employees, other agents, and independent contractors through training programs and publications.

    _____

5. Establish systems to monitor and audit misconduct and to allow employees and agents to report criminal activity.

    _____

6. Enforce standards and punishments consistently across all employees in the organization.

    _____

7. Respond immediately to misconduct and take reasonable steps to prevent further criminal conduct.

    _____

Notes

1. Wilfred Frost, "The Fed's Unprecedented Slap at Wells Fargo May Cost the Bank More than Just $400 Million This Year," *CNBC*, February 4, 2018, https://www.cnbc .com/2018/02/04/the-feds-unprecedented-slap-at-wells-fargo-may-cost-the-bank-more-than-400m-this-year.html (accessed March 30, 2018).

2. Benjamin West Janke, "Hershey's Protects Candy Bar Design," August 6, 2012, http://www.bakerdonelson .com/hersheys-protects-candy-bar-design-08-06-2012/ (accessed June 12, 2017); "Trade Dress," International Trademark Association, November 2015, http://www.inta .org/TrademarkBasics/FactSheets/Pages/Trade-Dress.aspx

(accessed June 12, 2017); Danielle Rubano, "Trade Dress: Who Should Bear the Burden of Proving or Disproving Functionality in a Section 43(a) Infringement Claim," *Fordham Intellectual Property, Media and Entertainment Law Journal* 6, no. 1 (1995), pp. 345–367.

3. "Market Down in Markham," *The Economist,* February 24–March 2, 2018, p. 28.

4. "FTC Challenges Proposed Merger of Staples, Inc. and Office Depot, Inc.," Federal Trade Commission, December 7, 2015, https://www.ftc.gov/news-events/press-releases/2015/12/ftc-challenges-proposed-merger-staples-inc-office-depot-inc (accessed February 26, 2016).

5. Maureen Dorney, "Congress Passes Federal Anti-Spam Law: Preempts Most State Anti-Spam Laws," *DLA Piper,* December 3, 2003, http://franchiseagreements.com/global/publications/detail.aspx?pub=622 (accessed April 7, 2014).

6. "CAN-SPAM Act: A Compliance Guide for Business," Federal Trade Commission, https://www.ftc.gov/tips-advice/business-center/guidance/can-spam-act-compliance-guide-business (accessed March 30, 2018).

7. Elizabeth Alterman, "As Kids Go Online, Identity Theft Claims More Victims," *CNBC,* October 10, 2011, http://www.cnbc.com/id/44583556 (accessed June 12, 2017); Ron Lieber, "Identify Theft Poses Extra Troubles for Children," *The New York Times,* April 17, 2015, http://www.nytimes.com/2015/04/18/your-money/a-childs-vulnerability-to-identity-theft.html?_r=0 (accessed June 12, 2017).

8. Federal Trade Commission, "Two App Developers Settle FTC Charges They Violated Children's Online Privacy Protection Act," December 17, 2015, https://www.ftc.gov/news-events/press-releases/2015/12/two-app-developers-settle-ftc-charges-they-violated-childrens (accessed June 12, 2017).

9. "FTC Cracking Down on Social Influencers' Labeling of Paid Promotions," *AdAge,* August 5, 2016, http://adage.com/article/digital/ftc-cracking-social-influencers-labeling-promotions/305345/ (accessed March 30, 2018).

10. Rachel Louise Ensign, "SEC Bolsters Whistleblowers," *The Wall Street Journal,* April 2, 2015, p. C1; Rachel Louise Ensign, "Treatment of Tipsters Is Focus of SEC," *The Wall Street Journal,* February 26, 2015, p. C1; Steve Goldstein, "Whistleblower on Credit Suisse Dark Pool Says He'll Ask SEC for Millions," *MarketWatch,* September 22, 2015, http://www.marketwatch.com/story/whistleblower-on-credit-suisse-dark-pool-says-hell-ask-sec-for-millions-2015-09-22 (accessed June 12, 2017); *National Business Ethics Survey of the U.S. Workforce* (Arlington, VA: Ethics Resource Center, 2014), p. 24.

11. Ashby Jones, "Nation's In-House Counsel Are Worried about Ethics, Data and 'Trolls,'" *The Wall Street Journal,* February 2, 2015, http://blogs.wsj.com/law/2015/02/02/nations-in-housecounsel-are-worried-about-ethics-data-and-trolls/ (accessed June 12, 2017).

12. Jean Eaglesham and Ashby Jones, "Whistle-blower Bounties Pose Challenges," *The Wall Street Journal,* December 13, 2010, pp. C1, C3.

13. "Office of Financial Research," U.S. Department of Treasury, www.treasury.gov/initiatives/Pages/ofr.aspx (accessed June 12, 2017); "Initiatives: Financial Stability Oversight Council," U.S. Department of Treasury, www.treasury.gov/initiatives/Pages/FSOC-index.aspx (accessed June 12, 2017).

Credits

B Personal Financial Planning*

©Slobodan Vasic/Getty Images

Chapter Outline

> **The Financial Planning Process**

> **Evaluate Your Financial Health**
> *The Personal Balance Sheet*
> *The Cash Flow Statement*

> **Set Short-Term and Long-Term Financial Goals**

> **Create and Adhere to a Budget**
> *Developing a Budget*
> *Tracking Your Budgeting Success*

> **Manage Credit Wisely**
> *Use and Abuse of Credit Cards*
> *Student Loans*

> **Develop a Savings and Investment Plan**
> *Understanding the Power of Compounded Returns*

The Link between Investment Choice and Savings Goals
Short-Term versus Long-Term Investment
Investment Choices
Planning for a Comfortable Retirement

> **Evaluate and Purchase Insurance**
> *Automobile Insurance*
> *Homeowners/Renters Insurance*
> *Life Insurance*
> *Health Insurance*
> *Disability Insurance*

> **Develop an Estate Plan**
> *The Importance of Having a Will*
> *Avoiding Estate Taxes*

> **Adjust Your Financial Plan to New Circumstances**

Learning Objectives

After reading this chapter, you will be able to:

LO B-1 Recall how to evaluate your financial situation.

LO B-2 Determine short-term and long-term personal financial goals.

LO B-3 Recall how to create and manage a personal financial budget.

LO B-4 Explain how to manage and use credit in your personal life.

LO B-5 Describe savings and investment choices.

LO B-6 Evaluate how insurance is a part of personal financial planning.

LO B-7 Explain the concept of estate planning.

* This bonus chapter was contributed by Dr. Vickie Bajtelsmit.

The Financial Planning Process

Personal financial planning is the process of managing your finances so that you can achieve your financial goals. By anticipating future needs and wants, you can take appropriate steps to prepare for them. Your needs and wants will undoubtedly change over time as you enter into various life circumstances. Although financial planning is not entirely about money management, a large part of this process is concerned with decisions related to expenditures, investments, and credit.

Although every person has unique needs, everyone can benefit from financial planning. Even if the entire financial plan is not implemented at once, the process itself will help you focus on what is important. With a little forethought and action, you may be able to achieve goals that you previously thought were unattainable.

The steps in development and implementation of an effective financial plan are:

- Evaluate your financial health.
- Set short-term and long-term financial goals.
- Create and adhere to a budget.
- Manage credit wisely.
- Develop a savings and investment plan.
- Evaluate and purchase insurance.
- Develop an estate plan.
- Adjust your financial plan to new circumstances.

LO B-1

Recall how to evaluate your financial situation.

personal financial planning the process of managing your finances so that you can achieve your financial goals.

Evaluate Your Financial Health

Just as businesses make use of financial reports to track their performance, good personal financial planning requires that individuals keep track of their income and expenses and their overall financial condition. Several software packages are readily available to help track personal finances (for example, Quicken and Microsoft Money), but all that is really needed is a simple spreadsheet program. This bonus chapter includes some simple worksheets that can be reproduced to provide a starting point for personal financial planning. Comprehensive financial planning sites are also available on the Internet. For example, http://money.msn.com/ and www.smartmoney.com both provide information and tools to simplify this process.

While it is possible to track all kinds of information over time, the two most critical elements of your finances are your personal net worth and your personal cash flow. The information necessary for these two measures is often required by lending institutions on loan applications, so keeping it up to date can save you time and effort later.

The Personal Balance Sheet

For businesses, net worth is usually defined as *assets minus liabilities,* and this is no different for individuals. **Personal net worth** is simply the total value of all personal assets less the total value of unpaid debts or liabilities. Although a business could not survive with a negative net worth since it would be technically insolvent, many students have negative net worth. As a student, you probably are not yet earning enough to have accumulated significant assets, such as a house or stock portfolio, but you are likely to have incurred various forms of debt, including student loans, car loans, and credit card debt.

personal net worth the total value of all personal assets less the total value of unpaid debts or liabilities.

At this stage in your life, negative net worth is not necessarily an indication of poor future financial prospects. Current investment in your "human capital" (education) is usually considered to have a resulting payoff in the form of better job opportunities and higher potential lifetime income, so this "upside-down" balance sheet should not stay that way forever. Unfortunately, there are many people in the United States who have negative net worth much later in their lives. This can result from unforeseen circumstances, like divorce, illness, or disability, but the easy availability of credit in the last couple of decades has also been blamed for the heavy debt loads of many American families. The most recent recession, caused partially by excessive risk-taking, has resulted in many bankruptcies and housing foreclosures. No matter the immediate trigger, it is usually poor financial planning—the failure to prepare in advance for those unforeseen circumstances—that makes the difference between those who fail and those who survive. It is interesting to note that we could say the exact same thing about business failures. Most are attributable to poor financial planning. If your net worth is negative, you should definitely include debt reduction on your list of short-and/or long-term goals.

You can use Table B.1 to estimate your net worth. On the left-hand side of the balance sheet, you should record the value of *assets,* all the things you own that have

TABLE B.1 Personal Net Worth

Assets	$	Liabilities	$
Checking accounts	___	Credit cards balances (list)	___
Savings accounts	___	1 ___	___
Money market accounts	___	2 ___	___
Other short-term investment	___	3 ___	___
	___	Personal loans	___
Market value of investments (stocks, bonds, mutual funds)	___	Student loans	___
	___	Car loans	___
Value of retirement funds	___	Home mortgage balance	___
College savings plan	___	Home equity loans	___
Other savings plans	___	Other real estate loans	___
Market value of real estate	___	Alimony/child support owed	___
Cars	___	Taxes owed (above withholding)	___
Home furnishings	___	Other investment loans	___
Jewelry/art/collectibles	___	Other liabilities/debts	___
Clothing/personal assets	___		___
Other assets	___		___
TOTAL ASSETS	___	TOTAL LIABILITIES	___

PERSONAL NET WORTH = TOTAL ASSETS - TOTAL LIABILITIES = $ ___

value. These include checking and savings account balances, investments, furniture, books, clothing, vehicles, houses, and the like. As with business balance sheets, assets are usually arranged from most liquid (easily convertible to cash) to least liquid. If you are a young student, it should not be surprising to find that you have little, if anything, to put on this side of your balance sheet. You should note that balance sheets are sensitive to the point in time chosen for evaluation. For example, if you always get paid on the first day of the month, your checking balance will be greatest at that point but will quickly be depleted as you pay for rent, food, and other needs. You may want to use your average daily balance in checking and savings accounts as a more accurate reflection of your financial condition. The right-hand side of the balance sheet is for recording *liabilities,* amounts of money that you owe to others. These include bank loans, mortgages, credit card debt, and other personal loans and are usually listed in order of how soon they must be paid back to the lender.

The Cash Flow Statement

Businesses forecast and track their regular inflows and outflows of cash with a cash budget and summarize annual cash flows on the statement of cash flows. Similarly, individuals should have a clear understanding of their flow of cash as they budget their expenditures and regularly check to be sure that they are sticking to their budget.

What is cash flow? Any time you receive cash or pay cash (including payments with checks), the dollar amount that is moving from one person to another is a cash flow. For students, the most likely cash inflows will be student loans, grants, and income from part-time jobs. Cash outflows will include rent, food, gas, car payments, books, tuition, and personal care expenses. Although it may seem obvious that you need to have enough inflows to cover the outflows, it is very common for people to estimate incorrectly and overspend. This may result in hefty bank overdraft charges or increasing debt as credit lines are used to make up the difference. Accurate forecasting of cash inflows and outflows allows you to make arrangements to cover estimated shortfalls before they occur. For students, this can be particularly valuable when cash inflows primarily occur at the beginning of the semester (for example, student loans) but outflows are spread over the semester.

How should you treat credit card purchases on your cash flow worksheet? Because credit purchases do not require payment of cash *now,* your cash flow statement should not reflect the value of the purchase as an outflow until you pay the bill. Take for example the purchase of a television set on credit. The $500 purchase will increase your assets and your liabilities by $500 but will only result in a negative cash flow of a few dollars per month because payments on credit cards are cash outflows when they are made. If you always pay your credit card balances in full each month, the purchases are really the same thing as cash, and your balance sheet will never reflect the debt. But if you purchase on credit and only pay minimum balances, you will be living beyond your means, and your balance sheet will get more and more "upside down." A further problem with using credit to purchase assets that decline in value is that the liability may still be there long after the asset you purchased has no value.

Table B.2 can be used to estimate your cash flow. The purpose of a cash flow worksheet for your financial plan is to heighten your awareness of where the cash is going. Many people are surprised to find that they are spending more than they make (by using too much credit) or that they have significant "cash leakage"—those little expenditures that add up to a lot without their even noticing. Examples include afternoon lattes or snacks, too many nights out at the local pub, eating lunch at the

TABLE B.2   Personal Cash Flow

Cash Inflows	Monthly	Annual
Salary/wage income (gross)	$ _____	$ _____
Interest/dividend income	_____	_____
Other income (self-employment)	_____	_____
Rental income (after expenses)	_____	_____
Capital gains	_____	_____
Other income	_____	_____
Total income	_____	_____
Cash Outflows	**Monthly**	**Annual**
Groceries	$ _____	$ _____
Housing	_____	_____
Mortgage or rent	_____	_____
House repairs/expenses	_____	_____
Property taxes	_____	_____
Utilities	_____	_____
Heating	_____	_____
Electric	_____	_____
Water and sewer	_____	_____
Cable/phone/satellite/Internet	_____	_____
Car loan payments	_____	_____
Car maintenance/gas	_____	_____
Credit card payments	_____	_____
Other loan payments	_____	_____
Income and payroll taxes	_____	_____
Other taxes	_____	_____
Insurance	_____	_____
Life	_____	_____
Health	_____	_____
Auto	_____	_____
Disability	_____	_____
Other insurance	_____	_____
Clothing	_____	_____
Gifts	_____	_____

TABLE B.2   Continued

Cash Inflows	Monthly	Annual
Other consumables (TVs, etc.)	____	____
Child care expenses	____	____
Sports-related expenses	____	____
Health club dues	____	____
Uninsured medical expenses	____	____
Education	____	____
Vacations	____	____
Entertainment	____	____
Alimony/child support	____	____
Charitable contributions	____	____
Required pension contributions	____	____
Magazine subscriptions/books	____	____
Other payments/expenses	____	____
Total Expenses	$ ____	$ ____

NET PERSONAL CASH FLOW = TOTAL INCOME - TOTAL EXPENSES = $ ____

Student Center instead of packing a bag, and regularly paying for parking (or parking tickets) instead of biking or riding the bus to school. In many cases, plugging the little leaks can free up enough cash to make a significant contribution toward achieving long-term savings goals.

Set Short-Term and Long-Term Financial Goals

Just as a business develops its vision and strategic plan, individuals should have a clear set of financial goals. This component of your financial plan is the roadmap that will lead you to achieving your short-term and long-term financial goals.

Short-term goals are those that can be achieved in two years or less. They may include saving for particular short-term objectives, such as a new car, a down payment for a home, a vacation, or other major consumer purchases. For many people, short-term financial goals should include tightening up on household spending patterns and reducing outstanding credit.

Long-term goals are those that require substantial time to achieve. Nearly everyone should include retirement planning as a long-term objective. Those who have or anticipate having children will probably consider college savings a priority. Protection of loved ones from the financial hazards of your unexpected death, illness, or disability is also a long-term objective for many individuals. If you have a spouse or other dependents, having adequate insurance and an estate plan in place should be part of your long-term goals.

Determine short-term and long-term personal financial goals.

short-term goals
goals that can be achieved in two years or less.

long-term goals
goals that require substantial time to achieve.

Recall how to create and manage a personal financial budget.

budget
an internal financial plan that forecasts expenses and income over a set period of time.

Create and Adhere to a Budget

Whereas the cash flow table you completed in the previous section tells you what you are doing with your money currently, a **budget** shows what you plan to do with it in the future. A budget can be for any period of time, but it is common to budget in monthly and/or annual intervals.

Developing a Budget

You can use the cash flow worksheet completed earlier to create a budget. Begin with the amount of income you have for the month. Enter your nondiscretionary expenditures (that is, bills you *must* pay, such as tuition, rent, and utilities) on the worksheet and determine the leftover amount. Next list your discretionary expenditures, such as entertainment and cable TV, in order of importance. You can then work down your discretionary list until your remaining available cash flow is zero.

An important component of your budget is the amount that you allocate to savings. If you put a high priority on saving and you do not use credit to spend beyond your income each month, you will be able to accumulate wealth that can be used to meet your short-term and long-term financial goals. In the bestseller *The Millionaire Next Door,* authors Thomas J. Stanley and William D. Danko point out that most millionaires have achieved financial success through hard work and thriftiness as opposed to luck or inheritance. You cannot achieve your financial goals unless your budget process places a high priority on saving and investing.

Tracking Your Budgeting Success

Businesses regularly identify budget items and track their variance from budget forecasts. People who follow a similar strategy in their personal finances are better able to meet their financial goals as well. If certain budgeted expenses routinely turn out to be under or over your previous estimates, then it is important to either revise the budget estimate or develop a strategy for reducing that expense.

College students commonly have trouble adhering to their budget for food and entertainment expenses. A strategy that works fairly well is to limit yourself to cash payments. At the beginning of the week, withdraw an amount from checking that will cover your weekly budgeted expenses. For the rest of the week, leave your checkbook, ATM card, and debit and credit cards at home. When the cash is gone, don't spend any more. While this is easier said than done, after a couple of weeks, you will learn to cut down on the cash leakage that inevitably occurs without careful cash management.

A debit card looks like a credit card but works like a check. For example, in the Netherlands almost no one writes a check, and everything is paid by debit card, which drafts directly from a checking account. You do not build up your credit rating when using a debit card. Figure B.1 indicates that the use of debit cards has become the predominant means of payment in the United States. On the other hand, credit cards allow you to promise

Do you know whether your expenses are going up or going down? Use a budget to track them.

©Blend Images - JGI/Jamie Grill/Getty Images

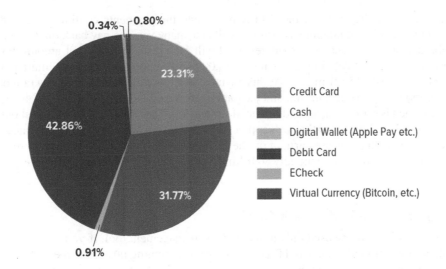

FIGURE B.1
Primary Payment Type

Sources: Mike Brown, "How Are People Making Payments in 2017?" *Lendedu*, September 25, 2017, https://lendedu.com/blog/how-are-people-making-payments/ (accessed May 9, 2018).

to pay for something at a later date by using preapproved lines of credit granted by a bank or finance company. Credit cards are easy to use and are accepted by most retailers today.

Manage Credit Wisely

Explain how to manage and use credit in your personal life.

One of the cornerstones of your financial plan should be to keep credit usage to a minimum and to work at reducing outstanding debt. The use of credit for consumer and home purchases is well entrenched in our culture and has arguably fueled our economy and enabled Americans to better their standard of living as compared to earlier generations. Nevertheless, credit abuse is a serious problem in this country, and the most recent economic downturn undoubtedly pushed many households over the edge into bankruptcy as a result.

To consider the pros and cons of credit usage, compare the following two scenarios. In the first case, Joel takes an 8 percent fixed-rate mortgage to purchase a house to live in while he is a college student. The mortgage payment is comparable to alternative monthly rental costs, and his house appreciates 20 percent in value over the four years he is in college. At the end of college, Joel will be able to sell his house and reap the return, having invested only a small amount of his own cash. For example, if he made an initial 5 percent down payment on a $100,000 home that is now worth $120,000 four years later, he has earned $12,800 (after a 6 percent commission to the real estate agent) on an investment of $5,000. This amounts to a sizable return on investment of more than 250 percent over four years. This example is oversimplified in that we did not take into account the principal that has been repaid over the four years, and we did not consider the mortgage payment costs or the tax deductibility of interest paid during that time. However, the point is still clear; borrowing money to buy an asset that appreciates in value by more than the cost of the debt is a terrific way to invest.

In the second case, Nicole uses her credit card to pay for some of her college expenses. Instead of paying off the balance each month, Nicole makes only the minimum payment and incurs 16 percent interest costs. Over the course of several years of college, Nicole's credit card debt is likely to amount to several thousand dollars, typical of college graduates in the United States. The beer and pizza Nicole purchased have long ago been digested, yet the debt remains, and the payments continue. If Nicole continues making minimum payments, it will take many years to pay back that original debt, and in the meantime the interest paid will far exceed the original amount borrowed. Credit card debt in the amount of $1,000 will usually require a minimum payment of at least $15 per month. At this payment level, it will take 166 months (almost 14 years) to pay the debt in full, and the total interest paid will be more than $1,400.

So when is borrowing a good financial strategy? A rule of thumb is that you should borrow only to buy assets that will appreciate in value or when your financing charges are less than what you are earning on the cash that you would otherwise use to make the purchase. This rule generally will limit your borrowing to home purchases and investments.

Use and Abuse of Credit Cards

Credit cards should be used only as a cash flow management tool. If you pay off your balance every month, you avoid financing charges (assuming no annual fee), you have proof of expenditures, which may be necessary for tax or business reasons, and you may be able to better match your cash inflows and outflows over the course of the month. There are several aspects of credit cards that you should be familiar with.

- *Finance charges.* Credit card companies make money by lending to you at a higher rate than it costs them to obtain financing. Because many of their customers don't pay back their debts in a timely fashion (default), they must charge enough to cover the risk of default as well. Interest is usually calculated on the average daily balance over the month, and payments are applied to old debts first. Although there are "teaser" rates that may be less than 5 percent, most credit cards regularly charge 13 to 24 percent annual interest. The low introductory rates are subject to time limitations (often six months or less), and they revert to the higher rates if you don't pay on time.

- *Annual fee.* Many credit cards assess an annual fee that may be as low as $25 or as much as $500 per year. If you regularly carry a very low balance, this amounts to the equivalent of a very high additional interest charge. For example, a $50 annual fee is the equivalent of an additional 5 percent on your annual interest rate if your balance is $1,000. Because the cards with fees do not generally provide you with different services, it is best to choose no-annual-fee credit cards.

- *Credit line.* The credit line is the maximum you are allowed to borrow. This may begin with a small amount for a new customer, perhaps as low as $300. As the customer shows the ability and intent to repay (by doing so in a timely fashion), the limit can increase to many thousands of dollars.

- *Grace period.* The grace period for most credit cards is 25 days. This may amount to twice as long a period of free credit depending on when your purchase date falls in the billing cycle. For example, if you used your card on January 1 and your billing cycle goes from the 1st to the 31st, then the bill for January purchases will arrive the first week in February and will be due on February 25. If you pay it in full on the last possible day, you will have had 55 days of free credit. Keep in mind that the lender considers the bill paid when the check is *received,* not when it is mailed.

- *Fees and penalties.* In addition to charging interest and annual fees, credit card companies charge extra for late payments and for going over the stated limit on the card. These fees have been on the rise in the past decade, and $25 or higher penalties are now fairly common.

- *ATM withdrawals.* Most credit cards can be used to obtain cash from ATMs. Although this may be convenient, it contributes to your increasing credit card balance and may result in extra expenditures that you would otherwise have avoided. In addition, these withdrawals may have hidden costs. Withdrawing cash from a machine that is not owned by your credit card lender will usually cause you to incur a fee of $1 or $1.50. The effective interest that this represents can be substantial if you are withdrawing small amounts of cash. A $1 charge on a withdrawal of $50 is the equivalent of 2 percent interest in addition to any interest you might pay to the credit card lender.
- *Perks.* Most credit cards provide a number of additional services. These may include a limitation on your potential liability in the event your card is lost or stolen or trip insurance. Some cards promise "cash back" in the form of a small rebate based on dollar volume of credit purchases. Many credit card companies offer the opportunity to participate in airline mileage programs. The general rule of thumb is that none of these perks is worth the credit card interest that is charged. If, however, you use your credit card as a cash management tool only, paying off your balance every month, then these perks are truly free to you.

Student Loans

Student loans are fairly common in today's environment of rising college tuition and costs. These loans can be a great deal, offering lower interest rates than other loans and terms that allow deferral of repayment until graduation. Furthermore, the money is being borrowed to pay for an asset that is expected to increase in value—your human capital. Don't underestimate, however, the monthly payments that will be required upon graduation. The rate of student debt is increasing at a rate higher than inflation.[1] For larger outstanding debt amounts, new college graduates in entry-level positions find that it is difficult to make the necessary payments without help.

Although the average student loan debt is $39,400, many students end up owing much more than that. The amount of student loan debt among Americans has reached more than $1.48 trillion across 44 million borrowers.[2] Both political parties agree that something must be done to curb this debt, although they appear split on how to do so. In 2013, Congress passed the Bipartisan Student Loan Certainty Act, which bases the interest rate on federal student loans according to market rate fluctuations. The year in which the loan is borrowed will determine the interest rate for the life of the loan.[3] In 2014, President Obama signed an executive order mandating that repayments not exceed 10 percent of the borrower's monthly income.[4]

Before borrowing for your education, check into federal student loans because they are less risky and less expensive than private loans. In order to see what kind of federal loans for which you qualify, fill out the Free Application for Federal Student Aid (FAFSA). It is important to keep track of the details of your loans, and keep your loan servicer updated on any changes in your information to avoid expensive late fees. For a list of all your federal student loans, frequently visit the National Student Loan Data System at www.nslds.ed.gov. Private loans can be accessed from your credit report, of which you can request a copy at www.annualcreditreport.com.

It is also helpful to understand your repayment plan options. The most common are standard repayment and income-driven repayment. Standard repayment is when the same sum is paid every month, and while the payment may be high, you will pay off your loans more quickly and pay fewer interest payments. Income-driven repayment bases your monthly payments on a percentage of income. They require annual income verification and other paperwork, and interest charges are high. Visit the Department of Education at www.studentaid.ed.gov/repay-loans for information and calculators regarding the different repayment options. Those who work in government, nonprofit, and other public service jobs may be eligible for student loan forgiveness after 10 years of faithful repayments. Other programs offer forgiveness of debt for teachers, military service members, or medical practitioners. More information can be found at www.studentloanborrowerassistance.org.[5]

Describe savings and investment choices.

Develop a Savings and Investment Plan

The next step to achieving your financial goals is to decide on a savings plan. A common recommendation of financial planners is to "pay yourself first." What this means is that you begin the month by setting aside an amount of money for your savings and investments, as compared to waiting until the end of the month and seeing what's left to save or invest. The budget is extremely important for deciding on a reasonable dollar amount to apply to this component of your financial plan.

As students, you might think that you cannot possibly find any extra dollars in your budget for saving, but, in fact, nearly everyone can stretch their budget a little. Some strategies for students might include taking public transportation several times a week and setting aside the gas or parking dollars you would have spent, buying regular coffees instead of Starbucks lattes, or eating at home one more night per week.

Understanding the Power of Compounded Returns

Even better, if you are a college student living on a typically small budget, you should be able to use this experience to help jump-start a viable savings program after graduation. If you currently live on $10,000 per year and your first job pays $30,000, it should be easy to "pay yourself" $2,000 or more per year. Putting the current maximum of $3,000 in an individual retirement account (IRA) will give you some tax advantages and can result in substantial wealth accumulation over time. An investment of only $2,000 per year from age 22 to retirement at 67 at 6 percent return per year will result in $425,487 at the retirement date. An annual contribution of $5,000 for 45 years will result in retirement wealth of about $1 million, not considering any additional tax benefits you might qualify for. If you invest that $5,000 per year for only 10 years and discontinue your contributions, you will still have about half a million dollars at age 67. And that assumes only a 6 percent return on investment.

What happens if you wait 10 years to start, beginning your $5,000 annual savings at age 32? By age 67, you will have only about a half million. Thirty-five years of investing instead of 45 doesn't sound like a big difference, but it cuts your retirement wealth in half. These examples illustrate an important point about long-term savings and wealth accumulation—the earlier you start, the better off you will be.

The Link between Investment Choice and Savings Goals

Once you have decided how much you can save, your choice of investment should be guided by your financial goals and the investment's risk and return and whether it will be long term or short term.

In general, investments differ in risk and return. The types of risk that you should be aware of are:

- *Liquidity risk*—How easy/costly is it to convert the investment to cash without loss of value?
- *Default risk*—How likely are you to receive the promised cash flows?
- *Inflation risk*—Will changes in purchasing power of the dollar over time erode the value of future cash flows your investment will generate?
- *Price risk*—How much might your investment fluctuate in value in the short run and the long run?

In general, the riskier an investment, the higher the return it will generate to you. Therefore, even though individuals differ in their willingness to take risk, it is important to invest in assets that expose you to at least moderate risk so that you can accumulate sufficient wealth to fund your long-term goals. To illustrate this more clearly, consider a $1 investment made in 1926. If this dollar had been invested in short-term Treasury bills, at the end of 2017 it would have grown to only $13.83. If the dollar had been invested in the S&P 500 index, which includes a diversified mix of stocks, the investment would be worth $2,393 in 2017, about 172 times more than an investment in Treasury bills. But this gain was not without risk. In some of those 70 years, the stock market lost money and your investment would have actually declined in value.

Short-Term versus Long-Term Investment

Given the differences in risk exposure across investments, your investment time horizon plays an important role in choice of investment vehicle. For example, suppose you borrow $5,000 on a student loan today but the money will be needed to pay tuition six months from now. Because you cannot afford to lose *any* of this principal in the short run, your investment should be in a low-risk security such as a bank certificate of deposit. These types of accounts promise that the original $5,000 principal plus promised interest will be available to you when your tuition is due. During the bull market of the 1990s, many students were tempted to take student loans and invest in the stock market in the hopes of doubling their money (although this undoubtedly violated their lender's rules). However, in the recent bear market, this strategy might have reduced the tuition funds by 20 percent or more.

In contrast to money that you are saving for near-term goals, your retirement is likely to be many decades away, so you can afford to take more risk for greater return. The average return on stocks over the past 10 years has been around 8.1 percent. In contrast, the average return on long-term corporate bonds, which offer regular payments of interest to investors, has been around 4.6 percent.[6] The differences in investment returns between these three categories is explainable based on the difference in risk imposed on the owners. Stock is the riskiest. Corporate bonds with their regular payments of

Want to buy a car? How you have handled your debts will help determine if you get a loan.

©maridav/123RF

interest are less risky to you since you do not have to wait until you sell your investment to get some of your return on the investment. Because they are less risky, investors expect a lower percentage return.

Investment Choices

There are numerous possible investments, both domestic and international. The difficulty lies in deciding which ones are most appropriate for your needs and risk tolerance.

Savings Accounts and Certificates of Deposit.   The easiest parking spot for your cash is in a savings account. Unfortunately, investments in these low-risk (FDIC-insured), low-return accounts will barely keep up with inflation. If you have a need for liquidity but not necessarily immediate access to cash, a certificate of deposit wherein you promise to leave the money in the bank for six months or more will give you a slightly higher rate of return.

Bonds.   Corporations regularly borrow money from investors and issue bonds, which are securities that contain the firm's promise to pay regular interest and to repay principal at the end of the loan period, often 20 or more years in the future. These investments provide higher return to investors than short-term, -interest-bearing accounts, but they also expose investors to price volatility, liquidity, and default risk.

A second group of bonds are those offered by government entities, commonly referred to as municipal bonds. These are typically issued to finance government projects, such as roads, airports, and bridges. Like corporate bonds, municipal bonds will pay interest on a regular basis, and the principal amount will be paid back to the investor at the end of a stated period of time, often 20 or more years. This type of bond has fewer interested investors and therefore has more liquidity risk.

Stocks.   A share of stock represents proportionate ownership interest in a business. Stockholders are thus exposed to all the risks that impact the business environment—interest rates, competition from other firms, input and output price risk, and others. In return for being willing to bear this risk, shareholders may receive dividends and/or capital appreciation in the value of their share(s). In any given year, stocks may fare better or worse than other investments, but there is substantial evidence that for long holding periods (20-plus years) stocks tend to outperform other investment choices.

Mutual Funds.   For the novice investor with a small amount of money to invest, the best choice is mutual funds. A mutual fund is a pool of funds from many investors that is managed by professionals who allocate the pooled dollars among various investments that meet the requirements of the mutual fund investors. There are literally thousands of these funds from which to choose, and they differ in type of investment (bonds, stocks, real estate, etc.), management style (active versus passive), and fee structure. Although even small investors have access to the market for individual securities, professional investors spend 100 percent of their time following the market and are likely to have more information at their disposal to aid in making buy and sell decisions.

Purchase of a Home.   For many people, one of the best investments is the purchase of a home. With a small up-front investment (your down payment) and relatively low borrowing costs, modest appreciation in the home's value can generate a large return on investment. This return benefits from the tax deductibility of home mortgage interest and capital gains tax relief at the point of sale. And to top it off, you have a place to live and thus save any additional rental costs you would incur if you invested your money elsewhere. There are many sources of information about home ownership for investors on the Internet. What type of home can you afford? What mortgage can you qualify for? How much difference does investment choice make?

Everyone needs to have a place to live, and two-thirds of Americans own their own homes. Nevertheless, owning a home is not necessarily the best choice for everyone. The decision on when and how to buy a house and how much to spend must be made based on a careful examination of your ability to pay the mortgage and to cover the time and expense of maintenance and repair. A home is probably the largest purchase you will ever make in your life. It is also one of the best investments you can make. As in the example given earlier, the ability to buy with a small down payment and to deduct the cost of interest paid from your taxable income provides financial benefits that are not available with any other investment type.

Few people could afford to buy homes at young ages if they were required to pay the full purchase price on their own. Instead, it is common for people to borrow most of the money from a financial institution and pay it back over time. The process of buying a home can begin with your search for the perfect home or it can begin with a visit to your local lender, who can give you an estimate of the amount of mortgage for which you can qualify. Mortgage companies and banks have specific guidelines that help them determine your creditworthiness. These include consideration of your ability and willingness to repay the loan in a timely fashion, as well as an estimate of the value of the house that will be the basis for the loan.

A **mortgage** is a special type of loan that commonly requires that you make a constant payment over time to repay the lender the original money you borrowed (**principal**) together with **interest**, the amount that the lender charges for your use of its money. In the event that you do not make timely payments, the lender has the right to sell your property to get its money back (a process called **foreclosure**).

Mortgage interest rates in the past decade have ranged from 5 to 10 percent per year, depending on the terms and creditworthiness of the borrower. There are many variations on mortgages—some that lock in an interest rate for the full term of the loan, often 30 years, and others that allow the rate to vary with market rates of interest. In low-interest-rate economic circumstances, it makes sense to lock in the mortgage at favorable low rates.

Several measures are commonly applied to assess your *ability to repay* the loan. In addition to requiring some work history, most lenders will apply two ratio tests. First, the ratio of your total mortgage payment (including principal, interest, property taxes, and homeowners insurance) to your gross monthly income can be no more than a prespecified percentage that varies from lender to lender but is rarely greater than 28 percent. Second, the ratio of your credit payments (including credit cards, car loan or lease payments, and mortgage payment) to your gross monthly income is limited to no more than 36 percent. More restrictive lenders will have lower limits on both of these ratios.

Lenders also consider your *willingness to repay* the loan by looking at how you have managed debt obligations in the past. The primary source of information will be a credit report provided by one of the large credit reporting agencies. Late payments and defaulted loans will appear on that report and may result in denial of the mortgage loan. Most lenders, however, will overlook previously poor credit if more recent credit management shows a change in behavior. This can be

mortgage
a special type of loan that commonly requires that you make a constant payment over time to repay the lender the original money you borrowed together with interest.

principal
The original money borrowed in the form of a loan.

interest
the amount that the lender charges for your use of the money.

foreclosure
when the lender gets the right to sell your property to get its money back if you cannot make timely payments.

collateral
assets pledged as security for a loan; if the loan is not repaid, the lender can use the assets pledged to settle the debt.

helpful to college students who had trouble paying bills before they were gainfully employed.

The value of the home is important to the lender since it is the **collateral** for the loan; that is, in the event that you default on the loan (don't pay), the lender has the right to take the home in payment of the loan. To ensure that they are adequately covered, lenders will rarely lend more than 95 percent of the appraised value of the home. If you borrow more than 80 percent of the value, you will usually be required to pay a mortgage insurance premium with your regular payments. This will effectively increase the financing costs by 0.5 percent per year.

To illustrate the process of buying a home and qualifying for a mortgage, consider the following example. Jennifer graduated from college two years ago and has saved $7,000. She intends to use some of her savings as a down payment on a home. Her current salary is $36,000. She has a car payment of $250 per month and credit card debt that requires a minimum monthly payment of $100 per month. Suppose that Jennifer has found her dream home, which has a price of $105,000. She intends to make a down payment of $5,000 and borrow the rest. Can she qualify for the $100,000 loan at a rate of 7 percent?

Using Table B.3, her payment of principal and interest on a loan of $100,000 at 7 percent annual interest will be $665. With an additional $150 per month for property taxes and insurance (which may vary substantially in different areas of the country), her total payment will be $815. Her gross monthly income is $3,000, so the ratio of her payment to her income is 27 percent. Unless her lender has fairly strict rules, this should be acceptable. Her ratio of total payments to income will be ($815 + $250 + $150)/$3,000 = 40.5 percent. Unfortunately, Jennifer will not be able to qualify for this loan in her current financial circumstances.

So what can she do? The simplest solution is to use some of her remaining savings to pay off her credit card debt. By doing this, her debt ratio will drop to 35.5 percent and she will be accomplishing another element of good financial planning—reducing credit card debt and investing in assets that increase in value.

TABLE B.3   Calculating Monthly Mortgage Payments (30-year loan, principal and interest only)

Annual Interest %	Amount Borrowed			
	$75,000	$100,000	$125,000	$150,000
4.0	$358	$477	$ 597	$ 716
5.0	$403	$537	$ 671	$ 805
6.0	$450	$600	$ 749	$ 899
6.5	$474	$632	$ 790	$ 948
7.0	$499	$665	$ 832	$ 998
7.5	$524	$699	$ 874	$1,049
8.0	$550	$734	$ 917	$ 1,101
8.5	$577	$769	$ 961	$ 1,153
9.0	$603	$805	$1,006	$1,207

Planning for a Comfortable Retirement

Although it may seem like it's too early to start thinking about retirement when you are still in college, this is actually the best time to do so. In the investment section, you learned about the power of compound interest over long periods of time. The earlier you start saving for long-term goals, the easier it will be to achieve them.

How Much to Save.   There is no "magic number" that will tell you how much to save. You must determine, based on budgeted income and expenses, what amount is realistic to set aside for this important goal. Several factors should help to guide this decision:

- Contributions to qualified retirement plans can be made before tax. This allows you to defer the payment of taxes until you retire many years from now.
- Earnings on retirement plan assets are tax deferred. If you have money in nonretirement vehicles, you will have to pay state and federal taxes on your earnings, which will significantly reduce your ending accumulation.
- If you need the money at some time before you reach age 59.5, you will be subject to a withdrawal penalty of 10 percent, and the distribution will also be subject to taxes at the time of withdrawal.

In planning for your retirement needs, keep in mind that inflation will erode the purchasing power of your money. You should consider your ability to replace preretirement income as a measure of your success in retirement preparation. You can use the Social Security Administration website (www.ssa.gov) to estimate your future benefits from that program. In addition, most financial websites provide calculators to aid you in forecasting the future accumulations of your savings.

Employer Retirement Plans.   Many employers offer retirement plans as part of their employee benefits package. **Defined benefit plans** promise a specific benefit at retirement (for example, 60 percent of final salary). More commonly, firms offer **defined contribution plans**, where they promise to put a certain amount of money into the plan in your name every pay period. The plan may also allow you to make additional contributions or the employer may base its contribution on your contribution (for example, by matching the first 3 percent of salary that you put in). Employers also may make it possible for their employees to contribute additional amounts toward retirement on a tax-deferred basis. Many plans now allow employees to specify the investment allocation of their plan contributions and to shift account balances between different investment choices.

> **defined benefit plan**
> a type of retirement plan that promises a specific benefit at retirement.
>
> **defined contribution plan**
> a type of retirement plan firms offer where they promise to put a certain amount of money into the plan in your name every pay period.

Some simple rules to follow with respect to employer plans include the following:

- If your employer offers you the opportunity to participate in a retirement plan, you should do so.
- If your employer offers to match your contributions, you should contribute as much as is necessary to get the maximum match, if you can afford to. Every dollar that the employer matches is like getting a 100 percent return on your investment in the first year.
- If your plan allows you to select your investment allocation, do not be too conservative in your choices if you still have many years until retirement.

Individual Retirement Accounts (IRAs).   Even if you do not have an employer-sponsored plan, you can contribute to retirement through an individual retirement account (IRA). There are two types of IRAs with distinctively different characteristics (which are summarized in Table B.4). Although previously subject to

TABLE B.4 Comparing Individual Retirement Account Options

	Roth IRA	Traditional IRA
2008–2010 allowable contribution	$5,000	$5,000
Contributions deductible from current taxable income	No	Yes
Current tax on annual investment earnings	No	No
Tax due on withdrawal in retirement	No	Yes
10% penalty for withdrawal before age 59½	Yes	Yes
Mandatory distribution before age 70½	No	Yes
Tax-free withdrawals allowed for first-time homebuyers	Yes	No

a $2,000 maximum annual contribution limit, tax reform increased that limit gradually to $5,500. The critical difference between Roth IRAs and traditional IRAs is the taxation of contributions and withdrawals. Roth IRA contributions are taxable, but the withdrawals are tax-free. Traditional IRAs are deductible, but the withdrawals are taxable. Both types impose a penalty of 10 percent for withdrawal before the qualified retirement age of 59.5, subject to a few exceptions.

Social Security. Social Security is a public pension plan sponsored by the federal government and paid for by payroll taxes equally split between employers and employees. In addition to funding the retirement portion of the plan, Social Security payroll taxes pay for Medicare insurance (an old-age health program), disability insurance, and survivors benefits for the families of those who die prematurely.

The aging of the U.S. population has created a problem for funding the current Social Security system. Whereas it has traditionally been a pay-as-you-go program, with current payroll taxes going out to pay current retiree benefits, the impending retirement of baby boomers is forecast to bankrupt the system early in this century if changes are not made in a timely fashion. To understand the problem, consider that when Social Security began, there were 17 workers for each retiree receiving benefits. There are currently fewer than four workers per beneficiary. After the baby boom retirement, there will be only two workers to pay for each retiree. Obviously, that equation cannot work.

Does that mean that Social Security will not be around when you retire? Contrary to popular belief, it is unlikely that this will happen. There are simply too many voters relying on the future of Social Security for Congress to ever take such a drastic action. Instead, it is likely that the current system will be revised to help it balance. Prior to the heavy declines in the stock market in 2008–2009, there was some general support for a plan that would divert some of the current payroll taxes to fund individual retirement accounts that could be invested in market assets. In addition, it seems likely that the retirement age will increase gradually to age 67. Other possible changes are to increase payroll taxes or to limit benefits payable to wealthier individuals. The proposed solutions are all complicated by the necessity of providing a transition program for those who are too old to save significant additional amounts toward their retirement. Figure B.2 indicates that most people are concerned about receiving fewer Social Security benefits when they retire.

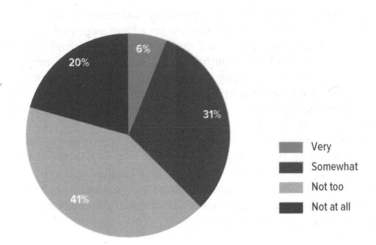

FIGURE B.2
**Worker Confidence
That Social Security
Will Continue to Provide
Benefits of at Least
Equal Value to Benefits
Received Today**

Source: *Statista,* "How Confident
Are You that Social Security System
Will Continue to Provide Benefits of
at Least Equal Value to the Benefits
Received by Retirees Today?," https://
www.statista.com/statistics/292108/
us-worker-confidence-social-security-
provisions/ (accessed May 9, 2018).

Very

Somewhat

Not too

Not at all

Evaluate and Purchase Insurance

LO B-6

Evaluate how insurance is
a part of personal financial
planning.

The next step in personal financial planning is the evaluation and purchase of insurance. Insurance policies are contracts between you and an insurance company wherein the insurer promises to pay you money in the event that a particular event occurs. Insurance is important, not only to protect your own assets from claims but also to protect your loved ones and dependents. The most common types of insurance for individuals are identified and briefly described next.

Automobile Insurance

In most states, drivers are required by law to carry a minimum amount of **auto liability insurance**. In the event that you are in a car accident, this coverage promises to pay claims against you for injuries to persons or property, up to a maximum per person and per accident. The basic liability policy will also cover your own medical costs. If you want to insure against damage to your own vehicle, you must purchase an additional type of coverage called **auto physical damage insurance**. If you have a car loan, the lender will require that you carry this type of insurance because the value of the car is the collateral for that loan and the lender wants to be sure that you can afford to fix any damage to the vehicle following an accident. The minimum limits in most states are too low to cover typical claim levels. Good financial planning requires that you pay for insurance coverage with much higher limits.

Auto physical damage insurance coverage is always subject to a **deductible**. A deductible is an amount that you must pay before the insurance company will pay. To illustrate this, suppose your policy has a $250 deductible. You back into your garage door and damage your bumper, which will cost $750 to fix. The insurer will only pay $500, because you are responsible for the first $250. Once you receive the check from the insurer, you are free to try to get it fixed for less than the full $750.

auto liability insurance
coverage that will pay claims
against you for injuries to
persons or property, up to a
maximum per person and per
accident.

**auto physical damage
insurance**
coverage that will insure
against damage to your own
vehicle.

deductible
an amount that you must
pay before the insurance
company will pay.

Homeowners/Renters Insurance

homeowners insurance
coverage for liability and property damage for your home.

renters insurance
similar to homeowners insurance in that it covers you for liability on your premises and for damage to your personal property.

life insurance
insurance that pays a benefit to your designated beneficiary in the event that you die during a coverage period.

term insurance
a type of life insurance that lasts usually for one year and the insurer promises to pay your designated beneficiary only the face amount of the policy in the event you die during the year of coverage.

permanent insurance
insurance designed to protect you with insurance protection over your lifetime.

health insurance
insurance that pays the cost of covered medical expenses during a policy period, which is usually six months to a year.

Homeowners insurance provides coverage for liability and property damage in your home. For example, if someone slips and falls on your front steps and sues you for medical expenses, this insurance policy will pay the claim (or defend you against the claim if the insurer thinks it is not justified). If your house and/or property are damaged in a fire, the insurance will pay for lost property and the costs of repair. It is a good idea to pay extra for replacement cost insurance because, otherwise, the insurance company is only obligated to pay you the depreciated value, which won't be enough to replace your belongings.

Renters insurance is similar to homeowners in that it covers you for liability on your premises (for example, if your dog bites someone) and for damage to your personal property. Because you do not own the house, your landlord needs to carry separate insurance for his building. This insurance is very cheap and is well worth the cost, since your landlord's insurance will not pay anything to you in the event that the house burns down and you lose all your belongings.

Life Insurance

As compared to other types of insurance, the primary purpose of life insurance is to provide protection for others. **Life insurance** pays a benefit to your designated beneficiary (usually your spouse or other family members) in the event that you die during the coverage period. Life insurance premiums will depend on the face amount of the policy, your age and health, your habits (smoker versus nonsmoker), and the type of policy (whether it includes an investment component in addition to the death benefit).

The simplest type of life insurance is **term insurance.** This policy is usually for one year and the insurer promises to pay your designated beneficiary only the face amount of the policy in the event that you die during the year of coverage. Because the probability of dying at a young age is very small, the cost of providing this promise to people in their 20s and 30s is very inexpensive, and premiums are fairly low. Term insurance becomes more expensive at older ages, since the probability of dying is much higher and insurers must charge more.

Other types of life insurance usually fall into a category often called **permanent insurance** because they are designed to provide you with insurance protection over your lifetime. To provide lifetime coverage at a reasonable cost, premiums will include an investment component. While there are many variations, typically in the early years of the policy you are paying a lot more than the actual cost of providing the death protection. The insurer takes that extra cost and invests it so that when you are older, the company has sufficient funds to cover your death risk. The primary difference between different types of permanent insurance is the way that they treat the investment component. Some policies allow the buyer to direct the investment choice and others do not.

Medical costs can be astronomical. Part of keeping your finances in order is making sure you have health insurance.

©numbeos/Getty Images

Health Insurance

Health insurance pays the cost of covered medical expenses during the policy period, which is usually six months or one year. Most health insurance is provided under group policies through employers, but it is possible to purchase an individual policy. Because those who want to buy individual insurance are likely to be people who anticipate medical expenses, individual policies can be very expensive and are usually subject to exclusions, high coinsurance (the percentage of each dollar of expenses that you must pay out of pocket), and deductibles (the amount you must pay in full before the insurance pays).

From a financial-planning perspective, the type of health coverage that is most important is that which will protect you and your family from unexpected large medical costs. The usual checkups, shots, and prescription drugs are all budgetable expenses and need not be insured. At a minimum, you should have a policy that covers hospitalization and care for major disease or injury. This can be accomplished at relatively low cost by contracting for a large deductible (for example, you pay the first $1,000 of costs per year).

The two main types of health insurance plans are *fee-for-service* and *managed care*. In a fee-for-service arrangement, the insurer simply pays for whatever covered medical costs you incur, subject to the deductible and coinsurance. Blue Cross and Blue Shield plans are the best known of this type. Managed care includes health maintenance organizations (HMOs) and preferred provider organizations (PPOs). In these health insurance arrangements, your health insurer pays all your costs (subject sometimes to small co-pays for office visits), but the care you receive is determined by your physician, who has contracted with the health insurer and has incentives to control overall costs. You are often limited in your choice of physician and your ability to seek specialist care under these plans.

Major changes in health insurance began to occur after the 2010 passage of the Patient Protection and Affordable Care Act. According to the law, individuals who are self-employed or who do not receive health insurance through their businesses can pay for insurance through state-based exchanges. Exchanges will also be created for small businesses to purchase health coverage, along with tax breaks for this purpose. Employers with more than 50 full-time employees must also pay for health care coverage. The purpose of this legislation is to provide health insurance for the more than 32 million Americans who were uninsured. The act also puts limits on insurers. For instance, insurers can no longer deny coverage or benefits based on a preexisting condition. The Patient Protection and Affordable Care Act is having a wide-ranging impact on the health care industry, including how much you will pay for future health care insurance.[7]

Disability Insurance

One of the most overlooked types of insurance is **disability insurance**, which pays replacement income to you in the event you are disabled under the definition in your policy. One in three people will be disabled for a period of three months or more during their lifetime, so disability insurance should be a component of the financial plan for anyone without sufficient financial resources to weather a period of loss of income.

disability insurance insurance that pays replacement income to you in the event you are disabled under the definition of your policy.

Develop an Estate Plan

As with retirement planning, it is difficult to think about estate planning when you are young. In fact, you probably don't need to think much about it yet. If you have no dependents, there is little point in doing so. However, if you are married or have other

>LO B-7

Explain the concept of estate planning.

estate plan
planning for the transfer of
your wealth and assets after
your death.

dependents, you should include this as a necessary part of your financial plan. The essential components of an **estate plan** are:

- Your will, including a plan for guardianship of your children.
- Minimization of taxes on your estate.
- Protection of estate assets.

Estate planning is a complicated subject that is mired in legal issues. As such, appropriate design and implementation of an estate plan requires the assistance of a qualified professional.

The Importance of Having a Will

There are several circumstances that necessitate having a will. If you have a spouse and/or dependent children, if you have substantial assets, or if you have specific assets that you would like to give to certain individuals in the event of your death, you *should* have a will. On the other hand, if you are single with no assets or obligations (like many students), a will is probably not necessary—yet.

Having a valid will makes the estate settlement simpler for your spouse. If your children are left parentless, will provisions specify who will take guardianship of the children and direct funds for their support. You might also like to include a *living will,* which gives your family directions for whether to keep you on life support in the event that an illness or injury makes it unlikely for you to survive without extraordinary interventions. Lastly, you may want to make a will so that you can give your iPad to your college roommate or Grandma's china to your daughter. Absent such provisions, relatives and friends have been known to take whatever they want without regard to your specific desires.

Gender Differences Create Special Financial Planning Concerns.   Although most people would agree that there are some essential differences between men and women, it is not as clear why their financial planning needs should be different. After all, people of both sexes need to invest for future financial goals like college educations for their children and retirement income for themselves. In the past few years, professionals have written articles considering this subject. The results are both controversial and eye-opening.

- Even though 75 percent of women in the United States are working, they still have greater responsibility for household chores, child care, and care of aging parents than their husbands. This leaves less time for household finances.
- Women still earn much less than men, on average.
- Women are much less likely to have a pension sponsored through their employer. Only one-third of all working women have one at their current employer.
- Women are more conservative investors than men. Although there is evidence that women are gradually getting smart about taking a little more risk in their portfolios, on average they allocate half as much as men do to stocks.
- Most women will someday be on their own, either divorced or widowed.

Because women live an average of five years longer than men, they actually need to have saved more to provide a comparable retirement income. The combined impact of these research findings makes it difficult but not impossible for women to save adequately for retirement. Much of the problem lies in education. Women need to be better informed about investing in order to make choices early in life that will pay off in the end. If they don't take the time to become informed about their finances or can't due to other obligations, in the end they will join the ranks of many women over age 65 who are living in poverty. But when women earn less, they don't have access to an employer pension, and they invest too conservatively, it is no surprise that women have so little wealth accumulation.

In her book, *The Busy Woman's Guide to Financial Freedom,* Dr. Vickie Bajtelsmit, academic director of the Master of Finance program at Colorado State University, provides a roadmap for women who are interested in taking charge of their financial future. With simple-to-follow instructions for all aspects of financial planning, from investing to insurance to home buying, the book provides information for women to get on the right financial track.

Avoiding Estate Taxes

As students, it will likely be many years before you will have accumulated a large enough estate (all your "worldly possessions") to have to worry about estate taxes. Currently, the first $11 million for an individual and the first $22 million for a married couple pays no estate taxes, so you probably don't have to worry about this for a while. Although federal tax law changes eliminated the estate tax in 2010, the tax was reinstated and raised to 35 percent in 2011. Today, the estate tax is at a maximum of 40 percent of assets upon death.[8] This is an area of law that frequently changes. Because no one can predict the date of his or her death, this implies that estate tax planning should be done assuming the worst-case scenario. Current estate taxes can take a big bite out of your family's inheritance for wealthy taxpayers. Thus, much of estate planning is actually tax-avoidance planning. Professionals can help set up trust arrangements that allow all or part of your estate to pass to your heirs without incurring taxes.

Adjust Your Financial Plan to New Circumstances

Finally, to ensure the success of your overall financial plan, it is vital that you evaluate it on a periodic basis and adjust it to accommodate changes in your life, such as marriage, children, or the addition or deletion of a second income from your spouse. You may be preparing income tax returns now, but as your income increases, you may have to make a decision about professional assistance. Figure B.3 indicates that most Americans prepare their own taxes, but many taxpayers use a professional service. Your plan also must be adjusted as your financial goals change (for example, desires to own a home, make a large purchase, or retire at an early age). Whatever your goals may be, the information and worksheets provided here will help with your personal financial planning.

FIGURE B.3

How Do Americans File Their Taxes?

Source: GOBankingRates.com.

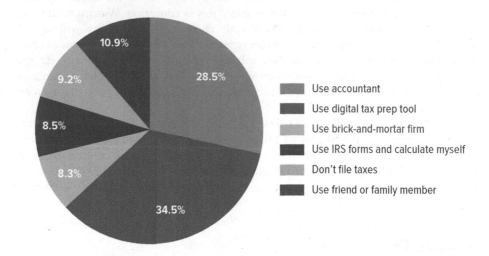

- Use accountant
- Use digital tax prep tool
- Use brick-and-mortar firm
- Use IRS forms and calculate myself
- Don't file taxes
- Use friend or family member

Review Your Understanding

Recall how to evaluate your financial situation.

Personal financial planning is the process of managing your finances so that you can achieve your financial goals. By anticipating future needs and wants, you can take appropriate steps to prepare for them. The two most critical elements of your finances you should track are your personal net worth, or the total value of all personal assets less the total value of unpaid debts or liabilities, and personal cash flow.

Determine short-term and long-term personal financial goals.

Short-term goals are those that can be achieved in two years or less. Long-term goals are those that require a substantial time to achieve. Short-term goals may include items such as a new car, a home, or a vacation. Long-term goals include retirement or college savings for children.

Recall how to create and manage a personal financial budget.

A budget shows what you plan to do with your cash in the future. It can be for any period of time but is most often done for monthly and/or annual intervals. A monthly budget would begin with the amount of income you have for the month. It is important to put high priority on investing and saving. It is also necessary to track how you are adhering to your budget.

Explain how to manage and use credit in your personal life.

Credit abuse is a serious problem in this country. A rule of thumb is that you should borrow only to buy assets that will appreciate in value or when your financing charges are less than what you are earning on the cash that you would otherwise use to make the purchase. It is important to be familiar with certain credit terms, such as finance charges, annual fees, credit lines, grace periods, fees and penalties, ATM withdrawals, and any perks associated with specific credit cards. An even bigger crisis than credit card debt, however, is student loans. Student loans are commonly used to pay for tuition costs, but it is important to manage student loan debt and look for favorable terms when selecting a loan. Students should also understand their repayment options.

Describe savings and investment choices.

Another major step is setting aside an amount of money for your savings and investments, as compared to waiting until the end of the month and seeing what's left to save or invest. Investing in individual retirement accounts gives you tax advantages and can result in substantial wealth accumulation over time. Once you have decided how much you can save, your choice of investment should be guided by your financial goals, the investment's risk and return, and whether it will be long term or short term. In general, the riskier an investment, the higher the return it will generate. Money invested for short-term needs should be invested in low-risk securities, while money invested for the long-term needs could be invested in riskier securities that offer a higher rate of return. Investment choices include savings accounts and certificates of deposit, bonds, stocks, mutual funds, and purchase of a home.

Purchasing a home requires a mortgage. A mortgage is a special type of loan that commonly requires that you make a constant payment over time to repay the lender the original money you borrowed (principal) together with interest. If you cannot make timely payments, the lender has the right to sell your property, known as a foreclosure. The value of the home is important to the lender because it can act as collateral for the loan.

In terms of saving for retirement, many employers offer defined benefit plans or defined contribution plans. Defined benefit plans promise a specific benefit at retirement. Defined contribution plans promise to put a certain amount of money into the plan in your name every pay period. For employees who do not have retirement benefit plans at their jobs, they can choose to contribute to retirement through a Roth IRA. Social security is a public pension plan sponsored by the federal government and paid for by payroll taxes equally split between employers and employees.

Evaluate how insurance is a part of personal financial planning.

In most states, drivers are required by law to carry a minimum amount of auto liability insurance, which is coverage that promises to pay claims against you for injuries to persons or property, up to a maximum per person and per accident. Ensuring against damage to your own vehicle requires auto physical damage insurance. Auto physical damage insurance coverage is always subject to a deductible, meaning a certain amount you must pay before an insurance company will pay. Homeowners insurance provides coverage for liability and property damage in your home. Renters insurance is similar to homeowners in that it covers you for liability on your premises and to your personal property. Life insurance pays a benefit to your designated beneficiary in the event that you die during the coverage period. The simplest type is known as term insurance. It is usually for one year, and the insurer promises to pay your designated beneficiary only the face amount of the policy in the event that you die during the year of coverage. Permanent insurance is designed to provide you with insurance protection over your lifetime. Health insurance pays the cost of covered medical expenses during the policy period. Finally, disability insurance pays replacement income to you in the event that you are disabled under the definition in your policy.

Explain the concept of estate planning.

Estate plans involve your will, protection of estate assets, and minimization of taxes on your estate. If you have assets that you would like to leave to certain individuals in case of death, you should have a will. It is important to plan for estate taxes as they can take a big bite out of a family's inheritance.

Learn the Terms

auto liability insurance 573
auto physical damage insurance 573
budget 562
collateral 570
deductible 573
defined benefit plan 571
defined contribution plan 571
disability insurance 575

estate plan 576
foreclosure 569
health insurance 574
homeowners insurance 574
interest 569
life insurance 574
long-term goals 561
mortgage 569

permanent insurance 574
personal financial planning 557
personal net worth 557
principal 569
renters insurance 574
short-term goals 561
term insurance 574

Check Your Progress

1. How do you determine your personal net worth?
2. What is the difference between short-term and long-term financial goals?
3. Why should you develop a budget?
4. What are some of the dangers or abuses that occur when using credit cards?
5. Why is it important even for entry-level employees to develop a savings plan?
6. What should you know about investment choices in determining the best alternative for your risk tolerance?
7. Why is it important to learn as much as possible about retirement plans?
8. What are the different types of insurance decisions that you will have to make?
9. What are some of the gender differences that create special financial planning concerns?
10. What is an estate plan, and why is it important to have a will?

Build Your Skills

Determining Net Worth

Task

On a separate sheet of paper, write down the information from Table B.I. Fill out the information with details about your personal assets and your personal liabilities. Subtract your net liabilities from your net assets to determine your net worth.

Notes

1. Mark Kantrowitz, "Why the Student Loan Crisis Is Even Worse Than You Think," *Time,* January 11, 2016, http://time.com/money/4168510/why-student-loan-crisis-is-worse-than-people-think/ (accessed May 24, 2016).

2. "A Look at the Shocking Student Loan Debt Statistics for 2018," *Student Loan Hero,* May 1, 2018, https://studentloanhero.com/student-loan-debt-statistics/ (accessed May 9, 2018).

3. "Interest Rates for New Direct Loans," Federal Student Aid an Office of the U.S. Department of Education, July 1, 2014, https://studentaid.ed.gov/About/announcements/interest-rate (accessed May 24, 2016); "Interest Rates and Fees on Federal Student Loans," Edvisors Network Inc., https://www.edvisors.com/college-loans/federal/stafford/interest-rates/ (accessed May 24, 2016).

4. Sam Frizzle, "Obama Looks to Reduce Student Loan Payments," *Time,* June 9, 2014, http://time.com/2847507/student-loan-debt-barack-obama/ (accessed June 16, 2014).

5. Maria Shriver, "Life Ed: How to Manage Student Loan Debt," *NBC News,* June 12, 2014, www.nbcnews.com/feature/maria-shriver/life-ed-how-manage-student-loan-debt-n129521 (accessed June 16, 2014).

6. Thomas Kenny, "Stocks vs. Bonds: The Long-Term Performance Data," *The Balance,* October 28, 2017, https://www.thebalance.com/stocks-vs-bonds-the-long-term-performance-data-416861 (accessed May 9, 2018).

7. Jill Jackson and John Nolen, "Health Care Reform Bill Summary: A Look at What's in the Bill," *CBS News,* March 21, 2010, www.cbsnews.com/8301-503544_162-20000846-503544.html (accessed May 30, 2012); " Patient Protection and Affordable Care Act," *Federal Register* 75, no 123 (June 28, 2010), www.gpo.gov/fdsys/pkg/FR-2010-06-28/html/2010-15278.htm (accessed June 18, 2014).

8. Ashlea Ebeling, "IRS Announces 2016 Estate and Gift Tax Limits: The $10.9 Million Tax Break," *Forbes,* October 22, 2015, http://www.forbes.com/sites/ashleaebeling/2015/10/22/irs-announces-2016-estate-and-gift-tax-limits-the-10-9-million-tax-break/#a2e69916a7c3 (accessed June 30, 2016).

Credits

Appendix A

Guidelines for the Development of the Business Plan

These guidelines are for students to create a hypothetical business plan for a good/service/business of their choice. Students should assume to have $25,000 to start this new business in their community.

At the end of every chapter, there is a section entitled "Build Your Business Plan" to assist you in the development of the business plan.

Phase 1: Development of the Business Proposal

You are encouraged to submit your idea for approval to your instructor as soon as possible. This will eliminate wasted effort on an idea that is not feasible in the instructor's view. Business plan proposals will be evaluated based on their thoroughness and your ability to provide support for the idea.

The business proposal consists of the following elements.

Business Description. This consists of an overview of the existing good/service or the good/service/business you will be starting (manufacturer, merchandiser, or service provider). This includes developing a mission (reason for existence; overall purpose of the firm) and a rationale for why you believe this business will be a success. What is your vision for this proposed product/business?

Brief Marketing Plan. (The marketing plan will be further developed as the plan evolves.) A description of your business/product is required. Identify the target market and develop a strategy for appealing to it. Justify your proposed location for this business. Describe how you will promote the new business and provide a rationale for your pricing strategy. Select a name for this business. The name should be catchy yet relate to the competencies of the business.

Competitive Analysis. Identify the competition as broadly as possible. Indicate why this business will be successful given the market.

Phase 2: Final Written Business Plan

Executive Summary. The executive summary appears first but should be written last.

Business Description. This section requires fleshing out the body of the business plan, including material from your revised preliminary proposal with more data, charts, and appendices. Include a description of the proposed form of organization, either a partnership or corporation, and the rationalization of the form chosen.

Industry and Market Analysis. An analysis of the industry including the growth rate of the industry and number of new entrants into this field is necessary. Identify uncontrollable variables within the industry. Determine an estimate of the proposed realistic size of the potential market. This will require interpretation of statistics from the U.S. census as well as from local sources such as the Chamber of Commerce.

Competitive Analysis. Include an exhaustive list of the primary and secondary competition, along with the competitive advantage of each.

Marketing Strategy. Target market specifics need to be developed.

Decisions on the marketing mix variables need to be made:

- Price (at the market, below market, above market).
- Promotion (sales associates, advertising budget, use of sales promotions, and publicity/goodwill).
- Distribution—Rationale of choice and level of distribution.
- Product—A detailed rationale of the perceived differential advantage of your product offering.

Operational Issues. How will you make or provide your product? Location rationale, facility type, leasing considerations, and sources of suppliers need to be detailed. Software/hardware requirements necessary to maintain operations must be determined.

Human Resources Requirement. Number and description of personnel needed, including realistic required education and skills.

Financial Projections. Statement of cash flows must be prepared for the first 12 months of the business. This must include startup costs, opening expenses, and estimation of cash inflows and outflows. A break-even analysis should be included and an explanation of your expected financial expenditures.

Appendixes

Phase 3: Oral Presentation

Specific separate guidelines on the oral presentation will be provided.

Appendix B

Personal Career Plan

The tools and techniques used in creating a business plan are just as useful in designing a plan to help sell yourself to potential employers. The outline in this appendix is designed to assist you in writing a personalized plan that will help you achieve your career goals. While this outline follows the same general format found in Appendix A, it has been adapted to be more relevant to career planning. Answering the questions presented in this outline will enable you to:

1. Organize and structure the data and information you collect about job prospects, the overall job market, and your competition.
2. Use this information to better understand your own personal strengths and weaknesses, as well as recognize the opportunities and threats that exist in your career development.
3. Develop goals and objectives that will capitalize on your strengths.
4. Develop a personalized strategy that will give you a competitive advantage.
5. Outline a plan for implementing your personalized strategy.

As you work through the following outline, it is very important that you be honest with yourself. If you do possess a strength in a given area, it is important to recognize that fact. Similarly, do not overlook your weaknesses. The viability of your SWOT analysis and your strategy depend on how well you have identified all of the relevant issues in an honest manner.

I. Summary
If you choose to write a summary, do so after you have written the entire plan. It should provide a brief overview of the strategy for your career. State your career objectives and what means you will use to achieve those objectives.

II. Situation Analysis
A. The External Environment
1. Competition
a. Who are your major competitors? What are their characteristics (number and growth in the number of graduates, skills, target employers)? Competitors to consider include peers at the same college or in the same degree field, peers at different colleges or in different degree fields, and graduates of trade, technical, or community colleges.
b. What are the key strengths and weaknesses of the total pool of potential employees (or recent college graduates)?
c. What are other college graduates doing in terms of developing skills, networking, showing a willingness to relocate, and promoting themselves to potential employers?
d. What are the current trends in terms of work experience versus getting an advanced degree?
e. Is your competitive skill set likely to change in the future? If so, how? Who are these new competitors likely to be?

2. Economic conditions
a. What are the general economic conditions of the country, region, state, and local area in which you live or in which you want to relocate?
b. Overall, are potential employers optimistic or pessimistic about the economy?
c. What is the overall outlook for major job/career categories Where do potential employers seem to be placing their recruitment and hiring emphasis?
d. What is the trend in terms of starting salaries for major job/career categories?

3. Political trends
a. Have recent elections changed the political landscape so that certain industries or companies are now more or less attractive as potential employers?

4. Legal and regulatory factors
a. What changes in international, federal, state, or local laws and regulations are being proposed that would affect your job/career prospects?
b. Have recent court decisions made it easier or harder for you to find employment?

c. Have global trade agreements changed in any way that makes certain industries or companies more or less attractive as potential employers?

5. Changes in technology

a. What impact has changing technology had on potential employers in terms of their need for employees?

b. What technological changes will affect the way you will have to work and compete for employment in the future?

c. What technological changes will affect the way you market your skills and abilities to potential employers?

d. How do technological advances threaten to make your skills and abilities obsolete?

6. Cultural trends

a. How are society's demographics and values changing? What effect will these changes have on your:
(1) Skills and abilities?
(2) Career/lifestyle choices?
(3) Ability to market yourself?
(4) Willingness to relocate?
(5) Required minimum salary?

b. What problems or opportunities are being created by changes in the cultural diversity of the labor pool and the requirements of potential employers?

c. What is the general attitude of society regarding the particular skills, abilities, and talents that you possess and the career/lifestyle choices that you have made?

B. The Employer Environment

1. Who are your potential employers?

a. Identify characteristics: industry, products, size, growth, profitability, hiring practices, union/nonunion, employee needs, etc.

b. Geographic characteristics: home office, local offices, global sites, expansion, etc.

c. Organizational culture: mission statement, values, priorities, employee training, etc.

d. In each organization, who is responsible for recruiting and selecting new employees?

2. What do your potential employers look for in new employees?

a. What are the basic or specific skills and abilities that employers are looking for in new employees?

b. What are the basic or specific needs and abilities that you currently possess and that other potential employees currently possess?

c. How well do your skills and abilities (and those of your competitors) currently meet the needs of potential employers?

d. How are the needs of potential employers expected to change in the future?

3. What are the recent hiring practices of your potential employers?

a. How many employees are being hired? What combination of skills and abilities do these new hires possess?

b. Is the growth or decline in hiring related to the recent expansion or downsizing of markets and/or territories? Changes in technology?

c. Are there major hiring differences between large and small companies? If so, why?

4. Where and how do your potential employers recruit new employees?

a. Where do employers make contact with potential employees?
(1) College placement offices
(2) Job/career fairs
(3) Internship programs
(4) Headhunting firms
(5) Unsolicited applications
(6) The Internet

b. Do potential employers place a premium on experience or are they willing to hire new graduates without experience?

5. When do your potential employers recruit new employees?

a. Does recruiting follow a seasonal pattern or do employers recruit new employees on an ongoing basis?

C. Personal Assessment

1. Review of personal goals, objectives, and performance

a. What are your personal goals and objectives in terms of employment, career, lifestyle, geographic preferences, etc.?

b. Are your personal goals and objectives consistent with the realities of the labor market? Why or why not?

c. Are your personal goals and objectives consistent with recent changes in the external or employer environments? Why or why not?

d. How are your current strategies for success working in areas such as course performance, internships, networking, job leads, career development, interviewing skills, etc.?

e. How does your current performance compare to that of your peers (competitors)? Are they performing well in terms of course performance, internships, networking, job leads, career development, interviewing skills, etc.?

f. If your performance is improving, what actions can you take to ensure that your performance continues in this direction?

2. Inventory of personal skills and resources

a. What do you consider to be your marketable skills? This list should be as comprehensive as possible and include areas such as interpersonal skills, organizational skills, technological skills, communication skills (oral and written), networking/team-building skills, etc.

b. Considering the current and future needs of your potential employers, what important skills are you lacking?

c. Other than personal skills, what do you consider to be your other career-enhancing resources? This list should be as comprehensive as possible and include areas such as financial resources (to pay for additional training, if necessary), personal contacts or "connections" with individuals who can assist your career development, specific degrees or certificates you hold, and intangible resources (family name, prestige of your educational institution, etc.).

d. Considering the current and future needs of your potential employers, what important resources are you lacking?

III. SWOT Analysis (your personal strengths and weaknesses and the opportunities and threats that may impact your career)

A. Personal Strengths

1. Three key strengths
   a. Strength 1
   b. Strength 2
   c. Strength 3

2. How do these strengths allow you to meet the needs of your potential employers?

3. How do these strengths compare to those of your peers/competitors? Do these strengths give you an advantage relative to your peers/competitors?

B. Personal Weaknesses

1. Three key weaknesses
   a. Weakness 1
   b. Weakness 2
   c. Weakness 3

2. How do these weaknesses cause you to fall short of meeting the needs of your potential employers?

3. How do these weaknesses compare to those of your peers/competitors? Do these weaknesses put you at a disadvantage relative to your peers/competitors?

C. Career Opportunities

1. Three key career opportunities
   a. Opportunity 1
   b. Opportunity 2
   c. Opportunity 3

2. How are these opportunities related to serving the needs of your potential employers?

3. What actions must be taken to capitalize on these opportunities in the short term? In the long term?

D. Career Threats

1. Three key career threats
   a. Threat 1
   b. Threat 2
   c. Threat 3

2. How are these threats related to serving the needs of your potential employers?

3. What actions must be taken to prevent these threats from limiting your capabilities in the short term? In the long term?

E. The SWOT Matrix

F. Matching, Converting, Minimizing, and Avoiding Strategies

1. How can you match your strengths to your opportunities to better serve the needs of your potential employers?

2. How can you convert your weaknesses into strengths?

3. How can you convert your threats into opportunities?

4. How can you minimize or avoid those weaknesses and threats that cannot be converted successfully?

IV. Resources

A. Financial

1. Do you have the financial resources necessary to undertake and successfully complete this plan (that is, preparation/duplication/mailing of a résumé; interviewing costs, including proper attire; etc.)?

B. Human

1. Is the industry in which you are interested currently hiring? Are companies in your area currently hiring?

C. Experience and Expertise

1. Do you have experience from either part-time or summer employment that could prove useful in your current plan?

2. Do you have the required expertise or skills to qualify for a job in your desired field? If not, do you have the resources to obtain them?

V. Strategies

A. Objective(s)

1. **Potential employer A:**
   a. Descriptive characteristics
   b. Geographic locations
   c. Culture/values/mission
   d. Basic employee needs
   e. Recruiting/hiring practices
   f. Employee training/compensation practices
   g. Justification for selection

2. **Potential employer B:**
   a. Descriptive characteristics

b. Geographic locations
c. Culture/values/mission
d. Basic employee needs
e. Recruiting/hiring practices
f. Employee training/compensation practices
g. Justification for selection

B. Strategy(ies) for Using Capabilities and Resources

1. Strategy A (to meet the needs of potential employer A)
   a. Personal skills, abilities, and resources
      (1) Description of your skills and abilities
      (2) Specific employer needs that your skills/abilities can fulfill
      (3) Differentiation relative to peers/competitors (why should you be hired?)
      (4) Additional resources that you have to offer
      (5) Needed or expected starting salary
      (6) Expected employee benefits
      (7) Additional employer-paid training that you require
      (8) Willingness to relocate
      (9) Geographic areas to target
      (10) Corporate divisions or offices to target
      (11) Summary of overall strategy
      (12) Tactics for standing out among the crowd of potential employees
      (13) Point of contact with potential employer
      (14) Specific elements
         (a) Résumé
         (b) Internships
         (c) Placement offices
         (d) Job fairs
         (e) Personal contacts
         (f) Unsolicited
      (15) Specific objectives and budget

2. Strategy B (to meet the needs of potential employer B)
   a. Personal skills, abilities, and resources
      (1) Description of your skills and abilities
      (2) Specific employer needs that your skills/abilities can fulfill

(3) Differentiation relative to peers/competitors (why should you be hired?)

(4) Additional resources that you have to offer

(5) Needed or expected starting salary

(6) Expected employee benefits

(7) Additional employer-paid training that you require

(8) Willingness to relocate

(9) Geographic areas to target

(10) Corporate divisions or offices to target

(11) Summary of overall strategy

(12) Tactics for standing out among the crowd of potential employees

(13) Point of contact with potential employer

(14) Specific elements

    (a) Résumé

    (b) Internships

    (c) Placement offices

    (d) Job fairs

    (e) Personal contacts

    (f) Unsolicited

(15) Specific objectives and budget

C. Strategy Summary

1. How does strategy A (B) give you a competitive advantage in serving the needs of potential employer A (B)?

2. Is this competitive advantage sustainable? Why or why not?

VI. Financial Projections and Budgets

A. Do you have a clear idea of your budgetary requirements (for example, housing, furnishings, clothing, transportation, food, other living expenses)?

B. Will the expected salaries/benefits from potential employers meet these requirements? If not, do you have an alternative plan (that is, a different job choice, a second job, requesting a higher salary)?

VII. Controls and Evaluation

A. Performance Standards

1. What do you have to offer? List corrective actions that can be taken if your skills, abilities, and resources do not match the needs of potential employers:

2. Are you worth it? Corrective actions that can be taken if potential employers do not think your skills/abilities are worth your asking price:

3. Where do you want to go? Corrective actions that can be taken if potential employers do not offer you a position in a preferred geographic location:

4. How will you stand out among the crowd? Corrective actions that can be taken if your message is not being heard by potential employers or is not reaching the right people:

B. Monitoring Procedures

1. What types and levels of formal control mechanisms are in place to ensure the proper implementation of your plan?

    a. Are your potential employers hiring?

    b. Do you need additional training/education?

    c. Have you allocated sufficient time to your career development?

    d. Are your investments in career development adequate?

        (1) Training/education

        (2) Networking/making contacts

        (3) Wardrobe/clothing

        (4) Development of interviewing skills

    e. Have you done your homework on potential employers?

    f. Have you been involved in an internship program?

    g. Have you attended job/career fairs?

    h. Are you using the resources of your placement center?

    i. Are you committed to your career development?

C. Performance Analysis

1. Number/quality/potential of all job contacts made

2. Number of job/career fairs attended and quality of the job leads generated

3. Number of résumés distributed

    a. Number of potential employers who responded

    b. Number of negative responses

4. Number of personal interviews

5. Number/quality of job offers

Appendix C

Risk: The Basics of Risk Management

Introduction

Risk is something we all face in our everyday lives. We live in a world where uncertainty is around the corner, from a terrorist attack at a train station to an automobile accident on our way to work. We do not often think about the risks of wandering into oncoming traffic as we walk down the street looking at our cell phones. Or we may light a candle for dinner and forget to put the flame out before we go to bed. What about the risk of outliving our retirement savings or not saving enough for retirement? Do you think about identity theft when you visit an unfamiliar website? Risk often presents itself in ways we do not think about.

So how do we protect ourselves from risk? Some risks are insurable, whereas others are not. One way to protect ourselves is to pay an insurance company such as State Farm or Allstate to cover any losses that are insurable. We can insure against the possibility of financial risk or loss due to personal injury, automobile crashes, house fires, floods, and other types of risk by buying casualty insurance. These risks are considered insurable risks. This type of risk is also called pure risk because without insurance, you have lost a car, a house, a boat, a business, or some other physical property without the ability to replace it.

Risks that no insurance company will write a policy for are considered uninsurable risks. Risks such as the economic risk of lost market share, currency fluctuations, interest rate risk, or the political risk of investing in foreign countries are considered uninsurable.

Speculative risk assumes the possibility of either making a profit or incurring a loss. For example, investing in stocks, bonds, oil, farmland, gold, silver, grains, and other financial products and commodities all have the possibility of a gain or a loss. However, although insurance companies do not cover losses on a speculative investment, products called financial derivatives, such as options and futures contracts, can be used to cover possible losses from fluctuating prices. When used properly, an investor can hedge (offset) either all or some of the losses by using these financial derivatives.

As individuals, we have our own set of personality traits. Some people are more risk averse—that is, they have an aversion or dislike for risk—than others, but we assume that most people are risk averse to some degree. Financial theory would say that in order to induce most people to take larger risks, there must be an increased possibility for return. However, taking more risk means that not only do you have the potential for higher returns, you also have the potential for higher losses. This is true even with property casualty insurance because you must decide whether to pay a minimum deductible of $0, $500, or $1,000 if your automobile is in a collision. The higher the amount you are willing to cover out of your own pocket, the lower the premium. Making the deductible choice is considered a risk–return trade-off, and your choice is most likely based on what you think the probability of having an accident is and how often you might have an accident. Insuring two automobiles might also impact your decision as well as where you live and how many miles you drive per year. All these factors could affect the probability of an accident.

Insurance Based on Probability Protects against Risk

The law of large numbers postulates that as a sample size grows, the mean of the sample will get closer to the mean of the total population. For an insurance company insuring a life against the probability of death, the more people who are insured, the higher the probability that the actual mean death rate will approximate the national mortality tables of life expectancy for men and women per age category.

Another example is insuring automobiles. If a company only insures one automobile, the risk of that automobile being in an accident might be lower or higher than the mean for the population. The price to repair an automobile that is in an accident could be much higher than the actual annual insurance premium the automobile owner pays. If the driver never has an accident, the insurance company makes a profit on that policy. However, if the driver has a serious accident in the first five years of the policy, the insurance company takes a large loss. If the company insures 100,000 cars in any one state, it knows that the accident rate for its group of policyholders will approximate the mean number of accidents in that state and will also approximate the average cost of repair per automobile. This information allows the company to set the price of its insurance so that it makes a reasonable profit. In reality, insurance companies annually adjust their rates on automobile insurance in every state to reflect the previous year's accident experience. The price of the policy also reflects the age of the automobile, the replacement value of the automobile, and the driver's driving record.

Insurance and the Law of Large Numbers

The law of large numbers gives us an idea as to why insurance companies like to insure a large number of people. The larger the number of lives, houses, cars, boats, and so on that the companies insure, the more closely the claims from losses will be predictable based on a normal statistical distribution. This allows insurance companies to price policies for life insurance and casualty insurance appropriately so they have a predictable return on their assets. Calculating the actual expected losses is a very intensive statistical process, and insurance companies employ actuaries to estimate losses and profits. The U.S. Bureau of Labor Statistics defines an actuary as follows:

> Actuaries use skills primarily in mathematics, particularly calculus-based probability and mathematical statistics, but also economics, computer science, finance, and business. For this reason, actuaries are essential to the insurance and reinsurance industries, either as staff employees or as consultants; to other businesses, including sponsors of pension plans; and to government agencies such as the Government's Actuary's Department in the United Kingdom or the Social Security Administration in the United States of America. *Actuaries assemble and analyze data to estimate the probability and likely cost of the -occurrence of an event such as death, sickness, injury, disability, or loss of property.* Actuaries also address financial questions, including those involving the level of pension contributions required to produce a certain retirement income and the way in which a company should invest resources to maximize its return on investments in light of potential risk. *Using their broad knowledge, actuaries help design and price insurance policies,* pension plans, and other financial strategies in a manner that will help ensure that the plans are maintained on a sound financial basis.
>
> Bureau of Labor Statistics 2015, Government
> Actuary's Department 2015.

Insurance Companies and Their Products

The most common types of insurance policies are health insurance, life insurance, and property and casualty insurance. Insurance companies can be either nonprofit mutual companies like State Farm that are owned by their policyholders or for-profit stock companies that are publicly traded on the New York Stock Exchange. (Examples include MetLife, Inc. [MET], Aflac Inc. [AFL], and The Travelers Companies [TRV]. The symbols in brackets are the ticker symbols you can use to look up information on these companies on Yahoo.com. finance or another financial website.)

Insurance companies create policies covering all the contractual terms of each policy and what is covered in case of loss. Once they calculate the expected loss, they can price the insurance policy to cover the expected loss, the cost of administration, and a target profit margin. Insurance companies guard against losses greater than expected with an investment portfolio. Because they have guaranteed payment, insurance companies need reserves to cover unexpected losses. Reserves are kept in financial assets such as stocks, bonds, and U.S. government securities. These assets are liquid, meaning they can be sold quickly and turned into cash within a few days. Life insurance companies tend to invest more conservatively with a larger percentage of their assets invested in bonds, whereas property casualty companies tend to have a larger percentage of stocks (equities) than life insurance companies. When losses occur and need to be paid, these assets can be sold and cash paid out to policyholders.

Life Insurance

A life insurance policy guarantees a payment upon death, and the price of the policy depends on how much money you want your beneficiary to receive on your death. Policies can be purchased for $10,000 or millions of dollars. You can name anyone as the beneficiary of the policy or even have multiple beneficiaries, such as your children. A beneficiary is the person, persons, or organization named in the life insurance policy who are to receive the payout upon your death. Your beneficiary can even be your university, church, or any other charitable organization. People buy life insurance for many reasons, but the two most common are to leave money to cover debts they want paid, such as the mortgage on a house or money to educate children or grandchildren, or to leave an estate for their loved ones. Wealthy people may use insurance proceeds to help pay estate taxes due after death. Some companies provide life insurance as part of their employment contract, with the amount either equal to the employee's salary or a multiple of the employee's salary. These corporate plans cover a wide swath of employees and would be called group life insurance policies.

Term Life Insurance

The simplest type of insurance is called term life insurance. The policy provides a death benefit based on your age, health, and life expectancy. The key again is the law of large numbers. If a company such as MetLife has millions of term life policyholders, they can be quite sure that the age of death will approximate the life expectancy (mortality) tables created by the U.S. government. Some people will die before their expected mortality and some will die later, but the average age of death will approximate the mean expected value. Unfortunately, these mortality tables are not updated every year, and as people live longer than they used to, the average age of death increases and may be above that listed in mortality tables. This is where actuaries use their statistical prowess to adjust the mortality tables. Each insurance company could have slightly different expectations of life expectancy that will affect the price of its term insurance policy. You can find the full mortality tables used by the U.S. Social Security Administration at the following website: https://www.ssa.gov/oact/STATS/table4c6.html. We have provided an abbreviated version here for discussion purposes labeled Table C.1.

The Social Security area population is comprised of (1) residents of the 50 States and the District of Columbia (adjusted for net census undercount); (2) civilian residents of Puerto Rico, the Virgin Islands, Guam, American Samoa and the Northern Mariana Islands; (3) federal civilian employees and persons in the U.S. Armed Forces abroad and their dependents; (4) noncitizens living abroad who are insured for Social Security benefits; and (5) all other U.S. citizens abroad.

Looking at Table C.1, at birth, a male is expected to live to be 76.18, whereas a female is expected to live to be 80.95. If you are reading this, most likely you are close to 20 years old and are expected to live another 57.07 or 61.63 years depending on whether you are male or female. Your retirement age is likely to be about 67 years old, and at that age, you would be expected to live another 16.21 or 18.63 years, respectively. Looking at a male who lives to be 76 years old, which is close to his life expectancy at birth, he would be expected to live another 10.34 years. Out of the 100,000 babies at birth, 61,195 males and 73,026 females would still be expected to be alive at age 76. So the tables indicate that the longer you live, the longer you are expected to live. Even a 100-year-old is expected to live a couple more years. The tables actually go up to 119 years of age, but at 111 years of age, only 1 male and 6 females are still expected to be alive out of 100,000 babies born in their birth year. Using these types of tables, actuaries can compute the expected price of a term life insurance policy.

TABLE C.1   Period Life Table, 2013

A period life table is based on the mortality experience of a population during a relatively short period of time. Here we present the 2014 period life table for the Social Security area population. For this table, the period life expectancy at a given age is the average remaining number of years expected prior to death for a person at that exact age, born on January 1, using the mortality rates for 2014 over the course of his or her remaining life.

	Period Life Table, 2013					
	Male			Female		
Exact Age	Death Probability[a]	Number of Lives[b]	Life Expectancy	Death Probability[a]	Number of Lives[b]	Life Expectancy
0	0.006322	100,000	76.33	0.005313	100,000	81.11
20	0.001019	98,802	57.18	0.000373	99,146	61.77
67	0.018295	77,977	16.40	0.011754	86,114	18.84
76	0.039882	61,377	10.51	0.028045	73,280	12.26
100	0.349027	1,001	2.15	0.301467	2,893	2.48

[a] Probability of dying within one year.

[b] Number of survivors out of 100,000 born alive.

Note: The period life expectancy at a given age for 2014 represents the average number of years of life remaining if a group of persons at that age were to experience the mortality rates for 2014 over the course of their remaining life.

Whole Life Insurance

An insurance policy that combines a term life policy and a savings account is called a whole life policy. This type of policy costs more than term life because the extra amount paid is invested by the insurance company and provides a guaranteed cash value to the policyholder. Many people use this type of policy as a forced savings plan that allows the policyholder to take out cash payments at retirement. In other words, you do not have to wait until you die to get some financial rewards—you can use the cash that has built up in the account. For example, the parent of one of the authors of this text took out a $10,000 whole life policy for their child shortly after birth, and after paying $99 per year for more than 50 years, the policy now has a $22,423 cash value that can be withdrawn.

Variable Life Insurance

Many people think that instead of buying whole life insurance, the smartest thing is to buy term insurance and invest the difference in the price of the policies into financial assets like stocks and bonds. The real question for most people is, "Would you be disciplined enough to invest the difference wisely?" To counter this argument, insurance companies have created a variable life policy that invests the savings amount into financial assets without guaranteeing the return. Thus, the policyholder does not get a guaranteed rate of return but has the opportunity to get a better return than the guaranteed rate. In the end, someone choosing a variable life policy is taking more risk than buying a whole life policy but is hoping to get a higher return. There are many variations on variable life policies. The warning here is not to buy something you do not understand. Know the risk–return trade-offs.

Annuities

An annuity is a guaranteed equal payment for a given number of periods. The word *annuity* might make you think it has to be an annual payment, but general interpretation is that the payments must be equally spaced—so monthly, quarterly, or annual payments all qualify as annuities. For example, a monthly car payment, a monthly mortgage payment, and a semi-annual bond interest payment are examples of annuities.

Insurance companies offer people the ability to buy annuities, with the insurance company guaranteeing the payments for a specific period of years, most often the life of the annuitant (the person receiving the annuity). The $22,423 cash value presented in the preceding whole life example could be turned into an annuity. If the policyholder had 30 years of expected life and the insurance company guarantees a 5 percent return, the annuitant could receive $1,460 per year for life. A guaranteed rate of 6 percent would increase that annuity to $1,640.

Now imagine that you just retired, have a retirement account worth $1 million, and are risk averse. You do not want to take a risk in the stock and bond markets, and you would like to know what your monthly income would be. Assume you retire at 67. Looking at Table C.1 you see that as a female you are expected to live another 18.63 years. If the life insurance company guarantees you a 5 percent return for life, you could receive an annual payment of $83,870. If you live longer than expected, you get more than bargained for, but given the law of large numbers, some will live longer than and some less than their life expectancy. In this example, you have passed the risk of generating a return to the insurance company. If they can earn more than 5 percent on your $1 million, they have an extra profit. If they earn less, they have a problem. Given that financial returns are volatile from one year to the next, you have exchanged variable returns for stable returns and have probably taken a lower rate of return than you could have gotten by having an investment advisor managing your money.

Property Casualty Insurance

We all own property of some sort—whether it be a computer, a smartphone, a car, a house, a painting left to us by our grandparents, an electric keyboard, or some other product. What happens if they are lost to a flood, a fire, a theft, or other calamity? As noted earlier, if we have no insurance, we have an uninsured loss and have taken pure risk. We can buy property casualty insurance for all types of property, including houses, automobiles, and physical possessions such as paintings, pianos, stamp collections, and jewelry, to name a few. Sometimes this type of insurance is called property and liability insurance because the policy also includes coverage for any liability the policyholder might suffer because of damage to other people or property. The insurance protects you up to a loss limit that you specify when you buy the insurance. Of course, the larger the dollar coverage you have, the more you pay.

For example, your automobile insurance covers medical expenses for you and other passengers in your automobile and for anyone you might have injured if the accident is your fault. The insurance also covers the damage to your automobile or another automobile if it is your fault. If you insure a business, you can buy business interruption insurance that will reimburse you for losses suffered due to fire, theft, or other calamities, and it will also cover lost income. Homeowner's insurance protects you from losses suffered to your structure and to the property within the house. Homeowner's insurance also protects you from liability in case of injury to others within your home. If someone slips on your stairs and ends up in the hospital, this may be covered by your homeowner's insurance. If you rent instead of own your home, you can buy renter's insurance to protect your belongings. This is wise because the landlord's insurance may not cover your belongings.

Sometimes, an insurance company insures a large piece of property but does not want to take on all the risk of loss, so it passes off some of the insurance to a reinsurance company. Reinsurance is the division of an insurance policy among many different insurance companies. Some companies specialize in reinsurance as their major line of business. Reinsurance allows the original writer of the policy to diversify the risk of its portfolio. A good example is insuring oil tankers. Would you like to insure 100 percent of a tanker filled with oil, or would you rather insure 1 percent of 100 oil tankers? If the insurance company is unlucky and the tanker it has insured by itself at 100 percent gets caught in a storm or hits a reef (as did the *Exxon Valdez* in Prince William Sound, Alaska, on March 24, 1989), it may be forced to satisfy a legal judgment and multibillion-dollar cleanup costs. This is a good example of why reinsurance exists. Other examples include large office buildings in Manhattan and high-rise apartments in earthquake-prone California.

Health Insurance

Ever since the Affordable Care Act was passed by Congress and signed by President Obama, health insurance has been a topic of discussion. Politicians have criticized the Affordable Care Act, and so have people who lost their previous insurance and had to pay more for a new policy. On the other hand, many have defended it on the basis that many people are now able to afford insurance who could not do so before. For example, although most people have health insurance through their employer's group health insurance, many self-employed individuals had trouble finding insurance

they could afford. Likewise, people who worked part-time were not able to get or afford health insurance because they did not qualify for their employer's insurance plan.

Health insurance is expensive and generally covers the two basic areas of hospitalization and medical. Hospitalization insurance covers most hospital expenses, but in many cases, there is a co-pay or a minimum payment the insured has to pay. For example, your insurance may cover 80 percent of the hospital cost, and after the insured pays $2,000, the coverage may go up to 100 percent. This can also be true of surgical medical insurance that pays for the doctor's fee, the anesthesiologist, and surgical expenses. Dental insurance is available but not always part of a corporation's group policy. Dental insurance usually covers two teeth-cleaning appointments per year and the partial cost of fillings and crowns up to a fixed yearly limit.

Most corporate insurance plans cover disability. Disability insurance covers a portion of your salary if you are unable to work due to an accident or illness and can include partial disability for several months or permanent disability if you are unable to ever work again. The Social Security Administration can also cover disability if the person was paying into the Social Security System prior to the accident. Once a person retires and receives Social Security benefits, he or she is eligible for Medicare health insurance parts A, B, and D. *Medicare Part A* covers hospital insurance, *Part B* covers doctor's fees, and *Part D* provides drug coverage. Medicare is not free, and the cost of the health insurance is subtracted from your monthly payment. Your health insurance fees are affected by your income. The more retirement income you make, the more you pay for insurance.

Corporate and Investment Risk Decisions

Corporations have their own unique risks, but they still need the same kind of insurance coverage mentioned earlier—especially health and disability insurance for their employees and life insurance for key executives where the company is the beneficiary of the policy. Companies also need property casualty insurance and liability insurance for injuries to employees and others. One critical insurance policy is directors and officers (D&O) liability insurance to protect the board of directors from lawsuits. Without D&O insurance, companies would have a difficult time attracting qualified directors because few people would be willing to put their own wealth at risk in case of a lawsuit against the company and its directors.

Risks that cannot be insured by companies include business risks. This is the risk related to the inability of the firm to hold its competitive position and maintain stability and growth in earnings. Sometimes, business risk and operating risk are confused, but they are not exactly the same. Business risk occurs because the company does not keep up with its competitors due to lack of research and development, lack of forward planning, poor management, and so on. For example, when it comes to predicting retail sales, it may be much easier to predict the industry trend than whether Abercrombie & Fitch will grow faster than Aeropostale or Eddie Bauer. Consumer tastes change, and that is a constant risk for retail stores. The competitive environment in any industry can cause changes in which a new company is winning the race. New players and new technology can change the landscape for the whole industry. For example, Amazon.com changed retail sales, Apple's iTunes changed the music industry, and eBay created an online marketplace. The risk is that any company can be left behind, such as once-great companies such as Eastman Kodak and Xerox. Kodak failed to move into digital photography and lost its market dominance. Xerox lost its patent protection when its patent expired, and it failed to compete with new entrants. Likewise, more and more textbooks are now available online. Thirty years ago, there were more than 30 textbook publishers specializing in business texts. Now there are three dominant publishers as technology and the costs of publishing forced many out of business or into mergers.

Operating risk focuses on the volatility of operating earnings. Given the cyclical nature of the economy and the stability of the industry, this risk can be measured by the standard deviation of operating earnings. One measure of operating risk is ratio of sales to assets. For example, airlines have a very large investment in planes. They also have high labor and fuel costs, and when fuel costs rise, their operating profit is negatively affected. When the economy slows or goes into a recession, fewer people fly, and profits fall. Because airlines have high-cost fixed assets (planes), they have a very high break-even point where revenues equal costs. For years after the terrorist attacks of 9/11, airlines lost money, and United Airlines, Continental, Delta, and American Airlines all went bankrupt. Electric utilities also have a heavy investment in fixed assets, but the difference is they have a stable revenue base.

One of the other problems that airlines had was that not only did they have high operating risk, they also had high financial risk because they borrowed money to buy their planes. They were heavily indebted. Financial risk occurs when a firm uses too much financial leverage as measured by the debt to asset ratio, the debt to equity ratio, and the times interest earned ratio. Too much debt can cause the firm to fail to pay its interest and principal on its debt obligations as they come due. When high financial risk is combined with high operating risk, double the problems ensue. This is what happened to the airlines in recent years and is currently happening to shale oil drillers, many of whom will go bankrupt as the price of oil falls and they cannot pay the interest or principal on their debts. Returning to the utility example, because utility companies have a relatively stable revenue and profit stream, they can afford to have high financial leverage because they do not have the risk of highly fluctuating operating earnings.

One risk that is always present for corporations with multinational operations is currency risk or currency fluctuation. Changes in the relative value of one currency to another can cause profits denominated in U.S. dollars to go up or down depending on the exchange rate. For example, the euro or yen may advance or decline in relation to the dollar. To the extent a foreign currency appreciates relative to the dollar, returns on foreign investments will increase in terms of dollars. The opposite would be true for declining foreign currencies. Consider companies such as Coca-Cola, Apple, Ford, and others with operations all over the world. In 2015, Microsoft stated in its annual report that foreign currency changes added $335 million to its income, whereas in 2014, the impact was a negative $165 million.

Companies such as Apple, Microsoft, and General Motors manage a great deal of cash. Just like mutual funds, pension funds, insurance companies, and individual investors, they own fixed income securities like bonds. All investors are subject to the interest rate risk caused by changing interest rates. Interest rate changes can affect both an investor's income stream and the value of the assets held by that investor. Rising interest rates cause bond prices to fall, and the prices of other assets will also fall if they are purchased with borrowed money. For example, housing prices fall when interest rates rise because the cost of the monthly payment rises, making it more difficult for people to buy houses.

When interest rates rise, bonds that pay a fixed interest payment fall because investors can buy new bonds with higher interest payments. This causes bonds with lower interest payments to fall to a price that causes their payments to have the same yield or interest rate as the new bond. When interest rates rise, it is also more expensive for companies to borrow money, and that can lower their returns.

Conversely, falling interest rates mean that investors who rely on fixed income payments such as bonds and certificates of deposits have less income. For example, a retired person with $1 million invested in long-term U.S. government securities could only earn about $30,000 per year, or 3 percent, at the beginning of 2018, whereas in previous times, he or she could earn closer to $70,000. Low interest rates penalize people investing in low-risk assets such as U.S. government securities. One basic rule of the interest rate risk is that long-term fixed income securities are more sensitive to a change in interest rates, whereas short-term securities are less volatile. Thus, when you expect interest rates to rise, own short-term securities, and when you expect them to fall, own long-term securities.

Political risk is the risk associated with investing in firms operating in foreign countries. Dangers include the nationalization of foreign firms, blockage of capital flows to investors, or a violent overthrow of the political party in power, with all the associated implications. Punitive legislation against foreign firms or investors is another political risk. Expropriation or confiscation of property for much less than market value has occurred many times in some of the more socialistic South American countries, such as Venezuela and Brazil. Companies making foreign investments must be continually aware of this risk.

Prices of financial securities and commodities fluctuate over the short term, and these price fluctuations create market risk. Consider the price of oil between 2014 and 2016, falling from more than $100 a barrel to less than $35 a barrel. Consider the Dow Jones Industrial Average falling from 13,930 in October 2007 to 7,062 in February 2009, and then rising back up over 25,000 in 2018. This volatility of oil and stock prices demonstrates why markets can be risky. However, despite these short-term price movements, over the long term, prices tend to follow supply and demand and trends related to gross domestic product growth rates and corporate earnings. It may seem to be an anomaly, but it is easier to predict long-term trends than short-term fluctuations. So when you think back to how an insurance company decides how much to charge for a guaranteed life insurance policy or an annuity, realize that it must rely on the statistical probability that long-term trends will continue to dominate short-term movements.

The final risk is liquidity risk, which is the ability to turn assets into cash quickly. If you have a debit card, you know that you can go to an ATM and withdraw cash immediately from your bank account. What if you own a house, a vacant piece of land, or an antique Chinese ceramic vase? How long would it take to get cash for these assets? Compare that to owning a U.S. government security or common stock in Johnson & Johnson. These financial assets can be sold in the financial markets within minutes, and you would have money credited to your account within a few days after the sale clears. For many investors, liquidity is very important, whereas for others it is not.

Managing Investment Portfolios

It would be safe to say that all the risks previously mentioned in this appendix apply to anyone who manages a portfolio of stocks, bonds, or commodities. Even if you are an investor in low-risk U.S. government securities, you are still subject to interest rate risk, but you would manage to escape the other risks associated with corporate securities. Anything that affects the profitability and growth of a corporation would have an impact on that corporation's securities. Thus, if you invest in airline stock, you need to know that you have a high degree of operating risk and financial risk. If you invest in Apple, Coca-Cola, Procter & Gamble, or any other multinational company, you have a currency risk and possibly a political risk.

In terms of managing a portfolio of assets, one of the basic lessons of risk management is diversification, or the lack of concentration in any similar security, industry, or country. A portfolio composed of many different securities is diversified. Investments that are negatively correlated or that have low positive correlation provide the best diversification benefits. Such benefits may be particularly evident in an internationally diversified portfolio. By diversifying their portfolio with different companies and different asset classes such as stocks and bonds, and foreign stocks as well as bonds and securities in developed and emerging markets, investors reduce the risk that one bad event like the decline in the price of oil will destroy their total investment.

If you are invested in a market, you are subject to the risks of that market. In other words, if you own common stocks in the United States, you are subject to the price volatility of that market. The same is true if you own bonds in the United States. You cannot escape market risk. One of the benefits of diversification is the ability to minimize price volatility or at least keep the volatility of returns equal to or less than the market you have invested in.

Using statistical measures such as the standard deviation or beta when looking at common stock portfolios is one way to measure our ability to minimize volatility. Beta is a measure of the volatility of returns on an individual stock relative to the market. Stocks with a beta of 1.0 are said to have risk equal to that of the market (equal volatility). Stocks with betas greater than 1.0 have more risk than the market, whereas those with betas of less than 1.0 have less risk than the market. Thus, in a diversified portfolio of common stocks, we can build a portfolio with betas greater than, equal to, or less than 1.0 depending on our risk preference. Of course, when we do this, we assume that our historical measures of beta for the portfolio remain relatively constant.

Reducing Risk with Derivatives

We have already presented how insurance can reduce risk for those events that are insurable. What about risks such as currency risk, interest rate risk, and others? Although we cannot buy insurance, investors of all kinds and especially sophisticated investors like insurance companies, hedge funds, investment banks, commercial banks, and others can resort to financial engineering to minimize risk. Financial engineering involves the use of derivative securities to reduce a specific risk. The prices of derivative securities are based on an underlying security such as a common stock or bond or on the price of a particular commodity such as wheat or gold. Options and futures contracts are examples of derivative securities, and while simple to understand in their basic form, they can be combined in various combinations to make complex financial products. The discussion of these instruments is beyond the scope of this text, but it is important to know that they can be used to mitigate some uninsurable risks.

One example of using derivatives is using futures contracts to hedge prices. Hedging engages in a transaction that partially or fully reduces a prior risk exposure by taking a position that is the opposite of your initial position. As an example, assume you are a farmer with a wheat crop and you want to guarantee the price per bushel today, even though you might not harvest the wheat for three months. You could engage in a contract to sell the wheat in the future at a set price on the Chicago Board of Trade, thereby guaranteeing you a price before harvest.

As an example of using derivative instruments, consider Apple's 2015 annual report stating that the company uses derivatives to hedge foreign currency and interest rate risk:

> The Company may use derivatives to partially offset its business exposure to foreign currency and interest rate risk on expected future cash flows, on net investments in certain foreign subsidiaries and on certain existing assets and liabilities. However, the Company may choose not to hedge certain exposures for a variety of reasons including, but not limited to, accounting considerations and the prohibitive economic cost of hedging particular exposures. There can be no assurance the hedges will offset more than a portion of the financial impact resulting from movements in foreign currency exchange or interest rates.
>
> Apple 2015 Annual Report 10 K, p. 50.

Conclusion

As this appendix has demonstrated, risk is an everpervasive fact of life. From the risk of death or disability to the risk of financial loss from market fluctuations, risk comes in various shapes and forms. Some risks are insurable, allowing individuals to purchase insurance for automobiles, houses, life, disability, and businesses from insurance companies. Purchasing insurance is a risk within itself. On the one hand, you may never experience a disaster or use very little of your insurance. In this case, you have forgone money through your insurance payments while the insurance company makes a profit. On the other hand, insurance can be a life-saver for catastrophic events. The insurance company also takes risks with the possibility of having to make large payouts should their insured clients experience a disaster. Insurance companies are able to mitigate this risk, however, with the law of large numbers. The more people the insurance companies insure, the more likely the mean of the sample will get closer to the mean of the total population. This helps insurance companies predict how much they will have to pay out and how much they will keep in profit.

However, even insurers can underestimate the risk of losses. For example, prior to the last recession, many financial companies used derivatives to transfer risk between parties. Companies like AIG guaranteed the creditworthiness of the financial products. Should the products fail, AIG promised to pay for the losses. This strategy worked well when economic times were good, but when the housing market fell, so many of these derivatives fell in value that AIG could not afford to pay all of them. Only a government bailout kept the institution from collapsing. This type of risk makes it all the more important to have people like actuaries and financial analysts who can accurately assess the probability of risk-taking, guarding against risks that may prove too costly.

The excessive risk taking that helped lead to the recession brings up another form of risk that is often not considered: *ethical risk*. Although businesses might anticipate the risks of market fluctuations or loss due to some disaster, they are much less likely to anticipate fallout from unethical behavior. Yet the large number of lawsuits and fines levied against firms that were caught in misconduct shows how damaging ignoring these risks can be. The risk of misconduct increases when a company becomes more globalized because it is more difficult to monitor all the different stakeholders involved in the firm. Ways to mitigate this type of risk include the creation of strong ethics programs, ethics training for employees and suppliers, disciplinary procedures for the violation of company policies, and reporting mechanisms that can be used to report questionable behavior.

With many risks also comes opportunity. In the financial industry, the higher the risks you take, the higher the possible return. It is up to you to determine the amount of risk you are willing to take.

Glossary

A

absolute advantage a monopoly that exists when a country is the only source of an item, the only producer of an item, or the most efficient producer of an item.

accountability the principle that employees who accept an assignment and the authority to carry it out are answerable to a superior for the outcome.

accounting the recording, measurement, and interpretation of financial information.

accounting cycle the four-step procedure of an accounting system: examining source documents, recording transactions in an accounting journal, posting recorded transactions, and preparing financial statements.

accounting equation assets equal liabilities plus owners' equity.

accounts payable the amount a company owes to suppliers for goods and services purchased with credit.

accounts receivable money owed a company by its clients or customers who have promised to pay for the products at a later date.

accrued expenses all unpaid financial obligations incurred by an organization.

acquisition the purchase of one company by another, usually by buying its stock.

administrative managers those who manage an entire business or a major segment of a business; they are not specialists but coordinate the activities of specialized managers.

advertising a paid form of nonpersonal communication transmitted through a mass medium, such as television commercials or magazine advertisements.

advertising campaign designing a series of advertisements and placing them in various media to reach a particular target market.

affirmative action programs legally mandated plans that try to increase job opportunities for minority groups by analyzing the current pool of workers, identifying areas where women and minorities are underrepresented, and establishing specific hiring and promotion goals, with target dates, for addressing the discrepancy.

agency a common business relationship created when one person acts on behalf of another and under that person's control.

agenda a calendar, containing both specific and vague items, that covers short-term goals and long-term objectives.

agent the one in an agency relationship who acts on behalf of the principal to accomplish the task.

analytical skills the ability to identify relevant issues, recognize their importance, understand the relationships between them, and perceive the underlying causes of a situation.

annual report summary of a firm's financial information, products, and growth plans for owners and potential investors.

appellate court a court that deals solely with appeals relating to the interpretation of law.

arbitration settlement of a labor/management dispute by a third party whose solution is legally binding and enforceable.

articles of partnership legal documents that set forth the basic agreement between partners.

Asia-Pacific Economic Cooperation (APEC) an international trade alliance that promotes open trade and economic and technical cooperation among member nations.

asset utilization ratios ratios that measure how well a firm uses its assets to generate each $1 of sales.

assets a firm's economic resources, or items of value that it owns, such as cash, inventory, land, equipment, buildings, and other tangible and intangible things.

Association of Southeast Asian Nations (ASEAN) a trade alliance that promotes trade and economic integration among member nations in Southeast Asia.

attitude knowledge and positive or negative feelings about something.

auto liability insurance coverage that will pay claims against you for injuries to persons or property, up to a maximum per person and per accident.

auto physical damage insurance coverage that will insure against damage to your own vehicle.

automated clearinghouses (ACHs) a system that permits payments such as deposits or withdrawals to be made to and from a bank account by magnetic computer tape.

automated teller machine (ATM) the most familiar form of electronic banking, which dispenses cash, accepts deposits, and allows balance inquiries and cash transfers from one account to another.

B

balance of payments the difference between the flow of money into and out of a country.

balance of trade the difference in value between a nation's exports and its imports.

balance sheet a "snapshot" of an organization's financial position at a given moment.

behavior modification changing behavior and encouraging appropriate actions by relating the consequences of behavior to the behavior itself.

benefits nonfinancial forms of compensation provided to employees, such as pension plans, health insurance, paid vacation and holidays, and the like.

blogs web-based journals in which writers can editorialize and interact with other Internet users.

board of directors a group of individuals, elected by the stockholders to oversee the general operation of the corporation, who set the corporation's long-range objectives.

bonds debt instruments that larger companies sell to raise long-term funds.

bonuses monetary rewards offered by companies for exceptional performance as incentives to further increase productivity.

boycott an attempt to keep people from purchasing the products of a company.

brainstorming a technique in which group members spontaneously suggest ideas to solve a problem.

branding the process of naming and identifying products.

breach of contract the failure or refusal of a party to a contract to live up to his or her promises.

bribes payments, gifts, or special favors intended to influence the outcome of a decision.

brokerage firms firms that buy and sell stocks, bonds, and other securities for their customers and provide other financial services.

budget an internal financial plan that forecasts expenses and income over a set period of time.

budget deficit the condition in which a nation spends more than it takes in from taxes.

business individuals or organizations who try to earn a profit by providing products that satisfy people's needs.

business ethics principles and standards that determine acceptable conduct in business.

business law the rules and regulations that govern the conduct of business.

business plan a precise statement of the rationale for a business and a step-by-step explanation of how it will achieve its goals.

business products products that are used directly or indirectly in the operation or manufacturing processes of businesses.

buying behavior the decision processes and actions of people who purchase and use products.

C

capacity the maximum load that an organizational unit can carry or operate.

capital budgeting the process of analyzing the needs of the business and selecting the assets that will maximize its value.

capitalism (free enterprise) an economic system in which individuals own and operate the majority of businesses that provide goods and services.

cartel a group of firms or nations that agrees to act as a monopoly and not compete with each other, in order to generate a competitive advantage in world markets.

cash flow the movement of money through an organization over a daily, weekly, monthly, or yearly basis.

centralized organization a structure in which authority is concentrated at the top, and very little decision-making authority is delegated to lower levels.

certificates of deposit (CDs) savings accounts that guarantee a depositor a set interest rate over a specified interval as long as the funds are not withdrawn before the end of the period—six months or one year, for example.

certified management accountants (CMAs) private accountants who, after rigorous examination, are certified by the National Association of Accountants and who have some managerial responsibility.

certified public accountant (CPA) an individual who has been state certified to provide accounting services ranging from the preparation of financial records and the filing of tax returns to complex audits of corporate financial records.

checking account money stored in an account at a bank or other financial institution that can be withdrawn without advance notice; also called a demand deposit.

classical theory of motivation theory suggesting that money is the sole motivator for workers.

Clayton Act prohibits price discrimination, tying and exclusive agreements, and the acquisition of stock in another corporation where the effect may be to substantially lessen competition or tend to create a monopoly.

codes of ethics formalized rules and standards that describe what a company expects of its employees.

collateral assets pledged as security for a loan; If the loan is not repaid, the lender can use the assets pledged to settle the debt.

collective bargaining the negotiation process through which management and unions reach an agreement about compensation, working hours, and working conditions for the bargaining unit.

commercial banks the largest and oldest of all financial institutions, relying mainly on checking and savings accounts as sources of funds for loans to businesses and individuals.

commercial certificates of deposit (CDs) certificates of deposit issued by commercial banks and brokerage companies, available in minimum amounts of $100,000, which may be traded prior to maturity.

commercial paper a written promise from one company to another to pay a specific amount of money.

commercialization the full introduction of a complete marketing strategy and the launch of the product for commercial success.

commission an incentive system that pays a fixed amount or a percentage of the employee's sales.

committee a permanent, formal group that performs a specific task.

common stock stock whose owners have voting rights in the corporation, yet do not receive preferential treatment regarding dividends.

communism first described by Karl Marx as a society in which the people, without regard to class, own all the nation's resources.

comparative advantage the basis of most international trade, when a country specializes in products that it can supply more efficiently or at a lower cost than it can produce other items.

competition the rivalry among businesses for consumers' dollars.

compressed workweek a four-day (or shorter) period during which an employee works 40 hours.

computer-assisted design (CAD) the design of components, products, and processes on computers instead of on paper.

computer-assisted manufacturing (CAM) manufacturing that employs specialized computer systems to actually guide and control the transformation processes.

computer-integrated manufacturing (CIM) a complete system that designs products, manages machines and materials, and controls the operations function.

concentration approach a market segmentation approach whereby a company develops one marketing strategy for a single market segment.

conceptual skills the ability to think in abstract terms and to see how parts fit together to form the whole.

conciliation a method of outside resolution of labor and management differences in which a third party is brought in to keep the two sides talking.

consumer products products intended for household or family use.

consumerism the activities that independent individuals, groups, and organizations undertake to protect their rights as consumers.

contingency planning an element in planning that deals with potential disasters such as product tampering, oil spills, fire, earthquake, computer virus, or airplane crash.

continuous manufacturing organizations companies that use continuously running assembly lines, creating products with many similar characteristics.

contract a mutual agreement between two or more parties that can be enforced in a court if one party chooses not to comply with the terms of the contract.

contract manufacturing the hiring of a foreign company to produce a specified volume of the initiating company's product to specification; the final product carries the domestic firm's name.

controlling the process of evaluating and correcting activities to keep the organization on course.

cooperative (co-op) an organization composed of individuals or small businesses that have banded together to reap the benefits of belonging to a larger organization.

corporate charter a legal document that the state issues to a company based on information the company provides in the articles of incorporation.

corporate citizenship the extent to which businesses meet the legal, ethical, economic, and voluntary responsibilities placed on them by their stakeholders.

corporation a legal entity, created by the state, whose assets and liabilities are separate from its owners.

cost of goods sold the amount of money a firm spent to buy or produce the products it sold during the period to which the income statement applies.

countertrade agreements foreign trade agreements that involve bartering products for other products instead of for currency.

credit cards means of access to preapproved lines of credit granted by a bank or finance company.

credit controls the authority to establish and enforce credit rules for financial institutions and some private investors.

credit union a financial institution owned and controlled by its depositors, who usually have a common employer, profession, trade group, or religion.

crisis management (contingency planning) an element in planning that deals with potential disasters such as product tampering, oil spills, fire, earthquake, computer virus, or airplane crash.

culture the integrated, accepted pattern of human behavior, including thought, speech, beliefs, actions, and artifacts.

current assets assets that are used or converted into cash within the course of a calendar year.

current liabilities a firm's financial obligations to short-term creditors, which must be repaid within one year.

current ratio current assets divided by current liabilities.

customer departmentalization the arrangement of jobs around the needs of various types of customers.

customization making products to meet a particular customer's needs or wants.

D

debit card a card that looks like a credit card but works like a check; using it results in a direct, immediate, electronic payment from the cardholder's checking account to a merchant or third party.

debt to total assets ratio a ratio indicating how much of the firm is financed by debt and how much by owners' equity.

debt utilization ratios ratios that measure how much debt an organization is using relative to other sources of capital, such as owners' equity.

decentralized organization an organization in which decision-making authority is delegated as far down the chain of command as possible.

deductible an amount that you must pay before the insurance company will pay.

defined benefit plan a type of retirement plan that promises a specific benefit at retirement.

defined contribution plan a type of retirement plan firms offer where they promise to put a certain amount of money into the plan in your name every pay period.

delegation of authority giving employees not only tasks, but also the power to make commitments, use resources, and take whatever actions are necessary to carry out those tasks.

demand the number of goods and services that consumers are willing to buy at different prices at a specific time.

departmentalization the grouping of jobs into working units usually called departments, units, groups, or divisions.

depreciation the process of spreading the costs of long-lived assets such as buildings and equipment over the total number of accounting periods in which they are expected to be used.

depression a condition of the economy in which unemployment is very high, consumer spending is low, and business output is sharply reduced.

development training that augments the skills and knowledge of managers and professionals.

digital marketing uses all digital media, including the Internet and mobile and interactive channels, to develop communication and exchanges with customers.

digital media electronic media that function using digital codes via computers, cellular phones, smartphones, and other digital devices that have been released in recent years.

direct investment the ownership of overseas facilities.

direct marketing the use of nonpersonal media to communicate products, information, and the opportunity to purchase via media such as mail, telephone, or the Internet.

direct selling the marketing of products to ultimate consumers through face-to-face sales presentations at home or in the workplace.

directing motivating and leading employees to achieve organizational objectives.

disability insurance coverage that pays replacement income to you in in the event you are disabled under the definition of your policy.

discount rate the rate of interest the Fed charges to loan money to any banking institution to meet reserve requirements.

discounts temporary price reductions, often employed to boost sales.

distribution making products available to customers in the quantities desired.

diversity the participation of different ages, genders, races, ethnicities, nationalities, and abilities in the workplace.

dividend yield the dividend per share divided by the stock price.

dividends profits of a corporation that are distributed in the form of cash payments to stockholders.

dividends per share the actual cash received for each share owned.

double-entry bookkeeping a system of recording and classifying business transactions that maintains the balance of the accounting equation.

downsizing the elimination of a significant number of employees from an organization.

dumping the act of a country or business selling products at less than what it costs to produce them.

E

e-business carrying out the goals of business through utilization of the Internet.

earnings per share net income or profit divided by the number of stock shares outstanding.

economic contraction a slowdown of the economy characterized by a decline in spending and during which businesses cut back on production and lay off workers.

economic expansion the situation that occurs when an economy is growing and people are spending more money; their purchases stimulate the production of goods and services, which in turn stimulates employment.

economic order quantity (EOQ) model a model that identifies the optimum number of items to order to minimize the costs of managing (ordering, storing, and using) them.

economic system a description of how a particular society distributes its resources to produce goods and services.

economics the study of how resources are distributed for the production of goods and services within a social system.

electronic funds transfer (EFT) any movement of funds by means of an electronic terminal, telephone, computer, or magnetic tape.

embargo a prohibition on trade in a particular product.

employee empowerment when employees are provided with the ability to take on responsibilities and make decisions about their jobs.

entrepreneur an individual who risks his or her wealth, time, and effort to develop for profit an innovative product or way of doing something.

entrepreneurship the process of creating and managing a business to achieve desired objectives.

equilibrium price the price at which the number of products that businesses are willing to supply equals the amount of products that consumers are willing to buy at a specific point in time.

equity theory an assumption that how much people are willing to contribute to an organization depends on their assessment of the fairness, or equity, of the rewards they will receive in exchange.

estate plan planning for the transfer of your wealth and assets after your death.

esteem needs the need for respect—both self-respect and respect from others.

ethical issue an identifiable problem, situation, or opportunity that requires a person to choose from among several actions that may be evaluated as right or wrong, ethical or unethical.

eurodollar market a market for trading U.S. dollars in foreign countries.

European Union (EU) a union of European nations established in 1958 to promote trade among its members; one of the largest single markets today.

exchange the act of giving up one thing (money, credit, labor, goods) in return for something else (goods, services, or ideas).

exchange controls regulations that restrict the amount of currency that can be bought or sold.

exchange rate the ratio at which one nation's currency can be exchanged for another nation's currency.

exclusive distribution the awarding by a manufacturer to an intermediary of the sole right to sell a product in a defined geographic territory.

expectancy theory the assumption that motivation depends not only on how much a person wants something but also on how likely he or she is to get it.

expenses the costs incurred in the day-to-day operations of an organization.

exporting the sale of goods and services to foreign markets.

express warranty stipulates the specific terms the seller will honor.

extrinsic rewards benefits and/or recognition received from someone else.

F

factor a finance company to which businesses sell their accounts receivable—usually for a percentage of the total face value.

Federal Deposit Insurance Corporation (FDIC) an insurance fund established in 1933 that insures individual bank accounts.

Federal Reserve Board an independent agency of the federal government established in 1913 to regulate the nation's banking and financial industry.

Federal Trade Commission (FTC) the federal regulatory unit that most influences business activities related to questionable practices that create disputes between businesses and their customers.

finance the study of how money is managed by individuals, companies, and governments.

finance companies businesses that offer short-term loans at substantially higher rates of interest than banks.

financial managers those who focus on obtaining needed funds for the successful operation of an organization and using those funds to further organizational goals.

financial resources (capital) the funds used to acquire the natural and human resources needed to provide products; also called capital.

first-line managers those who supervise both workers and the daily operations of an organization.

fixed-position layout a layout that brings all resources required to create the product to a central location.

flexible manufacturing the direction of machinery by computers to adapt to different versions of similar operations.

flextime a program that allows employees to choose their starting and ending times, provided that they are at work during a specified core period.

floating-rate bonds bonds with interest rates that change with current interest rates otherwise available in the economy.

foreclosure when the lender gets the right to sell your property to get its money back if you cannot make timely payments.

franchise a license to sell another's products or to use another's name in business, or both.

franchisee the purchaser of a franchise.

franchiser the company that sells a franchise.

franchising a form of licensing in which a company—the franchiser—agrees to provide a franchisee a name, logo, methods of operation, advertising, products, and other elements associated with a franchiser's business in return for a financial commitment and the agreement to conduct business in accordance with the franchiser's standard of operations.

fraud a purposefully unlawful act to deceive or manipulate in order to damage others.

free-market system pure capitalism, in which all economic decisions are made without government intervention.

functional departmentalization the grouping of jobs that perform similar functional activities, such as finance, manufacturing, marketing, and human resources.

G

General Agreement on Tariffs and Trade (GATT) a trade agreement, originally signed by 23 nations in 1947, that provided a forum for tariff negotiations and a place where international trade problems could be discussed and resolved.

general partnership a partnership that involves a complete sharing in both the management and the liability of the business.

generic products products with no brand name that often come in simple packages and carry only their generic name.

geographical departmentalization the grouping of jobs according to geographic location, such as state, region, country, or continent.

global strategy (globalization) a strategy that involves standardizing products (and, as much as possible, their promotion and distribution) for the whole world, as if it were a single entity.

goal-setting theory refers to the impact that setting goals has on performance.

grapevine an informal channel of communication, separate from management's formal, official communication channels.

gross domestic product (GDP) the sum of all goods and services produced in a country during a year.

gross income (or profit) revenues minus the cost of goods sold required to generate the revenues.

group two or more individuals who communicate with one another, share a common identity, and have a common goal.

H

health insurance insurance that pays the cost of covered medical expenses during a policy period, which is usually six months to a year.

homeowners insurance coverage for liability and property damage for your home.

human relations the study of the behavior of individuals and groups in organizational settings.

human relations skills the ability to deal with people, both inside and outside the organization.

human resources (labor) the physical and mental abilities that people use to produce goods and services; also called labor.

human resources management (HRM) all the activities involved in determining an organization's human resources needs, as well as acquiring, training, and compensating people to fill those needs.

human resources managers those who handle the staffing function and deal with employees in a formalized manner.

hygiene factors aspects of Herzberg's theory of motivation that focus on the work setting and not the content of the work; these aspects include adequate wages, comfortable and safe working conditions, fair company policies, and job security.

I

identity theft when criminals obtain personal information that allows them to impersonate someone else in order to use their credit to obtain financial accounts and make purchases.

implied warranty imposed on the producer or seller by law, although it may not be a written document provided at the time of sale.

import tariff a tax levied by a nation on goods imported into the country.

importing the purchase of goods and services from foreign sources.

income statement a financial report that shows an organization's profitability over a period of time—month, quarter, or year.

inflation a condition characterized by a continuing rise in prices.

information technology (IT) managers those who are responsible for implementing, maintaining, and controlling technology applications in business, such as computer networks.

infrastructure the physical facilities that support a country's economic activities, such as railroads, highways, ports, airfields, utilities and power plants, schools, hospitals, communication systems, and commercial distribution systems.

initial public offering (IPO) selling a corporation's stock on public markets for the first time.

inputs the resources—such as labor, money, materials, and energy—that are converted into outputs.

insurance companies businesses that protect their clients against financial losses from certain specified risks (death, accident, and theft, for example).

integrated marketing communications coordinating the promotion mix elements and synchronizing promotion as a unified effort.

intellectual property refers to property, such as musical works, artwork, books, and computer software, that is generated by a person's creative activities

intensive distribution a form of market coverage whereby a product is made available in as many outlets as possible.

interest the percentage that the lender charges for your use of the money.

intermittent organizations organizations that deal with products of a lesser magnitude than do project organizations; their products are not necessarily unique but possess a significant number of differences.

international business the buying, selling, and trading of goods and services across national boundaries.

International Monetary Fund (IMF) organization established in 1947 to promote trade among member nations by eliminating trade barriers and fostering financial cooperation.

intrapreneurs individuals in large firms who take responsibility for the development of innovations within the organizations.

intrinsic rewards the personal satisfaction and enjoyment felt after attaining a goal.

inventory all raw materials, components, completed or partially completed products, and pieces of equipment a firm uses.

inventory control the process of determining how many supplies and goods are needed and keeping track of quantities on hand, where each item is, and who is responsible for it.

inventory turnover sales divided by total inventory.

investment banker underwrites new issues of securities for corporations, states, and municipalities.

investment banking the sale of stocks and bonds for corporations.

ISO 9000 a series of quality assurance standards designed by the International Organization for Standardization (ISO) to ensure consistent product quality under many conditions.

ISO 14000 a comprehensive set of environmental standards that encourages a cleaner and safer world by promoting a more uniform approach to environmental management and helping companies attain and measure improvements in their environmental performance.

ISO 19600 a comprehensive set of guidelines for compliance management that address risks, legal requirements, and stakeholder needs.

J

job analysis the determination, through observation and study, of pertinent information about a job—including specific tasks and necessary abilities, knowledge, and skills.

job description a formal, written explanation of a specific job, usually including job title, tasks, relationship with other jobs, physical and mental skills required, duties, responsibilities, and working conditions.

job enlargement the addition of more tasks to a job instead of treating each task as separate.

job enrichment the incorporation of motivational factors, such as opportunity for achievement, recognition, responsibility, and advancement, into a job.

job rotation movement of employees from one job to another in an effort to relieve the boredom often associated with job specialization.

job sharing performance of one full-time job by two people on part-time hours.

job specification a description of the qualifications necessary for a specific job, in terms of education, experience, and personal and physical characteristics.

joint venture a partnership established for a specific project or for a limited time.

journal a time-ordered list of account transactions.

junk bonds a special type of high interest rate bond that carries higher inherent risks.

jurisdiction the legal power of a court, through a judge, to interpret and apply the law and make a binding decision in a particular case.

just-in-time (JIT) inventory management a technique using smaller quantities of materials that arrive "just in time" for use in the transformation process and therefore require less storage space and other inventory management expense.

L

labeling the presentation of important information on a package.

labor contract the formal, written document that spells out the relationship between the union and management for a specified period of time—usually two or three years.

labor unions employee organizations formed to deal with employers for achieving better pay, hours, and working conditions.

lawsuit where one individual or organization takes another to court using civil laws.

leadership the ability to influence employees to work toward organizational goals.

learning changes in a person's behavior based on information and experience.

ledger a book or computer file with separate sections for each account.

leveraged buyout (LBO) a purchase in which a group of investors borrows money from banks and other institutions to acquire a company (or a division of one), using the assets of the purchased company to guarantee repayment of the loan.

liabilities debts that a firm owes to others.

licensing a trade agreement in which one company—the licensor—allows another company—the licensee—to use its company name, products, patents, brands, trademarks, raw materials, and/or production processes in exchange for a fee or royalty.

life insurance insurance that pays a benefit to your designated beneficiary in the event that you die during a coverage period.

limited liability company (LLC) form of ownership that provides limited liability and taxation like a partnership but places fewer restrictions on members.

limited partnership a business organization that has at least one general partner, who assumes unlimited liability, and at least one limited partner, whose liability is limited to his or her investment in the business.

line of credit an arrangement by which a bank agrees to lend a specified amount of money to an organization upon request.

line-and-staff structure a structure having a traditional line relationship between superiors and subordinates and also specialized managers—called staff managers—who are available to assist line managers.

line structure the simplest organizational structure, in which direct lines of authority extend from the top manager to the lowest level of the organization.

liquidity ratios ratios that measure the speed with which a company can turn its assets into cash to meet short-term debt.

lockbox an address, usually a commercial bank, at which a company receives payments in order to speed collections from customers.

lockout management's version of a strike, wherein a work site is closed so that employees cannot go to work.

logistics the planning and coordination of inbound and outbound as well as third party services.

long-term (fixed) assets production facilities (plants), offices, and equipment—all of which are expected to last for many years.

long-term goals goals that require substantial time to achieve.

long-term liabilities debts that will be repaid over a number of years, such as long-term loans and bond issues.

M

management a process designed to achieve an organization's objectives by using its resources effectively and efficiently in a changing environment.

managerial accounting the internal use of accounting statements by managers in planning and directing the organization's activities.

managers those individuals in organizations who make decisions about the use of resources and who are concerned with planning, organizing, staffing, directing, and controlling the organization's activities to reach its objectives.

manufacturer brands brands initiated and owned by the manufacturer to identify products from the point of production to the point of purchase.

manufacturing the activities and processes used in making tangible products; also called production.

market a group of people who have a need, purchasing power, and the desire and authority to spend money on goods, services, and ideas.

market orientation an approach requiring organizations to gather information about customer needs, share that information throughout the firm, and use that information to help build long-term relationships with customers.

market segment a collection of individuals, groups, or organizations who share one or more characteristics and thus have relatively similar product needs and desires.

market segmentation a strategy whereby a firm divides the total market into groups of people who have relatively similar product needs.

marketable securities temporary investment of "extra" cash by organizations for up to one year in U.S. Treasury bills, certificates of deposit, commercial paper, or eurodollar loans.

marketing a group of activities designed to expedite transactions by creating, distributing, pricing, and promoting goods, services, and ideas.

marketing channel a group of organizations that moves products from their producer to customers; also called a channel of distribution.

marketing concept the idea that an organization should try to satisfy customers' needs through coordinated activities that also allow it to achieve its own goals.

marketing managers those who are responsible for planning, pricing, and promoting products and making them available to customers.

marketing mix the four marketing activities—product, price, promotion, and distribution—that the firm can control to achieve specific goals within a dynamic marketing environment.

marketing research a systematic, objective process of getting information about potential customers to guide marketing decisions.

marketing strategy a plan of action for developing, pricing, distributing, and promoting products that meet the needs of specific customers.

Maslow's hierarchy a theory that arranges the five basic needs of people—physiological, security, social, esteem, and self-actualization—into the order in which people strive to satisfy them.

material-requirements planning (MRP) a planning system that schedules the precise quantity of materials needed to make the product.

materials handling the physical handling and movement of products in warehousing and transportation.

matrix structure a structure that sets up teams from different departments, thereby creating two or more intersecting lines of authority; also called a project-management structure.

mediation a method of outside resolution of labor and management differences in which the third party's role is to suggest or propose a solution to the problem.

mentoring involves supporting, training, and guiding an employee in his or her professional development.

merger the combination of two companies (usually corporations) to form a new company.

microentrepreneur entrepreneurs who develop businesses with five or fewer employees.

middle managers those members of an organization responsible for the tactical planning that implements the general guidelines established by top management.

mini-trial a situation in which both parties agree to present a summarized version of their case to an independent third party; the third party advises them of his or her impression of the probable outcome if the case were to be tried.

mission the statement of an organization's fundamental purpose and basic philosophy.

mixed economies economies made up of elements from more than one economic system.

modular design the creation of an item in self-contained units, or modules, that can be combined or interchanged to create different products.

monetary policy means by which the Fed controls the amount of money available in the economy.

money anything generally accepted in exchange for goods and services.

money market accounts accounts that offer higher interest rates than standard bank rates but with greater restrictions.

monopolistic competition the market structure that exists when there are fewer businesses than in a pure-competition environment and the differences among the goods they sell are small.

monopoly the market structure that exists when there is only one business providing a product in a given market.

morale an employee's attitude toward his or her job, employer, and colleagues.

mortgage a special type of loan that commonly requires that you make a constant payment over time to repay the lender the original money you borrowed together with interest.

motivation an inner drive that directs a person's behavior toward goals.

motivational factors aspects of Herzberg's theory of motivation that focus on the content of the work itself; these aspects include achievement, recognition, involvement, responsibility, and advancement.

multidivisional structure a structure that organizes departments into larger groups called divisions.

multinational corporation (MNC) a corporation that operates on a worldwide scale, without significant ties to any one nation or region.

multinational strategy a plan, used by international companies, that involves customizing products, promotion, and distribution according to cultural, technological, regional, and national differences.

multisegment approach a market segmentation approach whereby the marketer aims its efforts at two or more segments, developing a marketing strategy for each.

mutual fund an investment company that pools individual investor dollars and invests them in large numbers of well-diversified securities.

mutual savings banks financial institutions that are similar to savings and loan associations but, like credit unions, are owned by their depositors.

N

National Credit Union Administration (NCUA) an agency that regulates and charters credit unions and insures their deposits through its National Credit Union Insurance Fund.

natural resources land, forests, minerals, water, and other things that are not made by people.

net income the total profit (or loss) after all expenses, including taxes, have been deducted from revenue; also called net earnings.

networking the building of relationships and sharing of information with colleagues who can help managers achieve the items on their agendas.

nonprofit corporations corporations that focus on providing a service rather than earning a profit but are not owned by a government entity.

nonprofit organizations organizations that may provide goods or services but do not have the fundamental purpose of earning profits.

North American Free Trade Agreement (NAFTA) agreement that eliminates most tariffs and trade restrictions on agricultural and manufactured products to encourage trade among Canada, the United States, and Mexico.

O

offshoring the relocation of business processes by a company or subsidiary to another country. Offshoring is different than outsourcing because the company retains control of the offshored processes.

oligopoly the market structure that exists when there are very few businesses selling a product.

omni-channel retailing a type of retail that integrates the different methods of shopping available to consumers (for example, online, in a physical store, or by phone).

online fraud any attempt to conduct fraudulent activities online.

open economy an economy in which economic activities occur between the country and the international community.

open market operations decisions to buy or sell U.S. Treasury bills (short-term debt issued by the U.S. government) and other investments in the open market.

operational plans very short-term plans that specify what actions individuals, work groups, or departments need to accomplish in order to achieve the tactical plan and ultimately the strategic plan.

operations the activities and processes used in making both tangible and intangible products.

operations management (OM) the development and administration of the activities involved in transforming resources into goods and services.

organizational chart a visual display of the organizational structure, lines of authority (chain of command), staff relationships, permanent committee arrangements, and lines of communication.

organizational culture a firm's shared values, beliefs, traditions, philosophies, rules, and role models for behavior.

organizational layers the levels of management in an organization.

organizing the structuring of resources and activities to accomplish objectives in an efficient and effective manner.

orientation familiarizing newly hired employees with fellow workers, company procedures, and the physical properties of the company.

outputs the goods, services, and ideas that result from the conversion of inputs.

outsourcing the transferring of manufacturing or other tasks—such as data processing—to countries where labor and supplies are less expensive.

over-the-counter (OTC) market a network of dealers all over the country linked by computers, telephones, and Teletype machines.

owners' equity equals assets minus liabilities and reflects historical values.

P

packaging the external container that holds and describes the product.

partnership a form of business organization defined by the Uniform Partnership Act as "an association of two or more persons who carry on as co-owners of a business for profit."

penetration price a low price designed to help a product enter the market and gain market share rapidly.

pension funds managed investment pools set aside by individuals, corporations, unions, and some nonprofit organizations to provide retirement income for members.

per share data data used by investors to compare the performance of one company with another on an equal, per share basis.

perception the process by which a person selects, organizes, and interprets information received from his or her senses.

permanent insurance insurance designed to protect you with insurance protection over your lifetime.

personal financial planning the process of managing your finances so that you can achieve your financial goals.

personal net worth the total value of all personal assets less the total value of unpaid debts or liabilities.

personal property all other property that is not real property.

personal selling direct, two-way communication with buyers and potential buyers.

personality the organization of an individual's distinguishing character traits, attitudes, or habits.

physical distribution all the activities necessary to move products from producers to customers—inventory control, transportation, warehousing, and materials handling.

physiological needs the most basic human needs to be satisfied—water, food, shelter, and clothing.

picketing a public protest against management practices that involves union members marching and carrying anti-management signs at the employer's plant.

plagiarism the act of taking someone else's work and presenting it as your own without mentioning the source.

planning the process of determining the organization's objectives and deciding how to accomplish them; the first function of management.

podcast audio or video file that can be downloaded from the Internet with a subscription that automatically delivers new content to listening devices or personal computers.

preferred stock a special type of stock whose owners, though not generally having a say in running the company, have a claim to profits before other stockholders do.

price a value placed on an object exchanged between a buyer and a seller.

price skimming charging the highest possible price that buyers who want the product will pay.

primary data marketing information that is observed, recorded, or collected directly from respondents.

primary market the market where firms raise financial capital.

prime rate the interest rate that commercial banks charge their best customers (usually large corporations) for short-term loans.

principal the original money borrowed in the form of a loan; the one in an agency relationship who wishes to have a specific task accomplished.

private accountants accountants employed by large corporations, government agencies, and other organizations to prepare and analyze their financial statements.

private corporation a corporation owned by just one or a few people who are closely involved in managing the business.

private court system similar to arbitration in that an independent third party resolves the case after hearing both sides of the story.

private distributor brands brands, which may cost less than manufacturer brands, that are owned and controlled by a wholesaler or retailer.

process layout a layout that organizes the transformation process into departments that group related processes.

product a good or service with tangible and intangible characteristics that provide satisfaction and benefits.

product departmentalization the organization of jobs in relation to the products of the firm.

product layout a layout requiring that production be broken down into relatively simple tasks assigned to workers, who are usually positioned along an assembly line.

product liability businesses' legal responsibility for any negligence in the design, production, sale, and consumption of products.

product line a group of closely related products that are treated as a unit because of similar marketing strategy, production, or end-use considerations.

product mix all the products offered by an organization.

product-development teams a specific type of project team formed to devise, design, and implement a new product.

production the activities and processes used in making tangible products; also called manufacturing.

production and operations managers those who develop and administer the activities involved in transforming resources into goods, services, and ideas ready for the marketplace.

profit the difference between what it costs to make and sell a product and what a customer pays for it.

profit margin net income divided by sales.

profit sharing a form of compensation whereby a percentage of company profits is distributed to the employees whose work helped to generate them.

profitability ratios ratios that measure the amount of operating income or net income an organization is able to generate relative to its assets, owners' equity, and sales.

project organization a company using a fixed-position layout because it is typically involved in large, complex projects such as construction or exploration.

project teams groups similar to task forces that normally run their operation and have total control of a specific work project.

promotion a persuasive form of communication that attempts to expedite a marketing exchange by influencing individuals, groups, and organizations to accept goods, services, and ideas.

promotional positioning the use of promotion to create and maintain an image of a product in buyers' minds.

psychological pricing encouraging purchases based on emotional rather than rational responses to the price.

public corporation a corporation whose stock anyone may buy, sell, or trade.

publicity nonpersonal communication transmitted through the mass media but not paid for directly by the firm.

pull strategy the use of promotion to create consumer demand for a product so that consumers exert pressure on marketing channel members to make it available.

purchasing the buying of all the materials needed by the organization; also called procurement.

pure competition the market structure that exists when there are many small businesses selling one standardized product.

push strategy an attempt to motivate intermediaries to push the product down to their customers.

Q

quality the degree to which a good, service, or idea meets the demands and requirements of customers.

quality control the processes an organization uses to maintain its established quality standards.

quality-assurance teams (or quality circles) small groups of workers brought together from throughout the organization to solve specific quality, productivity, or service problems.

quasi-public corporations corporations owned and operated by the federal, state, or local government.

quick ratio (acid test) a stringent measure of liquidity that eliminates inventory.

quota a restriction on the number of units of a particular product that can be imported into a country.

R

ratio analysis calculations that measure an organization's financial health.

real property consists of real estate and everything permanently attached to it

receivables turnover sales divided by accounts receivable.

recession a decline in production, employment, and income.

recruiting forming a pool of qualified applicants from which management can select employees.

reference groups groups with whom buyers identify and whose values or attitudes they adopt.

reference pricing a type of psychological pricing in which a lower-priced item is compared to a more expensive brand in hopes that the consumer will use the higher price as a comparison price.

reinforcement theory states that behavior can be strengthened or weakened through the use of rewards and punishment.

renters insurance similar to homeowners insurance in that it covers you for liability on your premises and for damage to your personal property

reserve requirement the percentage of deposits that banking institutions must hold in reserve.

responsibility the obligation, placed on employees through delegation, to perform assigned tasks satisfactorily and be held accountable for the proper execution of work.

restructure to change the basic structure of an organization.

retailers intermediaries who buy products from manufacturers (or other intermediaries) and sell them to consumers for home and household use rather than for resale or for use in producing other products.

retained earnings earnings after expenses and taxes that are reinvested in the assets of the firm and belong to the owners in the form of equity.

return on assets net income divided by assets.

return on equity net income divided by owners' equity; also called return on investment (ROI).

revenue the total amount of money received from the sale of goods or services, as well as from related business activities.

reward cards credit cards made available by stores that carry a benefit to the user.

routing the sequence of operations through which the product must pass.

S

S corporation corporation taxed as though it were a partnership with restrictions on shareholders.

salary a financial reward calculated on a weekly, monthly, or annual basis.

sales promotion direct inducements offering added value or some other incentive for buyers to enter into an exchange.

Sarbanes-Oxley Act a law that criminalized securities fraud and strengthened penalties for corporate fraud.

savings accounts accounts with funds that usually cannot be withdrawn without advance notice; also known as time deposits.

savings and loan associations (S&Ls) financial institutions that primarily offer savings accounts and make long-term loans for residential mortgages; also called "thrifts."

scheduling the assignment of required tasks to departments or even specific machines, workers, or teams.

secondary data information that is compiled inside or outside an organization for some purpose other than changing the current situation.

secondary markets stock exchanges and over-the-counter markets where investors can trade their securities with others.

secured bonds bonds that are backed by specific collateral that must be forfeited in the event that the issuing firm defaults.

secured loans loans backed by collateral that the bank can claim if the borrowers do not repay them.

securities markets the mechanism for buying and selling securities.

security needs the need to protect oneself from physical and economic harm.

selection the process of collecting information about applicants and using that information to make hiring decisions.

selective distribution a form of market coverage whereby only a small number of all available outlets are used to expose products.

self-actualization needs the need to be the best one can be; at the top of Maslow's hierarchy.

self-directed work team (SDWT) a group of employees responsible for an entire work process or segment that delivers a product to an internal or external customer.

separations employment changes involving resignation, retirement, termination, or layoff.

serial bonds a sequence of small bond issues of progressively longer maturity.

sharing economy an economic model involving the sharing of underutilized resources.

Sherman Antitrust Act passed in 1890 to prevent businesses from restraining trade and monopolizing markets.

short-term goals goals that can be achieved in two years or less.

small business any independently owned and operated business that is not dominant in its competitive area and does not employ more than 500 people.

Small Business Administration (SBA) an independent agency of the federal government that offers managerial and financial assistance to small businesses.

social classes a ranking of people into higher or lower positions of respect.

social entrepreneurs individuals who use entrepreneurship to address social problems.

social needs the need for love, companionship, and friendship—the desire for acceptance by others.

social network a web-based meeting place for friends, family, co-workers, and peers that lets users create a profile and connect with other users for a wide range of purposes.

social responsibility a business's obligation to maximize its positive impact and minimize its negative impact on society.

social roles a set of expectations for individuals based on some position they occupy.

socialism an economic system in which the government owns and operates basic industries but individuals own most businesses.

sole proprietorships businesses owned and operated by one individual; the most common form of business organization in the United States.

span of management the number of subordinates who report to a particular manager.

specialization the division of labor into small, specific tasks and the assignment of employees to do a single task.

staffing the hiring of people to carry out the work of the organization.

stakeholders groups that have a stake in the success and outcomes of a business.

standard of living refers to the level of wealth and material comfort that people have available to them.

standardization the making of identical interchangeable components or products.

statement of cash flows explains how the company's cash changed from the beginning of the accounting period to the end.

statistical process control a system in which management collects and analyzes information about the production process to pinpoint quality problems in the production system.

stock shares of a corporation that may be bought or sold.

strategic alliance a partnership formed to create competitive advantage on a worldwide basis.

strategic plans those plans that establish the long-range objectives and overall strategy or course of action by which a firm fulfills its mission.

strikebreakers people hired by management to replace striking employees; called "scabs" by striking union members.

strikes employee walkouts; one of the most effective weapons labor has.

structure the arrangement or relationship of positions within an organization.

supply the number of products—goods and services—that businesses are willing to sell at different prices at a specific time.

supply chain management connecting and integrating all parties or members of the distribution system in order to satisfy customers.

sustainability conducting activities in a way that allows for the long-term well-being of the natural environment, including all biological entities; involves the assessment and improvement of business strategies, economic sectors, work practices, technologies, and lifestyles so that they maintain the health of the natural environment.

T

tactical plans short-range plans designed to implement the activities and objectives specified in the strategic plan.

target market a specific group of consumers on whose needs and wants a company focuses its marketing efforts.

task force a temporary group of employees responsible for bringing about a particular change.

team a small group whose members have complementary skills; have a common purpose, goals, and approach; and hold themselves mutually accountable.

technical expertise the specialized knowledge and training needed to perform jobs that are related to particular areas of management.

term insurance a type of life insurance that lasts usually for one year and the insurer promises to pay your designated beneficiary only the face amount of the policy in the event you die during the year of coverage.

test marketing a trial minilaunch of a product in limited areas that represent the potential market.

Theory X McGregor's traditional view of management whereby it is assumed that workers generally dislike work and must be forced to do their jobs.

Theory Y McGregor's humanistic view of management whereby it is assumed that workers like to work and that under proper conditions employees will seek out responsibility in an attempt to satisfy their social, esteem, and self-actualization needs.

Theory Z a management philosophy that stresses employee participation in all aspects of company decision making.

times interest earned ratio operating income divided by interest expense.

Title VII of the Civil Rights Act prohibits discrimination in employment and created the Equal Employment Opportunity Commission.

top managers the president and other top executives of a business, such as the chief executive officer (CEO), chief financial officer (CFO), and chief operations officer (COO), who have overall responsibility for the organization.

tort a private or civil wrong other than breach of contract.

total asset turnover sales divided by total assets.

total quality management (TQM) a philosophy that uniform commitment to quality in all areas of an organization will promote a culture that meets customers' perceptions of quality.

total-market approach an approach whereby a firm tries to appeal to everyone and assumes that all buyers have similar needs.

trade credit credit extended by suppliers for the purchase of their goods and services.

trade deficit a nation's negative balance of trade, which exists when that country imports more products than it exports.

trademark a brand that is registered with the U.S. Patent and Trademark Office and is thus legally protected from use by any other firm.

trading company a firm that buys goods in one country and sells them to buyers in another country.

training teaching employees to do specific job tasks through either classroom development or on-the-job experience.

transaction balances cash kept on hand by a firm to pay normal daily expenses, such as employee wages and bills for supplies and utilities.

transfer a move to another job within the company at essentially the same level and wage.

transportation the shipment of products to buyers.

Treasury bills (T-bills) short-term debt obligations the U.S. government sells to raise money.

trial court when a court (acting through the judge or jury) must determine the facts of the case, decide which law or set of laws is pertinent to the case, and apply those laws to resolve the dispute.

turnover occurs when employees quit or are fired and must be replaced by new employees.

U

undercapitalization the lack of funds to operate a business normally.

unemployment the condition in which a percentage of the population wants to work but is unable to find jobs.

Uniform Commercial Code (UCC) set of statutory laws covering several business law topics.

unsecured bonds debentures or bonds that are not backed by specific collateral.

unsecured loans loans backed only by the borrower's good reputation and previous credit rating.

V

value a customer's subjective assessment of benefits relative to costs in determining the worth of a product.

venture capitalists persons or organizations that agree to provide some funds for a new business in exchange for an ownership interest or stock.

viral marketing a marketing tool that uses a networking effect to spread a message and create brand awareness. The purpose of this marketing technique is to encourage the consumer to share the message with friends, family, co-workers, and peers.

W

wage/salary survey a study that tells a company how much compensation comparable firms are paying for specific jobs that the firms have in common.

wages financial rewards based on the number of hours the employee works or the level of output achieved.

warehousing the design and operation of facilities to receive, store, and ship products.

whistleblowing the act of an employee exposing an employer's wrongdoing to outsiders, such as the media or government regulatory agencies.

wholesalers intermediaries who buy from producers or from other wholesalers and sell to retailers.

wiki software that creates an interface that enables users to add or edit the content of some types of websites.

working capital management the managing of short-term assets and liabilities.

World Bank an organization established by the industrialized nations in 1946 to loan money to underdeveloped and developing countries; formally known as the International Bank for Reconstruction and Development.

World Trade Organization (WTO) international organization dealing with the rules of trade between nations.

Name Index

Company Index

Subject Index

Credits

Online Supplements

Connect Online Access for Business Foundations, 12th Edition

McGraw-Hill Connect is a digital teaching and learning environment that improves performance over a variety of critical outcomes. With Connect, instructors can deliver assignments, quizzes and tests easily online. Students can practice important skills at their own pace and on their own schedule.

HOW TO REGISTER

Using a <u>Print Book</u>?
To register and activate your Connect account, simply follow these easy steps:
1. **Go to the Connect course web address provided by your instructor or visit the Connect link set up on your instructor's course within your campus learning management system.**
2. **Click on the link to register.**
3. **When prompted, enter the Connect code found on the inside back cover of your book and click Submit. Complete the brief registration form that follows to begin using Connect.**

Using an <u>eBook</u>?
To register and activate your Connect account, simply follow these easy steps:
1. **Upon purchase of your eBook, you will be granted automatic access to Connect.**
2. **Go to the Connect course web address provided by your instructor or visit the Connect link set up on your instructor's course within your campus learning management system.**
3. **Sign in using the same email address and password you used to register on the eBookstore. Complete your registration and begin using Connect.**

Note: Access Code is for one use only. If you did not purchase this book new, the access code included in this book is no longer valid.

Need help? Visit mhhe.com/support